Linear Programming

METHODS AND APPLICATIONS

Fifth Edition

Saul I. Gass

College of Business and Management
University of Maryland

boyd & fraser publishing company

I(T)P   An International Thomson Publishing Company

Danvers • Albany • Bonn • Boston • Cincinnati • Detroit • London • Madrid • Melbourne
Mexico City • New York • Paris • San Francisco • Singapore • Tokyo • Toronto • Washington

TO TRUDY

For her encouragement,
patience,
and lost weekends

Editor: DeVilla Williams
Designer: Naomi Auerbach

Linear Programming: Methods and Applications, 5th edition
by Saul I. Gass

Originally published by McGraw Hill, Inc. ©1958, 1964, 1969, 1975, 1985

 boyd & fraser publishing company
A division of International Thomson Publishing Inc.

I(T)P   The ITP™ logo is a trademark under license.

Printed in the United States of America

For more information, contact boyd & fraser publishing company:

boyd & fraser publishing company
One Corporate Place • Ferncroft Village
Danvers, Massachusetts 01923, USA

International Thomson Publishing Europe
Berkshire House 168-173
High Holborn
London, WC1V 7AA, England

Thomas Nelson Australia
102 Dodds Street
South Melbourne 3205
Victoria, Australia

Nelson Canada
1120 Birchmount Road
Scarborough, Ontario
Canada M1K 5G4

International Thomson Editores
Campose Eliseos 385, Piso 7
Col. Polanco
11560 Mexico D.F. Mexico

International Thomson Publishing GmbH
Konigswinterer Strasse 418
53227 Bonn, Germany

International Thomson Publishing Asia
221 Henderson Road
#05-10 Henderson Building
Singapore 0315

International Thomson Publishing Japan
Hirakawacho Kyowa Building, 3F
2-2-1 Hirakawacho
Chiyoda-ku, Tokyo 102, Japan

6 7 8 9 10 D 9 8 7 6 5

Contents

Preface

The material in this book was originally prepared for an introductory course in linear programming given at the Graduate School, U.S. Department of Agriculture, Washington, D.C. In developing and expanding the notes into a suitable text, I have attempted to pursue the same objectives that guided the presentation of the course material. These basic aims were to instill in the student an ability to recognize potential linear-programming problems, to formulate such problems as linear-programming models, to employ the proper computational techniques to solve these problems, and to understand the mathematical aspects that tie together these elements of linear programming.

It is very convenient to divide the subject matter of linear programming into three separate, but not distinct, areas: *theoretical, computational,* and *applied.* In teaching the course, I found it appropriate, instructive, and beneficial to the students to interlace material from all three areas as much as possible. Hence, after an introductory lecture on applications and the mathematical model of linear programming (Chap. 1), the mathematics of convex sets and linear inequalities were developed and followed by the computational aspects of the elimination method for solving linear equations (Chap. 2). The mathematical properties of a solution to the general linear-programming problem were next evolved. Then, in an attempt to explain fully the fundamentals of the simplex computational procedure, the ability to generate extreme-point solutions was shown to be a simple variation of the elimination technique of Gauss and Jordan (Chap. 3). The next set of lectures developed the theoretical and computational elements of the simplex method of G. B. Dantzig (Chap. 4). A discussion on the duality problems of linear programming (Chap. 6) was followed by lectures on the formulation of certain illustrative applications (Chaps. 10 and 11). The final lectures of the course[1] described the relationship between linear programming and the zero-sum two-person game.

The basic-course notes have been revised to include full discussions of the revised simplex method (Chap. 5), degeneracy procedures (Chap. 7), parametric programming (Chap. 8), further computational techniques (Chap. 9), and other topics and applications. As a result of a suggestion to the author, all the material has been gathered into three parts: an introduction, methods, both theoretical and computational, and applications.[2] This arrangement enhances the usefulness of the book for reference, and presents the three areas of linear programming in a related but separate manner. Consequently, the reader will find some relatively advanced topics, such as the decomposition method and parametric linear programming, appearing before the basic discussions of the transportation problem and general applications. It is suggested that, in order to motivate one's study of linear programming, the chapters should not be studied in numerical sequence. Instead, one should, as soon as possible (probably after Chap. 5 or 6), become acquainted with material in the applications sections.[3]

The material covered in this text is appropriate for courses at the upper-level undergraduate or first-year-graduate level. However, because of the interest in linear-programming methods outside the academic field, it seemed advisable to include sufficient material on matrices and vectors to make the work complete for all readers (Chap. 2). It should be noted that much of the mathematical notation used in subsequent chapters is developed in Chap. 2.[4]

The search for the best, the maximum, the minimum, or, generally speaking, the optimal solutions to a variety of problems has entertained and intrigued man throughout the ages. Euclid, in Book III, was concerned with finding the greatest and least straight lines that can be drawn from a point to the circumference of a circle, and in Book IV he described how to find the parallelogram of maximum area with a given perimeter. However, the rigorous approach to these and more sophisticated problems had to wait until the great mathematicians of the seventeenth and eighteenth centuries developed the powerful methods of the calculus and the calculus of varia-

---

[1] This one-semester course consisted of 16 evening lectures of $2\frac{1}{2}$ hours' duration.

[2] This edition includes an additional part on nonlinear programming, Chap. 12.

[3] In a course that meets three times a week, one lecture could be devoted to applications and/or reports of case studies cited in the Bibliography.

[4] The student will soon find that one of the main difficulties in understanding the mathematics of linear programming arises from the diverse and often intricate notation used in many of the source papers of this field. Wherever possible, I have employed "standard," consistent, and, I hope, explicit notation.

tions. With these techniques we can find the maximum and minimum solutions to a wide range of optimization problems. These and other mathematical optimization procedures were mainly concerned with the solutions to problems of a geometric, dynamic, or physical nature. Such problems as finding the minimum curves of revolution and the curve of quickest descent are resolved by these classical optimization methods.

Recently, a new class of optimization problems has originated out of the complex organizational structures that permeate modern society. Here we are concerned with such matters as the most efficient manner in which to run an economy, or the optimal deployment of aircraft that maximizes a country's chances of winning a war, or with such mundane tasks as mixing the ingredients of a fertilizer to meet agricultural specifications at a minimum cost. Research on how to formulate and solve such problems has led to the development of new and important optimization techniques. Among these we find the subject of this book—*linear programming.* The linear-programming model, i.e., the optimization of a linear function subject to linear constraints, is simple in its mathematical structure but powerful in its adaptability to a wide range of applications.

Historically, the general problem of linear programming was first developed and applied in 1947 by George B. Dantzig, Marshall Wood, and their associates of the U.S. Department of the Air Force.[1] At that time, this group was called on to investigate the feasibility of applying mathematical and related techniques to military programming and planning problems. This inquiry led Dantzig to propose "that interrelations between activities of a large organization be viewed as a linear programming type model and the optimizing program determined by minimizing a linear objective function." In order to develop and extend these ideas further, the Air Force organized a research group under the title of Project SCOOP (Scientific Computation of Optimum Programs). Besides putting the Air Force programming and budgeting problems on a more scientific basis, Project SCOOP's major contribution was the formal development and application of the linear-programming model. These early applications of linear-programming methods fell into three major categories: military applications generated by Project SCOOP, interindustry economics based on the Leontief input-output model, and problems involving the relationship between zero-sum two-person games and linear programming. In the past years these areas of applications have been extended and developed, but the main emphasis in linear-programming applications has shifted to the general industrial area, with many applications found in the social and urban fields.

The initial mathematical statement of the general problem of linear programming was made by Dantzig in 1947 along with the *simplex method,* a

---

[1] Acknowledgment should also be given to the Soviet mathematician and economist L. V. Kantorovich, who as early as 1939 formulated and solved a linear-programming problem dealing with the organization and planning of production.

systematic procedure for solving the problem. Prior to this, a number of problems (some unsolved) were recognized as being of the type that dealt with the optimization of a linear function subject to linear constraints. The more important examples include the transportation problem posed by Hitchcock (1941) and independently by Koopmans (1947) and the diet problem of Stigler (1945). The first successful solution of a linear-programming problem on a high-speed electronic computer occurred in January 1952, on the National Bureau of Standards SEAC machine. Since that time, the simplex algorithm, or variations of this procedure, has been coded for most intermediate and large general-purpose electronic computers.

Linear programming has become an important tool of modern theoretical and applied mathematics. This remarkable growth can be traced to the pioneering efforts of many individuals and research organizations. Specifically, I would like to make special mention of George B. Dantzig, Murray A. Geisler, Leon Goldstein, Julian L. Holley, Walter W. Jacobs, Alex Orden, Emil D. Schell, and Marshall K. Wood, all formerly with the U.S. Department of the Air Force; Leon Gainen, Alan J. Hoffman, and Solomon Pollack, formerly with the National Bureau of Standards; and the research groups of the Graduate School of Industrial Administration of the Carnegie Institute of Technology, The RAND Corporation, the Department of Mathematics of Princeton University, and the Cowles Commission for Research in Economics. I would like to thank the above named individuals and groups, other authors, and their publishers for their kind permission to use certain basic material contained in what might be considered "source documents" of linear programming. Appropriate references are given in the text.

This edition includes new sections on multiobjective linear programming, computational considerations, the lowest-index anticycling rules, the simplex algorithm and computational efficiency, and separable programming; an expanded chapter on applications; an expanded section on the solution of linear equations; and new explanatory material, numerical examples, and exercises with selected answers in all chapters. Many new publications have been included in the Bibliography of Linear-Programming Applications and References sections. In sum, this revision has continued my attempt to accomplish the somewhat conflicting (nonlinear) objectives of maintaining the book's status as an upper-level undergraduate and graduate text and a research source.

As most universities have computer-based systems for solving linear-programming problems, I encourage instructors and students to integrate the use of such systems into their courses, especially for solving formulation problems and performing sensitivity analyses. Computer scientists and numerical analysts will find the translation of the simplex algorithm into a working computer code a task that illustrates just about all of the computational subtleties encountered in more difficult endeavors.

I wish to acknowledge the initial encouragement I received to write this text from my former associates at the Directorate of Management Analysis of the Department of the Air Force, and to thank Moshe Dror, Jack Edmonds, Harold Fassberg, Stan Fromovitz, Tomas Gal, Gerald A. Gass, Bruce Golden, Harvey Greenberg, Donald Gross, Eli Hellerman, Richard H. F. Jackson, Walter W. Jacobs, Lambert Joel, Robert R. Meyer, Dianne P. O'Leary, William Orchard-Hays, Paul Rech, Thomas L. Saaty, Stephen P. Shao, Jr., Craig C. Sherbrooke, Kenneth Webb, Christoph Witzgall, and Sanjo Zlobec for their many valuable suggestions.

Saul I. Gass

Introduction

CHAPTER 1

General Discussion

1. LINEAR-PROGRAMMING PROBLEMS

Programming problems are concerned with the efficient use or allocation of limited resources to meet desired objectives.   These problems are characterized by the large number of solutions that satisfy the basic conditions of each problem.   The selection of a particular solution as the best solution to a problem depends on some aim or over-all objective that is implied in the statement of the problem.   A solution that satisfies both the conditions of the problem and the given objective is termed an *optimal solution*.   A typical example is that of the manufacturing company that must determine what combination of available resources will enable it to manufacture products in a way which not only satisfies its production schedule, but also maximizes its profit.   This problem has as its basic conditions the limitations of the available resources and the requirements of the production schedule, and as its objective the desire of the company to maximize its gain.

We shall mainly consider only a very special subclass of programming problems called *linear-programming problems*.   A linear-programming problem differs from the general variety in that a *mathematical model* or description of the problem can be stated using relationships which are called "straight-line" or linear. Mathematically, these relationships are of the form

$$a_1 x_1 + a_2 x_2 + \cdots + a_j x_j + \cdots + a_n x_n = b^{\dagger}$$

where the a_j's and b are known coefficients and the x_j's are unknown variables. The complete mathematical statement of a linear-programming problem includes a set of linear equations which represent the conditions of the problem and a linear function which expresses the objective of the problem. In Sec. 2 we shall state a number of programming problems and formulate them as linear-programming problems.

To solve a linear-programming problem, we must initially concern ourselves with the solution of the associated set of linear equations. There are various criteria which can be applied to a set of linear equations to reveal whether a solution or solutions to the problem exist (see Hadley [342]).[1] The set of two equations in two variables

$$2x_1 + 3x_2 = 8$$

$$x_1 + 2x_2 = 5$$

has the *unique* solution $x_1 = 1$ and $x_2 = 2$, while the single equation

$$x_1 + 2x_2 = 8 \tag{1.1}$$

has an *infinite* number[2] of solutions. From (1.1) we have

$$x_1 = 8 - 2x_2 \qquad \text{or} \qquad x_2 = 4 - \tfrac{1}{2}x_1$$

For every value of x_2 (or x_1) there is a corresponding value of x_1 (or x_2). If we further restrict the variables to be *nonnegative*, that is, $x_1 \geq 0$ and $x_2 \geq 0$, we limit the range of the variables, since

$$x_1 = 8 - 2x_2 \geq 0 \qquad \text{implies} \qquad 0 \leq x_2 \leq 4$$

and

$$x_2 = 4 - \tfrac{1}{2}x_1 \geq 0 \qquad \text{implies} \qquad 0 \leq x_1 \leq 8$$

We still have an infinite number of solutions, but the addition of further restrictions or constraints to (1.1) has resulted in less freedom of action. As we shall show, the condition of nonnegativity of the variables is an important requirement of linear-programming problems. Systems like (1.1) in which there are more variables than equations are called *underdetermined*. In general, underdetermined systems of linear equations have either no solution or an infinite number of solutions.

[†] Geometrically, these relationships are equivalent to straight lines in two dimensions, planes in three dimensions, and hyperplanes in higher dimensions.

[1] Numbers in brackets refer to the publications listed in the References at the back of the book.

[2] A more exact phrase would be "has an infinity of solutions." However, we shall use the common expression of "an infinite number."

One important method for finding solutions to underdetermined systems of equations is to reduce the system to a set containing just as many variables as equations, i.e., a *determined* set. This can be accomplished by letting the appropriate number of variables equal zero. For example, the underdetermined system

$$2x_1 + 3x_2 + x_3 = 8$$
$$x_1 + 2x_2 + 2x_3 = 5$$

(1.2)

has three such solutions:

$$x_1 = 0 \qquad x_2 = \tfrac{11}{4} \qquad x_3 = -\tfrac{1}{4}$$
$$x_1 = \tfrac{11}{3} \qquad x_2 = 0 \qquad x_3 = \tfrac{2}{3}$$

and

$$x_1 = 1 \qquad x_2 = 2 \qquad x_3 = 0^{\dagger}$$

Mathematically, linear programming deals with *nonnegative solutions* to underdetermined systems of linear equations. As we shall show in the succeeding chapters, the only solutions we have to be concerned with are those corresponding to determined subsets of equations that have been obtained in the manner described above.[1] If, for example, we let Eqs. (1.2) represent the conditions of a linear-programming problem, we need only to consider the two nonnegative solutions $x_1 = 1$, $x_2 = 2$, $x_3 = 0$ and $x_1 = \tfrac{11}{3}$, $x_2 = 0$, $x_3 = \tfrac{2}{3}$. The remaining solutions fail to satisfy either the nonnegativity requirements or other criteria to be discussed.

Just as the general-programming problem has some objective that guides the selection of the solution to be used, the linear-programming problem has a linear function of the variables to aid in choosing a solution to the problem. This linear combination of the variables, called the *objective function*, must be optimized by the selected solution. If for (1.2) we wished to maximize the objective function $x_1 + x_2 + x_3$, then, of the two nonnegative solutions, the solution $x_1 = \tfrac{11}{3}$, $x_2 = 0$, $x_3 = \tfrac{2}{3}$ is the optimum, as it yields a value of $\tfrac{13}{3}$ for the objective function compared with a value of 3 for the other nonnegative solution. If we wanted to minimize the objective function $x_1 - x_2$, then the solution $x_1 = 1$, $x_2 = 2$, $x_3 = 0$ would be the optimum, with a value of -1. As we have implied, the optimal solution either maximizes or minimizes some linear combination of the variables. Since the maximum of a linear function is equal to minus the minimum of the negative of the linear function, we lose no generality by considering only the minimization problem (see Exercise 15, Chap. 2).

$\dagger$ There are, of course, an infinite number of other solutions to (1.2) which can be obtained by arbitrarily setting one of the variables equal to a constant; e.g., with $x_1 = a$ we have $x_2 = (11 - 3a)/4$ and $x_3 = (-1 + a)/4$.

[1] An efficient, systematic procedure for finding nonnegative solutions to a set of linear equations is the *simplex method* described in Chap. 4.

With the added condition of optimizing an objective function, we are now able to select a single solution that satisfies all the conditions of the problem. There might be multiple solutions in that more than one nonnegative solution to the equations gives the same optimal value of the objective function. Generally speaking, combining the linear constraints of the programming problem with the optimization of a linear objective function transforms an underdetermined system of linear equations that describes a programming problem with many possible solutions to a system that can be solved for a solution that yields the unique optimal value of the objective function.

We next state the general *mathematical model of the linear-programming problem*: Minimize the objective function

$$c_1 x_1 + c_2 x_2 + \cdots + c_j x_j + \cdots + c_n x_n$$

subject to the conditions

$$a_{11} x_1 + a_{12} x_2 + \cdots + a_{1j} x_j + \cdots + a_{1n} x_n = b_1$$
$$a_{21} x_1 + a_{22} x_2 + \cdots + a_{2j} x_j + \cdots + a_{2n} x_n = b_2$$
$$\cdots\cdots\cdots\cdots\cdots\cdots\cdots\cdots\cdots\cdots\cdots\cdots\cdots\cdots$$
$$a_{i1} x_1 + a_{i2} x_2 + \cdots + a_{ij} x_j + \cdots + a_{in} x_n = b_i$$
$$\cdots\cdots\cdots\cdots\cdots\cdots\cdots\cdots\cdots\cdots\cdots\cdots\cdots\cdots$$
$$a_{m1} x_1 + a_{m2} x_2 + \cdots + a_{mj} x_j + \cdots + a_{mn} x_n = b_m$$

and

$$
\begin{aligned}
x_1 &\geq 0 \\
x_2 &\geq 0 \\
&\ \ \vdots \\
x_j &\geq 0 \\
&\ \ \vdots \\
x_n &\geq 0
\end{aligned}
$$

where the c_j for $j = 1, 2, \ldots, n$; b_i for $i = 1, 2, \ldots, m$; and a_{ij} are all constants, and $m < n$. The c_j are called *cost coefficients*.

As will be discussed in the following chapters, every linear-programming problem has either:

1. No solution in terms of nonnegative values of the variables
2. A nonnegative solution that yields an infinite value to the objective function
3. A nonnegative solution that yields a finite value to the objective function

A linear-programming problem that describes a valid, practical programming problem usually has a nonnegative solution with a corresponding finite value of the objective function.

2. EXAMPLES OF LINEAR-PROGRAMMING PROBLEMS

To illustrate the application of the above mathematical description of the linear-programming model, we shall next discuss the linear-programming formulation of three problems. More detailed discussions of these and other problems are given in Part 3.

The transportation problem. A manufacturer wishes to ship a number of units of an item from several warehouses to a number of retail stores. Each store requires a certain number of units of the item, while each warehouse can supply up to a certain amount. Let us define the following:

m = the number of warehouses

n = the number of stores

a_i = the total amount of the item available for shipment at warehouse i

b_j = the total requirement of the item by store j

x_{ij} = the amount of the item shipped from warehouse i to store j

We shall assume that the total amount available is equal to the total required, that is $\sum_i a_i = \sum_j b_j$. As will be shown later, this assumption is not a restrictive one.

The x_{ij} are the unknown shipments to be determined. If we form the array (from $m = 2$ and $n = 3$)

<table>
<tr><td rowspan="2"></td><td colspan="3" align="center">Stores</td><td></td></tr>
<tr><td align="center">1</td><td align="center">2</td><td align="center">3</td></tr>
<tr><td align="center">Warehouses 1</td><td>x_{11}</td><td>x_{12}</td><td>x_{13}</td><td>a_1</td></tr>
<tr><td align="center">2</td><td>x_{21}</td><td>x_{22}</td><td>x_{23}</td><td>a_2</td></tr>
<tr><td></td><td align="center">b_1</td><td align="center">b_2</td><td align="center">b_3</td><td></td></tr>
</table>

we see that the total amount shipped from warehouse 1 can be expressed by the linear equation

$$x_{11} + x_{12} + x_{13} = a_1 \tag{2.1}$$

For warehouse 2, we have

$$x_{21} + x_{22} + x_{23} = a_2 \tag{2.2}$$

We also note that the total amounts shipped to the three stores are expressed by the equations

$$
\begin{aligned}
x_{11} + x_{21} &= b_1 \\
x_{12} + x_{22} &= b_2 \\
x_{13} + x_{23} &= b_3
\end{aligned}
\tag{2.3}
$$

The manufacturer knows the cost c_{ij} of shipping one unit of the item from warehouse i to store j. We have the additional assumption that the cost relationship is linear; i.e., the cost of shipping x_{ij} units is $c_{ij}x_{ij}$.

The manufacturer wishes to determine how many units should be sent from each warehouse to each store so that the total shipping cost is a minimum. This objective of minimizing the cost is achieved by minimizing the linear cost function

$$c_{11}x_{11} + c_{12}x_{12} + c_{13}x_{13} + c_{21}x_{21} + c_{22}x_{22} + c_{23}x_{23} \tag{2.4}$$

Since a negative x_{ij} would represent a shipment from store j to warehouse i, we require that all the variables $x_{ij} \geq 0$.

By combining Eqs. (2.1) to (2.3), the objective function (2.4), and the condition of nonnegativity of the variables, the transportation problem for $m = 2$ and $n = 3$ can be formulated in terms of the following linear-programming problem: Minimize the cost function

$$c_{11}x_{11} + c_{12}x_{12} + c_{13}x_{13} + c_{21}x_{21} + c_{22}x_{22} + c_{23}x_{23}$$

subject to the conditions

$$\begin{aligned}
x_{11} + x_{12} + x_{13} && = a_1 \\
x_{21} + x_{22} + x_{23} &= a_2 \\
x_{11} \qquad\quad + x_{21} \qquad\qquad &= b_1 \\
x_{12} \qquad\quad + x_{22} \qquad &= b_2 \\
x_{13} \qquad\quad + x_{23} &= b_3
\end{aligned} \tag{2.5}$$

and

$$\begin{aligned}
x_{11} &\geq 0 \\
x_{12} &\geq 0 \\
x_{13} &\geq 0 \\
x_{21} &\geq 0 \\
x_{22} &\geq 0 \\
x_{23} &\geq 0
\end{aligned}$$

As a numerical example of the transportation problem, let us consider the two-warehouse, three-store problem with availabilities $a_1 = 5$, $a_2 = 10$ and requirements $b_1 = 8$, $b_2 = 5$, $b_3 = 2$. We note that $\sum_i a_i = \sum_j b_j = 15$. The corresponding costs c_{ij} are given in the table

$$(c_{ij}) = \begin{array}{|c|c|c|} \hline 1 & 2 & 4 \\ \hline 3 & 2 & 1 \\ \hline \end{array}$$

The linear-programming statement of this problem is given by: Minimize

$$x_{11} + 2x_{12} + 4x_{13} + 3x_{21} + 2x_{22} + x_{23}$$

subject to

$$
\begin{aligned}
x_{11} + x_{12} + x_{13} &&&= 5 \\
&x_{21} + x_{22} + x_{23} &&= 10 \\
x_{11} &+ x_{21} &&= 8 \\
x_{12} &+ x_{22} &&= 5 \\
x_{13} &+ x_{23} &&= 2
\end{aligned}
$$

with all $x_{ij} \geq 0$.

We note that the above equations represent a set of accounting statements which keep track of the flow of units between warehouses and stores. The equations of many linear-programming problems are similarly just no more than mathematical statements of the accounting process.

A trial solution is given by the array

$$
(x_{ij}) =
\begin{array}{|c|c|c|c}
\hline
0 & 5 & 0 & 5 \\
\hline
8 & 0 & 2 & 10 \\
\hline
8 & 5 & 2
\end{array}
$$

i.e., $x_{11} = 0$, $x_{12} = 5$, $x_{13} = 0$, $x_{21} = 8$, $x_{22} = 0$, $x_{23} = 2$. The corresponding value of the objective function is

$$
2x_{12} + 3x_{21} + x_{23} = 2 \cdot 5 + 3 \cdot 8 + 1 \cdot 2 = 36
$$

A second solution (out of the many possible) is

$$
(x_{ij}) =
\begin{array}{|c|c|c|c}
\hline
5 & 0 & 0 & 5 \\
\hline
3 & 5 & 2 & 10 \\
\hline
8 & 5 & 2
\end{array}
$$

with a value of the objective of 26. It can be shown by the method of Chap. 10 that this solution is also the minimum solution. (For this example, the reader should be able to demonstrate that any other solution will have a cost greater than 26; see Exercise 9.)

Activity-analysis problem. A manufacturing company has at its disposal fixed amounts of a number of different resources. These resources, such as raw material, labor, and equipment, can be combined to produce any one of several different commodities or combinations of commodities. The company knows how much of resource i it takes to produce one unit of commodity j. It also knows how much profit it makes for each unit of commodity j

produced.   The company desires to produce that combination of commodities which will maximize the total profit.   For this problem, we define the following:

m = the number of resources
n = the number of commodities
a_{ij} = the number of units of resource i required to produce one unit of commodity j
b_i = the total number of units of resource i available
c_j = profit per unit of commodity j produced
x_j = the level of activity (the amount produced) of the jth commodity

The a_{ij} are sometimes called *input-output* or *technological coefficients.*

The amount of the ith resource that is used is given by the linear expression

$$a_{i1}x_1 + a_{i2}x_2 + \cdots + a_{in}x_n$$

Since this amount must be less than or equal to the total number of units of the ith resource available, we then have, for each i, a linear inequality of the form

$$a_{i1}x_1 + a_{i2}x_2 + \cdots + a_{in}x_n \leq b_i$$

As a negative x_j has no appropriate interpretation, we require that all $x_j \geq 0$.   The profit derived from producing x_j units of the jth commodity is given by $c_j x_j$.   Our formulation has yielded the problem of maximizing the profit function

$$c_1x_1 + c_2x_2 + \cdots + c_nx_n$$

subject to the conditions

$$a_{11}x_1 + a_{12}x_2 + \cdots + a_{1n}x_n \leq b_1$$
$$a_{21}x_1 + a_{22}x_2 + \cdots + a_{2n}x_n \leq b_2$$
$$\cdots\cdots\cdots\cdots\cdots\cdots\cdots\cdots\cdots\cdots\cdots$$
$$a_{m1}x_1 + a_{m2}x_2 + \cdots + a_{mn}x_n \leq b_m$$

and

$$
\begin{aligned}
x_1 & & \geq 0 \\
& x_2 & \geq 0 \\
& & \vdots \\
& & x_n \geq 0
\end{aligned}
$$

Since, as will be discussed in Chap. 2, an inequality has an equivalent representation as an equality in nonnegative variables, the above problem is another statement of the general linear-programming problem.

To illustrate the above formulation, we consider the following example from Gass [287]. A furniture manufacturing company plans to make two products—chairs and tables—from its available resources, which consist of 400 board feet of mahogany lumber and 450 labor-hours.   It knows that to make a chair requires 5 board feet and 10 labor-hours and yields a profit of $45, while each table uses 20 board feet and 15 labor-hours and has a profit of $80.   The problem is to determine how many chairs and tables the company can make, keeping within its resource constraints, so that it maximizes its profit.   The activity of making a chair is equivalent to the expenditure of 5 board feet of lumber and 10 labor-hours; for a table the expenditure of 20 board feet of lumber and 15 labor-hours.   Letting x_1 be the level of activity (production) of chairs and x_2 the level of activity of tables, the above activity-analysis problem can be stated in linear-programming terms as that of maximizing the profit function

$$\$45x_1 + \$80x_2$$

subject to

$$5x_1 + 20x_2 \leq 400$$
$$10x_1 + 15x_2 \leq 450$$
$$x_1 \qquad\ \geq\ \ 0$$
$$x_2 \geq\ \ 0$$

There are, of course, many possible solutions to these constraints; for example, the make only chairs solution, $x_1 = 45$, $x_2 = 0$, which has a profit of $\$45 \cdot 45 = \$2{,}025$; and the make only tables solution, $x_1 = 0$, $x_2 = 20$, which has a profit of $\$80 \cdot 20 = \$1{,}600$.   The optimal solution is given by a production schedule which calls for making both chairs and tables, $x_1 = 24$, $x_2 = 14$, profit $= \$2{,}200$.

The diet problem.   Here we are given the nutrient content of a number of different foods.   For example, we might know how many milligrams of phosphorus or iron are contained in 1 oz of each food being considered.   We are also given the recommended daily allowance (RDA) for each nutrient.   Since we know the cost per ounce of food, the problem is to determine the diet that meets the RDAs and is also the minimum cost diet.   Define

m = the number of nutrients
n = the number of foods
a_{ij} = the number of milligrams of the ith nutrient in 1 oz of the jth food
b_i = the recommended number of milligrams of the ith nutrient
c_j = the cost per ounce of the jth food
x_j = the number of ounces of the jth food to be purchased ($x_j \geq 0$)

The total amount of the ith nutrient contained in all the purchased food is given by

$$a_{i1}x_1 + a_{i2}x_2 + \cdots + a_{in}x_n$$

Since this total amount must be greater than or equal to the daily requirement of the ith nutrient, this linear-programming problem involves minimizing the cost function

$$c_1 x_1 + c_2 x_2 + \cdots + c_n x_n$$

subject to the conditions

$$a_{11} x_1 + a_{12} x_2 + \cdots + a_{1n} x_n \geq b_1$$
$$a_{21} x_1 + a_{22} x_2 + \cdots + a_{2n} x_n \geq b_2$$
$$\cdots\cdots\cdots\cdots\cdots\cdots\cdots\cdots\cdots\cdots$$
$$a_{m1} x_1 + a_{m2} x_2 + \cdots + a_{mn} x_n \geq b_m$$

and

$$
\begin{aligned}
x_1 &\geq 0 \\
x_2 &\geq 0 \\
&\ \vdots \\
x_n &\geq 0
\end{aligned}
$$

A simple numerical example of a diet problem is as follows. Let us consider a diet of two foods, x_1 and x_2, and nutrient constraints involving thiamine, phosphorus, and iron. The amount of each nutrient in each of the foods (in milligrams per ounce) is given in the table:

	x_1	x_2
Thiamine	0.15 mg/oz	0.10 mg/oz
Phosphorus	0.75 mg/oz	1.70 mg/oz
Iron	1.30 mg/oz	1.10 mg/oz

For our simple, two-food diet, we want to obtain at least 1.00 mg of thiamine, 7.50 mg of phosphorus, and 10.00 mg of iron. The cost of x_1 is 2 cents/oz, while the cost of x_2 is $\frac{5}{3}$ cents/oz.

Following the general diet problem format above, we then have the corresponding linear program: Minimize

$$2x_1 + \tfrac{5}{3} x_2$$

subject to

$$
\begin{aligned}
0.15 x_1 + 0.10 x_2 &\geq 1.00 \\
0.75 x_1 + 1.70 x_2 &\geq 7.50 \\
1.30 x_1 + 1.10 x_2 &\geq 10.00 \\
x_1 &\geq 0 \\
x_2 &\geq 0
\end{aligned}
\tag{2.6}
$$

The optimum solution to this problem is a mixture of $x_1 = \frac{20}{7}$ oz and $x_2 = \frac{40}{7}$ oz with a cost of \$0.152. The reader will find it instructive to calculate the use only x_1 food solution and the use only x_2 food solution and compare costs.

For the above examples and for all problems cast into the general format of a linear-programming problem, we assume that certain basic relationships hold. The *proportionality* requirement of linear programming is illustrated by the activity-analysis and diet problems in which we assume a change in the level of an activity x_j causes a proportional change in the resources used or nutrients purchased and in the corresponding term of the objective function. Thus, for any $x_j \geq 0$, the associated constraint term is given by the direct relationship $a_{ij}x_j$; similarly, the objective function term is $c_j x_j$. The *additivity* requirement is demonstrated by the activity-analysis problem when we measure the amount of resource i used by all the activities by the sum $(a_{i1}x_1 + a_{i2}x_2 + \cdots + a_{in}x_n)$ and the total profit by the sum $(c_1x_1 + c_2x_2 + \cdots + c_nx_n)$; similarly, for the diet problem, the amount of the ith nutrient contained in all the foods purchased is given by the associated sum of the individual terms $a_{ij}x_j$, and the total cost is given by the sum of the individual cost terms $c_j x_j$. With respect to the possible values of the variables, not only must the *nonnegativity* requirements hold, but we assume that each variable x_j can take on any value allowed by the constraints of the problem. Thus, although we would like the optimal solution to an activity-analysis problem to be in terms of integer values to be produced, we might (and usually) find fractional solutions, e.g., produce $5\frac{2}{3}$ chairs. (We deal, of course, in the domain of rational numbers.) For problems of this sort we may be able to adjust (round) the fractional values without requiring additional resources or causing much of a change in the optimal value. As we shall discuss, there are some linear-programming problems (such as the transportation problem) with mathematical models that guarantee optimal solutions in integers. Also, there are optimization problems that require some or all variables to be integer-valued for which a rounded solution might not be very good or satisfy the constraints, e.g., assigning the minimum number of employees to meet the personnel requirements of overlapping work shifts. Computational procedures, i.e., *algorithms*, exist that handle nonlinear-integer conditions (see Chap. 9); however, they are not as versatile as the standard algorithm for solving linear-programming problems, the *simplex method*. Although one can rightly question the universality of the above requirements and assumptions of the linear-programming model, their judicious use or approximation has led to a remarkably wide range of important and practical real-world applications.

REMARKS

The early applications of linear-programming methods fell into three major categories: military applications generated by the Air Force's Project SCOOP, interindustry economics based on the Leontief input-output model, and problems involving the relationship between zero-sum two-person games and linear programming. These areas of application have been extended and developed, but the emphasis in linear-programming applications has shifted to the industrial area. In addition, linear-programming applications have been developed for a diverse set of problems arising in the social, urban, and policy analysis fields, e.g., education, law enforcement, health, energy planning, environmental quality. A number of applications are discussed in Chap. 11. The Bibliography of applications contains references classified by areas. Details of problem formulation are discussed in Chap. 11; see also Beale [59].

For additional introductory material, the reader is referred to Dorfman [212], Cooper and Charnes [137], Henderson and Schlaifer [365], Directorate of Management Analysis [206], and Wagner [684]. A complete set of linear-programming references to 1958, including applications, is given in Riley and Gass [567]. For a short history on the origins of linear programming, see Dantzig [178]. Linear programming is an important methodological tool in the fields of operations research and management science; see, for example, Wagner [684]. Also see Bland [81] and Koopmans [437].

EXERCISES

1. Write out the complete linear-programming formulation of the following transportation problem and attempt to determine the optimum solution by trial-and-error methods:

A manufacturing organization has distribution centers located at Atlanta, Chicago, and New York City. These centers have available 40, 20, and 40 units of its product, respectively, and its retail outlets require the following number of units: Cleveland, 25; Louisville, 10; Memphis, 20; Pittsburgh, 30; and Richmond, 15. The shipping cost per unit in dollars between each center and outlet is given in the following table:

	1 Cleveland	2 Louisville	3 Memphis	4 Pittsburgh	5 Richmond
1. Atlanta	55	30	40	50	40
2. Chicago	35	30	100	45	60
3. New York	40	60	95	35	30

The optimal solution is the following: (Atlanta to Louisville) $x_{12} = 10$, $x_{13} = 20$, $x_{15} = 10$, $x_{21} = 20$, $x_{31} = 5$, $x_{34} = 30$, $x_{35} = 5$; minimum cost = \$3,600.

2. A furniture manufacturing company wishes to determine how many tables, chairs, desks, or bookcases it should make in order to optimize the use of its available resources. These products utilize two different types of lumber, and the company has on hand 1,500 board feet of the first type and 1,000 board feet of the second. There are 800 labor-hours available for the total job. Sales forecast plus back orders require the company to make at least 40 tables, 130 chairs, 30 desks, and no more than 10 bookcases. Each table, chair, desk, and bookcase requires 5, 1, 9, and 12 board feet, respectively, of the first type of lumber and 2, 3, 4, and 1 board feet of the second type. A table requires 3 labor-hours to make, a chair 2, a desk 5, and a bookcase 10. The company makes a total of \$12 profit on a table, \$5 on a chair, \$15 on a desk, and \$10 on a bookcase. Write out the complete linear-programming formulation of this problem in terms of maximizing the profit.

The optimal solution is to produce 130 tables, 130 chairs, 30 desks, and no bookcases for a profit of \$2,660. For the optimal solution, determine if the manufacturer has any resources that are not used up. A resource that is completely used up represents a bottleneck to increasing production; the corresponding constraint is termed a *bottleneck constraint*.

3. A bakery starts the day with a certain supply of flour, shortening, eggs, sugar, milk, and yeast. It specializes in making bread, cakes, English muffins, and cookies. We wish to determine how much of each product the bakery should make so as to maximize profits. The recipes are given in the following table (we ignore such plentiful supplies as salt, water, etc.).

	Bread	Cake	Muffins	Cookies	Available resources
Flour	12 C	3 C	$\frac{9}{2}$ C	$\frac{3}{2}$ C	500 C
Shortening	2 tbsp	12 tbsp	3 tbsp	4 tbsp	350 tbsp
Eggs	0	3	1	1	120
Sugar	$\frac{1}{4}$ C	$\frac{3}{2}$ C	$\frac{1}{8}$ C	1 C	300 C
Milk	2 C	$\frac{3}{4}$ C	1 C	0	250 C
Yeast	1 cake	0	1 cake	0	100 cakes
Profit	15¢	45¢	33¢	42¢	

Write the corresponding linear-programming formulation and discuss the applicability of such a model to a real-life situation.

4. Describe the model of a transportation problem in which the total availabilities are greater than the total requirements. From the mathematical point of view, why must the sum of the availabilities and the sum of the requirements be equal in the standard transportation model? Discuss the linearity assumption of the transportation-problem cost relationship.

5. Formulate the following diet problem, Gass [287]. A mother wishes her children to obtain certain amounts of nutrients from their breakfast cereals. The children have the choice of eating Krunchies or Crispies or a mixture of the two. From their breakfast they should obtain at least 1 mg of thiamine, 5 mg of niacin, and 400 cal. One ounce of Krunchies contains 0.10 mg of thiamine, 1 mg of niacin, and 110 cal; 1 oz of Crispies contains 0.25 mg of thiamine, 0.25 mg of niacin, and 120 cal. One ounce of Krunchies costs 3.8 cents and 1 oz of Crispies 4.2 cents. Answer: Krunchies $4\frac{4}{9}$ oz; Crispies $2\frac{2}{9}$ oz; minimum cost $26\frac{2}{9}$ cents.

6. For the diet problem in Exercise 5, determine the solutions which use only Krunchies or Crispies. Why would you expect a solution which allowed the children to mix the breakfast foods to be less costly than the single food solutions?

7. Describe a four-ingredient five-nutrient diet problem for the making of dog food. Collect actual dietary data and cost information.

8. For the furniture-manufacturer problem of Sec. 2, express as a linear constraint that the ratio of tables to chairs must be $1:4$. There are a number of linear-programming applications with constraints that are formed by the ratio of two variables or the ratio of the two linear expressions in the variables. Discuss the conditions that are necessary to convert these nonlinear expressions to linear constraints.

9. For the numerical transportation problem of Sec. 2, prove that the optimal solution is the one that has the objective function value of 26. HINT: Any other solution requires shipping at least one unit from either warehouse 1 to store 2 or from warehouse 1 to store 3. Show that any such shipment increases the cost. Try to use this idea to show that your solution to Exercise 1 is optimal, and, if it is not, how to improve your solution.

10. Consider the following problem. The recently elected treasurer of the Junior Class must manage the Class' treasury of \$2,000 so as to have the most money for next year's Senior Class dance. Eight-hundred dollars must be placed in the Class' interest-bearing checking account to take care of expected expenditures for the coming year, leaving \$1,200 for investment. The campus bank offers three types of investment possibilities: a savings passbook account that pays 8 per cent interest, a six-month certificate that yields 12 per cent interest, and a twelve-month, 14 per cent certificate. There are interest penalties if the certificates are cashed in early, but there is no penalty if money is taken out of the passbook account. The bank will allow school organizations to open certificate accounts with no minimum balance requirement. Any funds invested in a six-month certificate would be reinvested for the next six months at the 12 per cent rate, unless the money was required to pay unexpected expenses.

The treasurer recognizes that not all of the money should be placed into certificates as this would require a penalty payment if the $800 in the checking account were not enough to pay for the current year's expenses. In the hope of avoiding penalties and maintaining a "liquid" cash position, the treasurer has decided to impose the following prudent investment condition: Some of the $1,200 should go into the passbook account and it should open with an amount equal to $2 for every dollar in the six-month certificate account and $4 for every dollar in the twelve-month certificate account. Assuming that things go well and no certificates have to be turned in early, how should the treasurer invest the $1,200?

Letting x_1 be the amount of money to be placed in the passbook account, x_2 the amount to be placed in the six-month certificate account, and x_3 the amount to be placed in the twelve-month certificate account, show that the optimal solution to the Junior Class treasurer's problem is $x_1 = 800$ and $x_2 = 400$.   What is the corresponding value of the objective function?

Reformulate and solve the problem with the following prudent investment conditions for the funds placed into the passbook account:

a.   The passbook account should open with an amount equal to $1 for every dollar in the six-month certificate account and $2 for every dollar in the twelve-month certificate account.

b.   Let the dollar values in *a* be $1 and $3, respectively.

c.   The ratio of the money placed in the passbook account to the sum of the monies placed in both certificate accounts must be $4 : 1$.

(HINT: To solve these problems you can substitute for x_1 the expression for x_1 obtained from the prudent investment condition. The substitution is allowed as we know that x_1 will be nonnegative in the final solution.   We then have a single equation in x_2 and x_3, with the objective function in only x_2 and x_3.)

CHAPTER 2

Mathematical Background

A complete development and understanding of the theoretical and computational aspects of linear programming requires the blending of the basic concepts and techniques of a number of mathematical topics. In this chapter we shall present and discuss only those elements of these topics that facilitate the discussion, are required in the succeeding chapters, or will aid the reader in applying the material to be described.

1. MATRICES

A *matrix* is a rectangular array of mn numbers arranged in m rows and n columns as follows:

$$\begin{matrix} a_{11} & a_{12} & \cdots & a_{1n} \\ a_{21} & a_{22} & \cdots & a_{2n} \\ \cdot & \cdot & \cdots & \cdot \\ a_{m1} & a_{m2} & \cdots & a_{mn} \end{matrix}$$

Such an array is usually enclosed in parentheses and is called matrix **A**. The individual a_{ij} are called *elements*. We sometimes denote the matrix **A** by

(a_{ij}). For any m or n and any elements a_{ij} we have

$$\mathbf{A} = \begin{pmatrix} a_{11} & a_{12} & \cdots & a_{1n} \\ a_{21} & a_{22} & \cdots & a_{2n} \\ \cdots\cdots\cdots\cdots\cdots\cdots\cdots \\ a_{m1} & a_{m2} & \cdots & a_{mn} \end{pmatrix} = (a_{ij})$$

Matrix $\mathbf{A}$ is called *square* if $m = n$ and is said to be of *order n*.

A *column vector* is a matrix with only one column, and a *row vector* is a matrix with only one row. A *diagonal matrix* is a square matrix whose off-diagonal elements, a_{ij} for $i \neq j$, are all equal to zero. We let $\mathbf{D}$ denote a diagonal matrix whose nonzero elements can take on any values. The a_{ii} form the *principal diagonal*.

A *unit matrix*, or *identity matrix*, is a diagonal matrix whose diagonal elements are 1. A unit matrix of order n is denoted by $\mathbf{I}_n$ or simply $\mathbf{I}$. For $n = 3$, we have

$$\mathbf{I}_3 = \begin{pmatrix} 1 & 0 & 0 \\ 0 & 1 & 0 \\ 0 & 0 & 1 \end{pmatrix}$$

The *transposed matrix*, $\mathbf{A}'$, of a matrix $\mathbf{A}$ is

$$\mathbf{A}' = \begin{pmatrix} a_{11} & a_{21} & \cdots & a_{m1} \\ a_{12} & a_{22} & \cdots & a_{m2} \\ \cdots\cdots\cdots\cdots\cdots\cdots\cdots \\ a_{1n} & a_{2n} & \cdots & a_{mn} \end{pmatrix}$$

i.e., the rows and columns are interchanged.

Two matrices are equal if and only if their corresponding elements are equal. Hence the corresponding dimensions, m and n, must also be equal.

A square matrix $\mathbf{A}$ is *upper triangular* if all $a_{ij} = 0$ for $i > j$ (all elements below the diagonal are zero) and is *lower triangular* if all $a_{ij} = 0$ for $i < j$ (all elements above the diagonal are zero). The matrix

$$\begin{pmatrix} a_{11} & a_{12} & a_{13} & a_{14} \\ 0 & a_{22} & a_{23} & a_{24} \\ 0 & 0 & a_{33} & a_{34} \\ 0 & 0 & 0 & a_{44} \end{pmatrix}$$

is upper triangular. An upper triangular matrix will be denoted by $\mathbf{U}$ and a lower triangular matrix by $\mathbf{L}$.[1]

A matrix $\mathbf{A}$ is *symmetric* if $\mathbf{A} = \mathbf{A}'$, that is, if $a_{ij} = a_{ji}$. A matrix is *skew-symmetric* if $\mathbf{A} = -\mathbf{A}'$, that is, if $a_{ij} = -a_{ji}$. This last condition implies that all $a_{ij} = 0$ for $i = j$.

---

[1] The reader should note that we have had to resort to using the same notation to express different mathematical quantities. A case in point is the letter $\mathbf{U}$. Besides representing an upper triangular matrix, it is also used to express a vector in Sec. 2 and a special matrix in Chap. 5. The exact meaning will be clear by the context.

A *null matrix* has all its elements equal to zero and is denoted by $\mathbf{0} = (0)$. In the succeeding chapters the matrix $\mathbf{0}$ will be used to denote either a null column vector or a null row vector.

We next define the operations of matrix addition and multiplication.

Given any scalar α (a real number α) and any matrix $\mathbf{A}$, the product $\alpha\mathbf{A}$ is given by

$$\alpha\mathbf{A} = \begin{pmatrix} \alpha a_{11} & \alpha a_{12} & \cdots & \alpha a_{1n} \\ \cdots\cdots\cdots\cdots\cdots\cdots\cdots \\ \alpha a_{m1} & \alpha a_{m2} & \cdots & \alpha a_{mn} \end{pmatrix} = (\alpha a_{ij})$$

Given any two $m \times n$ matrices $\mathbf{A}$ and $\mathbf{B}$, the sum $\mathbf{A} + \mathbf{B} = \mathbf{C}$ is given by

$$\mathbf{C} = \begin{pmatrix} a_{11} & a_{12} & \cdots & a_{1n} \\ \cdots\cdots\cdots\cdots\cdots\cdots \\ a_{m1} & a_{m2} & \cdots & a_{mn} \end{pmatrix} + \begin{pmatrix} b_{11} & b_{12} & \cdots & b_{1n} \\ \cdots\cdots\cdots\cdots\cdots\cdots \\ b_{m1} & b_{m2} & \cdots & b_{mn} \end{pmatrix}$$

$$= \begin{pmatrix} a_{11} + b_{11} & a_{12} + b_{12} & \cdots & a_{1n} + b_{1n} \\ \cdots\cdots\cdots\cdots\cdots\cdots\cdots\cdots\cdots \\ a_{m1} + b_{m1} & a_{m2} + b_{m2} & \cdots & a_{mn} + b_{mn} \end{pmatrix} = (a_{ij} + b_{ij})$$

The properties of scalar multiplication and matrix addition are:

(*a*) $(\mathbf{A} + \mathbf{B}) + \mathbf{C} = \mathbf{A} + (\mathbf{B} + \mathbf{C})$ (Associative law)

(*b*) $\mathbf{A} + \mathbf{B} = \mathbf{B} + \mathbf{A}$ (Commutative law)

(*c*) $(\alpha + \beta)\mathbf{A} = \alpha\mathbf{A} + \beta\mathbf{A}$

       $\alpha(\mathbf{A} + \mathbf{B}) = \alpha\mathbf{A} + \alpha\mathbf{B}$ (Distributive law)

(*d*) $\mathbf{A} + \mathbf{0} = \mathbf{A}$

where $\mathbf{A}$, $\mathbf{B}$, and $\mathbf{C}$ are all of dimensions $m \times n$ and α and β are scalars.

The multiplication of two matrices $\mathbf{A}$ and $\mathbf{B}$ is defined only on the assumption that the number of columns of $\mathbf{A}$ is equal to the number of rows of $\mathbf{B}$. On this assumption, the elements of the product $\mathbf{AB} = \mathbf{C}$ are defined as follows:

The element in the ith row and jth column of matrix $\mathbf{C}$ is equal to the sum of the products of the elements of the ith row of $\mathbf{A}$ multiplied by the corresponding elements of the jth column of $\mathbf{B}$. For example,

$$\mathbf{AB} = \begin{pmatrix} a_{11} & a_{12} & a_{13} \\ a_{21} & a_{22} & a_{23} \end{pmatrix} \begin{pmatrix} b_{11} & b_{12} \\ b_{21} & b_{22} \\ b_{31} & b_{32} \end{pmatrix} = \begin{pmatrix} c_{11} & c_{12} \\ c_{21} & c_{22} \end{pmatrix} = \mathbf{C}$$

where

$$c_{ij} = a_{i1}b_{1j} + a_{i2}b_{2j} + a_{i3}b_{3j}$$

The product of a matrix $\mathbf{A}$ of dimensions $m \times n$ and a matrix $\mathbf{B}$ of dimensions $n \times q$ yields a matrix $\mathbf{C}$ of dimensions $m \times q$. The reader can easily verify that matrix multiplication is not commutative; i.e., in general $\mathbf{AB} \neq \mathbf{BA}$.

Matrix multiplication has the following properties:

(*a*) $(AB)C = A(BC)$ (Associative law)

(*b*) $(A + B)C = AC + BC$

         $C(A + B) = CA + CB$ (Distributive law)

(*c*) $\alpha(AB) = (\alpha A)B = (A)(\alpha B)$

(*d*) $AI = IA = A$

where matrices A, B, C, and I are of the correct dimensions and α is a scalar. We note that $(AB)' = B'A'$.

Associated with every square matrix A is a number called the *determinant* of A. The determinant of A, denoted by $|A|$, is obtained as the sum of all possible products in each of which there appears one and only one element from each row and each column of A, each product of this sum being assigned a plus or minus sign according to the following rule: Let the elements in a given product be joined in pairs by line segments (see example below). If the total number of such segments sloping upward to the right is even, prefix a plus sign to the product; otherwise prefix a negative sign. (Each determinant has $n!$ such products where n is the order of the matrix.[1])

For $n = 3$, we have

$$A = \begin{pmatrix} a_{11} & a_{12} & a_{13} \\ a_{21} & a_{22} & a_{23} \\ a_{31} & a_{32} & a_{33} \end{pmatrix}$$

and

$$|A| = \begin{vmatrix} a_{11} & a_{12} & a_{13} \\ a_{21} & a_{22} & a_{23} \\ a_{31} & a_{32} & a_{33} \end{vmatrix}$$

$$= a_{11}a_{22}a_{33} + a_{12}a_{23}a_{31} + a_{13}a_{21}a_{32} - a_{11}a_{23}a_{32} - a_{12}a_{21}a_{33} - a_{13}a_{22}a_{31}$$

The lines for determining the sign of the third term have been drawn in the array for the determinant.

The following properties of determinants can be verified from the above definition:

1. If every element of a column or a row of a determinant is zero, then the value of the determinant is zero.

---

[1] The factorial notation of $n!$ is defined as $n! = n(n - 1)(n - 2) \cdots 2 \cdot 1$, where $0! = 1$. An alternate definition of $|A|$ is the sum of the $n!$ terms $\sum (\pm)a_{1i}a_{2j}a_{3k} \cdots a_{nr}$, where the second subscripts are a permutation of the numbers 1, 2, 3, ..., n. The even permutations, those that contain an even number of inversions, are given a plus sign and the odd ones a minus sign. The number of inversions in a permutation of n integers is the number of pairs of elements (not necessarily adjacent) in which a larger integer precedes a smaller one.

2. The value of a determinant is not changed if corresponding columns and rows are interchanged, i.e., $|\mathbf{A}| = |\mathbf{A}'|$.

3. If $|\mathbf{B}|$ is the determinant formed by interchanging two columns or rows in $|\mathbf{A}|$, then $|\mathbf{B}| = -|\mathbf{A}|$.

4. If two columns or rows of a determinant are identical, then the determinant has a value of zero.

5. If every element of a column or a row of a determinant is multiplied by a fixed number k, then the value of the determinant is multiplied by k.

6. The value of a determinant is not changed if, to every element of a column or row, we add k times the corresponding element of another column or row.

The *rank* of any matrix $\mathbf{A}$ is the order of the largest square array in $\mathbf{A}$ whose determinant does not vanish, i.e., does not have a value of zero.

A square matrix $\mathbf{A}$ is said to be *nonsingular* if its determinant does not equal zero. If $|\mathbf{A}| = 0$, the matrix is *singular*.

The following *elementary row or column transformations* on a matrix $\mathbf{A}$ preserve the rank of $\mathbf{A}$:

1. Interchange of two rows (columns)

2. Multiplication of a row (column) by any scalar $\alpha \neq 0$

3. Addition to the ith row (column) of α times the jth row (column), where α is any scalar and $i \neq j$

An elementary row transformation can be performed on an $m \times n$ matrix $\mathbf{A}$ by premultiplying $\mathbf{A}$ by an *elementary row transformation matrix* $\mathbf{E}$, where $\mathbf{E}$ is an $m \times m$ matrix and the transformed matrix $\mathbf{B}$ is given by $\mathbf{EA} = \mathbf{B}$. To interchange two rows i and j of $\mathbf{A}$, $\mathbf{E}$ is a unit matrix whose ith and jth rows have been interchanged. To multiply row i of $\mathbf{A}$ by the scalar α, $\mathbf{E}$ is a unit matrix whose ith row has been multiplied by α. To add to the ith row of $\mathbf{A}$ α times the jth row of $\mathbf{A}$, $\mathbf{E}$ is a unit matrix whose ith row has been replaced by the sum of its ith row and α times its jth row. We define *elementary column transformation matrices* in a similar fashion. As examples of 3×3 elementary row transformation matrices as described above, we have, respectively, the following:

$$\begin{pmatrix} 1 & 0 & 0 \\ 0 & 0 & 1 \\ 0 & 1 & 0 \end{pmatrix} \quad \begin{pmatrix} 1 & 0 & 0 \\ 0 & \alpha & 0 \\ 0 & 0 & 1 \end{pmatrix} \quad \begin{pmatrix} 1 & 0 & 0 \\ 0 & 1 & \alpha \\ 0 & 0 & 1 \end{pmatrix}$$

Note that these matrices can also serve as elementary column transformation matrices when postmultiplying a 3×3 matrix.

A matrix $\mathbf{E}^j$ that differs from a unit matrix in only the jth column and $a_{jj} \neq 0$ is termed an *elementary column matrix*; a matrix $\mathbf{E}_i$ that differs from a unit matrix in only the ith row and $a_{ii} \neq 0$ is termed an *elementary row matrix*. For example,

$$\mathbf{E}^2 = \begin{pmatrix} 1 & 2 & 0 \\ 0 & 3 & 0 \\ 0 & -1 & 1 \end{pmatrix} \qquad \mathbf{E}_3 = \begin{pmatrix} 1 & 0 & 0 \\ 0 & 1 & 0 \\ -2 & 3 & 1 \end{pmatrix}$$

A *permutation matrix* $\mathbf{P}$ is a matrix with exactly one 1 in each row and column and zeros elsewhere. The elementary row transformation matrix $\mathbf{E}$ that interchanges two rows is a permutation matrix. If an $m \times n$ matrix $\mathbf{A}$ is premultiplied by an $m \times m$ permutation matrix $\mathbf{P}$, then the product $\mathbf{PA} = \mathbf{B}$ is matrix $\mathbf{A}$ with its rows interchanged according to the locations of the 1's in $\mathbf{P}$. For example, for

$$\mathbf{P} = \begin{pmatrix} 0 & 0 & 1 \\ 1 & 0 & 0 \\ 0 & 1 & 0 \end{pmatrix} \qquad \mathbf{A} = \begin{pmatrix} 1 & 2 \\ 3 & 4 \\ 5 & 6 \end{pmatrix}$$

we have

$$\mathbf{PA} = \mathbf{B} = \begin{pmatrix} 5 & 6 \\ 1 & 2 \\ 3 & 4 \end{pmatrix}$$

Note that $\mathbf{P}$ is just a unit matrix whose rows have been permuted. The postmultiplication of a matrix $\mathbf{A}$ by a $\mathbf{P}$ interchanges the columns of $\mathbf{A}$. For the above $\mathbf{A}$ and

$$\mathbf{P} = \begin{pmatrix} 0 & 1 \\ 1 & 0 \end{pmatrix}$$

we have

$$\mathbf{AP} = \begin{pmatrix} 2 & 1 \\ 4 & 3 \\ 6 & 5 \end{pmatrix}$$

The *minor* $\mathbf{D}_{ij}$ of the element a_{ij} is the determinant obtained from the square matrix $\mathbf{A}$ by striking out the ith row and jth column.

The *cofactor* $\mathbf{A}_{ij}$ of the element a_{ij} equals $(-1)^{i+j} \mathbf{D}_{ij}$. $\mathbf{A}_{ij}$ is referred to as the signed minor of the element a_{ij}.

The *adjoint* of an $n \times n$ matrix $\mathbf{A}$ is another $n \times n$ matrix $\mathbf{J} = (\mathbf{A}_{ji})$ in which the element in the ith row and jth column is the cofactor of the element in the jth row and ith column of $\mathbf{A}$. We have

$$\mathbf{J} = \begin{pmatrix} \mathbf{A}_{11} & \mathbf{A}_{21} & \cdots & \mathbf{A}_{n1} \\ \mathbf{A}_{12} & \mathbf{A}_{22} & \cdots & \mathbf{A}_{n2} \\ \cdots\cdots\cdots\cdots\cdots\cdots \\ \mathbf{A}_{1n} & \mathbf{A}_{2n} & \cdots & \mathbf{A}_{nn} \end{pmatrix}$$

A matrix $\mathbf{B}$ is called the *inverse* of the square matrix $\mathbf{A}$ if $\mathbf{AB} = \mathbf{I}$. The inverse of $\mathbf{A}$ is denoted by $\mathbf{A}^{-1}$. For every nonsingular square matrix $\mathbf{A}$ there exists a unique $\mathbf{A}^{-1}$ such that

$$\mathbf{AA}^{-1} = \mathbf{A}^{-1}\mathbf{A} = \mathbf{I}$$

It can be shown that, if $|\mathbf{A}| \neq 0$,

$$\mathbf{A}^{-1} = \frac{1}{|\mathbf{A}|}\mathbf{J}$$

We see that only nonsingular square matrices have inverses. We also note that $(\mathbf{AB})^{-1} = \mathbf{B}^{-1}\mathbf{A}^{-1}$ and that $(\mathbf{A}')^{-1} = (\mathbf{A}^{-1})'$.

The determinant of a matrix product $\mathbf{AB}$ is the product of the determinants, i.e., $|\mathbf{AB}| = |\mathbf{A}| \cdot |\mathbf{B}|$.

If matrix $\mathbf{A}$ is triangular, then its determinant is the product of the diagonal elements a_{ii}. The inverse of a lower (upper) triangular matrix is a lower (upper) triangular matrix; the product of lower (upper) triangular matrices is a lower (upper) triangular matrix.

We next illustrate some of the above concepts using the following matrices:

$$\mathbf{A} = \begin{pmatrix} 1 & 1 & -1 \\ -2 & 1 & 1 \\ 1 & 1 & 1 \end{pmatrix} \quad \mathbf{B} = \begin{pmatrix} 1 & -1 & 0 \\ 2 & 1 & 1 \\ -1 & 0 & 1 \end{pmatrix} \quad \mathbf{C} = \begin{pmatrix} 1 \\ 2 \\ -1 \end{pmatrix}$$

Then

$$\mathbf{A}' = \begin{pmatrix} 1 & -2 & 1 \\ 1 & 1 & 1 \\ -1 & 1 & 1 \end{pmatrix}$$

$$\alpha\mathbf{A} = \begin{pmatrix} \alpha & \alpha & -\alpha \\ -2\alpha & \alpha & \alpha \\ \alpha & \alpha & \alpha \end{pmatrix}$$

$$\mathbf{A} + \mathbf{B} = \begin{pmatrix} 1 & 1 & -1 \\ -2 & 1 & 1 \\ 1 & 1 & 1 \end{pmatrix} + \begin{pmatrix} 1 & -1 & 0 \\ 2 & 1 & 1 \\ -1 & 0 & 1 \end{pmatrix} = \begin{pmatrix} 2 & 0 & -1 \\ 0 & 2 & 2 \\ 0 & 1 & 2 \end{pmatrix}$$

$$\mathbf{AB} = \begin{pmatrix} 1 & 1 & -1 \\ -2 & 1 & 1 \\ 1 & 1 & 1 \end{pmatrix}\begin{pmatrix} 1 & -1 & 0 \\ 2 & 1 & 1 \\ -1 & 0 & 1 \end{pmatrix} = \begin{pmatrix} 4 & 0 & 0 \\ -1 & 3 & 2 \\ 2 & 0 & 2 \end{pmatrix}$$

and

$$\mathbf{ABC} = \begin{pmatrix} 4 & 0 & 0 \\ -1 & 3 & 2 \\ 2 & 0 & 2 \end{pmatrix}\begin{pmatrix} 1 \\ 2 \\ -1 \end{pmatrix} = \begin{pmatrix} 4 \\ 3 \\ 0 \end{pmatrix}$$

The rank of $\mathbf{A}$ is 3 and $|\mathbf{A}| = 6$. For $\mathbf{A}$ the adjoint matrix $\mathbf{J}$ is [recalling that $\mathbf{A}_{ij}$ is $(-1)^{i+j}\mathbf{D}_{ij}$]

$$\mathbf{J} = \begin{pmatrix} 0 & -2 & 2 \\ 3 & 2 & 1 \\ -3 & 0 & 3 \end{pmatrix}$$

Then $\mathbf{A}^{-1}$ is given by

$$\mathbf{A}^{-1} = \frac{1}{|\mathbf{A}|}\,\mathbf{J} = \frac{1}{6}\begin{pmatrix} 0 & -2 & 2 \\ 3 & 2 & 1 \\ -3 & 0 & 3 \end{pmatrix} = \begin{pmatrix} 0 & -\frac{1}{3} & \frac{1}{3} \\ \frac{1}{2} & \frac{1}{3} & \frac{1}{6} \\ -\frac{1}{2} & 0 & \frac{1}{2} \end{pmatrix}$$

The determination of the inverse of a matrix can also be accomplished by other computational procedures. Two of these, the *Gauss–Jordan complete-elimination method* and the *elimination method of Gauss*, are discussed in Sec. 5 and illustrated for matrix **A** above. The value of $|\mathbf{A}|$ may also be obtained in terms of lower-order cofactors by expansion along the ith row or jth column by the following relationships, respectively:

$$|\mathbf{A}| = \sum_{j=1}^{n} a_{ij}\mathbf{A}_{ij} \qquad \text{for any } i$$

$$|\mathbf{A}| = \sum_{i=1}^{n} a_{ij}\mathbf{A}_{ij} \qquad \text{for any } j$$

A matrix **A** may be partitioned by grouping successive rows and columns into a number of submatrices. For example, the matrix

$$\mathbf{A} = \begin{pmatrix} a_{11} & a_{12} & a_{13} & a_{14} \\ a_{21} & a_{22} & a_{23} & a_{24} \\ a_{31} & a_{32} & a_{33} & a_{34} \end{pmatrix}$$

can be partitioned into

$$\mathbf{A} = \left(\begin{array}{c|ccc} a_{11} & a_{12} & a_{13} & a_{14} \\ \hline a_{21} & a_{22} & a_{23} & a_{24} \\ a_{31} & a_{32} & a_{33} & a_{34} \end{array}\right)$$

or

$$\mathbf{A} = \left(\begin{array}{cc|cc} a_{11} & a_{12} & a_{13} & a_{14} \\ \hline a_{21} & a_{22} & a_{23} & a_{24} \\ a_{31} & a_{32} & a_{33} & a_{34} \end{array}\right)$$

or other appropriate groupings. We can denote each submatrix by $\mathbf{A}_{ij}$, where i and j refer to the partition row and column number, respectively. For example, either of the two illustrations above can be represented by

$$\mathbf{A} = \begin{pmatrix} \mathbf{A}_{11} & \mathbf{A}_{12} \\ \mathbf{A}_{21} & \mathbf{A}_{22} \end{pmatrix}$$

where the dimensions of each $\mathbf{A}_{ij}$ are according to the corresponding partition. The rule for addition of partitioned matrices is the same as the rule for addition of matrices if the corresponding submatrices have the same dimensions; similarly for multiplication of partitioned matrices. For example, assuming the dimensions are correct, we have for

$$\mathbf{A} = \begin{pmatrix} \mathbf{A}_{11} & \mathbf{A}_{12} \\ \mathbf{A}_{21} & \mathbf{A}_{22} \end{pmatrix} \quad \text{and} \quad \mathbf{B} = \begin{pmatrix} \mathbf{B}_{11} & \mathbf{B}_{12} \\ \mathbf{B}_{21} & \mathbf{B}_{22} \end{pmatrix}$$

$$\mathbf{A} + \mathbf{B} = \begin{pmatrix} \mathbf{A}_{11} + \mathbf{B}_{11} & \mathbf{A}_{12} + \mathbf{B}_{12} \\ \mathbf{A}_{21} + \mathbf{B}_{21} & \mathbf{A}_{22} + \mathbf{B}_{22} \end{pmatrix}$$

and

$$\mathbf{AB} = \begin{pmatrix} \mathbf{A}_{11}\mathbf{B}_{11} + \mathbf{A}_{12}\mathbf{B}_{21} & \mathbf{A}_{11}\mathbf{B}_{12} + \mathbf{A}_{12}\mathbf{B}_{22} \\ \mathbf{A}_{21}\mathbf{B}_{11} + \mathbf{A}_{22}\mathbf{B}_{21} & \mathbf{A}_{21}\mathbf{B}_{12} + \mathbf{A}_{22}\mathbf{B}_{22} \end{pmatrix} = \begin{pmatrix} \mathbf{C}_{11} & \mathbf{C}_{12} \\ \mathbf{C}_{21} & \mathbf{C}_{22} \end{pmatrix}$$

Let an $n \times n$ nonsingular matrix $\mathbf{A}$ be partitioned as follows:

$$\mathbf{A} = \begin{pmatrix} \mathbf{K} & \mathbf{L} \\ \mathbf{M} & \mathbf{N} \end{pmatrix}$$

where $\mathbf{K}$ is a $k \times k$ submatrix, $\mathbf{L}$ is $k \times l$, $\mathbf{M}$ is $l \times k$, and $\mathbf{N}$ is $l \times l$, with $n = k + l$. Assume $\mathbf{N}^{-1}$ exists. Then $\mathbf{A}^{-1}$ is given by the similarly dimensioned partitioned matrix

$$\mathbf{A}^{-1} = \begin{pmatrix} \mathbf{P} & \mathbf{Q} \\ \mathbf{R} & \mathbf{S} \end{pmatrix} = \begin{pmatrix} (\mathbf{K} - \mathbf{L}\mathbf{N}^{-1}\mathbf{M})^{-1} & -\mathbf{P}\mathbf{L}\mathbf{N}^{-1} \\ -\mathbf{N}^{-1}\mathbf{M}\mathbf{P} & \mathbf{N}^{-1} - \mathbf{N}^{-1}\mathbf{M}\mathbf{Q} \end{pmatrix}$$

Given an $n \times n$ nonsingular matrix $\mathbf{A}$, then $\mathbf{A}$ can be expressed as the product $\mathbf{A} = \mathbf{LU}$, where $\mathbf{L}$ is a lower triangular matrix and $\mathbf{U}$ is an upper triangular matrix. This is termed the $\mathbf{LU}$ *decomposition of* $\mathbf{A}$. Conditions on $\mathbf{A}$ that must hold are that its upper-left (leading) principal submatrices (i.e., those formed down along the principal diagonal) must be nonsingular (see Exercise 25). Any nonsingular matrix can be made to satisfy these conditions by permutations of its rows and columns. If we let the diagonal elements of $\mathbf{U}$ (or $\mathbf{L}$) be all equal to 1, then the $\mathbf{LU}$ decomposition will be unique (see Exercise 24). For example, we have

$$\mathbf{A} = \begin{pmatrix} 1 & 1 & -1 \\ -2 & 1 & 1 \\ 1 & 1 & 1 \end{pmatrix} = \begin{pmatrix} 1 & 0 & 0 \\ -2 & 3 & 0 \\ 1 & 0 & 2 \end{pmatrix} \begin{pmatrix} 1 & 1 & -1 \\ 0 & 1 & -\frac{1}{3} \\ 0 & 0 & 1 \end{pmatrix} = \mathbf{LU} \tag{1.1}$$

A procedure for finding $\mathbf{L}$ and $\mathbf{U}$ is given in Sec. 5. A lower (upper) triangular matrix can be factored into the product of n lower (upper) triangular matrices. We illustrate such a factorization for a general 3×3 matrix $\mathbf{A}$ and then for the numerical $\mathbf{L}$ and $\mathbf{U}$ of (1.1). Let $\mathbf{A} = \mathbf{LU}$ with

$$\mathbf{L} = \begin{pmatrix} l_{11} & 0 & 0 \\ l_{21} & l_{22} & 0 \\ l_{31} & l_{32} & l_{33} \end{pmatrix} \quad \text{and} \quad \mathbf{U} = \begin{pmatrix} u_{11} & u_{12} & u_{13} \\ 0 & u_{22} & u_{23} \\ 0 & 0 & u_{33} \end{pmatrix} \tag{1.2}$$

$\mathbf{L}$ ($\mathbf{U}$) can be factored readily into the product of three lower (upper) triangular matrices $\mathbf{L}_k$ ($\mathbf{U}_k$) as follows:

$$\mathbf{L} = \begin{pmatrix} l_{11} & 0 & 0 \\ l_{21} & 1 & 0 \\ l_{31} & 0 & 1 \end{pmatrix}\begin{pmatrix} 1 & 0 & 0 \\ 0 & l_{22} & 0 \\ 0 & l_{32} & 1 \end{pmatrix}\begin{pmatrix} 1 & 0 & 0 \\ 0 & 1 & 0 \\ 0 & 0 & l_{33} \end{pmatrix} = \begin{pmatrix} l_{11} & 0 & 0 \\ l_{21} & l_{22} & 0 \\ l_{31} & l_{32} & l_{33} \end{pmatrix}$$
$$\qquad\quad \mathbf{L}_1 \qquad\qquad \mathbf{L}_2 \qquad\qquad \mathbf{L}_3$$

and

$$\mathbf{U} = \begin{pmatrix} 1 & 0 & u_{13} \\ 0 & 1 & u_{23} \\ 0 & 0 & u_{33} \end{pmatrix}\begin{pmatrix} 1 & u_{12} & 0 \\ 0 & u_{22} & 0 \\ 0 & 0 & 1 \end{pmatrix}\begin{pmatrix} u_{11} & 0 & 0 \\ 0 & 1 & 0 \\ 0 & 0 & 1 \end{pmatrix} = \begin{pmatrix} u_{11} & u_{12} & u_{13} \\ 0 & u_{22} & u_{23} \\ 0 & 0 & u_{33} \end{pmatrix}$$
$$\qquad\quad \mathbf{U}_3 \qquad\qquad \mathbf{U}_2 \qquad\qquad \mathbf{U}_1$$

Thus,

$$\mathbf{A} = \mathbf{LU} = \mathbf{L}_1\mathbf{L}_2\mathbf{L}_3\mathbf{U}_3\mathbf{U}_2\mathbf{U}_1$$

For the numerical $\mathbf{L}$ and $\mathbf{U}$ of (1.1), we have

$$\mathbf{L} = \begin{pmatrix} 1 & 0 & 0 \\ -2 & 1 & 0 \\ 1 & 0 & 1 \end{pmatrix}\begin{pmatrix} 1 & 0 & 0 \\ 0 & 3 & 0 \\ 0 & 0 & 1 \end{pmatrix}\begin{pmatrix} 1 & 0 & 0 \\ 0 & 1 & 0 \\ 0 & 0 & 2 \end{pmatrix} = \mathbf{L}_1\mathbf{L}_2\mathbf{L}_3$$

$$\mathbf{U} = \begin{pmatrix} 1 & 0 & -1 \\ 0 & 1 & -\frac{1}{3} \\ 0 & 0 & 1 \end{pmatrix}\begin{pmatrix} 1 & 1 & 0 \\ 0 & 1 & 0 \\ 0 & 0 & 1 \end{pmatrix}\begin{pmatrix} 1 & 0 & 0 \\ 0 & 1 & 0 \\ 0 & 0 & 1 \end{pmatrix} = \mathbf{U}_3\mathbf{U}_2\mathbf{U}_1$$

For a general $n \times n$ nonsingular matrix $\mathbf{A}$ [assuming that any necessary rearrangement (permutation) of its rows and columns has been accomplished], we have

$$\mathbf{A} = \mathbf{LU} = \mathbf{L}_1\mathbf{L}_2\cdots\mathbf{L}_n\mathbf{U}_n\mathbf{U}_{n-1}\cdots\mathbf{U}_1$$

where

$$\mathbf{L}_k = \begin{pmatrix} 1 & & & & \\ & \ddots & & & \\ & & l_{kk} & & \\ & & \vdots & 1 & \\ & & \vdots & & \ddots \\ & & l_{nk} & & & 1 \end{pmatrix} \quad \text{and} \quad \mathbf{U}_k = \begin{pmatrix} 1 & & u_{1k} & & \\ & \ddots & \vdots & & \\ & & u_{kk} & & \\ & & & 1 & \\ & & & & \ddots \\ & & & & & 1 \end{pmatrix}$$

$$(1.3)$$

We next determine the inverse of a matrix decomposed in terms of $\mathbf{L}$ and $\mathbf{U}$. Here, $\mathbf{A} = \mathbf{LU}$ and $\mathbf{A}^{-1} = (\mathbf{LU})^{-1} = \mathbf{U}^{-1}\mathbf{L}^{-1}$. If $\mathbf{A}$ is a 3×3 matrix, then

$$\mathbf{A} = \mathbf{LU} = \mathbf{L}_1\mathbf{L}_2\mathbf{L}_3\mathbf{U}_3\mathbf{U}_2\mathbf{U}_1 \qquad \text{and} \qquad \mathbf{A}^{-1} = (\mathbf{L}_1\mathbf{L}_2\mathbf{L}_3\mathbf{U}_3\mathbf{U}_2\mathbf{U}_1)^{-1}$$

or

$$\mathbf{A}^{-1} = \mathbf{U}_1^{-1}\mathbf{U}_2^{-1}\mathbf{U}_3^{-1}\mathbf{L}_3^{-1}\mathbf{L}_2^{-1}\mathbf{L}_1^{-1}$$

For a general $n \times n$ matrix $\mathbf{A}$, we have

$$\mathbf{A}^{-1} = \mathbf{U}^{-1}\mathbf{L}^{-1} = \mathbf{U}_1^{-1}\mathbf{U}_2^{-1} \cdots \mathbf{U}_n^{-1}\mathbf{L}_n^{-1}\mathbf{L}_{n-1}^{-1} \cdots \mathbf{L}_1^{-1}$$

Since each $\mathbf{L}_k$ and $\mathbf{U}_k$ is triangular, it is rather easy to determine $\mathbf{L}_k^{-1}$ and $\mathbf{U}_k^{-1}$. We have

$$\mathbf{L}_k^{-1} = \begin{pmatrix} 1 & & & & & & \\ & \ddots & & & & & \\ & & 1 & & & & \\ & & \frac{1}{l_{kk}} & 1 & & & \\ & & \vdots & & \ddots & & \\ & & \frac{-l_{ik}}{l_{kk}} & & & \ddots & \\ & & \vdots & & & & \\ & & \frac{-l_{nk}}{l_{kk}} & & & & 1 \end{pmatrix} \tag{1.4}$$

and

$$\mathbf{U}_k^{-1} = \begin{pmatrix} 1 & & & \frac{-u_{1k}}{u_{kk}} & & & \\ & \ddots & & \vdots & & & \\ & & & \frac{-u_{ik}}{u_{kk}} & & & \\ & & 1 & \vdots & & & \\ & & & \frac{1}{u_{kk}} & & & \\ & & & & 1 & & \\ & & & & & \ddots & \\ & & & & & & 1 \end{pmatrix} \tag{1.5}$$

For the general 3×3 system $\mathbf{A} = \mathbf{LU} = \mathbf{L}_1\mathbf{L}_2\mathbf{L}_3\mathbf{U}_3\mathbf{U}_2\mathbf{U}_1$ of (1.2), we have

$$\mathbf{A}^{-1} = \mathbf{U}^{-1}\mathbf{L}^{-1} = \begin{pmatrix} \frac{1}{u_{11}} & 0 & 0 \\ 0 & 1 & 0 \\ 0 & 0 & 1 \end{pmatrix}\begin{pmatrix} 1 & \frac{-u_{12}}{u_{22}} & 0 \\ 0 & \frac{1}{u_{22}} & 0 \\ 0 & 0 & 1 \end{pmatrix}\begin{pmatrix} 1 & 0 & \frac{-u_{13}}{u_{33}} \\ 0 & 1 & \frac{-u_{23}}{u_{33}} \\ 0 & 0 & \frac{1}{u_{33}} \end{pmatrix}\begin{pmatrix} 1 & 0 & 0 \\ 0 & 1 & 0 \\ 0 & 0 & \frac{1}{l_{33}} \end{pmatrix}\begin{pmatrix} 1 & 0 & 0 \\ 0 & \frac{1}{l_{22}} & 0 \\ 0 & \frac{-l_{32}}{l_{22}} & 1 \end{pmatrix}\begin{pmatrix} \frac{1}{l_{11}} & 0 & 0 \\ \frac{-l_{21}}{l_{11}} & 1 & 0 \\ \frac{-l_{31}}{l_{11}} & 0 & 1 \end{pmatrix}$$

$$= \mathbf{U}_1^{-1}\mathbf{U}_2^{-1}\mathbf{U}_3^{-1}\mathbf{L}_3^{-1}\mathbf{L}_2^{-1}\mathbf{L}_1^{-1}$$

For the 3×3 numerical example of (1.1), $\mathbf{A}^{-1}$ is given by the following product:

$$\mathbf{A}^{-1} = \begin{pmatrix} 1 & 0 & 0 \\ 0 & 1 & 0 \\ 0 & 0 & 1 \end{pmatrix}\begin{pmatrix} 1 & -1 & 0 \\ 0 & 1 & 0 \\ 0 & 0 & 1 \end{pmatrix}\begin{pmatrix} 1 & 0 & 1 \\ 0 & 1 & \frac{1}{3} \\ 0 & 0 & 1 \end{pmatrix}\begin{pmatrix} 1 & 0 & 0 \\ 0 & 1 & 0 \\ 0 & 0 & \frac{1}{2} \end{pmatrix}\begin{pmatrix} 1 & 0 & 0 \\ 0 & \frac{1}{3} & 0 \\ 0 & 0 & 1 \end{pmatrix}\begin{pmatrix} 1 & 0 & 0 \\ 2 & 1 & 0 \\ -1 & 0 & 1 \end{pmatrix}$$

$$= \mathbf{U}_1^{-1}\mathbf{U}_2^{-1}\mathbf{U}_3^{-1}\mathbf{L}_3^{-1}\mathbf{L}_2^{-1}\mathbf{L}_1^{-1} = \begin{pmatrix} 0 & -\frac{1}{3} & \frac{1}{3} \\ \frac{1}{2} & \frac{1}{3} & \frac{1}{6} \\ -\frac{1}{2} & 0 & \frac{1}{2} \end{pmatrix}$$

with

$$\mathbf{U}^{-1} = \begin{pmatrix} 1 & -1 & \frac{2}{3} \\ 0 & 1 & \frac{1}{3} \\ 0 & 0 & 1 \end{pmatrix} \quad \text{and} \quad \mathbf{L}^{-1} = \begin{pmatrix} 1 & 0 & 0 \\ \frac{2}{3} & \frac{1}{3} & 0 \\ -\frac{1}{2} & 0 & \frac{1}{2} \end{pmatrix} \tag{1.6}$$

2. VECTORS AND VECTOR SPACES

In the familiar two-dimensional Euclidean plane, points are represented by ordered pairs of numbers $\mathbf{U} = (u_1, u_2)$. $\mathbf{U}$ can be regarded as a point with coordinates (u_1, u_2) relative to a fixed origin $\mathbf{0} = (0, 0)$ or as a *vector*, i.e., a translation of the origin from $(0, 0)$ by the amounts u_1 and u_2 in two fixed coordinate directions (Fig. 2.1). These two notions are used interchangeably. We shall sometimes write $\mathbf{U} = \begin{pmatrix} u_1 \\ u_2 \end{pmatrix}$.

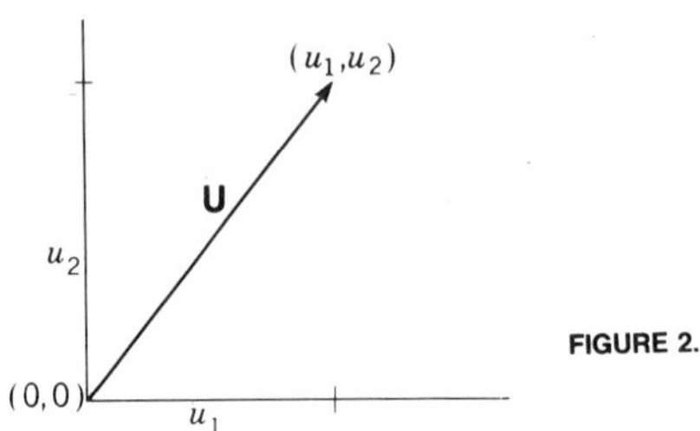

FIGURE 2.1

Let us call the two-dimensional Euclidean space $\mathbf{E}_2$ and note the important properties of vectors in $\mathbf{E}_2$.

a. Multiplication of vectors by scalars (Fig. 2.2). To every pair α and $\mathbf{U}$,

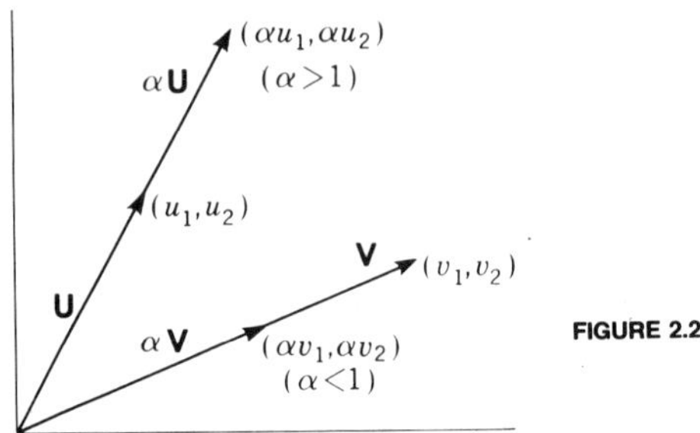

FIGURE 2.2

where α is a scalar and $\mathbf{U}$ is a vector, there corresponds a vector called the *scalar product* of α and $\mathbf{U}$, written $\alpha\mathbf{U} = (\alpha u_1, \alpha u_2)$, and such that

(a) $\qquad \alpha(\beta\mathbf{U}) = (\alpha\beta)\mathbf{U}$ $\qquad$ (Associative law)

(b) $\qquad \alpha(\mathbf{U} + \mathbf{V}) = \alpha\mathbf{U} + \alpha\mathbf{V}$ $\qquad$ (Distributive law)

$\qquad\qquad (\alpha + \beta)\mathbf{U} = \alpha\mathbf{U} + \beta\mathbf{U}$

(c) $\qquad\qquad 1\mathbf{U} = \mathbf{U}$

(d) $\qquad\qquad 0\mathbf{U} = \mathbf{0} = (0, 0)$

b. Addition of vectors (Fig. 2.3). To every pair $U = (u_1, u_2)$ and $V = (v_1, v_2)$ of vectors in E_2 there corresponds a vector called the *sum of* U and V, written $U + V = (u_1 + v_1, u_2 + v_2)$ and such that

(a)　　　　$U + V = V + U$　　　　(Commutative law)

(b)　　$(U + V) + W = U + (V + W)$　　(Associative law)

There exists in E_2 a unique vector 0, called the *origin*, such that

(c)　　$U + 0 = U$　　for all U in E_2

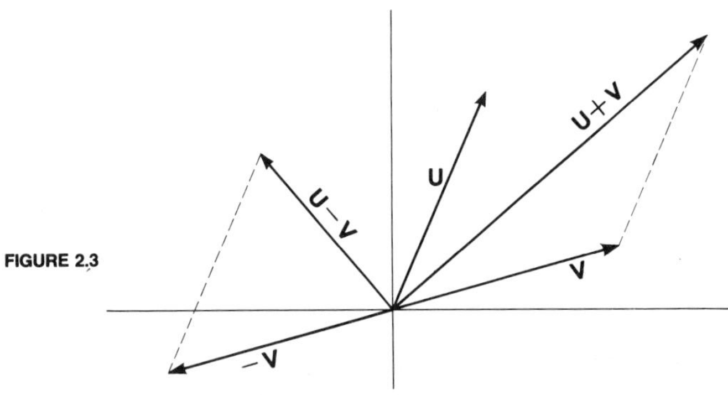

FIGURE 2.3

To each U in E_2 there corresponds a unique vector, called the *inverse* of U and denoted by $-U$, such that

(d)　　$U + (-U) = 0$

c. Inner product of vectors. To every pair of vectors U and V in E_2 there corresponds a real number called the *inner product* of U and V, written $U \cdot V = u_1 v_1 + u_2 v_2$ and such that

(a)　　　　$U \cdot V = V \cdot U$　　　　(Commutative law)

(b)　　$(\alpha U + \beta V) \cdot W = \alpha(U \cdot W) + \beta(V \cdot W)$

for all scalars α and β and all vectors U, V, W in E_2.

(c)　　$U \cdot U \geq 0$　　and　　$U \cdot U = 0$　　if and only if $U = 0$

Two vectors U and V are said to be *orthogonal* to one another if $U \cdot V = 0$.

d. Length of a vector (Fig. 2.4). To every vector U in E_2 there corresponds a real number, called the *length* of U, written

$$\|U\| = +\sqrt{u_1^2 + u_2^2}$$

and such that

(a)　　$\|U\| \geq 0$　　and　　$\|U\| = 0$　　if and only if $U = 0$

(b)　　$\|\alpha U\| = |\alpha| \cdot \|U\|$

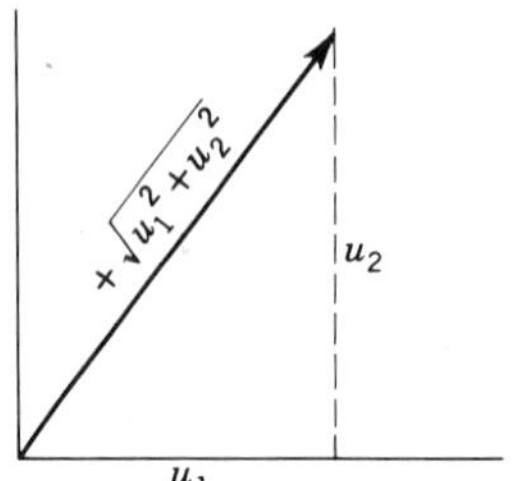

FIGURE 2.4

The length satisfies the triangle inequality

(c) $\qquad \|\mathbf{U} + \mathbf{V}\| \leq \|\mathbf{U}\| + \|\mathbf{V}\|$

for every $\mathbf{U}$ and $\mathbf{V}$ in $\mathbf{E}_2$.

(d) $\qquad \|\mathbf{U}\| = +\sqrt{\mathbf{U} \cdot \mathbf{U}}$

The number $\|\mathbf{U}\|$ is often called the *norm* of the vector $\mathbf{U}$.

e. Distance between two vectors. To every pair of vectors $\mathbf{U}$ and $\mathbf{V}$ in $\mathbf{E}_2$ there corresponds a real number, called the *distance between* $\mathbf{U}$ and $\mathbf{V}$, written $d(\mathbf{U}, \mathbf{V})$, such that

$$d(\mathbf{U}, \mathbf{V}) = \|\mathbf{U} - \mathbf{V}\| = +\sqrt{(u_1 - v_1)^2 + (u_2 - v_2)^2}$$

Given k vectors $\mathbf{U}_1, \mathbf{U}_2, \ldots, \mathbf{U}_k$, the vector

$$\mathbf{U} = \alpha_1 \mathbf{U}_1 + \alpha_2 \mathbf{U}_2 + \cdots + \alpha_k \mathbf{U}_k$$

is a *linear combination* of $\mathbf{U}_1, \mathbf{U}_2, \ldots, \mathbf{U}_k$ for any scalars α_i, $i = 1, \ldots, k$.

If $\mathbf{U}_1$ and $\mathbf{U}_2$ are different points, the set of points

$$\mathbf{U} = (1 - \lambda)\mathbf{U}_1 + \lambda \mathbf{U}_2 = \mathbf{U}_1 + \lambda(\mathbf{U}_2 - \mathbf{U}_1)$$

for all λ is called the *line through $\mathbf{U}_1$ and $\mathbf{U}_2$*. For $0 \leq \lambda \leq 1$, $\mathbf{U}$ defines the *line segment* joining $\mathbf{U}_1$ and $\mathbf{U}_2$.

Note that the above concepts and definitions, and those given below, generalize to higher dimensional spaces.

A set of vectors $\mathbf{U}_1, \mathbf{U}_2, \ldots, \mathbf{U}_k$ is called *linearly independent* if, for all numbers $\alpha_1, \alpha_2, \ldots, \alpha_k$,

$$\alpha_1 \mathbf{U}_1 + \alpha_2 \mathbf{U}_2 + \cdots + \alpha_k \mathbf{U}_k = \mathbf{0}$$

implies

$$\alpha_1 = \alpha_2 = \cdots = \alpha_k = 0$$

Otherwise the set is *linearly dependent*. For example, the set of two vectors

$$\mathbf{U}_1 = \begin{pmatrix} 1 \\ 0 \end{pmatrix} \qquad \text{and} \qquad \mathbf{U}_2 = \begin{pmatrix} 1 \\ 1 \end{pmatrix}$$

is linearly independent. To show this, we write

$$\alpha_1 \mathbf{U}_1 + \alpha_2 \mathbf{U}_2 = \alpha_1 \begin{pmatrix} 1 \\ 0 \end{pmatrix} + \alpha_2 \begin{pmatrix} 1 \\ 1 \end{pmatrix} = \begin{pmatrix} 0 \\ 0 \end{pmatrix}$$

or

$$\begin{pmatrix} \alpha_1 \\ 0 \end{pmatrix} + \begin{pmatrix} \alpha_2 \\ \alpha_2 \end{pmatrix} = \begin{pmatrix} 0 \\ 0 \end{pmatrix}$$

$$\begin{pmatrix} \alpha_1 + \alpha_2 \\ \alpha_2 \end{pmatrix} = \begin{pmatrix} 0 \\ 0 \end{pmatrix}$$

Since in the last equation the corresponding elements in the two column vectors must be equal, we have $\alpha_2 = 0$ and hence $\alpha_1 = 0$. In a similar fashion we can demonstrate that the set of vectors

$$\mathbf{U}_1 = \begin{pmatrix} 1 \\ 0 \end{pmatrix} \qquad \mathbf{U}_2 = \begin{pmatrix} 0 \\ 1 \end{pmatrix} \qquad \text{and} \qquad \mathbf{U}_3 = \begin{pmatrix} 1 \\ 1 \end{pmatrix}$$

is linearly dependent. Here we have

$$\alpha_1 \begin{pmatrix} 1 \\ 0 \end{pmatrix} + \alpha_2 \begin{pmatrix} 0 \\ 1 \end{pmatrix} + \alpha_3 \begin{pmatrix} 1 \\ 1 \end{pmatrix} = \begin{pmatrix} 0 \\ 0 \end{pmatrix} \tag{2.1}$$

or

$$\begin{pmatrix} \alpha_1 + \alpha_3 \\ \alpha_2 + \alpha_3 \end{pmatrix} = \begin{pmatrix} 0 \\ 0 \end{pmatrix}$$

We are then led to the conclusions that $\alpha_3 = -\alpha_1$ and $\alpha_3 = -\alpha_2$. Hence for $\alpha_1 = \alpha_2 = \alpha$ and $\alpha_3 = -\alpha$, where α can be *any* number, Eq. (2.1) holds.

This last example illustrates the property of $\mathbf{E}_2$ which makes it two-dimensional: that there exist two linearly independent vectors [e.g., the unit vectors $(1, 0)$ and $(0, 1)$], whereas every set of three vectors is linearly dependent. Generalizing this notion of dimensionality, we have the following definition:

An *n-dimensional Euclidean space* $\mathbf{E}_n$ is a set of objects called vectors which satisfy properties described above under subheads *a* to *c*, with the property that there exist n linearly independent vectors while every set of $n + 1$ vectors is linearly dependent.

In $\mathbf{E}_n$ we have $\mathbf{U} = (u_1, u_2, \ldots, u_n)$ and

$$\alpha(u_1, u_2, \ldots, u_n) = (\alpha u_1, \alpha u_2, \ldots, \alpha u_n)$$

$$(u_1, u_2, \ldots, u_n) + (v_1, v_2, \ldots, v_n) = (u_1 + v_1, u_2 + v_2, \ldots, u_n + v_n)$$

$$(u_1, u_2, \ldots, u_n) \cdot (v_1, v_2, \ldots, v_n) = u_1 v_1 + u_2 v_2 + \cdots + u_n v_n$$

A *basis* for $\mathbf{E}_n$ is a set of n linearly independent vectors. Any vector in $\mathbf{E}_n$ can be expressed uniquely as a linear combination of the vectors of a given basis. The set of n unit vectors $(1, 0, \ldots, 0)$, $(0, 1, \ldots, 0)$, $(0, 0, \ldots, 1)$ is a basis for $\mathbf{E}_n$.

If a set of vectors is linearly independent, then any subset of these vectors is also linearly independent. Similarly, if any set of vectors is linearly dependent, any larger set of vectors containing this set of vectors is also linearly dependent.

One linear condition in $\mathbf{E}_2$ (such as $a_1 u_1 + a_2 u_2 = b$, where a_1, a_2, and b are constants) defines a line; one linear condition in $\mathbf{E}_3$ defines a plane; the corresponding object defined by one linear condition in $\mathbf{E}_n$ is called a *hyperplane*. A hyperplane $\mathbf{H}(\mathbf{A}, b)$ in $\mathbf{E}_n$ is the set of all vectors $\mathbf{U}$ such that $\mathbf{A} \cdot \mathbf{U} = b$ for a given vector $\mathbf{A} \neq \mathbf{0}$ and a real number b.

A hyperplane divides $\mathbf{E}_n$ into two half spaces which we denote by

$$\mathbf{H}^+(\mathbf{A}, b) = (\mathbf{U} \mid \mathbf{A} \cdot \mathbf{U} \geq b) \qquad \mathbf{H}^-(\mathbf{A}, b) = (\mathbf{U} \mid \mathbf{A} \cdot \mathbf{U} \leq b)$$

$\mathbf{H}^+(\mathbf{A}, b)$ is the half space, i.e., that portion of $\mathbf{E}_n$ that contains the vectors $\mathbf{U}$ for which $\mathbf{A} \cdot \mathbf{U} \geq b$. $\mathbf{H}^-(\mathbf{A}, b)$ is the half space that contains the vectors $\mathbf{U}$ for which $\mathbf{A} \cdot \mathbf{U} \leq b$.

For $\mathbf{A} = (2, -1)$ and $b = 1$, we have the inner product

$$\mathbf{A} \cdot \mathbf{U} = 2u_1 - u_2$$

with the two half spaces

$$\mathbf{H}^+(\mathbf{A}, b) = 2u_1 - u_2 \geq 1 \qquad \text{and} \qquad \mathbf{H}^-(\mathbf{A}, b) = 2u_1 - u_2 \leq 1$$

as depicted in Fig. 2.5. The points on the line $2u_1 - u_2 = 1$ belong to both half spaces.

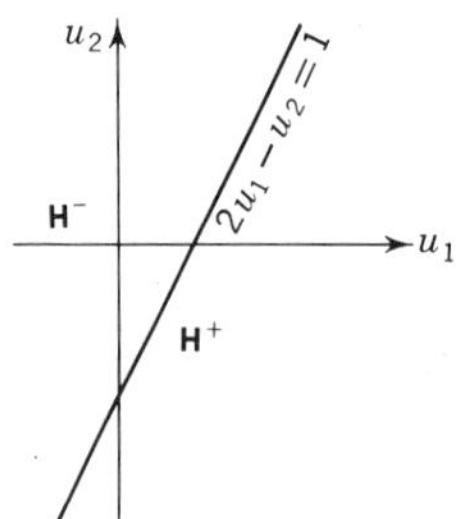

FIGURE 2.5

3. CONVEX SETS

A *convex combination* of the points $\mathbf{U}_1$, $\mathbf{U}_2$, ..., $\mathbf{U}_k$ is a point

$$\mathbf{U} = \alpha_1 \mathbf{U}_1 + \alpha_2 \mathbf{U}_2 + \cdots + \alpha_k \mathbf{U}_k$$

where the α_i are scalars, $\alpha_i \geq 0$, and $\sum_i \alpha_i = 1$. A subset $\mathbf{C}$ of $\mathbf{E}_n$ is *convex* if, for all pairs of points $\mathbf{U}_1$ and $\mathbf{U}_2$ in $\mathbf{C}$, any convex combination

$$\mathbf{U} = \alpha_1 \mathbf{U}_1 + \alpha_2 \mathbf{U}_2$$

is also in $\mathbf{C}$.

Examples of convex sets are the whole space $\mathbf{E}_n$, a circle, and a cube. The set of points that form the boundary of a circle is not a convex set.

We next show that, for a given convex set C, any convex combination of any number of points in C is also in C. Here we are given that the convex combination of any two points U_1 and U_2 in C is also in C. First we prove that

$$U = \alpha_1 U_1 + \alpha_2 U_2 + \alpha_3 U_3$$

for $\alpha_i \geq 0$, $\sum_i \alpha_i = 1$, is also in C. Let $\alpha_i' = \dfrac{\alpha_i}{\alpha_1 + \alpha_2}$ for $i = 1, 2$. We note that $\sum_i \alpha_i' = 1$ and $\alpha_i' \geq 0$. We have

$$U = \alpha_1 U_1 + \alpha_2 U_2 + \alpha_3 U_3 = (\alpha_1 + \alpha_2)(\alpha_1' U_1 + \alpha_2' U_2) + \alpha_3 U_3$$

By hypothesis $\alpha_1' U_1 + \alpha_2' U_2$ is in C. Let $U_4 = \alpha_1' U_1 + \alpha_2' U_2$. Then

$$U = (\alpha_1 + \alpha_2)U_4 + \alpha_3 U_3$$

where U is a convex combination of two points in C, and hence U is in C. This procedure can be generalized for any number of points.

Theorem 1. *Any point on the line segment joining two points in E_n can be expressed as a convex combination of the two points.*

Proof: Denote the two points by U and V and let W lie on the line segment joining U and V. This line segment is parallel to the line defined by

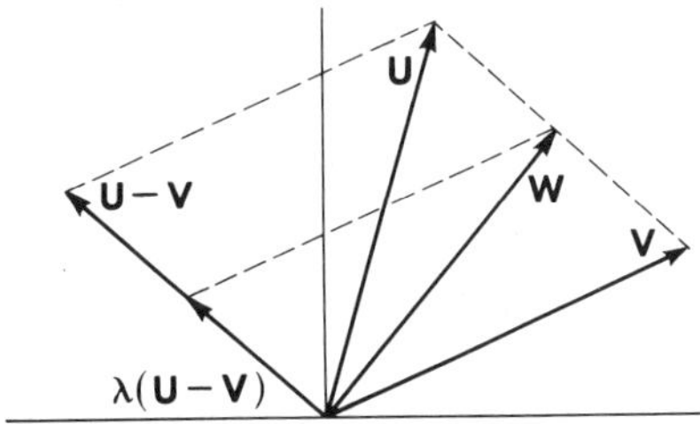

FIGURE 2.6

vector $U - V$ (Fig. 2.6). By the rules of vector addition we have, for some $0 \leq \lambda \leq 1$,

$$V + \lambda(U - V) = W \qquad \text{or} \qquad (1 - \lambda)V + \lambda U = W$$

which expresses the point W as a convex combination of V and U.

Theorem 2 (converse of Theorem 1). *Any point that can be expressed as a convex combination of two points in E_n lies on the line segment joining the two points.*

Proof: Here we are given

$$W = (1 - \lambda)V + \lambda U \qquad \text{or} \qquad W - V = \lambda(U - V)$$

where $0 \leq \lambda \leq 1$. Hence the vector $W - V$ is some positive multiple of vector $U - V$, and these vectors *cannot* have the configuration as pictured in Fig. 2.7.   Vector $\mathbf{W} - \mathbf{V}$ must coincide in direction with $\mathbf{U} - \mathbf{V}$.   Since the line segment joining $\mathbf{U}$ and $\mathbf{V}$ and the line segment joining $\mathbf{W}$ and $\mathbf{V}$ are parallel to the lines defined by $\mathbf{U} - \mathbf{V}$ and $\mathbf{W} - \mathbf{V}$, respectively, point $\mathbf{W}$ must lie on the line segment joining $\mathbf{U}$ and $\mathbf{V}$.

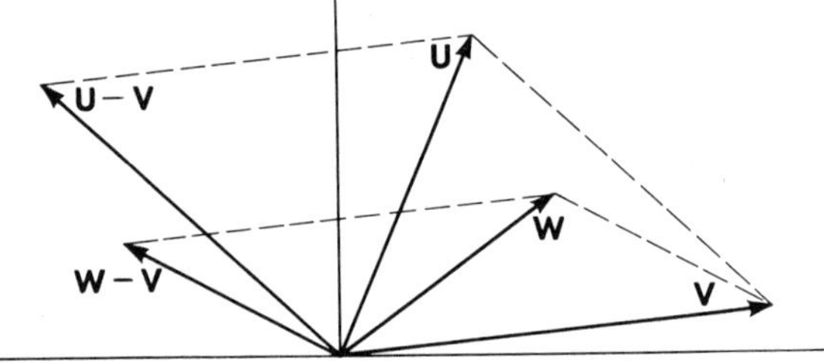

FIGURE 2.7

By Theorems 1 and 2 we see that geometrically a convex set is one which contains all line segments joining any two points in the set.   The sets shown in Fig. 2.8a and b are convex sets, while those of Fig. 2.8c and d are not.

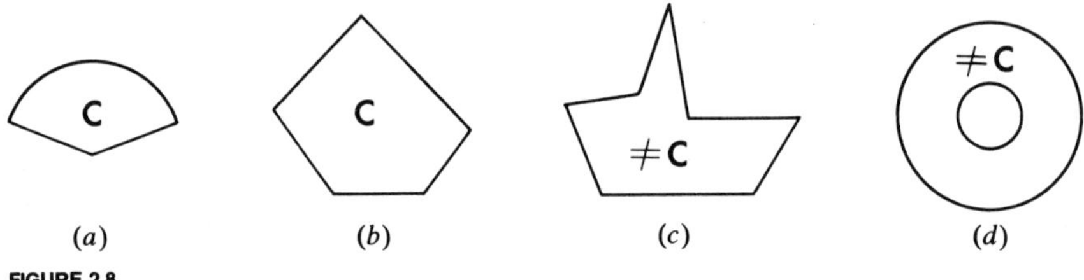

(a)   (b)   (c)   (d)

FIGURE 2.8

A point $\mathbf{U}$ in a convex set $\mathbf{C}$ is called an *extreme point* if $\mathbf{U}$ cannot be expressed as a convex combination of any other two distinct points in $\mathbf{C}$.   Every point on the boundary of a circle is an extreme point of the convex set which includes the boundary and the interior of the circle.   If the convex set did not include the boundary, then this set would contain no extreme points.   The extreme points of a triangle are its vertices.

The *convex hull* $\mathbf{C(S)}$ of any given set of points $\mathbf{S}$ is the set of all convex combinations of sets of points from $\mathbf{S}$. $\mathbf{C(S)}$ is the smallest convex set containing $\mathbf{S}$.   If $\mathbf{S}$ is just the eight vertices of a cube, then $\mathbf{C(S)}$ is the whole cube, or, if $\mathbf{S}$ is the boundary of a circle, then $\mathbf{C(S)}$ is the whole circle.

A *ray* or *half-line* generated by a vector $\mathbf{Y} \neq \mathbf{0}$ in $\mathbf{E}_n$ is the set of points $\theta\mathbf{Y}$ for all $\theta \geq 0$.   Here $\theta\mathbf{Y}$ originates at the origin of $\mathbf{E}_n$.   We can translate a ray $\theta\mathbf{Y}$ to another origin $\mathbf{U}_0$ in $\mathbf{E}_n$ and define the ray to be the set of points $\mathbf{U}_0 + \theta\mathbf{Y}$ (Fig. 2.9).   We see that the *direction* $\mathbf{Y}$ of the ray $\mathbf{U}_0 + \theta\mathbf{Y}$ is independent of its origin $\mathbf{U}_0$.   Two directions $\mathbf{Y}_1$ and $\mathbf{Y}_2$ are the same if $\mathbf{Y}_1 = \alpha\mathbf{Y}_2$ for $\alpha > 0$.

A nonempty closed convex set $\mathbf{C}$ is *unbounded* if for some $\mathbf{Y} \neq \mathbf{0}$ it contains the rays $\mathbf{U} + \theta\mathbf{Y}$ for all $\mathbf{U}$ in $\mathbf{C}$, i.e., the set $\mathbf{C}$ has a direction $\mathbf{Y}$ (Fig. 2.10). The convex set $\mathbf{C}$ is *bounded* if it contains no rays; it has no direc-

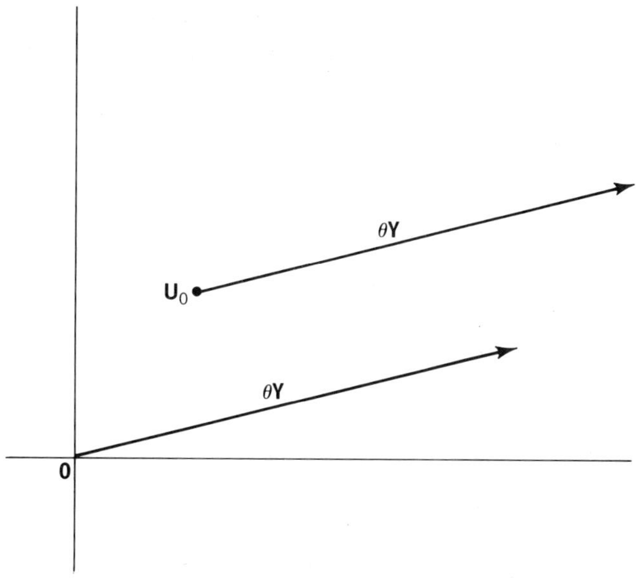

FIGURE 2.9

tions.[1]   The two-dimensional nonnegative quadrant is an unbounded convex set; the unit square is a bounded convex set.

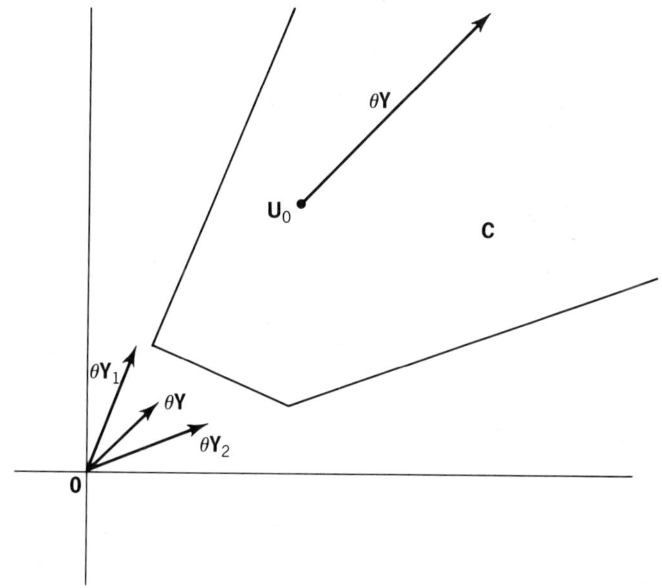

FIGURE 2.10

[1] This definition of boundedness coincides with the general definition of the term as applied to convex sets (see Stoer and Witzgall [632]).   The reader unfamiliar with general set concepts such as closure and boundedness is referred to Buck [99].

In Chap. 3 we prove that the solution space to a linear-programming problem is a convex set. In general, a set of points defined by the intersection of a finite number of linear equations and/or inequalities is termed a *convex polyhedron*. Such polyhedra can be bounded or unbounded. In either case, they have a finite number of extreme points. An example of a bounded convex polyhedron is the convex hull of a finite number of points, e.g., $C(S)$ of the eight vertices that define a cube. If a convex set C is closed and bounded, then any point of C can be expressed as a convex combination of the extreme points of C, i.e., C is the convex hull of its extreme points. Thus, if $U_1, \ldots, U_k$ are extreme points of a bounded convex polyhedron C, then any point U in C has a representation given by $U = \alpha_1 U_1 + \cdots + \alpha_k U_k$, with

$$\sum_{i=1}^{k} \alpha_i = 1 \qquad \alpha_i \geq 0$$

For unbounded convex sets, the above discussion can be extended as follows. A *face* of a convex set C is a convex subset F of C such that every line segment in C containing a point U in F, U not an end point of the line segment, has both end points in F. For example, the sides of a cube are two-dimensional faces, its edges are one-dimensional faces, and its vertices (extreme points) are zero-dimensional faces. If F is a half-line face of C, then the direction of F is called an *extreme direction*. An extreme direction Y of a convex set C cannot be expressed as a positive combination of two other distinct directions Y_1 and Y_2 of C, i.e., $Y \neq \theta_1 Y_1 + \theta_2 Y_2$ for $Y_1 \neq Y_2$ and $(\theta_1, \theta_2) > 0$. A convex polyhedron has a finite number of extreme directions. In Fig. 2.10, we illustrate the two extreme directions for the given C by the half-lines θY_1 and θY_2.

For an unbounded convex polyhedron C, a point U in C can be expressed as a convex combination of its extreme points plus a nonnegative combination of its extreme directions. Thus, if $U_1, \ldots, U_k$ are extreme points of C and $Y_1, \ldots, Y_l$ are its extreme directions, then any point U in C can be given by

$$U = \alpha_1 U_1 + \cdots + \alpha_k U_k + \beta_1 Y_1 + \cdots + \beta_l Y_l$$

with

$$\sum_{i=1}^{k} \alpha_i = 1 \qquad \alpha_i \geq 0 \qquad \beta_j \geq 0$$

where $j = 1, \ldots, l$. This statement is called the *representation theorem for convex polyhedra*. It can be generalized to arbitrary convex sets. (For additional aspects of the representation theorem, see Exercises 17 and 45; Chap. 3, Exercises 7, 8, and 13; and Chap. 9, Exercise 16.)

A set of vectors S is called a *cone* if, for every vector U in S, λU is in S for $\lambda \geq 0$; i.e., a cone contains the origin and all the rays generated by the nonzero vectors of S. Examples of a cone are the whole space, the origin, and set S of Fig. 2.11. An *extreme ray* of a cone is a half-line face that begins at the origin.

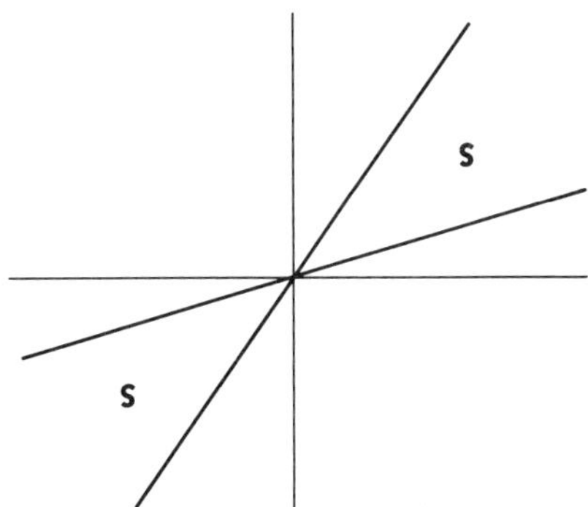

FIGURE 2.11

A *convex cone* is a cone that is convex. The cone of Fig. 2.11 is not convex. That part of S in the first quadrant is a convex cone. The non-negative orthant is a convex cone. If vectors U_1 and U_2 are in a convex cone, then the vectors $U = \alpha U_1 + \beta U_2$ are also in the cone for $(\alpha, \beta) > 0$.

A *simplex* is an n-dimensional convex polyhedron having exactly $n + 1$ vertices. The boundary of the simplex contains simplices of lower dimension which are called *simplical faces*. The number of such faces of dimension i is $\binom{n+1}{i+1}$.[†] A simplex in zero dimension is a point; in one dimension it is a line; in two dimensions it is a triangle; and in three dimensions it is a tetrahedron. The equation of a simplex with unit intercepts is

$$\sum_{i=1}^{m} x_i \leq 1$$

and $x_i \geq 0$

For $m = 3$ we have the tetrahedron with vertices $(0, 0, 0)$, $(1, 0, 0)$, $(0, 1, 0)$, and $(0, 0, 1)$.

4. LINEAR INEQUALITIES

Let us consider the following set of five linear inequalities in two dimensions:

$$
\begin{aligned}
(a) \quad & x_1 && \geq 0 \\
(b) \quad & & x_2 &\geq 0 \\
(c) \quad & x_1 + & x_2 &\geq 1 \\
(d) \quad & x_1 - & x_2 &\geq 1 \\
(e) \quad & -x_1 + & 2x_2 &\leq 0
\end{aligned}
\tag{4.1}
$$

We wish to determine which points (x_1, x_2) in two-dimensional Euclidean space satisfy the inequalities (4.1). We note (Fig. 2.12) that the first two inequalities

---

[†] The notation $\binom{n}{m}$ represents the number $n!/m!(n-m)!$.

limit our search to the nonnegative quadrant; i.e., only those points having nonnegative coordinates need be considered. Inequality (4.1a) is satisfied only by points lying on or to the right of the line $x_1 = 0$; inequality (4.1b) is satisfied only by points lying on or above the line $x_2 = 0$. The only points which simultaneously satisfy both inequalities are those lying in the first, or nonnegative, quadrant. Each inequality is satisfied by all the points lying in one of the half spaces that is generated when we treat the inequality as an equality. We note the solution half space by drawing arrows from the dividing hyperplane into the solution space. We next treat inequality (4.1c) as an equality and draw the line corresponding to $x_1 + x_2 = 1$. We determine that the half space which contains all the solutions to inequality (4.1c) lies upward to the right, by noting that the origin $(0, 0)$ does not satisfy the inequality; i.e., the solution space does not contain the origin. The solution space to inequalities (4.1a) to (4.1c) is that part of the nonnegative quadrant whose points simultaneously satisfy the three inequalities. For inequality (4.1d), we draw the line $x_1 - x_2 = 1$ and note that the half space whose points satisfy the inequality does not contain the origin. The line which corresponds to inequality (4.1e), $-x_1 + 2x_2 = 0$, contains the origin. To determine which of the associated half spaces represents the solution space for inequality (4.1e), we take any point, say $(1, 0)$, and note whether or not it satisfies (4.1e). Since $(1, 0)$ does satisfy (4.1e), the half space which contains $(1, 0)$ is the solution space for (4.1e). Only those points in region **C** described by the hatched lines simultaneously satisfy the five inequalities. The solution space **C** is an unbounded, polygonal convex region whose boundaries are portions of the lines which correspond to inequalities (4.1b), (4.1d), and (4.1e). Inequalities (4.1a) and (4.1c) do not enter into the determination of the solution space. They are redundant to the system (4.1), but only some type of analysis, such as the one above, will reveal this fact.

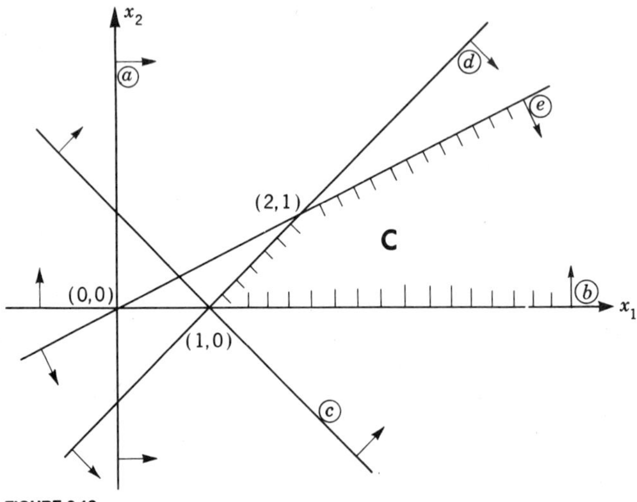

FIGURE 2.12

We have interpreted each of the inequalities of (4.1) as defining a hyperplane in two-dimensional space. An alternate interpretation is obtained by writing the matrix representation of the system (4.1). We let matrix

$$\mathbf{A} = \begin{pmatrix} 1 & 0 \\ 0 & 1 \\ 1 & 1 \\ 1 & -1 \\ 1 & -2 \end{pmatrix}$$

be the matrix of coefficients, and write the matrix inequality

$$\begin{pmatrix} 1 & 0 \\ 0 & 1 \\ 1 & 1 \\ 1 & -1 \\ 1 & -2 \end{pmatrix} \begin{pmatrix} x_1 \\ x_2 \end{pmatrix} \geq \begin{pmatrix} 0 \\ 0 \\ 1 \\ 1 \\ 0 \end{pmatrix} \tag{4.2}$$

By multiplying the matrix $\mathbf{A}$ by the column vector whose elements are x_1 and x_2, we have the five inequalities of the original system. Here, inequality (4.1e) has been multiplied by -1 to reverse the direction of the inequality. We denote the first column of $\mathbf{A}$ by $\mathbf{P}_1$ and the second column by $\mathbf{P}_2$; that is,

$$\mathbf{P}_1 = \begin{pmatrix} 1 \\ 0 \\ 1 \\ 1 \\ 1 \end{pmatrix} \quad \text{and} \quad \mathbf{P}_2 = \begin{pmatrix} 0 \\ 1 \\ 1 \\ -1 \\ -2 \end{pmatrix}$$

and let

$$\mathbf{P}_0 = \begin{pmatrix} 0 \\ 0 \\ 1 \\ 1 \\ 0 \end{pmatrix}$$

the column vector of constants. We can then write the system (4.2) as

$$x_1 \mathbf{P}_1 + x_2 \mathbf{P}_2 \geq \mathbf{P}_0 \tag{4.3}$$

Here we interpret the column vectors $\mathbf{P}_1$, $\mathbf{P}_2$, and $\mathbf{P}_0$ as points in five-dimensional space, and the problem is to determine the weights x_1 and x_2 such that all the conditions defined by (4.3) are satisfied.

In general, the points which satisfy a linear inequality of the type

$$a_{11}x_1 + a_{12}x_2 + \cdots + a_{1n}x_n \geq b_1$$

define a half space in n-dimensional space. The joint solution space to the set of inequalities

$$a_{11}x_1 + a_{12}x_2 + \cdots + a_{1n}x_n \geq b_1$$
$$a_{21}x_1 + a_{22}x_2 + \cdots + a_{2n}x_n \geq b_2$$
$$\cdots\cdots\cdots\cdots\cdots\cdots\cdots\cdots\cdots$$
$$a_{m1}x_1 + a_{m2}x_2 + \cdots + a_{mn}x_n \geq b_m$$

$$(4.4)$$

is a convex region in n-dimensional space. (We shall prove the convexity of the solution space in Chap. 3.) Since the direction of the inequality can be reversed by multiplication by -1, any set of inequalities can be written in the form of (4.4). The set of inequalities (4.4) can be transformed into a set of equalities by subtracting from each inequality an unknown nonnegative number. These numbers are termed *slack variables*. For (4.4), this transformation yields the set of equalities

$$a_{11}x_1 + a_{12}x_2 + \cdots + a_{1n}x_n - x_{n+1} \qquad\qquad = b_1$$
$$a_{21}x_1 + a_{22}x_2 + \cdots + a_{2n}x_n \qquad - x_{n+2} \qquad = b_2$$
$$\cdots\cdots\cdots\cdots\cdots\cdots\cdots\cdots\cdots\cdots\cdots\cdots\cdots$$
$$a_{m1}x_1 + a_{m2}x_2 + \cdots + a_{mn}x_n \qquad\qquad - x_{n+m} = b_m$$

where $x_{n+i} \geq 0$ $(i = 1, \ldots, m)$. Since we can represent each variable x_i as the difference of two nonnegative variables,[1] the system (4.4) has the following equivalent system in terms of nonnegative variables and equalities:

$$a_{11}(x_1' - x_1'') + a_{12}(x_2' - x_2'') + \cdots + a_{1n}(x_n' - x_n'') - x_{n+1} \qquad\qquad = b_1$$
$$a_{21}(x_1' - x_1'') + a_{22}(x_2' - x_2'') + \cdots + a_{2n}(x_n' - x_n'') \qquad - x_{n+2} \qquad = b_2$$
$$\cdots\cdots\cdots\cdots\cdots\cdots\cdots\cdots\cdots\cdots\cdots\cdots\cdots\cdots\cdots\cdots$$
$$a_{m1}(x_1' - x_1'') + a_{m2}(x_2' - x_2'') + \cdots + a_{mn}(x_n' - x_n'') \qquad\qquad - x_{n+m} = b_m$$

where

$$x_j = x_j' - x_j''$$
$$x_j' \geq 0 \qquad\qquad j = 1, \ldots, n$$
$$x_j'' \geq 0$$
$$x_{n+i} \geq 0 \qquad\qquad i = 1, \ldots, m$$

An inequality of the form

$$a_{11}x_1 + a_{12}x_2 + \cdots + a_{1n}x_n \leq b_1$$

can be transformed into an equality by adding to it an unknown nonnegative number (slack variable), e.g.,

$$a_{11}x_1 + a_{12}x_2 + \cdots + a_{1n}x_n + x_{n+1} = b_1$$

where $x_{n+1} \geq 0$.

From the above discussion we note that any set of linear expressions can

---

[1] The reader should note that any number can be written as the difference of two nonnegative numbers. A variable that can assume any value, i.e., $-\infty < x_j < +\infty$, is termed an **unrestricted** or *free variable*. See Exercise 19; also Exercise 34, Chap. 4.

be transformed into an equivalent set of linear equations in nonnegative variables.

We illustrate the above concepts with the following example. Consider the set of inequalities

$$
\begin{aligned}
x_1 - 2x_2 + 3x_3 &\geq 6 \\
2x_1 + x_2 - x_3 &\leq 3 \\
x_1 \quad\quad &\geq 0 \\
x_3 &\geq 0
\end{aligned}
$$

We note that x_2 is unrestricted as to sign and that the vector $\mathbf{X} = (x_1 = 2, x_2 = 1, x_3 = 2)$ does satisfy all the constraints; i.e., the solution space is not empty. Letting $x_2 = x_2' - x_2''$, with $x_2' \geq 0$, $x_2'' \geq 0$ and subtracting the nonnegative slack variable x_4 from the first constraint and adding the nonnegative slack variable x_5 to the second constraint, we have the equivalent system of equations in nonnegative variables of

$$
\begin{aligned}
x_1 - 2x_2' + 2x_2'' + 3x_3 - x_4 &= 6 \\
2x_1 + x_2' - x_2'' - x_3 \quad + x_5 &= 3
\end{aligned}
$$

and

$$
(x_1, x_2', x_2'', x_3, x_4, x_5) \geq 0
$$

It should be noted that the solution space for a set of inequalities might be void. There are no solutions to the set of inequalities

$$
\begin{aligned}
x_1 + x_2 &\leq 1 \\
2x_1 + 2x_2 &\geq 3
\end{aligned}
$$

As a mathematical problem, the general linear-programming problem can be described as follows: We are given a convex set which is defined by a set of linear constraints in n-dimensional Euclidean space. From all the points that belong to the convex set, we wish to determine a subset of points (which will contain either one or many points) for which a linear objective function is optimized. To illustrate how this can be done for two-dimensional constraints, we have represented in Fig. 2.13 the convex set $\mathbf{C}$, that is, the solution space, for Eqs. (4.1). Let us determine the subset of points for which the linear objective function $2x_1 + 2x_2$ is a minimum. In Fig. 2.13 we have also drawn the line

$$
2x_1 + 2x_2 = b
$$

for $b = 12, 6,$ and 2. The intersections of these lines with $\mathbf{C}$ have been represented by heavy lines, with the last intersection being a point. Every point (x_1, x_2) of the first intersection will yield a value of the objective function equal to 12; all points of the second intersection yield a value of 6; and finally we see that the last intersection, which consists of only the point $(1, 0)$, gives the minimum value of 2. Hence, out of the infinite number of points contained in

C only one point minimizes the objective function. This point is also an extreme point of **C**, and the full consequences of this will be discussed in Chap. 3. For any other objective function we can similarly determine the subset of points for which it is optimized. We need only to plot the objective function for some particular value and then move this line parallel to itself in the appropriate direction for minimizing or maximizing, as was done in Fig. 2.13. We move the line until the intersection of the line and convex set is reduced either to a point, a portion of the boundary of **C**, or until it is apparent that the maximum or minimum value is infinite. These last two situations are illustrated by minimizing $2x_1 - 2x_2$ and maximizing $2x_1 + 2x_2$ for **C** of Fig. 2.13. If our convex set were a bounded polyhedron, then all objective functions would have a finite minimum or maximum. We illustrate this case by solving the numerical activity-analysis (furniture manufacturer) example given in Chap. 1.

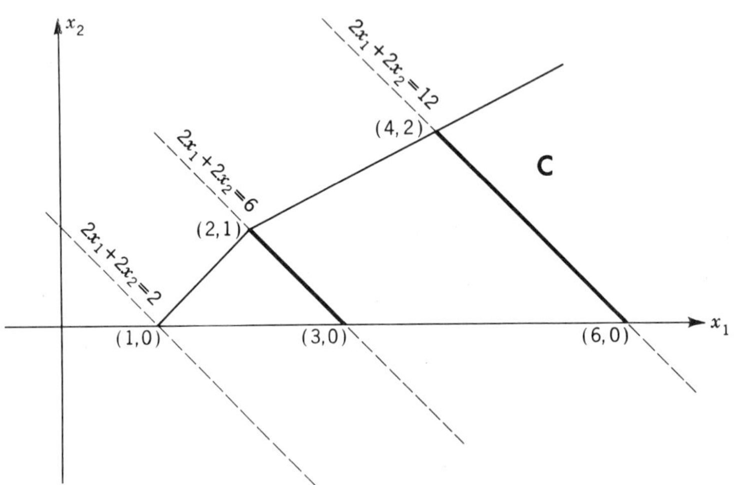

FIGURE 2.13

The problem is to maximize the profit function

$$45x_1 + 80x_2$$

subject to

$$5x_1 + 20x_2 \le 400$$
$$10x_1 + 15x_2 \le 450$$
$$x_1 \qquad \ge \quad 0$$
$$x_2 \ge \quad 0$$

where x_1 and x_2 represent the number of chairs and tables to be produced, respectively. We plot the constraint set as shown in Fig. 2.14.

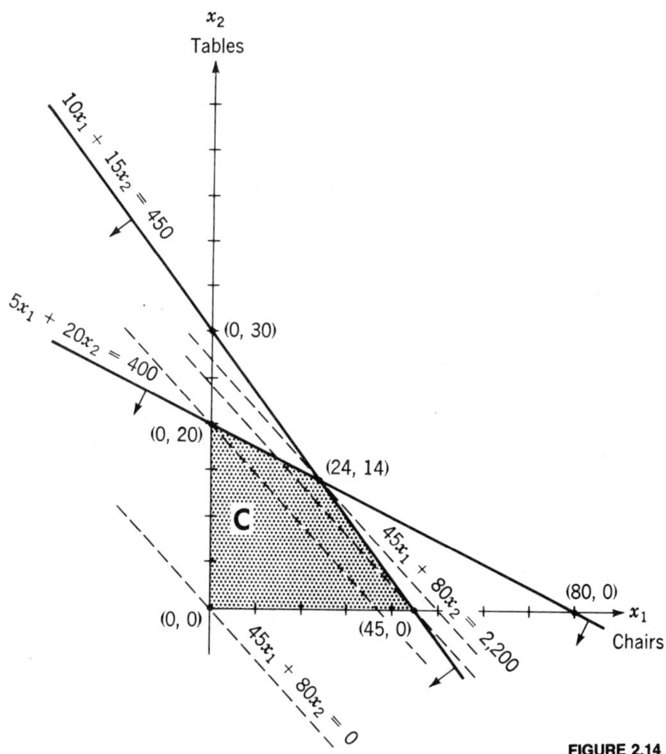

FIGURE 2.14

The joint solution space **C** to all the resource and nonnegativity constraints is denoted by the bounded shaded area. Any point in this area, which includes the boundary, represents a feasible manufacturing schedule, i.e., one which satisfies the constraint set. If the company does not make any chairs or tables, its profit function is zero, as denoted by the dashed line through the point $(0, 0)$. As we move the profit function $45x_1 + 80x_2$ parallel to itself from the point $(0, 0)$ through the feasible region **C**, it passes through the "all table" solution $(0, 20)$, which has a profit of \$1,600, then the "all chair" solution $(45, 0)$, which has a profit of \$2,025, until it reaches the extreme-point solution, which calls for the production of 24 chairs and 14 tables and has a maximum profit of \$2,200. This is the unique optimal solution.

5. SOLUTION OF A SET OF LINEAR EQUATIONS

Any set of linear equations has a convenient representation using matrix notation. The system

$$a_{11}x_1 + a_{12}x_2 + \cdots + a_{1n}x_n = b_1$$
$$a_{21}x_1 + a_{22}x_2 + \cdots + a_{2n}x_n = b_2$$
$$\cdots\cdots\cdots\cdots\cdots\cdots\cdots\cdots\cdots\cdots$$
$$a_{m1}x_1 + a_{m2}x_2 + \cdots + a_{mn}x_n = b_m$$

$$(5.1)$$

can be written as

$$\mathbf{AX} = \mathbf{b} \tag{5.1a}$$

where

$$\mathbf{A} = (a_{ij}) \qquad \mathbf{X} = \begin{pmatrix} x_1 \\ \vdots \\ x_n \end{pmatrix} \qquad \text{and} \qquad \mathbf{b} = \begin{pmatrix} b_1 \\ \vdots \\ b_m \end{pmatrix}^{\dagger}$$

If $\mathbf{A}$ is square (that is, if $m = n$) and nonsingular, the solution vector is given by

$$\mathbf{X} = \mathbf{A}^{-1}\mathbf{b} \tag{5.1b}$$

There is a simple computational scheme that can be applied to any nonsingular system to obtain the solution vector and/or the inverse. This procedure, the *complete elimination method of Gauss and Jordan*, has a finite number of steps, or iterations. In just $m(=n)$ iterations, the procedure multiplies the system (5.1) by $\mathbf{A}^{-1}$ to obtain $\mathbf{X} = \mathbf{A}^{-1}\mathbf{b}$. If it is so desired, $\mathbf{A}^{-1}$ may be explicitly developed at the same time. Before developing the method in general, we illustrate it by the following example.

Consider the set of three equations in three variables:

$$\begin{aligned} x_1 + x_2 - x_3 &= 2 \\ -2x_1 + x_2 + x_3 &= 3 \\ x_1 + x_2 + x_3 &= 6 \end{aligned} \tag{5.2}$$

Then

$$\mathbf{A} = \begin{pmatrix} 1 & 1 & -1 \\ -2 & 1 & 1 \\ 1 & 1 & 1 \end{pmatrix}$$

is the matrix of coefficients. It is easy to verify that $\mathbf{A}$ is nonsingular. We can rewrite (5.2) as the matrix equation

$$\begin{pmatrix} 1 & 1 & -1 \\ -2 & 1 & 1 \\ 1 & 1 & 1 \end{pmatrix} \begin{pmatrix} x_1 \\ x_2 \\ x_3 \end{pmatrix} = \begin{pmatrix} 2 \\ 3 \\ 6 \end{pmatrix}$$

or, by letting

$$\mathbf{P}_1 = \begin{pmatrix} 1 \\ -2 \\ 1 \end{pmatrix} \qquad \mathbf{P}_2 = \begin{pmatrix} 1 \\ 1 \\ 1 \end{pmatrix} \qquad \mathbf{P}_3 = \begin{pmatrix} -1 \\ 1 \\ 1 \end{pmatrix} \qquad \mathbf{P}_0 = \begin{pmatrix} 2 \\ 3 \\ 6 \end{pmatrix}$$

we can rewrite (5.2) as

$$x_1\mathbf{P}_1 + x_2\mathbf{P}_2 + x_3\mathbf{P}_3 = \mathbf{P}_0$$

---

† The reader is referred to the discussion of matrix multiplication in Sec. 1 for the validity of this representation.

Since $\mathbf{A}$ is nonsingular, the set of vectors $\mathbf{P}_1$, $\mathbf{P}_2$, and $\mathbf{P}_3$ is linearly independent and hence forms a basis in three-dimensional space.[1]   To solve (5.2), we need to find the unique linear combination (i.e., the numbers x_1, x_2, and x_3) of $\mathbf{P}_1$, $\mathbf{P}_2$ and $\mathbf{P}_3$ which equals vector $\mathbf{P}_0$.   The complete elimination procedure successively eliminates the first, the second, the third, etc., variable from all equations except the first, the second, the third, etc., respectively.   For (5.2), the first step is to eliminate the first variable from all the equations except the first.   This elimination is accomplished by adding and subtracting suitable multiples of the first equation to the second and third equations.

STEP 1

$$x_1 + x_2 - x_3 = 2$$
$$-2x_1 + x_2 + x_3 = 3$$
$$x_1 + x_2 + x_3 = 6$$

Since matrix $\mathbf{A}$ is nonsingular, Eqs. (5.2) can be listed in such an order that the coefficient of the first variable in the first equation does not equal zero.   Here $a_{11} = 1$.   To eliminate x_1 from the second equation, we multiply the first equation by 2 and add it to the second.   To eliminate the first variable from the third equation, we multiply the first equation by -1 and add it to the third.   These two substeps transform the original equations into the equations

$$x_1 + x_2 - x_3 = 2$$
$$3x_2 - x_3 = 7$$
$$2x_3 = 4$$

STEP 2   Here we eliminate the second variable from all the equations obtained as a result of Step 1 except the second.   Again the equations can be arranged to make the coefficient of x_2 in the second equation not equal to zero (this would have been done if the third equation had originally been the

---

[1] To prove that the vectors $\mathbf{P}_1$, $\mathbf{P}_2, \ldots, \mathbf{P}_n$ of a nonsingular matrix $\mathbf{A}$ form a linearly independent set, we first assume that the set is linearly dependent.   Hence, for some set of α_j we must have

$$\alpha_1 \mathbf{P}_1 + \alpha_2 \mathbf{P}_2 + \cdots + \alpha_n \mathbf{P}_n = \mathbf{0}$$

with at least one $\alpha_j \neq 0$.   We can then solve for some vector $\mathbf{P}_j$, for example, $\mathbf{P}_1$, in terms of the other vectors to obtain

$$\mathbf{P}_1 = \beta_2 \mathbf{P}_2 + \cdots + \beta_n \mathbf{P}_n$$

where $\beta_j = -\alpha_j / \alpha_1$.   We substitute this expression for vector $\mathbf{P}_1$ in matrix $\mathbf{A}$.   Then by successively adding to the first column $-\beta_j \mathbf{P}_j$ for $j = 2, 3, \ldots, n$, we reduce the first column to the zero vector.   From the properties of determinants given in Sec. 1, the addition of these vectors does not change the value of the determinant, and since a column is now equal to zero, the value of the determinant is zero.   However, this contradicts the given information that $\mathbf{A}$ is a nonsingular matrix.   Hence, our assumption of linear dependence has led to a contradiction, and therefore the set of vectors of $\mathbf{A}$ must be linearly independent.

second). Since at the start of an iteration it is convenient to make the coefficient of the variable being eliminated equal to 1, we divide the second equation by 3.   The coefficient 3 is referred to as the *pivot element.*   (In Step 1 the pivot element was $a_{11} = 1$.)   We then have to multiply this transformed second equation by -1 and add it to the first to obtain the equations

$$
\begin{aligned}
x_1 \quad\;\; - \tfrac{2}{3}x_3 &= -\tfrac{1}{3}\\
x_2 - \tfrac{1}{3}x_3 &= \tfrac{7}{3}\\
2x_3 &= 4
\end{aligned}
$$

STEP 3   Here the pivot element is 2.   We first divide the third equation obtained in Step 2 by 2, and then multiply it by $\tfrac{2}{3}$ and $\tfrac{1}{3}$ and add the corresponding results to the first and second equations, respectively.   The system (5.2) has been reduced to the solution

$$
\begin{aligned}
x_1 \quad\;\; &= 1\\
x_2 \quad &= 3\\
x_3 &= 2
\end{aligned}
\tag{5.3}
$$

The operations of the complete elimination procedure have transformed the matrix of coefficients

$$
\mathbf{A} = \begin{pmatrix} 1 & 1 & -1 \\ -2 & 1 & 1 \\ 1 & 1 & 1 \end{pmatrix}
$$

into the identity matrix [the matrix of coefficients for the transformed Eqs. (5.3)]

$$
\mathbf{I}_3 = \begin{pmatrix} 1 & 0 & 0 \\ 0 & 1 & 0 \\ 0 & 0 & 1 \end{pmatrix}
$$

We then see that the total transformation accomplished by the complete elimination procedure is equivalent to (5.1*b*), i.e., the multiplication of the system (5.1) by $\mathbf{A}^{-1}$.

It is very often necessary not only to solve a system of simultaneous linear equations, but also to obtain the inverse of the matrix of coefficients.   This is accomplished by attaching an $n \times n$ identity matrix to the right of the original matrix of coefficients and applying the elimination transformations to the extended matrix.   The inverse is generated in place of the identity matrix.   If we write the partitioned matrix

$$(\mathbf{A}|\mathbf{I}|\mathbf{P}_0)$$

(i.e., the matrix $\mathbf{A}$ with an $n \times n$ identity matrix written beside it and the $\mathbf{P}_0$ vector attached to the end) and apply the complete elimination transformation, we have

$$(\mathbf{A}^{-1}\mathbf{A}|\mathbf{A}^{-1}\mathbf{I}|\mathbf{A}^{-1}\mathbf{P}_0) = (\mathbf{I}|\mathbf{A}^{-1}|\mathbf{X})$$

We shall illustrate this procedure for the previous example. The partitioned matrix is

$$\left(\begin{array}{rrr|rrr|r} 1 & 1 & -1 & 1 & 0 & 0 & 2 \\ -2 & 1 & 1 & 0 & 1 & 0 & 3 \\ 1 & 1 & 1 & 0 & 0 & 1 & 6 \end{array}\right)$$

The successive steps, as before, yield

STEP 1

$$\left(\begin{array}{rrr|rrr|r} 1 & 1 & -1 & 1 & 0 & 0 & 2 \\ 0 & 3 & -1 & 2 & 1 & 0 & 7 \\ 0 & 0 & 2 & -1 & 0 & 1 & 4 \end{array}\right)$$

STEP 2

$$\left(\begin{array}{rrr|rrr|r} 1 & 0 & -\frac{2}{3} & \frac{1}{3} & -\frac{1}{3} & 0 & -\frac{1}{3} \\ 0 & 1 & -\frac{1}{3} & \frac{2}{3} & \frac{1}{3} & 0 & \frac{7}{3} \\ 0 & 0 & 2 & -1 & 0 & 1 & 4 \end{array}\right)$$

STEP 3

$$\left(\begin{array}{rrr|rrr|r} 1 & 0 & 0 & 0 & -\frac{1}{3} & \frac{1}{3} & 1 \\ 0 & 1 & 0 & \frac{1}{2} & \frac{1}{3} & \frac{1}{6} & 3 \\ 0 & 0 & 1 & -\frac{1}{2} & 0 & \frac{1}{2} & 2 \end{array}\right)$$

We then have

$$\mathbf{A}^{-1} = \left(\begin{array}{rrr} 0 & -\frac{1}{3} & \frac{1}{3} \\ \frac{1}{2} & \frac{1}{3} & \frac{1}{6} \\ -\frac{1}{2} & 0 & \frac{1}{2} \end{array}\right)$$

The unique linear combination of the vectors $\mathbf{P}_1$, $\mathbf{P}_2$, and $\mathbf{P}_3$ which equal $\mathbf{P}_0$ has been determined to be

$$1\mathbf{P}_1 + 3\mathbf{P}_2 + 2\mathbf{P}_3 = \mathbf{P}_0$$

Since the vectors $\mathbf{P}_1$, $\mathbf{P}_2$, $\mathbf{P}_3$ form a basis in three-dimensional space, any other vector in three-space can also be expressed as a unique linear combination of these three vectors. For any such vector, say $\mathbf{P}_4$, we wish to determine y_1, y_2, and y_3 such that

$$y_1\mathbf{P}_1 + y_2\mathbf{P}_2 + y_3\mathbf{P}_3 = \mathbf{P}_4$$

or

$$\mathbf{AY} = \mathbf{P}_4 \tag{5.4}$$

To solve for $\mathbf{Y}$, we must multiply (5.4) by $\mathbf{A}^{-1}$ to obtain

$$\mathbf{A}^{-1}\mathbf{AY} = \mathbf{A}^{-1}\mathbf{P}_4$$
$$\mathbf{Y} = \mathbf{A}^{-1}\mathbf{P}_4$$

For example, let

$$\mathbf{P}_4 = \begin{pmatrix} 4 \\ -2 \\ 4 \end{pmatrix}$$

$\mathbf{Y}$ is given by

$$\mathbf{Y} = \begin{pmatrix} 0 & -\frac{1}{3} & \frac{1}{3} \\ \frac{1}{2} & \frac{1}{3} & \frac{1}{6} \\ -\frac{1}{2} & 0 & \frac{1}{2} \end{pmatrix} \begin{pmatrix} 4 \\ -2 \\ 4 \end{pmatrix} = \begin{pmatrix} 2 \\ 2 \\ 0 \end{pmatrix}$$

$\mathbf{P}_4$ expressed as a linear combination of $\mathbf{P}_1$, $\mathbf{P}_2$, and $\mathbf{P}_3$ is then

$$2\mathbf{P}_1 + 2\mathbf{P}_2 + 0\mathbf{P}_3 = \mathbf{P}_4$$

Here $\mathbf{P}_4$ is equal to a linear combination of only two of the three basis vectors. The three-dimensional vectors $\mathbf{P}_1$, $\mathbf{P}_2$, $\mathbf{P}_4$ are linearly dependent since

$$2\mathbf{P}_1 + 2\mathbf{P}_2 - \mathbf{P}_4 = \mathbf{0}$$

with the coefficients of vectors $\mathbf{P}_1$, $\mathbf{P}_2$, and $\mathbf{P}_4$ not all zero. The situation where a vector can be expressed as a linear combination of less than n vectors from a given basis is termed *degenerate*.

We note that the accomplishment of the elimination procedure does not hinge on our being able to eliminate the first variable from the first equation, the second variable from the second equation, etc. Given that the matrix $\mathbf{A}$ of the system is nonsingular, we can select the pivot element in an arbitrary fashion. For example, let us return to the original partitioned matrix associated with Eqs. (5.2). We have

$$\begin{pmatrix} 1 & 1 & -1 & | & 1 & 0 & 0 & | & 2 \\ -2 & 1 & 1 & | & 0 & 1 & 0 & | & 3 \\ 1 & 1 & 1 & | & 0 & 0 & 1 & | & 6 \end{pmatrix}$$

We first eliminate x_1 by selecting the -2 element of the second row.

$$\begin{pmatrix} 0 & \frac{3}{2} & -\frac{1}{2} & | & 1 & \frac{1}{2} & 0 & | & \frac{7}{2} \\ 1 & -\frac{1}{2} & -\frac{1}{2} & | & 0 & -\frac{1}{2} & 0 & | & -\frac{3}{2} \\ 0 & \frac{3}{2} & \frac{3}{2} & | & 0 & \frac{1}{2} & 1 & | & \frac{15}{2} \end{pmatrix}$$

Next, we eliminate x_3 by pivoting on the $-\frac{1}{2}$ element of the first row.

$$\begin{pmatrix} 0 & -3 & 1 & | & -2 & -1 & 0 & | & -7 \\ 1 & -2 & 0 & | & -1 & -1 & 0 & | & -5 \\ 0 & 6 & 0 & | & 3 & 2 & 1 & | & 18 \end{pmatrix}$$

Finally we eliminate x_2 by pivoting on the 6 element of the third row.

$$\begin{pmatrix} 0 & 0 & 1 & | & -\frac{1}{2} & 0 & \frac{1}{2} & | & 2 \\ 1 & 0 & 0 & | & 0 & -\frac{1}{3} & \frac{1}{3} & | & 1 \\ 0 & 1 & 0 & | & \frac{1}{2} & \frac{1}{3} & \frac{1}{6} & | & 3 \end{pmatrix}$$

From the last tableau we have the solution of $x_3 = 2$, $x_1 = 1$, and $x_2 = 3$. We also note that the order of the unit elements has been permuted; similarly with the rows corresponding to $\mathbf{A}^{-1}$. We obtain the correct form of $\mathbf{A}^{-1}$ by interchanging rows to form a unit matrix in normal form. Here we shift the first row and make it the third row. The product of the pivot elements is $(-2)(-\frac{1}{2})(6) = 6 = |\mathbf{A}|$, the determinant value.

The reader should become familiar with the interpretation of elimination tableaus in which the order of the variables eliminated is not in sequence. This will be the usual situation when we apply the elimination procedure to solve linear-programming problems. We next develop the notational form of the Gauss–Jordan complete elimination method, followed by a discussion of Gauss' original elimination procedure. We then review some of the computational requirements and concerns that need to be addressed when solving large systems of linear equations by elimination methods.

For an $n \times n$ nonsingular system of equations $\mathbf{AX} = \mathbf{b}$, the Gauss–Jordan complete elimination method transforms the matrix $\mathbf{A}$ into a diagonal matrix. Assuming that the elimination of each variable can be done in sequence using the transformed principal-diagonal elements as pivots, we have the following representation of the Gauss–Jordan method. (Note that we may need to permute rows and columns to use the diagonal elements, i.e., for each step we have to make sure that the corresponding diagonal pivot element is nonzero.)

STEP 0   (the given system)

$$
\begin{aligned}
a_{11}x_1 + a_{12}x_2 + a_{13}x_3 + \cdots + a_{1n}x_n &= b_1 \\
a_{21}x_1 + a_{22}x_2 + a_{23}x_3 + \cdots + a_{2n}x_n &= b_2 \\
a_{31}x_1 + a_{32}x_2 + a_{33}x_3 + \cdots + a_{3n}x_n &= b_3 \\
&\,\,\vdots \\
a_{n1}x_1 + a_{n2}x_2 + a_{n3}x_3 + \cdots + a_{nn}x_n &= b_n
\end{aligned}
\tag{5.5}
$$

Starting with the Step 0 system, we perform n steps, each step eliminating a variable from all equations but one, as follows.

STEP 1

$$
\begin{aligned}
a_{11}^{(1)}x_1 + a_{12}^{(1)}x_2 + a_{13}^{(1)}x_3 + \cdots + a_{1n}^{(1)}x_n &= b_1^{(1)} \\
a_{22}^{(1)}x_2 + a_{23}^{(1)}x_3 + \cdots + a_{2n}^{(1)}x_n &= b_2^{(1)} \\
a_{32}^{(1)}x_2 + a_{33}^{(1)}x_3 + \cdots + a_{3n}^{(1)}x_n &= b_3^{(1)} \\
&\,\,\vdots \\
a_{n2}^{(1)}x_2 + a_{n3}^{(1)}x_3 + \cdots + a_{nn}^{(1)}x_n &= b_n^{(1)}
\end{aligned}
$$

STEP 2

$$
\begin{aligned}
a_{11}^{(1)}x_1 \qquad\quad + a_{13}^{(2)}x_3 + \cdots + a_{1n}^{(2)}x_n &= b_1^{(2)} \\
a_{22}^{(2)}x_2 + a_{23}^{(2)}x_3 + \cdots + a_{2n}^{(2)}x_n &= b_2^{(2)} \\
a_{33}^{(2)}x_3 + \cdots + a_{3n}^{(2)}x_n &= b_3^{(2)} \\
&\,\,\vdots \\
a_{n3}^{(2)}x_3 + \cdots + a_{nn}^{(2)}x_n &= b_n^{(2)}
\end{aligned}
$$

STEP $n-1$

$$a_{11}^{(1)}x_1 \qquad\qquad\qquad + a_{1n}^{(n-1)}x_n = b_1^{(n-1)}$$
$$a_{22}^{(2)}x_2 \qquad\qquad + a_{2n}^{(n-1)}x_n = b_2^{(n-1)}$$
$$a_{33}^{(3)}x_3 \quad + a_{3n}^{(n-1)}x_n = b_3^{(n-1)}$$
$$\cdots\cdots\cdots\cdots\cdots\cdots\cdots$$
$$a_{nn}^{(n-1)}x_n = b_n^{(n-1)}$$

STEP n

$$a_{11}^{(1)}x_1 \qquad\qquad\qquad\qquad = b_1^{(n)}$$
$$a_{22}^{(2)}x_2 \qquad\qquad\qquad = b_2^{(n)}$$
$$a_{33}^{(3)}x_3 \qquad\qquad = b_3^{(n)}$$
$$\cdots\cdots\cdots\cdots\cdots\cdots$$
$$a_{nn}^{(n)}x_n = b_n^{(n)}$$

The values of the x_i are obtained from the Step n equations directly and we have

$$x_i = \frac{b_i^{(n)}}{a_{ii}^{(i)}} \qquad i = 1, \ldots, n$$

Note that the elements of row k in step k are equal to the corresponding elements in step $k - 1$. Also, in the above general discussion of the Gauss–Jordan method, we do not divide by the pivot elements until the system of equations has been transformed into a diagonal structure. Row k is called the *pivot row* and column k the *pivot column.*

In Chaps. 3 and 4 we use the Gauss–Jordan elimination method as a means of determining solutions to a linear-programming problem. We will not be solving just a single square set of equations, but we will be finding a sequence of solutions that corresponds to square sets of equations formed from the original $m < n$ rectangular system of equations. There the representation of a solution requires each step of the Gauss–Jordan method to first divide the pivot row by the pivot element. For this situation, the reader should verify that the transformations from step $k - 1$ to step k are given by formulas (5.6) and (5.7). These formulas should be compared to Eqs. (2.7) of Chap. 4; also see the discussion of Sec. 3, Chap. 3.

$$a_{ij}^{(k)} = a_{ij}^{(k-1)} - \left(\frac{a_{ik}^{(k-1)}}{a_{kk}^{(k-1)}} \right) a_{kj}^{(k-1)} \qquad i \neq k$$
$$a_{kj}^{(k)} = \frac{a_{kj}^{(k-1)}}{a_{kk}^{(k-1)}} \tag{5.6}$$

$$b_i^{(k)} = b_i^{(k-1)} - \left(\frac{a_{ik}^{(k-1)}}{a_{kk}^{(k-1)}} \right) b_k^{(k-1)} \qquad i \neq k$$
$$b_k^{(k)} = \frac{b_k^{(k-1)}}{a_{kk}^{(k-1)}} \tag{5.7}$$

with the coefficients $a_{ij}^{(0)}$ and $b_i^{(0)}$ being the original data from Eqs. (5.5).

We next investigate the use of the Gauss–Jordan method to find the inverse of a matrix $\mathbf{A}$ given by

$$
\begin{pmatrix}
a_{11} & a_{12} & a_{13} & \cdots & a_{1n} \\
a_{21} & a_{22} & a_{23} & \cdots & a_{2n} \\
a_{31} & a_{32} & a_{33} & \cdots & a_{3n} \\
\cdots\cdots\cdots\cdots\cdots\cdots\cdots \\
a_{n1} & a_{n2} & a_{n3} & \cdots & a_{nn}
\end{pmatrix}
\tag{5.8}
$$

Assuming that the necessary row and column permutations take place so we may use the main diagonal elements as pivots, we see from formulas (5.6) that the first such pivot using a_{11} is equivalent to premultiplying (5.8) by the elementary column matrix

$$
\mathbf{E}^1 =
\begin{pmatrix}
\dfrac{1}{a_{11}} & 0 & 0 & \cdots & 0 \\[2mm]
\dfrac{-a_{21}}{a_{11}} & 1 & 0 & \cdots & 0 \\[2mm]
\dfrac{-a_{31}}{a_{11}} & 0 & 1 & \cdots & 0 \\[2mm]
\cdots\cdots\cdots\cdots\cdots\cdots \\[1mm]
\dfrac{-a_{n1}}{a_{11}} & 0 & 0 & \cdots & 1
\end{pmatrix}
$$

to yield

$$
\mathbf{E}^1\mathbf{A} =
\begin{pmatrix}
1 & a_{12}^{(1)} & a_{13}^{(1)} & \cdots & a_{1n}^{(1)} \\
0 & a_{22}^{(1)} & a_{23}^{(1)} & \cdots & a_{2n}^{(1)} \\
0 & a_{32}^{(1)} & a_{33}^{(1)} & \cdots & a_{3n}^{(1)} \\
\cdots\cdots\cdots\cdots\cdots\cdots\cdots\cdots \\
0 & a_{n2}^{(1)} & a_{n3}^{(1)} & \cdots & a_{nn}^{(1)}
\end{pmatrix}
\tag{5.9}
$$

Note that we are reducing the diagonal elements of the transformed matrix to unity and that $\mathbf{E}^1$ is just the inverse of the elementary column matrix

$$
\begin{pmatrix}
a_{11} & 0 & \cdots & 0 \\
a_{21} & 1 & \cdots & 0 \\
\cdots\cdots\cdots\cdots\cdots \\
a_{n1} & 0 & \cdots & 1
\end{pmatrix}
$$

The second step uses $a_{22}^{(1)}$ as the pivot element and is equivalent to

$$
\mathbf{E}^2\mathbf{E}^1\mathbf{A} =
\begin{pmatrix}
1 & 0 & a_{13}^{(2)} & \cdots & a_{1n}^{(2)} \\
0 & 1 & a_{23}^{(2)} & \cdots & a_{2n}^{(2)} \\
0 & 0 & a_{33}^{(2)} & \cdots & a_{3n}^{(2)} \\
\cdots\cdots\cdots\cdots\cdots\cdots\cdots\cdots \\
0 & 0 & a_{n3}^{(2)} & \cdots & a_{nn}^{(2)}
\end{pmatrix}
$$

where

$$\mathbf{E}^2 = \begin{pmatrix} 1 & \dfrac{-a_{12}^{(1)}}{a_{22}^{(1)}} & 0 & \cdots & 0 \\[2ex] 0 & \dfrac{1}{a_{22}^{(1)}} & 0 & \cdots & 0 \\[2ex] 0 & \dfrac{-a_{32}^{(1)}}{a_{22}^{(1)}} & 1 & \cdots & 0 \\[1ex] & \cdots\cdots\cdots\cdots\cdots \\[1ex] 0 & \dfrac{-a_{n2}^{(1)}}{a_{22}^{(1)}} & 0 & \cdots & 1 \end{pmatrix}$$

Continuing in this manner we perform n steps to obtain the transformation

$$\mathbf{E}^n\mathbf{E}^{n-1}\cdots\mathbf{E}^2\mathbf{E}^1\mathbf{A} = \mathbf{I} \tag{5.10}$$

Thus, $\mathbf{A}^{-1} = (\mathbf{E}^n\cdots\mathbf{E}^1)$ and it is given as the product of elementary column matrices. This representation is termed the *product form of the inverse*. Its application to linear programming is discussed in Chap. 6. Note from (5.10) that $\mathbf{A}$ can be represented as the product of elementary column matrices (Exercise 50).

To illustrate the product form of the inverse, consider the matrix

$$\mathbf{A} = \begin{pmatrix} 1 & 1 & -1 \\ -2 & 1 & 1 \\ 1 & 1 & 1 \end{pmatrix}$$

of Eqs. (5.2). Referring to the steps by which we found $\mathbf{A}^{-1}$ (p. 47), the reader can verify that

$$\mathbf{E}^3\mathbf{E}^2\mathbf{E}^1 = \begin{pmatrix} 1 & 0 & \frac{1}{3} \\ 0 & 1 & \frac{1}{6} \\ 0 & 0 & \frac{1}{2} \end{pmatrix} \begin{pmatrix} 1 & -\frac{1}{3} & 0 \\ 0 & \frac{1}{3} & 0 \\ 0 & 0 & 1 \end{pmatrix} \begin{pmatrix} 1 & 0 & 0 \\ 2 & 1 & 0 \\ -1 & 0 & 1 \end{pmatrix}$$

$$= \begin{pmatrix} 0 & -\frac{1}{3} & \frac{1}{3} \\ \frac{1}{2} & \frac{1}{3} & \frac{1}{6} \\ -\frac{1}{2} & 0 & \frac{1}{2} \end{pmatrix} = \mathbf{A}^{-1}$$

To solve the system $\mathbf{A}\mathbf{X} = \mathbf{b}$ by the Gauss–Jordan method requires the following number of arithmetic operations: n divisions, $\frac{1}{2}n^3 + n^2 - \frac{1}{2}n$ multiplications, and $\frac{1}{2}n^3 - \frac{1}{2}n$ additions and subtractions (Westlake [694]). To find $\mathbf{A}^{-1}$ by the Gauss–Jordan method, the corresponding number of arithmetic operations are, respectively, n, $n^3 - 1$, and $n^3 - 2n^2 + n$. We see that the amount of computational work to solve $\mathbf{A}\mathbf{X} = \mathbf{b}$ by the Gauss–Jordan method is for large values of n of the order of $n^3/2$. A more efficient scheme that is of the order $n^3/3$ is the original elimination method of Gauss.

Gaussian elimination first requires the carrying out of $n-1$ steps termed the *forward-course elimination*. The forward course applies a sequence of elementary row transformations to $\mathbf{A}$ that transforms $\mathbf{A}$ into an upper triangular matrix and the system $\mathbf{A}\mathbf{X} = \mathbf{b}$ into a corresponding triangular system of equations. These equations are then solved by the *return course* or *back-*

substitution procedure that produces values of the x_i. We illustrate the gaussian elimination method for the $n \times n$ nonsingular system $\mathbf{AX} = \mathbf{b}$ (we assume that the rows and columns are permuted as necessary to produce nonzero pivot elements along the principal diagonal).

STEP 0   (the given system)

$$a_{11}x_1 + a_{12}x_2 + a_{13}x_3 + \cdots + a_{1n}x_n = b_1$$
$$a_{21}x_1 + a_{22}x_2 + a_{23}x_3 + \cdots + a_{2n}x_n = b_2$$
$$a_{31}x_1 + a_{32}x_2 + a_{33}x_3 + \cdots + a_{3n}x_n = b_3$$
$$\cdots \cdots \cdots \cdots \cdots \cdots \cdots \cdots \cdots \cdots$$
$$a_{n1}x_1 + a_{n2}x_2 + a_{n3}x_3 + \cdots + a_{nn}x_n = b_n$$

The forward solution transforms these equations in $n - 1$ partial elimination steps as follows.

STEP 1

$$a_{11}x_1 + a_{12}\ x_2 + a_{13}\ x_3 + \cdots + a_{1n}\ x_n = b_1$$
$$a_{22}^{(1)}x_2 + a_{23}^{(1)}x_3 + \cdots + a_{2n}^{(1)}x_n = b_2^{(1)}$$
$$a_{32}^{(1)}x_3 + a_{33}^{(1)}x_3 + \cdots + a_{3n}^{(1)}x_n = b_3^{(1)}$$
$$\cdots \cdots \cdots \cdots \cdots \cdots \cdots \cdots \cdots \cdots$$
$$a_{n2}^{(1)}x_2 + a_{n3}^{(1)}x_3 + \cdots + a_{nn}^{(1)}x_n = b_n^{(1)}$$

STEP 2

$$a_{11}x_1 + a_{12}\ x_2 + a_{13}\ x_3 + \cdots + a_{1n}\ x_n = b_1$$
$$a_{22}^{(1)}x_2 + a_{23}^{(1)}x_3 + \cdots + a_{2n}^{(1)}x_n = b_2^{(1)}$$
$$a_{33}^{(2)}x_3 + \cdots + a_{3n}^{(2)}x_n = b_3^{(2)}$$
$$\cdots \cdots \cdots \cdots \cdots \cdots \cdots \cdots \cdots$$
$$a_{n3}^{(2)}x_3 + \cdots + a_{nn}^{(2)}x_n = b_n^{(2)}$$

STEP $n - 1$ (last step of the forward solution)

$$a_{11}x_1 + a_{12}\ x_2 + a_{13}\ x_3 + \cdots + a_{1n}\ x_n = b_1$$
$$a_{22}^{(1)}x_2 + a_{23}^{(1)}x_3 + \cdots + a_{2n}^{(1)}x_n = b_2^{(1)}$$
$$a_{33}^{(2)}x_3 + \cdots + a_{3n}^{(2)}x_n = b_3^{(2)}$$
$$\cdots \cdots \cdots \cdots$$
$$a_{nn}^{(n-1)}x_n = b_n^{(n-1)}$$

At the end of Step $n - 1$ the value of x_n is given by the nth equation, i.e.,

$$x_n = \frac{b_n^{(n-1)}}{a_{nn}^{(n-1)}}$$

The $(n - 1)$st equation only involves x_n and x_{n-1} and is given by

$$a_{n-1,n-1}^{(n-2)}x_{n-1} + a_{n-1,n}^{(n-2)}x_n = b_n^{(n-2)}$$

Thus, it is a simple matter to (back) substitute the value of x_n into the $(n-1)$st equation to find the value of x_{n-1}. The values of x_n and x_{n-1} are then substituted into the $(n-2)$nd equation to determine x_{n-2}, and so on, until all x_i are found. Note that the pivot row elements in each step are equal to the corresponding elements of the previous step and that final division by the pivot elements is accomplished during the back substitution.

The $n-1$ row transformation steps of the forward solution accomplish a **LU** decomposition of **A**. Let us denote the combined row transformations of the forward solution by the equivalent matrix $\mathbf{L}^{-1}$. Note that at the end of the forward solution, **A** has been transformed into an upper triangular matrix **U**. It is easy to verify that the row transformation applied in step k of the forward solution can be represented by a lower triangular matrix $\mathbf{L}_k^{-1}$, and that the product of these matrices is the lower triangular matrix $\mathbf{L}^{-1}$. If we denote the inverse of $\mathbf{L}_k^{-1}$ by $\mathbf{L}_k$ and the inverse of $\mathbf{L}^{-1}$ by $\mathbf{L}$, we have (with $\mathbf{L}_n^{-1} = \mathbf{I}_n$)

$$\mathbf{L}^{-1}\mathbf{A} = \mathbf{U} \qquad \text{or} \qquad \mathbf{L}_n^{-1}\cdots\mathbf{L}_1^{-1}\mathbf{A} = \mathbf{U}$$

and

$$\mathbf{A} = \mathbf{L}_1 \cdots \mathbf{L}_n\mathbf{U} = \mathbf{L}\mathbf{U}$$

and

$$\mathbf{A}^{-1} = \mathbf{U}^{-1}\mathbf{L}^{-1}$$

This representation is the *elimination form of the inverse.* The use of the **LU** decomposition in linear programming will be discussed in Chap. 6.

The matrix representations of $\mathbf{L}_k^{-1}$ and $\mathbf{L}_k$ are the following:

$$\mathbf{L}_k^{-1} = \begin{pmatrix} 1 & & & & & & \\ & \ddots & & & & & \\ & & 1 & & & & \\ & & & \dfrac{1}{a_{kk}^{(k-1)}} & & & \\ & & & -\dfrac{a_{k+1,k}^{(k-1)}}{a_{kk}^{(k-1)}} & 1 & & \\ & & & \vdots & & \ddots & \\ & & & -\dfrac{a_{nk}^{(k-1)}}{a_{kk}^{(k-1)}} & & & 1 \end{pmatrix}$$

and

$$\mathbf{L}_k = \begin{pmatrix} 1 & & & & & \\ & \ddots & & & & \\ & & 1 & & & \\ & & & a_{kk}^{(k-1)} & & \\ & & & a_{k+1,k}^{(k-1)} & 1 & \\ & & & \vdots & & \ddots \\ & & & a_{nk}^{(k-1)} & & 1 \end{pmatrix}$$

After the forward solution has been completed, the equations $\mathbf{AX} = \mathbf{b}$ have been transformed to $\mathbf{UX} = (\mathbf{L}^{-1}\mathbf{b})$.   The backward solution is then equivalent to finding the solution $\mathbf{X} = \mathbf{U}^{-1}\mathbf{L}^{-1}\mathbf{b} = \mathbf{U}_1^{-1} \cdots \mathbf{U}_n^{-1}\mathbf{L}_n^{-1} \cdots \mathbf{L}_1^{-1}\mathbf{b}$.

The total number of arithmetic operations required to solve $\mathbf{AX} = \mathbf{b}$ by gaussian elimination is n divisions, $\frac{1}{3}n^3 + n^2 - \frac{1}{3}n$ multiplications, and $\frac{1}{3}n^3 + n^2 - \frac{5}{6}n$ additions and subtractions (see Exercise 35).   The computation of $\mathbf{A}^{-1}$ takes as many operations as the Gauss–Jordan method.   (Note that it is more efficient to solve $\mathbf{AX} = \mathbf{b}$ for $\mathbf{X}$ directly than to first solve for $\mathbf{A}^{-1}$ and then $\mathbf{X} = \mathbf{A}^{-1}\mathbf{b}$.)   For large values of n, gaussian elimination takes on the order of $n^3/3$ operations, and, in general, is more efficient than the Gauss–Jordan method.

We illustrate the gaussian elimination method for Eqs. (5.2).

STEP 1   Starting with

$$x_1 + x_2 - x_3 = 2$$
$$-2x_1 + x_2 + x_3 = 3$$
$$x_1 + x_2 + x_3 = 6$$

we eliminate x_1 from the second and third equations to obtain

$$x_1 + x_2 - x_3 = 2$$
$$3x_2 - x_3 = 7$$
$$2x_3 = 4$$

Here

$$\mathbf{L}_1^{-1} = \begin{pmatrix} 1 & 0 & 0 \\ 2 & 1 & 0 \\ -1 & 0 & 1 \end{pmatrix}$$

Since the equations are now in the required upper triangular form (Step 1 also accomplished Step 2), we need go no further with the forward solution.   We can take

$$\mathbf{L}_2^{-1} = \mathbf{L}_3^{-1} = \begin{pmatrix} 1 & 0 & 0 \\ 0 & 1 & 0 \\ 0 & 0 & 1 \end{pmatrix}$$

We solve for the x_i by the back solution as follows:

$$x_3 = \tfrac{4}{2} = 2$$
$$x_2 = \frac{(7 + x_3)}{3} = 3$$
$$x_1 = 2 + x_3 - x_2 = 1$$

We also have that

$$
\mathbf{U} = \begin{pmatrix} 1 & 1 & -1 \\ 0 & 3 & -1 \\ 0 & 0 & 2 \end{pmatrix} = \mathbf{U}_3\mathbf{U}_2\mathbf{U}_1 = \begin{pmatrix} 1 & 0 & -1 \\ 0 & 1 & -1 \\ 0 & 0 & 2 \end{pmatrix}\begin{pmatrix} 1 & 1 & 0 \\ 0 & 3 & 0 \\ 0 & 0 & 1 \end{pmatrix}\begin{pmatrix} 1 & 0 & 0 \\ 0 & 1 & 0 \\ 0 & 0 & 1 \end{pmatrix}
$$

Thus,

$$
\mathbf{U}^{-1} = \mathbf{U}_1^{-1}\mathbf{U}_2^{-1}\mathbf{U}_3^{-1} = \begin{pmatrix} 1 & 0 & 0 \\ 0 & 1 & 0 \\ 0 & 0 & 1 \end{pmatrix}\begin{pmatrix} 1 & -\frac{1}{3} & 0 \\ 0 & \frac{1}{3} & 0 \\ 0 & 0 & 1 \end{pmatrix}\begin{pmatrix} 1 & 0 & \frac{1}{2} \\ 0 & 1 & \frac{1}{2} \\ 0 & 0 & \frac{1}{2} \end{pmatrix}
$$

$$
= \begin{pmatrix} 1 & -\frac{1}{3} & \frac{1}{3} \\ 0 & \frac{1}{3} & \frac{1}{6} \\ 0 & 0 & \frac{1}{2} \end{pmatrix}
$$

As

$$
\mathbf{L}^{-1} = \mathbf{L}_3^{-1}\mathbf{L}_2^{-1}\mathbf{L}_1^{-1} = \begin{pmatrix} 1 & 0 & 0 \\ 2 & 1 & 0 \\ -1 & 0 & 1 \end{pmatrix} \qquad \text{with} \qquad \mathbf{L} = \begin{pmatrix} 1 & 0 & 0 \\ -2 & 1 & 0 \\ 1 & 0 & 1 \end{pmatrix}
$$

we have

$$
\mathbf{A}^{-1} = \mathbf{U}^{-1}\mathbf{L}^{-1} = \begin{pmatrix} 1 & -\frac{1}{3} & \frac{1}{3} \\ 0 & \frac{1}{3} & \frac{1}{6} \\ 0 & 0 & \frac{1}{2} \end{pmatrix}\begin{pmatrix} 1 & 0 & 0 \\ 2 & 1 & 0 \\ -1 & 0 & 1 \end{pmatrix} = \begin{pmatrix} 0 & -\frac{1}{3} & \frac{1}{3} \\ \frac{1}{2} & \frac{1}{3} & \frac{1}{6} \\ -\frac{1}{2} & 0 & \frac{1}{2} \end{pmatrix}
$$

and $\mathbf{A} = \mathbf{LU}$. Note that we have determined a different $\mathbf{LU}$ decomposition of $\mathbf{A}$ [see (1.1)] as we did not reduce the diagonal elements of the gaussian elimination to unity (Exercise 24). Also, in solving the equations, only $\mathbf{U}$ is determined explicitly.

When carrying out any elimination-based method, we need to be concerned with the numerical accuracy of the process. This is especially true when solving large systems of equations on a digital computer. For elimination procedures, accuracy control centers about the size of the pivot elements, i.e., the absolute value of those elements that are positioned along the diagonal and used for division. The problem arises if a pivot element is close to zero; a division by a small number usually introduces numerical instability and causes severe roundoff errors. Thus, the pivoting selection strategy is to select at each step a pivot that has a large absolute value. Westlake [694] suggests three possible strategies, each having increasing accuracy accompanied by increased computation (see also Ralston and Rabinowitz [557]).

The easiest scheme is *no positioning for size* and the process just moves down the diagonal at each step. However, if this leads to a zero (or small) pivot, then some nonzero element in the same pivot column is found and it is moved to the diagonal by interchanging rows, i.e., making its row the new pivot row (the actual interchange does not have to be effected). This always can be done for a nonsingular matrix. This scheme is not recommended for large systems. The second strategy, termed *partial positioning for size* (partial pivoting), chooses the pivot for the kth step to be the element of largest absolute value in column k, rows k through n. The third pivot selection scheme is that of *complete positioning for size* (complete pivoting). Here, at the kth step, the pivot is chosen to be the element of largest absolute value in the remaining submatrix consisting of columns k through n and rows k through n. Both row and column interchanges may be required to place the pivot element on the diagonal. Complete positioning yields, in general, the most accurate results but requires the most computational effort. Westlake [694] notes that the number of comparisons needed to select pivots for partial positioning is $n(n + 1)/2$ and the number for complete positioning is $n(n + 1)(2n + 1)/6$. As a compromise between accuracy and computational effort, partial positioning is the recommended pivot strategy. Note that row and column interchanges can be accomplished by pre- and postmultiplication of permutation matrices applied to the original system. We often find that some systems of equations have many original $a_{ij} = 0$, with the zero elements falling into special patterns. Complete positioning tends to reduce this pattern, while partial positioning often maintains it. This is especially important when solving large-scale linear-programming problems, problems that always have structure (patterns) with many zeros (see Chap. 9, Sec. 4, on the decomposition of large-scale systems). In Sec. 3 of Chap. 5, we discuss elimination procedures that have been designed to take advantage of the *sparsity* of nonzero elements found in large problems.

Another method for solving square sets of simultaneous linear equations is given by Cramer's rule, Hadley [342]. We cite it here for reference only, as we will not employ it for computational purposes. Cramer's rule states that for the set of n equations $\mathbf{AX} = \mathbf{b}$, where $\mathbf{A} = (\mathbf{P}_1, \mathbf{P}_2, \ldots, \mathbf{P}_n)$, $\mathbf{X} = (x_1, x_2, \ldots, x_n)$ then the unique solution is given by

$$x_1 = \frac{|\mathbf{bP}_2 \cdots \mathbf{P}_n|}{|\mathbf{A}|}, \; x_2 = \frac{|\mathbf{P}_1\mathbf{b} \cdots \mathbf{P}_n|}{|\mathbf{A}|}, \ldots, \; x_n = \frac{|\mathbf{P}_1\mathbf{P}_2 \cdots \mathbf{b}|}{|\mathbf{A}|}$$

where $|\mathbf{A}| \neq 0$ and the numerator for variable x_i is the determinant formed from matrix $\mathbf{A}$ by replacing vector $\mathbf{P}_i$ by the right-hand-side vector $\mathbf{b}$.

A system of linear equations $\mathbf{AX} = \mathbf{b}$ is said to be *homogeneous* if $\mathbf{b} = \mathbf{0}$. Such a system always has the *trivial solution* $\mathbf{X} = \mathbf{0}$. (See Exercises 14, 17, 42, and 43.)

REMARKS

Much of the material in Sec. 1 is from Hildebrand [366], and much of that in Sec. 2 is from Kuhn [445]. Additional material in these and the other sections is from Ralston and Rabinowitz [557], Westlake [694], Tewarson [639], Dwyer [226], Faddeeva [244], Forrest and Tomlin [257], Koopmans [437], Kuhn and Tucker [449], Hadley [342], Nef [523], Munakata [519], Krekó [443], Mangasarian [489], Rockafellar [571], Gantmacher [279], Stoer and Witzgall [632], Goldman and Tucker [317, 318], and Tucker [659].

EXERCISES

1. Graph the convex hull of the points

    $(0, 0)$, $(1, 1)$, $(-1, -1)$, $(-2, 2)$, $(1, 4)$, $(0, 3)$, $(-1, 1)$, $(\tfrac{1}{2}, 4)$, $(-1, 2)$, $(2, 5)$

Find convex combinations of the extreme points which express those given points that are interior points of the convex hull.

2. Graph the following linear constraints, mark the area that satisfies the constraints, and determine the extreme points of this convex set:

$$-2x_1 + 5x_2 - 10 \leq 0$$
$$2x_1 + x_2 - 6 \leq 0$$
$$x_1 + 2x_2 - 2 \geq 0$$
$$-x_1 + 3x_2 - 3 \leq 0$$

3. Write the system of inequalities of Exercise 2 as a system of equalities in nonnegative variables.

4. Assume a given basis of the vectors

$$\mathbf{P}_1 = \begin{pmatrix} 1 \\ 0 \\ 1 \end{pmatrix} \qquad \mathbf{P}_2 = \begin{pmatrix} 1 \\ 1 \\ 1 \end{pmatrix} \qquad \mathbf{P}_3 = \begin{pmatrix} 2 \\ 1 \\ 1 \end{pmatrix}$$

Compute the inverse of the matrix associated with this basis and determine the linear combination of the basis vectors that equals vector $\mathbf{P}_4 = \begin{pmatrix} 1 \\ 3 \\ 4 \end{pmatrix}$.

Answer: $\mathbf{A}^{-1} = \begin{pmatrix} 0 & -1 & 1 \\ -1 & 1 & 1 \\ 1 & 0 & -1 \end{pmatrix}$, $\mathbf{A}^{-1}\mathbf{P}_4 = \begin{pmatrix} 1 \\ 6 \\ -3 \end{pmatrix}$

5. The convex region shown in Fig. E5 is the set of solutions determined by a set of linear inequalities. At what point or points are the following linear functions optimized?

 a. $x_1 + x_2 - 1$, to be a maximum
 b. $3x_1 - x_2 + 6$, to be a minimum
 c. $-2x_1 - 2x_2 + 2$, to be a maximum

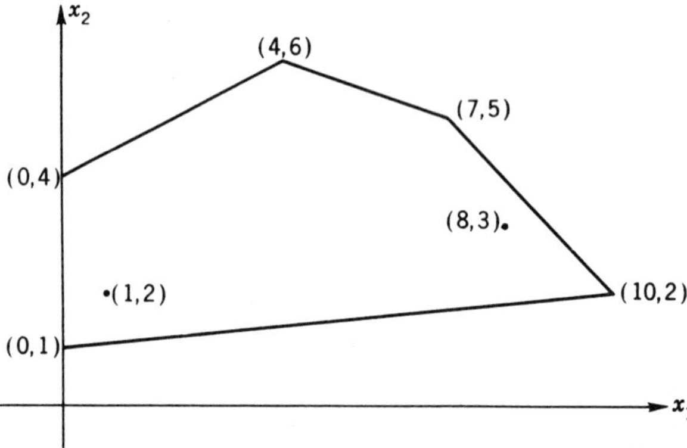

FIGURE E5

6. For the points (1, 2) and (8, 3) in Fig. E5, determine equivalent representations in terms of convex combinations of the extreme points.

7. Solve the following set of equations using both the gaussian and Gauss–Jordan elimination procedures. Determine an **LU** representation of **A**. Also compute $\mathbf{A}^{-1}$ by means of the adjoint matrix and by finding $(\mathbf{LU})^{-1}$. Check your results by computing $\mathbf{AA}^{-1}$.

$$2x_1 + 2x_2 - x_3 = 4$$
$$3x_1 - x_2 - 3x_3 = 7$$
$$x_1 + x_2 + 2x_3 = 3$$

Answer: $x_1 = \frac{13}{5}$, $x_2 = -\frac{2}{5}$, $x_3 = \frac{2}{5}$

8. Solve the following linear-programming problems by graphical methods:

a. Constraints:

$$2x_1 - x_2 \geq -2$$
$$x_1 + 2x_2 \leq 8$$
$$x_1 \qquad \geq 0$$
$$x_2 \geq 0$$

Objective functions: maximize x_2, maximize $3x_1 + 2x_2$, minimize $2x_1 - 2x_2$, maximize $2x_1 + 4x_2$, minimize $-3x_1 - 2x_2$.

b. Constraints:

$$3x_1 + 2x_2 \leq 6$$
$$x_1 - x_2 \geq -1$$
$$-x_1 - 2x_2 \geq 1$$
$$x_1 \qquad \geq 0$$
$$x_2 \geq 0$$

Objective function: maximize $2x_1 - 6x_2$.

c. Constraints:

$$x_1 - 3x_2 \leq 6$$
$$2x_1 + 4x_2 \geq 8$$
$$x_1 - 3x_2 \geq -6$$
$$x_1 \qquad \geq 0$$
$$x_2 \geq 0$$

Objective functions: maximize $2x_1 + 3x_2$, minimize $x_1 + 2x_2$, maximize $x_1 - 2x_2$, maximize $x_1 - 3x_2$, maximize $x_1 - 6x_2$.

d. Constraints:

$$x_1 + x_2 + x_3 \leq 2$$
$$x_1 + x_2 - x_3 \leq 1$$
$$x_1 \qquad\qquad \geq 0$$
$$x_2 \qquad \geq 0$$
$$x_3 \geq 0$$

Objective functions: maximize $x_1 + x_2 + x_3$, minimize $x_1 + x_3$, maximize $x_1 + x_2 - 3x_3$.

9. For a given set of inequalities

$$a_{i1}x_1 + \cdots + a_{in}x_n \leq 1 \qquad i = 1, \ldots, m$$
$$x_j \geq 0 \qquad j = 1, \ldots, n$$

discuss conditions that indicate that an inequality does not form part of the boundary of the solution space.

10. Define the intersection of two sets $\mathbf{K}_1$ and $\mathbf{K}_2$ to be the set of points belonging to both $\mathbf{K}_1$ and $\mathbf{K}_2$. If $\mathbf{K}_1$ and $\mathbf{K}_2$ are both convex, prove that their intersection is also convex.

11. For a convex cone, show that the positive sum of any two vectors in the cone is also in the cone.

12. Solve Exercise 5 of Chap. 1 graphically. Describe what happens to the optimal solution as the cost of 1 oz of Crispies changes from 0.0 to 10.0 cents. For what values of the costs do we obtain multiple, optimal solutions?

13. Discuss the geometric interpretation of degeneracy.

14. Given a solution to the equations $AX = b$, show that any other solution can be expressed as a vector sum of the given solution and a solution of the corresponding homogeneous system, and thus all solutions to $AX = b$ can be generated by knowing a single solution to $AX = b$ and all solutions to the homogeneous system (Hadley [342]).

15. Show that the maximum of a linear function $f(X)$ is equal to $-[\text{minimum} - f(X)]$.

16. Show for $b \geq 0$, that $X = 0$ is an extreme point of the solution space to $AX \leq b$, $X \geq 0$, Krekó [443].

17. Show that the set of solutions K to the constraints $AX \leq b$, $X \geq 0$, $b \geq 0$ is unbounded if and only if the corresponding homogeneous system $AX \leq 0$, $X \geq 0$ has a solution other than $X = 0$, Krekó [443].

18. Show how we can represent the general linear equation $a_{i1}x_1 + a_{i2}x_2 + \cdots + a_{in}x_n = b_i$ by two linear inequalities.

19. Show that a problem with all n variables x_j unrestricted can be restated in nonnegative variables $x_j = x_j' - x_0$, $x_j' \geq 0$, $x_0 \geq 0$, i.e., in terms of $(n + 1)$ nonnegative variables (Deutsch [201]).

20. Show that a half space is a convex set.

21. For Exercise 8c, graph the constraints and determine the solution set's extreme points and extreme directions. Determine representations of the following points in terms of convex combinations of the extreme points and nonnegative combinations of the extreme directions: $(3, \frac{1}{2})$, $(3, \frac{3}{4})$, $(3, 1)$, $(6, 4)$, $(8, 2)$, $(8, 3)$.

22. Show that the inverse of a lower (upper) triangular matrix is a lower (upper) triangular matrix, and that the product of lower (upper) triangular matrices is a lower (upper) triangular matrix. Also show that the inverse of a lower (upper) triangular matrix has nonzero elements only in the same positions as the original matrix.

23. Develop the formulas for the inverse of the block-partitioned matrix

$$A = \begin{pmatrix} K & L \\ M & N \end{pmatrix}$$

HINT: For

$$A^{-1} = \begin{pmatrix} P & Q \\ R & S \end{pmatrix}$$

form the product

$$AA^{-1} = \begin{pmatrix} K & L \\ M & N \end{pmatrix}\begin{pmatrix} P & Q \\ R & S \end{pmatrix} = \begin{pmatrix} I_k & 0 \\ 0 & I_l \end{pmatrix}$$

and solve the resulting four matrix equations (Hadley [342] and Faddeeva [244]). Assume the matrices have the proper dimensions. For $M = (0)$, show that

$$A^{-1} = \begin{pmatrix} K^{-1} & -K^{-1}LN^{-1} \\ 0 & N^{-1} \end{pmatrix}$$

(Stewart [627]).

24. Consider the following 3×3 nonsingular matrix $A = LU$:

$$A = \begin{pmatrix} a_{11} & a_{12} & a_{13} \\ a_{21} & a_{22} & a_{23} \\ a_{31} & a_{32} & a_{33} \end{pmatrix} = \begin{pmatrix} l_{11} & 0 & 0 \\ l_{21} & l_{22} & 0 \\ l_{31} & l_{32} & l_{33} \end{pmatrix}\begin{pmatrix} u_{11} & u_{12} & u_{13} \\ 0 & u_{22} & u_{23} \\ 0 & 0 & u_{33} \end{pmatrix}$$

Develop formulas for determining the l_{ij} and u_{ij} in terms of the a_{ij} and show that if all $l_{ii} = 1$ (or $u_{ii} = 1$) then the remaining l_{ij} and u_{ij} are uniquely determined. HINT: Set a_{ij} equal to the corresponding element of the product of LU. (See Faddeeva [244], Westlake [694], and Gantmacher [279].)

25. Show that the **LU** decomposition of a nonsingular matrix

$$\mathbf{A} = \begin{pmatrix} a_{11} & a_{12} & \cdots & a_{1n} \\ a_{21} & a_{22} & \cdots & a_{2n} \\ \hdotsfor{4} \\ a_{n1} & a_{n2} & \cdots & a_{nn} \end{pmatrix}$$

requires that all of its leading principal minors be nonzero, i.e.,

$$a_{11} \neq 0, \quad \begin{vmatrix} a_{11} & a_{12} \\ a_{21} & a_{22} \end{vmatrix} \neq 0, \ldots, |\mathbf{A}| \neq 0$$

(Faddeeva [244], Westlake [694], and Gantmacher [279]).   HINT: Use the structure of Exercise 24.

26. Based on the results of Exercise 24, show that a nonsingular matrix **A** can be expressed uniquely as $\mathbf{A} = \mathbf{LDU}$, where **L** is a unit lower triangular matrix (all $l_{ii} = 1$), **D** is a diagonal matrix, and **U** is a unit upper triangular matrix (all $u_{ii} = 1$).   Assume **A** satisfies the conditions given in Exercise 25.

27. Using the formulas found in Exercise 24, find **LU** decompositions of the following three matrices.   Also, factor the **LU** forms and determine the corresponding inverses.

$$\begin{pmatrix} 2 & 1 & 1 \\ 1 & 0 & 2 \\ 3 & 1 & 2 \end{pmatrix} \quad \begin{pmatrix} 1 & 2 & 1 \\ 2 & 1 & 0 \\ -1 & 0 & 1 \end{pmatrix} \quad \begin{pmatrix} 2 & -2 & 1 \\ 2 & -3 & 2 \\ -1 & 2 & 0 \end{pmatrix}$$

28. Show that the inverses of elementary row (column) matrices are elementary row (column) matrices and determine the forms of the inverses.

29. For the matrices of Exercise 27, find suitable block partitions and find their inverses using block partition formulas. (See Exercise 23.)

30. Solve Eqs. (5.2) by gaussian elimination using both partial and complete positioning for size.

31. Using the formulas given in Sec. 5, determine the exact number of arithmetic operations required to solve an $n = 100$ equation system by gaussian and Gauss–Jordan elimination.

32. Using the properties of matrix multiplication, explain how you can combine a sequence of elementary row (column) transformations into one composite transformation matrix.   How must the elementary row (column) transformations be restricted to make the composite matrix lower (upper) triangular?

33. Describe how you can carry out the Gauss–Jordan or the gaussian elimination procedure with pivoting on the principal diagonal.

34. For the gaussian elimination and the Gauss–Jordan methods, develop formulas that express the transformations from step $k - 1$ to step k.   For the original data let $a_{ij} = a_{ij}^{(0)}$ and $b_i = b_i^{(0)}$ (Ralston and Rabinowitz [557], Westlake [694], and Gantmacher [279]).

35. For the gaussian elimination method, show that the total number of multiplications and divisions required to solve a system of $n \times n$ equations is given by the formula

$$\sum_{k=1}^{n-1} (n - k)(n - k + 1) + (n - k)$$

From this, show that this total is dominated by the term $\frac{1}{3}n^3$ when n is large (Ralston and Rabinowitz [557]).   You will need to use the following relationships: $\sum_{k=1}^{n} k = n(n + 1)/2$ and $\sum_{k=1}^{n} k^2 = n(n + 1)(2n + 1)/6$.

36. Discuss the accuracy and roundoff aspects of solving the following set of equations by gaussian elimination with and without positioning for size (Ralston and Rabinowitz [557]):

$$0.3 \times 10^{-11} x_1 + x_2 = 0.7$$
$$x_1 + x_2 = 0.9$$

The true solution to 10 digits is $x_1 = 0.2000000000$, $x_2 = 0.7000000000$.

37. Show that partial pivoting requires $n(n + 1)/2$ comparisons and complete partitioning requires $n(n + 1)(2n + 1)/6$ comparisons.

38. Discuss how you would solve the system of equations $\mathbf{AX} = \mathbf{LUX} = \mathbf{b}$, given $\mathbf{L}$ and $\mathbf{U}$. HINT: Let $\mathbf{Y} = \mathbf{UX}$ and solve the triangular system $\mathbf{LY} = \mathbf{b}$ for $\mathbf{Y}$ by *forward substitution*, and then solve the triangular system $\mathbf{UX} = \mathbf{Y}$ by back substitution. Develop the explicit formulas for the individual y_i and x_i in terms of the elements of $\mathbf{L} = (l_{ij})$, $\mathbf{U} = (u_{ij})$ and $\mathbf{b} = (b_i)$. Note that determining $\mathbf{L}$ and $\mathbf{U}$ is independent of $\mathbf{b}$. (Ralston and Rabinowitz [557] and Westlake [694].)

39. Let $\mathbf{E}_{ij}$ be the elementary matrix that interchanges rows i and j when used as a left multiplier and $\mathbf{E}_{kl}$ be the elementary matrix that interchanges columns k and l when used as a right multiplier. Let $\mathbf{A}$ be a nonsingular matrix that has been permuted as follows:

$$\bar{\mathbf{A}} = \mathbf{E}_{i_r j_r} \cdots \mathbf{E}_{i_1 j_1}(\mathbf{A})\mathbf{E}_{k_1 l_1} \cdots \mathbf{E}_{k_s l_s}$$

Show that

$$\mathbf{A}^{-1} = \mathbf{E}_{k_1 l_1} \cdots \mathbf{E}_{k_s l_s}(\bar{\mathbf{A}}^{-1})\mathbf{E}_{i_r j_r} \cdots \mathbf{E}_{i_1 j_1}$$

(Westlake [694]).

40. For the $\mathbf{L}_k$ and $\mathbf{U}_k$ of (1.3), develop the $\mathbf{L}_k^{-1}$ and $\mathbf{U}_k^{-1}$ of (1.4) and (1.5).

41. For the gaussian elimination method, verify the form of the $\mathbf{L}_k^{-1}$.

42. A standard numerical analysis method for checking the accuracy of the computational steps in elimination methods is to add a column of n elements to the system of equations, with the elements being equal to $b_i + \sum_j a_{ij}$. This is equivalent to solving a new system formed by the substitution $\bar{x}_i = x_i + 1$. The elements of the added column are transformed as a second constant column. Each transformed element should be equal to the corresponding sums of the transformed b_i and a_{ij} after each step. Try this check scheme for the problem of Exercise 7. A basic means of determining the accuracy of your answers is to substitute them into the original equations.

43. Justify formulas (5.6) and (5.7). Develop similar formulas for the forward solution of gaussian elimination.

44. In our discussion of the elimination procedures, we assumed for the $n \times n$ system $\mathbf{AX} = \mathbf{b}$ that $|\mathbf{A}| \neq 0$, i.e., the equations are determinate and consistent (have a unique solution). If we apply an elimination method to a square system of equations for which $|\mathbf{A}| = 0$, then one of the following two cases can occur: (*i*) All elements of some row, including the right-hand-side term, transform to a zero during the (forward) solution process, or (*ii*) all elements of some row transform to zero, but the corresponding right-hand-side term is nonzero. The first case indicates that the equations are indeterminate and consistent (have an infinite number of solutions) and the original equations corresponding to the zero rows are linear combinations of the others, i.e., they can be obtained by multiplying the other equations by certain values and adding. The second case indicates that the equations are indeterminate and inconsistent (have no solution). When an equation (row) transforms to all zeros, that equation is called *redundant* (it is dependent on the other equations and can be derived from them) and can be deleted. Contrast the use of the term "redundant equation" here with the term "redundant constraint" (inequality) in describing the solution space of a convex polyhedron in Sec. 4, Chap. 2. Show how the above cases manifest themselves in solving the following problems (Dwyer [226]). (See Telgen [637] for a complete discussion on redundant constraints.)

a.

$$\begin{aligned} x_1 + x_2 - x_3 &= 1 \\ 2x_1 - x_2 + 2x_3 &= 2 \\ 3x_1 \quad\quad + x_3 &= 3 \end{aligned}$$

c.

$$\begin{aligned} x_1 + x_2 + x_3 &= 1 \\ x_1 + x_2 - 2x_3 &= 1 \\ 2x_1 + 2x_2 - x_3 &= 2 \end{aligned}$$

b.

$$\begin{aligned} 3x_1 - 2x_2 + x_3 &= -4 \\ x_1 - x_2 + 2x_3 &= -3 \\ 5x_1 - 4x_2 + 5x_3 &= 2 \end{aligned}$$

45. For the solution set to $\mathbf{AX} = \mathbf{b}$, $\mathbf{X} \geq \mathbf{0}$, show that $\mathbf{Y} \neq \mathbf{0}$ is a direction of the set if and only if $\mathbf{A}(\mathbf{X} + \theta\mathbf{Y}) = \mathbf{b}$ and $(\mathbf{X} + \theta\mathbf{Y}) \geq \mathbf{0}$, for all $\mathbf{X}$ in the set. (HINT: Show that $\mathbf{Y}$ is a nonzero, nonnegative solution to the corresponding homogeneous equations (Bazaraa and Jarvis [50], and Stoer and Witzgall [632]).

46. For the homogeneous equations $\mathbf{AX} = \mathbf{0}$, show that if the solution set has a nontrivial solution, then the set is a convex cone, Krekó [443].

47. For the solution set to $\mathbf{AX} = \mathbf{b}$, show that if it contains two different points, then the set is unbounded (Krekó [443]).

48. Transform the following set of constraints into a system of equations in nonnegative variables:

$$-2x_1 + 3x_2 - 4x_3 + 6x_4 \leq 3$$
$$x_1 - x_2 + 3x_3 - 2x_4 \geq 1$$
$$x_1 + x_2 + x_3 - x_4 \leq 3$$

where $(x_1, x_3) \geq 0$, $(x_2, x_4$ unrestricted).

49. Show that the inverse of an elementary column (row) matrix $\mathbf{E}^j$ ($\mathbf{E}_i$) is another elementary column (row) matrix $\mathbf{E}^j$ ($\mathbf{E}_i$). Note that an $\mathbf{E}^j$ or $\mathbf{E}_i$ is nonsingular.

50. Show how to express a matrix $\mathbf{A}$ as the product of elementary matrices. HINT: Use Eq. (5.10).

51. For the matrices of Exercise 27, determine both the product form and the elimination form of the corresponding inverses. Also express each matrix as the product of elementary column matrices that are derived from each representation.

52. Show that the inverse of a permutation matrix $\mathbf{P}$ is $\mathbf{P}'$.

PART 2

Methods: Theoretical and Computational

CHAPTER 3

The General Linear-Programming Problem

1. THE LINEAR-PROGRAMMING PROBLEM

The general linear-programming problem is to find a vector $(x_1, x_2, \ldots, x_j, \ldots, x_n)$ which minimizes the linear form (i.e., the objective function)

$$c_1 x_1 + c_2 x_2 + \cdots + c_j x_j + \cdots + c_n x_n \tag{1.1}$$

subject to the linear constraints

$$
\begin{aligned}
a_{11} x_1 + a_{12} x_2 + \cdots + a_{1j} x_j + \cdots + a_{1n} x_n &= b_1 \\
a_{21} x_1 + a_{22} x_2 + \cdots + a_{2j} x_j + \cdots + a_{2n} x_n &= b_2 \\
&\cdots \\
a_{i1} x_1 + a_{i2} x_2 + \cdots + a_{ij} x_j + \cdots + a_{in} x_n &= b_i \\
&\cdots \\
a_{m1} x_1 + a_{m2} x_2 + \cdots + a_{mj} x_j + \cdots + a_{mn} x_n &= b_m
\end{aligned}
\tag{1.2}
$$

and

$$x_j \geq 0 \qquad j = 1, 2, \ldots, n \tag{1.3}$$

where the a_{ij}, b_i, and c_j are given constants and $m < n$. We shall always assume that Eqs. (1.2) have been multiplied by -1 where necessary to make all $b_i \geq 0$. Because of the variety of notations in common use, one will find the general linear-programming problem stated in many forms. The more common are the following:

a. Minimize

$$\sum_{j=1}^{n} c_j x_j$$

subject to

$$\sum_{j=1}^{n} a_{ij} x_j = b_i \qquad i = 1, 2, \ldots, m$$

and

$$x_j \geq 0 \qquad j = 1, 2, \ldots, n$$

b. Minimize

$$f(\mathbf{X}) = \mathbf{cX}$$

subject to

$$\mathbf{AX} = \mathbf{b}$$

and

$$\mathbf{X} \geq \mathbf{0}$$

where $\mathbf{c} = (c_1, c_2, \ldots, c_n)$ is a row vector, $\mathbf{X} = (x_1, x_2, \ldots, x_n)$ is a column vector, $\mathbf{A} = (a_{ij})$, $\mathbf{b} = (b_1, b_2, \ldots, b_m)$ is a column vector, and $\mathbf{0}$ is an n-dimensional null column vector.

c. Minimize

$$\mathbf{cX}$$

subject to

$$x_1 \mathbf{P}_1 + x_2 \mathbf{P}_2 + \cdots + x_n \mathbf{P}_n = \mathbf{P}_0$$

and

$$\mathbf{X} \geq \mathbf{0}$$

where $\mathbf{P}_j$ for $j = 1, 2, \ldots, n$ is the jth column of the matrix $\mathbf{A}$ and $\mathbf{P}_0 = \mathbf{b}$.

In some situations, it is convenient to define a new, unrestricted variable, e.g., x_0 or z, which is equal to the value of the objective. We would then have the following representation for the linear-programming problem (see Exercise 8, Chap. 4):

d. Minimize

$$x_0$$

subject to

$$x_0 - \sum_{j=1}^{n} c_j x_j = 0$$

$$\sum_{j=1}^{n} a_{ij} x_j = b_i \qquad i = 1, 2, \ldots, m$$

$$x_j \geq 0 \qquad j = 1, 2, \ldots, n$$

When formulating a linear-programming model, we often find that the conditions of the problem can be represented by a mixture of linear equations and inequalities, usually in terms of nonnegative variables. As described in Sec. 4, Chap. 2, linear conditions can always be converted to equations in nonnegative variables by suitable additions of nonnegative slack variables and variable substitutions. It is convenient, and in most cases appropriate, to set the cost coefficients of the slack variables equal to zero. (There are circumstances, however, in which slack variable cost coefficients can be different from zero; see Sec. 4, Chap. 4.) To standardize our discussion, we shall assume the equation form (1.1), (1.2), and (1.3). The reader should recall that this form also includes the maximization of a linear form; see Exercise 15, Chap. 2. Although we refer to the c_j as cost coefficients, it should be emphasized that these coefficients can represent a variety of other measures to be optimized, e.g., net profit, distance, time; see Chap. 11 applications and exercises. For discussion purposes, we usually assume that the data of the linear-programming model—the a_{ij}, b_i, c_j—are known exactly, i.e., the problem is *determined*. This assumption is often not met in practice, and techniques such as sensitivity analysis are used to measure the impact of uncertainty in the data; see Chap. 8.

2. PROPERTIES OF A SOLUTION TO THE LINEAR-PROGRAMMING PROBLEM

In this section we shall state a number of standard definitions and describe the more important characteristics of a solution to the general linear-programming problem. Much of this material and that in Chap. 4 is contained in Dantzig [153] and Charnes, Cooper, and Henderson [118].

Definition 1. A *feasible solution* to the linear-programming problem is a vector $\mathbf{X} = (x_1, x_2, \ldots, x_n)$ which satisfies conditions (1.2) and (1.3).

Definition 2a. A *basic solution* to (1.2) is a solution obtained by setting $n - m$ variables equal to zero and solving for the remaining m variables, provided that the determinant of the coefficients of these m variables is nonzero. The m variables are called *basic variables*.

Definition 2b. A *basic feasible solution* is a basic solution which also satisfies (1.3); that is, all basic variables are nonnegative.

Definition 3. A *nondegenerate basic feasible solution* is a basic feasible solution with *exactly m* positive x_i; that is, all basic variables are positive.

Definition 4. A *minimum feasible solution* is a feasible solution which also minimizes (1.1).

Definition 5. An *optimal basic feasible solution* is a basic solution that satisfies conditions (1.1), (1.2), and (1.3).

Unless otherwise stated, when we refer to a solution, we shall mean any feasible solution.

Definition 6. A *linear functional* $f(\mathbf{X})$ is a real-valued function defined on an n-dimensional vector space such that, for every vector $\mathbf{X} = \alpha \mathbf{U} + \beta \mathbf{V}$, $f(\mathbf{X}) = f(\alpha \mathbf{U} + \beta \mathbf{V}) = \alpha f(\mathbf{U}) + \beta f(\mathbf{V})$ for all n-dimensional vectors $\mathbf{U}$ and $\mathbf{V}$ and all scalars α and β. For example, let $\mathbf{U} = (9, 3)$, $\mathbf{V} = (6, 6)$, $\alpha = \frac{1}{3}$, $\beta = -\frac{2}{3}$, and $f(\mathbf{X}) = 2x_1 + x_2$. We have $\alpha \mathbf{U} + \beta \mathbf{V} = (-1, -3)$ and $f(\alpha \mathbf{U} + \beta \mathbf{V}) = -5$; $\alpha f(\mathbf{U}) = 7$ and $\beta f(\mathbf{V}) = -12$.

We note that the objective function (1.1) is a linear functional for those $\mathbf{X}$ satisfying (1.2) and (1.3).

We illustrate some of the above definitions for the following linear-programming problem: Minimize

$$- x_2 \quad + 3x_4$$

subject to

$$-x_1 + 3x_2 \quad\quad + 6x_4 \quad\quad = 18$$
$$2x_2 + x_3 + 3x_4 \quad\quad = 24$$
$$x_2 \quad\quad - x_4 + x_5 = 4$$
$$x_j \geq 0$$

A feasible solution is $\mathbf{X} = (x_1, x_2, x_3, x_4, x_5) = (15, 5, 5, 3, 2)$; a basic solution is $\mathbf{X} = (-18, 0, 24, 0, 4)$; a nondegenerate basic feasible solution is $\mathbf{X} = (0, 0, 15, 3, 7)$; a minimum feasible solution is $\mathbf{X} = (0, \frac{14}{3}, \frac{38}{3}, \frac{2}{3}, 0)$. The minimum value of the objective function is $f(\mathbf{X}) = f(0, \frac{14}{3}, \frac{38}{3}, \frac{2}{3}, 0) = -\frac{14}{3} + 0(\frac{38}{3}) + 3(\frac{2}{3}) = -\frac{8}{3}$. The minimum feasible solution is also an optimal basic feasible solution. If a problem has a (optimal) feasible solution, it has a (optimal) basic feasible solution; see Exercise 7, Chap. 4.

Theorem 1. *The set of all feasible solutions to the linear-programming problem is a convex set.*

Proof: We need to show that every convex combination of any two feasible solutions is also a feasible solution. (The theorem is true, of course, if the set of solutions has only one element.)   Assume there are at least two solutions X_1 and X_2.   We have

$$AX_1 = b \qquad \text{for } X_1 \geq 0$$

and

$$AX_2 = b \qquad \text{for } X_2 \geq 0$$

For $0 \leq \alpha \leq 1$, let $X = \alpha X_1 + (1 - \alpha)X_2$ be any convex combination of X_1 and X_2.   We note all the elements of the vector X are nonnegative; that is, $X \geq 0$.   We then see that X is a feasible solution, for we have

$$AX = A[\alpha X_1 + (1 - \alpha)X_2] = \alpha AX_1 + (1 - \alpha)AX_2$$
$$= \alpha b + b - \alpha b = b$$

In a similar manner, one can prove that the sets of solutions to the inequalities (4.4) and the equalities (5.1) of Chap. 2 are convex sets.

We shall denote the convex set of solutions to the linear-programming problem by K.   Since K is determined by the intersection of the finite set of linear constraints (1.2) and (1.3), the boundary of K (if K is not void) will consist of sections of some of the corresponding hyperplanes. K will be a region of E_n and can either be void, a bounded convex polyhedron, or a convex polyhedron which is unbounded in some direction.   If K is void, then our problem does not have any solutions; if it is bounded, then our problem has a solution with a finite minimum value for the objective function; and if K is unbounded, the problem has a solution, but the minimum *might* be unbounded.   Valid linear-programming models should yield K's of the second or possibly the third type, as they are models of situations that have a number of possible solutions.   By Theorem 1, if a problem has more than one solution, it has, in reality, an infinite number of solutions.   Out of all these solutions, it is our task to determine the one which minimizes the corresponding objective function.   This work is somewhat simplified by the results of Theorem 2 below.   Before proceeding with this theorem, we recall the following: If K is a bounded convex polyhedron, then K is the convex hull of the extreme points of K.   That is, every feasible solution in K can be represented as a convex combination of the extreme feasible solutions in K.   An unbounded K also has a finite number of extreme points, but not all points of K can be represented as convex combinations of these extreme points; see the discussion on the representation theorem for convex polyhedra in Sec. 3, Chap. 2.   For ease in discussion, we can assume that all our problems have a K that is a bounded convex polyhedron.[1]   As will be shown in later chapters, computational devices exist that determine whether K is void or whether a problem has an unbounded minimum.   (For computational purposes, we always assume that K is a convex polyhedron.)

---

[1] See Exercises 7 and 8.

With the assumption that $\mathbf{K}$ is a convex polyhedron, we can surmise from the above discussion that we need only to look at the extreme points of the convex polyhedron in order to determine the minimum feasible solution. We prove this with the following theorem:

Theorem 2. *The objective function* (1.1) *assumes its minimum at an extreme point of the convex set* $\mathbf{K}$ *generated by the set of feasible solutions to the linear-programming problem. If it assumes its minimum at more than one extreme point, then it takes on the same value for every convex combination of those particular points.*

Proof: Since we have assumed $\mathbf{K}$ to be a bounded convex polyhedron, $\mathbf{K}$ has a finite number of extreme points. In two dimensions, $\mathbf{K}$ might look like Fig. 3.1. Let us denote the objective function by $f(\mathbf{X})$, the extreme points by $\bar{\mathbf{X}}_1, \bar{\mathbf{X}}_2, \ldots, \bar{\mathbf{X}}_p$, and the minimum feasible solution by $\mathbf{X}_0$. This means that $f(\mathbf{X}_0) \leq f(\mathbf{X})$ for all $\mathbf{X}$ in $\mathbf{K}$. If $\mathbf{X}_0$ is an extreme point, the first part of the theorem is true. Suppose $\mathbf{X}_0$ is not an extreme point (as indicated in Fig. 3.1). We can then write $\mathbf{X}_0$ as a convex combination of the extreme points of $\mathbf{K}$, that is,

$$\mathbf{X}_0 = \sum_{i=1}^{p} \alpha_i \bar{\mathbf{X}}_i$$

for $\alpha_i \geq 0$ and $\sum_i \alpha_i = 1$. Then, since $f(\mathbf{X})$ is a linear functional, we have

$$f(\mathbf{X}_0) = f\left(\sum_{i=1}^{p} \alpha_i \bar{\mathbf{X}}_i\right) = f(\alpha_1 \bar{\mathbf{X}}_1 + \alpha_2 \bar{\mathbf{X}}_2 + \cdots + \alpha_p \bar{\mathbf{X}}_p)$$

$$= \alpha_1 f(\bar{\mathbf{X}}_1) + \alpha_2 f(\bar{\mathbf{X}}_2) + \cdots + \alpha_p f(\bar{\mathbf{X}}_p) = m \qquad (2.1)$$

where m is the minimum of $f(\mathbf{X})$ for all $\mathbf{X}$ in $\mathbf{K}$.

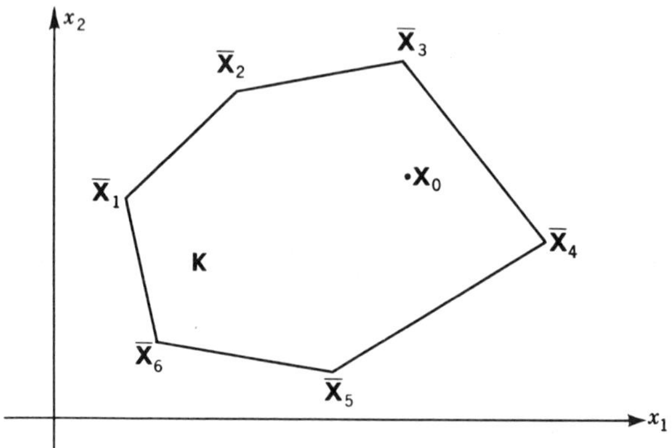

FIGURE 3.1

Since all $\alpha_i \geq 0$, we do not increase the sum (2.1) if we substitute for each $f(\bar{\mathbf{X}}_i)$ the minimum of the values $f(\bar{\mathbf{X}}_i)$. Let $f(\bar{\mathbf{X}}_m) = \min f(\bar{\mathbf{X}}_i)$. Substituting in (2.1) we have, since $\sum_i \alpha_i = 1$,

$$f(\mathbf{X}_0) \geq \alpha_1 f(\bar{\mathbf{X}}_m) + \alpha_2 f(\bar{\mathbf{X}}_m) + \cdots + \alpha_p f(\bar{\mathbf{X}}_m) = f(\bar{\mathbf{X}}_m)$$

Since we assumed $f(\mathbf{X}_0) \leq f(\mathbf{X})$ for all $\mathbf{X}$ in $\mathbf{K}$, we must have

$$f(\mathbf{X}_0) = f(\bar{\mathbf{X}}_m) = m$$

Therefore, there is an extreme point, $\bar{\mathbf{X}}_m$, at which the objective function assumes its minimum value.

To prove the second part of the theorem, let $f(\mathbf{X})$ assume its minimum at more than one extreme point, say at $\bar{\mathbf{X}}_1, \bar{\mathbf{X}}_2, \ldots, \bar{\mathbf{X}}_q$. Here we have $f(\bar{\mathbf{X}}_1) = f(\bar{\mathbf{X}}_2) = \cdots = f(\bar{\mathbf{X}}_q) = m$. If $\mathbf{X}$ is any convex combination of the above $\bar{\mathbf{X}}_i$, say

$$\mathbf{X} = \sum_{i=1}^{q} \alpha_i \bar{\mathbf{X}}_i$$

for $\alpha_i \geq 0$ and $\sum_i \alpha_i = 1$, then

$$f(\mathbf{X}) = f(\alpha_1 \bar{\mathbf{X}}_1 + \alpha_2 \bar{\mathbf{X}}_2 + \cdots + \alpha_q \bar{\mathbf{X}}_q)$$

$$= \alpha_1 f(\bar{\mathbf{X}}_1) + \alpha_2 f(\bar{\mathbf{X}}_2) + \cdots + \alpha_q f(\bar{\mathbf{X}}_q) = \sum_i \alpha_i m$$

$$= m$$

The proof is now complete. By making the obvious changes, the theorem can be proved for the case where (1.1) is to be maximized. By Theorem 2, we need only to consider the extreme points of $\mathbf{K}$ in our search for a minimum feasible solution to the linear-programming problem.

Recall that a feasible solution is a vector $\mathbf{X} = (x_1, x_2, \ldots, x_n)$, with all $x_i \geq 0$, such that

$$x_1 \mathbf{P}_1 + x_2 \mathbf{P}_2 + \cdots + x_n \mathbf{P}_n = \mathbf{P}_0$$

Assume we have found a set of k vectors that is linearly independent and that there exists a nonnegative combination of these vectors that is equal to $\mathbf{P}_0$. Let this set of vectors be $\mathbf{P}_1, \mathbf{P}_2, \ldots, \mathbf{P}_k$. We then have the following theorem:

Theorem 3. *If a set of $k \leq m$ vectors $\mathbf{P}_1, \mathbf{P}_2, \ldots, \mathbf{P}_k$ can be found that is linearly independent and such that*

$$x_1 \mathbf{P}_1 + x_2 \mathbf{P}_2 + \cdots + x_k \mathbf{P}_k = \mathbf{P}_0$$

and all $x_i \geq 0$, then the point $\mathbf{X} = (x_1, x_2, \ldots, x_k, 0, \ldots, 0)$ is an extreme *point of the convex set of feasible solutions. Here X is an n-dimensional vector whose last $n - k$ elements are zero.*

Proof: Suppose X is not an extreme point. Then, since X is a feasible solution, it can be written as a convex combination of two other points X_1 and X_2 in K. We have $X = \alpha X_1 + (1 - \alpha)X_2$ for $0 < \alpha < 1$. Since all the elements x_i of X are nonnegative and since $0 < \alpha < 1$, the last $n - k$ elements of X_1 and X_2 must also equal zero; that is,

$$X_1 = (x_1^{(1)}, x_2^{(1)}, \ldots, x_k^{(1)}, 0, \ldots, 0)$$
$$X_2 = (x_1^{(2)}, x_2^{(2)}, \ldots, x_k^{(2)}, 0, \ldots, 0)$$

Since X_1 and X_2 are feasible solutions, we have

$$AX_1 = b$$

and

$$AX_2 = b$$

Rewriting these equations in vector notation, we have

$$x_1^{(1)}P_1 + x_2^{(1)}P_2 + \cdots + x_k^{(1)}P_k = P_0$$

and

$$x_1^{(2)}P_1 + x_2^{(2)}P_2 + \cdots + x_k^{(2)}P_k = P_0$$

But $P_1, P_2, \ldots, P_k$ is a linearly independent set, and hence P_0 can be expressed as a *unique* linear combination in terms of $P_1, P_2, \ldots, P_k$.[†] This implies that $x_i = x_i^{(1)} = x_i^{(2)}$. Therefore, X cannot be expressed as a convex combination of two distinct points in K and must be an extreme point of K.

Theorem 4. *If $X = (x_1, x_2, \ldots, x_n)$ is an extreme point of K, then the vectors associated with positive x_i form a linearly independent set. From this it follows that, at most, m of the x_i are positive.*

[†] This can be shown as follows: Let $P_1, P_2, \ldots, P_k$ be a set of linearly independent vectors and assume that we can represent the vector P_0 in terms of these vectors by two different linear combinations, for example,

$$e_1P_1 + e_2P_2 + \cdots + e_kP_k = P_0$$

and

$$f_1P_1 + f_2P_2 + \cdots + f_kP_k = P_0$$

To show that these combinations have to be identical, we subtract the second from the first to obtain

$$(e_1 - f_1)P_1 + (e_2 - f_2)P_2 + \cdots + (e_k - f_k)P_k = 0$$

By the definition of linear independence we must have each $e_i - f_i = 0$, which implies that $e_i = f_i$.

Proof: Let the nonzero coefficients be the first k coefficients, so that $\sum_{i=1}^{k} x_i \mathbf{P}_i = \mathbf{P}_0$. We prove the main part of the theorem by contradiction. Assume that $\mathbf{P}_1, \mathbf{P}_2, \ldots, \mathbf{P}_k$ is a linearly dependent set. Then there exists a linear combination of these vectors which equals the zero vector,

$$d_1\mathbf{P}_1 + d_2\mathbf{P}_2 + \cdots + d_k\mathbf{P}_k = \mathbf{0} \tag{2.2}$$

with at least one $d_i \neq 0$. From the hypothesis of the theorem, we have

$$x_1\mathbf{P}_1 + x_2\mathbf{P}_2 + \cdots + x_k\mathbf{P}_k = \mathbf{P}_0 \tag{2.3}$$

For some $d > 0$, we multiply (2.2) by d and add and subtract the result from (2.3) to obtain the two equations

$$\sum_{i=1}^{k} x_i\mathbf{P}_i + d \sum_{i=1}^{k} d_i\mathbf{P}_i = \mathbf{P}_0$$

$$\sum_{i=1}^{k} x_i\mathbf{P}_i - d \sum_{i=1}^{k} d_i\mathbf{P}_i = \mathbf{P}_0$$

We then have the two solutions to (1.2) (note that they might not be feasible solutions):

$$\mathbf{X}_1 = (x_1 + dd_1, x_2 + dd_2, \ldots, x_k + dd_k, 0, \ldots, 0)$$

and

$$\mathbf{X}_2 = (x_1 - dd_1, x_2 - dd_2, \ldots, x_k - dd_k, 0, \ldots, 0)$$

Since all $x_i > 0$, we can let d be as small as necessary, but still positive, to make the first k components of both $\mathbf{X}_1$ and $\mathbf{X}_2$ positive. Then $\mathbf{X}_1$ and $\mathbf{X}_2$ are feasible solutions. But $\mathbf{X} = \frac{1}{2}\mathbf{X}_1 + \frac{1}{2}\mathbf{X}_2$, which contradicts the hypothesis that $\mathbf{X}$ is an extreme point. The assumption of linear dependence for the vectors $\mathbf{P}_1, \mathbf{P}_2, \ldots, \mathbf{P}_k$ has thus led to a contradiction and hence must be false; i.e., the set of vectors $\mathbf{P}_1, \mathbf{P}_2, \ldots, \mathbf{P}_k$ is linearly independent.

Since every set of $m + 1$ vectors in m-dimensional space is necessarily linearly dependent, we cannot have more than m positive x_i. For assume that we did. Then the above proof of the main part of the theorem would imply that there exist vectors $\mathbf{P}_1, \ldots, \mathbf{P}_m, \mathbf{P}_{m+1}$ that are linearly independent.

We can, without any loss of generality, assume that the set of vectors $\mathbf{P}_1, \mathbf{P}_2, \ldots, \mathbf{P}_n$ of the linear-programming problem always contains a set of m linearly independent vectors. If this property is not evident when a particular problem is being solved, the original set of vectors is augmented by a set of m linearly independent vectors, and we then seek a solution to the extended problem. This procedure will be explained in detail in the succeeding chapters.

Corollary 1. *Associated with every extreme point of* $\mathbf{K}$ *is a set of m linearly independent vectors from the given set* $\mathbf{P}_1, \mathbf{P}_2, \ldots, \mathbf{P}_n$.

Proof: Theorem 4 has shown that there are $k \le m$ such vectors. For $k = m$, the corollary is proved. Assume that $k < m$ and that we can find only additional vectors $\mathbf{P}_{k+1}, \ldots, \mathbf{P}_r$ such that the set

$$\mathbf{P}_1, \ldots, \mathbf{P}_k, \mathbf{P}_{k+1}, \ldots, \mathbf{P}_r$$

for $r < m$ is linearly independent. Then this implies that the remaining $n - r$ vectors are dependent on $\mathbf{P}_1, \ldots, \mathbf{P}_r$. But this contradicts the assumption that we always have a set of m linearly independent vectors in the given set of $\mathbf{P}_1, \ldots, \mathbf{P}_n$. Therefore, there must be m linearly independent vectors $\mathbf{P}_1, \ldots, \mathbf{P}_m$ associated with every extreme point, such that

$$\sum_{i=1}^{k} x_i \mathbf{P}_i + \sum_{i=k+1}^{m} 0 \mathbf{P}_i = \mathbf{P}_0$$

We can sum up the preceding theorems by the following:

Theorem 5. $\mathbf{X} = (x_1, x_2, \ldots, x_n)$ *is an extreme point of* $\mathbf{K}$ *if and only if the positive x_j are coefficients of linearly independent vectors $\mathbf{P}_j$ in*

$$\sum_{j=1}^{n} x_j \mathbf{P}_j = \mathbf{P}_0$$

As a result of the assumptions and theorems of this section, we have:

1. There is an extreme point of $\mathbf{K}$ at which the objective function takes on its minimum.
2. Every basic feasible solution corresponds to an extreme point of $\mathbf{K}$.
3. Every extreme point of $\mathbf{K}$ has m linearly independent vectors of the given set of n associated with it.

From the above we can conclude that we need only investigate extreme-point solutions and hence only those feasible solutions generated by m linearly independent vectors. Since there are at most $\binom{n}{m}$ sets of m linearly independent vectors from the given set of n, the value $\binom{n}{m}$ is an upper bound to the number of possible solutions to the problem.[1] For large n and m it would be an impossible task to evaluate all the possible solutions and select one that minimizes the objective function. What is required is a computational scheme that selects, in an orderly fashion, a small subset of the possible solutions which converges to a minimum solution. The *simplex procedure*, devised by G. B. Dantzig, is such a scheme.[2] This procedure finds an extreme point and

---

[1] See Saaty [583] and Quandt and Kuhn [553] for discussions of the upper bound to the number of possible solutions.

[2] The name simplex method is due to the use of the equation $\sum_j x_j = 1$ as a constraint in a geometric interpretation of this procedure, as described in Dantzig [153].

determines whether it is the minimum. If it is not, the procedure finds an adjacent extreme point[1] whose corresponding value of the objective function is less than or equal to the preceding value. In a finite number of such steps (usually between m and $3m$), a minimum feasible solution is found. The simplex method makes it possible to discover whether the problem has no finite minimum solutions or no feasible solutions. It is a powerful scheme for solving *any* linear-programming problem.

Before going into the validity and full computational aspects of the simplex procedure, we introduce the following computational element of the procedure.

3. GENERATING EXTREME-POINT SOLUTIONS

Here we assume that an extreme-point solution in terms of m vectors $\mathbf{P}_j$ of the original set of n vectors is known. We can let this set of m linearly independent vectors be the first m, and let

$$\mathbf{X} = (x_1, x_2, \ldots, x_m, 0, \ldots, 0)$$

be the solution vector. We then have

$$x_1\mathbf{P}_1 + x_2\mathbf{P}_2 + \cdots + x_m\mathbf{P}_m = \mathbf{P}_0 \tag{3.1}$$

where all $x_i \geq 0$. With this information, the problem is to determine, in a computationally efficient manner, a new extreme-point solution. (We shall, of course, assume that a different extreme solution exists.) Since the vectors $\mathbf{P}_1$, $\mathbf{P}_2, \ldots, \mathbf{P}_m$ are linearly independent, they form a basis in m-dimensional vector space. We can then express every vector of the given n as a linear combination of these basis vectors. We can write

$$\sum_{i=1}^{m} x_{ij}\mathbf{P}_i = \mathbf{P}_j \qquad j = 1, \ldots, n$$

Assume that some vector not in the given basis, say $\mathbf{P}_{m+1}$, has at least one element $x_{i,m+1} > 0$ in the expression

$$x_{1,m+1}\mathbf{P}_1 + x_{2,m+1}\mathbf{P}_2 + \cdots + x_{m,m+1}\mathbf{P}_m = \mathbf{P}_{m+1} \tag{3.2}$$

Let θ be any number, and multiply (3.2) by θ and subtract the result from (3.1) to obtain

$$(x_1 - \theta x_{1,m+1})\mathbf{P}_1 + (x_2 - \theta x_{2,m+1})\mathbf{P}_2 + \cdots + (x_m - \theta x_{m,m+1})\mathbf{P}_m + \theta\mathbf{P}_{m+1} = \mathbf{P}_0 \tag{3.3}$$

The vector $\mathbf{X}' = (x_1 - \theta x_{1,m+1}, x_2 - \theta x_{2,m+1}, \ldots, x_m - \theta x_{m,m+1}, \theta)$ is a solution to the problem, and if all the elements of $\mathbf{X}'$ are nonnegative, $\mathbf{X}'$ is a feasible solution. Since we want $\mathbf{X}'$ to be a feasible solution different from $\mathbf{X}$, we

---

[1] Two extreme points are said to be adjacent if they are joined by a boundary of the convex polyhedron; see Exercise 9.

restrict θ to be greater than zero.[1]   With this restriction, all elements of $\mathbf{X}'$ that have a negative or zero $x_{i,m+1}$ will also have a nonnegative $x_i - \theta x_{i,m+1}$. We need only concern ourselves with those elements having a positive $x_{i,m+1}$. We wish to find a $\theta > 0$ such that

$$x_i - \theta x_{i,m+1} \geq 0 \tag{3.4}$$

for all $x_{i,m+1} > 0$.

From (3.4) we have

$$\frac{x_i}{x_{i,m+1}} \geq \theta$$

and hence any θ for which

$$0 < \theta \leq \min_i \frac{x_i}{x_{i,m+1}}$$

will give a feasible solution for (3.3).   However, as we are looking for an extreme-point solution, we know by our theorems in Sec. 2 that we cannot have all the $m+1$ elements of $\mathbf{X}'$ positive.   We then must force at least one of the elements of $\mathbf{X}'$ to be exactly equal to zero.   We see that, if we let

$$\theta = \theta_0 = \min_i \frac{x_i}{x_{i,m+1}}$$

for $x_{i,m+1} > 0$, then the element in $\mathbf{X}'$ for which this minimum is attained will reduce to zero.   Let this element be the first, that is,

$$\theta_0 = \min_i \frac{x_i}{x_{i,m+1}} = \frac{x_1}{x_{1,m+1}}$$

We have now obtained a new feasible solution

$$x_2'\mathbf{P}_2 + x_3'\mathbf{P}_3 + \cdots + x_m'\mathbf{P}_m + x_{m+1}'\mathbf{P}_{m+1} = \mathbf{P}_0$$

where

$$x_i' = x_i - \theta_0 x_{i,m+1} \qquad i = 2, \ldots, m$$
$$x_{m+1}' = \theta_0$$

If all the $x_{i,m+1}$ had been equal to or less than zero, then we would not

---

[1] The reader should note that this restriction causes us to assume (for illustrative purposes) that associated with an $x_{i,m+1} > 0$ is an $x_i > 0$. This, in general, is not the case and will be true only if nondegeneracy is assumed for all basic feasible solutions. As discussed in Chap. 4, a value of $\theta = 0$ is an acceptable one; i.e., some $x_i = 0$ for an $x_{i,m+1} > 0$. With $\theta = 0$, the transformation retains the old extreme point but selects a new feasible basis. In sum, the simplex procedure allows for all values of $\theta \geq 0$. The reader should review the applicability of the above discussion to the case $\theta = \min_i(x_i/x_{i,m+1}) = 0$ for $x_{i,m+1} > 0$.

have been able to select a positive θ that would have eliminated at least one of the vectors $\mathbf{P}_1, \ldots, \mathbf{P}_m$ from the basis. For this situation we obtain for *any* $\theta > 0$ a non-extreme-point feasible solution associated with the $m + 1$ vectors $\mathbf{P}_1, \ldots, \mathbf{P}_m, \mathbf{P}_{m+1}$. Here we have found an extreme direction[1] $\mathbf{Y} = (y_i)$ of the defining convex polyhedron. For $\theta > 0$, we let $y_i = -\theta x_{i,m+1}$ for $i = 1, \ldots, m$, $y_{m+1} = \theta$, and $y_i = 0$ for $i = m + 2, \ldots, n$; see Exercise 13. Note that $\mathbf{Y}$ satisfies the associated homogeneous equations $\mathbf{AX} = \mathbf{0}, \mathbf{X} \geq \mathbf{0}$; see Exercise 45, Chap. 2. As will be shown in Chap. 4, the generation of an extreme direction by the simplex optimization procedure indicates that the problem does not have a finite minimum solution.

To show that $\mathbf{X}' = (x_2', \ldots, x_m', x_{m+1}')$ is an extreme point, we have to prove that the set $\mathbf{P}_2, \ldots, \mathbf{P}_m, \mathbf{P}_{m+1}$ is linearly independent. Assume it is linearly dependent. We can then find (from the definition of linear dependence)

$$d_2\mathbf{P}_2 + d_3\mathbf{P}_3 + \cdots + d_m\mathbf{P}_m + d_{m+1}\mathbf{P}_{m+1} = \mathbf{0} \tag{3.5}$$

where not all $d_i = 0$. Since any subset of a set of linearly independent vectors is also a set of linearly independent vectors, then the set $\mathbf{P}_2, \ldots, \mathbf{P}_m$ is linearly independent. This implies that $d_{m+1} \neq 0$. From (3.5) we have

$$e_2\mathbf{P}_2 + e_3\mathbf{P}_3 + \cdots + e_m\mathbf{P}_m = \mathbf{P}_{m+1} \tag{3.6}$$

where

$$e_i = \frac{-d_i}{d_{m+1}} \qquad i = 2, \ldots, m$$

Subtracting (3.6) from (3.2), we obtain

$$x_{1,m+1}\mathbf{P}_1 + (x_{2,m+1} - e_2)\mathbf{P}_2 + (x_{3,m+1} - e_3)\mathbf{P}_3 + \cdots + (x_{m,m+1} - e_m)\mathbf{P}_m = \mathbf{0} \tag{3.7}$$

Since $\mathbf{P}_1, \mathbf{P}_2, \ldots, \mathbf{P}_m$ are linearly independent, all the coefficients in (3.7) must equal zero. But $x_{1,m+1}$ was assumed to be positive. Hence the assumption of linear dependence for $\mathbf{P}_2, \ldots, \mathbf{P}_{m+1}$ has led to a contradiction, and these vectors must be linearly independent.

In order to continue this process of obtaining new extreme feasible solutions, we need the representation of any vector not in the new basis $\mathbf{P}_2, \mathbf{P}_3, \ldots, \mathbf{P}_{m+1}$ in terms of this basis. From (3.2) we have

$$\mathbf{P}_1 = \frac{1}{x_{1,m+1}} (\mathbf{P}_{m+1} - x_{2,m+1}\mathbf{P}_2 - \cdots - x_{m,m+1}\mathbf{P}_m) \tag{3.8}$$

Let

$$\mathbf{P}_j = x_{1j}\mathbf{P}_1 + x_{2j}\mathbf{P}_2 + \cdots + x_{mj}\mathbf{P}_m \tag{3.9}$$

be any vector not in the new basis. Substitute the expression (3.8) for $\mathbf{P}_1$ in (3.9) to obtain

---

[1] See Sec. 3 of Chap. 2 for the definition of "extreme direction."

$$\mathbf{P}_j = \left(x_{2j} - \frac{x_{1j}}{x_{1,\,m+1}}\, x_{2,\,m+1}\right)\mathbf{P}_2 + \left(x_{3j} - \frac{x_{1j}}{x_{1,\,m+1}}\, x_{3,\,m+1}\right)\mathbf{P}_3 + \cdots$$

$$+ \left(x_{mj} - \frac{x_{1j}}{x_{1,\,m+1}}\, x_{m,\,m+1}\right)\mathbf{P}_m + \frac{x_{1j}}{x_{1,\,m+1}}\, \mathbf{P}_{m+1}$$

The reader will note that the formulas for complete elimination required in Exercise 34, Chap. 2, are equivalent to the transformation that describes $\mathbf{P}_0$ and $\mathbf{P}_j$ in terms of the new basis. The procedure for obtaining new extreme-point solutions is that of selecting a new variable to be introduced into the system, determining which variable has to be removed from the solution in order to preserve feasibility, and applying the complete elimination formulas to obtain the new solution and the new representations of the vectors not in the basis. The criterion used to determine which variable is to be introduced into the solution is a feature of the simplex procedure and will be considered in Chap. 4.

Example 3.1. We are given the following set of equations:

$\mathbf{P}_1$	$\mathbf{P}_2$	$\mathbf{P}_3$	$\mathbf{P}_4$	$\mathbf{P}_5$	$\mathbf{P}_6$	$\mathbf{P}_0$

$$3x_1 - x_2 + 2x_3 + x_4 \qquad\qquad = 7$$
$$2x_1 - 4x_2 \qquad\qquad + x_5 \qquad = 12$$
$$-4x_1 - 3x_2 + 8x_3 \qquad\qquad + x_6 = 10$$

We have as an initial extreme-point solution $x_1 = 0$, $x_2 = 0$, $x_3 = 0$, $x_4 = 7$, $x_5 = 12$, $x_6 = 10$, which in vector notation is given by

$$7\mathbf{P}_4 + 12\mathbf{P}_5 + 10\mathbf{P}_6 = \mathbf{P}_0 \tag{3.10}$$

Here the basis vectors $\mathbf{P}_4$, $\mathbf{P}_5$, $\mathbf{P}_6$ are unit vectors. We wish to introduce vector $\mathbf{P}_1$ to obtain another extreme-point solution. The representation of $\mathbf{P}_1$ in terms of the basis vectors is simply

$$3\mathbf{P}_4 + 2\mathbf{P}_5 - 4\mathbf{P}_6 = \mathbf{P}_1 \tag{3.11}$$

that is,

$$x_{41} = 3 \qquad x_{51} = 2 \qquad x_{61} = -4$$

If we multiply (3.11) by θ and subtract the result from (3.10), we have

$$(7 - 3\theta)\mathbf{P}_4 + (12 - 2\theta)\mathbf{P}_5 + (10 + 4\theta)\mathbf{P}_6 + \theta\mathbf{P}_1 = \mathbf{P}_0 \tag{3.12}$$

Since $x_{41} = 3$ and $x_{51} = 2$ are both positive, we determine θ_0 by evaluating, for these positive x_{i1},

$$\theta = \theta_0 = \min \frac{x_i}{x_{i1}} = \frac{7}{3}^{\dagger}$$

---

† We form the ratios for those i in the current solution. Here $i = 4, 5, 6$.

Substituting this value in (3.12), we eliminate $\mathbf{P}_4$ from the basis to obtain

$$\tfrac{22}{3}\mathbf{P}_5 + \tfrac{58}{3}\mathbf{P}_6 + \tfrac{7}{3}\mathbf{P}_1 = \mathbf{P}_0$$

or the extreme-point solution $x_1 = \tfrac{7}{3}$, $x_2 = 0$, $x_3 = 0$, $x_4 = 0$, $x_5 = \tfrac{22}{3}$, $x_6 = \tfrac{58}{3}$.

If, instead of $\mathbf{P}_1$, we tried in a similar manner to obtain an extreme solution with $\mathbf{P}_2$, where

$$-\mathbf{P}_4 - 4\mathbf{P}_5 - 3\mathbf{P}_6 = \mathbf{P}_2$$

we would have developed the following expression for $\mathbf{P}_0$ in terms of $\mathbf{P}_4$, $\mathbf{P}_5$, $\mathbf{P}_6$, $\mathbf{P}_2$:

$$(7 + \theta)\mathbf{P}_4 + (12 + 4\theta)\mathbf{P}_5 + (10 + 3\theta)\mathbf{P}_6 + \theta\mathbf{P}_2 = \mathbf{P}_0 \tag{3.13}$$

From (3.13) we see that any $\theta > 0$ yields a feasible solution $x_1 = 0$, $x_2 = \theta$, $x_3 = 0$, $x_4 = 7 + \theta$, $x_5 = 12 + 4\theta$, $x_6 = 10 + 3\theta$. Here, since all $x_{i2} < 0$, we do not obtain a new extreme-point solution.

A more efficient way of interpreting the problem is as a transformation accomplished by complete elimination. Here we detach the coefficients of the equations and set up the following tableau:

$\mathbf{P}_1$	$\mathbf{P}_2$	$\mathbf{P}_3$	$\mathbf{P}_4$	$\mathbf{P}_5$	$\mathbf{P}_6$	$\mathbf{P}_0$	θ
③	-1	2	1	0	0	7	$\tfrac{7}{3} = \theta_0$
2	-4	0	0	1	0	12	$\tfrac{12}{2} = 6$
-4	-3	8	0	0	1	10	

As we want to introduce $\mathbf{P}_1$ into the basis, we again form the ratios x_i/x_{i1} for $x_{i1} > 0$. Since $\theta_0 = \tfrac{7}{3}$ is the minimum of these ratios, we let the element 3 of $\mathbf{P}_1$ be the pivot element of the elimination procedure, as denoted by the circle. That is, we eliminate x_1 from all the equations except the first. If we carry out the elimination transformation, we obtain a new tableau. Here $x_1 = \tfrac{7}{3}$, $x_5 = \tfrac{22}{3}$, $x_6 = \tfrac{58}{3}$, and $x_2 = x_3 = x_4 = 0$.

$\mathbf{P}'_1$	$\mathbf{P}'_2$	$\mathbf{P}'_3$	$\mathbf{P}'_4$	$\mathbf{P}'_5$	$\mathbf{P}'_6$	$\mathbf{P}'_0$	θ
1	$-\tfrac{1}{3}$	$\tfrac{2}{3}$	$\tfrac{1}{3}$	0	0	$\tfrac{7}{3}$	
0	$-\tfrac{10}{3}$	$-\tfrac{4}{3}$	$-\tfrac{2}{3}$	1	0	$\tfrac{22}{3}$	
0	$-\tfrac{13}{3}$	$\tfrac{32}{3}$	$\tfrac{4}{3}$	0	1	$\tfrac{58}{3}$	

We now have a basis of $\mathbf{P}_1$, $\mathbf{P}_5$, $\mathbf{P}_6$, with $\mathbf{P}_2$, $\mathbf{P}_3$, $\mathbf{P}_4$ explicitly given in terms of these basis vectors, that is,

$$-\tfrac{1}{3}\mathbf{P}_1 - \tfrac{10}{3}\mathbf{P}_5 - \tfrac{13}{3}\mathbf{P}_6 = \mathbf{P}_2$$

$$\tfrac{2}{3}\mathbf{P}_1 - \tfrac{4}{3}\mathbf{P}_5 + \tfrac{32}{3}\mathbf{P}_6 = \mathbf{P}_3$$

$$\tfrac{1}{3}\mathbf{P}_1 - \tfrac{2}{3}\mathbf{P}_5 + \tfrac{4}{3}\mathbf{P}_6 = \mathbf{P}_4$$

Hence if we wanted to obtain an extreme-point solution with $\mathbf{P}_3$ in the basis, we could start with the second tableau and determine θ_0 as before and transform this tableau by the elimination formulas. The resulting tableau will yield the representation of the vectors not in the basis in terms of the new basis vectors.

Example 3.2. We next analyze the extreme points and corresponding bases of the furniture manufacturer activity-analysis problem of Chap. 1. The mathematical statement is the following: Maximize

$$x_0 = 45x_1 + 80x_2$$

subject to

$$5x_1 + 20x_2 \le 400$$
$$10x_1 + 15x_2 \le 450$$
$$x_1 \qquad \ge \quad 0$$
$$x_2 \ge \quad 0$$

The two-dimensional (x_1, x_2) solution space and its extreme points are shown in Fig. 3.2. The problem is converted to equation form by adding slack variables x_3 (that measures unused lumber) and x_4 (that measures unused labor-hours) as follows: Maximize

$$x_0 = 45x_1 + 80x_2$$

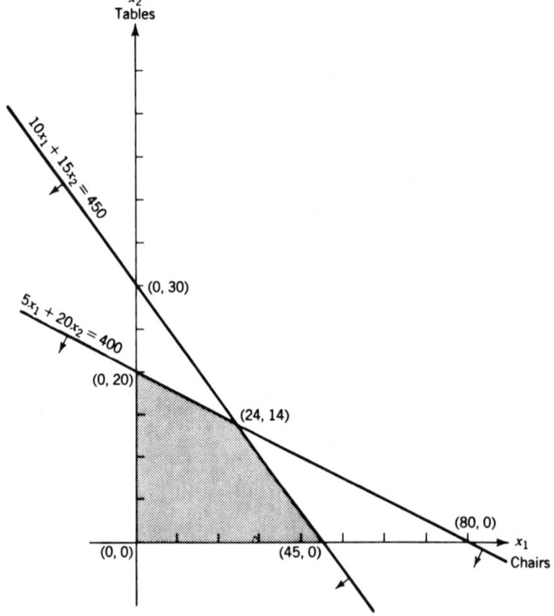

FIGURE 3.2

subject to

$$5x_1 + 20x_2 + x_3 \qquad = 400$$
$$10x_1 + 15x_2 \qquad + x_4 = 450$$
$$x_j \geq \quad 0$$

The two-dimensional extreme points, corresponding solutions, bases, and resultant square system of equations are given below.

Extreme point	Solution	Basis (equations)
$(0, 0)$	$(x_1 = \ 0, x_2 = \ 0, x_3 = 400, x_4 = 450)$	$\begin{pmatrix} 1 & 0 \\ 0 & 1 \end{pmatrix}\begin{pmatrix} x_3 \\ x_4 \end{pmatrix} = \begin{pmatrix} 400 \\ 450 \end{pmatrix}$
$(45, 0)$	$(x_1 = 45, x_2 = \ 0, x_3 = 175, x_4 = \ \ 0)$	$\begin{pmatrix} 5 & 1 \\ 10 & 0 \end{pmatrix}\begin{pmatrix} x_1 \\ x_3 \end{pmatrix} = \begin{pmatrix} 400 \\ 450 \end{pmatrix}$
$(0, 20)$	$(x_1 = \ 0, x_2 = 20, x_3 = \ \ 0, x_4 = 150)$	$\begin{pmatrix} 20 & 0 \\ 15 & 1 \end{pmatrix}\begin{pmatrix} x_2 \\ x_4 \end{pmatrix} = \begin{pmatrix} 400 \\ 450 \end{pmatrix}$
$(24, 14)$	$(x_1 = 24, x_2 = 14, x_3 = \ \ 0, x_4 = \ \ 0)$	$\begin{pmatrix} 5 & 20 \\ 10 & 15 \end{pmatrix}\begin{pmatrix} x_1 \\ x_2 \end{pmatrix} = \begin{pmatrix} 400 \\ 450 \end{pmatrix}$

To follow the basis changes for this problem in terms of its extreme points, we start with the basis $(\mathbf{P}_3, \mathbf{P}_4)$ associated with variables x_3 and x_4 (extreme point with $x_1 = 0$, $x_2 = 0$) and the tableau

$\mathbf{P}_1$	$\mathbf{P}_2$	$\mathbf{P}_3$	$\mathbf{P}_4$	$\mathbf{P}_0$	θ
5	(20)	1	0	400	$\frac{400}{5} = 80$
(10)	15	0	1	450	$\frac{450}{10} = 45 = \theta_0$

By introducing $\mathbf{P}_1$ into the basis, i.e., pivoting x_1 into the basic solution using coefficient 10 as the pivot element, we obtain for the basis $(\mathbf{P}_1\mathbf{P}_3)$ the adjacent extreme-point solution of $x_1 = 45$, $x_2 = 0$, $x_3 = 175$, $x_4 = 0$, and the corresponding transformed tableau

$\mathbf{P}_1'$	$\mathbf{P}_2'$	$\mathbf{P}_3'$	$\mathbf{P}_4'$	$\mathbf{P}_0'$	θ
0	$\left(\frac{25}{2}\right)$	1	$-\frac{1}{2}$	175	$\frac{350}{25} = 14 = \theta_0$
1	$\frac{3}{2}$	0	$\frac{1}{10}$	45	$\frac{90}{3} = 30$

The reader can verify that if x_2 was first introduced as indicated by the dashed circle of the first tableau, we would have moved to the adjacent extreme-point solution $x_1 = 0$, $x_2 = 20$, $x_3 = 0$, $x_4 = 150$. To find the remaining extreme point, we introduce x_2 into the basic solution by pivoting on coefficient $\frac{25}{2}$ of the second tableau. This step moves the solution to the adjacent extreme point $x_1 = 24$, $x_2 = 14$, $x_3 = 0$, $x_4 = 0$ corresponding to the basis $(\mathbf{P}_1\mathbf{P}_2)$. The transformation is left as an exercise. How the simplex algorithm selects the adjacent extreme points for this problem is shown in Sec. 2 of Chap. 4.

EXERCISES

1. Find all basic feasible solutions for the equations

$$2x_1 + 6x_2 + 2x_3 + x_4 = 3$$
$$6x_1 + 4x_2 + 4x_3 + 6x_4 = 2$$

and determine the associated general convex combination of extreme-point solutions.

2. Construct and graph linear-programming problems in three variables with a unique extreme-point optimum solution, three extreme-point optimum solutions, and four extreme-point optimum solutions. Must the constraints be inequalities?

3. In the discussion in Sec. 3, the extreme-point solution (3.1) can be degenerate. Discuss the computation of θ and the transformation to a new basic feasible solution if some $x_i = 0$. Can there be any assurance that $\theta_0 > 0$? Also discuss and interpret the situation where the selection of $\min_i (x_i/x_{ij})$ for $x_{ij} > 0$ is not unique.

4. The following set of equations has a given extreme-point solution $\mathbf{X} = (x_1, x_2, x_3) = (4, 3, 6)$. By the algebraic procedure of Sec. 3, obtain two basic solutions for the bases $\mathbf{P}_1, \mathbf{P}_2, \mathbf{P}_4$ and $\mathbf{P}_2, \mathbf{P}_3, \mathbf{P}_4$. In each case, determine the expression for $\mathbf{P}_5$ and the vector eliminated in terms of the new basis.

$\mathbf{P}_1$	$\mathbf{P}_2$	$\mathbf{P}_3$	$\mathbf{P}_4$	$\mathbf{P}_5$	$\mathbf{P}_0$
x_1			$+2x_4$	$-x_5$	$= 4$
	x_2		$-x_4$	$+x_5$	$= 3$
		x_3	$+3x_4$	$-2x_5$	$= 6$

5. Given the following set of equations:

$$x_1 + 4x_2 - x_3 = 3$$
$$5x_1 + 2x_2 + 3x_3 = 4$$

Determine the basic feasible solution involving x_1 and x_2. Do basic feasible solutions exist for x_1 and x_3 and for x_2 and x_3?

Discuss the graphical solution to this problem if each column of the matrix is assumed to be a vector in two-dimensional space.

6. Give an algebraic explanation why the point $\mathbf{X}_0$ of Fig. 3.1 cannot be in the interior of $\mathbf{K}$, as shown.

7. Prove that if the set of solutions $\mathbf{K}$ to a linear-programming problem is bounded, then $\mathbf{K}$ is the set of all convex combinations of the extreme points of $\mathbf{K}$. Hence, since the number of extreme points is finite, then $\mathbf{K}$ is a convex polyhedron, Krekó [443]. We can always bound $\mathbf{K}$ by adding the constraint $\sum_1^n x_j \leq M$ to any linear-programming problem, where M is a very large positive number. If, in the optimal solution, we find $\sum_1^n x_j = M$, then we know that the original problem has an unbounded optimum.

8. Show that the set of all feasible solutions $\mathbf{K}$ to the constraints $\mathbf{AX} \leq \mathbf{b}$, $\mathbf{X} \geq \mathbf{0}$ can be represented as a convex combination of extreme points of $\mathbf{K}$ and a nonnegative linear combination of the extreme points of the solution space $\mathbf{AX} \leq \mathbf{0}$, $\sum_1^n x_j = 1$, $\mathbf{X} \geq \mathbf{0}$, Krekó [443], Tucker [659].

9. We define an *edge* of a convex polyhedron $\mathbf{K}$ as the line segment joining two extreme points such that no point on the segment is the midpoint of two other points in $\mathbf{K}$ not on the segment. The extreme points are called adjacent. Prove that a simplex iteration selects adjacent extreme points, i.e., moves along an edge, and this movement causes the greatest decrease in $f(\mathbf{X})$ per unit change in the new variable being introduced, Dantzig [152]. (See Sec. 2 of Chap. 4.)

10. We have shown that a basic feasible solution corresponds to the optimum solution of a linear-programming problem, assuming a bounded optimum. What does this mean, for example, to a manufacturing company in terms of the number of products it should make as described by the optimum solution to the activity-analysis problem? Also for the number of foods in the optimum solution to a diet problem?

11. Graph the following set of constraints and compute all extreme-point solutions. Indicate the corresponding basic feasible solutions obtained after converting the inequalities to equalities.

$$3x_1 + 2x_2 \le 18$$
$$x_1 - x_2 \ge -6$$
$$5x_1 + 2x_2 \le 20$$
$$x_1 \qquad \ge 0$$
$$x_2 \ge 0$$

12. State a general convex combination of the extreme points of Exercise 11. Find a convex combination which is equal to the point (2, 3).

13. Consider the convex polyhedron $\mathbf{AX} = \mathbf{b}$, $\mathbf{X} \ge \mathbf{0}$, with $\mathbf{A}$ of rank m. Let $\mathbf{B}$ be a basis consisting of m vectors of $\mathbf{A}$ and let $\mathbf{P}_j$ be a nonbasic vector of $\mathbf{A}$. Then a vector $\mathbf{Y}$ is an extreme direction of the polyhedron if and only if we have a $\mathbf{B}$ and $\mathbf{P}_j$ such that $\mathbf{B}^{-1}\mathbf{P}_j \le \mathbf{0}$ and $\mathbf{Y}$ is a positive multiple of the vector $\begin{pmatrix} -\mathbf{B}^{-1}\mathbf{P}_j \\ \mathbf{e}_j \end{pmatrix}$, where $\mathbf{e}_j$ is an $n - m$ component unit vector with unity in position j (Bazaraa and Shetty [51]). Apply the results of this theorem to Example 3.1 and determine the extreme direction associated with Eq. (3.13).

14. For Example 3.2, determine the tableaus corresponding to bases $(\mathbf{P}_2\mathbf{P}_4)$ and $(\mathbf{P}_1\mathbf{P}_2)$. Express the nonbasic vectors as linear combinations of the corresponding basis.

CHAPTER 4

The Simplex Computational Procedure

We shall next discuss and prove the validity of the basic elements of the simplex procedure and related computational algorithms. By the simplex procedure, we can, once any basic (extreme-point) feasible solution has been determined, obtain a minimum feasible solution in a finite number of steps. These steps, or iterations, consist in finding a new feasible solution whose corresponding value of the objective function is less than the value of the objective function for the preceding solution. This process is continued until a minimum solution has been reached. From the discussion of Chap. 3, we have that all extreme-point solutions, and especially the minimum solution, have m linearly independent vectors associated with them. We then limit our search to those solutions that are generated by m linearly independent vectors. We note that there are a finite number of such solutions.

Before we formalize the simplex procedure by Theorems 1 and 2 below, we first examine a numerical linear-programming problem, keeping in mind the extreme-point and feasible-basis discussions of Chap. 3. We now wish to investigate the relationship between a basic feasible solution of the linear-programming problem and the associated objective function. We illustrate this relationship by considering the following example: Minimize

$$z = 2x_1 - x_2 + x_3 - 5x_4 + 22x_5$$

subject to

$$x_1 \qquad\qquad -2x_4 + \ x_5 = \ 6$$
$$x_2 \quad\ \ + \ x_4 - 4x_5 = \ 3$$
$$x_3 + 3x_4 + 2x_5 = 10$$
$$x_j \geq \ 0$$

The initial basic feasible solution is $x_1 = 6$, $x_2 = 3$, $x_3 = 10$, $x_4 = 0$, $x_5 = 0$, with the value of the objective function for this solution given by the unrestricted variable $z = 2x_1 - x_2 + x_3 = 19$.   We would like to determine if a different basic feasible solution will yield a smaller value of the objective function or whether the current solution is the optimum.   Note that this is equivalent to asking whether one of the nonbasic variables, here x_4 and x_5, which are now set equal to zero, should be allowed to take on a positive value, if possible.   From the above equations we solve for the current basic variables in terms of the nonbasic variables to obtain

$$x_1 \quad = \ 6 + 2x_4 - \ x_5$$
$$x_2 \ = \ 3 - \ x_4 + 4x_5$$
$$x_3 = 10 - 3x_4 - 2x_5$$

We next rewrite the objective function in terms of only the nonbasic variables by substituting for x_1, x_2, and x_3 the corresponding right-hand-side expressions above to obtain

$$z = 2(6 + 2x_4 - x_5) - (3 - x_4 + 4x_5) + (10 - 3x_4 - 2x_5) - 5x_4 + 22x_5$$

or

$$z = 19 - 3x_4 + 14x_5$$

or

$$z = 19 - (3x_4) - (-14x_5)$$

(The reason for rewriting z with the terms in parentheses is to put the expression for z in the form treated below.)   With $x_4 = 0$ and $x_5 = 0$, $z = 19$, which is the value for the current basic feasible solution.   We see from the last transformed expression of z that if x_4 can be made positive, the objective will *decrease* 3 units for each unit increase of x_4; while any positive unit increase to x_5 will *increase* the value of the objective function by 14 units.   Since we are minimizing, it would appear to be appropriate to determine a new basic feasible solution, i.e., an extreme-point solution, involving x_4 at a positive level, if possible.

Utilizing the procedure of Chap. 3, Sec. 3, we next generate a new extreme-point solution by replacing x_2 by x_4 to obtain the solution

$$x_1 + 2x_2 \qquad\qquad - \ 7x_5 = 12$$
$$x_2 \quad\ + x_4 - \ 4x_5 = \ 3$$
$$-3x_2 + x_3 \quad\ + 14x_5 = \ 1$$

or

$$x_1 \quad = 12 - 2x_2 + 7x_5$$
$$x_4 \quad = 3 - x_2 + 4x_5$$
$$x_3 = 1 + 3x_2 - 14x_5$$

(The reader should verify the pivot selection and elimination process.) The new basic feasible solution is $x_1 = 12$, $x_2 = 0$, $x_3 = 1$, $x_4 = 3$, $x_5 = 0$. Substituting the expressions of the basic variables x_1, x_3, x_4 in terms of the nonbasic variables x_2 and x_5 into the objective function, we now have

$$z = 2(12 - 2x_2 + 7x_5) - x_2 + (1 + 3x_2 - 14x_5) - 5(3 - x_2 + 4x_5) + 22x_5$$

or

$$z = 10 - (-3x_2) - (-2x_5)$$

From this last expression we see that any increase in the values of the nonbasic variables x_2 and x_5 would increase the value of the objective. We thus conclude that the new basic feasible solution is an optimum, with the value of the objective function of $z = 10$.

The above process is just the direct application of the elimination procedure on the linear-programming system of equations which now includes an explicit expression of the objective function. We will restate the above analysis using general notation after we introduce some necessary definitions and concepts associated with the simplex procedure.

1. DEVELOPMENT OF A MINIMUM FEASIBLE SOLUTION

We assume that the linear-programming problem is feasible, that every basic feasible solution is nondegenerate, and that we are given a basic feasible solution.[1] These assumptions, as will be discussed later, are made without any loss in generality. Let the given solution be $\mathbf{X}_0 = (x_{10}, x_{20}, \ldots, x_{m0})^{\dagger}$ and the associated set of linearly independent vectors be $\mathbf{P}_1, \mathbf{P}_2, \ldots, \mathbf{P}_m$. We then have

$$x_{10}\mathbf{P}_1 + x_{20}\mathbf{P}_2 + \cdots + x_{m0}\mathbf{P}_m = \mathbf{P}_0 \tag{1.1}$$

$$x_{10}c_1 + x_{20}c_2 + \cdots + x_{m0}c_m = z_0 \tag{1.2}$$

where all $x_{i0} > 0$, the c_i are the cost coefficients of the objective function, and z_0 is the corresponding value of the objective function for the given solution. Since the set $\mathbf{P}_1, \mathbf{P}_2, \ldots, \mathbf{P}_m$ is linearly independent and thus forms a basis, we can express any vector from the set $\mathbf{P}_1, \mathbf{P}_2, \ldots, \mathbf{P}_n$ in terms of $\mathbf{P}_1, \mathbf{P}_2,$

---

[1] The definitions of these terms are given in Sec. 2 of Chap. 3.

$\dagger$ In order to generalize the simplex transformations we now denote the solution vector $\mathbf{X} = (x_1, x_2, \ldots, x_m)$ by the vector $\mathbf{X}_0 = (x_{10}, x_{20}, \ldots, x_{m0})$. We should note that the remaining $n - m$ values of the solution vector have arbitrarily been set equal to zero.

$\ldots$, $\mathbf{P}_m$. Let $\mathbf{P}_j$ be given by

$$x_{1j}\mathbf{P}_1 + x_{2j}\mathbf{P}_2 + \cdots + x_{mj}\mathbf{P}_m = \mathbf{P}_j \qquad j = 1, \ldots, n \tag{1.3}$$

and *define*

$$x_{1j}c_1 + x_{2j}c_2 + \cdots + x_{mj}c_m = z_j \qquad j = 1, \ldots, n \tag{1.4}$$

where the c_i are the cost coefficients corresponding to the $\mathbf{P}_i$.

Theorem 1. *If, for any fixed j, the condition $z_j - c_j > 0$ holds, then a set of feasible solutions can be constructed such that $z < z_0$ for any member of the set, where the lower bound of z is either finite or infinite. (z is the value of the objective function for a particular member of the set of feasible solutions.)*

Case I. If the lower bound is finite, a new feasible solution consisting of exactly m positive variables can be constructed whose value of the objective function is less than the value for the preceding solution.

Case II. If the lower bound is infinite, a new feasible solution consisting of exactly $m + 1$ positive variables can be constructed whose value of the objective function can be made arbitrarily small.

The following analysis applies to the proof of both cases:

Multiplying (1.3) by some number θ and subtracting from (1.1), and similarly multiplying (1.4) by the same θ and subtracting from (1.2), for $j = 1, 2, \ldots, n$ we get

$$(x_{10} - \theta x_{1j})\mathbf{P}_1 + (x_{20} - \theta x_{2j})\mathbf{P}_2 + \cdots + (x_{m0} - \theta x_{mj})\mathbf{P}_m + \theta \mathbf{P}_j = \mathbf{P}_0 \tag{1.5}$$

$$(x_{10} - \theta x_{1j})c_1 + (x_{20} - \theta x_{2j})c_2 + \cdots + (x_{m0} - \theta x_{mj})c_m + \theta c_j = z_0 - \theta(z_j - c_j)$$

$$\tag{1.6}$$

where θc_j has been added to both sides of (1.6). If all the coefficients of the vectors $\mathbf{P}_1$, $\mathbf{P}_2$, $\ldots$, $\mathbf{P}_m$, $\mathbf{P}_j$ in (1.5) are nonnegative, then we have determined a new feasible solution whose value of the objective function is, by (1.6), $z = z_0 - \theta(z_j - c_j)$. Since the variables $x_{10}, x_{20}, \ldots, x_{m0}$ in (1.5) are all positive, it is clear, from our discussion in Sec. 3 of Chap. 3, that there is a value of $\theta > 0$ (either finite or infinite) for which the coefficients of the vectors in (1.5) remain positive. From the assumption that, for a fixed j, $z_j - c_j > 0$, we have

$$z = z_0 - \theta(z_j - c_j) < z_0$$

for $\theta > 0$. We see that in either event a new feasible solution can be obtained whose corresponding value of the objective function is less than the value for the preceding solution.

The proof of Case I follows:

If, for the fixed j, at least one $x_{ij} > 0$ in (1.3) for $i = 1, 2, \ldots, m$, the largest value of θ for which all coefficients of (1.5) remain nonnegative is given by

$$\theta_0 = \min_{i} \frac{x_{i0}}{x_{ij}} > 0 \tag{1.7}$$

for $x_{ij} > 0$ (see Sec. 3 of Chap. 3). Since we assumed that the problem is nondegenerate, i.e., that all basic feasible solutions have m positive elements, the minimum in (1.7) will be obtained for a unique i. If θ_0 is substituted for θ in (1.5) and (1.6), the coefficient corresponding to this unique i will vanish. We have then constructed a new basic feasible solution consisting of $\mathbf{P}_j$ and $m - 1$ vectors of the original basis. This new basis can be used as the previous one. If a new $z_j - c_j > 0$ and a corresponding $x_{ij} > 0$, another solution can be obtained which has a smaller value of the objective function. This process will continue either until all $z_j - c_j \leq 0$, or until, for some $z_j - c_j > 0$, all $x_{ij} \leq 0$. If all $z_j - c_j \leq 0$, the process terminates.

For Case II:

If at any stage we have, for some j, $z_j - c_j > 0$ and all $x_{ij} \leq 0$, then there is no upper bound to θ and the objective function has a lower bound of $-\infty$. We see for this case that, for any $\theta > 0$, all the coefficients of (1.5) are positive. We then have a feasible solution consisting of $m + 1$ positive elements. Hence, by taking θ large enough, the corresponding value of the objective function given by the right-hand side of (1.6) can be made arbitrarily small.

Theorem 2. *If for any basic feasible solution* $\mathbf{X} = (x_{10}, \; x_{20}, \; \ldots, \; x_{m0})$ *the conditions* $z_j - c_j \leq 0$ *hold for all* $j = 1, 2, \ldots, n$, *then* (1.1) *and* (1.2) *constitute a minimum feasible solution.*[1]

Proof: Let

$$y_{10}\mathbf{P}_1 + y_{20}\mathbf{P}_2 + \cdots + y_{n0}\mathbf{P}_n = \mathbf{P}_0 \tag{1.8}$$

and

$$y_{10}c_1 + y_{20}c_2 + \cdots + y_{n0}c_n = z^* \tag{1.9}$$

be any other feasible solution with z^* the corresponding value of the

---

[1] This optimality criterion is sometimes varied. For a minimization problem we could have computed, instead of the $z_j - c_j$, the numbers $c_j - z_j$ and selected as the vector to be introduced into the basis the one corresponding to the min $(c_j - z_j)$. An optimum has been reached when all $c_j - z_j \geq 0$. If the problem was originally to be maximized, we could use the following criterion instead of changing to a minimization problem: Compute the $z_j - c_j$ and select a new vector corresponding to min $(z_j - c_j)$; an optimum solution has been found when all $z_j - c_j \geq 0$. Or compute the $c_j - z_j$ elements with the new vector corresponding to max $(c_j - z_j)$, the procedure stopping when all $c_j - z_j \leq 0$. It is much more efficient, however, especially in developing a computational procedure for an electronic computer, to select one criterion to be used in solving all problems. See page 96 for the discussion on the criterion used to select a vector to be introduced into the basis.

objective function. We shall show that $z_0 \leq z^*$. (Note that the non-degeneracy assumption is not required for this theorem.)

By hypothesis, $z_j - c_j \leq 0$ for all j, so that replacing c_j by z_j in (1.9) yields

$$y_{10}z_1 + y_{20}z_2 + \cdots + y_{n0}z_n \leq z^* \tag{1.10}$$

For each j we substitute the expression for $\mathbf{P}_j$ given by (1.3) into (1.8), to obtain

$$y_{10}\left(\sum_{i=1}^{m} x_{i1}\mathbf{P}_i\right) + y_{20}\left(\sum_{i=1}^{m} x_{i2}\mathbf{P}_i\right) + \cdots + y_{n0}\left(\sum_{i=1}^{m} x_{in}\mathbf{P}_i\right) = \mathbf{P}_0$$

or, by regrouping terms,

$$\left(\sum_{j=1}^{n} y_{j0}x_{1j}\right)\mathbf{P}_1 + \left(\sum_{j=1}^{n} y_{j0}x_{2j}\right)\mathbf{P}_2 + \cdots + \left(\sum_{j=1}^{n} y_{j0}x_{mj}\right)\mathbf{P}_m = \mathbf{P}_0 \tag{1.11}$$

Similarly, for each j we substitute the expression for z_j given by (1.4) into (1.10) to obtain

$$\left(\sum_{j=1}^{n} y_{j0}x_{1j}\right)c_1 + \left(\sum_{j=1}^{n} y_{j0}x_{2j}\right)c_2 + \cdots + \left(\sum_{j=1}^{n} y_{j0}x_{mj}\right)c_m \leq z^* \tag{1.12}$$

Since the set of vectors $\mathbf{P}_1$, $\mathbf{P}_2$, ..., $\mathbf{P}_m$ is linearly independent, the coefficients of the corresponding vectors in (1.1) and (1.11) must be equal,[1] and hence (1.12) becomes

$$x_{10}c_1 + x_{20}c_2 + \cdots + x_{m0}c_m \leq z^*$$

or, by (1.2), $z_0 \leq z^*$.

The results of Theorems 1 and 2 enable us to start with a basic feasible solution and generate a set of new basic feasible solutions that converge to the minimum solution or determine that a finite solution does not exist.[2]

---

[1] See footnote of Theorem 3, Chap. 3, for a proof of the equality of the coefficients of (1.1) and (1.11).

[2] We refer to the z_j as *indirect costs*, while the cost coefficients c_j are *direct costs*. The z_j can be interpreted as follows. One unit of the nonbasic activity $\mathbf{P}_j$ (associated with variable x_j) consumes the vector of resources (a_{ij}). For the basis $(\mathbf{P}_1\mathbf{P}_2 \cdots \mathbf{P}_m)$, the $\mathbf{P}_j$ can be expressed (indirectly) by the linear combination

$$\mathbf{P}_j = x_{1j}\mathbf{P}_1 + x_{2j}\mathbf{P}_2 + \cdots + x_{mj}\mathbf{P}_m$$

The x_{ij} (which can be positive, negative, or zero) indicate how much of the corresponding activity $\mathbf{P}_i$, at cost c_i, is required to represent the resources of $\mathbf{P}_j$ in the linear combination of the basis vectors. The linear combination represents another way—an indirect way—of accomplishing the activity $\mathbf{P}_j$ as a resource user. (They are not equivalent from a technological point of view.) The associated indirect cost for the resources is then given by

$$z_j = x_{1j}c_1 + x_{2j}c_2 + \cdots + x_{mj}c_m$$

The (minimizing) optimality condition of the simplex method states that we have an

We can summarize the above discussion of the optimality of a basic feasible solution in the following manner, recalling the explanation of the numerical example which opened this chapter. Define the linear-programming problem by: Minimize

$$z = c_1 x_1 + \cdots + c_m x_m + c_{m+1} x_{m+1} + \cdots + c_n x_n \tag{1.13}$$

subject to

$$a_{11} x_1 + \cdots + a_{1m} x_m + a_{1,m+1} x_{m+1} + \cdots + a_{1n} x_n = b_1$$
$$\tag{1.14}$$
$$a_{m1} x_1 + \cdots + a_{mm} x_m + a_{m,m+1} x_{m+1} + \cdots + a_{mn} x_n = b_m$$

and

$$x_j \geq 0 \tag{1.15}$$

Note that z is an unrestricted variable which measures the value of the given objective function. For discussion purposes, assume the first m variables $(x_1, \ldots, x_m)$ form a basic feasible solution and thus we can solve (1.14) for these basic variables by the elimination procedure to obtain

$$x_1 = x_{10} - x_{1,m+1} x_{m+1} - \cdots - x_{1n} x_n$$
$$\tag{1.16}$$
$$x_m = x_{m_0} - x_{m,m+1} x_{m+1} - \cdots - x_{mn} x_n$$

Equations (1.16) are solved for the basic variables in terms of the solution $(x_{10}, \ldots, x_{m0})$ and the nonbasic variables $(x_{m+1}, \ldots, x_n)$. The x_{ij} for $j = m + 1, \ldots, n$ are the x_{ij} of (1.3). By letting all the nonbasic variables equal zero we obtain the usual solution for the basic variables $\mathbf{X}_0 = (x_{10}, \ldots, x_{m0})$. We next substitute for $(x_1, \ldots, x_m)$ the corresponding right-hand sides of (1.16), i.e., $x_i = x_{i0} - x_{i,m+1} x_{m+1} - \cdots - x_{in} x_n$, in the objective function (1.13) to obtain [remembering the definition of z_0 and z; see (1.2) and (1.4)]

$$z = z_0 - \sum_{j=m+1}^{n} (z_j - c_j) x_j \tag{1.17}$$

When all the nonbasic variables x_j are set equal to zero, we have the $z = z_0$ of (1.2).

Since all $x_j \geq 0$, by (1.17) we see that, if for the given basic feasible solution all $(z_j - c_j) \leq 0$, then an increase above zero of any nonbasic x_j with $z_j - c_j < 0$ would *increase* z; while if some $(z_j - c_j) > 0$, an increase of the corresponding

---

optimal basis if all the indirect costs are less than or equal to the direct costs, i.e., $z_j \leq c_j$. (A basis activity has $z_j = c_j$.) Note that a nonbasic $\mathbf{P}_j$ has a different indirect representation for each basis. For this reason, the z_j are sometimes called *relative costs* (Dantzig [152]); also see the discussion of Sec. 3, Chap. 6 and Gale [277]. The individual $z_j - c_j$ are termed *reduced costs*.

nonbasic variable would *decrease z.*   Thus, in the former situation (all $z_j - c_j \leq 0$) we have an optimal solution, and in the latter we can improve the solution.

The nondegeneracy assumption was invoked to ensure the convergence to the minimum solution.   If we did not make this assumption, it would be possible to have at least one of the m x_{i0} of the given solution equal to zero.   If this were the case, then θ_0 could equal zero and the value of the objective function for the new solution would then equal the value of the old solution.   This lack of improvement in the solution could continue for a number of successive steps.   For this situation, the procedure could conceivably repeat a basis and hence keep returning to this basis.   The simplex procedure is then said to have *cycled*, and the computational routine for determining the minimum solution breaks down.   During actual computation, the phenomenon of degeneracy is reflected by basic solutions with less than m positive x_{i0} and/or by more than one i yielding $\theta_0 = \min_i (x_{i0}/x_{ij})$ for $x_{ij} > 0$.   If the i is not unique, some of the x_{i0} in the new solution will equal zero.

Dantzig, Orden, and Wolfe [191], Dantzig [79], Wolfe [715], Charnes, Cooper, and Henderson [118], and Bland [80] have resolved degeneracy from both the theoretical and computational points of view.   As noted in Benichou et al. [70] cycling is rarely detected in practice, but degeneracy is a common phenomenon which may significantly increase the number of basis changes required to solve a problem.   In the past, when digital computers were relatively slow and computational costs high, the burdensome antidegeneracy and anticycling techniques were usually not included in the large-scale computer codes (called *mathematical programming systems*)[1] used to solve linear-programming models.   Now, however, the designers of many of the mathematical programming systems find it best to include an antidegeneracy scheme, as the trade-off between added computation and reduced number of iterations is cost-effective (see Benichou et al. [70]).   There have been reports of practical linear-programming applications that have cycled.   Whether such occurrences were due to numerical instability of the problems, the design features of the mathematical programming systems, or the arithmetic characteristics of the digital computer are not clear.   Problems that appear to cycle on one system may converge without difficulty on another (see Benichou et al. [70], Orchard-Hays [532], Beale [58], and Gass [289]).   There are some artificially constructed linear-programming problems that are known to cycle when solved by the simplex algorithm as described in this chapter (Hoffman [373] and Beale [52]).   When solving a problem by hand, what is normally done is to treat degenerate solutions as nothing unusual and to compute with $\theta_0 = 0$ whenever the procedure yields such a value for θ.   If ties occur when determining θ_0, the usual rule is to select $\theta_0 = \min_i (x_{i0}/x_{ij})$ for the smallest index i.   Degeneracy techniques and a cycling problem of Beale are discussed in Chap. 7.

---

[1] See, for example, IBM [387] and Sperry Univac [619].

2. COMPUTATIONAL PROCEDURE

In this section we assume either that (1) we have selected a set of m linearly independent vectors that yield a feasible solution and have expressed all other vectors in terms of this basis or that (2) our problem matrix contains m vectors that can be explicitly arranged to form a unit matrix of order m.

For the first case let the set of m linearly independent vectors be $\mathbf{P}_1$, $\mathbf{P}_2$, ..., $\mathbf{P}_m$ and denote the $m \times m$ matrix $(\mathbf{P}_1 \mathbf{P}_2 \cdots \mathbf{P}_m)$ by $\mathbf{B}$. The matrix $\mathbf{B}$ is termed an *admissible basis*. To compute the corresponding solution vector $\mathbf{X}$ and the representation of the other vectors in terms of the basis, we must first compute $\mathbf{B}^{-1}$. Since

$$\mathbf{BX}_0 = \mathbf{P}_0$$

we have

$$\mathbf{X}_0 = \mathbf{B}^{-1}\mathbf{P}_0$$

and

$$\mathbf{X}_j = \mathbf{B}^{-1}\mathbf{P}_j$$

where

$$\mathbf{X}_0 = (x_{10}, x_{20}, \ldots, x_{m0}) \geq \mathbf{0}$$

and

$$\mathbf{X}_j = (x_{1j}, x_{2j}, \ldots, x_{mj})$$

are both column vectors.

To start the simplex process, we group the vectors of the problem matrix as follows:

$$(\mathbf{P}_0 | \mathbf{P}_1 \mathbf{P}_2 \cdots \mathbf{P}_m | \mathbf{P}_{m+1} \cdots \mathbf{P}_n)$$

or

$$(\mathbf{P}_0 | \mathbf{B} | \mathbf{P}_{m+1} \cdots \mathbf{P}_n) \tag{2.1}$$

By multiplying the elements in the partitioned matrix (2.1) by $\mathbf{B}^{-1}$, we obtain

$$(\mathbf{X}_0 | \mathbf{I}_m | \mathbf{X}_{m+1} \cdots \mathbf{X}_n)^{\dagger}$$

Since we know the c_j, we next compute the $z_j - c_j$ and determine whether, for any j, the corresponding $z_j - c_j > 0$. If so, we carry out the computational procedure described in Theorem 1. If not, we have found the minimum feasible solution. In general, since we have no assurance that an arbitrary set of m vectors from the given set will be linearly independent, let alone yield a feasible solution, our assumption in the first case is not the usual situation

---

† Note that this transformation can be obtained by applying complete elimination to the columns of $\mathbf{B}$ in (2.1).

encountered in practice. (There have been computational procedures proposed that do start with an arbitrary, but educated, selection of m vectors. Some of these procedures are described in Sec. 1 of Chap. 9.) The condition of the second case is quite common, and we shall describe it in great detail.

In the second case we assume that the given set of n vectors $\mathbf{P}_1, \ldots, \mathbf{P}_n$ contains m unit vectors that can be grouped together to form an $m \times m$ unit matrix. It will be shown in Sec. 3 below that this assumption is not restrictive. We let these vectors be $\mathbf{P}_1, \mathbf{P}_2, \ldots, \mathbf{P}_m$ and take as our admissible basis

$$\mathbf{B} = (\mathbf{P}_1 \mathbf{P}_2 \cdots \mathbf{P}_m) = \mathbf{I}_m$$

Since $\mathbf{B}^{-1} = \mathbf{I}_m$ and since all the elements of $\mathbf{P}_0$ were originally assumed to be nonnegative, we have the initial extreme-point solution

$$\mathbf{X}_0 = \mathbf{P}_0$$

and

$$\mathbf{X}_j = \mathbf{P}_j$$

where

$$\mathbf{X}_0 = (x_{10}, x_{20}, \ldots, x_{m0}) \geq \mathbf{0}$$

and

$$\mathbf{X}_j = (x_{1j}, x_{2j}, \ldots, x_{mj})$$

To start the simplex procedure, we arrange the problem matrix as shown in Tableau 4.1. (In practice, one does not have to group the unit vectors together or even to keep track of them, but we shall do so for illustrative purposes.) From the original equations of the problem given by $\mathbf{AX} = \mathbf{b}$, we have let $x_{i0} = b_i$ and $x_{ij} = a_{ij}$. z_j for $j = 0, 1, \ldots, n$ is obtained by taking the

TABLEAU 4.1   Initial Step of Computational Procedure

i	Basis	$\mathbf{c}$		c_1	c_2	·	c_l	·	c_m	c_{m+1}	·	c_j	·	c_k	·	c_n
			$\mathbf{P}_0$	$\mathbf{P}_1$	$\mathbf{P}_2$	·	$\mathbf{P}_l$	·	$\mathbf{P}_m$	$\mathbf{P}_{m+1}$	·	$\mathbf{P}_j$	·	$\mathbf{P}_k$	·	$\mathbf{P}_n$
1	$\mathbf{P}_1$	c_1	x_{10}	1	0	·	0	·	0	$x_{1,m+1}$	·	x_{1j}	·	x_{1k}	·	x_{1n}
2	$\mathbf{P}_2$	c_2	x_{20}	0	1	·	0	·	0	$x_{2,m+1}$	·	x_{2j}	·	x_{2k}	·	x_{2n}
·	·	·	·	·	·	·	·	·	·	·	·	·	·	·	·	·
·	·	·	·	·	·	·	·	·	·	·	·	·	·	·	·	·
l	$\mathbf{P}_l$	c_l	x_{l0}	0	0	·	1	·	0	$x_{l,m+1}$	·	x_{lj}	·	x_{lk}	·	x_{ln}
·	·	·	·	·	·	·	·	·	·	·	·	·	·	·	·	·
·	·	·	·	·	·	·	·	·	·	·	·	·	·	·	·	·
m	$\mathbf{P}_m$	c_m	x_{m0}	0	0	·	0	·	1	$x_{m,m+1}$	·	x_{mj}	·	x_{mk}	·	x_{mn}
$m+1$			z_0	0	0	·	0	·	0	$z_{m+1} - c_{m+1}$	·	$z_j - c_j$	·	$z_k - c_k$	·	$z_n - c_n$

inner product of the jth vector with the column vector labeled $\mathbf{c}$, that is,

$$z_0 = \sum_{i=1}^{m} c_i x_{i0}$$

$$z_j = \sum_{i=1}^{m} c_i x_{ij} \qquad j = 1, 2, \ldots, n$$

The elements z_0 and $z_j - c_j$ are entered in the $(m+1)$st row of their respective columns. The $z_j - c_j$ for those vectors in the basis will always equal zero. If all the numbers $z_j - c_j \leq 0$ for $j = 1, 2, \ldots, n$, then the solution $\mathbf{X}_0 = (x_{10}, x_{20}, \ldots, x_{m0}) = (b_1, b_2, \ldots, b_m)$ is a minimum feasible solution, and the corresponding value of the objective function is z_0. We shall assume at least one $z_j - c_j > 0$ and compute a new feasible solution whose basis contains $m - 1$ vectors of the original basis $\mathbf{P}_1, \mathbf{P}_2, \ldots, \mathbf{P}_m$. In searching for a new vector to enter the basis, we can theoretically select any vector whose corresponding $z_j - c_j > 0$. As Dantzig [153] points out, the number of *iterations*, i.e., the number of basis changes, necessary to obtain a minimum solution can, in general, be greatly reduced by not selecting at random any vector $\mathbf{P}_j$ with its $z_j - c_j > 0$, but by selecting the one which gives the greatest immediate decrease in the value of the objective function. The vector $\mathbf{P}_j$ should then be the one which corresponds to the

$$\max_{j} \ \theta_0(z_j - c_j) \qquad\qquad (2.1a)$$

where, for each j, θ_0 is given by (1.7). If there are a number of j for which $z_j - c_j > 0$, the above rule is rather complicated to apply. A much simpler criterion for selecting the vector to be introduced is to select the one which corresponds to the

$$\max_{j} \ (z_j - c_j) \qquad\qquad (2.1b)$$

If there are ties, the rule is to select the vector with the lowest (or the highest) index j. This selection criterion has proved to be effective and it or variations are employed in most mathematical programming systems. When using this type of rule, empirical evidence indicates that, in general, approximately m to $3m$ changes of basis are required to go from the first feasible solution to the minimizing solution.[1] We shall employ the second criterion $(2.1b)$.

[1] The work of Cutler and Wolfe [149] and Quandt and Kuhn [553] describes and evaluates alternate selection rules; see also the selection rules of Dickson and Frederick [202] and Harris [353, 354]. McCall [481] describes the results of using Sperry Univac's Functional Mathematical Programming System (FMPS) to solve a set of 16 real-world problems and reports that the total number of iterations ranged from $0.03m$ to $6.6m$, with the average being $2.4m$. The smallest problem had $m = 355$ and the largest $m = 8952$. Benichou et al. [70] describe alternate selection rules based on the Devex method of Harris [353] and their efficacy in IBM's Mathematical Programming System Extended/370 (MPSX-370). Note that the number of iterations is dependent on the selection scheme and rules more complex than $(2.1b)$ require more computation to effect, but they tend to reduce the total number of basis changes to find the optimal solution. See Sec. 3, Chap. 5, for further discussion.

In our example, let

$$\max_{j} \ (z_j - c_j) = z_k - c_k > 0$$

The vector $\mathbf{P}_k$ is to be introduced into the basis. We next compute

$$\theta_0 = \min_{i} \frac{x_{i0}}{x_{ik}}$$

for $x_{ik} > 0.$[†] If all $x_{ik} \leq 0$, we can then find a feasible solution whose value of the objective function can be made arbitrarily small (Theorem 1, Case II). Our computation is then complete. Assume, however, some $x_{ik} > 0$ and

$$\theta_0 = \min_{i} \frac{x_{i0}}{x_{ik}} = \frac{x_{l0}}{x_{lk}}$$

Vector $\mathbf{P}_l$ will be the one removed from the basis. Our new feasible solution will have a new basis consisting of $\mathbf{P}_1, \ldots, \mathbf{P}_{l-1}, \mathbf{P}_k, \mathbf{P}_{l+1}, \ldots, \mathbf{P}_m$. We next wish to compute the new solution explicitly and to express each vector not in the basis in terms of the new basis.

Since our initial basis is $(\mathbf{P}_1 \mathbf{P}_2 \cdots \mathbf{P}_m) = \mathbf{I}_m$, we can readily express all the vectors $\mathbf{P}_j$ in terms of this basis. We then have

$$\mathbf{P}_0 = x_{10}\mathbf{P}_1 + \cdots + x_{l0}\mathbf{P}_l + \cdots + x_{m0}\mathbf{P}_m \tag{2.2}$$

$$\mathbf{P}_k = x_{1k}\mathbf{P}_1 + \cdots + x_{lk}\mathbf{P}_l + \cdots + x_{mk}\mathbf{P}_m \tag{2.3}$$

and

$$\mathbf{P}_j = x_{1j}\mathbf{P}_1 + \cdots + x_{lj}\mathbf{P}_l + \cdots + x_{mj}\mathbf{P}_m \tag{2.4}$$

From (2.3)

$$\mathbf{P}_l = \frac{1}{x_{lk}} \left(\mathbf{P}_k - x_{1k}\mathbf{P}_1 - \cdots - x_{mk}\mathbf{P}_m \right) \tag{2.5}$$

Substituting the above expression for $\mathbf{P}_l$ into (2.2), we obtain

$$\mathbf{P}_0 = x_{10}\mathbf{P}_1 + \cdots + x_{l0}\left[\frac{1}{x_{lk}} \left(\mathbf{P}_k - x_{1k}\mathbf{P}_1 - \cdots - x_{mk}\mathbf{P}_m \right) \right] + \cdots + x_{m0}\mathbf{P}_m$$

or

$$\mathbf{P}_0 = \left(x_{10} - \frac{x_{l0}}{x_{lk}} x_{1k} \right)\mathbf{P}_1 + \cdots + \frac{x_{l0}}{x_{lk}} \mathbf{P}_k + \cdots + \left(x_{m0} - \frac{x_{l0}}{x_{lk}} x_{mk} \right)\mathbf{P}_m{}^{‡}$$

The new feasible solution $\mathbf{X}_0' = (x_{10}', \ldots, x_{k0}', \ldots, x_{m0}')$, $x_{i0}' \geq 0$, is given by

$$\mathbf{P}_0 = x_{10}'\mathbf{P}_1 + \cdots + x_{k0}'\mathbf{P}_k + \cdots + x_{m0}'\mathbf{P}_m$$

[†] For the general situation the index i ranges over those i whose corresponding variables are in the basic solution.

[‡] This expression is equivalent to Eq. (1.5) with $j = k$ and $\theta = x_{l0}/x_{lk} = \theta_0$.

where

$$x'_{i0} = x_{i0} - \frac{x_{l0}}{x_{lk}} x_{ik} \qquad \text{for } i = 1, 2, \ldots, l-1, l+1, \ldots, m$$

$$x'_{k0} = \frac{x_{l0}}{x_{lk}}$$

(2.6)

Similarly, by substituting (2.5) into (2.4), we can obtain the expression for each $\mathbf{P}_j$ not in the new basis in terms of this basis. This yields

$$\mathbf{P}_j = x'_{1j}\mathbf{P}_1 + \cdots + x'_{kj}\mathbf{P}_k + \cdots + x'_{mj}\mathbf{P}_m$$

where

$$x'_{ij} = x_{ij} - \frac{x_{lj}}{x_{lk}} x_{ik} \qquad \text{for } i \neq l$$

$$x'_{kj} = \frac{x_{lj}}{x_{lk}}$$

(2.7)

Since

$$z'_j - c_j = x'_{1j}c_1 + \cdots + x'_{kj}c_k + \cdots + x'_{mj}c_m - c_j$$

one can readily verify by substituting the values (2.7) for x'_{ij} that

$$z'_j - c_j = z_j - c_j - \frac{x_{lj}}{x_{lk}}(z_k - c_k)$$

(2.7a)

and by substituting the values (2.6) for x'_{i0} into

$$z'_0 = c_1 x'_{10} + \cdots + c_k x'_{k0} + \cdots + c_m x'_{m0}$$

that

$$z'_0 = z_0 - \frac{x_{l0}}{x_{lk}}(z_k - c_k)$$

(2.7b)

We then note that, in order to obtain the new solution $\mathbf{X}'_0$, the new vectors $\mathbf{X}'_j$, and the corresponding $z'_j - c_j$, all elements in Tableau 4.1 for rows $i = 1, \ldots, m+1$ and columns $j = 0, 1, \ldots, n$ are transformed by the formulas

$$x'_{ij} = x_{ij} - \frac{x_{lj}}{x_{lk}} x_{ik} \qquad \text{for } i \neq l$$

$$x'_{lj} = \frac{x_{lj}}{x_{lk}}$$

(2.8)

where

$$z'_0 = x'_{m+1,0} \qquad z'_j - c_j = x'_{m+1,j}$$

Here we are letting the general formulas (2.8) apply to all elements of the computational tableau including the $\mathbf{P}_0$ column and the $(m+1)$st row. The transformation defined by (2.8) is equivalent to the complete elimination formulas when the pivot element is x_{lk}.[†]

Once an initial computational tableau has been constructed, the simplex procedure calls for the successive application (i.e., an iteration) of:

1. The testing of the $z_j - c_j$ elements to determine whether a minimum solution has been found, i.e., whether $z_j - c_j \leq 0$ for all j.
2. The selection of the vector to be introduced into the basis if some $z_j - c_j > 0$, i.e., selection of the vector with maximum $z_j - c_j$.
3. The selection of the vector to be removed from the basis to ensure feasibility of the new solution, i.e., the vector with $\min_i (x_{i0}/x_{ik})$ for those $x_{ik} > 0$, where k corresponds to the vector selected in step 2. If all $x_{ik} \leq 0$, then the solution is unbounded.
4. The transformation of the tableau by the complete elimination procedure to obtain the new solution and associated elements.

Each such iteration produces a new basic feasible solution, and by the discussions of Theorems 1 and 2 we eventually obtain a minimum solution or determine an unbounded solution. Simple tie-breaking rules as previously discussed are used in steps 2 and 3.

An application of the simplex procedure to the initial tableau yields the transformed values of Tableau 4.2.[‡]

Example 2.1. As an example, let us solve the following linear-programming problem by means of the simplex procedure: Minimize

$$x_2 - 3x_3 \quad\quad + 2x_5$$

subject to

$$
\begin{aligned}
x_1 + 3x_2 - x_3 + 2x_5 &= 7 \\
- 2x_2 + 4x_3 + x_4 &= 12 \\
- 4x_2 + 3x_3 + 8x_5 + x_6 &= 10 \\
x_j &\geq 0
\end{aligned}
$$

[†] Note that we are now letting the indices i and l refer to the corresponding rows in the computational tableau. We keep track of which variables are in the current basic solution by means of the column labeled Basis. Hence, the general elimination formulas given by Eqs. (2.8) are applied to each element x_{ij}, where i refers to the row and j to the column of the tableau. Element x_{lk} is the pivot element of the elimination transformation, with row l being the pivot row and column k being the pivot column.

In applying Eqs. (2.8) for manual computation, it is best to compute first the new elements of the pivot row and add suitable multiples of this row to the other rows of the tableau in order to carry out the elimination of x_k from all rows except the lth. The reader should note that a computer-based simplex algorithm should not use the tableau format described here; see Sec. 3 of Chap. 5 on computational considerations.

[‡] A condensed version of the tableau, one in which the unit vectors are not carried, can also be used; see Exercise 19.

TABLEAU 4.2. Second Step of Computational Procedure

i	Basis	$\mathbf{c}$	$\mathbf{P}_0$	c_1 / $\mathbf{P}_1$	c_2 / $\mathbf{P}_2$	·	c_l / $\mathbf{P}_l$	·	c_m / $\mathbf{P}_m$	c_{m+1} / $\mathbf{P}_{m+1}$	·	c_j / $\mathbf{P}_j$	·	c_k / $\mathbf{P}_k$	·	c_n / $\mathbf{P}_n$
1	$\mathbf{P}_1$	c_1	x'_{10}	1	0	·	x'_{11}	·	0	$x'_{1,m+1}$	·	x'_{1j}	·	0	·	x'_{1n}
2	$\mathbf{P}_2$	c_2	x'_{20}	0	1	·	x'_{21}	·	0	$x'_{2,m+1}$	·	x'_{2j}	·	0	·	x'_{2n}
·	·	·	·	·	·	·	·	·	·	·	·	·	·	·	·	·
·	·	·	·	·	·	·	·	·	·	·	·	·	·	·	·	·
l	$\mathbf{P}_k$	c_k	x'_{k0}	0	0	·	x'_{ll}	·	0	$x'_{l,m+1}$	·	x'_{lj}	·	1	·	x'_{ln}
·	·	·	·	·	·	·	·	·	·	·	·	·	·	·	·	·
·	·	·	·	·	·	·	·	·	·	·	·	·	·	·	·	·
m	$\mathbf{P}_m$	c_m	x'_{m0}	0	0	·	x'_{ml}	·	1	$x'_{m,m+1}$	·	x'_{mj}	·	0	·	x'_{mn}
$m+1$			z'_0	0	0	·	$z'_l - c_l$	·	0	$z'_{m+1} - c_{m+1}$	·	$z'_j + c_j$	·	0	·	$z'_n - c_n$

Our initial basis (see Tableau 4.3) consists of $\mathbf{P}_1$, $\mathbf{P}_4$, $\mathbf{P}_6$, and the corresponding solution is $\mathbf{X}_0 = (x_1, x_4, x_6) = (7, 12, 10)$. Since

$$c_1 = c_4 = c_6 = 0$$

the corresponding value of the objective function, z_0, equals zero. $\mathbf{P}_3$ is selected to go into the basis, since

$$\max_j \, (z_j - c_j) = z_3 - c_3 = 3 > 0$$

θ_0 is the minimum of x_{i0}/x_{i3} for $x_{i3} > 0$, that is,

$$\min \left(\tfrac{12}{4}, \tfrac{10}{3} \right) = \tfrac{12}{4} = \theta_0{}^\dagger$$

and hence $\mathbf{P}_4$ is removed. We transform the tableau (see Second Step, Tableau 4.3) and obtain a new solution

$$\mathbf{X}'_0 = (x_1, x_3, x_6) = (10, 3, 1)$$

and the value of the objective function is -9. In the second step, since

$$\max_j \, (z'_j - c_j) = z'_2 - c_2 = \tfrac{1}{2} > 0 \qquad \text{and} \qquad \theta_0 = \frac{10}{\frac{5}{2}}$$

$\mathbf{P}_2$ is introduced into the basis and $\mathbf{P}_1$ is removed. We transform the second-step values of Tableau 4.3 and obtain the third solution

$$\mathbf{X}''_0 = (x_2, x_3, x_6) = (4, 5, 11)$$

with a value of the objective function equal to -11. Since

$$\max \, (z''_j - c_j) = 0$$

$\dagger$ For computational convenience, we can add a column as in Tableau 4.3 for the calculation of θ_0.

TABLEAU 4.3

Initial Step

i	Basis	c		0	1	-3	0	2	0	
			P_0	P_1	P_2	P_3	P_4	P_5	P_6	θ
1	P_1	0	7	1	3	-1	0	2	0	—
2	P_4	0	12	0	-2	④	1	0	0	$12/4 = 3$
3	P_6	0	10	0	-4	3	0	8	1	$10/3 = 3\frac{1}{3}$
4			0	0	-1	3	0	-2	0	

Second Step

i	Basis	c	P_0	P_1	P_2	P_3	P_4	P_5	P_6	
1	P_1	0	10	1	$\left(\frac{5}{2}\right)$	0	$\frac{1}{4}$	2	0	$10/\frac{5}{2} = 4$
2	P_3	-3	3	0	$-\frac{1}{2}$	1	$\frac{1}{4}$	0	0	—
3	P_6	0	1	0	$-\frac{5}{2}$	0	$-\frac{3}{4}$	8	1	—
4			-9	0	$\frac{1}{2}$	0	$-\frac{3}{4}$	-2	0	

Third Step

i	Basis	c	P_0	P_1	P_2	P_3	P_4	P_5	P_6
1	P_2	1	4	$\frac{2}{5}$	1	0	$\frac{1}{10}$	$\frac{4}{5}$	0
2	P_3	-3	5	$\frac{1}{5}$	0	1	$\frac{3}{10}$	$\frac{2}{5}$	0
3	P_6	0	11	1	0	0	$-\frac{1}{2}$	10	1
4			-11	$-\frac{1}{5}$	0	0	$-\frac{4}{5}$	$-\frac{12}{5}$	0

this solution is a minimum feasible solution.   As a check on each complete elimination transformation, one should explicitly compute the individual z_0 and $z_j - c_j$ and compare them with the transformed values of z_0 and $z_j - c_j$.   If for the minimum feasible solution some $z'_j - c_j = 0$ for a vector P_j not in the final basis, then this vector can be introduced into the basis without changing the final value of the objective function.   The resulting solution will also be a minimum feasible solution, and hence we have determined *multiple minimum solutions*.   Any convex combination of these minimum solutions will also be a minimum solution.

Example 2.2.   To further illustrate the simplex procedure, we next solve the furniture-manufacturer problem shown in Fig. 2.14: The problem is to maximize

$$45x_1 + 80x_2$$

subject to

$$5x_1 + 20x_2 \leq 400$$
$$10x_1 + 15x_2 \leq 450$$
$$x_1 \qquad \geq 0$$
$$x_2 \geq 0$$

Since we want to standardize the computations and use the optimality conditions for minimization, we multiply the objective function by -1. Adding on slack variables, the problem is now to minimize

$$-45x_1 - 80x_2$$

subject to

$$5x_1 + 20x_2 + x_3 \qquad = 400$$
$$10x_1 + 15x_2 \qquad + x_4 = 450$$
$$x_j \geq 0$$

The simplex computations are given in Tableau 4.4. The optimal solution

TABLEAU 4.4

Initial Step

				-45	-80	0	0	
i	Basis	c	$\mathbf{P}_0$	$\mathbf{P}_1$	$\mathbf{P}_2$	$\mathbf{P}_3$	$\mathbf{P}_4$	θ
1	$\mathbf{P}_3$	0	400	5	⑳	1	0	$400/20 = 20$
2	$\mathbf{P}_4$	0	450	10	15	0	1	$450/15 = 30$
3			0	45	80	0	0	

Second Step

1	$\mathbf{P}_2$	-80	20	$\frac{5}{20}$	1	$\frac{1}{20}$	0	$20/\frac{5}{20} = 80$
2	$\mathbf{P}_4$	0	150	$\frac{125}{20}$	0	$-\frac{15}{20}$	1	$150/\frac{125}{20} = 24$
3			$-1{,}600$	25	0	-4	0	

Third Step

1	$\mathbf{P}_2$	-80	14	0	1	$\frac{2}{25}$	$-\frac{1}{25}$	
2	$\mathbf{P}_1$	-45	24	1	0	$-\frac{3}{25}$	$\frac{4}{25}$	
3			$-2{,}200$	0	0	-1	-4	

is $x_1 = 24$, $x_2 = 14$, and remembering that we are really maximizing, the value of the optimal solution is 2,200. The reader should trace the extreme-point path to the simplex procedure on Fig. 2.14.

Note that when solving linear-programming problems by hand, it is usually easier to work with fractions than with decimal expansions, given that the original data are integers or fractions. The use of hand calculators will, of course, introduce finite decimal representations of the transformed data; see Exercise 34.

Example 2.3.   We next develop the simplex tableaus for the numerical example that introduced this chapter.   The problem is to minimize

$$z = 2x_1 - x_2 + x_3 - 5x_4 + 22x_5$$

subject to

$$x_1 \qquad\qquad - 2x_4 + \quad x_5 = 6$$
$$x_2 \quad + \quad x_4 - \quad 4x_5 = 3$$
$$x_3 + 3x_4 + \quad 2x_5 = 10$$
$$x_j \geq 0$$

Using x_1, x_2, x_3 as the first basic feasible solution, we form the initial step of Tableau 4.5. Note that these basic variables have cost coefficients $c_j \neq 0$, while the associated basis $(\mathbf{P}_1\mathbf{P}_2\mathbf{P}_3)$ is an identity matrix. Structuring the initial simplex tableau step is equivalent to eliminating x_1, x_2, x_3 from

TABLEAU 4.5

Initial Step

i	Basis	c	$\mathbf{P}_0$	$\mathbf{P}_1$ (2)	$\mathbf{P}_2$ (−1)	$\mathbf{P}_3$ (1)	$\mathbf{P}_4$ (−5)	$\mathbf{P}_5$ (22)	θ
1	$\mathbf{P}_1$	2	6	1	0	0	−2	1	—
2	$\mathbf{P}_2$	−1	3	0	1	0	①	−4	$\frac{3}{1}$
3	$\mathbf{P}_3$	1	10	0	0	1	3	2	$\frac{10}{3}$
4			19	0	0	0	3	−14	

Second Step

i	Basis	c	$\mathbf{P}_0$	$\mathbf{P}_1$	$\mathbf{P}_2$	$\mathbf{P}_3$	$\mathbf{P}_4$	$\mathbf{P}_5$
1	$\mathbf{P}_1$	2	12	1	2	0	0	−7
2	$\mathbf{P}_4$	−5	3	0	1	0	1	−4
3	$\mathbf{P}_3$	1	1	0	−3	1	0	14
4			10	0	−3	0	0	−2

the equations

$$z - 2x_1 + x_2 - x_3 + 5x_4 - 22x_5 = 0$$
$$x_1 \qquad\qquad\quad - 2x_4 + x_5 = 6$$
$$x_2 \qquad + x_4 - 4x_5 = 3$$
$$x_3 + 3x_4 + 2x_5 = 10$$

to obtain

$$z \qquad\qquad\qquad + 3x_4 - 14x_5 = 19$$
$$x_1 \qquad\qquad - 2x_4 + x_5 = 6$$
$$x_2 \qquad + x_4 - 4x_5 = 3$$
$$x_3 + 3x_4 + 2x_5 = 10$$

In the initial step, x_4 is the only nonbasic variable that can be introduced into the solution to decrease the value of the objective function ($z_4 - c_4 = 3 > 0$). Introducing $\mathbf{P}_4$ into the basis requires the removal of $\mathbf{P}_2$ as $\theta_0 = \min\ x_{i0}/x_{i4} = x_{20}/x_{24}$ for $x_{i4} > 0$, that is

$$\min \left(\tfrac{3}{1}, \tfrac{10}{3}\right) = 3 = \theta_0$$

We pivot x_4 into the solution by the complete elimination method to obtain the Second Step shown in Tableau 4.5. Here all $z_j - c_j \leq 0$ and thus the optimal solution is $\mathbf{X}_0' = (x_1, x_3, x_4) = (12, 1, 3)$ and minimum $z = 10$.

Example 2.4. Consider the following problem: Minimize

$$x_0 = -6x_1 - 12x_2 - 4x_3$$

subject to

$$x_1 - 2x_2 + x_3 \leq 5$$
$$2x_1 - x_2 + 3x_3 \leq 2$$
$$x_j \geq 0$$

By adding slack variables x_4 and x_5, we form the Initial Step of Tableau 4.6. The simplex algorithm calls for the introduction of x_2 into the solution as $\max\ (z_j - c_j) = z_2 - c_2 = 12 > 0$. But, all $x_{i2} \leq 0$. This implies that the problem has an unbounded optimum. Here we have a family of feasible solutions given by

$$x_4 = 5 + 2x_2$$
$$x_5 = 2 + x_2$$

for any $x_2 \geq 0$, with $x_0 = -12x_2$ (see Exercise 45, Chap. 2, and Exercise 13, Chap. 3, on extreme directions). Note that if any nonbasic transformed $\mathbf{P}_j$ has all components $x_{ij} \leq 0$ with its $(z_j - c_j) > 0$, then we have an unbounded situation.

TABLEAU 4.6

Initial Step

				−6	−12	−4	0	0
i	Basis	c	P_0	P_1	P_2	P_3	P_4	P_5
1	P_4	0	5	1	−2	1	1	0
2	P_5	0	2	2	−1	3	0	1
3			0	6	12	4	0	0

Example 2.5. If the objective function of Example 2.2 was to maximize $x_0 = 30x_1 + 45x_2$, then the final step of Tableau 4.4 would be the following:

Basis	c	P_0	P_1	P_2	P_3	P_4
P_2	−45	14	0	1	$\left(\frac{2}{25}\right)$	$-\frac{1}{25}$
P_1	−30	24	1	0	$-\frac{3}{25}$	$\frac{4}{25}$
		−1,350	0	0	0	−3

Here all $z_j - c_j \leq 0$, but for the nonbasic variable x_3 we have $z_3 - c_3 = 0$. This means that we can introduce x_3 into the solution to obtain a new basic feasible solution that has the same optimal value of 1,350. Doing so removes P_2 from the basis and yields the solution of $x_1 = 45$, $x_3 = 175$, and $x_0 = 1,350$. The reader should verify the multiple solution for the new objective function by graphical technique (see Fig. 2.14). Any convex combination of the two extreme-point solutions $X_1 = (24, 14, 0, 0)$ and $X_2 = (45, 0, 175, 0)$ is also an optimal solution.

The determinants of the bases used in the simplex computation can be readily obtained as a by-product of the computation. Let the basis of the pth iteration be

$$B_p = (P_1 P_2 \cdots P_l \cdots P_m)$$

and the $(p + 1)$st basis be

$$B_{p+1} = (P_1 P_2 \cdots P_k \cdots P_m)$$

where

$$P_k = \sum_{i=1}^{m} x_{ik} P_i$$

We see that

$$
\mathbf{B}_{p+1} = \mathbf{B}_p \mathbf{C}_p = \mathbf{B}_p
\begin{pmatrix}
1 & \cdots & x_{1k} & \cdots & 0 \\
& \cdots \cdots \cdots \cdots \cdots & & \\
0 & \cdots & x_{lk} & \cdots & 0 \\
& \cdots \cdots \cdots \cdots \cdots & & \\
0 & \cdots & x_{mk} & \cdots & 1
\end{pmatrix}
$$

where $\mathbf{C}_p$ differs from the identity matrix in the lth column. We then have

$$
|\mathbf{B}_{p+1}| = |\mathbf{B}_p| \cdot |\mathbf{C}_p| = x_{lk}|\mathbf{B}_p|
$$

In the simplex process for $p = 0$, $\mathbf{B}_0 = \mathbf{I}$; hence the determinant of the pth basis of the simplex method is given by the product

$$
|\mathbf{B}_p| = \prod_{q=0}^{p} x_{lk}^{(q)}
$$

where $x_{lk}^{(0)} = 1$, and $x_{lk}^{(q)}$ are the pivot elements of the successive bases. We note that $|\mathbf{B}_p| > 0$.

The determinant of the final basis in Tableau 4.3 is $(4)(\frac{5}{2}) = 10$; while the determinant of the final basis $(\mathbf{P}_2\mathbf{P}_1)$ of Tableau 4.4 is $(20)(\frac{125}{20}) = 125$.

3. THE ARTIFICIAL-BASIS TECHNIQUE

Up to this point, we have always assumed that the given linear-programming problem was feasible and contained a unit matrix that could be used for the initial basis. Although a correct formulation of a problem will usually guarantee that the problem will be feasible, many problems do not contain a unit matrix. For such problems, the method of the *artificial basis* (see Orden [535]) is a satisfactory way to start the simplex process. This procedure also determines whether or not the problem has any feasible solutions.

The general linear-programming problem is to minimize

$$
c_1 x_1 + \cdots + c_n x_n
$$

subject to

$$
a_{11} x_1 + \cdots + a_{1n} x_n = b_1
$$
$$
a_{21} x_1 + \cdots + a_{2n} x_n = b_2
$$
$$
\cdots \cdots \cdots \cdots \cdots \cdots
$$
$$
a_{m1} x_1 + \cdots + a_{mn} x_n = b_m
$$

and

$$
x_j \geq 0
$$

For the method of the artificial basis we augment the above system as follows: Minimize

$$
c_1 x_1 + \cdots + c_n x_n + w x_{n+1} + w x_{n+2} + \cdots + w x_{n+m}
$$

subject to

$$a_{11}x_1 + \cdots + a_{1n}x_n + x_{n+1} \qquad\qquad\quad = b_1$$
$$a_{21}x_1 + \cdots + a_{2n}x_n \qquad\quad + x_{n+2} \qquad = b_2$$
$$\cdots\cdots\cdots\cdots\cdots\cdots\cdots\cdots\cdots\cdots\cdots\cdots\cdots\cdots\cdots$$
$$a_{m1}x_1 + \cdots + a_{mn}x_n \qquad\qquad\quad + x_{n+m} = b_m$$

and $x_j \geq 0$ for $j = 1, \ldots, n, n + 1, \ldots, n + m$. The quantity w is taken to be an unspecified large positive number. The vectors $\mathbf{P}_{n+1}, \mathbf{P}_{n+2}, \ldots, \mathbf{P}_{n+m}$ form a basis—an artificial basis—for the augmented system. If there is at least one feasible solution to the original problem, then this solution is also a feasible one for the augmented system. The simplex procedure will then ensure our obtaining the minimum solution, in which it is impossible for one of the artificial variables, x_{n+i}, to appear with a positive value. If the original problem is not feasible, then the minimum feasible solution to the augmented problem will contain at least one $x_{n+i} > 0$. As we shall show below, it is not necessary to assign a specific value to w.

For the augmented problem, the first feasible solution is

$$\mathbf{X}_0 = (x_{n+1,0},\ x_{n+2,0},\ \ldots,\ x_{n+m,0}) = (b_1,\ b_2,\ \ldots,\ b_m) \geq \mathbf{0}$$

with a corresponding value of the objective function $z_0 = w \sum\limits_{i=1}^{m} b_i$. Since the basis is a unit matrix, $\mathbf{X}_j = (x_{1j},\ x_{2j},\ \ldots,\ x_{mj}) = (a_{1j},\ a_{2j},\ \ldots,\ a_{mj})$, and $z_j = w \sum\limits_{i=1}^{m} x_{ij}$. As long as there are artificial vectors in the basis, each $z_j - c_j$ will be a linear function of w. For the first solution,

$$z_j - c_j = w \sum_{i=1}^{m} x_{ij} - c_j$$

Each $z_j - c_j$ will then have a w coefficient and a non-w coefficient which are independent of each other. We next set up the associated computational procedure as Tableau 4.7. For each j, the non-w component and the w component of $z_j - c_j$ have been placed in the $(m + 1)$st and $(m + 2)$nd rows, respectively, of that column.

We treat this tableau exactly like the original simplex tableau (Tableau 4.1), except that the vector introduced into the basis is associated with the largest positive element in the $(m + 2)$nd row. For the first iteration, the vector corresponding to $\max\limits_{j} \sum\limits_{i=1}^{m} x_{ij}$ is introduced into the basis. The elements in the $(m + 2)$nd row are also transformed by the usual elimination procedure. Once an artificial vector is removed from the basis, it is never selected to reenter the basis. Hence we do not have to transform the last m columns of the tableau. If we are interested in the inverse of the final basis, then these last m vectors should be transformed as usual.[1] It should be noted that, even if there are artificial vectors in the basis, the iteration may not

TABLEAU 4.7　First Step of Artificial-Basis Computational Procedure

i	Basis	c		c_1	c_2	$\cdot$	c_k	$\cdot$	c_n	w	$\cdot$	w	$\cdot$	w
			P_0	P_1	P_2	$\cdot$	P_k	$\cdot$	P_n	P_{n+1}	$\cdot$	P_{n+l}	$\cdot$	P_{n+m}
1	P_{n+1}	w	$x_{n+1,0}$	x_{11}	x_{12}	$\cdot$	x_{1k}	$\cdot$	x_{1n}	1	$\cdot$	0	$\cdot$	0
2	P_{n+2}	w	$x_{n+2,0}$	x_{21}	x_{22}	$\cdot$	x_{2k}	$\cdot$	x_{2n}	0	$\cdot$	0	$\cdot$	0
$\cdot$		$\cdot$	$\cdot$	$\cdot$	$\cdot$	$\cdot$	$\cdot$	$\cdot$	$\cdot$	$\cdot$	$\cdot$	$\cdot$	$\cdot$	$\cdot$
$\cdot$		$\cdot$	$\cdot$	$\cdot$	$\cdot$	$\cdot$	$\cdot$	$\cdot$	$\cdot$	$\cdot$	$\cdot$	$\cdot$	$\cdot$	$\cdot$
l	P_{n+l}	w	$x_{n+l,0}$	x_{l1}	x_{l2}	$\cdot$	x_{lk}	$\cdot$	x_{ln}	0	$\cdot$	1	$\cdot$	0
$\cdot$		$\cdot$	$\cdot$	$\cdot$	$\cdot$	$\cdot$	$\cdot$	$\cdot$	$\cdot$	$\cdot$	$\cdot$	$\cdot$	$\cdot$	$\cdot$
$\cdot$		$\cdot$	$\cdot$	$\cdot$	$\cdot$	$\cdot$	$\cdot$	$\cdot$	$\cdot$	$\cdot$	$\cdot$	$\cdot$	$\cdot$	$\cdot$
m	P_{n+m}	w	$x_{n+m,0}$	x_{m1}	x_{m2}	$\cdot$	x_{mk}	$\cdot$	x_{mn}	0	$\cdot$	0	$\cdot$	1
$m+1$			0	$-c_1$	$-c_2$	$\cdot$	$-c_k$	$\cdot$	$-c_n$	0	$\cdot$	0	$\cdot$	0
$m+2$			$\Sigma x_{n+i,0}$	Σx_{i1}	Σx_{i2}	$\cdot$	Σx_{ik}	$\cdot$	Σx_{in}	0	$\cdot$	0	$\cdot$	0

remove one of them. When using a full artificial basis, at least m iterations are required to exchange them with columns of the given problem, although fewer iterations may be necessary if the problem is degenerate.

We continue to select a vector to be introduced into the basis, using the element in the $(m+2)$nd row as criterion, until either (1) all the artificial vectors are removed from the basis or (2) no positive $(m+2)$nd element exists. The first alternative implies that all the elements in the $(m+2)$nd row equal zero and that the corresponding basis is a feasible basis for the original problem. We then apply the regular simplex algorithm to determine the minimum feasible solution. In the second alternative, if the $(m+2, 0)$ element, i.e., the artificial part of the corresponding value of the objective function, is greater than zero, then the original problem is not feasible. Theorem 2 tells us that there are no other feasible solutions whose value of the objective function is smaller than this final solution. If the $(m+2, 0)$ element is equal to zero, then we have a degenerate feasible solution to the original problem which contains at least one artificial vector. The artificial variables have values of zero. We have not, however, reached the minimum feasible solution. We continue the iterations by introducing a vector that corresponds to the maximum positive element in the $(m+1)$st row which is above a zero element in the $(m+2)$nd row.[2] This criterion is used until the minimum solution has been reached, i.e., until there are no more positive $(m+1)$st elements over a zero in the $(m+2)$nd row. The final solution may or may not contain artificial variables with values equal to zero. When we use

---

[1] See Orden [535] for a discussion of the above and the application of the simplex procedure to related matrix problems.

[2] This procedure forces the artificial variables still in the solution always to maintain a value of zero and also eliminates the selection of certain vectors that would not reduce the value of the objective function.

this criterion, the elements in the $(m + 2)$nd row are never transformed, since the $z_j - c_j$ of the vector being introduced is equal to $0 \cdot w + (z_j - c_j)$.

For both alternatives (1) and (2), all the $(m + 2, j)$ elements are less than or equal to zero, with the possible exception of the $(m + 2, 0)$ element. The latter element is always nonnegative, and its value is nonincreasing. As Orden [535] points out, if the original problem is feasible but has an unbounded minimum, the method of the artificial basis will determine the feasibility before the unboundedness. See Exercises 29 and 33 for further aspects of the artificial-basis technique.

Whenever the original problem contains some unit vectors, these vectors along with the necessary artificial ones should be used for the initial basis. Doing this will tend to decrease the total number of iterations. The following examples illustrate the artificial-basis technique.

Example 3.1. Minimize

$$2x_1 + x_2$$

subject to

$$3x_1 + x_2 - x_3 \qquad = 3$$
$$4x_1 + 3x_2 \qquad - x_4 \qquad = 6$$
$$x_1 + 2x_2 \qquad\qquad - x_5 = 2$$

and

$$x_j \geq 0$$

Since the given system does not contain any unit vectors and as all elements of $\mathbf{P}_0$ are strictly positive, we need to augment the system with a full artificial basis of three vectors. We denote these vectors as $\mathbf{P}_6, \mathbf{P}_7, \mathbf{P}_8$ and develop the first solution (see Tableau 4.8) according to the general formulation of Tableau 4.7. The first (artificial) solution is then $\mathbf{X}_0 = (x_6, x_7, x_8) = (3, 6, 2)$ with a value of the objective function equal to $11w$. The z_j elements are obtained by taking the inner product of $\mathbf{P}_j$ with the cost column $\mathbf{c}$. For example, $z_1 - c_1 = \mathbf{c}\mathbf{P}_1 - c_1 = 3w + 4w + 1w - 2 = 8w - 2$, with the coefficient of w noted in the fifth $(m + 2)$ row of the tableau and the -2 in the fourth $(m + 1)$ row under column $\mathbf{P}_1$. Vector $\mathbf{P}_1$ is selected to enter the basis as the maximum $(m + 2, j)$ element is $(m + 2, 1) = 8$. The corresponding $\theta_0 = \frac{3}{3} = 1$, and the artificial vector $\mathbf{P}_6$ is removed. We next transform all the elements of the initial step (Tableau 4.8), except the artificial vector columns, by the elimination formulas. The new solution, as shown in the second step, is $\mathbf{X}_0' = (x_1, x_7, x_8) = (1, 2, 1)$, with a value of the objective function of $3w + 2$. Vector $\mathbf{P}_2$ goes into the basis next, and the artificial vector $\mathbf{P}_8$ is removed. The new solution of the third step is $\mathbf{X}_0'' = (x_1, x_7, x_2) = (\frac{4}{5}, 1, \frac{3}{5})$ with the objective function equal to $w + \frac{11}{5}$. Both the elements in the positions $(m + 2, 3)$

Initial Step

i	Basis	c	P_0	2 P_1	1 P_2	0 P_3	0 P_4	0 P_5	w P_6	w P_7	w P_8	θ
1	P_6	w	3	③	1	-1	0	0	1	0	0	$3/3 = 1$
2	P_7	w	6	4	3	0	-1	0	0	1	0	$6/4 = 1\frac{1}{2}$
3	P_8	w	2	1	2	0	0	-1	0	0	1	$2/1 = 2$
4			0	-2	-1	0	0	0	0	0	0	
5			11	8	6	-1	-1	-1	0	0	0	

Second Step

Basis	c	P_0	P_1	P_2	P_3	P_4	P_5	P_6	P_7	P_8	θ
P_1	2	1	1	$\frac{1}{3}$	$-\frac{1}{3}$	0	0		0	0	$1/\frac{1}{3} = 3$
P_7	w	2	0	$\frac{5}{3}$	$\frac{4}{3}$	-1	0		1	0	$2/\frac{5}{3} = 1\frac{1}{5}$
P_8	w	1	0	⑤⁄③	$\frac{1}{3}$	0	-1		0	1	$1/\frac{5}{3} = \frac{3}{5}$
		2	0	$-\frac{1}{3}$	$-\frac{2}{3}$	0	0		0	0	
		3	0	$\frac{10}{3}$	$\frac{5}{3}$	-1	-1		0	0	

Third Step

Basis	c	P_0	P_1	P_2	P_3	P_4	P_5	P_7	θ
P_1	2	$\frac{4}{5}$	1	0	$-\frac{2}{5}$	0	$\frac{1}{5}$	0	—
P_7	w	1	0	0	①	-1	1	1	$1/1 = 1$
P_2	1	$\frac{3}{5}$	0	1	$\frac{1}{5}$	0	$-\frac{3}{5}$	0	$\frac{3}{5}/\frac{1}{5} = 3$
		$\frac{11}{5}$	0	0	$-\frac{3}{5}$	0	$-\frac{1}{5}$	0	
		1	0	0	1	-1	1	0	

Fourth Step

Basis	c	P_0	P_1	P_2	P_3	P_4	P_5	θ
P_1	2	$\frac{6}{5}$	1	0	0	$-\frac{2}{5}$	$\frac{3}{5}$	$\frac{6}{5}/\frac{3}{5} = 2$
P_3	0	1	0	0	1	-1	①	$1/1 = 1$
P_2	1	$\frac{2}{5}$	0	1	0	$\frac{1}{5}$	$-\frac{4}{5}$	—
		$\frac{14}{5}$	0	0	0	$-\frac{3}{5}$	$\frac{2}{5}$	
		0	0	0	0	0	0	

Fifth Step

Basis	c	P_0	P_1	P_2	P_3	P_4	P_5
P_1	2	$\frac{3}{5}$	1	0	$-\frac{3}{5}$	$\frac{1}{5}$	0
P_5	0	1	0	0	1	-1	1
P_2	1	$\frac{6}{5}$	0	1	$\frac{4}{5}$	$-\frac{3}{5}$	0
		$\frac{12}{5}$	0	0	$-\frac{2}{5}$	$-\frac{1}{5}$	0

and $(m+2, 5)$ equal 1; i.e., $z_3 - c_3 = w - \frac{3}{5}$ and $z_j - c_j = w - \frac{1}{5}$. We can choose either vector to enter the basis, and we arbitrarily choose $\mathbf{P}_3$, removing the last artificial vector $\mathbf{P}_7$ to obtain the fourth step solution $\mathbf{X}_0''' = (x_1, x_3, x_2) = (\frac{6}{5}, 1, \frac{2}{5})$, with the objective function equal to $\frac{14}{5}$. We have now found our first feasible solution, i.e., a solution which involves only the variables $(x_1, x_2, x_3, x_4, x_5)$ and satisfies the given constraints.

In the fourth step, as all the artificial variables have been removed, we use the usual simplex criterion to select the vector to enter the basis, i.e., maximum $(m+1, j)$ element. Thus, $\mathbf{P}_5$ is introduced and $\mathbf{P}_3$ removed. As the new basis of $(\mathbf{P}_1, \mathbf{P}_5, \mathbf{P}_2)$ causes all the $(m+1, j)$ elements to be nonpositive, i.e., all $z_j - c_j \le 0$, the fifth step contains the optimal feasible solution of $\mathbf{X}_0'''' = (x_1, x_5, x_2) = (\frac{3}{5}, 1, \frac{6}{5})$ with the minimum of the objective function $2x_1 + x_2$ equal to $\frac{12}{5}$. The reader will note that if $\mathbf{P}_5$ had been introduced into the basis in going from the third to the fourth step, we would have saved one iteration.

Example 3.2.[†] Maximize

$$x_1 + 2x_2 + 3x_3 - x_4$$

subject to

$$x_1 + 2x_2 + 3x_3 \qquad = 15$$
$$2x_1 + x_2 + 5x_3 \qquad = 20$$
$$x_1 + 2x_2 + x_3 + x_4 = 10$$

and $x_j \ge 0$. Since we are always minimizing, the corresponding objective is to minimize $-x_1 - 2x_2 - 3x_3 + x_4$.

Since the given system contains a unit vector $\mathbf{P}_4$, we need only two artificial vectors $\mathbf{P}_5$ and $\mathbf{P}_6$. The first solution (see Tableau 4.9) is $\mathbf{X}_0 = (x_5, x_6, x_4) = (15, 20, 10)$, with a value of the objective function equal to $10 + 35w$. Each z_j is computed by taking the inner product of $\mathbf{P}_j$ with the column vector $\mathbf{c}$. For example,

$$z_1 - c_1 = w + 2w + 1 - (-1) = 2 + 3w$$

Vector $\mathbf{P}_3$ is introduced into the basis because the maximum $(m+2, j)$ element is $(m+2, 3) = 8$. The corresponding $\theta_0 = \frac{20}{5}$, and the artificial vector $\mathbf{P}_6$ is removed from the basis. We transform all the elements of the initial step of Tableau 4.9 by the elimination formulas. The new solution is $\mathbf{X}_0' = (x_5, x_3, x_4) = (3, 4, 6)$, with a value of the objective function of $-6 + 3w$. $\mathbf{P}_2$ goes into the basis, and $\mathbf{P}_5$ is removed. In the third

---

[†] This simple example used to illustrate the computational technique also points out the advantages of applying a bit of analysis to the equations of a linear-programming model before solving it. Here, since $x_1 + 2x_2 + 3x_3 = 15$, the problem is really to maximize the objective function $15 - x_4$. Hence, for $x_4 \ge 0$, the maximum basic feasible solution involves x_1, x_2, and x_3 with a maximum value of 15.

TABLEAU 4.9

Initial Step

i	Basis	$\mathbf{c}$	$\mathbf{P}_0$	-1 $\mathbf{P}_1$	-2 $\mathbf{P}_2$	-3 $\mathbf{P}_3$	1 $\mathbf{P}_4$	w $\mathbf{P}_5$	w $\mathbf{P}_6$	θ
1	$\mathbf{P}_5$	w	15	1	2	3	0	1	0	$15/3 = 5$
2	$\mathbf{P}_6$	w	20	2	1	⑤	0	0	1	$20/5 = 4$
3	$\mathbf{P}_4$	1	10	1	2	1	1	0	0	$10/1 = 10$
4			10	2	4	4	0	0	0	
5			35	3	3	8	0	0	0	

Second Step

$\mathbf{P}_5$	w	3	$-\frac{1}{5}$	⑦⁄₅	0	0	1	$3/\frac{7}{5} = 2\frac{1}{7}$		
$\mathbf{P}_3$	-3	4	$\frac{2}{5}$	$\frac{1}{5}$	1	0	0	$4/\frac{1}{5} = 20$		
$\mathbf{P}_4$	1	6	$\frac{3}{5}$	$\frac{9}{5}$	0	1	0	$6/\frac{9}{5} = 3\frac{1}{3}$		
		-6	$\frac{2}{5}$	$\frac{16}{5}$	0	0	0			
		3	$-\frac{1}{5}$	$\frac{7}{5}$	0	0	0			

Third Step

$\mathbf{P}_2$	-2	$\frac{15}{7}$	$-\frac{1}{7}$	1	0	0	$-$
$\mathbf{P}_3$	-3	$\frac{25}{7}$	$\frac{3}{7}$	0	1	0	$25/7 / 3/7 = 8\frac{1}{3}$
$\mathbf{P}_4$	1	$\frac{15}{7}$	⑥⁄₇	0	0	1	$15/7 / 6/7 = 2\frac{1}{2}$
		$-\frac{90}{7}$	$\frac{6}{7}$	0	0	0	
		0	0	0	0	0	

Fourth Step

$\mathbf{P}_2$	-2	$\frac{5}{2}$	0	1	0	$\frac{1}{6}$
$\mathbf{P}_3$	-3	$\frac{5}{2}$	0	0	1	$-\frac{3}{6}$
$\mathbf{P}_1$	-1	$\frac{5}{2}$	1	0	0	$\frac{7}{6}$
		-15	0	0	0	-1

step, since all the $(m + 2, j) \leq 0$ and $(m + 2, 0) = 0$, we have a feasible solution to the original problem; this solution is

$$\mathbf{X}_0'' = (x_2, x_3, x_4) = \left(\tfrac{15}{7}, \tfrac{25}{7}, \tfrac{15}{7}\right)$$

with the objective function equal to $-\frac{90}{7}$. The fourth step yields a minimum feasible solution

$$\mathbf{X}_0''' = (x_2, x_3, x_1) = \left(\tfrac{5}{2}, \tfrac{5}{2}, \tfrac{5}{2}\right)$$

with $x_4 = 0$ and the objective function equal to -15. The correct value of the objective function is $+15$, since we were originally dealing with a maximization problem.

4. A FIRST FEASIBLE SOLUTION USING SLACK VARIABLES[1]

If the linear-programming problem is originally of the form $\mathbf{AX} \le \mathbf{b}$, its corresponding initial computational tableau will contain an $m \times m$ unit matrix. This matrix appears when we write each inequality as an equality by adding a nonnegative variable called a *slack variable*. These slack variables have associated cost coefficients that are usually set equal to zero.

For example, given the linear-programming problem: Maximize

$$2x_1 + 3x_2$$

subject to

$$-x_1 + 2x_2 \le 4$$
$$x_1 + x_2 \le 6$$
$$x_1 + 3x_2 \le 9$$
$$x_j \ge 0$$

Transforming to equations by adding a new nonnegative slack variable to all inequalities except the nonnegativity inequalities, we obtain the equivalent linear-programming problem: Maximize

$$2x_1 + 3x_2$$

subject to

$$-x_1 + 2x_2 + x_3 \qquad\qquad = 4$$
$$x_1 + x_2 \qquad + x_4 \qquad = 6$$
$$x_1 + 3x_2 \qquad\qquad + x_5 = 9$$
$$x_j \ge 0$$

We see that these equations contain a starting basis of the *slack vectors* $(\mathbf{P}_3\mathbf{P}_4\mathbf{P}_5)$ with the associated first feasible solution of $x_1 = 0$, $x_2 = 0$, $x_3 = 4$, $x_4 = 6$, $x_5 = 9$. The corresponding value of the objective function is zero, as all cost coefficients for the slack variables are here assumed to be zero. Some applications might call for these cost coefficients to be other than zero, e.g., penalty costs for not using raw materials, storage costs.

If the problem was originally $\mathbf{AX} \ge \mathbf{b}$, the equivalent set of equations is obtained by the subtraction of a new nonnegative slack variable from each inequality. Here the equations contain a negative unit basis which, since we assume $\mathbf{b} \ge \mathbf{0}$, cannot be used to initiate the computational procedure.[2] For

---

[1] See Sec. 4 of Chap. 2 for previous discussion of slack variables.
[2] See Example 3.1.

this case the problem, in general, must employ at least one artificial vector. This is further described in Sec. 1 of Chap. 9. A problem can have a mixture of $\leq$, $\geq$, and $=$ constraints and is put into equation form by suitable additions or subtractions of slack variables. If a unit vector is used in a starting basis and its $c_j \neq 0$, then we must eliminate the corresponding variable from the objective function, as was done in Example 2.3; see also Exercises 1*b* and 1*f*. Slack variables are sometimes referred to as *logical variables*, while the problem's original variables are termed *structural variables*; similar terms are used for the associated column vectors.

5. GEOMETRIC INTERPRETATION OF THE SIMPLEX PROCEDURE

The preceding algebraic description of the simplex procedure has the following geometric interpretation, described in Hoffman, Mannos, Sokolowsky, and Wiegmann [377]. It was first given in Dantzig [153].

Let the $(m + 1)$-dimensional vectors $\mathbf{P}'_1$, $\mathbf{P}'_2$, $\ldots$, $\mathbf{P}'_n$ be obtained by appending the cost coefficients c_1, c_2, $\ldots$, c_n, respectively, to the m-dimensional column vectors $\mathbf{P}_1$, $\mathbf{P}_2$, $\ldots$, $\mathbf{P}_n$; that is, for $\mathbf{P}'_j$ the corresponding c_j becomes the $(m + 1)$st coordinate. Let $\mathbf{C}$ be the convex cone in $(m + 1)$-space determined by $\mathbf{P}'_1$, $\mathbf{P}'_2$, $\ldots$, $\mathbf{P}'_n$. Let $\mathbf{L}$ be the line in $(m + 1)$-space consisting of all points whose first m coordinates are b_1, b_2, $\ldots$, b_m. This line can be thought of as being generated by an $(m + 1)$-dimensional vector $\mathbf{P}'_0$ whose first m coordinates are b_1, b_2, $\ldots$, b_m and whose $(m + 1)$st coordinate takes on all values. The object of the computation is to find the lowest point of $\mathbf{L}$ which is also in $\mathbf{C}$, that is, the point of $\mathbf{L}$ in $\mathbf{C}$ whose $(m + 1)$st coordinate is a minimum.

The computational technique is as follows: Assume that m of the vectors $\mathbf{P}'_1$, $\mathbf{P}'_2$, $\ldots$, $\mathbf{P}'_n$, say the first m, form a linearly independent set and have the property that the m-dimensional cone $\mathbf{D}$ which they generate contains a point of $\mathbf{L}$. This property is equivalent to saying that the m vectors are an admissible basis, i.e., correspond to a feasible solution. These vectors can either be given, as in Sec. 2, or be artificial, as in Sec. 3. The hyperplane containing $\mathbf{D}$ divides the remaining vectors $\mathbf{P}'_j$ into two groups. One group is on the side of the hyperplane which contains the positive $(m + 1)$st coordinate axis. The second group contains all the vectors which are below the hyperplane and are eligible to be selected as new basis vectors. Each vector of this second group can be joined to the hyperplane containing $\mathbf{D}$ by a line segment parallel to the $(m + 1)$st coordinate axis. Let $\mathbf{P}'_k$ be the vector with the property that its line segment is the longest. This corresponds to selecting the vector with $\max_j (z_j - c_j) > 0$. Then $\mathbf{P}'_k$ and a certain set of $m - 1$ of the vectors $\mathbf{P}'_1$, $\ldots$, $\mathbf{P}'_m$ have the property that the m-dimensional cone they form contains a point of $\mathbf{L}$, and this point will be lower than the intersection of $\mathbf{D}$ with $\mathbf{L}$. We replace the eliminated vector by $\mathbf{P}'_k$ and continue the process until a minimum solution has been found.

We next illustrate the above with an example. Consider the problem of minimizing

$$5x_1 + 3x_2 + 4x_3 + 6x_4$$

subject to the conditions

$$x_1 + 3x_2 + 5x_3 + 6x_4 = 3$$
$$6x_1 + 4x_2 + 2x_3 + x_4 = 2$$

and

$$x_j \geq 0$$

We have

$$\mathbf{P}_1' = \begin{pmatrix} 1 \\ 6 \\ 5 \end{pmatrix} \qquad \mathbf{P}_2' = \begin{pmatrix} 3 \\ 4 \\ 3 \end{pmatrix} \qquad \mathbf{P}_3' = \begin{pmatrix} 5 \\ 2 \\ 4 \end{pmatrix} \qquad \mathbf{P}_4' = \begin{pmatrix} 6 \\ 1 \\ 6 \end{pmatrix}$$

We plot these points as shown in Fig. 4.1. Since the line **L** intersects the cone **C**, feasible solutions to the problems exist. We can readily show that $\mathbf{P}_1'$ and $\mathbf{P}_4'$

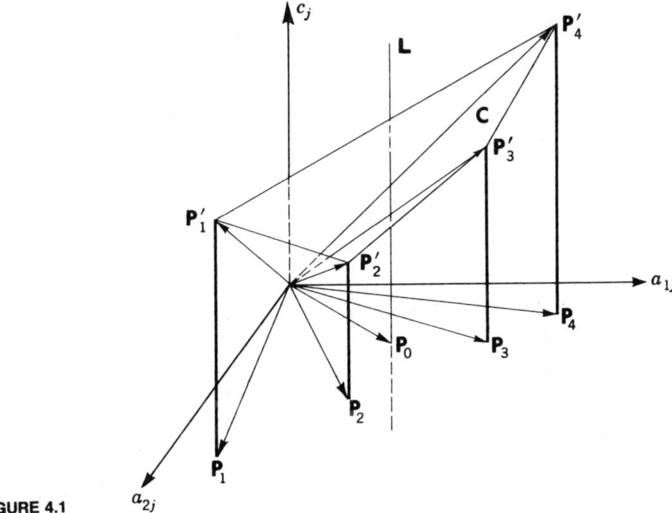

FIGURE 4.1

are linearly independent and that $\mathbf{P}_0$ can be expressed as a positive combination of $\mathbf{P}_1$ and $\mathbf{P}_4$. We then take as our first basis vectors $\mathbf{P}_1$ and $\mathbf{P}_4$, and we determine the two-dimensional cone formed by vectors $\mathbf{P}_1'$ and $\mathbf{P}_4'$ (Fig. 4.2). The point $\mathbf{P}_2$ is below the hyperplane containing **D** by a distance of z_2, where z_2 is defined in Eq. (1.4). $\mathbf{P}_3$ is a distance z_3 below this hyperplane. Point $\mathbf{P}_2'$ is below the hyperplane containing **D** by a distance of $z_2 - c_2$. The corresponding value of the objective function is given by z. Since $z_2 - c_2 > z_3 - c_3$, we introduce $\mathbf{P}_2$ into the basis. From Fig. 4.2 we see that $\mathbf{P}_0$ can be expressed as a positive combination of vectors $\mathbf{P}_2$ and $\mathbf{P}_4$ and

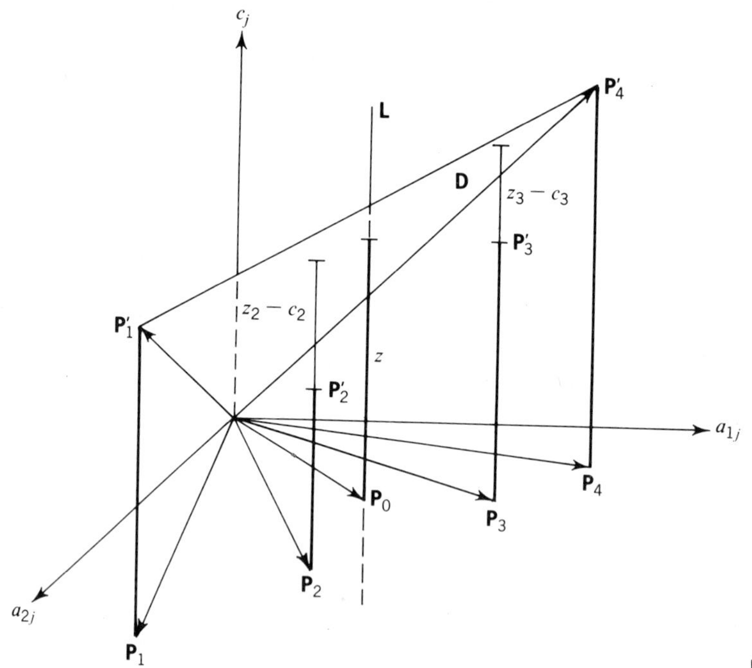

FIGURE 4.2

cannot be expressed as a positive combination of P_2 and P_1; that is, P_0 is not contained in the cone generated by vectors P_1 and P_2. Hence P_2 is introduced into the basis and removes P_1, giving a new basis of P_2 and P_4. For this basis we can perform a similar analysis that introduces P_3 into the basis and removes P_4. This basis of vectors P_2 and P_3 corresponds to the minimum solution.

We may also interpret the simplex procedure in terms of moving from one extreme point to an adjacent extreme point (see Sec. 4 of Chap. 2 and Saaty [583]). The procedure starts with the linear form passing through an extreme point of K (Fig. 4.3). Here we have pictured an arbitrary two-dimensional set of linear inequalities and its corresponding region of solution.

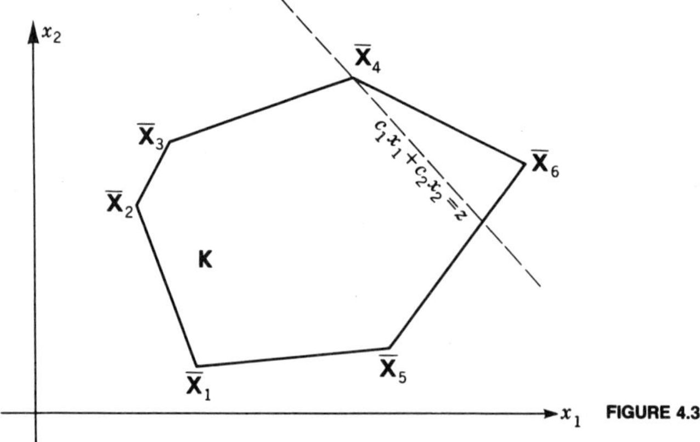

FIGURE 4.3

The next iteration would advance the linear form parallel to itself until it passes through $\bar{\mathbf{X}}_3$. In two additional iterations the linear form would pass through $\bar{\mathbf{X}}_1$, and we would have the minimum solution. The total number of iterations necessary to reach the minimum depends on which feasible solution is used to initiate the computation. If, as above, we started with the solution that corresponds to the extreme point $\bar{\mathbf{X}}_4$, then it would take us three iterations. If the solution associated with $\bar{\mathbf{X}}_5$ was taken as the first, then it would require only one iteration. If the coefficients of the linear form and the equations are normalized, then z is the distance of the linear form from the origin.[1]

We illustrate the above interpretation for Example 3.1, treating the constraints as inequalities. Thus, we want to solve the problem of minimizing

$$2x_1 + x_2$$

subject to

(a) $3x_1 + x_2 \geq 3$

(b) $4x_1 + 3x_2 \geq 6$

(c) $x_1 + 2x_2 \geq 2$

(d) $x_1 \quad\quad \geq 0$

(e) $x_2 \geq 0$

The joint-solution space $\mathbf{K}$ of inequalities (a) to (e) is shown in Fig. 4.4. It is an unbounded, convex region with four extreme points. The objective function is shown passing through the optimum solution point $x_1 = \frac{3}{5}$, $x_2 = \frac{6}{5}$. This corresponds to the fifth step of Tableau 4.8. The point $x_1 = \frac{6}{5}$, $x_2 = \frac{2}{5}$ is the first feasible solution of the fourth step.

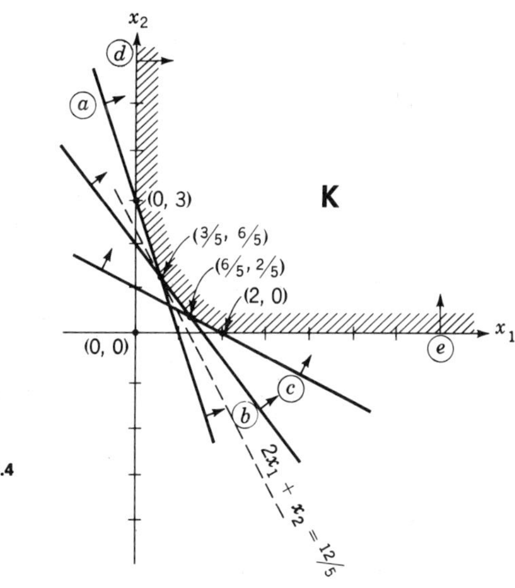

FIGURE 4.4

[1] See Exercise 18.

EXERCISES

1. Solve the following linear-programming problems by the simplex method:

a. Minimize

$$x_1 + x_2 + x_3$$

subject to

$$x_1 \qquad\quad - x_4 \qquad\quad - 2x_6 = 5$$
$$x_2 \quad + 2x_4 - 3x_5 + \; x_6 = 3$$
$$x_3 + 2x_4 - 5x_5 + 6x_6 = 5$$

and $x_j \geq 0$.

Answer: $x_1 = \frac{71}{10}$, $x_4 = \frac{13}{10}$, $x_6 = \frac{4}{10}$; $x_2 = x_3 = x_5 = 0$; minimum $z = \frac{71}{10}$.

b. Minimize

$$x_1 - x_2 + \; x_3 + x_4 + x_5 - \; x_6$$

subject to

$$x_1 \qquad\qquad + x_4 \qquad + 6x_6 = 9$$
$$3x_1 + x_2 - 4x_3 \qquad\qquad + 2x_6 = 2$$
$$x_1 \qquad + 2x_3 \qquad + x_5 + 2x_6 = 6$$

and $x_j \geq 0$.

Answer: $x_2 = 5$, $x_3 = \frac{3}{2}$, $x_6 = \frac{3}{2}$; $x_1 = x_4 = x_5 = 0$; minimum $z = -5$.

c. Minimize

$$2x_1 + x_2 - \; x_3 - x_4$$

subject to

$$x_1 - x_2 + 2x_3 - x_4 = 2$$
$$2x_1 + x_2 - 3x_3 + x_4 = 6$$
$$x_1 + x_2 + \; x_3 + x_4 = 7$$

and $x_j \geq 0$.

Answer: $x_1 = 3$, $x_3 = 1$, $x_4 = 3$; $x_2 = 0$; minimum $z = 2$.

d. Minimize

$$-3x_1 + \; x_2 + 3x_3 - \; x_4$$

subject to

$$x_1 + 2x_2 - \; x_3 + \; x_4 = 0$$
$$2x_1 - 2x_2 + 3x_3 + 3x_4 = 9$$
$$x_1 - \; x_2 + 2x_3 - \; x_4 = 6$$

and $x_j \geq 0$.

Answer: $x_1 = 1$, $x_2 = 1$, $x_3 = 3$; $x_4 = 0$; minimum $z = 7$.

e. Maximize

$$x_4 - x_5$$

subject to

$$2x_2 - \ x_3 - x_4 + x_5 \geq 0$$
$$-2x_1 \qquad + 2x_3 - x_4 + x_5 \geq 0$$
$$x_1 - 2x_2 \qquad - x_4 + x_5 \geq 0$$
$$x_1 + \ x_2 + \ x_3 \qquad\ = 1$$

and $x_j \geq 0$.

Answer: $x_1 = \frac{2}{5}$, $x_2 = \frac{1}{5}$, $x_3 = \frac{2}{5}$; $x_4 = x_5 = 0$; maximum $z = 0$.

f. Maximize

$$x_1 - x_2 + \ x_3 - 3x_4 + x_5 - x_6 - 3x_7$$

subject to

$$3x_3 \qquad + x_5 + x_6 \qquad = \ 6$$
$$x_2 + 2x_3 - \ x_4 \qquad = 10$$
$$-x_1 \qquad + x_6 \qquad = \ 0$$
$$x_3 \qquad + x_6 + \ x_7 = \ 6$$

and $x_j \geq 0$.

Answer: $x_1 = 6$, $x_2 = 10$, $x_3 = 0$; $x_4 = 0$, $x_5 = 0$, $x_6 = 6$, $x_7 = 0$; maximum $z = -10$.

g. Minimize

$$x_1 - 2x_2 + 3x_3$$

subject to

$$-2x_1 + \ x_2 + 3x_3 = 2$$
$$2x_1 + 3x_2 + 4x_3 = 1$$

and $x_j \geq 0$.

h. Maximize

$$3x_1 - \ x_2$$

subject to

$$2x_1 + \ x_2 \geq 2$$
$$x_1 + 3x_2 \leq 3$$
$$x_2 \leq 4$$

and $x_j \geq 0$.

i. Minimize

$$2x_1 - 3x_2 + 6x_3$$

subject to

$$3x_1 - 4x_2 - 6x_3 \leq \ 2$$
$$2x_1 + \ x_2 + 2x_3 \geq 11$$
$$x_1 + 3x_2 - 2x_3 \leq \ 5$$

and $x_j \geq 0$.

Answer: $x_1 = 0$, $x_2 = 4$, $x_3 = \frac{7}{2}$; minimum $z = 9$.

j. Maximize

$$x_1 + x_2$$

subject to

$$x_1 + x_2 \geq 1$$
$$x_1 - x_2 \leq 1$$
$$-x_1 + x_2 \leq 1$$

and $x_j \geq 0$.

  k. Maximize

$$4x_1 + x_2$$

subject to

$$-x_1 + 2x_2 \leq 4$$
$$2x_1 + 3x_2 \leq 12$$
$$4x_1 - 4x_2 \leq 12$$

and $x_j \geq 0$.

Answer: $x_1 = 4\frac{1}{5}$, $x_2 = 1\frac{1}{5}$; maximum $z = 18$.

2. Given a linear-programming problem that has a finite minimum feasible solution. Let the minimum value of the objective function be z_0. The problem is being solved by the simplex procedure, and in the kth iteration a degenerate basic feasible solution with exactly one $x_i = 0$ appears. The corresponding value of the objective function is $\bar{z}$ with $\bar{z} > z_0$. Prove that this kth basis cannot reappear in the succeeding iterations.

3. (A test for a near-optimal solution in the simplex procedure due to L. Goldstein.) Given a linear-programming problem with a known upper bound S to the sum of the variables, e.g., the transportation problem. Let $\bar{z}$ be the value of the objective function for a given basic feasible solution and z_0 be the unknown minimum value of the objective function. Let $m = z_k - c_k = \max (z_j - c_j)$ for the given basic solution. Show that $\bar{z} - z_0 \leq \epsilon$ is a predetermined tolerance limit if $m \leq \epsilon/S$.

4. Solve Exercise 1b, 1d, and 1i using the criterion (2.1a) for selecting the variable to be introduced into the solution.

5. Prove the validity of the transformation formulas (2.7a) and (2.7b).

6. Prove the following statement: If the original problem is feasible but has an unbounded minimum, the method of the artificial basis will determine the feasibility before the unboundedness.

7. Prove that if a set of m linear equations in n unknowns, where $m \leq n$, has a feasible solution, there exists a basic feasible solution. (Assume that the rank of the system is m.) Also, given an optimal feasible solution with more than m positive variables, show how it can be reduced to an optimal basic feasible solution (Gale [277], Danø [151], and Lawler [462]).

8. For the standard linear-programming problem, let

$$x_0 = c_1x_1 + c_2x_2 + \cdots + c_nx_n$$

Write an equivalent linear-programming problem which uses this equation as a constraint and discuss the application of the simplex algorithm to the new problem.

9. For Fig. 4.2 show that for some $\mathbf{P}_j$ not in the basis, e.g., vector $\mathbf{P}_2$, the corresponding z_j does measure the vertical distance of the point $\mathbf{P}_j$ from the plane $\mathbf{D}$.

10. In the system $\mathbf{AX} = \mathbf{b}$ let some x_j be unrestricted as to sign. If we substitute $x_j = x_j' - x_j''$, explain why both x_j' and x_j'' cannot be in the same basic feasible solution; i.e., if x_j is nonzero for some basic feasible solution, then it is represented in the basis by either x_j' or x_j'', but not both.

11. Under what conditions can a variable be eliminated from a problem by using one equation to solve for it in terms of the remaining variables?

12. Solve Exercise 2 of Chap. 1.

13. Solve Exercises 1g, 1h, 1j, and 1k by graphical procedures.

14. Develop a computer flow chart of the simplex procedure.

15. Given that the system $\mathbf{AX} = \mathbf{b}$, $\mathbf{X} \geq \mathbf{0}$ has a feasible solution, show that the results of Exercise 17, Chap. 2, apply. (HINT: The artificial-basis technique will find an equivalent basic feasible solution.)

16.   Describe how you would change (scale) the coefficients of a column to make them smaller or larger.   (NOTE: Scaling is an important concept which must be considered, especially when solving a problem on a digital computer.   The ability of the simplex algorithm to converge and the avoidance of numerical difficulties are dependent upon the coefficients in the simplex tableaus, e.g., ensuring against division by zero, determining if a $z_j - c_j < 0$.   Many computer codes have built-in scaling procedures.)

17.   Let x_{n+i} be the ith slack variable for the system $\mathbf{AX} \le \mathbf{b}, \mathbf{X} \ge 0$; i.e., the ith constraint can be written as $a_{i1}x_1 + \cdots + a_{in}x_n + x_{n+i} = b_i$.   Show for any nonnegative solution that $x_{n+i} = h_i \sqrt{a_{i1}^2 + \cdots + a_{in}^2}$, where h_i is the perpendicular distance from the solution point to the ith hyperplane (Barsov [45]).

18.   Show that the linear functional $c_1x_1 + \cdots + c_nx_n$ is perpendicular to the vector from the origin to the point $(c_1, \ldots, c_n)$.   Also, if the linear functional is transformed to $z = d_1x_1 + \cdots + d_nx_n$, where

$$d_j = \frac{c_j}{\sqrt{c_1^2 + \cdots + c_n^2}}$$

then z is the distance of the linear functional from the origin, Hadley [342].

19.   Develop a condensed simplex computational tableau and associated computational procedure which does not require the unit vectors corresponding to the vectors in the basis; i.e., the tableau will have only $(n - m)$ columns.   (HINT: The elimination procedure formulas still apply but the pivot column, instead of reducing to a unit vector, becomes the representation of the eliminated vector in terms of the new basis vectors.)

20.   If an optimal solution is degenerate and some nonbasic variable x_j has its $z_j - c_j = 0$, under what conditions will the introduction of x_j into the solution not produce a new and different optimal solution?

21.   Solve the following problem by the simplex method and by graphical techniques.   What are the solution set's extreme points and extreme directions?   Use the representation theorem (Sec. 3, Chap. 2) to express any feasible solution in terms of the extreme points and directions.

Maximize

$$x_0 = x_1 + x_2$$

subject to

$$-x_1 + x_2 \le 1 \qquad x_j \ge 0$$

22.   Solve the problem of Example 2.5 starting with the slack variable, first basic feasible solution, and calculate the tableaus for both optimal solutions.

23.   Based on the definition of a convex cone (see also Exercise 11, Chap. 2), explain why, in Fig. 4.1, the bases $(\mathbf{P_1P_2})$ and $(\mathbf{P_3P_4})$ are not admissible (feasible) bases for $\mathbf{P_0}$, while $(\mathbf{P_2P_3})$ and $(\mathbf{P_2P_4})$ are admissible.   What about the other possible bases?

24.   Show how the elimination computational check scheme of Exercise 42, Chap. 2, can be used in the simplex tableau computations.

25.   Discuss why the simplex tableau scheme always has the unrestricted objective function variable z (or x_0) in the basis.

26.   Assume that a standard linear-programming problem is feasible but the $\mathbf{A}$ matrix has rank $r < m$.   Describe what happens if a full artificial basis is used to find an initial feasible solution.   What can you say about the artificial-basis technique and the finding of redundant equations?   (See Exercise 44, Chap. 2.)

27.   Solve the system (4.1) of Chap. 2 using the three objective functions minimize $2x_1 + 2x_2$, minimize $2x_1 - 2x_2$, and maximize $2x_1 + 2x_2$.   Describe how you could save some work by using the first feasible solution for all problems and the carrying along of all the objective functions in the simplex tableaus.

28.   Discuss why you really do not need to specify a particular value of w when using the artificial basis technique.   Note that you are, in effect, minimizing the sum of the artificial variables.   If you do give the artificial weight a specific value, you can then consider the problem with artificial variables as an equivalent numerical problem and solve it by the simplex algorithm.   As the algorithm is an optimizing one, it should force the artificial variables out of the basic solution, if the weight was selected properly.   Discuss the difficulties that might arise when using this approach, e.g., you do not know the exact value of w in that a value that is not large enough could give a false signal that a problem is infeasible, while a value that is too large might cause numerical problems.   Solve Exercises $1d$ and $1f$ using a numerical w.   Construct an example with a "suitable" value of w that does not work.   See Chap. 5 and the two-phase method for handling an artificial basis.

29.   Solve the problem in Sec. 5 associated with Fig. 4.1 with $(\mathbf{P}_1\mathbf{P}_4)$ the starting basis.

30.   Assume you have found a first feasible solution using an artificial basis and artificial variables are still in the basis.   Describe ways you can remove them and/or how to make sure their values do not increase above zero (Dantzig [152]).   (HINT: Any number can be used as a pivot in a degenerate row; the remaining artificial variables can be constrained to equal zero by the addition of a suitable constraint; redundant rows can be dropped.)

31.   Kaplan [411] describes a way to implement the successive basis transformations so that if an initial simplex tableau has all integer elements, the succeeding tableaus will also be all integer, i.e., no fractions (decimals) are introduced.   Can you discover how to do it?   Of course, you have to take care of the possibility that the final solution is noninteger.

32.   For the $m \times n$ system $\mathbf{AX} = \mathbf{b}$, assume $\mathbf{A}$ is of rank m and all a_{ij} and b_i are integers.   Show that for any basic solution $|x_j| \le m!\,ba^{m-1}$, where $b = \max |b_i|$ and $a = \max |a_{ij}|$.   HINT: Recall Cramer's rule, Sec. 5, Chap. 2 (Papadimitriou and Steiglitz [541]).

33.   Show that the same sequence of vectors will enter the basis if the linear-programming problem is solved either as a maximization or minimization problem (Hadley [343]).

34.   Some linear-programming applications have free variables that can be equal to either a positive, negative, or zero value (see Secs. 1 and 6 of Chap. 11).   A free (or unrestricted) variable is denoted by $-\infty < x_j < +\infty$.   As noted in Sec. 4 of Chap. 2, any free variable can be defined as the difference of two nonnegative variables.   From a computational point of view, such a transformation is not efficient as it increases the number of variables [although only one of the defining nonnegative variables can be in a basis (see Exercise 10)].   Show that a free variable must be in a finite optimal (or multiple optimal) solution; that at the start of the simplex process, we can force a free variable into the basis and it must remain in the basis; and that if a problem has free variables, a basic solution may not be an extreme point (Dantzig [152], Gass [290], and Collatz and Wetterling [131]).   Discuss the above when there are many free variables in the problem, especially if there are at least m of them.   Note that commercial mathematical programming systems for solving linear-programming problems on a computer force and keep free variables in the basis, instead of defining them by the difference of two nonnegative variables.

35.   Given a nonbasic, optimal feasible solution to a linear-programming problem, can you prove that it is optimal without comparing it to an optimal basic feasible solution found by direct application of the simplex method?

CHAPTER 5

The Revised Simplex Method

1. THE GENERAL FORM OF THE INVERSE

Upon critically analyzing the simplex computational procedure (see Chaps. 3 and 4), we see that the essential element which enables us to progress from each basic feasible solution to the optimal solution is the explicit knowledge of the representation of the vectors not in the current basis in terms of the basis vectors.   Given this information, we can do the following:

1. Calculate the $z_j - c_j$ elements to determine which vector to introduce into the basis or to determine whether the current solution is optimal.
2. Determine which vector to remove from the basis.
3. Transform the basis and obtain the new solution.

In Sec. 5 of Chap. 2 we showed that, given a basis $\mathbf{B}$ of m-dimensional vectors $(\mathbf{P}_1 \ \mathbf{P}_2 \ \cdots \ \mathbf{P}_m)$, the linear combination that expresses any other m-dimensional vector $\mathbf{P}_j$ in terms of $\mathbf{B}$ is determined by computing the vector

$$\mathbf{X}_j = \mathbf{B}^{-1}\mathbf{P}_j \tag{1.1}$$

where $\mathbf{X}_j = (x_{1j}, x_{2j}, \ldots, x_{mj})$ is a column vector.   We then have

$$\mathbf{P}_j = x_{1j}\mathbf{P}_1 + x_{2j}\mathbf{P}_2 + \cdots + x_{mj}\mathbf{P}_m$$

which is the desired linear combination. Consider the general linear-programming problem of minimizing

$$\mathbf{cX}$$

subject to

$$\mathbf{AX} = \mathbf{b}$$

and

$$\mathbf{X} \geq \mathbf{0}$$

If we let the matrix $\mathbf{B}$ correspond to the first m vectors of $\mathbf{A}$, such that

$$\mathbf{BX}_0 = \mathbf{b}$$

$$\mathbf{X}_0 \geq \mathbf{0}$$

where $\mathbf{X}_0 = (x_{10}, x_{20}, \ldots, x_{m0})$, we have a basic feasible solution given by

$$\mathbf{X}_0 = \mathbf{B}^{-1}\mathbf{b} \tag{1.2}$$

The linear combinations of all the vectors of $\mathbf{A}$ in terms of $\mathbf{B}$ can be determined by (1.1) for $j = 1, 2, \ldots, n$. In Sec. 1 of Chap. 4 we defined for any feasible basis the quantities

$$z_j = c_1 x_{1j} + c_2 x_{2j} + \cdots + c_m x_{mj} \tag{1.3}$$

for $j = 1, 2, \ldots, n$, where the c_i correspond to the cost coefficients of those vectors in the basis. For any j we have from (1.1) that (1.3) can be rewritten to read

$$z_j = \mathbf{c}_0 \mathbf{X}_j = \mathbf{c}_0 \mathbf{B}^{-1} \mathbf{P}_j \qquad j = 1, 2, \ldots, n$$

where $\mathbf{c}_0 = (c_1, c_2, \ldots, c_m)$ is a row vector. Hence, given the m-dimensional row vector

$$\boldsymbol{\pi} = \mathbf{c}_0 \mathbf{B}^{-1} \tag{1.4}$$

for a feasible basis $\mathbf{B}$, we can compute the corresponding z_j. The vector $\boldsymbol{\pi} = (\pi_1, \pi_2, \ldots, \pi_m)$ is called the *pricing* or *multiplier vector* and the individual π_i are the *pricing* or *simplex multipliers*. We *price out* a vector $\mathbf{P}_j$ not in the basis by computing $\boldsymbol{\pi}\mathbf{P}_j - c_j = z_j - c_j$. From (1.1), (1.2), and (1.4) we note that the information necessary to proceed from feasible solution to feasible solution can be obtained if, for each feasible basis $\mathbf{B}$, we have explicit knowledge of $\mathbf{B}^{-1}$ and the original data consisting of $\mathbf{A}$, $\mathbf{b}$, and $\mathbf{c}$. This fact has led to the development of a computational procedure to solve the general linear-programming problem termed the *revised simplex method* (Dantzig, Orden, and Wolfe [191] and Dantzig [156, 157]).

The main difference between the original simplex method and the revised procedure is that in the former we transform all the elements of the simplex tableau by means of the elimination formulas, while in the latter we need transform only the elements of an inverse matrix by means of the same formulas.

Since its development the revised procedure and especially its variation which employs the product form of the inverse have been selected for use on most electronic computers.[1]   The two reasons for this are:

1. For problems whose matrix of coefficients contains a large number of zero elements the total amount of computation is reduced.[2]   The revised procedure always deals with the original coefficients, and because the computer codes can be developed to multiply only nonzero elements, the total processing time is greatly reduced.   Also, the original nonzero elements can be compactly stored in the computer memory.   The original simplex procedure transforms the zero elements to nonzeros as the computation progresses.   The total number of computations in the revised method is, in general, less than the number in the original method.[3]

2. The amount of new information the computer is required to record is, in general, reduced, since in the revised procedure we need to record only the inverse and the solution vector, while in the original method the complete simplex tableau has to be recorded.   The recording is even further reduced if the product form of the inverse is used.

Before describing the systematic computational rules for the revised procedure, we wish to show how the inverse of each new basis can be obtained from the preceding basis by application of the elimination formulas.   Here we are given an old basis $\mathbf{B} = (\mathbf{P}_1\mathbf{P}_2 \cdots \mathbf{P}_l \cdots \mathbf{P}_m)$, which differs from the new one in that vector $\mathbf{P}_k$ has replaced vector $\mathbf{P}_l$.   Let $\bar{\mathbf{B}} = (\mathbf{P}_1\mathbf{P}_2 \cdots \mathbf{P}_k \cdots \mathbf{P}_m)$ represent the new basis.   We then have

$$\mathbf{B}^{-1}\mathbf{B} = \mathbf{B}^{-1}(\mathbf{P}_1\mathbf{P}_2 \cdots \mathbf{P}_l \cdots \mathbf{P}_m) = \begin{pmatrix} 1 & 0 & \cdots & 0 & \cdots & 0 \\ 0 & 1 & \cdots & 0 & \cdots & 0 \\ \multicolumn{6}{c}{\dotfill} \\ 0 & 0 & \cdots & 1 & \cdots & 0 \\ \multicolumn{6}{c}{\dotfill} \\ 0 & 0 & \cdots & 0 & \cdots & 1 \end{pmatrix} \tag{1.5}$$

and from (1.1)

---

[1] See Dantzig and Orchard-Hays [187] and Sec. 2 of this chapter.

[2] Problems such as the caterer problem, production-scheduling problem, and inter-industry problem described in Chap. 11 and most large-scale linear-programming problems are of this nature.   A matrix is said to be *sparse* if the number of nonzero coefficients is small, i.e., its *density* is low, where *density* is measured by the ratio of the numbers of nonzero coefficients divided by the total number of coefficients of the matrix.   Most large-scale problems have densities between 1 and 5 per cent and often much less (Murtagh [521] and Greenberg [333]).   Also see Rose and Willoughby [574] relative to computational procedures involving sparse systems of equations.

[3] Wagner [678] gives a complete comparison of the original and revised simplex methods.   There he shows that, for $n > 3m$, the revised method is better in terms of the number of computations.

$$\mathbf{B}^{-1}\bar{\mathbf{B}} = \mathbf{B}^{-1}(\mathbf{P}_1\mathbf{P}_2 \cdots \mathbf{P}_k \cdots \mathbf{P}_m) = \begin{pmatrix} 1 & 0 & \cdots & x_{1k} & \cdots & 0 \\ 0 & 1 & \cdots & x_{2k} & \cdots & 0 \\ & & \cdots\cdots\cdots\cdots & & \\ 0 & 0 & \cdots & x_{lk} & \cdots & 0 \\ & & \cdots\cdots\cdots\cdots & & \\ 0 & 0 & \cdots & x_{mk} & \cdots & 1 \end{pmatrix} \qquad (1.5a)$$

If we let b_{ij} equal the element in the ith row and jth column of $\mathbf{B}^{-1}$ and let $\bar{b}_{ij}$ be the corresponding element of $\bar{\mathbf{B}}^{-1}$, we have the $\bar{b}_{ij}$ given by the elimination formulas

$$\bar{b}_{ij} = b_{ij} - \frac{b_{lj}}{x_{lk}}\, x_{ik} \qquad \text{for } i \neq l$$

$$\bar{b}_{ij} = \frac{b_{lj}}{x_{lk}} \tag{1.6}$$

The validity of this transformation can be verified by direct multiplication of $(\bar{\mathbf{B}}^{-1}\bar{\mathbf{B}})$ to determine whether the product yields the identity matrix.[1] For example, we have the inner product of the first row of $\bar{\mathbf{B}}^{-1}$ and the first column of $\bar{\mathbf{B}}$ given by[2]

$$\left(b_{11} - \frac{b_{l1}}{x_{lk}}\, x_{1k}\right)a_{11} + \left(b_{12} - \frac{b_{l2}}{x_{lk}}\, x_{1k}\right)a_{21} + \cdots + \left(b_{1m} - \frac{b_{lm}}{x_{lk}}\, x_{1k}\right)a_{m1}$$

or

$$(b_{11}a_{11} + b_{12}a_{21} + \cdots + b_{1m}a_{m1}) - (b_{l1}a_{11} + b_{l2}a_{21} + \cdots + b_{lm}a_{m1})\frac{x_{1k}}{x_{lk}} \tag{1.7}$$

The first term of (1.7) is the inner product of the first row of $\mathbf{B}^{-1}$ and $\mathbf{P}_1$, which from (1.5) is equal to 1; and the second term of (1.7) contains the inner product of the lth row of $\mathbf{B}^{-1}$ and $\mathbf{P}_1$, which equals 0. Hence (1.7) is equal to 1. In a similar manner we can show that $(\bar{\mathbf{B}}^{-1}\bar{\mathbf{B}}) = \mathbf{I}$, and hence formulas (1.6) do generate $\bar{\mathbf{B}}^{-1}$.[†]

To illustrate the concepts of pricing and the use of the inverse, we next analyze the furniture-manufacturer problem as solved in Tableau 4.4. The problem was to minimize

$$-45x_1 - 80x_2$$

subject to

$$5x_1 + 20x_2 + x_3 \quad\;\; = 400$$
$$10x_1 + 15x_2 \quad\;\; + x_4 = 450$$
$$x_j \geq 0$$

[1] From (1.5a), $\bar{\mathbf{B}}^{-1}$ in terms of $\mathbf{B}^{-1}$ is given by $\mathbf{B}^{-1}\bar{\mathbf{B}} = \mathbf{I}_l$, where $\mathbf{I}_l$ is the right-hand side of (1.5a). Then $\bar{\mathbf{B}} = \mathbf{B}\mathbf{I}_l$, $\bar{\mathbf{B}}^{-1} = (\mathbf{B}\mathbf{I}_l)^{-1}$, and finally, $\bar{\mathbf{B}}^{-1} = \mathbf{I}_l^{-1}\mathbf{B}^{-1}$. In Sec. 2, we show that $\mathbf{I}_l^{-1} = \mathbf{E}^l$, where $\mathbf{E}^l$ is an identity matrix except for the lth column.

[2] This product is obtained by the rule of matrix multiplication described in Sec. 1 of Chap. 2.

[†] The matrices formed by vectors $\mathbf{P}_1$, $\mathbf{P}_4$, and $\mathbf{P}_6$ in the example of Sec. 2, Chap. 4, illustrate this transformation.

Here

$$\mathbf{c} = (-45, -80, 0, 0)$$

$$\mathbf{A} = (\mathbf{P_1 P_2 P_3 P_4}) = \begin{pmatrix} 5 & 20 & 1 & 0 \\ 10 & 15 & 0 & 1 \end{pmatrix}$$

$$\mathbf{b} = \begin{pmatrix} 400 \\ 450 \end{pmatrix}$$

The first feasible basis $\mathbf{B}$ is given by the unit vectors $\mathbf{P_3}$ and $\mathbf{P_4}$; thus

$$\mathbf{B} = \begin{pmatrix} 1 & 0 \\ 0 & 1 \end{pmatrix} \qquad \mathbf{B}^{-1} = \begin{pmatrix} 1 & 0 \\ 0 & 1 \end{pmatrix} \qquad \mathbf{c_0} = (0, 0)$$

and

$$\mathbf{X_0} = \mathbf{B}^{-1}\mathbf{b} = \begin{pmatrix} 1 & 0 \\ 0 & 1 \end{pmatrix}\begin{pmatrix} 400 \\ 450 \end{pmatrix} = \begin{pmatrix} 400 \\ 450 \end{pmatrix} = (x_3, x_4)$$

The corresponding pricing vector is

$$\boldsymbol{\pi} = \mathbf{c_0}\mathbf{B}^{-1} = (0 \quad 0)\begin{pmatrix} 1 & 0 \\ 0 & 1 \end{pmatrix} = (0, 0)$$

We price out the vectors $\mathbf{P_1}$ and $\mathbf{P_2}$ not in the basis by computing the $z_j - c_j$ using

$$\boldsymbol{\pi}\mathbf{P_1} - c_1 = (0, 0)\begin{pmatrix} 5 \\ 10 \end{pmatrix} + 45 = 45$$

$$\boldsymbol{\pi}\mathbf{P_2} - c_2 = (0, 0)\begin{pmatrix} 20 \\ 15 \end{pmatrix} + 80 = 80$$

We then select vector $\mathbf{P_2}$ to enter the basis as it corresponds to the maximum $z_j - c_j$. As we need the vector $\mathbf{X_2}$, which is the representation of $\mathbf{P_2}$ in terms of the current basis $\mathbf{B}$, we compute

$$\mathbf{X_2} = \mathbf{B}^{-1}\mathbf{P_2} = \begin{pmatrix} 1 & 0 \\ 0 & 1 \end{pmatrix}\begin{pmatrix} 20 \\ 15 \end{pmatrix} = \begin{pmatrix} 20 \\ 15 \end{pmatrix}$$

Next, we determine the vector to be removed by calculating the corresponding θ ratios of $400:20$ and $450:15$. As the former ratio is the minimum and corresponds to x_3, vector $\mathbf{P_3}$ is removed from the basis and is replaced by vector $\mathbf{P_2}$. To compute the inverse of the new basis $(\mathbf{P_2 P_4})$ we need to apply the elimination formulas to the matrix

$$\begin{array}{c|cc} \mathbf{X_2} & \multicolumn{2}{c}{\mathbf{B}^{-1}} \\ \hline \boxed{20} & 1 & 0 \\ 15 & 0 & 1 \end{array}$$

to obtain the new inverse

$$(\mathbf{P_2 P_4})^{-1} = \begin{pmatrix} \frac{1}{20} & 0 \\ -\frac{15}{20} & 1 \end{pmatrix}$$

(See Second Step of Tableau 4.4.)

We repeat the process and denote the new basis by $\mathbf{B} = (\mathbf{P_2 P_4})$ with

$$\mathbf{B} = \begin{pmatrix} 20 & 0 \\ 15 & 1 \end{pmatrix} \qquad \mathbf{B}^{-1} = \begin{pmatrix} \frac{1}{20} & 0 \\ -\frac{15}{20} & 1 \end{pmatrix} \qquad \mathbf{c}_0 = (-80, 0)$$

and

$$\mathbf{X}_0 = \mathbf{B}^{-1}\mathbf{b} = \begin{pmatrix} \frac{1}{20} & 0 \\ -\frac{15}{20} & 1 \end{pmatrix} \begin{pmatrix} 400 \\ 450 \end{pmatrix} = \begin{pmatrix} 20 \\ 150 \end{pmatrix} = (x_2, x_4)$$

The corresponding pricing vector is

$$\pi = \mathbf{c}_0 \mathbf{B}^{-1} = (-80, 0) \begin{pmatrix} \frac{1}{20} & 0 \\ -\frac{15}{20} & 1 \end{pmatrix} = (-4, 0)$$

We price out vectors $\mathbf{P}_1$ and $\mathbf{P}_3$ which are not in the current basis and obtain

$$\pi \mathbf{P}_1 - c_1 = (-4, 0) \begin{pmatrix} 5 \\ 10 \end{pmatrix} + 45 = 25$$

$$\pi \mathbf{P}_3 - c_3 = (-4, 0) \begin{pmatrix} 1 \\ 0 \end{pmatrix} - 0 = -4$$

Thus, vector $\mathbf{P}_1$ is selected to enter the basis and we compute

$$\mathbf{X}_1 = \mathbf{B}^{-1}\mathbf{P}_1 = \begin{pmatrix} \frac{1}{20} & 0 \\ -\frac{15}{20} & 1 \end{pmatrix} \begin{pmatrix} 5 \\ 10 \end{pmatrix} = \begin{pmatrix} \frac{5}{20} \\ \frac{125}{20} \end{pmatrix}$$

with the corresponding θ ratios of $20 : \frac{5}{20}$ and $150 : \frac{125}{20}$; vector $\mathbf{P}_4$ is removed from the basis.   The new basis is $\mathbf{B} = (\mathbf{P_2 P_1})$, and the inverse is obtained by applying the elimination formulas to the matrix

$$
\begin{array}{c|cc}
\mathbf{X}_1 & \multicolumn{2}{c}{\mathbf{B}^{-1}} \\
\hline
\end{array}
$$

$$\begin{pmatrix} \frac{5}{20} & \vline & \frac{1}{20} & 0 \\ \boxed{\frac{125}{20}} & \vline & -\frac{15}{20} & 1 \end{pmatrix}$$

i.e., the new $\mathbf{B}^{-1} = (\mathbf{P_2 P_1})^{-1} = \begin{pmatrix} \frac{2}{25} & -\frac{1}{25} \\ -\frac{3}{25} & \frac{4}{25} \end{pmatrix}$.   (See Third Step of Tableau 4.4.)

Finally, using the new inverse we see that the vectors $\mathbf{P}_3$ and $\mathbf{P}_4$ not in the basis price out negatively and the optimal solution is

$$\mathbf{X}_0 = \mathbf{B}^{-1}\mathbf{b} = \begin{pmatrix} \frac{2}{25} & -\frac{1}{25} \\ -\frac{3}{25} & \frac{4}{25} \end{pmatrix} \begin{pmatrix} 400 \\ 450 \end{pmatrix} = \begin{pmatrix} 14 \\ 24 \end{pmatrix} = (x_2, x_1)$$

with the minimum value of the objective function

$$z_0 = \mathbf{c}_0 \mathbf{X}_0 = \mathbf{c}_0(\mathbf{B}^{-1}\mathbf{b}) = (\mathbf{c}_0 \mathbf{B}^{-1})\, \mathbf{b}$$

$$= (-80, -45) \begin{pmatrix} \frac{2}{25} & -\frac{1}{25} \\ -\frac{3}{25} & \frac{4}{25} \end{pmatrix} \begin{pmatrix} 400 \\ 450 \end{pmatrix} = \pi \mathbf{b} = (-1, -4) \begin{pmatrix} 400 \\ 450 \end{pmatrix} = -2{,}200$$

We next describe the computational format for the revised simplex method as developed by Dantzig [157] and Orchard-Hays [529]. We shall not attempt to develop the revised method fully, in the way the original simplex method was presented in Chaps. 3 and 4, but the reader will be able to determine the relationship between these methods.

In order to facilitate the use of the revised procedure, we shall introduce and define below two new variables x_{n+m+1} and x_{n+m+2}.

For the revised procedure we write the general linear-programming problem as follows: Maximize

$$x_{n+m+1} \tag{1.8}$$

subject to

$$
\begin{aligned}
a_{11}x_1 + a_{12}x_2 + \cdots + a_{1n}x_n &= b_1 \\
&\,\cdots\cdots \\
a_{m1}x_1 + a_{m2}x_2 + \cdots + a_{mn}x_n &= b_m \\
c_1x_1 + c_2x_2 + \cdots + c_nx_n + x_{n+m+1} &= 0
\end{aligned}
\tag{1.9}
$$

$$
\begin{aligned}
x_1 &\geq 0 \\
x_2 &\geq 0 \\
&\;\;\vdots \\
x_n &\geq 0
\end{aligned}
\tag{1.10}
$$

Here we assume all $b_i \geq 0$. Since

$$x_{n+m+1} = -c_1x_1 - c_2x_2 - \cdots - c_nx_n$$

and since the linear-programming problem that minimizes an objective function is basically the same as the one which maximizes the negative of this objective function subject to the same constraints, the above problem is equivalent to the standard form of the minimizing linear program treated in Chaps. 3 and 4, i.e., minimize

$$c_1x_1 + c_2x_2 + \cdots + c_nx_n \tag{1.8a}$$

subject to

$$
\begin{aligned}
a_{11}x_1 + a_{12}x_2 + \cdots + a_{1n}x_n &= b_1 \\
&\,\cdots\cdots \\
a_{m1}x_1 + a_{m2}x_2 + \cdots + a_{mn}x_n &= b_m
\end{aligned}
\tag{1.9a}
$$

$$
\begin{aligned}
x_1 &\geq 0 \\
x_2 &\geq 0 \\
&\;\;\vdots \\
x_n &\geq 0
\end{aligned}
\tag{1.10a}
$$

Since the objective function in the revised problem consists of only x_{n+m+1}, that is, the value of x_{n+m+1} is the negative value of the minimizing objective function, we do not restrict the sign of x_{n+m+1} in (1.8) and (1.9).

As in the original method, the revised procedure has to begin with a basis consisting of an identity matrix whose unit vectors correspond to either real or artificial vectors. If the revised procedure starts with artificial vectors, we again have the problem of determining a first basic feasible solution (if one exists) and then continuing on to the optimal solution. We shall let Phase I refer to the part of the computation that determines feasibility and Phase II to the part that develops the optimal solution.   In order to generalize the concept of artificial vectors and to facilitate the computation of Phase I, we calculate a "redundant" equation defined as

$$a_{m+2,1}x_1 + a_{m+2,2}x_2 + \cdots + a_{m+2,n}x_n + x_{m+n+2} = b_{m+2}$$

where

$$a_{m+2,j} = -\sum_{i=1}^{m} a_{ij} \qquad j = 1, 2, \ldots, n$$

$$b_{m+2} = -\sum_{i=1}^{m} b_i$$

$$(1.11)$$

We see that $a_{m+2,j}$ is just the negative sum of the coefficients in the jth column of the coefficient matrix $\mathbf{A}$ and that b_{m+2} is the negative sum of all the b_i. The significance of the "redundant" variable x_{m+n+2} will be described below. If we let $c_j = a_{m+1,j}$, we then rewrite the revised problem (1.8) to (1.10) to maximize

$$x_{n+m+1} \tag{1.12}$$

subject to

$$
\begin{aligned}
a_{11}x_1 + \quad a_{12}x_2 + \cdots + \quad a_{1n}x_n + x_{n+1} &= b_1 \\
a_{21}x_1 + \quad a_{22}x_2 + \cdots + \quad a_{2n}x_n \quad\quad + x_{n+2} &= b_2 \\
\cdots\cdots\cdots\cdots\cdots\cdots\cdots\cdots\cdots\cdots\cdots\cdots \\
a_{m1}x_1 + \quad a_{m2}x_2 + \cdots + \quad a_{mn}x_n \quad\quad\quad + x_{n+m} &= b_m \\
a_{m+1,1}x_1 + a_{m+1,2}x_2 + \cdots + a_{m+1,n}x_n \quad\quad\quad + x_{n+m+1} &= 0 \\
a_{m+2,1}x_1 + a_{m+2,2}x_2 + \cdots + a_{m+2,n}x_n \quad\quad\quad + x_{n+m+2} &= b_{m+2}
\end{aligned}
\tag{1.13}
$$

$$
\begin{aligned}
x_1 &\geq 0 \\
x_2 &\geq 0 \\
&\ \ \vdots \\
x_n &\geq 0 \\
x_{n+1} &\geq 0 \\
&\ \ \vdots \\
x_{n+m} &\geq 0
\end{aligned}
\tag{1.14}
$$

where the nonnegative artificial variables x_{n+i} for $i = 1, 2, \ldots, m$ have been added to the corresponding equations. The artificial variables can be interpreted as the measure of the error between the sides of the original equations of (1.9) evaluated for a vector $\mathbf{X} = (x_1, x_2, \ldots, x_n)$ which does not satisfy all the constraints of (1.9).   From (1.11) and (1.13) we have

$$x_{n+1} + x_{n+2} + \cdots + x_{n+m} + x_{n+m+2} = 0$$

The variable x_{n+m+2} can then be thought of as representing the negative of the absolute sum of errors of an approximate nonnegative solution to the equations.   Since $x_{n+i} \geq 0$ for $i = 1, 2, \ldots, m$, it is clear that x_{n+m+2} cannot be positive.

As written in (1.12) to (1.14), the revised problem consists of $m + 2$ equations in $n + m + 2$ variables.   Here a basic feasible solution will contain $m + 2$ variables from the set $(x_1, x_2, \ldots, x_n, x_{n+m+1}, x_{n+m+2})$.   The signs of the last two variables are unrestricted, and they will always be in the solution.   The variables from the set $(x_1, x_2, \ldots, x_n)$ that are in the optimal solution represent a corresponding basic maximum feasible solution to the problem of (1.8) to (1.10) with x_{n+m+1} the value of the objective function.   This optimal solution is also a basic minimum feasible solution to (1.8a) to (1.10a), where

$$-x_{n+m+1} = c_1 x_1 + \cdots + c_n x_n$$

is the corresponding value of the objective function.   The computational procedure for the revised problem starts out with the m artificial variables $x_{n+1}, \ldots, x_{n+m}$ and variables x_{n+m+1} and x_{n+m+2} in the solution.   We must employ the Phase I procedure to find a first basic feasible solution.

In Phase I we first consider solving the problem of maximizing x_{n+m+2}, subject to (1.13) and (1.14), with both x_{n+m+1} and x_{n+m+2} unrestricted as to sign.   If the maximum value of x_{n+m+2} is zero, then all the x_{n+i} for $i = 1, 2, \ldots, m$ must also equal zero, and the x_j for $j = 1, 2, \ldots, n$ of this "preliminary maximum solution" represent a basic feasible solution to the problems (1.9), (1.10) and (1.9a), (1.10a).   If the maximum of x_{n+m+2} is less than zero, then this implies that at least one artificial variable is in the Phase I solution with a nonzero value and hence that no feasible solutions exist for the original problem.   In the former case we then go on to Phase II, where we maximize x_{n+m+1}, subject to (1.13) and (1.14), while keeping x_{n+m+2} equal to zero.   It should be noted that the set of basic feasible solutions found in Phase II might contain artificial variables with zero values.   The final Phase II solution is the desired optimal solution.

We shall next describe and illustrate the complete computational procedure for the revised problem which starts with a full artificial basis.   In the computational tableau for the procedure we need to keep record of only the following information: the variables in the solution, the corresponding values of the variables, and the inverse of the current basis.   In developing new solutions, we must have a means of calculating the z_j terms.   We shall do this by having the pricing vector defined by (1.4) available for the current basis.   We can then determine which vector is to be introduced into the basis and calculate (1.1), where $\mathbf{P}_j$ corresponds to the vector being introduced.   We next find out which vector is to be removed, using the method described in Chap. 4, and then transform the inverse of the current basis by the elimination formulas to obtain the inverse of the new basis.

Let us detach the required coefficients from (1.13) and arrange them in an $m + 2 \times n$ matrix as follows:

$$
\bar{\mathbf{A}} = \begin{pmatrix}
a_{11} & a_{12} & \cdots & a_{1k} & \cdots & a_{1n} \\
a_{21} & a_{22} & \cdots & a_{2k} & \cdots & a_{2n} \\
\cdots & \cdots & \cdots & \cdots & \cdots & \cdots \\
a_{l1} & a_{l2} & \cdots & a_{lk} & \cdots & a_{ln} \\
\cdots & \cdots & \cdots & \cdots & \cdots & \cdots \\
a_{m1} & a_{m2} & \cdots & a_{mk} & \cdots & a_{mn} \\
a_{m+1,1} & a_{m+1,2} & \cdots & a_{m+1,k} & \cdots & a_{m+1,n} \\
a_{m+2,1} & a_{m+2,2} & \cdots & a_{m+2,k} & \cdots & a_{m+2,n}
\end{pmatrix}
$$

We shall denote each column vector of $\bar{\mathbf{A}}$ by $\bar{\mathbf{A}}_j$. Let us compare the information in matrix $\bar{\mathbf{A}}$ with the initial artificial-basis computational tableau for the original procedure for the minimization problem described in Chap. 4 (see Tableau 4.7). We note that the matrix formed by rows 1, 2, ..., m, $m + 1$, $m + 2$ and the columns $\mathbf{P}_1$, $\mathbf{P}_2$, ..., $\mathbf{P}_n$ of the tableau is equal to matrix $\bar{\mathbf{A}}$, except that rows $m + 1$ and $m + 2$ of $\bar{\mathbf{A}}$ are the negative of the corresponding rows of the tableau. As in the original method, row $m + 2$ of $\bar{\mathbf{A}}$ will be used in computing the $z_j - c_j$ elements when artificial variables are still in the solution, and row $m + 1$ will be used when they have been removed.[1]

Since our starting basis for the revised procedure consists of an identity matrix, its inverse is also an identity matrix, and this information is recorded in the following $m + 2 \times m + 2$ matrix:

$$
\mathbf{U} = \begin{pmatrix}
1 & 0 & \cdots & 0 & \cdots & 0 & 0 & 0 \\
0 & 1 & \cdots & 0 & \cdots & 0 & 0 & 0 \\
\cdots & \cdots & \cdots & \cdots & \cdots & \cdots & \cdots & \cdots \\
0 & 0 & \cdots & 1 & \cdots & 0 & 0 & 0 \\
\cdots & \cdots & \cdots & \cdots & \cdots & \cdots & \cdots & \cdots \\
0 & 0 & \cdots & 0 & \cdots & 1 & 0 & 0 \\
0 & 0 & \cdots & 0 & \cdots & 0 & 1 & 0 \\
0 & 0 & \cdots & 0 & \cdots & 0 & 0 & 1
\end{pmatrix}
$$

The first m rows and m columns of $\mathbf{U}$ represent the inverse of the starting basis $\mathbf{B}$. The last two rows of $\mathbf{U}$ will be used to determine which vector should be introduced into the basis, with row $m + 2$ generating the information in Phase I and row $m + 1$ in Phase II. We will embed the matrix $\mathbf{U}$ in the starting computational tableau and transform its elements as described below. Let us designate the elements of $\mathbf{U}$ by u_{ij} and its rows by the row vectors $\mathbf{U}_i = (u_{i1}, u_{i2}, \ldots, u_{i,m+2})$. For convenience in describing the computational procedure, we

---

[1] Dantzig and Orchard-Hays [187] write the revised problem with the equations containing variables x_{n+m+1} and x_{n+m+2} as the zeroth and first equations, respectively. They renumber the equations to make $x_{n+m+1} = x_0$ and $x_{n+m+2} = x_{n+1}$.

shall let U_i represent both the original and transformed versions of the ith row of U. The matrix U will be the inverse of an $m + 2 \times m + 2$ matrix whose columns are vectors of (1.13).

We arrange the matrix U and the initial solution of $x_{n+i} = b_i$ for $i = 1, 2, \ldots, m$; $x_{n+m+1} = 0$; and $x_{n+m+2} = b_{m+2}$ as shown in the starting tableau of Tableau 5.1. The computational steps for the phases are as follows:

Phase I. Artificial variables in the solution with positive values

STEP 1   If $x_{m+n+2} < 0$, compute

$$\delta_j = U_{m+2}\bar{A}_j^\dagger$$
$$= u_{m+2,1}a_{1j} + u_{m+2,2}a_{2j} + \cdots + u_{m+2,m+2}a_{m+2,j} \qquad j = 1, 2, \ldots, n$$

and continue to Step 2.

If $x_{m+n+2} = 0$, go to Step 1 of Phase II.

STEP 2   If all $\delta_j \geq 0$, then x_{n+m+2} is at its maximum, and hence no feasible solution exists for the problem (1.8) to (1.10).

If at least one $\delta_j < 0$, then the variable to be introduced into the solution, x_k, corresponds to

$$\delta_k = \min \delta_j$$

If more than one of the δ_j are equal to the minimum, select δ_k such that k is the smallest index. This is an arbitrary, but computationally sound, rule for breaking ties.

STEP 3   Compute

$$x_{ik} = U_i\bar{A}_k$$
$$= u_{i1}a_{1k} + u_{i2}a_{2k} + \cdots + u_{i,m+2}a_{m+2,k}$$

for $i = 1, 2, \ldots, m, m+1, m+2$. The variable x_l to be removed from the solution corresponds to the ratio

$$\theta_0 = \min_i \frac{x_{i0}}{x_{ik}} = \frac{x_{l0}}{x_{lk}} \qquad i = 1, 2, \ldots, m$$

[†] In Phase I for the revised problem the objective function to be maximized is x_{m+n+2}. Hence, all cost coefficients except $c_{m+n+2} = 1$ for this maximizing problem are equal to zero. Similarly, in Phase II the objective function to be maximized is x_{m+n+1}, and the cost coefficients are all equal to zero except $c_{m+n+1} = 1$. The quantities δ_j of Phase I and γ_j defined in Phase II can be interpreted in two ways. They are equal to, respectively, the artificial and real $z_j - c_j$ elements of the original simplex procedure adapted to maximizing the objective functions of the revised problem. The criterion then states that the optimal solution has been reached when all $z_j - c_j \geq 0$. For the corresponding iteration, the quantities δ_j and γ_j are also equivalent to the $c_j - z_j$ elements that would be computed in solving the minimization problem defined by (1.8a) to (1.10a). Here the criterion states that the optimal solution has been reached when all $c_j - z_j \geq 0$. In setting up the revised problem, we shall always write the original problem as one to be minimized and then determine matrix $\bar{A}$, where $a_{m+1,j} = c_j$ of the minimization problem.

where the ratio is formed only for those $x_{ik} > 0$. If there are ties for the minimum, we select l to be the smallest index.[1] [For discussion purposes we have taken the ratios for the values of index $i = 1, 2, \ldots, m$. The ratios x_{i0}/x_{ik} are actually taken for those i whose corresponding x_{i0} are in the basic solution to (1.13).]

STEP 4   The new values of the variables in the basic solution are obtained by the formulas

$$x'_{i0} = x_{i0} - \frac{x_{l0}}{x_{lk}} x_{ik} \qquad \text{for } i \neq k$$

$$x'_{k0} = \frac{x_{l0}}{x_{lk}}$$

The new elements of the matrix $\mathbf{U}$ are transformed by

$$u'_{ij} = u_{ij} - \frac{u_{lj}}{x_{lk}} x_{ik} \qquad \text{for } i \neq l$$

$$u'_{lj} = \frac{u_{lj}}{x_{lk}}$$

Under this transformation the $(m + 1)$st and $(m + 2)$nd columns of $\mathbf{U}$ will never change. These unit vectors enable us to add the correct c_j when we compute the $c_j - z_j$. Using the transformed rows of $\mathbf{U}$ as shown in the transformed section of Tableau 5.1, the steps of Phase I are repeated until it is determined either that no feasible solutions exist or that the value of $x_{n+m+2} = 0$. In the latter case we go on to Phase II.

Phase II.   No artificial variables in the solution with positive values[2]

STEP 1   Here $x_{m+n+2} = 0$. Compute

$$\gamma_j = \mathbf{U}_{m+1}\bar{\mathbf{A}}_j$$
$$= u_{m+1,1}a_{1j} + u_{m+1,2}a_{2j} + \cdots + u_{m+1,m+2}a_{m+2,j} \qquad j = 1, 2, \ldots, n$$

STEP 2   If all $\gamma_j \geq 0$, then x_{n+m+1} is at its maximum value and the corresponding basic feasible solution is an optimal solution. The negative value of x_{n+m+1} is the true value of the objective function that was to be minimized.

If at least one $\gamma_j < 0$, let

$$\gamma_k = \min \gamma_j$$

Variable x_k is selected to be introduced into the solution. Ties in selecting γ_k are broken by selecting the variable with the smallest index.

---

[1] In this situation the resultant basic feasible solution will be degenerate. In Chap. 7 we discuss computational rules for avoiding cycling (see Sec. 1, Chap. 4) due to degeneracy. Their use in computer-based mathematical programming systems is discussed there and in Sec. 3. Such rules are usually not employed in hand computations unless it is obvious that a cycle has occurred.

[2] See Exercise 7.

STEP 3   Compute

$$x_{ik} = \mathbf{U}_i \bar{\mathbf{A}}_k$$

$$= u_{i1} a_{1k} + u_{i2} a_{2k} + \cdots + u_{i,\,m+2} a_{m+2,\,k}$$

for $i = 1, 2, \ldots, m, m+1, m+2$.   The variable x_{l0} to be eliminated from the solution is determined by finding the ratio

$$\theta_0 = \min_i \frac{x_{i0}}{x_{ik}} = \frac{x_{l0}}{x_{lk}} \qquad i = 1, 2, \ldots, m$$

for those $x_{lk} > 0$.   (See Phase I, Step 3, for actual range of index i.)   Ties are broken in the usual fashion.   If all $x_{ik} \leq 0$, then the procedure has yielded a solution whose value of the objective function can be made arbitrarily large.

STEP 4   The new values of the variables are obtained by the formulas

$$x'_{i0} = x_{i0} - \frac{x_{l0}}{x_{lk}} x_{ik} \qquad \text{for } i \neq k$$

$$x'_{k0} = \frac{x_{l0}}{x_{lk}}$$

The elements of the matrix $\mathbf{U}$ are transformed by

$$u'_{ij} = u_{ij} - \frac{u_{lj}}{x_{lk}} x_{ik} \qquad \text{for } i \neq l$$

$$u'_{lj} = \frac{u_{lj}}{x_{lk}}$$

The steps are repeated until an optimal solution with either a finite or an infinite value of the objective function is determined.

The initial solution to (1.12) to (1.14) and the above steps can be arranged and carried out by the computational scheme shown in Tableau 5.1.   A column has been set aside for the recording of the x_{ik} associated with each inverse.   The inverse matrix $\mathbf{U}$ consists of the $m + 2 \times m + 2$ array enclosed in heavy lines.   The last two columns of the transformed $\mathbf{U}$ do not change.   The inverse of the m-dimensional basis selected from the $\mathbf{P}_j$ of (1.13) without the $(m + 1)$st and $(m + 2)$nd rows is enclosed in heavy and dashed lines.

If the revised problem has unit vectors in its explicit mathematical statement, these vectors can be used in the starting basis for the revised procedure if their cost coefficients are equal to zero; for example, slack vectors.[1]   If these vectors are used, formulas (1.11) must be changed to read, respectively,

$$a_{m+2,\,j} = -\sum_{i \notin \mathbf{B}} a_{ij} \tag{1.15}$$

and

---

[1] If the cost coefficients are not equal to zero, then the original objective function must be transformed to make them zero; see Example 2.3, Chap. 4.

TABLEAU 5.1

Row index of tableau	Index of variables in solution	Value of variables	Matrix $\mathbf{U}$							$\mathbf{U}_i \overline{\mathbf{A}}_k$ $(\mathbf{B}^{-1}\mathbf{P}_k)$
			Inverse of basis, $\mathbf{B}^{-1}$							
			$\mathbf{U}_{m+1}$ $\mathbf{U}_{m+2}$							$\mathbf{U}_{m+1}\overline{\mathbf{A}}_k$ $\mathbf{U}_{m+2}\overline{\mathbf{A}}_k$

1. Initial Step

Row index of tableau	Index of variables in solution	Value of variables									$\mathbf{U}_i\overline{\mathbf{A}}_k$
1	$n+1$	$x_{n+1}=b_1$	1	0	$\cdots$	0	$\cdots$	0	0	0	x_{1k}
2	$n+2$	$x_{n+2}=b_2$	0	1	$\cdots$	0	$\cdots$	0	0	0	x_{2k}
.	.	.	.	.	$\cdots$	.	$\cdots$	.	.	.	.
.	.	.	.	.	$\cdots$	.	$\cdots$	.	.	.	.
.	.	.	.	.	$\cdots$	.	$\cdots$	.	.	.	.
l	$n+l$	$x_{n+l}=b_l$	0	0	$\cdots$	1	$\cdots$	0	0	0	x_{lk}
.	.	.	.	.	$\cdots$	.	$\cdots$	.	.	.	.
.	.	.	.	.	$\cdots$	.	$\cdots$	.	.	.	.
.	.	.	.	.	$\cdots$	.	$\cdots$	.	.	.	.
m	$n+m$	$x_{n+m}=b_m$	0	0	$\cdots$	0	$\cdots$	1	0	0	x_{mk}
$m+1$	$n+m+1$	$x_{n+m+1}=0$	0	0	$\cdots$	0	$\cdots$	0	1	0	$x_{m+1,k}$
$m+2$	$n+m+2$	$x_{n+m+2}=b_{m+2}$	0	0	$\cdots$	0	$\cdots$	0	0	1	$x_{m+2,k}$

2. Transformed Step

Row index of tableau	Index of variables in solution	Value of variables									
1	$n+1$	x'_{n+1}	u'_{11}	u'_{12}	$\cdots$	u'_{1l}	$\cdots$	u'_{1m}	0	0	
2	$n+2$	x'_{n+2}	u'_{21}	u'_{22}	$\cdots$	u'_{2l}	$\cdots$	u'_{2m}	0	0	
.	.	.	.	.	$\cdots$	.	$\cdots$	.	.	.	
.	.	.	.	.	$\cdots$	.	$\cdots$	.	.	.	
.	.	.	.	.	$\cdots$	.	$\cdots$	.	.	.	
l	k	x'_k	u'_{l1}	u'_{l2}	$\cdots$	u'_{ll}	$\cdots$	u'_{lm}	0	0	
.	.	.	.	.	$\cdots$	.	$\cdots$	.	.	.	
.	.	.	.	.	$\cdots$	.	$\cdots$	.	.	.	
.	.	.	.	.	$\cdots$	.	$\cdots$	.	.	.	
m	$n+m$	x'_{n+m}	u'_{m1}	u'_{m2}	$\cdots$	u'_{ml}	$\cdots$	u'_{mn}	0	0	
$m+1$	$n+m+1$	x'_{n+m+1}	$u'_{m+1,1}$	$u'_{m+1,2}$	$\cdots$	$u'_{m+1,l}$	$\cdots$	$u'_{m+1,m}$	1	0	
$m+2$	$n+m+2$	x'_{n+m+2}	$u'_{m+2,1}$	$u'_{m+2,2}$	$\cdots$	$u'_{m+2,l}$	$\cdots$	$u'_{m+2,m}$	0	1	

$$b_{m+2} = -\sum_{i\notin\mathbf{B}} b_i \qquad (1.16)$$

where $i \notin \mathbf{B}$ represents the set of indices of the rows that still require an artificial vector. If m appropriate natural unit vectors are available, then $b_{m+2} = 0$ and the procedure starts with Phase II.[1]

[1] See Sec. 1, Chap. 9, on how to find a starting solution that requires only one artificial variable.

To illustrate the revised procedure and to help point out the relationships between the two simplex methods, we shall next solve Example 3.2 of Chap. 4, using the revised procedure.

Example 1.1. Maximize

$$x_1 + 2x_2 + 3x_3 - x_4$$

subject to

$$x_1 + 2x_2 + 3x_3 \quad\;\; = 15$$
$$2x_1 + \;\; x_2 + 5x_3 \quad\;\; = 20$$
$$x_1 + 2x_2 + \;\; x_3 + x_4 = 10$$

and $x_j \geq 0$.

Here $m = 3$ and $n = 4$.   The objective function for the corresponding minimization problem is $-x_1 - 2x_2 - 3x_3 + x_4$.

For the revised procedure with a full artificial basis[1] we rewrite the example to read: Maximize

$$x_8$$

subject to

$$x_1 + 2x_2 + 3x_3 \quad\;\; + x_5 \qquad\qquad\qquad\qquad = 15$$
$$2x_1 + \;\; x_2 + 5x_3 \quad\qquad + x_6 \qquad\qquad\qquad = 20$$
$$x_1 + 2x_2 + \;\; x_3 + x_4 \qquad\qquad + x_7 \qquad\qquad = 10$$
$$-x_1 - 2x_2 - 3x_3 + x_4 \qquad\qquad\qquad + x_8 \qquad = 0$$
$$-4x_1 - 5x_2 - 9x_3 - x_4 \qquad\qquad\qquad\qquad\quad + x_9 = -45$$

The coefficients of the first four variables for the fourth row are equal to the corresponding c_j as written in the original objective function to be minimized.   The coefficients of variables x_1, x_2, x_3, x_4 and the right side of the fifth equation were obtained by formulas (1.11).

The $\bar{\mathbf{A}}$ and $\mathbf{U}$ matrices for the example are

$$\bar{\mathbf{A}} = \begin{pmatrix} 1 & 2 & 3 & 0 \\ 2 & 1 & 5 & 0 \\ 1 & 2 & 1 & 1 \\ -1 & -2 & -3 & 1 \\ -4 & -5 & -9 & -1 \end{pmatrix} \quad \text{and} \quad \mathbf{U} = \begin{pmatrix} 1 & 0 & 0 & 0 & 0 \\ 0 & 1 & 0 & 0 & 0 \\ 0 & 0 & 1 & 0 & 0 \\ 0 & 0 & 0 & 1 & 0 \\ 0 & 0 & 0 & 0 & 1 \end{pmatrix}$$

The initial step and the sequence of iterations are shown in Tableau 5.2.

---

[1] See Exercise 9 on using variable x_4 in the starting solution.

TABLEAU 5.2

Row index of tableau	Index of variables in solution	Value of variables	Matrix U					x_{ik}

1. Initial Step   *Phase I*

1	5	15	1	0	0	0	0	3
2	6	20	0	1	0	0	0	⑤
3	7	10	0	0	1	0	0	1
4	8	0	0	0	0	1	0	-3
5	9	-45	0	0	0	0	1	-9

$$\delta_k = \delta_3 = U_{m+2}\,\overline{A}_3 = -9$$
$$\theta_0 = \frac{x_6}{x_{63}} = \frac{20}{5} = 4$$

2. Second Step

1	5	3	1	$-\frac{3}{5}$	0	0	0	⑦⁄₅
2	3	4	0	$\frac{1}{5}$	0	0	0	$\frac{1}{5}$
3	7	6	0	$-\frac{1}{5}$	1	0	0	$\frac{9}{5}$
4	8	$+12$	0	$\frac{3}{5}$	0	1	0	$-\frac{7}{5}$
5	9	-9	0	$\frac{9}{5}$	0	0	1	$-\frac{16}{5}$

$$\delta_k = \delta_2 = U_{m+2}\,\overline{A}_2 = -\frac{16}{5}$$
$$\theta_0 = \frac{x_5}{x_{52}} = \frac{15}{7}$$

3. Third Step

1	2	$\frac{15}{7}$	$\frac{5}{7}$	$-\frac{3}{7}$	0	0	0	0
2	3	$\frac{25}{7}$	$-\frac{1}{7}$	$\frac{2}{7}$	0	0	0	0
3	7	$\frac{15}{7}$	$-\frac{9}{7}$	$\frac{4}{7}$	1	0	0	①
4	8	$+15$	1	0	0	1	0	1
5	9	$-\frac{15}{7}$	$\frac{16}{7}$	$\frac{3}{7}$	0	0	1	-1

$$\delta_k = \delta_4 = U_{m+2}\,\overline{A}_4 = -1$$
$$\theta_0 = \frac{x_7}{x_{74}} = \frac{15}{7}$$

4. Fourth Step   *Phase II*

1	2	$\frac{15}{7}$	$\frac{5}{7}$	$-\frac{3}{7}$	0	0	0	$-\frac{1}{7}$
2	3	$\frac{25}{7}$	$-\frac{1}{7}$	$\frac{2}{7}$	0	0	0	$\frac{3}{7}$
3	4	$\frac{15}{7}$	$-\frac{9}{7}$	$\frac{4}{7}$	1	0	0	⑥⁄₇
4	8	$+\frac{90}{7}$	$\frac{16}{7}$	$-\frac{4}{7}$	-1	1	0	$-\frac{6}{7}$
5	9	0	1	1	1	0	1	0

$$x_9 = 0$$
$$\gamma_k = \gamma_1 = U_{m+1}\,\overline{A}_1 = -\frac{6}{7}$$
$$\theta_0 = \frac{x_4}{x_{41}} = \frac{5}{2}$$

5. Fifth Step

1	2	$\frac{5}{2}$	$\frac{3}{6}$	$-\frac{2}{6}$	$\frac{1}{6}$	0	0	
2	3	$\frac{5}{2}$	$\frac{3}{6}$	0	$-\frac{3}{6}$	0	0	
3	1	$\frac{5}{2}$	$-\frac{9}{6}$	$\frac{4}{6}$	$\frac{7}{6}$	0	0	
4	8	$+15$	1	0	0	1	0	
5	9	0	1	1	1	0	1	

All $\gamma_i \geq 0$

Optimal solution:

$$x_1 = x_2 = x_3 = \tfrac{5}{2},\ x_4 = 0$$

Value of objective function:

$$x_8 = 15$$

2. THE PRODUCT FORM OF THE INVERSE

In Sec. 1 we developed the computational formulas (1.6) that enabled us to compute the inverse of a basis $\bar{\mathbf{B}}$ which differs by one vector from a basis $\mathbf{B}$ with known inverse $\mathbf{B}^{-1}$.   Making use of the notation of Sec. 1, we have

$$\mathbf{B} = (\mathbf{P}_1\mathbf{P}_2 \cdots \mathbf{P}_l \cdots \mathbf{P}_m)$$

$$\bar{\mathbf{B}} = (\mathbf{P}_1\mathbf{P}_2 \cdots \mathbf{P}_k \cdots \mathbf{P}_m)$$

$$\mathbf{B}^{-1} = \begin{pmatrix} b_{11} & b_{12} & \cdots & b_{1l} & \cdots & b_{1m} \\ \cdots\cdots\cdots\cdots\cdots\cdots\cdots\cdots\cdots\cdots \\ b_{l1} & b_{l2} & \cdots & b_{ll} & \cdots & b_{lm} \\ \cdots\cdots\cdots\cdots\cdots\cdots\cdots\cdots\cdots\cdots \\ b_{m1} & b_{m2} & \cdots & b_{ml} & \cdots & b_{mm} \end{pmatrix}$$

and

$$\mathbf{B}^{-1}\mathbf{P}_k = \mathbf{X}_k = \begin{pmatrix} x_{1k} \\ \cdot \\ \cdot \\ \cdot \\ x_{lk} \\ \cdot \\ \cdot \\ \cdot \\ x_{mk} \end{pmatrix}$$

Also

$$\bar{\mathbf{B}}^{-1} = \begin{pmatrix} \bar{b}_{11} & \bar{b}_{12} & \cdots & \bar{b}_{1l} & \cdots & \bar{b}_{1m} \\ \cdots\cdots\cdots\cdots\cdots\cdots\cdots\cdots\cdots\cdots \\ \bar{b}_{l1} & \bar{b}_{l2} & \cdots & \bar{b}_{ll} & \cdots & \bar{b}_{lm} \\ \cdots\cdots\cdots\cdots\cdots\cdots\cdots\cdots\cdots\cdots \\ \bar{b}_{m1} & \bar{b}_{m2} & \cdots & \bar{b}_{ml} & \cdots & \bar{b}_{mm} \end{pmatrix}$$

where

$$\bar{b}_{ij} = b_{ij} - \frac{x_{ik}}{x_{lk}} b_{lj} \qquad \text{for } i \neq l$$

$$\bar{b}_{lj} = b_{lj} \frac{1}{x_{lk}}$$

Let us form the $m \times m$ matrix

$$\mathbf{E}^l = \begin{pmatrix} 1 & 0 & \cdots & -\dfrac{x_{1k}}{x_{lk}} & \cdots & 0 \\ 0 & 1 & \cdots & -\dfrac{x_{2k}}{x_{lk}} & \cdots & 0 \\ \cdots\cdots\cdots\cdots\cdots\cdots\cdots\cdots \\ 0 & 0 & \cdots & \dfrac{1}{x_{lk}} & \cdots & 0 \\ \cdots\cdots\cdots\cdots\cdots\cdots\cdots\cdots \\ 0 & 0 & \cdots & -\dfrac{x_{mk}}{x_{lk}} & \cdots & 1 \end{pmatrix}$$

The matrix $\mathbf{E}^l$ is called an elementary column matrix, i.e., an identity matrix except for the lth column. The reader can readily verify by direct multiplication that the matrix product

$$\mathbf{E}^l \mathbf{B}^{-1} = \bar{\mathbf{B}}^{-1}$$

Hence the premultiplication of $\mathbf{B}^{-1}$ by $\mathbf{E}^l$ is equivalent to the application of formulas (1.6).[1] As in (1.6), this is accomplished with the knowledge of only $\mathbf{X}_k$ and $\mathbf{B}^{-1}$. If we let

$$\eta_{il} = -\frac{x_{ik}}{x_{lk}} \qquad \text{for } i \neq l \tag{2.1}$$

$$\eta_{ll} = \frac{1}{x_{lk}} \tag{2.2}$$

then $\mathbf{E}^l$ can be written as

$$\mathbf{E}^l = \begin{pmatrix} 1 & \cdots & \eta_{1l} & \cdots & 0 \\ & & & & \\ 0 & \cdots & \eta_{ll} & \cdots & 0 \\ & & & & \\ 0 & \cdots & \eta_{ml} & \cdots & 1 \end{pmatrix}$$

In fact, to write out the explicit representation of the elementary matrix, we need to know only the column index l and the set of m elements η_{il}. Using a shorthand notation, we define and denote the *eta* (η) *vector* by

$$\mathbf{E}^l = (l; \eta_{1l}, \ldots, \eta_{ll}, \ldots, \eta_{ml})$$

In the original simplex method we usually initiate the computation with either an artificial or a real basis $\mathbf{B}_0$ which consists of an m-dimensional unit matrix. The inverse of $\mathbf{B}_0$ is, of course, an identity matrix; let us denote $\mathbf{B}_0^{-1} = \mathbf{I} = \mathbf{E}_0$. The inverse of the next basis $\mathbf{B}_1$, which differs from $\mathbf{B}_0$ in that vector $\mathbf{P}_k$ has replaced vector $\mathbf{P}_l$, can be computed by multiplying

$$\mathbf{E}_1^l \mathbf{E}_0 = \mathbf{B}_1^{-1}$$

where $\mathbf{E}_1^l$ is the elementary matrix obtained in making the first change of basis. In general, given the associated set of eta vectors, the inverse of the pth basis can be obtained from *the product form of the inverse* (PFI)

$$\mathbf{E}_p^l \mathbf{E}_{p-1}^l \cdots \mathbf{E}_1^l \mathbf{E}_0 = \mathbf{B}_p^{-1} \tag{2.3}$$

We should note that, in each $\mathbf{E}_p^l$, the index l corresponds to the positional index of the column in the preceding basis that is being replaced. We see that, for any basis $\mathbf{B}_p$, given the corresponding set of elementary matrices, we can obtain $\mathbf{B}_p^{-1}$, and hence the methodology of the revised simplex procedure applies.

In the product form of the revised simplex method the computation begins

---

[1] $\mathbf{E}^l$ is the inverse of the elementary column matrix given by the right-hand side of (1.5a). See footnote 1, page 126 and Exercise 49, Chap. 2.

with the $m + 2 \times m + 2$ identity matrix $\mathbf{U}$. $\mathbf{U}$ is interpreted as being the inverse of the initial basis to the problem stated by (1.12) to (1.14).   The associated elementary matrices are of order $m + 2$, and the rules of Phase I and Phase II of Sec. 1 apply, with the following modifications: Since, in the general form of the revised method, we always have the explicit knowledge of the "inverse matrix" $\mathbf{U}$, we can immediately calculate in Phase I, Step 1, the necessary $\delta_j = \mathbf{U}_{m+2}\bar{\mathbf{A}}_j$.   In the product form, however, we have to determine first the current $\mathbf{U}_{m+2}$, and this is best done by computing

$$\mathbf{V}_{m+2}\mathbf{E}_p^l\mathbf{E}_{p-1}^l \cdots \mathbf{E}_1^l = \mathbf{U}_{m+2}$$

where

$$\mathbf{V}_{m+2} = (0, 0, \ldots, 0, 1)$$

is a unit row vector of dimension $m + 2$.   The calculation can be efficiently carried out by interpreting the product to be

$$\{[(\mathbf{V}_{m+2}\mathbf{E}_p^l)\mathbf{E}_{p-1}^l] \cdots \mathbf{E}_1^l\} \tag{2.4}$$

We should note that, in the product of any row vector

$$\mathbf{V} = (v_1, \ldots, v_{m+2})$$

and an elementary matrix $\mathbf{E}_p^l$, that is, in the product $\mathbf{W} = \mathbf{V}\mathbf{E}_p^l$, the elements can be computed by the formulas

$$w_i = v_i \qquad \text{for } i \neq l$$

$$w_l = \sum_i v_i \eta_{il}$$

In Step 3, the vector $\mathbf{X}_k$ is best computed by

$$\mathbf{X}_k = \{\mathbf{E}_p^l[\mathbf{E}_{p-1}^l \cdots (\mathbf{E}_1^l\mathbf{A}_k)]\} \tag{2.5}$$

In computing these products, we note that, in the product of a column vector $\mathbf{A}_j = (a_{1j}, \ldots, a_{m+2,j})$ and an elementary matrix $\mathbf{E}_p^l$, that is, in $\mathbf{E}_p^l\mathbf{A}_j = \mathbf{D}_j$, the elements can be computed by the formulas

$$d_{ij} = a_{ij} + \eta_{il}a_{lj} \qquad \text{for } i \neq l$$
$$d_{lj} = \eta_{ll}a_{lj}$$

The new elementary matrix and the current solution are developed in Step 4 by formulas (2.1) and (2.2), and

$$x'_{i0} = x_{i0} + \eta_{il}x_{l0} \qquad \text{for } i \neq k$$

$$x'_{k0} = \eta_{ll}x_{l0}$$

respectively.

The above modifications apply to Phase II, except that in Step 1 we compute

$$\{[(\mathbf{V}_{m+1}\mathbf{E}_p^l)\mathbf{E}_{p-1}^l] \cdots \mathbf{E}_1^l\} = \mathbf{U}_{m+1}$$

where

$$\mathbf{V}_{m+1} = (0, 0, \ldots, 0, 1, 0)$$

is a unit row vector of dimension $m + 2$.

The major advantage in generating the inverse for each basis by means of elementary matrices is that only a minimum amount of information, namely, $\mathbf{E}_p^l$, need be recorded. This is extremely important when one uses a computer that has limited high-speed storage. Here we need only record in the short-hand notation that $\mathbf{E}_p^l = (l; \eta_{1l}, \ldots, \eta_{m+2,l})_p$ instead of recording the full inverse $\mathbf{U}$ after each iteration. The sequence of eta vectors in (2.3) defines the current basis inverse $\mathbf{B}_p^{-1}$ and is stored in an *eta file*.

Just about all computer codes for solving linear-programming problems use the revised simplex method and some variation of the product form of the inverse. We next discuss some of the computational aspects of the simplex method.

3. COMPUTATIONAL CONSIDERATIONS

The simplex method has proven to be a most versatile procedure for solving a remarkable range of linear-programming applications.[1] This has been due to the adaptability of the linear-programming model and to the wide availability of simplex method computer codes. Many of these codes, especially those used on large-scale computer systems, are very sophisticated computational procedures termed mathematical programming systems (MPS). They involve state-of-the-art ideas from many fields: linear-programming theory, numerical analysis, computer programming, data preparation and storage procedures, report writers and displays, and computational aids for problem formulation and analysis; see, for example, the MPS user manuals for IBM's MPSX/370 [387] and Sperry Univac's FMPS [619]. The design, implementation, updating, and maintenance of a MPS is a major undertaking that requires technical expertise in all the fields noted above. The proper use of a MPS is not a casual affair. The analyst with a problem to be solved—large-scale or not—is well advised to read and understand the details of the MPS involved in order to take advantage of the many practical features that have been incorporated into it. The reader interested in the use and design of a MPS should consult the manuals cited above (among others) and Benichou et al. [70], Orchard-Hays [532], Murtagh [521], Ketron [428], Greenberg [330], Forrest and Tomlin [257], Hellerman and Rarick [363, 364], Beale [58, 60], and Driebeek [221]. For our purposes, we shall briefly describe some of the ideas and features that are usually incorporated into a MPS. The MPS technology and the high-speed computer have been integrated to form a powerful tool for the

---

[1] See the applications given in Part 3 and the Bibliography.

efficient solution of very large-scale problems—problems having 10,000 to 20,000 constraints and 100,000 to a million variables.[1]

A major concern of those who develop mathematical programming systems has been how to produce efficient computational procedures for solving large-scale problems. The key to MPS efficiency is the fact that most large-scale problems are sparse, and computational and numerical techniques have been developed to take advantage of the original sparseness and to preserve it.[2] Sparse computational procedures allow for the storage of large problems and the speeding up of the calculations. With respect to the revised simplex algorithm, sparseness considerations center about the manner by which the basis inverse is developed and maintained.

First, let us consider the problem data in the form $\begin{pmatrix} \mathbf{c} & \mathbf{0} \\ \mathbf{A} & \mathbf{b} \end{pmatrix}$. The nonzero coefficients (c_j, a_{ij}, b_i) are usually stored column by column. By suitable blocking and noting the number of nonzero elements in each column, the nonzero data can be packed so only a row index is required. Further savings in storage and processing time can be accomplished by taking advantage of a characteristic found in most large-scale problems known as *super-sparsity*; i.e., the number of distinct numerical values in the problem matrix is usually much smaller than the number of nonzero coefficients (Kalan [408]).[3] In a MPS, super-sparsity can be handled in the following manner (Ketron [428]). Each distinct numerical value is recorded once in a value table that is stored in main (core) memory. Each nonzero coefficient is recorded in a main memory index array by means of a number triple: row index, column index, and a pointer. The pointer locates the coefficient's numerical value in the value table. The +1 and −1 coefficients are recorded with a special indicator that does not reference the value table. The pointer occupies less space than a

---

[1] Of course, most linear-programming problems solved in practice are not that large. McCall [481] analyzed 16 "real-world" problems whose average dimensions were $1,800 \times 2,603$, with one having 8,952 constraints and another having 7,907 variables. The reader and student should note that the classroom solution of "toy" examples by hand or by using the linear-programming codes that are available at most university computing centers does not impart the complexities of formulation, data collection, computation, and analysis that are necessary to solve most real-world problems.

[2] See, for example, the problem structure of Fig. 9.3. Beale [60] notes that in most practical problems a typical variable occurs in only about six constraints. Also, the columns of such problems usually average less than 10 nonzero elements, regardless of dimensions (Saunders [597], Greenberg [333]). The books by Rose and Willoughby [574], Tewarson [639], and Bunch and Rose [101] discuss most aspects of sparse matrix computation, including applications to linear programming. See footnote 2, page 125, for a definition of sparseness.

[3] Super-sparsity is illustrated by the following problems. A 6,908 row by 41,801 column problem had 167,561 nonzero coefficients, with only 6,584 distinct values; a $589 \times 1,285$ problem had 14,978 nonzeros and 3,599 distinct values (Ketron [428]). Note the low densities of these problems.

coefficient and the ± 1's do not require pointers (there is usually a high percentage of ± 1's).

Although we might start with a sparse problem matrix, the simplex method pivot transformations that take us from basis to basis can cause the resultant structure to "fill in" with nonzero elements. It is clear that the full tableau procedure used to introduce the simplex method in Chap. 4 can destroy any given sparse structure. This is also true if we want to maintain an explicit representation of the full basis inverse for the revised simplex method. For example, consider the following sparse basis (Orchard-Hays [532])[1]:

$$\mathbf{B} = \begin{pmatrix} 1 & 0 & 0 & 0 & 1 \\ 1 & 1 & 0 & 0 & 0 \\ 0 & 1 & 1 & 0 & 0 \\ 0 & 0 & 1 & 1 & 0 \\ 0 & 0 & 0 & 1 & 1 \end{pmatrix} \tag{3.1}$$

Its explicit inverse is completely dense:

$$\mathbf{B}^{-1} = \begin{pmatrix} \frac{1}{2} & \frac{1}{2} & -\frac{1}{2} & \frac{1}{2} & -\frac{1}{2} \\ -\frac{1}{2} & \frac{1}{2} & \frac{1}{2} & -\frac{1}{2} & \frac{1}{2} \\ \frac{1}{2} & -\frac{1}{2} & \frac{1}{2} & \frac{1}{2} & -\frac{1}{2} \\ -\frac{1}{2} & \frac{1}{2} & -\frac{1}{2} & \frac{1}{2} & \frac{1}{2} \\ \frac{1}{2} & -\frac{1}{2} & \frac{1}{2} & -\frac{1}{2} & \frac{1}{2} \end{pmatrix}$$

The selection of a computational strategy for maintaining a sparse and accurate representation of the current basis inverse is essential to a successful MPS. For the product form of the inverse (PFI), the eta vectors may be sparse, but they are often about 50 percent dense, (Murtagh [521], Saunders [596]). Even the order in which the eta vectors are generated can lead to different densities of the eta file. For the matrix $\mathbf{B}$ above, the eta file obtained by pivoting down the main diagonal selecting columns 1 to 5 in order contains 13 nonzero elements, while the eta file developed by pivoting up the diagonal selecting columns 1 to 5 in reverse order has 19 nonzero elements (see Exercise 10). As Orchard-Hays [532] notes, inverse methods applied to large and sparse matrices can produce pronounced differences in their nonzero counts. Such considerations have led to a number of procedures designed to control the growth of nonzeros in the PFI implementation of the simplex method.

For the PFI, there are two occasions during which we can employ nonzero control strategies. The first is at the beginning of the computation and assumes we are given a starting basis $\mathbf{B} \neq \mathbf{I}$. The second can occur often during the course of the computation. It is due to the fact that each iteration of the simplex procedure adds a new eta vector to the ever-lengthening eta

---

[1] The general form of this matrix occurs in a number of personnel scheduling problems; see, for example, Exercise 31*f*, Chap. 11.

file. The processing of a large number of eta vectors eventually becomes inefficient compared with reinitializing the file with m eta vectors that represent the PFI of the current basis. Also, as the eta file increases, rounding error increases. Most mathematical programming systems automatically call for a reinversion to take place based on some function of the number of iterations since the last reinversion, accuracy, storage requirements, etc. A summary of the basic steps for forming the eta file for a given basis is as follows (Hellerman and Rarick [364]):

STEP a Select a column of the basis not already used for pivoting.

STEP b Transform the column by applying the current set of eta vectors in the order of their generation.

STEP c Choose a pivot element for the transformed column in a row where no column has pivoted previously.

STEP d Form an eta vector from the transformed column and add it to the eta file.

STEP e Repeat steps a through d until all columns have been used as pivot columns.

Note that there is some flexibility in selecting the pivot elements; they can be any appropriate nonzero element. As discussed in Sec. 5, Chap. 2, positioning for size of the pivot element aids in reducing rounding error. Here we address what can be done to reduce the number of nonzero elements in the PFI representation of the starting or reinitialized inverse.

An early proposal for controlling the growth of nonzeros was due to Markowitz [495]. His inversion scheme selects pivot elements that reduce the number of new nonzero coefficients created at each step. For a given basis $\mathbf{B}$, let R_i and C_j be equal to the initial count of the nonzero elements in row i and column j, respectively. We then form the eta file as follows. The first pivot operation selects an $a_{ij} \neq 0$ that corresponds to the position with the minimum value of $(R_i - 1)(C_j - 1)$. This value indicates the total number of new non-zero coefficients that could occur (the R_i and C_j are reduced by 1 to account for the pivot position). This operation produces the first eta vector. The remaining columns of $\mathbf{B}$ are updated (transformed) by the eta vector, the row and column counts are also updated, and the process continues until the m eta vectors that represent $\mathbf{B}^{-1}$ are determined. Although the procedure is laborious, the reduction in nonzero eta vector elements is remarkable, with the succeeding simplex iterations being accomplished much more quickly and with greater accuracy (Orchard-Hays [532]). A simpler rule that has proved effective, due to Larsen [458], is to select for column j an $a_{ij} \neq 0$ whose current row count is minimum over those rows that have not been used as pivot rows. The columns of $\mathbf{B}$ are processed in order (which can be stipulated) and the row counts are based on the original nonzero elements in each row. After the eta vector is determined for column j, the row counts are reduced according to the nonzero coefficients in the original column j. Note that in these schemes sparsity concerns are given prominence over round-off error.

However, an a_{ij} selected to be a pivot that is too small (e.g., is less than 0.001) would not be used.

A more complicated but very effective sparse inversion scheme is the *partitioned preassigned pivot procedure* (P^4) of Hellerman and Rarick [363, 364]. The key to P^4 is that most linear-programming problems, especially large-scale ones, have sparse structures that more often than not allow the basis vectors to be arranged in a near triangular matrix form, i.e., the basis would be a lower triangular matrix except for a few "bumps" along the main diagonal. The reader will recall from the discussion of triangular matrices and the **LU** decomposition (Sec. 1, Chap. 2) that the inverse of a lower triangular matrix is lower triangular and the PFI is readily available. Also, a lower triangular matrix **L** and its inverse $\mathbf{L}^{-1}$ have the same number of nonzero elements. [Compare (1.1) and (1.6) and see Exercise 22 of Chap. 2.] P^4 takes advantage of these facts by rearranging the basis in a lower triangular matrix, if possible, or a near triangular matrix. If the latter is the case, then various procedures are employed to obtain sparse PFI eta vectors or sparse factors of the **LU** inverse. The P^4 approach is an efficient and effective way for generating the starting or reinitialized inverse and has been incorporated into many mathematical programming systems. Note that in practice the rows and columns of the basis are not actually rearranged by P^4 or similar schemes. (See Hellerman and Rarick [364] and Murtagh [521] for details of P^4; also see Orchard-Hays [532], Dantzig [169], Perold and Dantzig [545], and McBride [478] for other triangularization inverse procedures.)

The PFI method forms and maintains the eta file used to carry out the operations of the revised simplex method. The basic steps can be summarized as follows (Forrest and Tomlin [257], Murtagh [521]). We assume that an initial feasible basis **B** is given and its inverse is known in product form, i.e., $\mathbf{B}^{-1} = \mathbf{E}_p^l \mathbf{E}_{p-1}^l \cdots \mathbf{E}_1^l \mathbf{E}_0$.

STEP 1   Determine the pricing vector

$$\boldsymbol{\pi} = \mathbf{c}_0 \mathbf{B}^{-1} = \mathbf{c}_0 \mathbf{E}_p^l \mathbf{E}_{p-1}^l \cdots \mathbf{E}_1^l \mathbf{E}_0$$

where $\mathbf{c}_0$ is a row vector whose components c_i correspond to the basis vectors. The calculations are done by multiplying the eta file with $\mathbf{c}_0$ in reverse order as described in Sec. 2.4. In mathematical programming systems this process is termed BTRAN, for "backward transformation."

STEP 2   Find the reduced costs of the nonbasic columns, i.e.,

$$d_j = z_j - c_j = \boldsymbol{\pi} \mathbf{A}_j - c_j$$

This is known as the PRICE operation. [In many mathematical programming systems the objective function (cost row) is imbedded in the problem matrix as described in Eqs. (1.8) to (1.10). For that form, the $\mathbf{c}_0$ in Step 1 is a unit vector (see footnote on page 133).]

STEP 3   Select a vector $\mathbf{A}_k$ to enter the basis with $d_k < 0$ and calculate the updated form

$$\mathbf{X}_k = \mathbf{B}^{-1}\mathbf{A}_k = \mathbf{E}_p^l \mathbf{E}_{p-1}^l \cdots \mathbf{E}_1^l \mathbf{E}_0 \mathbf{A}_k$$

Here the eta file is multiplied with $\mathbf{A}_k$ in a forward manner; this operation is termed FTRAN. (Note that the selection criterion is for a maximization problem. The usual maximization rule is to select a vector $\mathbf{A}_j$ corresponding to the minimum $d_j < 0$. Mathematical programming systems use a variety of selection rules, some of which are discussed below.)

STEP 4   Determine the solution vector

$$\mathbf{X}_0 = \mathbf{B}^{-1}\mathbf{b} = \mathbf{E}_p^p \mathbf{E}_{p-1}^l \cdots \mathbf{E}_1^l \mathbf{E}_0 \mathbf{b}$$

and the pivot row l corresponding to

$$\frac{x_{l0}}{x_{lk}} = \min_i \frac{x_{i0}}{x_{ik}} \qquad \text{for } x_{ik} > 0$$

This is the usual ratio test to preserve feasibility. The x_{i0} are the components of $\mathbf{X}_0$ and the x_{ik} are the components of $\mathbf{X}_k$. This operation is called CHUZR, for "choose-the-row."

STEP 5   Update the basis inverse and $\mathbf{X}_0$ by forming the eta vector

$$\mathbf{E}_{p+1}^l = \begin{pmatrix} 1 & \cdots & \eta_{1l} & \cdots & 0 \\ & & \cdots & & \\ 0 & \cdots & \eta_{ll} & \cdots & 0 \\ & & \cdots & & \\ 0 & \cdots & \eta_{ml} & \cdots & 1 \end{pmatrix}$$

where

$$\eta_{il} = -\frac{x_{ik}}{x_{lk}} \qquad \text{for } i \neq l$$

$$\eta_{ll} = \frac{1}{x_{lk}}$$

The inverse of the new basis $\bar{\mathbf{B}}$ is given by

$$(\bar{\mathbf{B}})^{-1} = \mathbf{E}_{p+1}^l \mathbf{E}_p^l \mathbf{E}_{p-1}^l \cdots \mathbf{E}_1^l \mathbf{E}_0$$

and the updated new solution is

$$\bar{\mathbf{X}}_0 = \mathbf{E}_{p+1}^l \mathbf{X}_0$$

The operations of Step 5 are termed WRETA, for "write eta."

The usual tie-breaking and stopping rules of the simplex algorithm are applied in the above steps.

For an $m \times m$ basis $\mathbf{B}$, P^4 or similar procedures generate a sparse eta file containing $p = m$ eta vectors. We have up to now considered the eta file to be in standard PFI form, i.e.,

$$\mathbf{B}^{-1} = \mathbf{E}_m^l \mathbf{E}_{m-1}^l \cdots \mathbf{E}_1^l \mathbf{E}_0$$

An alternative form that has proved to be effective is based on the $\mathbf{B} = \mathbf{LU}$ decomposition and the gaussian elimination form of the inverse (EFI) described in Chap. 2. $\mathbf{P}^4$ can be used to develop a sparse $\mathbf{LU}$ representation and inverse in the factored form

$$\mathbf{B}^{-1} = \mathbf{U}^{-1}\mathbf{L}^{-1} = \mathbf{U}_1^{-1} \cdots \mathbf{U}_m^{-1}\mathbf{L}_m^{-1} \cdots \mathbf{L}_1^{-1}$$

Note that the $\mathbf{U}_i^{-1}$ and $\mathbf{L}_i^{-1}$ can be represented as eta vectors, and can be stored efficiently and processed as an eta file.   Either form of the inverse can be used to initiate the simplex method or when reinversion is required. (IBM's mathematical programming system MPSX/370 selects either the PFI or EFI form at reinversion time based on the problem's characteristics; see Benichou et al. [70].)   The standard PFI method calls for the transformations of subsequent iterations to be recorded as eta vectors and added to the PFI or EFI eta file.   For example, the next iteration produces a basis $\bar{\mathbf{B}}$ whose inverse is given by

$$(\bar{\mathbf{B}})^{-1} = \mathbf{E}_{m+1}^l (\mathbf{E}_m^l \cdots \mathbf{E}_1^l \mathbf{E}_0)$$

or

$$(\bar{\mathbf{B}})^{-1} = \mathbf{E}_{m+1}^l (\mathbf{U}_1^{-1} \cdots \mathbf{U}_m^{-1}\mathbf{L}_1^{-1} \cdots \mathbf{L}_m^{-1})$$

Instead of maintaining a PFI eta file, an alternative procedure is to start with a sparse EFI representation of $\mathbf{B} = \mathbf{LU}$ and to transform subsequent bases into the same form in a manner that exploits sparsity.[1]   Here the pivot operation that replaces a vector in $\mathbf{B}$ is chosen so as to keep $\mathbf{L}$ the same (except for possible permutations) and to minimize the fill-in of the new $\mathbf{U}$.   Let us assume an initial basis $\mathbf{B} = (\mathbf{P}_1 \cdots \mathbf{P}_l \cdots \mathbf{P}_m) = \mathbf{LU}$ and a new basis $\bar{\mathbf{B}} = (\mathbf{P}_1 \cdots \mathbf{P}_k \cdots \mathbf{P}_m)$, where vector $\mathbf{P}_k$ has replaced vector $\mathbf{P}_l$.   We have $\mathbf{L}^{-1}\mathbf{B} = \mathbf{U} = (\mathbf{U}_1 \cdots \mathbf{U}_l \cdots \mathbf{U}_m)$ and $\mathbf{L}^{-1}\bar{\mathbf{B}} = (\mathbf{U}_1 \cdots \mathbf{Y}_k \cdots \mathbf{U}_m)$, where $\mathbf{Y}_k = \mathbf{L}^{-1}\mathbf{P}_k$.   The vector $\mathbf{Y}_k$ is the "partially updated" form of the vector $\mathbf{P}_k$.   Note that the system of equations $\mathbf{LY}_k = \mathbf{P}_k$ is triangular and can be readily solved without explicit knowledge of $\mathbf{L}^{-1}$.   The product $\mathbf{L}^{-1}\bar{\mathbf{B}}$ is not upper triangular and, pictorially, can be represented by

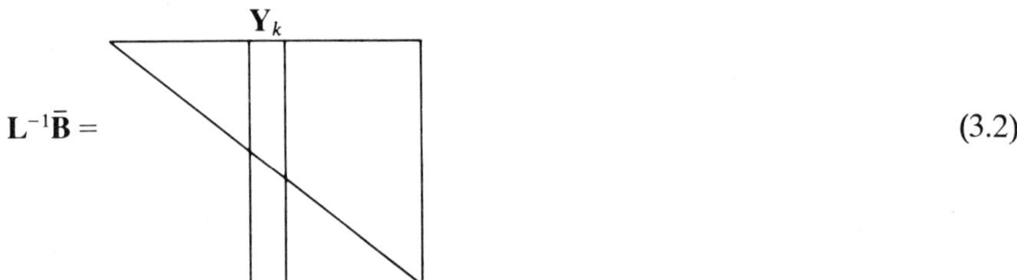

$$(3.2)$$

---

[1] The EFI eta file is usually more sparse than the PFI eta file (Tomlin [649]).

The problem now is to reduce $\mathbf{L}^{-1}\bar{\mathbf{B}}$ by gaussian elimination to an upper triangular form $\bar{\mathbf{U}}$ while causing little fill-in of nonzero elements. We can, of course, permute the rows and columns of (3.2) as required.

We shall review two procedures for producing $\bar{\mathbf{U}}$. The first, by Bartels and Golub [46], moves column $\mathbf{Y}_k$ to the last position of (3.2) and compensates by moving columns $l + 1$ to m one place to the left to obtain the following form:

$$\mathbf{H} = \mathbf{L}^{-1}\bar{\mathbf{B}}\mathbf{Q} =$$

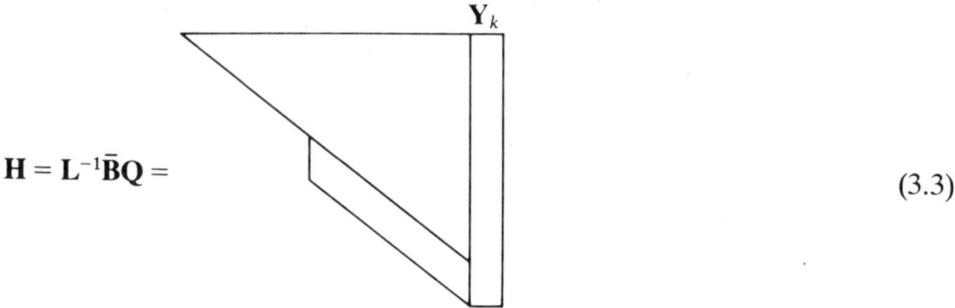

$$(3.3)$$

where $\mathbf{Q}$ is the appropriate column permutation matrix. $\mathbf{H}$ is known as a Hessenberg matrix and is upper triangular except for nonzero elements immediately below the main diagonal in rows l to $m - 1$. Elimination transformations must be applied to (3.3) that reduce the subdiagonal elements to zero, and this is done in order from column l to column $m - 1$. The pivot for column j is either the diagonal element or the subdiagonal element, whichever is the largest in absolute value. If the subdiagonal element is chosen, then rows j and $j + 1$ are first interchanged to place the subdiagonal element on the diagonal. The transformation of $\mathbf{H}$ into the upper triangular form $\bar{\mathbf{U}}$ is accomplished by

$$\mathbf{E}_{m-1}^{-1} \cdots \mathbf{E}_{l}^{-1}\mathbf{L}^{-1}\bar{\mathbf{B}}\mathbf{Q} = \bar{\mathbf{U}} \tag{3.4}$$

where $\mathbf{E}_j$ is an elementary matrix that performs the pivot operation in row j and interchanges rows if required; see Exercise 16. We then have

$$\bar{\mathbf{B}}^{-1} = \mathbf{Q}\bar{\mathbf{U}}^{-1}\mathbf{E}_{m-1}^{-1} \cdots \mathbf{E}_{l}^{-1}\mathbf{L}^{-1} \tag{3.5}$$

The Bartels and Golub **LU** basis updating scheme tends to be numerically stable because of the manner in which the pivot elements are selected. However, it can cause fill-in of $\bar{\mathbf{U}}$ in rows $l + 1$ to m.[1]

We illustrate the structure of (3.2) and (3.3) by the following numerical example. Let

---

[1] Saunders [596] discusses an implementation of the Bartels and Golub procedure that is fast and efficient; see also Reid [565].

$$\mathbf{L}^{-1}\bar{\mathbf{B}} = \begin{array}{c} \mathbf{Y}_k \\ \begin{bmatrix} 1 & 2 & 0 & 0 & 0 & 1 \\ 0 & 2 & 5 & 0 & 3 & 0 \\ 0 & 0 & 4 & 4 & 0 & 4 \\ 0 & 0 & 0 & 5 & 7 & 0 \\ 0 & 0 & 0 & 0 & 6 & 6 \\ 0 & 2 & 0 & 0 & 0 & 5 \end{bmatrix} \end{array} \qquad (3.2a)$$

where $\mathbf{Y}_k$ corresponds to the second column. We then permute the columns to obtain[1]

$$\mathbf{H} = \begin{array}{c} \mathbf{Y}_k \\ \begin{bmatrix} 1 & 0 & 0 & 0 & 1 & 2 \\ 0 & 5 & 0 & 3 & 0 & 2 \\ 0 & 4 & 4 & 0 & 4 & 0 \\ 0 & 0 & 5 & 7 & 0 & 0 \\ 0 & 0 & 0 & 6 & 6 & 0 \\ 0 & 0 & 0 & 0 & 5 & 2 \end{bmatrix} \end{array} \qquad (3.3a)$$

A second scheme for transforming (3.2) into an upper triangular form, one that adjusts for fill-in, is the procedure of Forrest and Tomlin [257]. Here we start with the EFI given by

$$\mathbf{B}^{-1} = \mathbf{U}_1^{-1} \cdots \mathbf{U}_m^{-1} \mathbf{L}_m^{-1} \cdots \mathbf{L}_1^{-1}$$

with the $\mathbf{U}_i^{-1}$ and $\mathbf{L}_i^{-1}$ represented as eta vectors. The set of $\mathbf{U}_i^{-1}$ is one eta file and the set of $\mathbf{L}_i^{-1}$ is another. We form $\mathbf{H}$ as before, but shift row l to position m and move rows $l+1$ to m to rows l to $m-1$, respectively. Letting $\mathbf{P}$ denote the permutation matrix for the row exchanges, we then have

$$\bar{\mathbf{H}} = \mathbf{P}\mathbf{L}^{-1}\bar{\mathbf{B}}\mathbf{Q} = \qquad (3.6)$$

[1] The reader should carry out the steps that transform (3.3a) into an upper triangular matrix $\mathbf{U}$ (Exercise 16).

The last row of (3.6) (originally row l) is reduced to zero, except for its diagonal element, by selecting pivot elements coming down the diagonal, starting with the element in row l, column l. These pivots will be nonzero as they were diagonal elements of **U**. Thus, in transforming $\tilde{\mathbf{H}}$ to $\bar{\mathbf{U}}$, no fill-in occurs and new nonzero elements can appear only in the last column. The pivot operations are represented by a row transformation matrix $\mathbf{R}^{-1}$ which is stored as an eta vector; see Exercise 17. We then have

$$\mathbf{R}^{-1}\tilde{\mathbf{H}} = \mathbf{R}^{-1}\mathbf{P}\mathbf{L}^{-1}\bar{\mathbf{B}}\mathbf{Q} = \bar{\mathbf{U}} \tag{3.7}$$

and

$$\bar{\mathbf{B}}^{-1} = \mathbf{Q}\bar{\mathbf{U}}^{-1}\mathbf{R}^{-1}\mathbf{P}\mathbf{L}^{-1} \tag{3.8}$$

To illustrate the form of $\tilde{\mathbf{H}}$, we apply row permutations to (3.3a) to obtain

$$
\tilde{\mathbf{H}} =
\begin{array}{c}
 \\

\end{array}
\left[
\begin{array}{ccccc|c}
1 & 0 & 0 & 0 & 1 & 2 \\
0 & 4 & 4 & 0 & 4 & 0 \\
0 & 0 & 5 & 7 & 0 & 0 \\
0 & 0 & 0 & 6 & 6 & 0 \\
0 & 0 & 0 & 0 & 5 & 2 \\
\hline
0 & 5 & 0 & 3 & 0 & 2 \\
\end{array}
\right]
\tag{3.6a}
$$

where the last column is headed $\mathbf{Y}_k$.

[The reader should determine the transformations and carry out the steps that make (3.6a) upper triangular.]

The Forrest and Tomlin procedure is not as numerically stable as the method of Bartels and Golub because of the restricted choice of pivot elements. The reader is referred to Tomlin [649] for details on how the eta files of both these procedures are updated; also see Benichou et al. [70], Saunders [596], and Tewarson [639].

Mathematical programming systems include a variety of options that help reduce the number of iterations and speed the convergence of the simplex algorithm. As described in Secs. 1 and 2 and Chap. 4, the simplex method selects a vector $\mathbf{A}_k$ to enter the basis as a function of the reduced costs d_j; $\mathbf{A}_k$ corresponds to min $d_k < 0$ (maximization problem) or max $d_k > 0$ (minimization problem). To find such an $\mathbf{A}_k$, the full set of nonbasic vectors has to be searched and priced out. This can be a fairly time-consuming process, especially when the matrix **A** has to be stored outside of the main memory of the computer. One way to reduce this search time is to first price out all the nonbasic vectors and select p vectors as candidates to enter the basis. The selection is based on the values of the d_j's. The simplex algorithm continues by introducing the *best* of the p vectors and then prices out *only* the remaining $p-1$ candidates. If any are favorable (in terms of their d_j's), the *best* is introduced into the basis, and so on, until all of the p vectors have been used as pivot columns, or until none of the remaining ones will improve the value of the objective function.   Here the *best vector* is the one that gives the greatest change $\theta_0(z_j - c_j) = \theta_0 d_j$ in the objective function; see $(2.1a)$ of Chap. 4.   The process continues by again pricing out the full set of nonbasic vectors and selecting another set of p candidates.   This variation is termed *multiple pricing*.   The pricing out of the full set of nonbasic variables and the selection of the p vectors is called a *major iteration*; the subsequent pricing of the vectors remaining from the set of p is called a *minor iteration*.   The selection of the best vector is termed *suboptimization*.   The parameter p is usually set to a value of 5 or less (IBM [387]).   A modification of this process, known as *partial pricing*, is the following.   The nonbasic vector file is searched until a set of p vectors that price out properly is found.   These vectors are then processed by minor iterations.   A new set of p vectors is selected by searching the nonbasic vector file starting from the point where the first search stopped, and so on.   A variation of this scheme is to search a fixed number of the nonbasic vectors and select a set of p candidates.   The next iteration begins its search where the first ended, and so on.   Of course, if p vectors cannot be found, these processes continue with whatever number can be selected until the usual optimality condition arises.   The reader is referred to Orchard-Hays [532], Murtagh [521], IBM [387], and Sperry Univac [619] for further details; see Cutler and Wolfe [149] and Quandt and Kuhn [553] for discussions on the effectiveness of different pricing strategies.

The manner in which the pricing function is defined is crucial to the efficiency of the simplex algorithm.   In the multiple pricing schemes described above, the major iteration selects the p candidates based on the values of the d_j's; the minor iteration selects the entering vector from the set of candidates corresponding to the one with the $\theta_0 d_j$ term that makes the largest improvement in the objective function.   These selection rules are functions of the standard pricing formula πA_j.   Other formulas have been suggested and, although they require more computation, as does the suboptimization scheme, they tend to reduce the number of iterations.   Most MPS use the standard pricing procedures described above, but have options to use others.

The original reduced-cost selection rule of Dantzig (i.e., selection based on max d_j or min d_j) is termed *steepest unit ascent* and is equivalent to maximizing the gradient[1] (rate of change of the objective function) in the space of the nonbasic variables (Dantzig [152], Quandt and Kuhn [553]).   Using this rule, the simplex method moves from extreme point to a neighboring extreme point along an edge that joins these points (see Exercise 9, Chap. 3).   Movement along the edge gives the best rate of improvement of the objective function with respect to the change in the out-of-basis variable that characterizes that edge (Goldfarb and Reid [315]).   In defining new pricing functions, emphasis has been placed on respecifying the gradient and the selection of the steepest

---

[1] See Chap. 12.

edge.   In particular, we cite the procedures of Harris [353, 354], Goldfarb and Reid [315], Dickson and Frederick [202], and Crowder and Hattingh [143].   We next review some of the basic ideas behind these schemes.

First we note that the $d_j = \pi \mathbf{A}_j - c_j$ changes in magnitude if the original column $\begin{pmatrix} c_j \\ \mathbf{A}_j \end{pmatrix}$ is multiplied by a column scale s_j; we would have $\bar{d}_j = s_j d_j$.   Thus, a selection rule based on the d_j may be misleading (Greenberg and Kalan [334]).   (Scaling of columns and rows is often done and is a MPS feature; proper scaling improves the stability and accuracy of the computations; see Exercise 18 in this chapter and Exercise 16, Chap. 4.)   Harris' Devex[1] is a dynamic scaling procedure that modifies the d_j's so they can be interpreted as gradients in the reference space of the initial nonbasic variables.   The rationale for this idea is the maintaining of a consistent reference space and, thus, a consistent set of selection decisions from iteration to iteration; also, the initial set of nonbasic variables tends to include the more important structural variables, as the first feasible basis usually consists of logical (slack) and artificial variables (Murtagh [521]).   Devex is applied in the following manner.   After several iterations, let C denote the set of vectors from the initial basis that are now nonbasic, and let R denote the set of row positions corresponding to those vectors that were originally nonbasic but are now basic.   For $d_j < 0$ (or $d_j > 0$ depending on the objective function), we base the selection of the new vector to enter the basis on the reference space gradients d_j / T_j where

$$T_j = \sqrt{\delta_j + \sum_{i \in R} x_{ij}^2}$$

$$\mathbf{X}_j = (x_{ij}) = \mathbf{B}^{-1}\mathbf{A}_j$$

$$\delta_j = \begin{cases} 1 & j \notin C \\ 0 & j \in C \end{cases}$$

As T_j is rather costly to compute, various approximations have been suggested (Harris [353], Benichou et al. [70]).

The procedure of Goldfarb and Reid [315] moves along the *steepest edge* leading from the current extreme point to a neighboring extreme point.[2]   It is based on normalized reduced costs; see Exercise 21.

Devex has proved to be quite effective in terms of requiring fewer iterations to solve a wide class of linear-programming problems, but the steepest-edge method has been shown to be better for a set of test problems.   Pricing procedures such as Devex and steepest edge have been incorporated into MPS and brought into play depending on the characteristics of the problem (Benichou et al. [70]).   They may be used in conjunction with partial and multiple pricing strategies.

---

[1] Devex is from the Latin *devexus*, meaning "steep."
[2] See Exercise 9, Chap. 3.

REMARKS

The material in Secs. 1 and 2 is from Orchard-Hays [528] and Dantzig [169]. For additional remarks on the computational elements of the general form of the revised procedure, the reader is referred to Dantzig [157]; for the product form of the inverse, see Dantzig, Orchard-Hays, and Waters [188], Hellerman and Rarick [364], and, in general, Orchard-Hays [531, 532]. The material in Sec. 3 is based on Orchard-Hays [532], Murtagh [521], Forrest and Tomlin [257], Hellerman and Rarick [363, 364], Benichou et al. [70], Magnanti [488], Tomlin [646, 649], Saunders [596], Harris [353], Quandt and Kuhn [553], Markowitz [495], and Greenberg and Kalan [334]. Also see Reid [565] and Gill, Murray, and Wright [303].

EXERCISES

    1. Solve the example of Sec. 1 by means of the product form of the inverse.

    2. Solve the following problems by both forms of the revised procedure:

    a. Maximize

$$5x_1 - x_2 + x_3 - 10x_4 + 7x_5$$

subject to

$$3x_1 - x_2 - x_3 \qquad\qquad = 4$$
$$x_1 - x_2 + x_3 + x_4 \qquad = 1$$
$$2x_1 + x_2 + 2x_3 \qquad + x_5 = 7$$
$$x_j \geq 0$$

    b. Minimize

$$-x_1 + 2x_2$$

subject to

$$5x_1 - 2x_2 \leq 3$$
$$x_1 + x_2 \geq 1$$
$$-3x_1 + x_2 \leq 3$$
$$-3x_1 - 3x_2 \leq 2$$
$$x_1 \qquad \geq 0$$
$$x_2 \geq 0$$

    3. Solve Exercise 2*b* by graphical methods.

    4. Solve the following problem by the original simplex method and by both forms of the revised procedure: Minimize

$$x_1 + x_2 + x_3$$

subject to

$$x_1 \qquad\qquad + x_4 \qquad - 2x_6 = 5$$
$$x_2 \qquad + 2x_4 - 3x_5 + x_6 = 3$$
$$x_3 + 2x_4 - 5x_5 + 6x_6 = 5$$
$$x_j \geq 0$$

    5. Solve Example 3.1 of Chap. 4 by the revised simplex method.

    6. Compare the number of multiplications required in performing one iteration of the standard and revised simplex algorithms, Dantzig [152] and Wagner [678].

    7. In Phase I, we maximize the variable x_{m+n+2}, which is equal to the negative sum of the artificial variables used in the first basic solution. If Phase I ends with maximum $x_{m+n+2} < 0$, explain why we know that the original problem is infeasible.

8. If Phase I ends with $x_{m+n+2} = 0$ and some artificial variables are still in the basis (their values must be zero), discuss what you must do in Phase II to ensure that these artificial variables never become positive. (See Exercise 44, Chap. 2, and Dantzig [152] and Murtagh [521].)

9. Solve Example 1.1 by the revised simplex method using variable x_4 and two artificial variables as the starting solution.

10. For the matrix (3.1), determine product form inverses by first pivoting down the main diagonal and then pivoting up the main diagonal. Also determine product form inverses by applying Markowitz's and Larsen's rules for pivot selection. Form the eta files and compare the number of nonzero elements (Orchard-Hays [532]).

11. Consider the following basis:

$$\begin{pmatrix} 1 & 0 & 2 \\ 0 & 0 & 6 \\ 2 & 4 & 2 \end{pmatrix}$$

Compute product form inverses using pivot positions (3, 3), (2, 1), (1, 2) and (3, 2), (1, 1), (2, 3). Compare the number of nonzero elements in each representation. Apply each PFI to the vector $\begin{pmatrix} 1 \\ 1 \\ 1 \end{pmatrix}$ and note the permuted results (Hellerman and Rarick [363]).

12. Under what conditions are the Bartels-Golub and Forrest-Tomlin updating procedures equivalent?

13. Rewrite Steps 1 through 5, Sec. 3, with the necessary simplex method tie-breaking and stopping rules.

14. Let $\mathbf{B}$ be a feasible basis and π the corresponding pricing vector, with $d_j = \pi \mathbf{A}_j - c_j$. A basis change is made with vector $\mathbf{A}_k$ replacing vector $\mathbf{A}_l$. Denote the new basis by $\bar{\mathbf{B}}$, the new pricing vector by $\bar{\pi}$, and the new reduced costs by $\bar{d}_j$. Show that $\bar{d}_j = d_j - \bar{\mathbf{B}}_l \mathbf{A}_j d_k$ and $\bar{\pi} = \pi - d_k \bar{\mathbf{B}}_l$, where $\bar{\mathbf{B}}_l$ is the lth row of $\bar{\mathbf{B}}^{-1}$. Express $\bar{\pi}$ in terms of the lth row of $\mathbf{B}^{-1}$. {See Eq. (2.7a), Chap. 4, Dantzig [152], and Tomlin [649].}

15. Show that $\bar{\mathbf{B}}$ of Eq. (1.5a) can be expressed as $\bar{\mathbf{B}} = \mathbf{B} + (\mathbf{P}_k - \mathbf{B}\mathbf{e}_l)\mathbf{e}_l'$, where $\mathbf{e}_l$ is the lth column of a unit matrix (Murtagh [521]). Note that we can express $\bar{\mathbf{B}} = \mathbf{B}(\mathbf{e}_1, \mathbf{e}_2, \ldots, \mathbf{e}_{l-1}, \mathbf{X}_k, \mathbf{e}_{l+1}, \ldots, \mathbf{e}_m)$; see footnote 1, page 126.

16. For the Bartels and Golub updating procedure, develop the explicit expressions for the $\mathbf{E}_j$ transformations in Eq. (3.5). Each $\mathbf{E}_j$ must select either the diagonal element in row j as the pivot element or the subdiagonal element in row $j + 1$, whichever is the largest. The $\mathbf{E}_j$ are $m \times m$ unit matrices except for the 2×2 submatrix formed by the row and column elements in positions j and $j + 1$. Let $\mathbf{U}_1^{-1} \cdots \mathbf{U}_m^{-1}$ be the backward eta file and $\mathbf{L}_m^{-1} \cdots \mathbf{L}_1^{-1}$ be the forward eta file. The $\mathbf{E}_{m-1} \cdots \mathbf{E}_l$ are added to the beginning of the forward eta file. What must be added to the end of the backward eta file to adjust for the $\bar{\mathbf{U}}^{-1}$ term in Eq. (3.5)? Determine the $\mathbf{E}_j$ for matrix (3.3a). (See Tomlin [649] for a discussion of the updating procedure; also see Murtagh [521] and Tewarson [639].)

17. For the Forrest and Tomlin updating scheme, develop the explicit form of the transformation $\mathbf{R}^{-1}$ in Eq. (3.8). Show that $\mathbf{R} = \mathbf{I} + \mathbf{e}_l r$, $\mathbf{R}^{-1} = \mathbf{I} - \mathbf{e}_l r$, where $r = (0, \ldots, r_{l+1}, \ldots, r_m)$, $r = \bar{u}\mathbf{U}^{-1}$, and $\bar{u} = (0, \ldots, 0, u_{l,l+1}, \ldots, u_{lm})$. In updating the eta files, $\mathbf{R}^{-1}$ is added to the left of the forward eta file (see Exercise 16). What modifications and additions are made to the backward eta file? (See Tomlin [649] and Forrest and Tomlin [257] for a discussion of the updating procedure; also see Murtagh [521], Tewarson [639], and Benichou et al. [70].)

18. *Scaling of a linear-programming problem.* Tewarson [639] notes that it is generally recommended that the rows and columns of a matrix (linear-programming problem) be transformed in such a manner that they have magnitudes of similar order. This is called *scaling*. Mathematical programming systems have a SCALE command that can apply several options. In MPSX/370 (IBM [387]), these include scaling the rows and columns according to their maximum (absolute value) magnitudes (this is termed *equilibration*), and scaling the rows and columns according to the square root of the product of their elements of maximum and minimum magnitudes (*geometric mean scaling*). The basic process in FMPS (Sperry Univac [619]) is equilibration in which the rows are scaled first by the largest coefficient (absolute value) and then the columns. Other scaling options are available. Tomlin [648] notes that the reasons for scaling linear-programming problems are (a) to make all variables with a finite upper bound have the same bound, e.g., all have unit upper bounds (see Orchard-Hays [532]), (b) to reduce the number of iterations, and (c) to improve the numerical behavior of the simplex algorithm. Given the linear-programming problem $(\mathbf{I},\ \mathbf{A})\mathbf{X} = \mathbf{b}$, $l_j \le x_j \le u_j$, describe how you would transform the problem so that the lower bounds are zeros and a unit basis is maintained; i.e., what are the relationships between the row and column scales? Determine the scaled problem $\mathbf{R}(\mathbf{I},\ \mathbf{A})\mathbf{C}\bar{\mathbf{X}} = \mathbf{Rb}$, where $\mathbf{R}$ is an $m \times m$ diagonal matrix and $\mathbf{C}$ is an $n \times n$ diagonal matrix. Apply equilibration to the problems of Exercises 2 and 4. (Also see Westlake [694] and Ralston and Rabinowitz [557].)

19. Using gaussian elimination, show that

$$\mathbf{E}_p \mathbf{E}_{p-1} \cdots \mathbf{E}_1 \mathbf{B} = \mathbf{PUQ}$$

where $\mathbf{E}_i$ is a matrix that differs from $\mathbf{I}$ in just one off-diagonal element (it is an elementary row transformation matrix), $\mathbf{P}$ and $\mathbf{Q}$ are permutation matrices, $\mathbf{U}$ is upper triangular, and $\mathbf{B}$ is a basis, (Reid [565]). What is $\mathbf{B}^{-1}$? HINT: See Exercise 52, Chap. 2. Show how the $\mathbf{L}_k^{-1}$ of Eq. (1.4), Chap. 2, can be factored into elementary row transformation matrices.

20. For a minimization problem, if a vector $\mathbf{P}_k$ not in the basis has its reduced cost $z_k - c_k > 0$, then by proper scaling of the kth column the simplex algorithm will select $\mathbf{P}_k$ to enter the basis (Dantzig [152]). How would you determine the proper scaling factors?

21. Given a basic feasible solution with basis $\mathbf{B}$ consisting of the first m vectors $\mathbf{P}_1, \ldots, \mathbf{P}_m$, show that moving along the edge of steepest descent leading from the current extreme point requires a basis change where the new variable corresponds to

$$\max_{j \notin \mathbf{B}} \mathbf{c}\left(\frac{\tilde{\mathbf{X}}_j}{\|\tilde{\mathbf{X}}_j\|}\right)$$

where $\tilde{\mathbf{X}}_j = \begin{pmatrix} \mathbf{X}_j \\ -\mathbf{e}_j \end{pmatrix}$, $\mathbf{e}_j$ is an $n - m$ unit vector with unity in position j of $\tilde{\mathbf{X}}_j$, and $\mathbf{X}_j = \mathbf{B}^{-1}\mathbf{P}_j$ (Goldfarb and Reid [315]).

CHAPTER 6

The Duality Problems of Linear Programming

Associated with every linear-programming problem as defined in Sec. 1 of Chap. 4 is a corresponding optimization problem called the *dual problem.* The original problem is termed the *primal.* The optimal solution of either problem reveals information concerning the optimal solution of the other. In fact, if the initial simplex tableau for the primal problem contains an $m \times m$ unit matrix, then the solution of either problem by the simplex procedure yields an explicit solution to the other. We shall state and prove the theorems associated with the dual problems as given by Dantzig and Orden [190]. For additional reading, the reader is referred to Gale, Kuhn, and Tucker [278], Vajda [661], Kuhn and Tucker [449], and Balinski and Tucker [33].

1. THE UNSYMMETRIC PRIMAL-DUAL PROBLEMS

The primal problem.[1] Find a column vector $\mathbf{X} = (x_1, x_2, \ldots, x_n)$ which minimizes the linear functional

$$f(\mathbf{X}) = \mathbf{c}\mathbf{X} \tag{1.1}$$

[1] For discussion purposes we have designated the minimization problem as the primal. In general, either problem can be considered the primal, with the remaining problem the dual.

subject to the conditions

$$\mathbf{AX} = \mathbf{b} \tag{1.2}$$

and

$$\mathbf{X} \geq 0 \tag{1.3}$$

This statement of the general linear-programming problem assumes that the number of rows of $\mathbf{A}$ is less than the number of columns of $\mathbf{A}$.

The dual problem. Find a row vector $\mathbf{W} = (w_1, w_2, \ldots, w_m)$ which maximizes the linear functional

$$g(\mathbf{W}) = \mathbf{Wb} \tag{1.4}$$

subject to the conditions

$$\mathbf{WA} \leq \mathbf{c} \tag{1.5}$$

In the dual problem, the variables w_i are not restricted to be non-negative. For both problems we have that $\mathbf{c} = (c_1, c_2, \ldots, c_n)$ is a row vector, $\mathbf{b} = (b_1, b_2, \ldots, b_m)$ is a column vector, and $\mathbf{A} = (a_{ij})$ is the matrix of coefficients. Multiplying the $1 \times m$ row vector $\mathbf{W}$ by the $m \times n$ matrix $\mathbf{A}$, we have the following explicit representation of (1.5):

$$a_{11}w_1 + a_{21}w_2 + \cdots + a_{m1}w_m \leq c_1$$
$$a_{12}w_1 + a_{22}w_2 + \cdots + a_{m2}w_m \leq c_2$$
$$\cdots\cdots\cdots\cdots\cdots\cdots\cdots\cdots\cdots\cdots\cdots\cdots$$
$$a_{1n}w_1 + a_{2n}w_2 + \cdots + a_{mn}w_m \leq c_n$$

The matrix of coefficients for the above inequalities is given by $\mathbf{A}'$. Note that there is a dual variable for each equation of (1.2).

Associated with these problems is the following important theorem:

Theorem 1: The duality theorem. *If either the primal or the dual problem has a finite optimal solution, then the other problem has a finite optimal solution and the extremes of the linear functions are equal, that is,* $\min f(\mathbf{X}) = \max g(\mathbf{W})$.

If either problem has an unbounded optimal solution, then the other problem has no feasible solutions.[1]

Proof: We first assume that the primal is feasible and that a minimum feasible solution has been obtained by the simplex procedure. For discussion purposes, let the first m vectors $\mathbf{P}_1, \mathbf{P}_2, \ldots, \mathbf{P}_m$ be the final basis vectors. Let $\mathbf{B}$ equal the $m \times m$ matrix $(\mathbf{P}_1\mathbf{P}_2 \cdots \mathbf{P}_m)$. The final computational tableau contains the vectors of the original system, $\mathbf{P}_1, \mathbf{P}_2, \ldots, \mathbf{P}_m, \mathbf{P}_{m+1},$

---

[1] For the dual problem, a solution is feasible if it satisfies (1.5). The variables are not required to be nonnegative.

$\ldots$, $\mathbf{P}_n$, in terms of the final basis vectors; i.e., for each vector $\mathbf{P}_j$ the final tableau contains the vector $\mathbf{X}_j$, where $\mathbf{P}_j = \mathbf{B}\mathbf{X}_j$. Let the $m \times n$ matrix

$$\bar{\mathbf{X}} = (\mathbf{X}_1 \mathbf{X}_2 \cdots \mathbf{X}_m \mathbf{X}_{m+1} \cdots \mathbf{X}_n)$$

be the matrix of coefficients contained in the final simplex tableau. Since we assumed that the final basis contained the first m vectors, we have

$$\bar{\mathbf{X}} = \begin{pmatrix} 1 & 0 & \cdots & 0 & x_{1,m+1} & \cdots & x_{1n} \\ 0 & 1 & \cdots & 0 & x_{2,m+1} & \cdots & x_{2n} \\ \multicolumn{7}{c}{\dotfill} \\ 0 & 0 & \cdots & 1 & x_{m,m+1} & \cdots & x_{mn} \end{pmatrix}$$

The minimum-solution vector is given by $\mathbf{X}^0 = \mathbf{B}^{-1}\mathbf{b}$. For this final solution we then have the following relationships:

$$\mathbf{B}^{-1}\mathbf{A} = \bar{\mathbf{X}} \qquad \mathbf{A} = \mathbf{B}\bar{\mathbf{X}} \tag{1.6}$$

$$\mathbf{B}^{-1}\mathbf{b} = \mathbf{X}^0 \qquad \mathbf{b} = \mathbf{B}\mathbf{X}^0 \tag{1.7}$$

$$\min f(\mathbf{X}) = \mathbf{c}^0\mathbf{X}^0 \tag{1.8}$$

$$\mathbf{Z} = \mathbf{c}^0\bar{\mathbf{X}} - \mathbf{c} \leq 0 \tag{1.9}$$

where $\mathbf{c}^0 = (c_1, c_2, \ldots, c_m)$ is a row vector and, by (1.4) of Chap. 4,

$$\mathbf{Z} = (\mathbf{c}^0\mathbf{X}_1 - c_1, \mathbf{c}^0\mathbf{X}_2 - c_2, \ldots, \mathbf{c}^0\mathbf{X}_n - c_n)$$
$$= (z_1 - c_1, z_2 - c_2, \ldots, z_n - c_n)$$

is a row vector whose elements are nonpositive, as they are the $z_j - c_j$ elements corresponding to an optimal solution. Here the vector $\mathbf{0}$ is a null n-dimensional row vector.

Let $\mathbf{W}^0 = (w_1^0, w_2^0, \ldots, w_m^0)$ be defined by

$$\mathbf{W}^0 = \mathbf{c}^0\mathbf{B}^{-1} \tag{1.10}$$

Then, by (1.6) and (1.9), we have

$$\mathbf{W}^0\mathbf{A} - \mathbf{c} = \mathbf{c}^0\mathbf{B}^{-1}\mathbf{A} - \mathbf{c} = \mathbf{c}^0\bar{\mathbf{X}} - \mathbf{c} \leq 0$$

or

$$\mathbf{W}^0\mathbf{A} \leq \mathbf{c}$$

The vector $\mathbf{W}^0$ is a solution to the dual problem, since it satisfies the dual constraints (1.5). For this solution the corresponding value of the dual objective function (1.4) is given by $\mathbf{W}^0\mathbf{b}$, or, from (1.7) and (1.8), we have

$$\mathbf{W}^0\mathbf{b} = \mathbf{c}^0\mathbf{B}^{-1}\mathbf{b} = \mathbf{c}^0\mathbf{X}^0 = \min f(\mathbf{X})$$

Hence, for the solution $\mathbf{W}^0$, the value of the dual objective function is equal to the minimum value of the primal objective function. Now we need only to show that $\mathbf{W}^0$ is also an optimal solution for the dual problem.

For any $n \times 1$ vector $\mathbf{X}$ which satisfies (1.2) and (1.3) and any $1 \times m$ vector $\mathbf{W}$ satisfying (1.5), we have

$$\mathbf{WAX} = \mathbf{Wb} = g(\mathbf{W}) \tag{1.11}$$

and

$$\mathbf{WAX} \le \mathbf{cX} = f(\mathbf{X}) \tag{1.12}$$

By (1.11) and (1.12) we obtain the important relationship

$$g(\mathbf{W}) \le f(\mathbf{X}) \tag{1.13}$$

for *all* feasible $\mathbf{W}$ and $\mathbf{X}$. The extreme values of (1.1) and (1.4) are related by

$$\max g(\mathbf{W}) \le \min f(\mathbf{X}) \tag{1.14}$$

For the dual solution vector $\mathbf{W}^0$, we have from (1.10)

$$g(\mathbf{W}^0) = \mathbf{W}^0\mathbf{b} = \min f(\mathbf{X})$$

and hence, for the optimal solution to the primal problem, $\mathbf{X}^0$, and the dual solution $\mathbf{W}^0$, (1.14) becomes

$$g(\mathbf{W}^0) = f(\mathbf{X}^0)$$

i.e., for the solution $\mathbf{W}^0 = \mathbf{c}^0\mathbf{B}^{-1}$, the dual objective function takes on its maximum value. Therefore, for $\mathbf{W}^0$ and $\mathbf{X}^0$ we have the corresponding values of the objective functions related by

$$\max g(\mathbf{W}) = \min f(\mathbf{X}) \tag{1.15}$$

We stress that an optimal solution $\mathbf{W}^0$ to the dual problem is the final pricing vector $\boldsymbol{\pi}$ of the revised simplex method; see Eq. (1.4) of Chap. 5.

The above results have been shown to hold whenever the primal has a finite optimal solution. In a similar fashion, we can show that, when the dual problem has a finite optimal solution, the primal is feasible, and (1.15) holds.

To do this, transform the dual problem

$$\max g(\mathbf{W}) = \max \mathbf{Wb}$$
$$\mathbf{WA} \le \mathbf{c}$$

to the primal format and show that its dual (which will be feasible by the above theorem) is just the original primal. We have

$$\max \mathbf{Wb} = -\min (-\mathbf{Wb})$$

or

$$-\max \mathbf{Wb} = \min (-\mathbf{Wb})$$

subject to

$$\mathbf{WA} + \mathbf{W}_3\mathbf{I} = \mathbf{c}$$
$$\mathbf{W}_3 \ge \mathbf{0}$$

where the W_3 is a set of nonnegative slack variables.   We next transform $W = W_1 - W_2$, where W_1 and W_2 are nonnegative variables.   Our problem is now

$$\min\ (-W_1 + W_2)b + W_3 0 \qquad (1.15a)$$

subject to

$$(W_1 - W_2)A + W_3 I = c \qquad (1.15b)$$
$$W_1 \geq 0$$
$$W_2 \geq 0$$
$$W_3 \geq 0$$

and is in the standard format.

The duality theorem holds for this problem.   To determine its dual in the proper format, it is convenient to multiply the constraint equation $(1.15b)$ by -1 to obtain

$$-W_1 A + W_2 A - W_3 I = -c \qquad (1.15c)$$

The dual to $(1.15a)$ and $(1.15c)$ is then

$$\max\ (-cX) = -\min\ cX$$

subject to

$$(-A \quad A \quad -I)X \leq (-b \quad b \quad 0)$$

which is equivalent to

$$-\max\ (-cX) = \min\ cX$$

subject to

$$-AX \leq -b$$
$$AX \leq b$$
$$-IX \leq 0$$

or

$$AX \geq b$$
$$AX \leq b$$
$$X \geq 0$$

or

$$AX = b$$
$$X \geq 0$$

which is the original primal problem.

This completes the proof of the first part of the theorem.

To prove the second part, we note that, if the primal is unbounded, then we have by (1.13)

$$g(\mathbf{W}) \leq -\infty \tag{1.16}$$

Any solution to the dual inequalities (1.5) must have a corresponding value for the dual objective function (1.4) which is a lower bound for the primal objective function. Since this contradicts the assumption of unboundedness, we must conclude that there are no solutions to the dual problem and hence that the dual inequalities are inconsistent. A similar argument will show that, when the dual has an unbounded solution, the primal has no solutions.

In a slightly different form, the basic duality theorem can be restated as follows (Dantzig [152]).

Theorem 1a. *If feasible solutions to both the primal and dual systems exist, there exists an optimal solution to both systems and*

$$\min \ f(\mathbf{X}) = \max \ g(\mathbf{W})$$

Example 1.1: Unsymmetric dual problems. Using Example 2.1, Chap. 4, as the primal problem, we construct the associated dual problem and analyze the optimal simplex tableau to determine the optimal solution to the dual problem.

The primal problem. Minimize

$$x_2 - 3x_3 \quad\quad + 2x_5$$

subject to

$$x_1 + 3x_2 - x_3 \quad\quad + 2x_5 \quad\quad = 7$$
$$- 2x_2 + 4x_3 + x_4 \quad\quad\quad = 12$$
$$- 4x_2 + 3x_3 \quad\quad + 8x_5 + x_6 = 10$$

and

$$x_j \geq 0$$

Here we have $\mathbf{c} = (0, 1, -3, 0, 2, 0)$, $\mathbf{b} = (7, 12, 10)$,

$$\mathbf{A} = \begin{pmatrix} 1 & 3 & -1 & 0 & 2 & 0 \\ 0 & -2 & 4 & 1 & 0 & 0 \\ 0 & -4 & 3 & 0 & 8 & 1 \end{pmatrix}$$

and

$$\mathbf{A}' = \begin{pmatrix} 1 & 0 & 0 \\ 3 & -2 & -4 \\ -1 & 4 & 3 \\ 0 & 1 & 0 \\ 2 & 0 & 8 \\ 0 & 0 & 1 \end{pmatrix}$$

The dual problem.   Maximize

$$7w_1 + 12w_2 + 10w_3$$

subject to

$$
\begin{aligned}
w_1 && &\le 0 \\
3w_1 - 2w_2 - 4w_3 &\le 1 \\
-w_1 + 4w_2 + 3w_3 &\le -3 \\
w_2 && &\le 0 \\
2w_1 + 8w_3 &\le 2 \\
w_3 &\le 0
\end{aligned}
$$

The final basis corresponding to an optimal solution to the primal problem, as computed by the simplex method in Sec. 2 of Chap. 4, is given by the vectors $\mathbf{P}_2$, $\mathbf{P}_3$, $\mathbf{P}_6$, that is,

$$\mathbf{B} = (\mathbf{P}_2\mathbf{P}_3\mathbf{P}_6) = \begin{pmatrix} 3 & -1 & 0 \\ -2 & 4 & 0 \\ -4 & 3 & 1 \end{pmatrix}$$

The corresponding optimal solution $\mathbf{X}^0$ is

$$\mathbf{X}^0 = \mathbf{B}^{-1}\mathbf{b} = (x_2^0, x_3^0, x_6^0) = (4, 5, 11)$$
$$\mathbf{c}^0 = (c_2, c_3, c_6) = (1, -3, 0)$$

and the minimum value of the objective function is

$$\mathbf{c}^0\mathbf{X}^0 = (1, -3, 0)\begin{pmatrix} 4 \\ 5 \\ 11 \end{pmatrix} = -11$$

From the final simplex tableau (see Tableau 4.3) we have

$$\bar{\mathbf{X}} = \begin{pmatrix} \frac{2}{5} & 1 & 0 & \frac{1}{10} & \frac{4}{5} & 0 \\ \frac{1}{5} & 0 & 1 & \frac{3}{10} & \frac{2}{5} & 0 \\ 1 & 0 & 0 & -\frac{1}{2} & 10 & 1 \end{pmatrix}$$

The vector $\mathbf{Z}$ for the optimal solution is given by

$$\mathbf{Z} = \mathbf{c}^0\bar{\mathbf{X}} - \mathbf{c} = (1, -3, 0)\begin{pmatrix} \frac{2}{5} & 1 & 0 & \frac{1}{10} & \frac{4}{5} & 0 \\ \frac{1}{5} & 0 & 1 & \frac{3}{10} & \frac{2}{5} & 0 \\ 1 & 0 & 0 & -\frac{1}{2} & 10 & 1 \end{pmatrix} - (0, 1, -3, 0, 2, 0)$$

or

$$\mathbf{Z} = (-\tfrac{1}{5}, 0, 0, -\tfrac{4}{5}, -\tfrac{12}{5}, 0) \le \mathbf{0}$$

The elements of $\mathbf{Z}$ are just those elements contained in the $(m + 1)$st row of the final tableau, i.e., the $z_j - c_j$ elements.

In the discussion of the elimination method described in Sec. 5 of

Chap. 2, we showed that, if the original matrix of coefficients for the equations contained a unit matrix or was augmented by a unit matrix, then the solution of the equations by the elimination method also yielded the inverse of the basis that corresponds to the solution. The elimination procedure transformed the unit matrix into the desired inverse. The same property is true for a linear-programming problem that has been solved by the simplex procedure. If the original matrix of coefficients, $\mathbf{A}$, contains a unit matrix or is augmented by a unit matrix, then the inverse of the basis at each step is developed in the columns that correspond to the unit matrix. In our example, we see that $\mathbf{A}$ does contain a unit matrix whose columns correspond to $\mathbf{P}_1$, $\mathbf{P}_4$, $\mathbf{P}_6$. Hence, in our final tableau, which is the third step of the example, the columns which correspond to $\mathbf{P}_1$, $\mathbf{P}_4$, $\mathbf{P}_6$ have been transformed to the inverse of the final basis $\mathbf{B}$. These columns form the set of vectors $\mathbf{X}_1$, $\mathbf{X}_4$, $\mathbf{X}_6$, and hence

$$\mathbf{B}^{-1} = (\mathbf{X}_1\mathbf{X}_4\mathbf{X}_6) = \begin{pmatrix} \frac{2}{5} & \frac{1}{10} & 0 \\ \frac{1}{5} & \frac{3}{10} & 0 \\ 1 & -\frac{1}{2} & 1 \end{pmatrix}$$

As we have shown in the proof of the duality theorem, the optimal solution $\mathbf{W}^0$ to the dual is given by

$$\mathbf{W}^0 = \mathbf{c}^0\mathbf{B}^{-1} = (1,\ -3,\ 0)\begin{pmatrix} \frac{2}{5} & \frac{1}{10} & 0 \\ \frac{1}{5} & \frac{3}{10} & 0 \\ 1 & -\frac{1}{2} & 1 \end{pmatrix} = (-\tfrac{1}{5},\ -\tfrac{4}{5},\ 0)$$

We check this solution by substituting it in the dual constraints, and we have

$$\mathbf{W}^0\mathbf{A} \leq \mathbf{c}$$

$$(-\tfrac{1}{5},\ -\tfrac{4}{5},\ 0)\begin{pmatrix} 1 & 3 & -1 & 0 & 2 & 0 \\ 0 & -2 & 4 & 1 & 0 & 0 \\ 0 & 4 & 3 & 0 & 8 & 1 \end{pmatrix} \leq \begin{pmatrix} 0 \\ 1 \\ -3 \\ 0 \\ 2 \\ 0 \end{pmatrix}$$

$$-\tfrac{1}{5} \leq\ 0$$
$$1 \leq\ 1$$
$$-3 \leq -3$$
$$-\tfrac{4}{5} \leq\ 0$$
$$-\tfrac{2}{5} \leq\ 2$$
$$0 \leq\ 0$$

We have as the value of the dual objective function

$$\mathbf{W}^0\mathbf{b} = (-\tfrac{1}{5}, -\tfrac{4}{5}, 0)\begin{pmatrix} 7 \\ 12 \\ 10 \end{pmatrix} = -11$$

The values of the variables for the optimal solution to the dual do not have to be obtained by multiplying $\mathbf{c}^0\mathbf{B}^{-1}$ if the matrix $\mathbf{A}$ contains a unit matrix. We have

$$\mathbf{W}^0 = \mathbf{c}^0\mathbf{B}^{-1} = \mathbf{c}^0(\mathbf{X}_1\mathbf{X}_4\mathbf{X}_6) = (w_1^0, w_2^0, w_3^0)$$

By definition of z_j we have $\mathbf{c}^0\mathbf{X}_1 = z_1$, $\mathbf{c}^0\mathbf{X}_4 = z_4$, and $\mathbf{c}^0\mathbf{X}_6 = z_6$. In the $(m+1)$st row of the final tableau we have for each $\mathbf{X}_j$ the corresponding $z_j - c_j$ element. For $j = 1, 4, 6$ we note that the corresponding $c_j = 0$, and hence the elements in the $(m+1)$st row corresponding to $j = 1, 4, 6$ are equal to the corresponding values of the dual variables, that is, $w_1^0 = z_1$, $w_2^0 = z_4$, and $w_3^0 = z_6$. If a vector which formed the unit matrix had a $c_j \neq 0$, then the value of this c_j would have to be added back to the corresponding $z_j - c_j$ in the final tableau in order to obtain the correct value for the w_i^0. We note that w_i^0 is equal to the z_j which has, for its corresponding unit vector in the initial simplex tableau, the vector whose unit element is in position i. In our example, $w_2^0 = z_4$, since $\mathbf{P}_4$ is a unit vector with its unit element in position $i = 2$.

Example 1.2. Here we consider the revised simplex solution of the furniture-manufacturer problem (Tableau 4.4) as discussed in Chap. 5. The problem is to minimize

$$-45x_1 - 80x_2$$

subject to

$$\begin{aligned}
5x_1 + 20x_2 + x_3 \quad\;\;\; &= 400 \\
10x_1 + 15x_2 \quad\;\; + x_4 &= 450 \\
x_j &\geq \;\; 0
\end{aligned}$$

The dual problem is to maximize

$$400w_1 + 450w_2$$

subject to

$$\begin{aligned}
5w_1 + \;\; 10w_2 &\leq -45 \\
20w_1 + \;\; 15w_2 &\leq -80 \\
w_1 \qquad\quad &\leq \;\;\; 0 \\
w_2 &\leq \;\;\; 0
\end{aligned}$$

Note that having a unit basis in the primal with zero cost coefficients imposes nonpositive conditions on the dual variables.

The optimal basis of the primal problem (Third Step of Tableau 4.4) is

$\mathbf{B} = (\mathbf{P}_2\mathbf{P}_1)$, with $\mathbf{B}^{-1} = \begin{pmatrix} \frac{2}{25} & -\frac{1}{25} \\ -\frac{3}{25} & \frac{4}{25} \end{pmatrix}$. The corresponding pricing vector (optimal solution to the dual) is

$$\boldsymbol{\pi} = (-80, -45)\begin{pmatrix} \frac{2}{25} & -\frac{1}{25} \\ -\frac{3}{25} & \frac{4}{25} \end{pmatrix} = (-1, -4) = \mathbf{W}^0$$

The value of the dual objective function is given by

$$\mathbf{W}^0\mathbf{b} = (-1, -4)\begin{pmatrix} 400 \\ 450 \end{pmatrix} = -2{,}200$$

A variation of the unsymmetric dual problems is given in Sec. 2. For both of these new problems, the *symmetric dual problems*, the constraints are inequalities and the variables are restricted to be nonnegative. We show that the duality theorem holds for the symmetric dual problems by transforming them to their equivalent unsymmetric primal and dual problems.

2. THE SYMMETRIC PRIMAL-DUAL PROBLEMS

The primal problem. Find a vector $\mathbf{X} = (x_1, x_2, \ldots, x_n)$ which *minimizes* the linear functional

$$f(\mathbf{X}) = \mathbf{c}\mathbf{X} \tag{2.1}$$

subject to the conditions

$$\mathbf{A}\mathbf{X} \geq \mathbf{b} \tag{2.2}$$

and

$$\mathbf{X} \geq \mathbf{0} \tag{2.3}$$

The dual problem. Find a vector $\mathbf{W} = (w_1, w_2, \ldots, w_m)$ which *maximizes* the linear functional

$$g(\mathbf{W}) = \mathbf{W}\mathbf{b} \tag{2.4}$$

subject to the conditions

$$\mathbf{W}\mathbf{A} \leq \mathbf{c} \tag{2.5}$$

and

$$\mathbf{W} \geq \mathbf{0} \tag{2.6}$$

We next show that the duality theorem of Sec. 1 also applies to the symmetric dual problems. Let the m nonnegative elements of the column vector $\mathbf{Y} = (y_1, y_2, \ldots, y_m)$ be the slack variables which transform the primal into a set of equations. The equivalent linear-programming problem in terms of partitioned matrices[1] is to *minimize*

$$f(\mathbf{X}, \mathbf{Y}) = (\mathbf{c}|\bar{\mathbf{0}})\left(\frac{\mathbf{X}}{\mathbf{Y}}\right) \tag{2.7}$$

subject to the constraints

$$(\mathbf{A}|-\mathbf{I})\left(\frac{\mathbf{X}}{\mathbf{Y}}\right) = \mathbf{b} \tag{2.8}$$

and

$$\mathbf{X} \geq \mathbf{0} \quad \text{and} \quad \mathbf{Y} \geq \mathbf{0} \tag{2.9}$$

Here $\bar{\mathbf{0}} = (0, \ldots, 0)$ is an m-component row vector and $\mathbf{I}$ is an $m \times m$ identity matrix.

The dual of this transformed primal is to find a vector $\mathbf{W}$ which maximizes

$$g(\mathbf{W}) = \mathbf{W}\mathbf{b} \tag{2.10}$$

subject to the constraints

$$\mathbf{W}(\mathbf{A}|-\mathbf{I}) \leq (\mathbf{c}|\bar{\mathbf{0}}) \tag{2.11}$$

We see that (2.11) decomposes to (2.5) and (2.6), respectively, i.e.,

$$\mathbf{W}\mathbf{A} \leq \mathbf{c}$$

and

$$-\mathbf{W}\mathbf{I} \leq \bar{\mathbf{0}}$$

The last expression is equivalent to $\mathbf{W} \geq \bar{\mathbf{0}}$.

The original problems are now given as unsymmetric dual problems, and the duality theorem holds.

As an example of symmetric primal-dual problems, we consider the following.

The primal problem.   Minimize

$$2x_1 + x_2$$

subject to

$$3x_1 + x_2 \geq 3$$
$$4x_1 + 3x_2 \geq 6$$
$$x_1 + 2x_2 \geq 2$$
$$x_1 \qquad \geq 0$$
$$x_2 \geq 0$$

This is the inequality form of Example 3.1, Chap. 4, prior to the subtraction of

---

[1] See Secs. 2 and 5 of Chap. 2 for discussions on the use of partitioned matrices.

nonnegative slack variables and solution by the use of artificial variables (see Tableau 4.8).

The dual problem. Maximize

$$3w_1 + 6w_2 + 2w_3$$

subject to

$$3w_1 + 4w_2 + w_3 \leq 2$$
$$w_1 + 3w_2 + 2w_3 \leq 1$$
$$w_1 \qquad\qquad \geq 0$$
$$w_2 \qquad \geq 0$$
$$w_3 \geq 0$$

From the Fifth Step of Tableau 4.8, we note that the optimal solution of the primal problem is $\mathbf{X}^0 = (x_1,\ x_2) = (\tfrac{3}{5},\ \tfrac{6}{5})$ with $f(\mathbf{X}^0) = \tfrac{12}{5}$. As the primal simplex tableau starts with a full negative unit matrix (see the Initial Step of Tableau 4.8), the elements in row $(m + 1)$ in the corresponding columns of the Fifth Step (the optimal solution of the primal) are the negative of the optimal values of the dual variables. From the Fifth Step, we note that the optimal basis is $\mathbf{B} = (\mathbf{P}_1\mathbf{P}_5\mathbf{P}_2)$, and by taking the negative of the last three columns,

$$\mathbf{B}^{-1} = \begin{pmatrix} \tfrac{3}{5} & -\tfrac{1}{5} & 0 \\ -1 & 1 & -1 \\ -\tfrac{4}{5} & \tfrac{3}{5} & 0 \end{pmatrix}$$

with $\mathbf{c}^0 = (2, 0, 1)$. Thus,

$$\mathbf{W}^0 = \mathbf{c}^0\mathbf{B}^{-1} = (2, 0, 1)\begin{pmatrix} \tfrac{3}{5} & -\tfrac{1}{5} & 0 \\ -1 & 1 & -1 \\ -\tfrac{4}{5} & \tfrac{3}{5} & 0 \end{pmatrix}$$

$$\mathbf{W}^0 = (\tfrac{2}{5}, \tfrac{1}{5}, 0)$$

Of course, $g(\mathbf{W}^0) = 3(\tfrac{2}{5}) + 6(\tfrac{1}{5}) + 2(0) = \tfrac{12}{5}$.

For the primal, the slack variables $(x_3, x_4, x_5) = (0, 0, 1)$; while for the dual, the slack variables $(w_4, w_5) = (0, 0)$.

We next prove, for the symmetric dual problems, the following theorem which can be interpreted as establishing necessary and sufficient conditions for the optimality of feasible solutions to the primal and dual problems.

Theorem 2: Complementary slackness theorem (Dantzig and Orden [190] and Dantzig [152]). *For optimal feasible solutions of the primal and dual systems [(2.1) to (2.3) and (2.4) to (2.6), respectively] whenever inequality occurs in the kth relation of either system (the corresponding slack variable is* *positive*), *then the kth variable of its dual vanishes; if the kth variable is positive in either system, the kth relation of its dual is equality* (*the corresponding slack variable is zero*).

Proof: We rewrite the inequalities (2.2) of the primal system as equalities by subtracting a nonnegative slack variable x_{n+i} from the ith inequality to obtain

$$a_{11}x_1 + \cdots + a_{1n}x_n - x_{n+1} \qquad\qquad = b_1$$

$$\vdots \qquad\qquad \vdots \qquad\qquad \vdots \qquad\qquad \vdots \qquad\qquad (2.2')$$

$$a_{m1}x_1 + \cdots + a_{mn}x_n \qquad\qquad - x_{n+m} = b_m$$

Similarly, for the dual inequalities (2.5) we add a nonnegative slack variable w_{m+j} to the jth inequality to obtain

$$a_{11}w_1 + \cdots + a_{m1}w_m + w_{m+1} \qquad\qquad = c_1$$

$$\vdots \qquad\qquad \vdots \qquad\qquad \vdots \qquad\qquad (2.5')$$

$$a_{1n}w_1 + \cdots + a_{mn}w_m \qquad\qquad w_{m+n} = c_n$$

In explicit form, the primal objective function is

$$c_1x_1 + \cdots + c_nx_n = f(\mathbf{X}) \qquad\qquad (2.1')$$

and the dual objective function is

$$b_1w_1 + \cdots + b_mw_m = g(\mathbf{W}) \qquad\qquad (2.4')$$

We next multiply the ith equation of (2.2') by the corresponding dual variable w_i, $i = 1, 2, \ldots, m$, add the resulting set of equations, and subtract this sum from (2.1') to obtain

$$\left(c_1 - \sum_{i=1}^{m} a_{i1}w_i\right)x_1 + \left(c_2 - \sum_{i=1}^{m} a_{i2}w_i\right)x_2 + \cdots + \left(c_n - \sum_{i=1}^{m} a_{in}w_i\right)x_n$$

$$+ w_1x_{n+1} + w_2x_{n+2} + \cdots + w_mx_{n+m} = f(\mathbf{X}) - \sum_{i=1}^{m} w_ib_i$$

Noting that $w_{m+j} = c_j - \sum_{i=1}^{m} a_{ij}w_i$ and $g(\mathbf{W}) = \sum_{i=1}^{m} w_ib_i$, we have

$$w_{m+1}x_1 + w_{m+2}x_2 + \cdots + w_{m+n}x_n + w_1x_{n+1} + w_2x_{n+2}$$

$$+ \cdots + w_mx_{n+m} = f(\mathbf{X}) - g(\mathbf{W}) \qquad (2.12)$$

From the duality theorem, we have for an optimal solution $\mathbf{X}^0 = (x_1^0, x_2^0, \ldots, x_n^0)$ to the primal problem and $\mathbf{W}^0 = (w_1^0, w_2^0, \ldots, w_m^0)$ an optimal solution to the dual problem that $f(\mathbf{X}^0) - g(\mathbf{W}^0) = 0$. Thus, for these optimal solutions and corresponding slack variables $x_{n+i}^0 \geq 0$ and $w_{m+j}^0 \geq 0$, (2.12) becomes

$$w^0_{m+1}x^0_1 + w^0_{m+2}x^0_2 + \cdots + w^0_{m+n}x^0_n + w^0_1 x^0_{n+1} + w^0_2 x^0_{n+2}$$

$$+ \cdots + w^0_m x^0_{n+m} = 0 \qquad (2.13)$$

We note that the terms $w^0_{m+j}x^0_j$ are the product of the jth slack variable of the dual and the jth variable of the original primal; while the terms $w^0_i x^0_{n+i}$ are the product of the ith variable of the original dual and the ith slack variable of the primal. Since all variables are restricted to be non-negative, all the product terms of (2.13) are nonnegative and, as the sum of these terms must be equal to zero, they individually must be equal to zero. Thus $w^0_{m+j}x^0_j = 0$ for all j and $w^0_i x^0_{n+i} = 0$ for all i. If $w^0_{m+k} > 0$, we must have $x^0_k = 0$; if $x^0_{n+k} > 0$, then $w^0_k = 0$, which establishes the first part of the theorem. If $x^0_k > 0$, then $w^0_{m+k} = 0$, if $w^0_k > 0$, then $x^0_{n+k} = 0$, which completes the proof.

Noting the definitions of the slack variables in (2.2′) and (2.5′), the individual terms of the complementary slackness condition (2.13) must satisfy the following conditions for optimal $\mathbf{X}^0$ and $\mathbf{W}^0$:

$$\left(c_j - \sum_{i=1}^{m} a_{ij}w^0_i \right)x^0_j = 0 \qquad j = 1, \ldots, n$$

$$w^0_i \left(\sum_{j=1}^{n} a_{ij}x^0_j - b_i \right) = 0 \qquad i = 1, \ldots, m$$

We can conclude that if $x^0_j > 0$, then for the corresponding dual constraint we must have $\sum_{i=1}^{m} a_{ij}w^0_i = c_j$; if $\sum_{i=1}^{m} a_{ij}w^0_i < c_j$, then $x^0_j = 0$; if $w^0_i > 0$, then for the corresponding primal constraint we must have $\sum_{j=1}^{n} a_{ij}x^0_j = b_i$; and if $\sum_{j=1}^{n} a_{ij}x^0_j > b_i$, then $w^0_i = 0$.

We illustrate the complementary slackness theorem for the previous example (from Example 3.1, Chap. 4). The equation form of the primal, equivalent to (2.1′) and (2.2′), and the equation form of the dual, equivalent to (2.4′) and (2.5′), are, respectively, the following problems:

The primal problem. Minimize

$$2x_1 + x_2$$

subject to

$$3x_1 + x_2 - x_3 \qquad\quad = 3$$
$$4x_1 + 3x_2 \quad - x_4 \qquad = 6$$
$$x_1 + 2x_2 \qquad\quad - x_5 = 2$$
$$x_j \geq 0$$

The variables x_3, x_4, and x_5 are slack variables.

The dual problem.   Maximize

$$3w_1 + 6w_2 + 2w_3$$

subject to

$$3w_1 + 4w_2 + w_3 + w_4 \quad\quad = 2$$
$$w_1 + 3w_2 + 2w_3 \quad\quad + w_5 = 1$$
$$w_i \geq 0$$

The variables w_4 and w_5 are slack variables.

For these problems Eq. (2.12) becomes

$$w_4 x_1 + w_5 x_2 + w_1 x_3 + w_2 x_4 + w_3 x_5 = f(\mathbf{X}) - g(\mathbf{W})$$

For the optimal solutions $\mathbf{X}^0 = (\tfrac{3}{5}, \tfrac{6}{5}, 0, 0, 1)$ and $\mathbf{W}^0 = (\tfrac{2}{5}, \tfrac{1}{5}, 0, 0, 0)$, this becomes Eq. (2.13), i.e.,

$$0(\tfrac{3}{5}) + 0(\tfrac{6}{5}) + \tfrac{2}{5}(0) + \tfrac{1}{5}(0) + 0(1) = \tfrac{12}{5} - \tfrac{12}{5} = 0$$

For the unsymmetric primal-dual problems of (1.1) to (1.3) and (1.4) to (1.5) we can develop a relationship similar to (2.12). We rewrite these problems in explicit form as: Minimize

$$c_1 x_1 + \cdots + c_n x_n = f(\mathbf{X}) \tag{1.1'}$$

subject to

$$a_{11} x_1 + \cdots + a_{1n} x_n = b_1$$
$$\vdots \quad\quad\quad\quad \vdots \tag{1.2'}$$
$$a_{m1} x_1 + \cdots + a_{mn} x_n = b_m$$
$$x_j \geq 0 \tag{1.3'}$$

and maximize

$$b_1 w_1 + \cdots + b_m w_m = g(\mathbf{W}) \tag{1.4'}$$

subject to

$$a_{11} w_1 + \cdots + a_{m1} w_m \leq c_1$$
$$\vdots \quad\quad\quad\quad \vdots \tag{1.5'}$$
$$a_{1n} w_1 + \cdots + a_{mn} w_m \leq c_n$$

with the w_i unrestricted.

We multiply the ith equation of (1.2') by the corresponding dual variable w_i, $i = 1, 2, \ldots, m$, add the resulting set of equations, and subtract this sum from (1.1') to obtain

$$\left(c_1 - \sum_{i=1}^{m} a_{i1} w_i\right) x_1 + \left(c_2 - \sum_{i=1}^{m} a_{i2} w_i\right) x_2$$
$$+ \cdots + \left(c_n - \sum_{i=1}^{m} a_{in} w_i\right) x_n = f(\mathbf{X}) - g(\mathbf{W}) \tag{2.14}$$

From the proof of the duality theorem, we have the right-hand side of (2.14), $f(\mathbf{X}) - g(\mathbf{W}) \geq 0$, with the relationship equal to zero only if $f(\mathbf{X}^0) = g(\mathbf{W}^0)$; i.e., $\mathbf{X}^0$ and $\mathbf{W}^0$ are finite optimal feasible solutions to their respective problems. We also note that for feasible $\mathbf{W}$ each term $\left(c_j - \sum_{i=1}^{m} a_{ij} w_i \right)$ is nonnegative and that the left-hand side of (2.14) is the sum of products of nonnegative numbers. In particular, for optimal $\mathbf{X}^0$ and $\mathbf{W}^0$, the left-hand side of (2.14) sums to zero, and for each j, either the term $\left(c_j - \sum_{i=1}^{m} a_{ij} w_i^0 \right)$ is zero or the corresponding x_j^0 is zero or both. For the optimal $\mathbf{W}^0 = (w_1^0, w_2^0, \ldots, w_m^0) = \mathbf{c}^0 \mathbf{B}^{-1}$, as defined by (1.10), the terms $\left(c_j - \sum_{i=1}^{m} a_{ij} w_i^0 \right) = c_j - z_j \geq 0$. Thus, the above discussion restates the validity of the optimality conditions of the simplex algorithm; i.e., for a primal minimization problem an optimal solution $\mathbf{X}^0 = (x_1^0, x_2^0, \ldots, x_n^0)$ requires those $x_j^0 > 0$ to have the corresponding $z_j - c_j = 0$ and those $x_j = 0$ to have the corresponding $z_j - c_j \leq 0$.

We will use the above relationships in developing optimality arguments for the transportation-problem algorithm in Chap. 10. Similar arguments have been used for constructing computational approaches for other problems with special structures, e.g., bounded-variable problems (Chap. 9) and network problems (Chap. 11).

We note that each member of a pair of primal and dual problems can have finite optimal solutions; that one problem can have an unbounded optimal feasible solution while the other has no feasible solutions; and that both problems can have no feasible solutions.

As noted in Goldman and Tucker [318], the symmetric dual problems can be conveniently represented by the following tableau:

(≥ 0)	x_1	$\cdots$	x_j	$\cdots$	x_n	$\geq$
w_1	a_{11}	$\cdots$	a_{1j}	$\cdots$	a_{1n}	b_1
$\cdot$	$\cdot$	$\cdots$	$\cdot$	$\cdots$	$\cdot$	$\cdot$
w_i	a_{i1}	$\cdots$	a_{ij}	$\cdots$	a_{in}	b_i
$\cdot$	$\cdot$	$\cdots$	$\cdot$	$\cdots$	$\cdot$	$\cdot$
w_m	a_{m1}	$\cdots$	a_{mj}	$\cdots$	a_{mn}	b_m
$\leq$	c_1	$\cdots$	c_j	$\cdots$	c_n	min / max

By taking inner products of the rows of $\mathbf{A}$ with the row of x's, we obtain the constraints of the primal; the inner products of the columns of $\mathbf{A}$ with the column of w's yield the dual constraints.

Many linear-programming problems are originally stated in the form of either the symmetric primal or the symmetric dual. In either case, the problem is usually rewritten in equation form by adding or subtracting the necessary nonnegative slack variables. For the symmetric primal we subtract a set of slack vectors and solve the corresponding system, while for the symmetric dual we add a set of slack vectors. These slack vectors are given associated cost coefficients that are equal to zero. The resulting tableau for the primal contains a negative unit matrix, while the dual tableau contains a positive unit matrix. Hence the final simplex tableau of either will contain the optimal solution of the other. If the primal was solved to obtain the dual solution, then, since we start out with a negative unit matrix in the primal, the corresponding z_j values in the final tableau have to have their signs reversed.

As the size of a linear-programming problem that can be solved by hand or computer is usually limited by the number of rows involved, a problem that is too large when stated in terms of the primal may be of the right dimensions when written as the corresponding dual problem, and vice versa. In either situation the optimal solution to the original problem is contained in the final tableau. In mathematical programming systems, as they use some form of the revised simplex method, the optimal dual variables are calculated using the inverse of the optimal primal basis and the associated cost coefficient vector, i.e., by Eq. (1.10).

By means of the symmetric dual relationships the general linear-programming problem can be represented as a problem that is concerned with the solution of a set of inequalities in nonnegative variables. A vector $\mathbf{X}$ which satisfies the constraints

$$\mathbf{AX} \geq \mathbf{b}$$

$$\mathbf{WA} \leq \mathbf{c}$$

$$\mathbf{Wb} \geq \mathbf{cX} \tag{2.15}$$

$$\mathbf{X} \geq \mathbf{0}$$

$$\mathbf{W} \geq \mathbf{0}$$

will be an optimal solution to the problem of minimizing

$$\mathbf{cX}$$

subject to

$$\mathbf{AX} \geq \mathbf{b} \tag{2.16}$$

$$\mathbf{X} \geq \mathbf{0}$$

See Exercise 16 for other primal-dual problem relationships.

3. ECONOMIC INTERPRETATION OF THE PRIMAL-DUAL PROBLEMS

We next investigate the economic interpretation of the activity-analysis problem of Sec. 2, Chap. 1, and its dual.[1] The primal problem can be written

[1] See Baumol [48], Dorfman [213], Dorfman, Samuelson, and Solow [214], Gale [277], Harrison [355], and Tucker [657] for further discussion. For an interpretation of the primal and dual problems (and linear programming in general) in terms of classical Lagrange multipliers, see Tucker [658].

so as the maximize

$$\mathbf{cX}$$

subject to

$$\mathbf{AX} \leq \mathbf{b}$$
$$\mathbf{X} \geq \mathbf{0}$$

where the a_{ij} represent the number of units of resource i required to produce one unit of commodity j, the b_i represent the maximum number of units of resource i available, and the c_j represent the value (profit) per unit of commodity j produced. The corresponding dual problem is to minimize

$$\mathbf{Wb}$$

subject to

$$\mathbf{WA} \geq \mathbf{c}$$
$$\mathbf{W} \geq \mathbf{0}$$

Whereas the physical interpretation of the primal is straightforward, the corresponding interpretation of the dual is not so evident. Questions arise as to the meaning of the dual objective function and inequalities. These can best be answered by interpreting the primal and dual problems in terms of physical units in order to determine the meaning of the dual variables (Harrison [355]).

The primal problem is to maximize

$$\sum_{j=1}^{n} \left(\frac{\text{value}}{\text{output } j} \right) (\text{output } j) = (\text{value})$$

subject to

$$\sum_{j=1}^{n} \left(\frac{\text{input } i}{\text{output } j} \right) (\text{output } j) \leq (\text{input } i) \qquad i = 1, 2, \ldots, m$$
$$(\text{output } j) \geq 0 \qquad\qquad\qquad j = 1, 2, \ldots, n$$

and the dual is to minimize

$$\sum_{i=1}^{m} (\text{input } i) w_i = (?)$$

subject to

$$\sum_{i=1}^{m} \left(\frac{\text{input } i}{\text{output } j} \right) w_i \geq \left(\frac{\text{value}}{\text{output } j} \right) \qquad j = 1, 2, \ldots, n$$
$$w_i \geq 0 \qquad\qquad\qquad i = 1, 2, \ldots, m$$

We see that the dual constraints will be consistent if the w_i are in units of value per unit of input i. The dual problem would then be to minimize

$$\sum_{i=1}^{m} (\text{input } i)\left(\frac{\text{value}}{\text{input } i}\right) = (\text{value})$$

subject to

$$\sum_{i=1}^{m} \left(\frac{\text{input } i}{\text{output } j}\right)\left(\frac{\text{value}}{\text{input } i}\right) \geq \left(\frac{\text{value}}{\text{output } j}\right) \qquad j = 1, 2, \ldots, n$$

$$\left(\frac{\text{value}}{\text{input } i}\right) \geq 0 \qquad\qquad i = 1, 2, \ldots, m$$

Verbal descriptions of the primal and dual problems can then be stated as follows:

The primal problem.   With a given unit of value of each output (c_j) and a given upper limit for the availability of each input (b_i), how much of each output (x_j) should be produced in order to maximize the value of the total output?

The dual problem.   With a given availability of each input (b_i) and a given lower limit of unit value for each output (c_j), what unit values should be assigned to each input (w_i) in order to minimize the value of the total input?   The variables w_i are referred to by various names, e.g., *shadow prices*, *accounting* or *marginal values.*

An econometric interpretation of the furniture-manufacturer's activity-analysis problem (Example 1.2) is the following.   Here the primal problem is to maximize

$$45x_1 + 80x_2$$

subject to

$$5x_1 + 20x_2 \leq 400 \qquad \text{(labor-hours)}$$
$$10x_1 + 15x_2 \leq 450 \qquad \text{(board-feet of lumber)} \qquad\qquad (3.1)$$
$$x_j \geq \quad 0$$

and the dual problem is to minimize

$$400w_1 + 450w_2$$

subject to

$$5w_1 + 10w_2 \geq 45$$
$$20w_1 + 15w_2 \geq 80 \qquad\qquad (3.2)$$
$$w_i \geq \quad 0$$

Considering each w_i as the value of a unit of the corresponding resource, constraints (3.2) state that for each product the total value of the resources used to make one unit of a product must be at least as great as the unit's profit. For example, the first product (chairs) uses 5 units of labor and 10 units of lumber and makes \$45 profit per unit of production x_1; the constraint for the valuation of a chair's resources is $5w_1 + 10w_2 \geq 45$. At optimality of the primal, a product being produced has, by complementary slackness, equality holding for the corresponding value constraint, e.g., for $x_1 > 0$ we have $5w_1 + 10w_2 = 45$ ($z_1 - c_1 = 0$ for the basic variable x_1). For a nonoptimal basis, a product would be a candidate to enter the basis if its value constraint was not satisfied. Here the current valuation of the resources is less than the profit and, thus, the manufacturer would have an incentive to make the product. For example, if x_1 was nonbasic and $z_1 - c_1 < 0$ (which is the nonoptimal condition for a maximizing problem), we would have values of w_1 and w_2 such that $5w_1 + 10w_2 < 45$. Note that when we have a nonoptimal solution to the primal, the pricing vector π [see Eq. (1.4) of Chap. 5] is taken as a "trial" solution vector $\mathbf{W}$ to the dual problem. At optimality, the pricing vector is equal to the optimal dual solution $\mathbf{W}^0$ and the dual constraints are satisfied. A nonbasic variable would not be a candidate to enter the basis if the valuation of its resources is greater than the value of making a unit of the product. If x_1 were nonbasic in an optimal solution, then its corresponding resource valuation constraint would have $5w_1 + 10w_2 \geq 45$, which is equivalent to $z_1 - c_1 \geq 0$. If x_1 could enter the basis to form a multiple optimal solution to the primal, we would have $5w_1 + 10w_2 = 45$.

The optimal primal and dual solutions yield, in econometric terms, an *equilibrium solution* to the problem faced by the manufacturer competing against other firms in the "market place." In a sense, the manufacturer and the market place set internal prices (*accounting values*) on the manufacturer's resources, so that at equilibrium, the maximum profit to be made with the resources is equal to the minimum value of the resources (primal-dual theorem). If a resource owned by the manufacturer is not fully utilized in the optimal solution, then the market place, working against the manufacturer, sets a value of zero on the excess resource (complementary slackness). An extra unit of an excess resource is of no value to the manufacturer. On the other hand, if a resource is completely utilized in the optimal solution, i.e., is a scarce resource for the manufacturer, then having an extra unit should be of positive value to the manufacturer. The value of an extra unit of a resource is its *marginal value.* Using the results of Exercise 11, Chap. 8, we can show that the optimal dual variables represent the marginal values (profit) to the manufacturer of additional units of the corresponding resources, given that the units can be utilized by the production activities. For the furniture-manufacturer's problem (3.1) and its dual (3.2), both resources are scarce. The corresponding dual variables are $w_1 = 1$ (the marginal value of an extra labor-hour) and $w_2 = 5$ (the marginal value of an extra unit of lumber). The marginal values serve as upper bounds on the prices the manufacturer would want to pay for extra units of the scarce resources.

When formulating any linear-programming problem, the analyst should investigate the meaning and interpretation of the associated dual problem. In fact, for some applications, (e.g., the expenditure of funds to improve current facilities, Exercise 29), we might really be interested in the marginal values.   It is sometimes difficult to form an exact definition and meaning for a dual variable (recall that a dual variable is associated with each primal constraint).   As marginal values, the optimal dual variables may need to be adjusted for variable costs (see Murtagh [521], Perry and Crellin [546], and Ronen [572]).   If the dual problem has multiple optimal solutions (Exercise 19), care must be taken in interpreting the dual solutions as marginal values (Exercise 30 of this chapter, and Exercise 45, Chap. 8) (Shapiro [608], Williams [702], and Bradley, Hax, and Magnanti [92]; also see Dantzig [176]).   The reader should investigate the dual interpretation of the diet problem (Exercise 11) and of the transportation problem (Exercise 19 of Chap. 9).   See Exercises 8 and 9 of this chapter for further economic ramifications of the primal-dual problems.

In line with the above, let us assume that we have determined by means of the simplex method an optimal solution to the activity-analysis problem.   As we are maximizing the objective function, all vectors both in and out of the basis will have an associated $z_j - c_j \geq 0$.   Because of some external conditions, e.g., government regulations or marketing conditions, the problem might stipulate that the kth activity must be in the final solution at a positive level.   However, the optimal basis does not include vector $\mathbf{P}_k$.   We must then alter our maximum solution by forcing vector $\mathbf{P}_k$ into the basis and removing some other activity.   Assuming that $\mathbf{P}_k$ does not generate a multiple solution, i.e., that we have $z_k - c_k > 0$, the introduction of $\mathbf{P}_k$ reduces the maximum value of the objective function by the amount $(z_k - c_k)\theta_0$.   This value is the loss in profit forced upon the manufacturer by the external condition.   The value of $z_k - c_k$ is the net cost of introducing one unit of the kth activity, where the z_k is termed the indirect cost and the c_k the direct cost.   The relationship between the net costs and accounting prices can be seen by analyzing the simplex tableau corresponding to the optimal solution of the activity-analysis problem.   The set of indirect costs, i.e., the $z_j - c_j$ elements that correspond to the slack vectors, represents a set of accounting prices that solve the dual problem.   If all the slack variables are in the final basis, then, in terms of the primal, no outputs are generated, no inputs are used, and the maximum profit is zero.   For the dual this means that the minimum value of the total input is also zero.   This is evidenced by all the indirect costs corresponding to the slack vectors being equal to zero.   If the optimal solution of the primal does not involve any of the slack variables at a positive level, then the primal problem yields a positive profit while utilizing all the available inputs.   For this case none of the indirect costs associated with the slack vectors is equal to zero, and the minimum value of the total input is positive and equal to the maximum profit.

Other econometric discussions of the activity-analysis problem and the general linear-programming model include the allocation of resources in a competitive economy and equilibrium theory (Gale [277]), and managerial and entrepreneurial implications of the simplex method (Dantzig [152]; also see Koopmans [437, 438], Dorfman, Samuelson, and Solow [214], and Bradley, Hax, and Magnanti [92]).

EXERCISES

1. Write out the corresponding primal problem to the following symmetric dual problem and solve both problems by the simplex procedure: Maximize

$$w_1 + w_2 + w_3$$

subject to

$$2w_1 + w_2 + 2w_3 \leq 2$$
$$4w_1 + 2w_2 + w_3 \leq 2$$
$$w_1 \geq 0$$
$$ w_2 \geq 0$$
$$ w_3 \geq 0$$

Answer: $w_1 = 0$, $w_2 = \frac{2}{3}$, $w_3 = \frac{2}{3}$; $g(\mathbf{W}) = \frac{4}{3}$; $x_1 = \frac{1}{3}$, $x_2 = \frac{1}{3}$, $f(\mathbf{X}) = \frac{4}{3}$.

2. Solve the following problem by the simplex method: Minimize

$$2x_1 - 3x_2$$

subject to

$$2x_1 - x_2 - x_3 \geq 3$$
$$x_1 - x_2 + x_3 \geq 2$$
$$x_1 \geq 0$$
$$ x_2 \geq 0$$
$$ x_3 \geq 0$$

Answer: Solution unbounded.

3. Write out the dual problem to the problem given in Exercise 2 and solve it by graphing the constraints.

4. Graph the constraints of the following problem and its corresponding dual problem: Minimize

$$x_1 - x_2$$

subject to

$$2x_1 + x_2 \geq 2$$
$$-x_1 - x_2 \geq 1$$
$$x_1 \geq 0$$
$$ x_2 \geq 0$$

5. Write out the dual to the transportation problem of Sec. 2, Chap. 1.

6. Construct a two-variable, two-inequality primal problem which has no feasible solutions whose corresponding dual problem also has no feasible solutions.

7. *Farkas' Lemma* (Charnes and Cooper [112], Dantzig [152], and Hadley [344]). Prove the following: A vector $\mathbf{b}$ will satisfy $\mathbf{Wb} \geq 0$ for all $\mathbf{W}$ satisfying $\mathbf{WA} \geq 0$ if and only if there exists an $\mathbf{X} \geq 0$ such that $\mathbf{AX} = \mathbf{b}$. HINT: To prove necessity, define appropriate primal and dual problems and apply the duality theorem. (Also see Balinski and Tucker [33] for discussion on related linear inequality theorems which can be proved by application of duality theory.)

8. Given the optimal solution to an activity analysis problem, how would you determine whether or not it would be profitable to make a new commodity defined by a vector P_k, assuming the same level of available resources, i.e., how would you compute the "opportunity cost" for a new product? For Eq. (3.1), determine if the manufacturer can increase the company's profits by making a new product (desks) which requires 15 board feet of mahogany and 25 labor-hours and has a profit of \$110.

9. For the activity analysis problem, we say that the ith resource is a bottleneck if the corresponding ith constraint is an equality for the optimal solution, i.e., the resource is used to capacity and thus the slack variable x_{n+i} is zero. The $(z_{n+i} - c_{n+i})$ for such a slack variable will be positive (using the maximizing criteria and assuming no multiple solutions) and represents the optimum value of the corresponding dual variable w_i. Show that this value of w_i represents the extra (marginal) profit which could be added to the total profits if resource i is increased by one unit. (HINT: See Chap. 8, Exercise 11, and let $\theta = 1$.) Note that the extra profit may not be obtained fully as the other constraints may restrict our being able to use the complete additional unit.

10. For Eqs. (3.1) and (3.2), graph (3.2) and solve for the optimal dual solution. Solve the second part of Exercise 8 by this graphical method. Discuss what happens to both problems as the resource vector **b** or the cost vector **c** is varied (see Fig. 2.12 for the graph of the primal problem).

11. State the dual to the standard diet problem given in Chap. 1 and develop an economic interpretation of the dual variables. HINT: Consider a manufacturer who wants to substitute nutrient pills for the foods and needs to establish prices for the nutrients in the pills (Gale [277]).

12. Prove that if the problem Min $\mathbf{cX}$, $\mathbf{AX} = \mathbf{b}$, $\mathbf{X} \geq \mathbf{0}$ has a finite optimal solution, then the optimal solution to Min $\mathbf{cX}$, $\mathbf{AX} = \bar{\mathbf{b}}$, $\mathbf{X} \geq \mathbf{0}$ cannot be unbounded, where $\bar{\mathbf{b}}$ is any right-hand-side vector for which feasible solutions exist. From a geometric point of view, what happens to the dual solution as **b** is varied? HINT: Set up dual problem.

13. Write out the dual to the following problems:

a. Minimize

$$3x_1 - 2x_2 + x_3$$

subject to

$$2x_1 - 3x_2 + x_3 = 1$$
$$2x_1 + 3x_2 \quad\;\; \geq 8$$
$$x_j \geq 0$$

b. Exercise 1*e*, Chap. 4.
c. Exercise 1*g*, Chap. 4.
d. Exercise 1*i*, Chap. 4.
e. Maximize

$$x_1 - \;\; x_2 + 3x_3 + 2x_4$$

subject to

$$x_1 + \;\; x_2 \qquad\qquad \geq -1$$
$$x_1 - 3x_2 - \;\; x_3 \qquad \leq \;\; 7$$
$$x_1 \qquad + \;\; x_3 - 3x_4 = -2$$
$$x_1 \qquad\qquad\qquad \geq \;\; 0$$
$$x_4 \geq \;\; 0$$

14. Discuss the relationship between multiple optimal solutions in the primal and the optimal solution of the dual problem. HINT: Does $\mathbf{W}^0 = \mathbf{c}^0\mathbf{B}^{-1}$ correspond to a basic solution to the dual problem? Set up the dual problem for the final primal solution tableau.

15. Develop the condensed simplex tableau approach (Exercise 19, Chap. 4), so that both the optimal primal and dual solutions can be obtained from the final tableau, Balinski and Tucker [33].

16. Demonstrate the following:

a. Show that if a variable x_j in the primal problem is unrestricted as to sign, then the corresponding dual constraint is an equation.

b. Show that if the ith constraint of the primal problem is an equation, then the corresponding dual variable is unrestricted as to sign.

HINT: In both instances, consider the symmetric primal-dual problems. Also see Exercises 13*a* and 13*e* above.

17. Demonstrate the equivalence between Theorems 1 and 1*a*.

18. For the activity-analysis problem, interpret the dual problem from the point of view of an entrepreneur who wishes to buy the firm and its resources (Dantzig [152]).

19. Under what conditions will the optimal dual problem have multiple optimal solutions? (Dantzig [152].)

20. Construct two-variable, two-inequality examples that illustrate the following situations: (*a*) The primal and dual problems both have finite optimal solutions, (*b*) the primal (dual) problem is unbounded and the dual (primal) is infeasible, (*c*) both the primal and dual problems are infeasible.

21. Write out the dual to Example 2.3 of Chap. 4 and determine the optimal dual solution.

22. Show how any linear-programming problem can be transformed into either the symmetric or unsymmetric primal form.

23. Discuss why the solution to Eq. (2.15) yields an optimal solution to Eq. (2.16). Note that Eq. (2.15) is not stated as an optimization problem.

24. Consider the unsymmetric primal and dual problems and let **B** be a feasible basis to the primal. What would be the corresponding (complementary) dual basis? (Bazaraa and Jarvis [50].)

25. Write out the dual determine the optimal solution of Example 2.4, Chap. 4.

26. Write out the dual to the primal diet problem (2.6) of Chap. 1 and solve the dual by the simplex algorithm. Determine the solution to the primal *from the simplex tableau.*

27. Let $\bar{\mathbf{X}}$ be the solution set to $\mathbf{AX} \le \mathbf{b}$, $\mathbf{X} \ge 0$ and $\bar{\mathbf{W}}$ be the solution set to $\mathbf{WA} \ge \mathbf{c}$, $\mathbf{W} \ge 0$. Prove that if $\bar{\mathbf{X}}$ is nonempty and bounded, then $\bar{\mathbf{W}}$ is nonempty and unbounded. Also, if $\bar{\mathbf{W}}$ is nonempty and bounded, then $\bar{\mathbf{X}}$ is nonempty and unbounded (Charnes, Cooper and Thompson [120]). HINT: Consider the primal problem that maximizes $\sum_{j=1}^{n} x_j$ and the corresponding dual. (Clark [129] proves that if a linear-programming problem has a solution, then either the primal constraint set or the dual constraint set is unbounded [symmetric primal-dual problems].)

28. How would you explain the following statement to an entrepreneur unfamiliar with linear programming? Reduced costs are opportunity costs (Exercise 8) and optimal dual variables are marginal values.

29. A university has \$100,000 to spend improving its six student/faculty parking lots. Describe how you would use linear programming marginal analysis to determine which lots should be improved.

30. Given the following primal linear-programming problem, maximize

$$x_0 = 3x_1 + 2x_2$$

subject to

$$\begin{aligned}
x_1 + x_2 &\le 3 \\
2x_1 + x_2 &\le 4 \qquad (x_j \ge 0) \\
4x_1 + 3x_2 &\le 10
\end{aligned}$$

Show that the primal problem has the unique optimal solution of $x_0 = 7$, $x_1 = 1$, $x_2 = 2$. Using the optimal simplex tableau for the primal, show that the dual has the two optimal solutions $y_0 = 7$, $y_1 = 1$, $y_2 = 1$, $y_3 = 0$ and $y_0 = 7$, $y_1 = 0$, $y_2 = \frac{1}{2}$, $y_3 = \frac{1}{2}$. Use the graphical solution to the primal to explain why we have "ambiguity" in the marginal values (Williams [702]). (See Exercise 45, Chap. 8, for further discussion.)

CHAPTER 7

Degeneracy and Anticycling Procedures

A basic feasible solution to a linear-programming problem is degenerate if at least one basic variable equals zero. That is, if the vectors $(\mathbf{P}_1\mathbf{P}_2 \cdots \mathbf{P}_m)$ form an admissible basis, then, in the degenerate case, the corresponding non-negative linear combination

$$x_{10}\mathbf{P}_1 + x_{20}\mathbf{P}_2 + \cdots + x_{m0}\mathbf{P}_m = \mathbf{P}_0$$

has at least one $x_{i0} = 0$. In proving the validity of the simplex algorithm (Chap. 4), we assumed that all basic feasible solutions were non-degenerate. This assumption was necessary to demonstrate that, for each successive admissible basis, the associated value of the objective function is smaller than those that precede it. Hence, we will reach the minimum solution in a finite number of iterations, since there are only a finite number of possible bases. This proof breaks down if we admit the existence of degenerate basic feasible solutions. This is, of course, the more realistic situation. For a degenerate solution, we have the possibility of computing a $\theta_0 = x_{l0}/x_{lk}$ [see Eqs. (1.6) and (1.7) of Chap. 4] for which $\theta_0 = 0$. This choice of vector $\mathbf{P}_l$ to be removed and $\mathbf{P}_k$ to be introduced into the basis will give a new feasible solution whose value of the objective function is equal to the preceding one.[1] It is then theoretically possible to select a sequence of admissible bases

---

[1] Here we should distinguish between a multiple solution and a degenerate solution. A multiple solution has the same value of the objective function as the preceding solution because $z_k - c_k = 0$ with $\mathbf{P}_k$ not in the basis and $\theta_0 > 0$, while a degenerate solution will yield a new solution with the same value of the objective function because $\theta_0 = 0$.

that cycles, i.e., a sequence of bases that is repeatedly selected without ever satisfying the optimality criteria and, hence, never reaches a minimum solution. The possibility of cycling is crucial only if the current basic feasible solution has more than one $x_{i0} = 0$ (see Exercise 2, Chap. 4). With at least two $x_{i0} = 0$, it could be possible to have a tie with $\theta_0 = 0$ in the selection of the vector to be eliminated from the basis. It is also possible to have a tie in computing θ_0 even if the current solution is not degenerate. Here, of course, $\theta_0 > 0$, and the new solution will have an improved value for the objective function. However, because of this tie in the old solution, the new solution will be degenerate. To illustrate these situations, let us consider the following examples.

Example 1.1. Minimize

$$- x_4 - x_5$$

subject to

$$
\begin{aligned}
x_1 \quad\quad + 3x_4 + 2x_5 &= 6 \\
x_2 + x_4 - x_5 &= 0 \\
x_3 - 2x_4 + 2x_5 &= 4 \\
x_j &\geq 0
\end{aligned}
$$

This problem has an initial, degenerate, basic feasible solution of $x_1 = 6$, $x_2 = 0$, and $x_3 = 4$, with the objective function equal to zero. For the next solution, the simplex algorithm could select either x_4 or x_5 to enter the basis[1] as both have a $z_j - c_j$ value of $+1$. If x_4 is selected, then it comes into the basis with the value of $\theta_0 = \frac{0}{1} = 0$ and the objective function remains at zero. Here the degenerate solution has not changed. We have generated another basis that represents the same degenerate extreme point. If x_5 is selected, then it comes into the basis with the value of $\theta_0 = \frac{4}{2} = 2$ and the objective function decreases from zero to -2.

Example 1.2. Minimize

$$-6x_1 - 3x_2 + 4x_3$$

subject to

$$
\begin{aligned}
x_1 + 4x_2 - 2x_3 + x_4 \quad\quad\quad &= 1 \\
2x_1 - 2x_2 + 6x_3 \quad\quad + x_5 \quad\quad &= 2 \\
-2x_1 + 3x_2 + x_3 \quad\quad\quad + x_6 &= 5 \\
x_j &\geq 0
\end{aligned}
$$

We start with the first basic feasible solution (nondegenerate) of $x_4 = 1$, $x_5 = 2$, and $x_6 = 5$, with the objective function equal to zero. The simplex

---

[1] It is convenient to describe the variables as entering the basis, but, of course, we mean the corresponding vector.

algorithm would select x_1 to enter the basis ($z_1 - c_1 = 6$).   As $\theta_0 = \min(\frac{1}{1}, \frac{2}{2})$, we have a tie in selecting the variable to leave the basis; either x_4 or x_5 can be removed.   Selecting x_4, we have the new degenerate basic feasible solution of $x_1 = 1$, $x_5 = 0$, and $x_6 = 7$, with the objective function decreasing from zero to -6.   If x_5 were selected to leave the basis, we would have the same solution, but a different basis to represent it.

From the above discussion, we see that any scheme for relaxing the nondegeneracy assumption must be concerned with the determination of θ_0, i.e., the index l of the vector to be removed from the basis.   A number of *anticycling procedures* have been developed (Dantzig [153], Charnes [110], Wolfe [715], Dantzig, Orden, and Wolfe [191], and Bland [80]).   Computational experience on digital computers has minimized the importance of these procedures, as examples of practical problems that have truly cycled are rare.[1]   The successful solution of thousands of problems has not hinged on the development of these techniques.   As Beale [58] notes: "They are not used in practice—partly because they would slow the computer program down, unnecessarily in the vast majority of cases, and partly because of the difficulty of recognizing a true zero numerically."   What is important, however, is that these schemes "make the simplex method available, without blemish, as a crisp tool for proving pure theorems" (Hoffman [374]).

1. PERTURBATION TECHNIQUES

Geometrically, a degenerate situation is one in which the vector $\mathbf{P}_0$ lies on a bounding hyperplane or edge of the convex cone determined by its basis

---

[1] We need to differentiate between *classical cycling* and *computer cycling*.   Classical cycling can arise when applying the simplex algorithm, as described in Chap. 4, using rational arithmetic without loss of accuracy.   Computer cycling can arise when solving a problem using a computer-based version of the simplex algorithm, e.g., mathematical programming systems (MPS).   Such systems incorporate various numerical and computational devices that cause the resultant computer-based simplex algorithm to differ from the classical one.   Specially constructed problems that exhibit classical cycling are known (see Sec. 3 and Exercises 2 and 6), but they are readily solved by most MPS.   Problems that exhibit computer cycling are also known.   Whether such cycling is due to the numerical transformations and data truncations of the computer system or the original structure of the problem is unclear.   One type of cycling does not imply the other.   A practical problem that exhibits classical cycling has not been found as yet.   We note that most MPS employ an *antidegeneracy procedure*, not so much as a guard against cycling, but to decrease the number of iterations and speed up the computational process.   Given a degenerate solution, one approach is to change the basic variable zeros to 10^{-2} (or 10^{-3}), solve the perturbed problem, and then correct the optimal solution.   Another approach is to track the value of the objective function and if it does not change after a specified number of iterations (say $2m$ or 100), then do not allow any pivot steps on degenerate rows or initiate a perturbation scheme or special pivot selection rules.   For further discussion see Gass [289], IBM [387], Sperry Univac [619], and Benichou et al. [70].

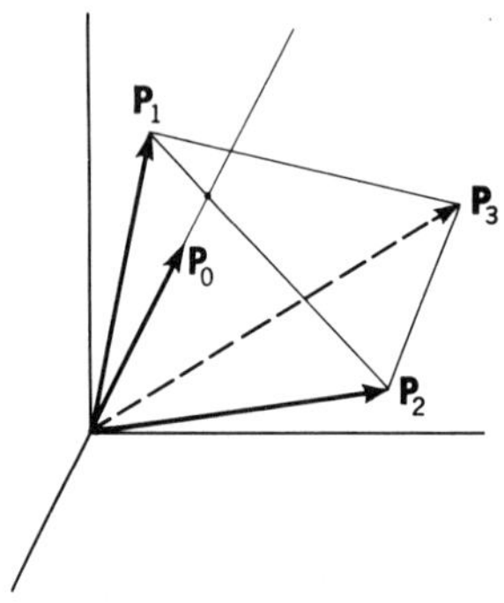

FIGURE 7.1

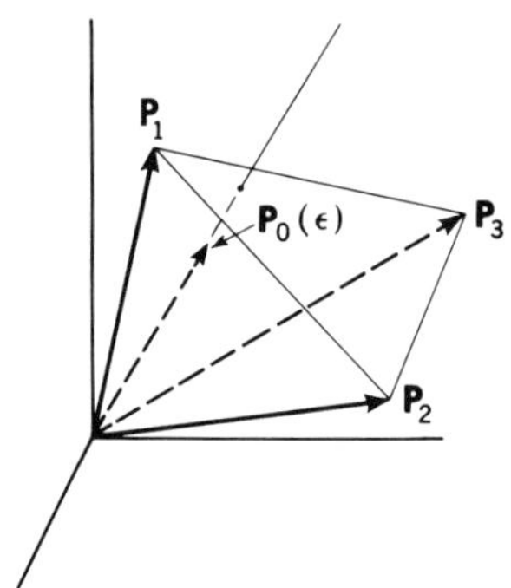

FIGURE 7.2

vectors. For example, in Fig. 7.1 vector $\mathbf{P}_0$ can be expressed as a positive combination of $\mathbf{P}_1$ and $\mathbf{P}_2$ but as a nonnegative combination of $\mathbf{P}_1$, $\mathbf{P}_2$, and $\mathbf{P}_3$ with $x_{30} = 0$. However, if we move, i.e., perturb, the vector $\mathbf{P}_0$ in such a manner that it lies inside the convex cone determined by vectors $\mathbf{P}_1$, $\mathbf{P}_2$, and $\mathbf{P}_3$, then the corresponding solution will be nondegenerate. We can do this by taking a positive linear combination of these vectors and adding it to $\mathbf{P}_0$. However, we wish to do this so as not to destroy or hide the original problem. Hence we make our positive combination small, and, as suggested by Charnes [110], we would let the positive combination for our example be

$$\epsilon\mathbf{P}_1 + \epsilon^2\mathbf{P}_2 + \epsilon^3\mathbf{P}_3$$

where ϵ is some small positive number. The constraints of the new problem would be

$$x_{10}\mathbf{P}_1 + x_{20}\mathbf{P}_2 + x_{30}\mathbf{P}_3 = \mathbf{P}_0 + \epsilon\mathbf{P}_1 + \epsilon^2\mathbf{P}_2 + \epsilon^3\mathbf{P}_3 = \mathbf{P}_0(\epsilon)$$

Its geometric interpretation is given in Fig. 7.2.

For the general linear-programming problem we wish to perturb the problem in a similar manner so as to ensure that, for any possible admissible basis, the corresponding solution will be nondegenerate. To accomplish this for the general problem, we rewrite the constraints

$$x_1\mathbf{P}_1 + x_2\mathbf{P}_2 + \cdots + x_m\mathbf{P}_m + \cdots + x_n\mathbf{P}_n = \mathbf{P}_0 \tag{1.1}$$

to read

$$x_1\mathbf{P}_1 + x_2\mathbf{P}_2 + \cdots + x_m\mathbf{P}_m + \cdots + x_n\mathbf{P}_n = \mathbf{P}_0 + \epsilon\mathbf{P}_1 + \epsilon^2\mathbf{P}_2$$
$$+ \cdots + \epsilon^m\mathbf{P}_m + \cdots + \epsilon^n\mathbf{P}_n = \mathbf{P}_0(\epsilon) \tag{1.2}$$

Let us assume that the vectors $\mathbf{P}_1$, $\mathbf{P}_2$, ..., $\mathbf{P}_m$ form an admissible basis **B**. Then a solution to (1.1) is

$$\bar{\mathbf{X}}_0 = \mathbf{B}^{-1}\mathbf{P}_0 \geq \mathbf{0} \tag{1.3}$$

and a solution to (1.2) is

$$\bar{\mathbf{X}}_0(\epsilon) = \mathbf{B}^{-1}\mathbf{P}_0(\epsilon) \tag{1.4}$$

Let

$$\mathbf{X}_j = \mathbf{B}^{-1}\mathbf{P}_j \tag{1.5}$$

Then (1.4) is given by

$$\bar{\mathbf{X}}_0(\epsilon) = \mathbf{B}^{-1}\mathbf{P}_0 + \epsilon\mathbf{B}^{-1}\mathbf{P}_1 + \epsilon^2\mathbf{B}^{-1}\mathbf{P}_2 + \cdots + \epsilon^m\mathbf{B}^{-1}\mathbf{P}_m + \cdots + \epsilon^n\mathbf{B}^{-1}\mathbf{P}_n$$
$$= \bar{\mathbf{X}}_0 + \epsilon\mathbf{X}_1 + \epsilon^2\mathbf{X}_2 + \cdots + \epsilon^m\mathbf{X}_m + \cdots + \epsilon^n\mathbf{X}_n \tag{1.6}$$

We note that, since $\mathbf{B}$ consists of vectors $\mathbf{P}_1$, $\mathbf{P}_2$, $\ldots$, $\mathbf{P}_m$, $\mathbf{X}_j = \mathbf{B}^{-1}\mathbf{P}_j$ is an identity vector with the unit element in position j for $j = 1, 2, \ldots, m$. Hence the $\bar{x}_{i0}(\epsilon)$ are given by

$$\bar{x}_{i0}(\epsilon) = \bar{x}_{i0} + \sum_{j=1}^{n} \epsilon^j x_{ij} \tag{1.7}$$

or

$$\bar{x}_{i0}(\epsilon) = \bar{x}_{i0} + \epsilon^i + \sum_{j=m+1}^{n} \epsilon^j x_{ij} \tag{1.8}$$

From (1.8), and by taking $\epsilon > 0$ but sufficiently small, we can make all $\bar{x}_{i0}(\epsilon) > 0$ for $i = 1, 2, \ldots, m$. (We can always rearrange the problem to make the basis vectors the first m.) In determining the vector $\mathbf{P}_l$ to be removed from the basis for the problem of (1.2), the criterion that ensures the feasibility of the new solution is the selection of

$$\theta_0 = \frac{\bar{x}_{l0}(\epsilon)}{x_{lk}} = \min_i \frac{\bar{x}_{i0}(\epsilon)}{x_{ik}} = \min_i \frac{\bar{x}_{i0} + \epsilon^i + \sum\limits_{j=m+1}^{n} \epsilon^j x_{ij}}{x_{ik}} > 0 \tag{1.9}$$

for $x_{ik} > 0$. We see from (1.8) that ties cannot occur since $\bar{x}_{l0}(\epsilon)$ is the only variable that contains ϵ^l.

Once a solution to the linear-programming problem with the constraints (1.2) has been obtained, by letting ϵ equal zero, we have the corresponding extreme-point solution to (1.1). In actual practice, it is not necessary to select an ϵ and rewrite the problem as was done in (1.2). From (1.6) and (1.9) we note that the information required to determine θ_0 is the $\bar{x}_{i0}$ of problem (1.1) and the coefficients of the ϵ^j. This information is all contained in the simplex tableau of the unperturbed problem. For ϵ small enough, we note that the significant coefficients of ϵ are the first ones, starting with $j = 1, 2, \ldots, n$. The procedure can then be routinized as follows: If, for all $\bar{x}_{i0}$ in the basic solution, the ratios $\bar{x}_{i0}/x_{ik}$ for $x_{ik} > 0$ yield a unique $\theta_0 = \min_i (\bar{x}_{i0}/x_{ik})$, then vector $\mathbf{P}_l$ is uniquely determined. If, however, there are ties for the minimum for some set of indices, we compute starting with $j = 1$ the ratios x_{ij}/x_{ik} for all rows i in this set and compare these ratios. The index that corresponds to the algebraically smallest ratio determines the index of the vector to be removed. If there are still ties in computing the minimum, we form the next ratios for the tied set of indices for column $j + 1$ and repeat the analysis. For example, if for the admissible basis $(\mathbf{P}_1 \mathbf{P}_2 \cdots \mathbf{P}_m)$ we have

$$\theta_0 = \frac{\bar{x}_{10}}{x_{1k}} = \frac{\bar{x}_{20}}{x_{2k}}$$

we then compute x_{11}/x_{1k} and x_{21}/x_{2k} and compare. If

$$\min_i \frac{x_{i1}}{x_{ik}} = \frac{x_{11}}{x_{1k}} \qquad \text{for } i = 1, 2$$

then vector $\mathbf{P}_1$ is removed. If

$$\min_i \frac{x_{i1}}{x_{ik}} = \frac{x_{21}}{x_{2k}} \qquad \text{for } i = 1, 2$$

then $\mathbf{P}_2$ is removed. In both cases $\mathbf{P}_k$ is the vector introduced. (The selection of the new vector does not depend in any way on the degenerate situation.) If $x_{11}/x_{1k} = x_{21}/x_{2k}$, we form the ratios x_{12}/x_{1k} and x_{22}/x_{2k} and repeat the comparison until the tie situation is broken. We known from (1.8) that this will happen for some j. The simplex tableau is transformed in the usual manner with the selected x_{lk} as pivot element. The new $\bar{\mathbf{X}}_0'(\epsilon)$ solution to (1.2) will be nondegenerate for ϵ sufficiently small, and the above process is repeated until a minimum solution is found.

The above perturbation scheme is ideally adapted to the original simplex method since all the necessary information to execute the scheme is readily available. For the revised procedure a different approach to the problem also yields an efficient technique that requires knowledge of the inverse of the current basis (Dantzig, Orden, and Wolfe [191]). We shall not discuss the development of the method except to describe the rule to be used when breaking ties in the computation of θ_0. This rule is given in Dantzig [157] and is: "If two or more indices l_1, l_2, ... are tied for the minimum, divide the corresponding entries in the *first column of the inverse* by $x_{l_1,k}$, $x_{l_2,k}$, ..., respectively, and take the index of the row with the minimizing ratio for l. If there still remain ties (for the minimum), repeat for those indices that are still tied, using as ratios the corresponding entries in the *second column of the inverse* divided by their respective $x_{l_i,k}$. Ratios formed from successive columns of the inverse are used until all ties are resolved. (Since no two columns of an inverse can be proportional, a *unique l* will be chosen by the last column.)"

2. THE LOWEST-INDEX ANTICYCLING RULES

As presented in Chap. 4, the simplex method requires us to make a choice when there are ties in selecting the variable x_k to enter the basic solution (i.e., when determining $z_k - c_k = \max_j z_j - c_j > 0$) and when ties occur in selecting the variable x_l to be removed from the basic solution (i.e., when determining $\theta_0 = x_{l0}/x_{lk} = \min_i x_{i0}/x_{ik} \geq 0$). In selecting x_k, we can choose any variable that will improve the value of the objective function, while the choice of x_l must correspond to a basis change that preserves feasibility. The perturbation anticycling procedure (described in Sec. 1) does not modify the rule for selecting x_k, but it does require a different and more computationally involved process for determining x_l. In contrast, a very-easy-to-use anticycling procedure, due to Bland [80], requires the selection of both x_k and x_l to be made under modified, but simple, decision rules. While perturbation methods (conceptually) make the degenerate basic variables positive, Bland's approach considers the zero values explicitly. The modified rules are the following:

The lowest-index anticycling rules
a. Among all candidates to enter the basic solution, select the variable x_k having the lowest index; i.e., $k = \min \{j$ for which $z_j - c_j > 0\}$. (The index j is associated with the nonbasic variables.)
b. Among all candidates to leave the basic solution, select variable x_l having the lowest index; i.e.,

$$l = \min \{i \text{ for which } x_{ik} > 0 \text{ and } x_{i0}/x_{ik} = \min_{x_{ik}>0} x_{i0}/x_{ik}\}$$

(The index i is associated with the basic variables.)

To prove that the use of these rules will prevent cycling, we require the following lemma.

Lemma 1. *Consider the linear-programming problem of minimizing $\mathbf{cX}$, $\mathbf{AX} = \mathbf{b}$, $\mathbf{X} \geq 0$. Let $\mathbf{B}$ be any basis formed from columns of $\mathbf{A}$ and denote the $m \times 1$ basic solution vector by $\mathbf{X}_0 = \mathbf{B}^{-1}\mathbf{b}$. Here $\mathbf{X}_0$ may not satisfy the non-negativity constraints. The corresponding objective function value is given by $x_0 = \mathbf{c}_0\mathbf{X}_0$, where $\mathbf{c}_0$ is the $1 \times m$ row vector of costs associated with the basis vectors. Let $\mathbf{Y}_0$ be any $n \times 1$ solution vector to the constraints $\mathbf{AY} = \mathbf{b}$; $\mathbf{Y}_0$ may have some negative components. The objective function value for $\mathbf{Y}_0$ is $y_0 = \mathbf{cY}_0$. For basis $\mathbf{B}$, we have $z_j = \mathbf{c}_0\mathbf{B}^{-1}\mathbf{P}_j$, for $\mathbf{P}_j$ any column of $\mathbf{A}$, and $\mathbf{z} = (z_1, \ldots, z_n) = \mathbf{c}_0\mathbf{B}^{-1}\mathbf{A}$. Let $\bar{\mathbf{c}} = \mathbf{z} - \mathbf{c} = (z_1 - c_1, \ldots, z_n - c_n) = (\bar{c}_1, \ldots, \bar{c}_n)$ be the ordered row vector of reduced costs. Then $\bar{\mathbf{c}}\mathbf{Y}_0 = x_0 - y_0$.*

Proof: By direct evaluation we have

$$\bar{\mathbf{c}}\mathbf{Y}_0 = (\mathbf{z} - \mathbf{c})\mathbf{Y}_0 = \mathbf{z}\mathbf{Y}_0 - \mathbf{c}\mathbf{Y}_0 = (\mathbf{c}_0\mathbf{B}^{-1})\mathbf{AY}_0 - \mathbf{c}\mathbf{Y}_0$$
$$= \mathbf{c}_0(\mathbf{B}^{-1}\mathbf{b}) - \mathbf{c}\mathbf{Y}_0 = \mathbf{c}_0\mathbf{X}_0 - \mathbf{c}\mathbf{Y}_0 = x_0 - y_0$$

Note that this lemma demonstrates a relationship between *any basic solution* to a linear-programming problem and *any other solution* to the main constraints $\mathbf{AX} = \mathbf{b}$.

Theorem 1. *The simplex method modified to include the lowest-index rules cannot cycle and, hence, it is finite.*[1]

Proof: We assume that a cycle has occurred while solving the linear-programming problem of minimizing cX, $AX = b$, $X \geq 0$ using the simplex algorithm with the lowest-index rules. We shall show that this assumption leads to a contradiction.

By the definition of a cycle, we have encountered a finite series of basis changes that has returned to a previous basic feasible solution. When the simplex algorithm is in the cycle, the pivot ratio $\theta_0 = 0$ for all basis changes, and the value of the objective function x_0 does not change. Also, although the basis is changing, the basic solution vector $X_0 = (x_{10}, \ldots, x_{m0})$ remains the same. (The indices of the x_{i0} are, of course, updated to reflect the current basis.) Let us consider the reduced set of $\bar{m} \times \bar{n}$ equations $\bar{A}\bar{X} = 0$ that contains only those columns of A chosen to enter some basis of the cycle and only those rows in which pivots take place during the cycle. As a linear-programming problem, this reduced problem has the same cycle as the original problem and has a constant objective function value of x_0 (which we can let equal zero). For those variables that enter a basis during the cycle, let q be the largest index. We need to consider, for the reduced problem, two simplex tableaus: The first, denoted by $\bar{X}_0$, corresponds to the tableau when x_q is about to enter the basis; the second $\bar{X}_1$, corresponds to the tableau as x_q is about to leave the basis, as illustrated in Tableau 7.1. Denote the basis of

TABLEAU 7.1 Tableau $\bar{X}_1$ (vector P_p to replace P_q)

Row index i	B_1 basis	c	P_0	$\cdots$	P_f	$\cdots$	P_p	$\cdots$	P_g	$\cdots$	P_h	$\cdots$	P_q
				$\cdots$	y_f	$\cdots$	$y_p = 1$	$\cdots$	y_g	$\cdots$	y_h	$\cdots$	y_q
				$\cdots$	c_f	$\cdots$	c_p	$\cdots$	c_g	$\cdots$	c_h	$\cdots$	c_q
1	P_f	c_f	0	$\cdots$	1	$\cdots$	$x_{1p}^{(1)}$	$\cdots$	0	$\cdots$	0	$\cdots$	0
2	P_h	c_h	0	$\cdots$	0	$\cdots$	$x_{2p}^{(1)}$	$\cdots$	0	$\cdots$	1	$\cdots$	0
.	.	.	.	$\cdots$	.	$\cdots$	.	$\cdots$	.	$\cdots$	.	$\cdots$	.
.	.	.	.	$\cdots$	.	$\cdots$	.	$\cdots$	.	$\cdots$	.	$\cdots$	.
.	.	.	.	$\cdots$	.	$\cdots$	.	$\cdots$	.	$\cdots$	.	$\cdots$	.
l	P_q	c_q	0	$\cdots$	0	$\cdots$	$x_{lp}^{(1)}$	$\cdots$	0	$\cdots$	0	$\cdots$	1
.	.	.	.	$\cdots$	.	$\cdots$	.	$\cdots$	.	$\cdots$	.	$\cdots$	.
.	.	.	.	$\cdots$	.	$\cdots$	.	$\cdots$	.	$\cdots$	.	$\cdots$	.
.	.	.	.	$\cdots$	.	$\cdots$	.	$\cdots$	.	$\cdots$	.	$\cdots$	.
$\bar{m}$	P_g	c_g	0	$\cdots$	0	$\cdots$	$x_{\bar{m}p}^{(1)}$	$\cdots$	1	$\cdots$	0	$\cdots$	0
			x_0	$\cdots$	0		$x_{\bar{m}+1,p}^{(1)}$	$\cdots$	0	$\cdots$	0	$\cdots$	0

[1] The proof given here is a simplification of the original by Bland [80] and is due to H. W. Kuhn as presented in Papadimtriou and Steiglitz [541].

tableau $\bar{\mathbf{X}}_0$ by $\mathbf{B}_0$ and that of $\bar{\mathbf{X}}_1$ by $\mathbf{B}_1$. Also, denote the elements of $\bar{\mathbf{X}}_0$ by $x_{ij}^{(0)}$ and of $\bar{\mathbf{X}}_1$ by $x_{ij}^{(1)}$. Let the column with index p be the one about to enter the basis in tableau $\bar{\mathbf{X}}_1$ and cause x_q to leave the basic solution. For column p, we have $x_{\bar{m}+1,p}^{(1)} = z_p - c_p > 0$ and the pivot element $x_{lp}^{(1)} > 0$. Using these tableaus, we identify two solutions for the reduced system $\tilde{\mathbf{A}}\tilde{\mathbf{X}} = \mathbf{0}$ as follows. For $\bar{\mathbf{X}}_0$, we take the solution corresponding to the basic feasible solution $\mathbf{X}_0$ with objective function value x_0. For $\bar{\mathbf{X}}_1$, we form a solution vector $\mathbf{Y}_0$ with components $y_j = 1$ if $j = p$, $y_j = -x_{ip}^{(1)}$ if column j is in $\mathbf{B}_1$, and $y_j = 0$ otherwise. We see that $\mathbf{Y}_0$ has at most $\bar{m} + 1$ nonzero elements, is nonbasic, has a negative component $-x_{lp}^{(1)}$, and $\tilde{\mathbf{A}}\tilde{\mathbf{Y}}_0 = \mathbf{0}$. By direct substitution in the objective function row of Tableau 7.1 [see Eq. (1.17), Chap. 4], the objective function value for solution $\mathbf{Y}_0$ is $y_0 = x_0 - x_{\bar{m}+1,p}^{(1)}$. Applying Lemma 1 to the reduced problem with basic solution $\mathbf{X}_0$ and other solution $\mathbf{Y}_0$, we have

$$\bar{\mathbf{c}}\mathbf{Y}_0 = x_0 - y_0 = x_0 - x_0 + x_{\bar{m}+1,p}^{(1)} > 0 \tag{2.1}$$

For tableau $\bar{\mathbf{X}}_0$ (where x_q is about to enter the basis), we must have (because of the lowest-index rule a)

$$\bar{c}_j = z_j - c_j \begin{cases} \leq 0 & \text{for } j < q \\ > 0 & \text{for } j = q \end{cases}$$

For tableau $\bar{\mathbf{X}}_1$ (where the selected pivot row removes x_q from the basic solution), we must have

$$y_j = \begin{cases} -x_{lp}^{(1)} < 0 & \text{for } j = q \\ 0,\ 1,\ \text{or} - x_{ip}^{(1)} \geq 0 & \text{for } j < q \end{cases}$$

This is because the first and only pivot element in column p of $\bar{\mathbf{X}}_1$ is $x_{lp}^{(1)} > 0$; the other elements of lower index must satisfy $x_{ip}^{(1)} \leq 0$ (by decision rule b). We then have

$$\bar{\mathbf{c}}\mathbf{Y}_0 = \sum_{j<q} \bar{c}_j y_j + \bar{c}_q y_q \leq \bar{c}_q y_q < 0$$

which contradicts (2.1).

Although the lowest-index rules are easy to apply, they have proved inefficient from a total computational point of view. For the simplex method, the basic measure of computational cost (time) is the number of iterations required to solve a problem. Tests by Avis and Chvátal [15] in which the max $(z_j - c_j)$ rule and the lowest-index rules were compared showed that the max $(z_j - c_j)$ rule required about one-fourth the number of iterations for their test problems. This result might have been expected since the lowest-index rules cause the column selection process of the simplex method to concentrate much of its search in the front end of the tableau, while ignoring the magnitudes of the $(z_j - c_j)$ terms. One possible use of the lowest-index rules is as an antidegeneracy procedure; that is, apply the rules to a problem after a number of iterations (say $2m$) have been completed without a change in the value of the objective function. {The Avis and Chvátal tests showed that the max $(z_j - c_j)$ rule was still much better on highly degenerate problems, e.g., assignment problems. In degenerate problems we may stay at the same solution for a number of iterations, changing only the bases, and eventually moving to a better solution without cycling. This phenomenon, termed *stalling* by Avis and Chvátal, has been a long recognized unpleasant feature of the simplex method. Perold [544] notes that it is typical for a problem with an average of 20 per cent degenerate basic variables to result in 50 per cent of its iterations showing no improvement in the objective function value. An adaptation of the simplex method to reduce the number of such *degenerate iterations* is described by Perold [544].}

3. EXAMPLE OF CYCLING

The literature contains few examples of problems that cycle when they are solved by the original simplex method (Hoffman [373] and Beale [52]). The problem of Beale [52] is discussed in terms of its dual. We shall next illustrate the cycling phenomenon in terms of the primal by a previously unpublished example due to Beale.[1]

The problem is to minimize

$$-\tfrac{3}{4}x_1 + 150x_2 - \tfrac{1}{50}x_3 + 6x_4$$

$$
\begin{aligned}
\tfrac{1}{4}x_1 - 60x_2 - \tfrac{1}{25}x_3 + 9x_4 + x_5 &= 0 \\
\tfrac{1}{2}x_1 - 90x_2 - \tfrac{1}{50}x_3 + 3x_4 \quad + x_6 &= 0 \\
x_3 \qquad\qquad + x_7 &= 1
\end{aligned}
$$

and

$$x_j \geq 0$$

TABLEAU 7.2

		P_0	P_1	P_2	P_3	P_4	P_5	P_6	P_7
I. Basis	**c**		$-\tfrac{3}{4}$	150	$-\tfrac{1}{50}$	6	0	0	0
P_5	0	0	$\left(\tfrac{1}{4}\right)$	-60	$-\tfrac{1}{25}$	9	1	0	0
P_6	0	0	$\tfrac{1}{2}$	-90	$-\tfrac{1}{50}$	3	0	1	0
P_7	0	1	0	0	1	0	0	0	1
		0	$\tfrac{3}{4}$	-150	$\tfrac{1}{50}$	-6	0	0	0

[1] For other cycling examples, see Exercise 20, Chap. 10 (transportation/assignment problem). Exercise 61, Chap. 12 (network problem), and Exercises 2, 6, and 7 of this chapter.

TABLEAU 7.2 (Continued)

		c								
II.	P_1	$-\frac{3}{4}$	0	1	-240	$-\frac{4}{25}$	36	4	0	0
	P_6	0	0	0	(30)	$\frac{3}{50}$	-15	-2	1	0
	P_7	0	1	0	0	1	0	0	0	1
			0	0	30	$\frac{7}{50}$	-33	-3	0	0

		c								
III.	P_1	$-\frac{3}{4}$	0	1	0	$\left(\frac{8}{25}\right)$	-84	-12	8	0
	P_2	150	0	0	1	$\frac{1}{500}$	$-\frac{1}{2}$	$-\frac{1}{15}$	$\frac{1}{30}$	0
	P_7	0	1	0	0	1	0	0	0	1
			0	0	0	$\frac{2}{25}$	-18	-1	-1	0

		c								
IV.	P_3	$-\frac{1}{50}$	0	$\frac{25}{8}$	0	1	$-\frac{525}{2}$	$-\frac{75}{2}$	25	0
	P_2	150	0	$-\frac{1}{160}$	1	0	$\left(\frac{1}{40}\right)$	$\frac{1}{120}$	$-\frac{1}{60}$	0
	P_7	0	1	$-\frac{25}{8}$	0	0	$\frac{525}{2}$	$\frac{75}{2}$	-25	1
			0	$-\frac{1}{4}$	0	0	3	2	-3	0

		c								
V.	P_3	$-\frac{1}{50}$	0	$-\frac{125}{2}$	10,500	1	0	(50)	-150	0
	P_4	6	0	$-\frac{1}{4}$	40	0	1	$\frac{1}{3}$	$-\frac{2}{3}$	0
	P_7	0	1	$\frac{125}{2}$	$-10,500$	0	0	-50	150	1
			0	$\frac{1}{2}$	-120	0	0	1	-1	0

		c								
VI.	P_5	0	0	$-\frac{5}{4}$	210	$\frac{1}{50}$	0	1	-3	0
	P_4	6	0	$\frac{1}{6}$	-30	$-\frac{1}{150}$	1	0	$\left(\frac{1}{3}\right)$	0
	P_7	0	1	0	0	1	0	0	0	1
			0	$\frac{7}{4}$	-330	$-\frac{1}{50}$	0	0	2	0

		c								
VII.	P_5	0	0	$\left(\frac{1}{4}\right)$	-60	$-\frac{1}{25}$	9	1	0	0
	P_6	0	0	$\frac{1}{2}$	-90	$-\frac{1}{50}$	3	0	1	0
	P_7	0	1	0	0	1	0	0	0	1
			0	$\frac{3}{4}$	-150	$\frac{1}{50}$	-6	0	0	0

We first demonstrate that this problem has a cycle, and then show how the perturbation technique and lowest-index rules break the cycle.   The iterations of the cycle are show in Tableau 7.2.   There we use the standard max $(z_j - c_j)$ rule for selecting a vector to enter the basis and, when ties occur in determining the vector to leave the basis, we choose the one with the lowest index.   (This is

TABLEAU 7.3

			P_0	P_1	P_2	P_3	P_4	P_5	P_6	P_7
	Basis	$\mathbf{c}$		$-\frac{3}{4}$	150	$-\frac{1}{50}$	6	0	0	0
I.	P_5	0	0	$\frac{1}{4}$	-60	$-\frac{1}{25}$	9	1	0	0
	P_6	0	0	$\frac{1}{2}$	-90	$-\frac{1}{50}$	3	0	1	0
	P_7	0	1	0	0	1	0	0	0	1
			0	$\frac{3}{4}$	-150	$\frac{1}{50}$	-6	0	0	0
II.	P_1	$-\frac{3}{4}$	0	1	-240	$-\frac{4}{25}$	36	4	0	0
	P_6	0	0	0	30	$\frac{3}{50}$	-15	-2	1	0
	P_7	0	1	0	0	1	0	0	0	1
			0	0	30	$\frac{7}{50}$	-33	-3	0	0
III.	P_1	$-\frac{3}{4}$	0	1	0	$\frac{8}{25}$	-84	-12	8	0
	P_2	150	0	0	1	$\frac{1}{500}$	$-\frac{1}{2}$	$-\frac{1}{15}$	$\frac{1}{30}$	0
	P_7	0	1	0	0	1	0	0	0	1
			0	0	0	$\frac{2}{25}$	-18	-1	-1	0
IV.	P_1	$-\frac{3}{4}$	0	1	-160	0	-4	$-\frac{4}{3}$	$\frac{8}{3}$	0
	P_3	$-\frac{1}{50}$	0	0	500	1	-250	$-\frac{100}{3}$	$\frac{50}{3}$	0
	P_7	0	1	0	-500	0	250	$\frac{100}{3}$	$-\frac{50}{3}$	1
			0	0	-40	0	2	$\frac{5}{3}$	$-\frac{7}{3}$	0
V.	P_1	$-\frac{3}{4}$	$\frac{2}{125}$	1	-168	0	0	$-\frac{4}{5}$	$\frac{12}{5}$	$\frac{2}{125}$
	P_3	$-\frac{1}{50}$	1	0	0	1	0	0	0	1
	P_4	6	$\frac{1}{250}$	0	-2	0	1	$\frac{2}{15}$	$-\frac{1}{15}$	$\frac{1}{250}$
			$-\frac{1}{125}$	0	-36	0	0	$\frac{7}{5}$	$-\frac{11}{5}$	$-\frac{1}{125}$
VI.	P_1	$-\frac{3}{4}$	$\frac{1}{25}$	1	-180	0	6	0	2	$\frac{1}{25}$
	P_3	$-\frac{1}{50}$	1	0	0	1	0	0	0	1
	P_5	0	$\frac{3}{100}$	0	-15	0	$\frac{15}{2}$	1	$-\frac{1}{2}$	$\frac{3}{100}$
			$-\frac{1}{20}$	0	-15	0	$-\frac{21}{2}$	0	$-\frac{3}{2}$	$-\frac{1}{20}$

the rule used in Chap. 4.) Such ties arose in the first, third, and fifth solutions. In Tableau 7.2, we see that the seventh solution is identical to the first, and an optimal solution has not been reached. If we continued the process with the seventh step, we would only repeat the previous solutions and never reach the minimum.

However, by applying the degeneracy procedure described in Sec. 1, we select a different sequence of solutions and do determine the minimum solution. The steps in this computation are shown in Tableau 7.3, where the sixth solution gives the final answer. There is no recurrence of any solution. Ties occur in computing θ_0 in the first and third solutions. The first three solutions in both Tableaus 7.2 and 7.3 are the same. The main difference occurs in going from the third to the fourth solution, where in the second case we removed $\mathbf{P}_2$ instead of $\mathbf{P}_1$. The minimum value of the objective function is $-\frac{1}{20}$.

In applying the lowest-index rules, we would generate the same solutions given in Steps I to V of Tableau 7.2. But in going from Step V to the Step VI we would select $\mathbf{P}_1$ to enter the basis (instead of $\mathbf{P}_5$). The resulting basis is the same as the one in Step V of Tableau 7.3 and leads to the optimal solution in the next iteration.

EXERCISES

1. Solve the cycling example of Sec. 2 using the revised simplex method and the corresponding rule for breaking ties.

2. Starting with vectors $\mathbf{P}_1$, $\mathbf{P}_2$, $\mathbf{P}_7$ as the initial basis, will the following problem cycle when the standard simplex rules are applied (Yudin and Gol'shtein [723])? Maximize

$$x_3 - x_4 + x_5 - x_6$$

subject to

$$x_1 \quad + 2x_3 - 3x_4 - 5x_5 + 6x_6 \quad = 0$$
$$x_2 + 6x_3 - 5x_4 - 3x_5 + 2x_6 \quad = 0$$
$$3x_3 + \ x_4 + 2x_5 + 4x_6 + x_7 = 1$$
$$x_j \geq 0$$

3. Prove that the minimum length of a cycle in a linear-programming problem is six iterations (Yudin and Gol'shtein [723]).

4. If you were attempting to construct a linear-programming problem which will cycle, what is the implication of Exercise 2, Chap. 4, and the above Exercise 3 on the dimensions of the problem?

5. Show that a cycling example must have at least two equations, at least six variables, and at least three nonbasic variables (Marshall and Suurballe [496]).

6. Show that the following problem due to H. W. Kuhn, as given in Balinski and Tucker [33], will cycle in six iterations: Minimize

$$-2x_4 - 3x_5 + x_6 + 12x_7$$

subject to

$$x_1 \qquad\qquad -2x_4 - 9x_5 + x_6 + 9x_7 = 0$$

$$x_2 \qquad +\tfrac{1}{3}x_4 + x_5 - \tfrac{1}{3}x_6 - 2x_7 = 0$$

$$x_3 + 2x_4 + 3x_5 - x_6 - 12x_7 = 2$$

$$x_j \geq 0$$

7. Using the standard simplex algorithm, show that the following problems will cycle (Marshall and Suurballe [496]):

a. Minimize

$$2x_2 + \qquad 4x_4 + \qquad 4x_6$$

subject to

$$x_1 - 3x_2 - x_3 - x_4 - x_5 + 6x_6 = 0$$

$$2x_2 + x_3 - 3x_4 - x_5 + 2x_6 = 0$$

$$x_j \geq 0$$

b. Minimize

$$-\tfrac{2}{5}x_5 - \tfrac{2}{5}x_6 + \tfrac{9}{5}x_7$$

subject to

$$x_1 \qquad\qquad +\tfrac{3}{5}x_5 - \tfrac{32}{5}x_6 + \tfrac{24}{5}x_7 = 0$$

$$x_2 \qquad +\tfrac{1}{5}x_5 - \tfrac{9}{5}x_6 + \tfrac{3}{5}x_7 = 0$$

$$x_3 \qquad +\tfrac{2}{5}x_5 - \tfrac{8}{5}x_6 + \tfrac{1}{5}x_7 = 0$$

$$x_4 \qquad + x_6 \qquad = 1$$

$$x_j \geq 0$$

8. Solve Exercises 2, 6, and 7 by the perturbation technique and the lowest-index rules.

9. Show that if rule *a* of the lowest-index rules is dropped then cycling can occur (Bland [80]).

10. Attempt to solve Exercises 2, 6, and 7 using a computer-based version of the simplex algorithm.

11. An anticycling rule of Azpeitia and Dickinson [16] is the following: At each iteration, solve the tie (for the vector to leave the basis) by making a choice that minimizes the number of times the selected basis has appeared in the previous iterations. Apply this rule to the example of Sec. 3 and discuss the computational implications of using it in a MPS.

12. Discuss the geometry of degeneracy when more than m constraints intersect to form an extreme point; see Exercise 30, Chap. 6, and Telgen [637].

13. In Lemma 1, let $\mathbf{X}_0$ be a basic feasible solution with corresponding $\mathbf{c} \leq 0$ and $\mathbf{Y}_0$ any feasible solution. Use the modified lemma to prove that $\mathbf{X}_0$ is optimal. Discuss the relationship between the modified lemma and Theorem 2, Chap. 4.

CHAPTER 8

Parametric Linear Programming and Sensitivity Analysis

A major task in the development of realistic linear-programming models is the collection of accurate and reliable numerical values for the coefficients. In some instances only estimates or average values of the coefficients are available, e.g., the cost of resources in a future period or the amount of inventory stored in a warehouse, and we are forced to use these point estimates and consider the problem to be deterministic. Hence, it is important to study the behavior of solutions to a linear-programming problem when the coefficients of that problem are allowed to vary; i.e., for what ranges of coefficient values will the deterministic solution remain optimal? Here we include variations in the coefficients of the $\mathbf{A}$ matrix (the a_{ij}), the objective-function cost coefficients (the c_j), and the constants of the right-hand side of the equations (the b_i). In this chapter we shall investigate changes of the coefficients which can occur in a known fashion, i.e., where some of the coefficients are assumed to be linear functions of a parameter, i.e., *parametric linear programming* (Gass and Saaty [293] and Manne [492]) and changes in individual coefficients which cause an optimal solution to become nonoptimal, i.e., *sensitivity analysis*. Other approaches to uncertainty in the values of the coefficients come under the headings of *stochastic, probabilistic,* or *chance-constrained programming*; i.e., some elements of the problem are assumed to be random variables with known probability distribution functions (see Dantzig [152], Vajda [665], Charnes and Cooper [114], Hansotia [352], Sengupta [603, 604], and Stancu-Minasian and Wets [622]). In Sec. 4, we extend the basic linear-programming model to include multiobjectives and discuss the related multiparameter problem.

1. THE PARAMETRIC OBJECTIVE FUNCTION[1]

The investigation of parametric programming as applied to the variation of the cost coefficients originated in the study of the production-scheduling problem described in Sec. 1 of Chap. 11.[2] There we introduced the parameter λ, which measures the cost of a unit increase in output relative to that of storing a unit for 1 month. Before selecting a particular production schedule that is optimal for some specified value of λ, the manufacturing company might find it advantageous to investigate a set of optimal solutions corresponding to ranges of values of λ. In this manner it can more accurately select the solution that not only best fits its production and storage capabilities, but also suits those factors that have not been incorporated into the linear-programming model. The procedure that enables one to compute efficiently the associated optimal solutions for ranges of λ is described below. This computational scheme is basically a variation of the standard simplex method.

To illustrate how such problems may arise in practice, we consider the furniture-manufacturer problem of Example 2.2, Chap. 4. The problem is to determine the number of chairs (x_1) and the number of tables (x_2) to be produced, subject to resource constraints, that maximize the profit function $\$45x_1 + \$80x_2$. The manufacturer has some flexibility in setting the profit for each product, depending on competition and overall marketing strategy. Thus, there would be interest in knowing for what range of the profit coefficients does the solution that optimizes $\$45x_1 + \$80x_2$ remain optimal, and what other solutions become optimal as the coefficients vary.

One way of structuring such an analysis is to assume that the profit coefficients are linear functions of a parameter and allow the parameter to take on positive or negative values, as appropriate. For example, the manufacturer may want to change the profit coefficient of the chairs and tables by the same amount, i.e., an equal dollar discount or increase. We would then consider the profit function $(45 + \lambda)x_1 + (80 + \lambda)x_2$, where λ is an unrestricted parameter. Other relationships could be considered, e.g., $(45 + \lambda)x_1 + (80 - \lambda)x_2$ or $(45 + \lambda)x_1 + (80 + 2\lambda)x_2$. The first makes compensating changes to the profits of the products, while the second doubles the change in profit for tables with respect to the change in chairs. We shall analyze the situation for the function $(45 + \lambda)x_1 + (80 + \lambda)x_2$.

The manufacturer's parametric-objective function problem is to maximize

$$z = (45 + \lambda)x_1 + (80 + \lambda)x_2$$

[1] Credit should be given to W. W. Jacobs for initiating the early work in parametric programming and establishing the underlying concepts.

[2] The reader not familiar with this problem should first review the cited section.

subject to

$$5x_1 + 20x_2 \leq 400$$

$$10x_1 + 15x_2 \leq 450$$

$$x_j \geq \quad 0$$

for all values of λ.

For the original problem, when $\lambda = 0$, the optimal solution, shown in Fig. 8.1, is $x_1 = 24$, $x_2 = 14$, and $z_0 = \$2{,}200$. We now need to consider what happens as λ is increased or decreased. Referencing Fig. 8.1, we see that the objective function z passes through the optimal extreme point (24, 14) of the feasible solution set. As λ varies, the slope of z changes. It is clear that the extreme point (24, 14) will remain optimal for some values of λ. As λ increases, the slope becomes steeper; as λ decreases, the slope becomes less inclined. The line defined by z rotates about the point (24, 14). This point remains optimal until a value of λ is chosen that causes z to be parallel to one of the boundary constraints that intersect at (24, 14). Thus, as λ increases, (24, 14) will remain optimal until z is parallel to the line defined by equation $10x_1 + 15x_2 = 450$; as λ decreases, (24, 14) will remain optimal until z is parallel to the line $5x_1 + 20x_2 = 400$. When z is parallel to either line, a multiple optimal solution is obtained. Parametric programming enables us to determine the range of λ for which (24, 14) remains optimal, and what other solutions are optimal (if any) for the full range of $-\infty < \lambda < +\infty$. Before developing the parametric computational procedure, we next determine, using the simplex algorithm, the range of λ for which (24, 14) stays optimal.

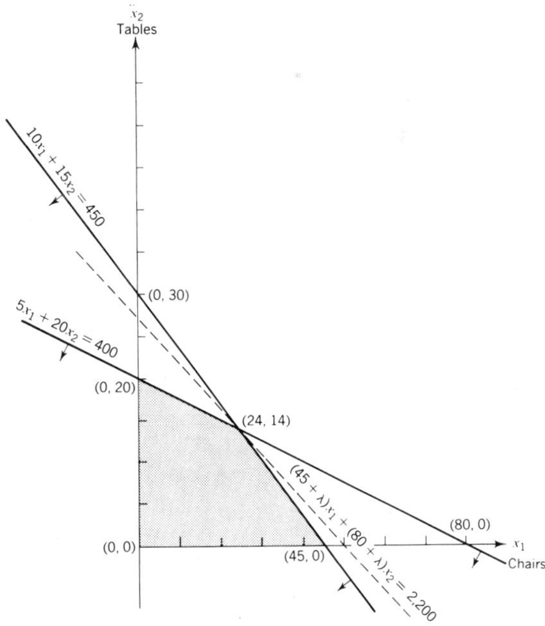

FIGURE 8.1

The solution of the parametrized chair-table problem for $\lambda = 0$ is given in Tableau 4.4 of Chap. 4. The solution step of that tableau is reproduced in Tableau 8.1, modified to include the parametric cost coefficients.

TABLEAU 8.1

	Basis	c	$\mathbf{P}_0$	$-45-\lambda$ $\mathbf{P}_1$	$-80-\lambda$ $\mathbf{P}_2$	0 $\mathbf{P}_3$	0 $\mathbf{P}_4$
1	$\mathbf{P}_2$	$-80-\lambda$	14	0	1	$\frac{2}{25}$	$-\frac{1}{25}$
2	$\mathbf{P}_1$	$-45-\lambda$	24	1	0	$-\frac{3}{25}$	$\frac{4}{25}$
3			$-2{,}200-38\lambda$	0	0	$-1+\frac{1}{25}\lambda$	$-4-\frac{3}{25}\lambda$

Note that the nonbasic vectors $\mathbf{P}_3$ and $\mathbf{P}_4$ have been priced out using the parametrized costs. Thus,

$$z_3 - c_3 = (-80-\lambda)(\tfrac{2}{25}) + (-45-\lambda)(-\tfrac{3}{25}) - 0$$
$$= -1 + \tfrac{1}{25}\lambda$$

and

$$z_4 - c_4 = (-80-\lambda)(-\tfrac{1}{25}) + (-45-\lambda)(\tfrac{4}{25}) - 0$$
$$= -4 - \tfrac{3}{25}\lambda$$

Of course, for $\lambda = 0$, we have $z_3 - c_3 = -1 \leq 0$ and $z_4 - c_4 = -4 \leq 0$, and the optimal solution $x_1 = 24$ and $x_2 = 14$. For this solution to remain optimal, the nonbasic $z_j - c_j$, which are now functions of λ, must be nonpositive. The conditions for optimality for the current basic solution (x_1, x_2) are that

$$z_3 - c_3 = -1 + \tfrac{1}{25}\lambda \leq 0$$

and

$$z_4 - c_4 = -4 - \tfrac{3}{25}\lambda \leq 0$$

These two inequalities (there is one inequality for each nonbasic variable) must be satisfied simultaneously. They are consistent in that they have a solution for $\lambda = 0$. Solving them for other values of λ, we have from the first condition that $\lambda \leq 25$ and from the second that $-\frac{100}{3} \leq \lambda$. Taken together, we see that the current solution will remain optimal for $-\frac{100}{3} \leq \lambda \leq 25$. The reader will note that when $\lambda = 25$, $z = 70x_1 + 105x_2$, which is parallel to $10x_1 + 15x_2 = 400$; when $\lambda = -\frac{100}{3}$, $z = \frac{35}{3}x_1 + \frac{140}{3}x_2$, which is parallel to $5x_1 + 20x_2 = 400$. For $\lambda = 25$, both $(24, 14)$ and $(45, 0)$ are optimal solutions with $z_0 = \$3{,}150$. For $\lambda = -\frac{100}{3}$, both $(24, 14)$ and $(0, 20)$ are optimal solutions with $z_0 = \$2{,}800/3$.

We are also interested in determining optimal solutions for $\lambda > 25$ and $\lambda < -\frac{100}{3}$. As one might expect, the analysis calls for a change of basis. We describe why this is so and how the new solutions can be found systematically by developing the parametric objective-function procedure for the general case. (See Exercise 18 for the complete solution to the chair-table parametric cost-coefficient problem.)

Mathematically, the problem can be stated as follows: Let $\delta \leq \lambda \leq \phi$, where δ may be an arbitrary, algebraically small, but finite number and ϕ may be an arbitrary, algebraically large, but finite number. For each λ in this interval, find a vector $\mathbf{X} = (x_1, x_2, \ldots, x_n)$ which minimizes

$$\sum_{j=1}^{n} (d_j + \lambda d_j')x_j \tag{1.1}$$

subject to

$$\sum_{j=1}^{n} a_{ij}x_j = b_i \qquad i = 1, \ldots, m$$
$$x_j \geq 0 \qquad j = 1, \ldots, n \tag{1.2}$$

where d_j, d_j', a_{ij}, and b_i are given constants.

We assume that our problem is nondegenerate and that we have available a basic feasible solution of (1.2). Using the simplex method, we can solve our problem for $\lambda = \delta$ and obtain either a solution (Case A) or the information that the objective function (1.1) with $\lambda = \delta$ has no finite minimum on the convex set defined by (1.2) (Case B).

Case A. Since in our problem the cost coefficients $c_j = d_j + \lambda d_j'$, we can for any basis represent the terms $z_j - c_j$ as a linear function of λ. Let us write these linear functions as $z_j - c_j = \alpha_j + \lambda\beta_j$. Then, since we are in Case A,

$$\alpha_j + \delta\beta_j \leq 0 \qquad j = 1, \ldots, n$$

which means that the inequalities

$$\alpha_j + \lambda\beta_j \leq 0 \qquad j = 1, \ldots, n \tag{1.3}$$

are consistent. For all $\beta_j < 0$, we have

$$\lambda \geq -\frac{\alpha_j}{\beta_j}$$

and for all $\beta_j > 0$,

$$\lambda \leq -\frac{\alpha_j}{\beta_j}$$

Let us define

$$\underline{\lambda} = \begin{cases} \max\limits_{\beta_j < 0} -\dfrac{\alpha_j}{\beta_j} \\ \text{or} \\ -\infty \quad \text{if all } \beta_j \geq 0 \end{cases} \qquad\qquad \bar{\lambda} = \begin{cases} \min\limits_{\beta_j > 0} -\dfrac{\alpha_j}{\beta_j} \\ \text{or} \\ +\infty \quad \text{if all } \beta_j \leq 0 \end{cases}$$

Hence, by the simplex method, the current solution will yield the minimum for all λ satisfying (1.3), i.e., for all λ such that

$$\underline{\lambda} \leq \lambda \leq \bar{\lambda}$$

If $\bar{\lambda} = +\infty$, our problem is over. Assume then that $\bar{\lambda}$ is finite, $\bar{\lambda} = -\alpha_k/\beta_k$ for $\beta_k > 0$. If all the corresponding $x_{ik} \leq 0$, then we know from the simplex method and the definition of $\bar{\lambda}$ that our problem has no minimum for $\lambda > \bar{\lambda}$; thus we are finished.[1] If at least one $x_{ik} > 0$, then the simplex process introduces the vector $\mathbf{P}_k$ into the basis and removes a vector $\mathbf{P}_l$ in the usual manner. Note that

$$x_{lk} > 0 \tag{1.4}$$

Theorem 1. *The new basis yields a minimum for at least one value of λ. If $\underline{\lambda}' \leq \lambda \leq \bar{\lambda}'$ is the entire set of values of λ for which the new basis yields a minimum, then $\underline{\lambda}' = \bar{\lambda}$.*

Proof: The basis is certainly feasible, since we have followed the simplex prescription. Further, the inequalities (for the new basis)

$$\alpha_j' + \lambda\beta_j' \leq 0 \qquad j = 1, \ldots, n \tag{1.5}$$

are consistent. For if we imagine $\lambda = \bar{\lambda}$, we have inserted into the basis the vector $\mathbf{P}_k$ whose $z_k - c_k = \alpha_k + \bar{\lambda}\beta_k = 0$. The new basis will still be a minimum for $\lambda = \bar{\lambda}$, since $\bar{\lambda}$ satisfies (1.5).

All that remains to be shown is that $\lambda < \bar{\lambda}$ does not satisfy (1.5). We have

$$\alpha_l' = -\frac{\alpha_k}{x_{lk}}$$

$$\beta_l' = -\frac{\beta_k}{x_{lk}} \tag{1.6}$$

In order to satisfy the inequality from the set (1.5) which corresponds to the eliminated vector $\mathbf{P}_l$, we must have

$$\alpha_l' + \lambda\beta_l' \leq 0$$

Or, by (1.4) and (1.6),

$$-\alpha_k - \lambda\beta_k \leq 0$$

Since $\beta_k > 0$, this last inequality forces the new range of λ to be

$$\lambda \geq -\frac{\alpha_k}{\beta_k} = \bar{\lambda}$$

In this manner, we proceed from one range of values of λ to the next

---

[1] As in the general simplex method, the column vector $\mathbf{X}_k = (x_{1k}, \ldots, x_{mk})$, where $\mathbf{X}_k = \mathbf{B}^{-1}\mathbf{P}_k$ for the current basis $\mathbf{B}$.

until we include $\lambda = \phi$. To prove that this process is valid, we have only to assure ourselves that no basis is repeated. This assurance is provided by the following remark: If $\bar{\lambda}$ is replaced by $\bar{\lambda} + \epsilon$, where ϵ is any (conceptually small) positive number, the steps followed would be precisely the same as we have already described. The nondegeneracy assumption guarantees that we will either solve the problem for $\lambda = \bar{\lambda} + \epsilon$ or obtain the information that there is no minimum for that value of λ. Hence we cannot remain indefinitely at a value λ such that $\underline{\lambda} = \lambda = \bar{\lambda}$. After we leave a basis, we cannot return to it or to any basis corresponding to lower values of λ. By our theorem, these intervals of λ will not overlap. It is entirely possible (indeed, it has frequently occurred) that, corresponding to some solutions, $\underline{\lambda} = \bar{\lambda}$. But, as we have seen, this cannot persist indefinitely through the various changes of bases. The various $\underline{\lambda}$ and $\bar{\lambda}$ that arise are called *characteristic values of* λ, and the minimizing solutions corresponding to the various values of λ are called *characteristic solutions*.

Case B. Here, as we attempt to find a minimum feasible solution for $\lambda = \delta$, a vector chosen to go into the basis cannot do so because all of its elements x_{ik} are nonpositive. There are two possible situations:

1. Let $\mathbf{P}_k$ be the vector that cannot be introduced into the basis. Here we are given $\alpha_k + \delta\beta_k > 0$ and all $x_{ik} \leq 0$. If $\beta_k \geq 0$, then the problem has no finite minimum solutions for any λ.
2. If $\beta_k < 0$, the inequality $\alpha_k + \lambda\beta_k > 0$ will hold for all

$$\lambda < \lambda_1' = -(\alpha_k/\beta_k)$$

and hence there will be no finite minimum feasible solutions for λ in the region $\delta \leq \lambda < \lambda_1'$. At this stage, we do not know whether a finite minimum solution exists for λ_1'. If all $\alpha_j + \lambda_1'\beta_j \leq 0$, we have a minimum feasible solution for λ_1', and λ_1 can be determined by

$$\lambda_1 = \min_{\beta_j > 0} \; (-\alpha_j/\beta_j)$$

The characteristic solution holds for $\lambda_1' \leq \lambda \leq \lambda_1$, and the procedure of Case A can now be applied. If not all $\alpha_j + \lambda_1'\beta_j \leq 0$, any vector having $\alpha_j + \lambda_1'\beta_j > 0$ can be introduced into the basis. This criterion is used in the following iterations until all the transformed $\alpha_j + \lambda_1'\beta_j \leq 0$ or until a vector with $\alpha_t + \lambda_1'\beta_t > 0$ cannot be introduced because all its transformed elements are nonpositive. The former condition can be handled by the method of Case A. In the latter condition, if $\beta_t \geq 0$, no finite minimum solutions exist. For $\beta_t < 0$, we know there are no finite solutions for $\lambda < \lambda_2' = -(\alpha_t/\beta_t)$, where $\lambda_2' > \lambda_1'$. We now attempt to determine whether a finite solution exists for $\lambda = \lambda_2'$. Successive applications of the above procedure either will lead us to a finite minimum for some λ (and then Case A can be used) or will show that there are no λ for which a finite minimum exists.

If we are given a characteristic solution for an arbitrary range

$$\lambda_i \leq \lambda \leq \lambda_{i+1}$$

we can proceed to the right of λ_{i+1} by the procedure of Case A. We can proceed to the left of λ_i by introducing the vector $\mathbf{P}_q$ which has

$$\lambda_i = \max_{\beta_j < 0} -\frac{\alpha_j}{\beta_j} = -\frac{\alpha_q}{\beta_q}$$

If all $x_{iq} \leq 0$, then there are no finite solutions for $\lambda < \lambda_i$.

Summarizing, we have seen that

1. By a modification of the general simplex procedure, it is possible to investigate systematically and solve the one-parameter objective-function problem.
2. Given any finite minimum solution, we can determine a set of characteristic solutions and the associated characteristic values for all possible values of the parameter.
3. A solution is minimum over a closed interval of λ.
4. The set of λ for which minimum solutions exist is closed and connected.

We next illustrate the parametric objective-function computational procedure by the following examples.

Example 1.1. Minimize

$$x_1 + x_2 - \lambda x_3 + 2\lambda x_4$$

subject to

$$
\begin{aligned}
x_1 \quad\quad + \;\; x_3 + 2x_4 &= 2 \\
2x_1 + x_2 \quad\quad + 3x_4 &= 5 \\
x_j &\geq 0
\end{aligned}
\tag{1.7}
$$

for all values of λ.

The iterative steps and corresponding solutions are given in Tableau 8.2.

As given, the problem (1.7) has the feasible basis $(\mathbf{P}_3\,\mathbf{P}_2)$. The associated simplex tableau is stated in Step I of Tableau 8.2. Here the $(z_j - c_j)$ terms are written as $\alpha_j + \lambda\beta_j$, with the α_j entered in the $(m+1)$st row and the β_j in the $(m+2)$nd row. For example, $z_4 - c_4 = 3 - 4\lambda$, with a 3, -4 entered in the corresponding rows under $\mathbf{P}_4$. The basic feasible solution is $x_2 = 5$, $x_3 = 2$, $x_1 = x_4 = 0$, with the value of the objective function $5 - 2\lambda$ noted in the $\mathbf{P}_0$ column. For the Step I basic feasible solution, we need to determine the values of the parameter λ, if any, for which this solution will remain optimal.

The $z_j - c_j = \alpha_j + \lambda\beta_j$ terms for the vector $\mathbf{P}_1$ and $\mathbf{P}_4$ which are not in the basis form a set of linear inequalities in λ. This set must be analyzed to determine if there are values of λ for which the set will be nonpositive; i.e., we solve the inequalities $\alpha_j + \lambda\beta_j \leq 0$ for each $\mathbf{P}_j$ not in the basis. For Step I we have the inequalities $1 - \lambda \leq 0$ and $3 - 4\lambda \leq 0$ or $1 \leq \lambda$ and $\frac{3}{4} \leq \lambda$. Based on the definition of $\underline{\lambda}$ and $\bar{\lambda}$, we see that for Step I the basis $(\mathbf{P}_3\mathbf{P}_2)$ will be optimal whenever $\underline{\lambda} = 1 \leq \lambda \leq +\infty = \bar{\lambda}$.

TABLEAU 8.2.

I. Problem (1.7) in Simplex Tableau

				1	1	$-\lambda$	2λ
i	Basis	$\mathbf{c}$	$\mathbf{P}_0$	$\mathbf{P}_1$	$\mathbf{P}_2$	$\mathbf{P}_3$	$\mathbf{P}_4$
1	$\mathbf{P}_3$	$-\lambda$	2	①	0	1	2
2	$\mathbf{P}_2$	1	5	2	1	0	3
$m+1$			5	1	0	0	3
$m+2$			-2	-1	0	0	-4

II. Vector $\mathbf{P}_1$ Replaces Vector $\mathbf{P}_3$

$\mathbf{P}_1$	1	2	1	0	1	②
$\mathbf{P}_2$	1	1	0	1	-2	-1
		3	0	0	-1	1
		0	0	0	1	-2

III. Vector $\mathbf{P}_4$ Replaces Vector $\mathbf{P}_1$

$\mathbf{P}_4$	2λ	1	$\frac{1}{2}$	0	$\frac{1}{2}$	1
$\mathbf{P}_2$	1	2	$\frac{1}{2}$	1	$-\frac{3}{2}$	0
		2	$-\frac{1}{2}$	0	$-\frac{3}{2}$	0
		2	1	0	2	0

To determine if there are any optimal basic solutions for $\lambda < 1$, we introduce vector $\mathbf{P}_1$ into the basis as shown in Step II. The new basis $(\mathbf{P}_1\mathbf{P}_2)$ will be optimal for $\lambda = 1$. Applying the parametric-programming algorithm, we note that for $-1 + \lambda \leq 0$ and $1 - 2\lambda \leq 0$ (the $\alpha_j + \lambda\beta_j$ terms of the nonbasic vectors $\mathbf{P}_3$ and $\mathbf{P}_4$) the new solution $x_1 = 2$, $x_2 = 1$, $x_3 = x_4 = 0$ will be optimal for $\underline{\lambda} = \frac{1}{2} \leq \lambda \leq 1 = \bar{\lambda}$. The value of the objective function is 3 and is independent of the value of λ.

In Step III we attempt to determine optimal solutions for $\lambda < \frac{1}{2}$ and introduce $\mathbf{P}_4$ into the basis. The corresponding solution is $x_2 = 2$, $x_4 = 1$, $x_1 = x_3 = 0$, with objective-function value of $2 + 2\lambda$. The solution will be optimal for all λ which satisfy $-\frac{1}{2} + \lambda \leq 0$ and $-\frac{3}{2} + 2\lambda \leq 0$, or

$$\underline{\lambda} = -\infty \leq \lambda \leq \frac{1}{2} = \bar{\lambda}$$

We can summarize the optimal solutions as follows:

(I) $x_2 = 5, x_3 = 2, x_1 = x_4 = 0; z_0 = 5 - 2\lambda$ whenever $1 \le \lambda \le +\infty$

(II) $x_1 = 2, x_2 = 1, x_3 = x_4 = 0; z_0 = 3$ whenever $\frac{1}{2} \le \lambda \le 1$

(III) $x_2 = 2, x_4 = 1, x_1 = x_3 = 0; z_0 = 2 + 2\lambda$ whenever $-\infty \le \lambda \le \frac{1}{2}$

For this problem an optimal solution exists for any λ.

The graph of the value of the objective function in terms of λ is shown in Fig. 8.2. The general shape of the resultant curve is discussed in Exercise 12. From a geometric point of view, the parametrization of the objective function can be considered as changing the slope of the corresponding hyperplane. The movement from one extreme point to another is due to the rotation of the hyperplane about the current optimal extreme point until it lies on a face of the convex set of solutions, which includes the old and new extreme points. For these extreme points the optimal value of the objective function is the same for the corresponding characteristic value. We illustrate this description of the parametric process by the following example.

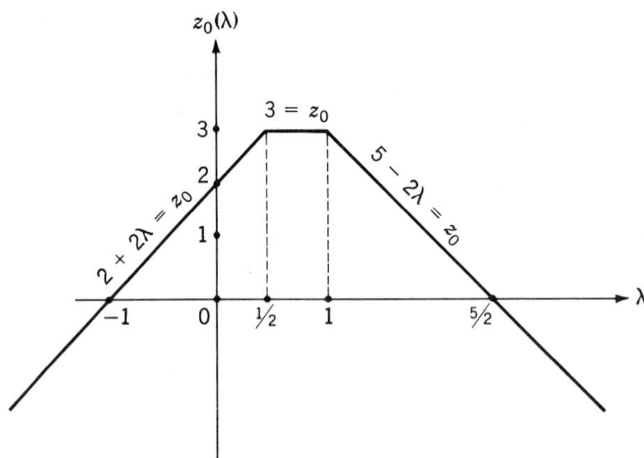

FIGURE 8.2

Example 1.2. Maximize

$$x_1 + \lambda x_2$$

subject to

$$x_1 + 3x_2 \le 3$$
$$2x_1 - 3x_2 \le 3 \tag{1.8}$$
$$x_j \ge 0$$

for all values of λ.

The solution space of inequalities (1.8) is shown by the shaded area of Fig. 8.3. The objective function is shown passing through the extreme point for which it assumes its maximum value for a specific value of λ, e.g., for $\lambda = 1$, $x_1 + x_2$ is maximum at the point $(2, \frac{1}{3})$. Also shown is the range of λ for which the corresponding extreme point remains optimal. Thus, the point $(\frac{3}{2}, 0)$ is optimal for $-\infty \le \lambda \le -\frac{3}{2}$, the point $(2, \frac{1}{3})$ is optimal for $-\frac{3}{2} \le \lambda \le 3$, and the point $(0, 1)$ is optimal for $3 \le \lambda \le +\infty$. We leave as an exercise for the reader the verification of the given characteristic values.

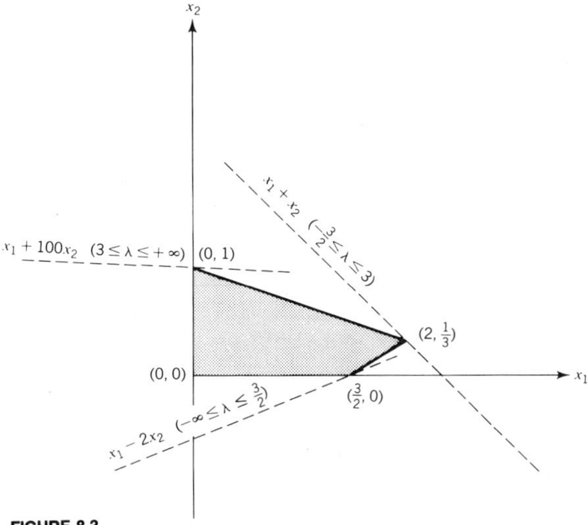

FIGURE 8.3

Example 1.3.   Minimize

$$\lambda x - y$$

subject to

$$3x - y \ge 5$$
$$2x + y \le 3 \tag{1.9}$$

where

$$-\infty < \delta \le \lambda \le \phi < +\infty$$

We transform system (1.9) to a set of equalities with nonnegative variables, with one artificial variable v with associated large positive cost coefficient w, to obtain the system (1.10):

$$3(x_1 - x_2) - (y_1 - y_2) - u_1 \qquad + v = 5$$
$$2(x_1 - x_2) + (y_1 - y_2) \qquad + u_2 \qquad = 3 \tag{1.10}$$

with the corresponding objective function of

$$\lambda(x_1 - x_2) - (y_1 - y_2) \qquad\qquad + wv$$

Step I of Tableau 8.3 is the system (1.10) in the usual simplex tableau. Since there is a natural unit vector, P_6, in the system, we need to employ only one artificial vector, P_7, with large positive weight, w. The first basis is $(P_7 P_6)$. The $z_j - c_j$ elements are written as $\alpha_j + \lambda\beta_j + w\gamma_j$ and are entered in the $(m+1)$st, $(m+2)$nd, and $(m+3)$rd rows, respectively. For example, vector P_1 has $z_1 - c_1 = -\lambda + 3w$; hence we enter 0, -1, 3 in the corresponding rows under P_1. The value of the objective function is $z = 5w$. Since P_1 has the largest positive value in the $(m+3)$rd

TABLEAU 8.3.

I. Problem (1.10) in Simplex Tableau (with Artificial Vector P_7)

i	Basis	c	P_0	λ	$-\lambda$	-1	1	0	0	w
				P_1	P_2	P_3	P_4	P_5	P_6	P_7
1	P_7	w	5	3	-3	-1	1	-1	0	1
2	P_6	0	3	②	-2	1	-1	0	1	0
$m+1$			0	0	0	1	-1	0	0	0
$m+2$			0	-1	1	0	0	0	0	0
$m+3$			5	3	-3	-1	1	-1	0	0

II. Vector P_1 Replaces Vector P_6

Basis	c	P_0	P_1	P_2	P_3	P_4	P_5	P_6	P_7
P_7	w	$\frac{1}{2}$	0	0	$-\frac{5}{2}$	⑤⁄②	-1	$-\frac{3}{2}$	1
P_1	λ	$\frac{3}{2}$	1	-1	$\frac{1}{2}$	$-\frac{1}{2}$	0	$\frac{1}{2}$	0
		0	0	0	1	-1	0	0	0
		$\frac{3}{2}$	0	0	$\frac{1}{2}$	$-\frac{1}{2}$	0	$\frac{1}{2}$	0
		$\frac{1}{2}$	0	0	$-\frac{5}{2}$	$\frac{5}{2}$	-1	$-\frac{3}{2}$	0

III. Vector P_4 Replaces the Artificial Vector P_7

Basis	c	P_0	P_1	P_2	P_3	P_4	P_5	P_6	P_7
P_4	1	$\frac{1}{5}$	0	0	-1	1	$-\frac{2}{5}$	$-\frac{3}{5}$	
P_1	λ	$\frac{8}{5}$	1	-1	0	0	$-\frac{1}{5}$	①⁄⑤	
		$\frac{1}{5}$	0	0	0	0	$-\frac{2}{5}$	$-\frac{3}{5}$	
		$\frac{8}{5}$	0	0	0	0	$-\frac{1}{5}$	$\frac{1}{5}$	
		0	0	0	0	0	0	0	

IV. Vector P_6 Replaces Vector P_1

Basis	c	P_0	P_1	P_2	P_3	P_4	P_5	P_6	P_7
P_4	1	5	3	-3	-1	1	-1	0	
P_6	0	8	5	-5	0	0	-1	1	
		5	3	-3	0	0	-1	0	
		0	-1	1	0	0	0	0	

row, it is introduced into the basis, and $\mathbf{P}_6$ is removed (Step II). Similarly we remove $\mathbf{P}_7$ by $\mathbf{P}_4$ to reach Step III. Now, since all the elements in the $(m+3)$rd row equal zero, we have our first feasible solution. The basis vectors are $\mathbf{P}_1$ and $\mathbf{P}_4$ and the solution is $x_1 = \frac{8}{5}$, $y_2 = \frac{1}{5}$; $x_2 = y_1 = u_1 = u_2 = 0$. The value of the objective function is $z_0 = \frac{1}{5} + \frac{8}{5}\lambda$.

Since we first wish to determine whether a finite solution for $\delta = \lambda$ exists, we should next introduce a vector with a *negative* element in row $m+2$. The only one is $\mathbf{P}_5$. However, all $x_{i5} \leq 0$, and hence there are no finite solutions for $\lambda < \lambda_1' = -\alpha_5/\beta_5 = -2$. Applying the procedure of Case B, we see that all $\alpha_j + \lambda_1'\beta_j \leq 0$, and Step III gives a minimum for $-2 \leq \lambda \leq 3$ ($\lambda_1 = -\alpha_6/\beta_6$). Introducing $\mathbf{P}_k = \mathbf{P}_6$ into the basis (we should expect a solution for $\lambda \geq 3$), we remove $\mathbf{P}_1$ and have a new feasible solution with $\mathbf{P}_4$ and $\mathbf{P}_6$ as basis vectors. The solution is $y_2 = 5$, $u_2 = 8$, $z_0 = 5$; $x_1 = x_2 = y_1 = u_1 = 0$. (Note that here the value of the objective function is independent of λ.) Applying the algorithm of Case A, we have $\mathbf{P}_k = \mathbf{P}_2$ and $\lambda_1 = \lambda_2 = 3$. $\mathbf{P}_2$ cannot be introduced, since all $x_{i2} \leq 0$, and we have no bounded solutions for $\lambda > 3$. Therefore, the problem has only one characteristic solution for $-2 \leq \lambda \leq 3$ and a multiple solution for $\lambda = 3$.

2. THE PARAMETRIC DUAL PROBLEM

For any parametric objective-function problem there is an associated dual problem whose right-hand-side elements are linear functions of the parameter. Such problems also arise naturally in that we often want to investigate how an optimal solution changes as a function of the right-hand-side elements. For example, the furniture manufacturer may be interested in knowing how the production schedule and profits vary as the available number of labor-hours is increased or decreased about the given 400; similarly for changes in the 450 units of lumber. A reasonable parametric representation of the manufacturer's concerns might be to let $b_1 = 400 + \theta$ and $b_2 = 450 + 5\theta$; i.e., for every 5 units change in lumber, a corresponding unit change in labor-hours is required. Here we assume that the parameter θ can take on any value. The parametric dual problem for this chair-table example is to maximize

$$z = 45x_1 + 80x_2$$

subject to

$$5x_1 + 20x_2 \leq 400 + \theta$$
$$10x_1 + 15x_2 \leq 450 + 5\theta$$
$$x_i \geq 0$$

for all values of θ.

We see from Fig. 8.1 that as θ takes on values other than $\theta = 0$, the extreme point formed by the intersection of the constraint lines will be shifted to the right for $\theta > 0$ and to the left for $\theta < 0$. Also, as $-\infty < \theta < \infty$, there may be values of θ for which the problem is infeasible.

Where in the parametric objective-function problem we were concerned with the conditions of optimality as the parameter changed, the parametric dual problem requires an analysis of the conditions of feasibility. We illustrate this for the furniture-manufacturer problem with right-hand sides $b_1 = 400 + \theta$ and $b_2 = 450 + 5\theta$. The optimal solution for $\theta = 0$ is given in Tableau 8.1 (we ignore the cost-coefficient parameter λ). Also from Tableau 8.1, we have that the inverse for the optimal basis $\mathbf{B} = (\mathbf{P}_2, \mathbf{P}_1)$ is given by

$$\mathbf{B}^{-1} = \begin{pmatrix} \frac{2}{25} & -\frac{1}{25} \\ -\frac{3}{25} & \frac{4}{25} \end{pmatrix}$$

For the parametric right-hand-side vector

$$\mathbf{b} = \begin{pmatrix} 400 + \theta \\ 450 + 5\theta \end{pmatrix}$$

we have the solution

$$\begin{pmatrix} x_2 \\ x_1 \end{pmatrix} = \begin{pmatrix} \frac{2}{25} & -\frac{1}{25} \\ -\frac{3}{25} & \frac{4}{25} \end{pmatrix} \begin{pmatrix} 400 + \theta \\ 450 + 5\theta \end{pmatrix} = \begin{pmatrix} 14 - \frac{3}{25}\theta \\ 24 + \frac{17}{25}\theta \end{pmatrix}$$

and

$$z_0 = c_1 x_1 + c_2 x_2 = 45(24 + \tfrac{17}{25}\theta) + 80(14 - \tfrac{3}{25}\theta)$$
$$= 2{,}200 + \tfrac{14}{25}\theta$$

For $\theta = 0$, the unparametrized optimal solution is $x_1 = 24$, $x_2 = 14$, and $z_0 = 2{,}200$. The corresponding basis $(\mathbf{P}_2, \mathbf{P}_1)$ will remain optimal for other values of θ as long as the values of the variables stay nonnegative, i.e., as θ changes, we must have

$$x_1 = 24 + \tfrac{17}{25}\theta \geq 0$$

and

$$x_2 = 14 - \tfrac{3}{25}\theta \geq 0$$

These inequalities (one for every constraint of the problem) are consistent as they are satisfied by $\theta = 0$. Solving them for θ, we have from the first condition that $\theta \geq -\frac{600}{17}$ and from the second condition that $\theta \leq \frac{350}{3}$. Taken together they tell us that the basis $(\mathbf{P}_2, \mathbf{P}_1)$ will remain optimal and feasible for $-\frac{600}{17} \leq \theta \leq \frac{350}{3}$. That is, $x_1 = 24 + \frac{17}{25}\theta$, $x_2 = 14 - \frac{3}{25}\theta$ and $z_0 = 2{,}200 + \frac{14}{25}\theta$ will be optimal and feasible for that range of θ. We note that, when $\theta = \frac{350}{3}$, then $x_2 = 0$ and any value of $\theta > \frac{350}{3}$ will cause x_2 to go negative. Similarly, for $\theta = -\frac{600}{17}$, $x_1 = 0$ and goes negative for $\theta < -\frac{600}{17}$. This implies that we must change bases or determine that there are no feasible solutions outside the range $-\frac{600}{17} \leq \theta \leq \frac{350}{3}$. Why this is so and how it is done are described by the following discussion of the computational procedure for the parametric dual problem. (See Exercise 19 for the complete solution to the chair-table parametric dual problem.)

The mathematical statement of the parametric dual problem is as follows. Let $\sigma \leq \theta \leq \rho$. For each θ in this interval find a vector

$$\mathbf{X} = (x_1, x_2, \ldots, x_n)$$

which minimizes

$$\sum_{j=1}^{n} c_j x_j$$

subject to

$$\sum_{j=1}^{n} a_{ij} x_j = b_i + \theta b_i' \qquad i = 1, 2, \ldots, m$$

and

$$x_j \geq 0 \qquad j = 1, 2, \ldots, n$$

Assume that we have found a minimum basic feasible solution $\mathbf{X}^0 = (x_{10}, x_{20}, \ldots, x_{m0})$ for the problem when $\theta = \sigma$. We can write each x_{i0} as a linear combination of σ and have

$$x_{i0} = q_i + \sigma p_i \geq 0$$

which implies that the inequalities

$$x_{i0} = q_i + \theta p_i \geq 0 \tag{2.1}$$

are consistent.

If all $p_i = 0$, then the solution $\mathbf{X}^0$ is a minimum feasible solution for all θ; if all $p_i \geq 0$, then the solution $\mathbf{X}^0$ is a minimum feasible solution for all $\theta \geq \sigma$; and if all $p_i \leq 0$, the solution $\mathbf{X}^0$ is a minimum feasible solution for all $\theta \leq \sigma$. In general, however, the p_i will be both positive and negative, and in order to determine for what range of values of θ the given solution $\mathbf{X}^0$ is a minimum, we must perform an analysis similar to that performed for the parametric objective-function problem.

For $p_i > 0$, we have

$$\theta \geq -\frac{q_i}{p_i}$$

and define

$$\underline{\theta} = \begin{cases} \max_{p_i > 0} -\dfrac{q_i}{p_i} \\ \text{or} \\ -\infty \qquad \text{if all } p_i \leq 0 \end{cases}$$

For $p_i < 0$, we have

$$\theta \leq -\frac{q_i}{p_i}$$

and define

$$\bar{\theta} = \begin{cases} \min_{p_i < 0} -\dfrac{q_i}{p_i} \\ \text{or} \\ +\infty & \text{if all } p_i \geq 0 \end{cases}$$

The given solution (2.1) is then a minimum for

$$\underline{\theta} \leq \theta \leq \bar{\theta}$$

We shall assume that the upper bound θ is not $+\infty$. As θ is increased, the solution remains optimal, i.e., each $z_j - c_j \leq 0$, but it need not stay feasible.

For a sufficiently large increase in θ, one of the values

$$x_{i0} = q_i + \theta p_i$$

will be forced negative. If x_{l0} is the first of these variables to go negative, then

$$\bar{\theta} = -\frac{q_l}{p_l}$$

where $p_l < 0$. We next wish to determine a new minimum feasible solution for a range of $\theta \geq \bar{\theta}$. We have to determine a vector to be introduced into the basis and a vector to be removed that will keep the right-hand elements of the transformed equations nonnegative and the transformed $z_j - c_j$ nonpositive. The method of selection is summarized as follows:

Theorem 2. *If the vector $\mathbf{P}_l$ corresponding to $\bar{\theta} = -q_l/p_l$ is removed from the basis and a vector $\mathbf{P}_k$ with*

$$\frac{z_k - c_k}{x_{lk}} = \min_{x_{lj} < 0} \frac{z_j - c_j}{x_{lj}}$$

is introduced into the basis, the new basic solution is a minimum feasible solution for at least one value of θ. If $\underline{\theta}' \leq \theta \leq \bar{\theta}'$ is the entire set of θ for which the new basis yields a minimum, then $\underline{\theta}' = \bar{\theta}$.

Proof: The new solution $\bar{\mathbf{X}}^0 = (\bar{x}^0_{10}, \bar{x}^0_{20}, \ldots, \bar{x}^0_{m0})$ is given by

$$\bar{x}^0_{i0} = q_i' + \theta p_i' = q_i + \theta p_i - \frac{x_{ik}}{x_{lk}}(q_l + \theta p_l) \qquad \text{for } i \neq l$$

$$\bar{x}^0_{k0} = q_l' + \theta p_l' = \frac{q_l + \theta p_l}{x_{lk}}$$

The new solution is certainly feasible for $\theta = \bar{\theta} = -q_l/p_l$. If $\bar{\mathbf{X}}^0$ is feasible for any other θ, then it must be for $\theta \geq \bar{\theta}$, since $x_{lk} < 0$, $p_l < 0$, and

$$\frac{q_l + \theta p_l}{x_{lk}} \geq 0$$

implies

$$\theta \geq -\frac{q_l}{p_l} = \bar{\theta}$$

To show that the new basis is optimal, we have

$$(z_j - c_j)' = z_j - c_j - \frac{x_{lj}}{x_{lk}}(z_k - c_k)$$

Since all $z_j - c_j \leq 0$ and $x_{lk} < 0$, all $(z_j - c_j)'$ with an $x_{lj} \geq 0$ are also non-positive. In order to have

$$z_j - c_j - \frac{x_{lj}}{x_{lk}}(z_k - c_k) \leq 0$$

for $x_{lj} < 0$, we must have

$$z_j - c_j \leq \frac{x_{lj}}{x_{lk}}(z_k - c_k)$$

or

$$\frac{z_j - c_j}{x_{lj}} \geq \frac{z_k - c_k}{x_{lk}}$$

Hence $\mathbf{P}_k$ must be the vector not in the basis that corresponds to

$$\min_{x_{lj} < 0} \frac{z_j - c_j}{x_{lj}}$$

If all $x_{lj} \geq 0$, then there are no feasible solutions for $\theta > \bar{\theta}$. The proof is left to the reader (Exercise 26).

A similar discussion holds, with appropriate changes, for finding solutions for values of $\theta \leq \underline{\theta}$.

To illustrate the parametric right-hand-side computational procedure we solve the following example:

Example 2.1. Minimize

$$x_1 + x_2 + 2x_3 + x_4$$

subject to

$$x_1 \qquad - 2x_3 - x_4 = 2 - \theta$$
$$x_2 - \ x_3 + x_4 = -1 + \theta \qquad\qquad (2.2)$$
$$x_j \geq 0$$

We develop the simplex tableau as shown in Tableau 8.4. In Step I, we see that the basis $(\mathbf{P}_1 \mathbf{P}_2)$ has all $z_j - c_j \leq 0$ with $x_1 = q_1 + \theta p_1 = 2 - \theta$ and $x_2 = q_2 - \theta p_2 = -1 + \theta$. (The values of the basis x_i are split into q_i and p_i

terms as shown in column $\mathbf{P}_0$.) If a value of θ can be found so that both x_1 and x_2 are nonnegative, then the basis will also be an optimal feasible basis. Thus, we need to solve the inequalities $2 - \theta \geq 0$ and $-1 + \theta \geq 0$ to determine if they are consistent, i.e., if a θ exists which satisfies both simultaneously. The procedure above shows that this will be true if $1 \leq \theta \leq 2$. The step I optimal, feasible solution is then $x_1 = 2 - \theta$, $x_2 = -1 + \theta$, $x_3 = x_4 = 0$, $z_0 = 1$ for $1 \leq \theta \leq 2$. To determine a solution for $\theta > 2$, the procedure requires vector $\mathbf{P}_4$ to replace vector $\mathbf{P}_1$. The reader will note that this calls for a negative pivot element. The resultant solution is shown in Step II. This solution with $x_2 = 1$, $x_4 = -2 + \theta$, $x_1 = x_3 = 0$, and $z_0 = -1 + \theta$ is optimal for $2 \leq \theta \leq +\infty$. To find an optimal feasible solution for $\theta < 1$, we need to go back to Step I and pivot on the second row. Doing this we find vector $\mathbf{P}_3$ replacing vector $\mathbf{P}_2$, with the optimal solution for $-\infty \leq \theta \leq 1$ being $x_1 = 4 - 3\theta$, $x_3 = 1 - \theta$, and $z_0 = 6 - 5\theta$, Step III. The graph of the objective function against values of the parameter θ is shown in Fig. 8.4. (See Exercise 12 for a statement of the general form of the graph.)

TABLEAU 8.4

I. Problem (2.2) in Simplex Tableau

Basis	c	$\mathbf{P}_0$	1	1	2	1
			$\mathbf{P}_1$	$\mathbf{P}_2$	$\mathbf{P}_3$	$\mathbf{P}_4$
$\mathbf{P}_1$	1	2	-1	1	0	-2	$\boxed{-1}$
$\mathbf{P}_2$	1	-1	1	0	1	(-1)	1
		1	0	0	0	-5	-1

II. Vector $\mathbf{P}_4$ Replaces Vector $\mathbf{P}_1$ in Step I

$\mathbf{P}_4$	1	-2	1	-1	0	2	1
$\mathbf{P}_2$	1	1	0	1	1	-3	0
		-1	1	-1	0	-3	0

III. Vector $\mathbf{P}_3$ Replaces Vector $\mathbf{P}_2$ in Step I

$\mathbf{P}_1$	1	4	-3	1	-2	0	-3
$\mathbf{P}_3$	2	1	-1	0	-1	1	-1
		6	-5	0	-5	0	-6

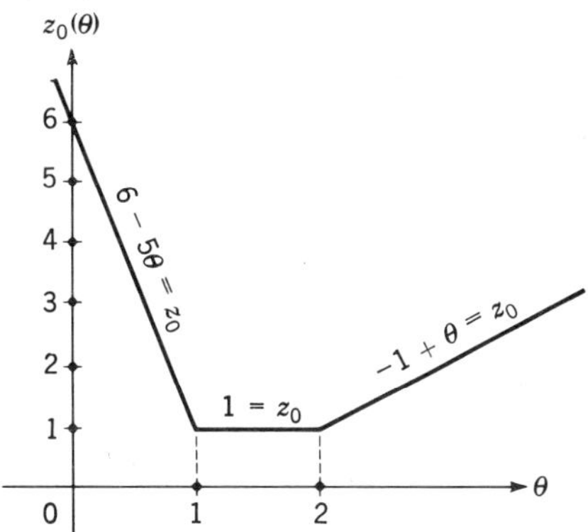

FIGURE 8.4

The ability to carry out a full range of parametric analyses is a standard feature of mathematical programming systems (see, for example, IBM [387] and Sperry Univac [619]). MPS have the following parametric options: PARAOBJ (parametric objective function), PARARHS (parametric dual— right-hand side), PARARIM in which both the objective function and right-hand-side column are functions of the same parameter, PARACOL in which the coefficients of a designated column are functions of a parameter, and PARAROW in which the coefficients of a given row are functions of a parameter. For each of these options, the MPS determines a series of optimal solutions for the specified range of the parameter. See Orchard-Hays [532] and Gal [274] for details of PARARIM, PARACOL, and PARAROW.

3. SENSITIVITY ANALYSIS

Except for the discussions in Secs. 1 and 2, we have assumed that all the coefficients of a linear-programming problem are given. However, for many problems either these constants, i.e., the a_{ij}, b_i, and c_j, are estimates or they vary over time. For example, in a diet problem such as the chicken-feed problem, the cost of any individual grain will vary from week to week, and it is quite important to know the cost range for which the solution remains optimal. Similar situations may arise for other elements of a problem. Investigations that deal with the changes in the optimal solution due to changes in the data are termed *sensitivity (postoptimal) analyses*. In this section, we shall be concerned with the analysis that determines the range of a given element for which the optimal basis remains optimal. Additional work in sensitivity and related areas is given in Shetty [610], Courtillot [142], Garvin [282], Madansky [486], Wagner [679], Saaty [584], Webb [691], and Gal [274].

Although sensitivity analysis can be treated as a special case of parametric programming (see Exercise 46), we shall develop it as a direct extension of the usual simplex computations. We illustrate this derivation by analysis of the furniture-manufacturer problem given in Sec. 1. The problem is to maximize

$$z = 45x_1 + 80x_2$$

subject to

$$5x_1 + 20x_2 \leq 400 \qquad \text{(labor-hours)}$$
$$10x_1 + 15x_2 \leq 450 \qquad \text{(units of lumber)}$$
$$x_j \geq \quad 0$$

Treating the problem as a minimization problem and adding slack variables x_3 and x_4, the optimal solution (as calculated in Sec. 2, Chap. 4) is given in Tableau 8.5. Figure 8.5 describes the problem geometrically and shows the objective function passing through the optimal extreme point (24, 14). For some problem coefficient, we now want to determine the range of its values that keeps this extreme point optimal. We first concentrate on the cost coefficients c_j, varying them one at a time.

TABLEAU 8.5

	Basis	c	$\mathbf{P}_0$	-45 $\mathbf{P}_1$	-80 $\mathbf{P}_2$	0 $\mathbf{P}_3$	0 $\mathbf{P}_4$
1	$\mathbf{P}_2$	-80	14	0	1	$\frac{2}{25}$	$-\frac{1}{25}$
2	$\mathbf{P}_1$	-45	24	1	0	$-\frac{3}{25}$	$\frac{4}{25}$
3			$-2{,}200$	0	0	-1	-4

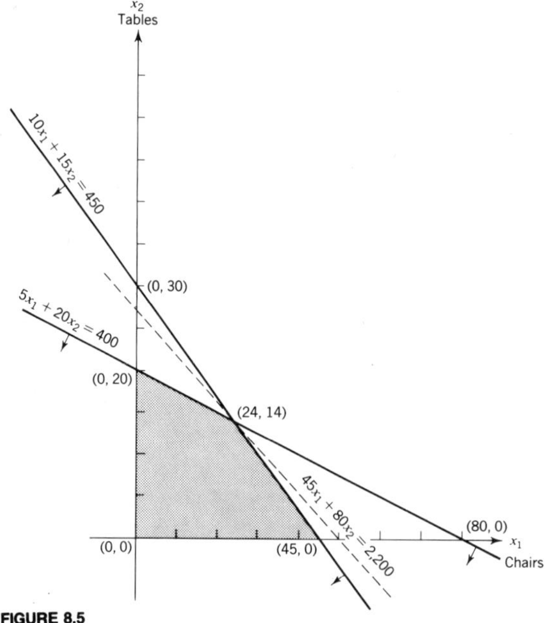

FIGURE 8.5

Let c_1 be changed to $c_1 + \Delta c_1 = (45 + \Delta c_1)$, where Δc_1 is some positive or negative increment to c_1. From Fig. 8.5 we see that as Δc_1 increases, the objective function z rotates about the extreme point $(24, 14)$ until z becomes parallel to the constraint $10x_1 + 15x_2 = 450$. For that value of Δc_1 (we shall show that it is $\frac{25}{3}$), the extreme points $(24, 14)$ and $(45, 0)$ are multiple optimal solutions. If $\Delta c_1 > \frac{25}{3}$, then the current basic solution $(24, 14)$ will no longer be optimal. Similarly, if Δc_1 takes on negative values, then eventually the objective function will be parallel to the constraint $5x_1 + 20x_2 = 400$. This happens when $\Delta c_1 = -25$. For that value both $(24, 14)$ and $(0, 20)$ will be multiple optimal solutions. If $\Delta c_1 < -25$, then the basic solution $(24, 14)$ will no longer be optimal. Thus, if $-25 \leq \Delta c_1 \leq \frac{25}{3}$, then the original optimal basic solution will remain optimal. The optimal value of the objective function would change to

$$z_0 = (45 + \Delta c_1)x_1 + 80x_2 = (45 + \Delta c_1)(24) + 80(14) = 2{,}200 + 24\Delta c_1$$

Computationally, we can perform a sensitivity analysis of a c_j by just knowing the information in the final simplex tableau. To demonstrate this, we consider the minimization form of the furniture-manufacturer problem that has the objective function of $z = (-45 + \Delta c_1)x_1 - 80x_2$. If we treat the Δc_1 as an algebraic quantity (but ignore it when carrying out the simplex computations), the final simplex tableau would be that shown as Tableau 8.6.

TABLEAU 8.6

	Basis	c	$\mathbf{P}_0$	$-45 + \Delta c_1$ $\mathbf{P}_1$	-80 $\mathbf{P}_2$	0 $\mathbf{P}_3$	0 $\mathbf{P}_4$
1	$\mathbf{P}_2$	-80	14	0	1	$\frac{2}{25}$	$-\frac{1}{25}$
2	$\mathbf{P}_1$	$-45 + \Delta c_1$	24	1	0	$-\frac{3}{25}$	$\frac{4}{25}$
3			$-2{,}200 + 24\Delta c_1$	0	0	$-1 - \frac{3}{25}\Delta c_1$	$-4 + \frac{4}{25}\Delta c_1$

For $\Delta c_1 = 0$, Tableau 8.6 reduces to the original optimal tableau. When $\Delta c_1 \neq 0$, the basis $(\mathbf{P}_2\mathbf{P}_1)$ will remain optimal only if the $z_j - c_j$ terms of the nonbasic vectors $\mathbf{P}_3$ and $\mathbf{P}_4$ are nonpositive. Thus, the conditions on Δc_1 that allow $(\mathbf{P}_2\mathbf{P}_1)$ to stay optimal are given by

$$z_3 - c_3 = -1 - \tfrac{3}{25}\Delta c_1 \leq 0$$

and

$$z_4 - c_4 = -4 + \tfrac{4}{25}\Delta c_1 \leq 0$$

Solving for Δc_1, we have $-\frac{25}{3} \leq \Delta c_1 \leq 25$. (This range is the negative of the one previously determined as here we are dealing directly with the negative of the original objective function.) The associated range for c_1 is $-\frac{160}{3} \leq c_1 \leq -20$. The reader should note that the conditions on Δc_1 were found by using the coefficients of the nonbasic columns that are in the row corresponding to the basis representation of P_1 (here row 2). For the sensitivity analysis of c_2, the reader can then verify that the conditions on Δc_2 are

$$-1 + \tfrac{2}{25}\Delta c_2 \le 0 \qquad \text{and} \qquad -4 - \tfrac{1}{25}\Delta c_2 \le 0$$

which yields the ranges $-100 \le \Delta c_2 \le \tfrac{25}{2}$ and $-180 \le c_2 \le -\tfrac{135}{2}$, with $z_0 = 2{,}200 + 14\Delta c_2$. As will be discussed below, the sensitivity analysis for a variable not in the optimal basic solution (here x_3 and x_4) is straightforward; the change Δc_j is restricted by $\Delta c_j \ge z_j - c_j$.

We next investigate the impact of changes in the right-hand-side elements b_i. For the furniture-manufacturer problem, we are interested in how the optimal solution changes as we vary the given resource levels, i.e., vary the 400 labor-hours or the 450 units of lumber. As such resource information is often inexact, it is important to determine the conditions under which the optimal production activities (the basis) remain an optimal set. Sensitivity analysis does this by examining each b_i one at a time. We let the b_i be changed to $b_i + \Delta b_i$, where Δb_i represents a positive or negative increment. For example, let us assume labor-hours are given as $400 + \Delta b_1$. Carrying out the simplex computations with this expression, we find that the optimal values of the basic variables are functions of Δb_1, i.e., $x_1 = 24 - \tfrac{3}{25}\Delta b_1$ and $x_2 = 14 + \tfrac{2}{25}\Delta b_1$, with $z_0 = 2{,}200 + \Delta b_1$. Values of Δb_1 must be such to ensure that x_1 and x_2 remain nonnegative. [The optimality conditions for the basis $(P_2\,P_1)$ will be satisfied for all values of Δb_1.] Thus, to determine the range of Δb_1 for which the optimal feasible basis $(P_2\,P_1)$ stays optimal and feasible, we must have

$$x_1 = 24 - \tfrac{3}{25}\Delta b_1 \ge 0 \qquad \text{and} \qquad x_2 = 14 + \tfrac{2}{25}\Delta b_1 \ge 0$$

or $-175 \le \Delta b_1 \le 200$ and $225 \le b_1 \le 600$. When Δb_1 assumes its lower and upper bounds, the optimal extreme points are $(45, 0)$ and $(0, 30)$, respectively.

As discussed below, the expressions of x_1, x_2, and z in terms of the Δb_i can be determined readily if the optimal simplex tableau contains an explicit representation of the inverse of the optimal basis. This is the case for the furniture-manufacturer problem (Tableau 8.6). The coefficients of Δb_1 come from column P_3, the transformed slack column for the labor-hour resource. The analysis for the lumber resource b_2 yields the functions

$$x_1 = 24 + \tfrac{4}{25}\Delta b_2 \ge 0 \qquad \text{and} \qquad x_2 = 14 - \tfrac{1}{25}\Delta b_2 \ge 0$$

and, thus, $-150 \le \Delta b_2 \le 350$, $300 \le b_2 \le 800$, and $z_0 = 2{,}200 + 4\Delta b_2$. The reader should investigate the geometric aspects of varying a b_i and note the conditions that force a basis change (see Fig. 8.5).

We next develop the theoretical basis of sensitivity analysis for linear programming and treat in turn the c_j, b_i, and a_{ij}. Elements of the theory are illustrated for the numerical problems of Tableau 4.3 and Tableau 8.5.

  a. Variation of the c_j. For an optimal solution we have the set of inequalities $z_j - c_j \le 0$ holding for all j.

Let Δc_j be the amount to be added to the corresponding c_j. For those variables not in the final basis, we must have $z_j - (c_j + \Delta c_j) \le 0$; hence, $z_j - c_j \le \Delta c_j$, with Δc_j having no upper bound. An appropriate change in c_j does not change the value of the objective function since $x_j = 0$. For the example of Tableau 4.3, we have for those variables not in the final solution that $-\frac{1}{5} \le \Delta c_1$, $-\frac{4}{5} \le \Delta c_4$, and $-\frac{12}{5} \le \Delta c_5$. For Tableau 8.5, changes in the nonbasic c_j's are restricted by $-1 \le \Delta c_3$ and $-4 \le \Delta c_4$.

For a variable in the final basis, a Δc_j affects all z_j, for all j not in the basis, since

$$z_j - c_j = \sum_{i \in \mathbf{B}} x_{ij} c_i - c_j \le 0^{\dagger}$$

Let the change Δc_k occur for some basis variable x_k. Then

$$\sum_{i \in \mathbf{B}} x_{ij} c_i + x_{kj} \Delta c_k - c_j \le 0$$

or

$$x_{kj}\, \Delta c_k \le -(z_j - c_j)$$

For those $x_{kj} > 0$

$$\Delta c_k \le \frac{-(z_j - c_j)}{x_{kj}}$$

and for those $x_{kj} < 0$

$$\Delta c_k \ge \frac{-(z_j - c_j)}{x_{kj}}$$

Hence

$$\max_{x_{kj} < 0} \frac{-(z_j - c_j)}{x_{kj}} \le \Delta c_k \le \min_{x_{kj} > 0} \frac{-(z_j - c_j)}{x_{kj}}$$

for all j not in the basis. If no $x_{kj} > 0$, there is no upper bound, and if no $x_{kj} < 0$, there is no lower bound. For the example of Tableau 4.3, let $\Delta c_k = \Delta c_6$. We have

$$\left(\frac{\frac{4}{5}}{-\frac{1}{2}}\right) \le \Delta c_6 \le \min \left(\frac{\frac{1}{5}}{1}, \frac{\frac{12}{5}}{10}\right)$$

$$-\frac{8}{5} \le \Delta c_6 \le \frac{1}{5}$$

For $\Delta c_k = \Delta c_2$, we have

$$-\infty \le \Delta c_2 \le \min \left(\frac{\frac{1}{5}}{\frac{2}{5}}, \frac{\frac{4}{5}}{\frac{1}{10}}, \frac{\frac{12}{5}}{\frac{4}{5}}\right)$$

$$-\infty \le \Delta c_2 \le \frac{1}{2}$$

---

† The notation $\displaystyle\sum_{i \in \mathbf{B}}$ means that the sum is taken for the set of indices i of those vectors in the basis $\mathbf{B}$.

For Tableau 8.5, we have for $\Delta c_k = c_1$

$$\left(\frac{1}{-\frac{3}{25}}\right) \leq \Delta c_1 \leq \left(\frac{4}{\frac{4}{25}}\right)$$

$$-\frac{25}{3} \leq \Delta c_1 \leq 25$$

and for $\Delta c_k = \Delta c_2$

$$\left(\frac{4}{-\frac{1}{25}}\right) \leq \Delta c_2 \leq \left(\frac{1}{\frac{2}{25}}\right)$$

$$-100 \leq \Delta c_2 \leq \frac{25}{2}$$

With Δc_k restricted by the appropriate bounds, the basis remains optimal but the value of the objective function changes by $\Delta c_k x_{k0}$.

b. Variation of the b_i. A change in a b_i must be of the magnitude that preserves the feasibility of the basis. For the optimal solution, $\mathbf{X}^0 = \mathbf{B}^{-1}\mathbf{b} \geq \mathbf{0}$, and for a change Δb_l in b_l, we must have, letting $\bar{\mathbf{b}}$ be the new right-hand-side vector,

$$\bar{\mathbf{X}}^0 = \mathbf{B}^{-1}\bar{\mathbf{b}} = (x_{i0} + b_{il}\,\Delta b_l) \geq \mathbf{0} \tag{3.1}$$

for all i in the basis and where b_{il} is the element in the ith row and lth column of $\mathbf{B}^{-1}$.

For $b_{il} > 0$, we have

$$\Delta b_l \geq \frac{-x_{i0}}{b_{il}}$$

and for $b_{il} < 0$

$$\Delta b_l \leq \frac{-x_{i0}}{b_{il}}$$

Hence,

$$\max_{b_{il}>0} \frac{-x_{i0}}{b_{il}} \leq \Delta b_l \leq \min_{b_{il}<0} \frac{-x_{i0}}{b_{il}}$$

For the example of Tableau 4.3, $\mathbf{B} = (\mathbf{P}_2\mathbf{P}_3\mathbf{P}_6)$, $\mathbf{X}^0 = (4, 5, 11)$, and

$$\mathbf{B}^{-1} = \begin{pmatrix} \frac{2}{5} & \frac{1}{10} & 0 \\ \frac{1}{5} & \frac{3}{10} & 0 \\ 1 & -\frac{1}{2} & 1 \end{pmatrix}$$

Letting $\Delta b_l = \Delta b_2$ yields

$$\max \left(\frac{-4}{\frac{1}{10}}, \frac{-5}{\frac{3}{10}}\right) \leq \Delta b_2 \leq \left(\frac{-11}{-\frac{1}{2}}\right)$$

$$\frac{-50}{3} \leq \Delta b_2 \leq 22$$

For Tableau 8.5, $\mathbf{B} = (\mathbf{P}_2 \mathbf{P}_1)$, $\mathbf{X}^0 = (x_{20}, x_{10}) = (24, 14)$, and

$$\mathbf{B}^{-1} = \begin{pmatrix} \frac{2}{25} & -\frac{1}{25} \\ -\frac{3}{25} & \frac{4}{25} \end{pmatrix}$$

For $\Delta b_l = \Delta b_1$, we have

$$\left(\frac{-14}{\frac{2}{25}} \right) \leq \Delta b_1 \leq \left(\frac{-24}{-\frac{3}{25}} \right)$$

$$-175 \leq \Delta b_1 \leq 200$$

and for $\Delta b_l = \Delta b_2$, we have

$$\left(\frac{-24}{\frac{4}{25}} \right) \leq \Delta b_2 \leq \left(\frac{-14}{-\frac{1}{25}} \right)$$

$$-150 = \Delta b_2 = 350$$

The new solution is given by (3.1), and the change in the value of the objective function is given by $\sum_{i \in \mathbf{B}} b_{il} \, \Delta b_l c_i$.

c. Variation in a_{ij}. Let a_{lk}, the element to be varied, be the element of the lth row of column $\mathbf{P}_k$. Assume that $\mathbf{P}_k$ is a vector of the optimal basis and that the optimal solution is $\mathbf{X}^0 = \mathbf{B}^{-1} \mathbf{b} = (x_{10}, \ldots, x_{m0})$. The new matrix is given by $\bar{\mathbf{B}} = \mathbf{B} + \Delta a_{lk} \mathbf{0}_{lk}$, where $\mathbf{0}_{lk}$ is a null matrix except for element (l, k), which equals unity. To preserve feasibility (assuming that $\bar{\mathbf{B}}$ is a nonsingular matrix), we must have

$$\bar{\mathbf{B}}^{-1} \mathbf{b} = (\mathbf{B} + \Delta a_{lk} \mathbf{0}_{lk})^{-1} \mathbf{b} \geq 0 \tag{3.2}$$

and to preserve optimality

$$z_j - c_j = \mathbf{c}_0 \bar{\mathbf{B}}^{-1} \mathbf{P}_j - c_j \leq 0 \tag{3.3}$$

for all j, where $\mathbf{c}_0$ is the set of cost coefficients for the basis variables. We have

$$\bar{\mathbf{B}} = \mathbf{B}(\mathbf{I} + \mathbf{B}^{-1} \Delta a_{lk} \mathbf{0}_{lk})$$

and, by Sec. 1 of Chap. 2,

$$\bar{\mathbf{B}}^{-1} = (\mathbf{I} + \mathbf{B}^{-1} \Delta a_{lk} \mathbf{0}_{lk})^{-1} \mathbf{B}^{-1}$$

Equation (3.2) can then be rewritten as

$$\bar{\mathbf{B}}^{-1} \mathbf{b} = (\mathbf{I} + \mathbf{B}^{-1} \Delta a_{lk} \mathbf{0}_{lk})^{-1} \mathbf{B}^{-1} \mathbf{b}$$

$$= (\mathbf{I} + \mathbf{B}^{-1} \Delta a_{lk} \mathbf{0}_{lk})^{-1} \mathbf{X}^0 \tag{3.4}$$

We now must determine for what conditions on Δa_{lk} the inverse

$$(\mathbf{I} + \mathbf{B}^{-1} \Delta a_{lk} \mathbf{0}_{lk})^{-1}$$

exists and restricts (3.4) to be nonnegative.

The term $\mathbf{B}^{-1}\Delta a_{lk}\mathbf{0}_{lk}$ can be written as

$$\mathbf{B}^{-1}\,\Delta a_{lk}\mathbf{0}_{lk} = \mathbf{B}^{-1}\begin{pmatrix} 0 & \cdots & 0 & \cdots & 0 \\ \cdots & \cdots & \cdots & \cdots & \cdots \\ 0 & \cdots & \Delta a_{lk} & \cdots & 0 \\ \cdots & \cdots & \cdots & \cdots & \cdots \\ 0 & \cdots & 0 & \cdots & 0 \end{pmatrix} = \begin{pmatrix} 0 & \cdots & b_{1l}\,\Delta a_{lk} & \cdots & 0 \\ \cdots & \cdots & \cdots & \cdots & \cdots \\ 0 & \cdots & b_{kl}\,\Delta a_{lk} & \cdots & 0 \\ \cdots & \cdots & \cdots & \cdots & \cdots \\ 0 & \cdots & b_{ml}\,\Delta a_{lk} & \cdots & 0 \end{pmatrix}$$

↑   ↑
kth column kth column

where the b_{il} are the corresponding elements of the lth column of $\mathbf{B}^{-1}$.   Hence,

$$(\mathbf{I}+\mathbf{B}^{-1}\,\Delta a_{lk}\mathbf{0}_{lk}) = \begin{pmatrix} 1 & \cdots & b_{1l}\,\Delta a_{lk} & \cdots & 0 \\ \cdots & \cdots & \cdots & \cdots & \cdots \\ 0 & \cdots & 1+b_{kl}\,\Delta a_{lk} & \cdots & 0 \\ \cdots & \cdots & \cdots & \cdots & \cdots \\ 0 & \cdots & b_{ml}\,\Delta a_{lk} & \cdots & 1 \end{pmatrix} \leftarrow k\text{th row} \qquad (3.5)$$

The inverse of (3.5) is given by

$$(\mathbf{I}+\mathbf{B}^{-1}\,\Delta a_{lk}\mathbf{0}_{lk})^{-1} = \begin{pmatrix} 1 & \cdots & \dfrac{-b_{1l}\,\Delta a_{lk}}{1+b_{kl}\,\Delta a_{lk}} & \cdots & 0 \\ & & \cdots & & \\ 0 & \cdots & \dfrac{1}{1+b_{kl}\,\Delta a_{lk}} & \cdots & 0 \\ & & \cdots & & \\ 0 & \cdots & \dfrac{-b_{ml}\,\Delta a_{lk}}{1+b_{kl}\,\Delta a_{lk}} & \cdots & 1 \end{pmatrix} \qquad (3.6)$$

Substituting (3.6) in (3.4), we obtain

$$\bar{\mathbf{B}}^{-1}\mathbf{b} = (\mathbf{I}+\mathbf{B}^{-1}\,\Delta a_{lk}\mathbf{0}_{lk})^{-1}\mathbf{X}^0$$

$$\bar{\mathbf{B}}^{-1}\mathbf{b} = \begin{pmatrix} x_{10} - \dfrac{b_{1l}\,\Delta a_{lk}}{1+b_{kl}\,\Delta a_{lk}}\,x_{k0} \\ \vdots \\ \dfrac{1}{1+b_{kl}\,\Delta a_{lk}}\,x_{k0} \\ \vdots \\ x_{m0} - \dfrac{b_{ml}\,\Delta a_{lk}}{1+b_{kl}\,\Delta a_{lk}}\,x_{k0} \end{pmatrix} \qquad (3.7)$$

In order for (3.7) to exist and to be nonnegative, we must have the following:

$$1+b_{kl}\,\Delta a_{lk} > 0 \qquad (3.8)$$

$$x_{i0} - \frac{b_{il}\,\Delta a_{lk}}{1+b_{kl}\,\Delta a_{lk}}\,x_{k0} \geq 0 \qquad (3.9)$$

Solving for Δa_{lk}, (3.9) yields (assuming that $1+b_{kl}\,\Delta a_{lk} > 0$) the conditions for all $i \neq k$:

$$\Delta a_{lk} \le \frac{x_{i0}}{b_{il}x_{k0} - b_{kl}x_{i0}} \qquad \text{for } b_{il}x_{k0} - b_{kl}x_{i0} > 0$$

$$\frac{x_{i0}}{b_{il}x_{k0} - b_{kl}x_{i0}} \le \Delta a_{lk} \qquad \text{for } b_{il}x_{k0} - b_{kl}x_{i0} < 0$$

The upper and lower bounds do not exist if corresponding denominators do not exist. If the change Δa_{lk} is for a vector in the basis, the feasibility conditions impose the restriction (3.8) and

$$\max_{i \ne k} \frac{x_{i0}}{(b_{il}x_{k0} - b_{kl}x_{i0}) < 0} \le \Delta a_{lk} \le \min_{i \ne k} \frac{x_{i0}}{(b_{il}x_{k0} - b_{kl}x_{i0}) > 0} \tag{3.10}$$

For those vectors $\mathbf{P}_j$ not in the basis, the optimality conditions (3.3) must be simultaneously satisfied. Since $\bar{\mathbf{B}}^{-1} = (\mathbf{I} + \mathbf{B}^{-1}\Delta a_{lk}\mathbf{0}_{lk})^{-1}\mathbf{B}^{-1}$, the above analysis yields for (3.3)

$$\mathbf{c}_0(\mathbf{I} + \mathbf{B}^{-1}\Delta a_{lk}\mathbf{0}_{lk})^{-1}\mathbf{B}^{-1}\mathbf{P}_j - c_j \le 0 \tag{3.11}$$

or

$$\mathbf{c}_0(\mathbf{I} + \mathbf{B}^{-1}\Delta a_{lk}\mathbf{0}_{lk})^{-1}\mathbf{X}_j - c_j \le 0 \tag{3.12}$$

where $\mathbf{X}_j = \mathbf{B}^{-1}\mathbf{P}_j$. From (3.6), inequalities (3.12) become

$$\mathbf{c}_0 \begin{pmatrix} x_{1j} - \dfrac{b_{1l}\,\Delta a_{lk}}{1 + b_{kl}\,\Delta a_{lk}}\, x_{kj} \\ \vdots \\ \dfrac{1}{1 + b_{kl}\,\Delta a_{lk}}\, x_{kj} \\ \vdots \\ x_{mj} - \dfrac{b_{ml}\,\Delta a_{lk}}{1 + b_{kl}\,\Delta a_{lk}}\, x_{kj} \end{pmatrix} - c_j \le 0$$

This yields

$$c_1\left(x_{1j} - \frac{b_{1l}\,\Delta a_{lk}}{1 + b_{kl}\,\Delta a_{lk}}\, x_{kj}\right) + \cdots + c_k \frac{1}{1 + b_{kl}\,\Delta a_{lk}}\, x_{kj}$$

$$+ \cdots + c_m\left(x_{mj} - \frac{b_{ml}\,\Delta a_{lk}}{1 + b_{kl}\,\Delta a_{lk}}\, x_{kj}\right) + c_k x_{kj} - c_k x_{kj} - c_j \le 0$$

where the term $c_k x_{kj} - c_k x_{kj}$ has been added to the left-hand side.

Collecting terms, we have

$$\sum_{i \in \mathbf{B}} c_i x_{ij} - \frac{b_{1l}\,\Delta a_{lk}}{1 + b_{kl}\,\Delta a_{lk}}\, c_1 x_{kj} - \cdots + \frac{1}{1 + b_{kl}\,\Delta a_{lk}}\, c_k x_{kj}$$

$$- \cdots - \frac{b_{ml}\,\Delta a_{lk}}{1 + b_{kl}\,\Delta a_{lk}}\, c_m x_{kj} - c_k x_{kj} - c_j \le 0 \tag{3.13}$$

By multiplying $c_k x_{kj}$ by $1 + b_{kl}\,\Delta a_{lk}/(1 + b_{kl}\,\Delta a_{lk})$, substituting for z_j, and collecting terms in (3.13), we have

$$(z_j - c_j) - \sum_{i \in \mathbf{B}} \frac{b_{il} \, \Delta a_{lk}}{1 + b_{kl} \, \Delta a_{lk}} \, c_i x_{kj} \leq 0$$

and from (3.8),

$$(z_j - c_j)(1 + b_{kl} \, \Delta_{lk}) - \Delta a_{lk} \sum_i b_{il} c_i x_{kj} \leq 0$$

$$(z_j - c_j) + (z_j - c_j) b_{kl} \, \Delta a_{lk} - \Delta a_{lk} \sum_i b_{il} c_i x_{kj} \leq 0$$

$$(z_j - c_j) \leq \Delta a_{lk} \left[\sum_i b_{il} c_i x_{kj} - (z_j - c_j) b_{kl} \right]$$

If the term in brackets is positive,

$$\frac{z_j - c_j}{\sum\limits_i b_{il} c_i x_{kj} - (z_j - c_j) b_{kl}} \leq \Delta a_{lk}$$

and if it is negative,

$$\Delta a_{lk} \leq \frac{z_j - c_j}{\sum\limits_i b_{il} c_i x_{kj} - (z_j - c_j) b_{kl}}$$

or, for those j not in the basis,

$$\max_j \frac{z_j - c_j}{\left[x_{kj} \sum\limits_i b_{il} c_i - (z_j - c_j) b_{kl} \right] > 0} \leq \Delta a_{lk}$$

$$\leq \min_j \frac{z_j - c_j}{\left[x_{kj} \sum\limits_i b_{il} c_i - (z_j - c_j) b_{kl} \right] < 0} \tag{3.14}$$

The Δa_{lk} is unbounded if corresponding denominators do not exist. We note that the summation term is the lth element of the row vector $\boldsymbol{\pi} = \mathbf{c}_0 \mathbf{B}^{-1}$. In order to apply a change Δa_{lk} to the lth component of a vector $\mathbf{P}_k$ in the basis, Δa_{lk} must satisfy (3.8), (3.10), and (3.14). The corresponding changes in the value of the variables are given by (3.7). The total change in the value of the objective function is determined by subtracting the value of the old solution $\mathbf{X}^0$ from the value of the solution (3.7), i.e.,

$$\left(x_{10} - \frac{b_{1l} \, \Delta a_{lk}}{1 + b_{kl} \, \Delta a_{lk}} x_{k0} \right) c_1 + \cdots + \frac{c_k x_{k0}}{1 + b_{kl} \, \Delta a_{lk}}$$

$$+ \cdots + \left(x_{m0} - \frac{b_{ml} \, \Delta a_{lk}}{1 + b_{kl} \, \Delta a_{lk}} x_{k0} \right) c_m - c_1 x_{10} - \cdots - c_k x_{k0} - \cdots - c_m x_{m0} = z$$

Adding the term

$$\frac{b_{kl} \, \Delta a_{lk}}{1 + b_{kl} \, \Delta a_{lk}} c_k x_{k0} - \frac{b_{kl} \, \Delta a_{lk}}{1 + b_{kl} \, \Delta a_{lk}} c_k x_{k0}$$

and collecting terms, we have

$$z = -\sum_i b_{il} c_i \frac{\Delta a_{lk} x_{k0}}{1 + b_{kl}\, \Delta a_{lk}}$$

For a change to an element a_{lj} of a vector $\mathbf{P}_j$ not in the basis, a_{lj} must be such that

$$\mathbf{c}_0 \mathbf{B}^{-1} \mathbf{P}_j - c_j \leq 0$$

For a given Δa_{lj}, we have

$$\mathbf{c}_0(\mathbf{X}_j + \Delta a_{lj} \mathbf{B}^l) - c_j \leq 0 \tag{3.15}$$

where $\mathbf{B}^l$ is the lth column of $\mathbf{B}^{-1}$.   Then (3.15) is

$$(z_j - c_j) + \Delta a_{lj} \sum_i b_{il} c_i \leq 0$$

Hence, for the vector $\mathbf{P}_j$ not in the basis, Δa_{lj} must lie between the limits

$$\frac{-(z_j - c_j)}{\left(\sum_i b_{il} c_i\right) < 0} \leq \Delta a_{lj} \leq \frac{-(z_j - c_j)}{\left(\sum_i b_{il} c_i\right) > 0}$$

That is,

$$\Delta a_{lj} \leq \begin{cases} \dfrac{-(z_j - c_j)}{\sum_i b_{il} c_i} & \text{if } \sum_i b_{il} c_i > 0 \\[2em] +\infty & \text{if } \sum_i b_{il} c_i = 0 \end{cases}$$

$$\Delta a_{lj} \geq \begin{cases} \dfrac{-(z_j - c_j)}{\sum_i b_{il} c_i} & \text{if } \sum_i b_{il} c_i < 0 \\[2em] -\infty & \text{if } \sum_i b_{il} c_i = 0 \end{cases}$$

As $\sum_i b_{il} c_i$ is the lth element of the optimal pricing vector (optimal dual variables) $\boldsymbol{\pi} = \mathbf{c}_0 \mathbf{B}^{-1}$, the above conditions can be rewritten as

$$\Delta a_{lj} \leq \frac{-(z_j - c_j)}{\pi_l} \qquad \text{if } \pi_l > 0$$

$$\Delta a_{lj} \geq \frac{-(z_j - c_j)}{\pi_l} \qquad \text{if } \pi_l < 0$$

Sensitivity analysis carried out by mathematical programming systems is termed *ranging* (RANGE Procedure).   The ranging output includes the bounds on the costs and right-hand-side elements and a listing of the vectors that would enter and leave the basis at the bounds (see Orchard-Hays [532] and Gal [274]).

4. MULTIOBJECTIVE LINEAR PROGRAMMING

In defining a linear-programming model of a decision problem, we have, up to now, assumed that we could characterize the optimizing aspects of the situation by a single linear objective function. However, many (if not most) decision problems have multiple objectives to be optimized simultaneously. Such objectives often conflict with one another and it is difficult, if not impossible, to find a solution that accomplishes the simultaneous optimization. For example, a transportation problem might require the minimization of the total transportation cost and the minimization of the total transit time of the goods being shipped. A production problem may require the meeting of demand while minimizing the use of a particular resource and maximizing profit; or the minimization of the fluctuations in production and the minimization of storage costs (see Sec. 1, Chap. 11). Problems in urban and social systems usually have conflicting goals, e.g., minimizing the cost of a sewage treatment plant while maximizing water quality (Beltrami [65]). When the constraints of a problem and its multiple objectives can be expressed as linear functions, we have a *multiobjective or multicriteria linear-programming problem.*

The basic approach to solving multiobjective problems is to determine a solution that represents an acceptable trade-off or compromise between the objectives, or to determine a set of such solutions and to allow the decision maker to choose among them. In general, new formulations and computational procedures are required to produce these solutions. In this section, we describe the multiparametric formulation of the multiobjective linear-programming problem and some of the related theoretical aspects. Complete discussions, including algorithms for solving the multiparametric problem, can be found in the references (see, for example, Gal [274]). An alternative approach for dealing with multiple objectives, called *goal programming,* is discussed in Chap. 11, Sec. 5.

We define the *linear multiobjective function problem* as follows. We are given p cost vectors denoted by $c_k = (c_{k1}, \ldots, c_{kn})$, $k = 1, \ldots, p$, and the usual set of $m \times n$ constraints $AX = b$ and $X \geq 0$. Each $c_k X$ defines a linear objective function to be optimized subject to the constraints; assume they are to be maximized.[1] We denote the simultaneous optimization of the p objective functions by the yet-to-be-defined concept of "maximization." The problem can now be stated as: "Maximize"

$$\begin{pmatrix} c_1 X \\ c_2 X \\ \vdots \\ c_p X \end{pmatrix}$$

subject to

$$AX = b$$
$$X \geq 0$$

---

[1] We can always take the negative of those objective functions to be minimized.

Letting

$$
\mathbf{C} = \begin{pmatrix} c_1 \\ c_2 \\ \cdot \\ \cdot \\ \cdot \\ c_p \end{pmatrix} = \begin{pmatrix} c_{11} & \cdots & c_{1n} \\ c_{21} & \cdots & c_{2n} \\ \cdots\cdots\cdots\cdots \\ \cdots\cdots\cdots\cdots \\ \cdots\cdots\cdots\cdots \\ c_{p1} & \cdots & c_{pn} \end{pmatrix}
$$

the problem can be stated in the more concise form as the following *linear vector maximum problem*: "Maximize"

$$\mathbf{CX}$$

subject to

$$
\begin{aligned}
\mathbf{AX} &= \mathbf{b} \\
\mathbf{X} &\geq \mathbf{0}
\end{aligned}
\tag{4.1}
$$

We next clarify what is meant by vector "maximization."

As a numerical example, consider the original furniture-manufacturer problem that required the maximization of the profit function

$$\mathbf{c}_1\mathbf{X} = 45x_1 + 80x_2$$

subject to the resource constraints

$$
\begin{aligned}
5x_1 + 20x_2 &\leq 400 \\
10x_1 + 15x_2 &\leq 450
\end{aligned}
$$

and $x_j \geq 0$. The optimal solution is $x_1 = 24$, $x_2 = 14$, and $\mathbf{c}_1\mathbf{X} = \$2{,}200$. If, instead, the manufacturer were interested in determining a solution that maximized total units produced, i.e., maximize $\mathbf{c}_2\mathbf{X} = x_1 + x_2$, we have from Fig. 8.6 that the optimal solution would then be $x_1 = 45$, $x_2 = 0$, and $\mathbf{c}_2\mathbf{X} = 45$ units. The attempt to "maximize" both objectives leads to the following linear vector maximum problem: "Maximize"

$$\mathbf{CX} = \begin{pmatrix} 45x_1 + 80x_2 \\ x_1 + x_2 \end{pmatrix}$$

subject to

$$
\begin{aligned}
5x_1 + 20x_2 &\leq 400 \\
10x_1 + 15x_2 &\leq 450 \\
x_j &\geq 0
\end{aligned}
\tag{4.1a}
$$

For $\mathbf{X}_1 = (24, 14)$ and $\mathbf{X}_2 = (45, 0)$, we have, respectively,

$$\mathbf{CX}_1 = \begin{pmatrix} \mathbf{c}_1\mathbf{X}_1 \\ \mathbf{c}_2\mathbf{X}_1 \end{pmatrix} = \begin{pmatrix} 2{,}200 \\ 38 \end{pmatrix} \quad \text{and} \quad \mathbf{CX}_2 = \begin{pmatrix} \mathbf{c}_1\mathbf{X}_2 \\ \mathbf{c}_2\mathbf{X}_2 \end{pmatrix} = \begin{pmatrix} 2{,}025 \\ 45 \end{pmatrix}$$

We see that these two solutions conflict with one another; we cannot choose either one without decreasing the value of an objective-function value with respect to the remaining solution. Also, any other feasible solution, extreme point or otherwise, would yield values of c_1X and c_2X that reduce at least one value, if not both.

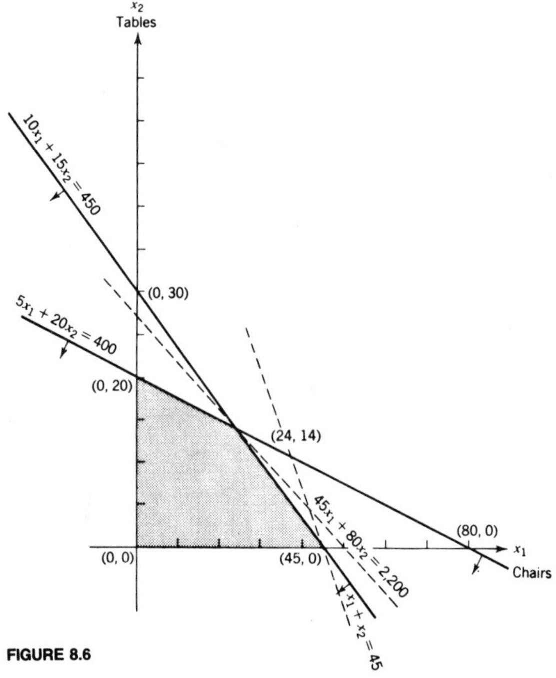

FIGURE 8.6

To resolve the question of vector "maximization," we need to redefine our concept of optimality. We do this in a fashion similar to the economic definition of efficiency or Pareto optimum. In an economic system, a *Pareto optimum* is a state in which no person's satisfaction can be raised without lowering someone else's satisfaction (Koopmans [438]). For the vector maximum problem (4.1), we define an efficient solution as follows (Charnes and Cooper [112]):

Definition. X^0 is said to be an *efficient solution* to (4.1) if X^0 is feasible and there exists no other feasible solution X such that $CX \geq CX^0$ and $CX \neq CX^0$.

X^0 is also termed a *nondominated solution* or *Pareto-optimal solution*. An equivalent definition is that a feasible X^0 is efficient if and only if there exists no other feasible X such that $c_kX \geq c_kX^0$ for all $k = 1, \ldots, p$ and $c_kX > c_kX^0$ for at least one k. We are at an efficient point when moving away from it lowers the value of at least one objective function. From these definitions, we see that the solutions (24, 14) and (45, 0) to the furniture-manufacturer problem are efficient. The definition of an efficient point yields the following theorem (Philip [549]).

Theorem 3. *Let $c_k X$ be one of the objective functions of the linear vector maximum problem (4.1). Consider the linear-programming problem of maximizing $c_k X$ subject to $AX = b$ and $X \geq 0$ and let X^0 be its unique solution. Then X^0 is an efficient solution of (4.1).* (The proof follows directly from the definition of an efficient solution.)

One way of trying to achieve a balance or suitable trade-off between multiple objectives is to assign positive weights to each one, where a weight reflects the importance of the corresponding objective function. For example, in the furniture-manufacturer problem let λ_1 be the weight assigned to the profit function $c_1 X = 45x_1 + 80x_2$ and let λ_2 be the weight assigned to the total production function $c_2 X = x_1 + x_2$. We can then consider the trade-off problem to be the two-parameter linear program: Maximize

$$z = \lambda_1(45x_1 + 80x_2) + \lambda_2(x_1 + x_2)$$

subject to

$$5x_1 + 20x_2 \leq 400$$
$$10x_1 + 15x_2 \leq 450 \tag{4.2}$$
$$x_j \geq \quad 0$$

As the weights are often not exact and are indeterminate, the problem of interest is the finding of all solutions of (4.2) for all positive values of the parameters and relating these solutions to the concept of efficient solutions.

To generalize the situation, we consider the *multiobjective, multiparameter linear-programming problem*: Maximize

$$z = \sum_{k=1}^{p} \lambda_k (c_k X) = \lambda (CX)$$

subject to

$$AX = b$$
$$X \geq 0 \tag{4.3}$$

where $\lambda = (\lambda_1, \ldots, \lambda_p)$ and all $\lambda_k > 0$. (Although we do not do so, we can restrict $\sum \lambda_k = 1$.) The vector maximum problem (4.1) and the multiparameter problem (4.3) are related by the following theorem.

Theorem 4. *A vector X^0 that solves the multiparameter problem (4.3) for fixed $\lambda_k > 0$ ($k = 1, \ldots, p$) is an efficient solution to the vector maximum problem (4.1).*

Proof: Suppose that X^0 is not efficient. Then by the definition of an efficient solution there exists a feasible X to (4.1) and (4.3) such that

$$c_k X \geq c_k X^0 \quad \text{for all } k$$
$$c_k X > c_k X^0 \quad \text{for at least one } k$$

We multiply each inequality k by the corresponding $\lambda_k > 0$ and add to obtain

$$\sum_{k=1}^{p} \lambda_k(\mathbf{c}_k\mathbf{X}) > \sum_{k=1}^{p} \lambda_k(\mathbf{c}_k\mathbf{X}^0) \tag{4.4}$$

As the right-hand side of (4.4) is the maximum value of (4.3), we have a contradiction.

Based on Theorem 4, if we can establish a set of weights $\lambda_k > 0$, we need only solve the resulting linear-programming problem to obtain an efficient solution to the associated vector maximum problem. However, as the concept of efficiency broadens our choice of a solution, we should be interested in finding a set of efficient solutions (if not all) and then selecting one of them. The following theorem, which incorporates Theorem 4, establishes the important fact that the set of efficient solutions to the linear vector maximum problem (4.1) is identical to the set of optimal solutions to the multiparameter linear-programming problem (4.3); see Evans and Steuer [239], Gal [274], Geoffrion [296], Isermann [394, 395], and Philip [549] for proofs and further discussion.

Theorem 5. *Let* $\mathbf{X}^0$ *satisfy the constraints* $\mathbf{AX} = \mathbf{b}$, $\mathbf{X} \geq \mathbf{0}$. *Then* $\mathbf{X}^0$ *is an efficient solution to the linear vector maximum problem* (4.1) *if and only if there exists a* $\boldsymbol{\lambda}^0 = (\lambda_1^0, \ldots, \lambda_p^0) > 0$ *such that* $\mathbf{X}^0$ *solves the multiparameter problem* (4.3) *for* $\boldsymbol{\lambda} = \boldsymbol{\lambda}^0$.

A number of computational procedures exist for finding all efficient solutions or selected subsets to problems (4.1) and (4.3); see Gal [274, 275], Steuer [625], Gal and Nedoma [276], Evans and Steuer [239], Ecker and Kouada [232], Zionts and Wallenius [732], Zeleny [729], Ecker, Hegner, and Kouada [230], Frank and Nygard [261], Isermann [395], Benayoun et al. [66], Chankong and Haimes [109], and Philip [549]. For some of these procedures, the technical approach can be viewed as an extension of the single-parameter objective-function algorithm developed in Sec. 1. For example, the (λ_1, λ_2) two-parameter linear-programming problem requires the finding of regions in the (λ_1, λ_2)-plane that correspond to optimal basic feasible solutions that are efficient and the generation of adjacent efficient basic solutions (see Exercise 5). We note that if the linear vector maximum problem (4.1) has an efficient solution, at least one basic feasible solution is efficient (Gal [275] and Isermann [395]). Also, two basic efficient solutions $\mathbf{X}^0$ and $\mathbf{X}^1$ are said to be adjacent if and only if they have $m - 1$ basic variables in common and the points on the line segment $\mathbf{X} = \alpha\mathbf{X}^0 + (1 - \alpha)\mathbf{X}^1$, $(0 \leq \alpha \leq 1)$, are efficient for (4.1). Thus, for the convex polyhedral set of solutions $\mathbf{AX} = \mathbf{b}$, $\mathbf{X} \geq \mathbf{0}$, which is common to both problems (4.1) and (4.3), we would need to determine the set of efficient extreme points and that part of the boundary containing efficient solutions.

For the furniture-manufacturer problem, the points (24, 14), (45, 0) and

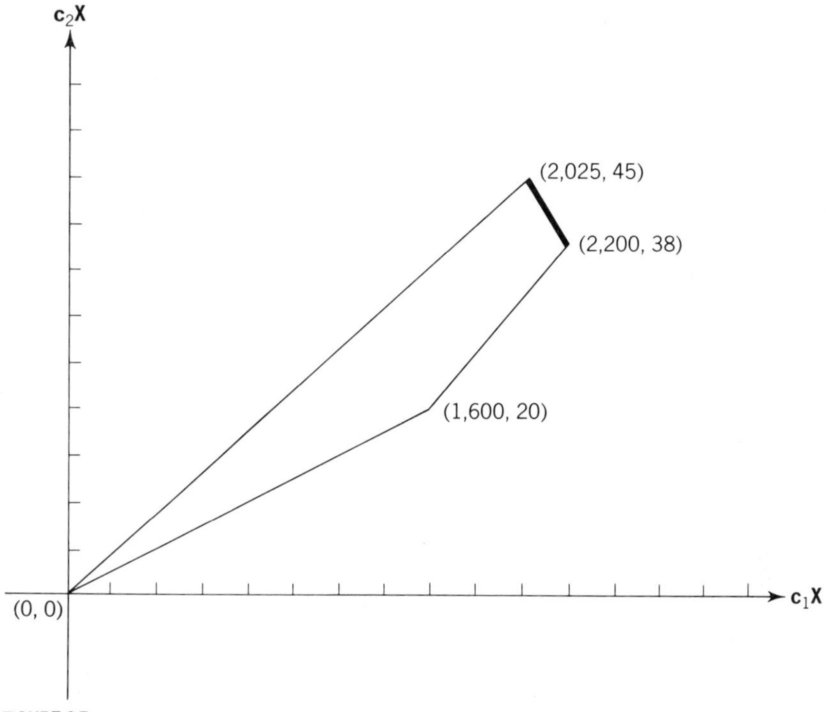

FIGURE 8.7

the points on the line segment joining them form the complete set of efficient solutions (Exercise 36).   This can be seen by reference to Fig. 8.7.   There we have plotted the values of c_1X against c_2X (note that the scales have been adjusted).   In particular, the labeled points of Fig. 8.7 correspond to the extreme points of the solution space shown in Fig. 8.6, as given in the following table.

Fig. 8.6	Fig. 8.7	
Extreme point $X = (x_1, x_2)$	c_1X	c_2X
0, 0	0	0
0, 20	1,600	20
45, 0	2,025	45
24, 14	2,200	38

Any feasible solution to the problem of Fig. 8.6 will yield a corresponding point that lies on the boundary or interior to the objective-function plot of Fig. 8.7.   If a feasible solution does not yield a point that lies on the heavy line joining (2,025, 45) and (2,200, 38) in Fig. 8.7, then that solution is not efficient.

Given a feasible solution $\mathbf{X}^0$ to the linear vector maximum problem (4.1), the following theorems yield tests that enable us to determine if $\mathbf{X}^0$ is efficient and, if not, to find an efficient solution or show that the problem has no efficient solutions (see Charnes and Cooper [112], Ecker and Kouada [231], and Exercise 40).

Theorem 6. *Let $\mathbf{X}^0$ be a feasible solution to the linear vector maximum problem (4.1). Consider the following problem: Maximize*

$$y_0 = \sum_{k=1}^{p} y_k$$

subject to

$$\mathbf{c}_k \mathbf{X} - y_k = \mathbf{c}_k \mathbf{X}^0 \qquad k = 1, \ldots, p$$
$$\mathbf{AX} = \mathbf{b} \tag{4.5}$$
$$\mathbf{X} \geq 0 \qquad \text{and} \qquad \text{all } y_k \geq 0$$

$\mathbf{X}^0$ *is efficient if and only if maximum* $y_0 = 0$. (Note that the p objective function constraints are based on the definition of an efficient solution and that the y_k are slack variables.)

Theorem 7. *If problem (4.5) of Theorem 6 has a finite nonzero maximum value for a solution $\mathbf{X}^1$, then $\mathbf{X}^1$ is efficient. If problem (4.5) does not have a finite maximum value, then there are no efficient solutions to the linear vector maximum problem (4.1).*

Decision problems with multiobjectives are described in Cohon [130] and Zeleny [728]. A procedure for solving linear-programming problems with *multiparametric right-hand sides* is given in Gal [274] and Gal and Nedoma [276]. Algorithms for vector optimization problems and applications are surveyed in French et al. [265]; also see Raff and Zeleny [555] and Chankong and Haimes [109]. Although intimately related in terms of their computational requirements, the reader should note the differences between the multiobjective, multiparameter problem and the general multiparameter problem in which the problem coefficients are functions of more than one parameter. Exercise 38 and the furniture-manufacturer problem described above illustrate the two-objective, two-parameter problem, while Exercises 5a and 5b illustrate the general two-parameter problem.

REMARKS

For additional reading the reader is referred to Saaty and Gass [587], Müller-Merbach [512, 513], Barnett [39], Kim [430], Van de Panne [668], Courtillot [142], Saaty [584], and Willner [705]. For the extension of parametric programming to the multiparameter case see Gass and Saaty [292], Gal [274], Gal and Nedoma [276], and Exercise 5. For nonlinear parameters see Graves [327]. The books by Gal [274] and Dinkelbach [204] cover most aspects of sensitivity analysis and parametric programming as applied to the general linear-programming problem. Kornbluth [440] and Hannan [350] discuss sensitivity analysis in multiple objective linear programming.

EXERCISES

1. Solve the following parametric objective-function problems for all values of λ and plot z as a function of λ:

   a. Minimize

$$z = (3 - \lambda)x_1 - (2 + \lambda)x_2$$

subject to

$$2x_1 + 5x_2 \le 10$$
$$6x_1 + x_2 \le 12$$
$$x_1 - x_2 \le 1$$
$$x_j \ge 0$$

   b. Minimize

$$z = 2\lambda x_1 + (1 - \lambda)x_2 - 3x_3 + \lambda x_4 + 2x_5 - 3\lambda x_6$$

subject to

$$x_1 + 3x_2 - x_3 + 2x_5 = 7$$
$$-2x_2 + 4x_3 + x_4 = 12$$
$$-4x_2 + 3x_3 + 8x_5 + x_6 = 10$$

and

$$x_j \ge 0$$

Answer: Optimal basic feasible solutions can be found for the following ranges of λ: $-\infty \le \lambda \le -\frac{1}{10}$, $-\frac{1}{10} \le \lambda \le -\frac{1}{27}$, $-\frac{1}{27} \le \lambda \le 2$, $2 \le \lambda \le +\infty$. For the last range, the optimal solution is $x_2 = \frac{7}{3}$, $x_4 = \frac{50}{3}$, $x_6 = \frac{58}{3}$, $x_1 = x_3 = x_5 = 0$, and minimum $z = \frac{7}{3} - \frac{131}{3}\lambda$.

   2. Solve Exercise 1*a* graphically.

   3. Solve the following parametric right-hand-side problem for all values of θ: Minimize

$$z = 2x_4 + 8x_5$$

subject to

$$x_1 + 3x_4 - x_5 = 3 - \theta$$
$$x_2 - 4x_4 + 2x_5 = 1 + 2\theta$$
$$x_3 - x_4 + 3x_5 = -1 + \theta$$
$$x_j \ge 0$$

Plot z as a function of θ.

Answer: Optimal basic feasible solutions exist for the following ranges of θ: $0 \le \theta \le 1$, $1 \le \theta \le 3$, $3 \le \theta \le 4$. For the first range, the optimal solution is $x_1 = 2\theta$, $x_2 = 5 - 2\theta$, $x_4 = 1 - \theta$, $x_3 = x_5 = 0$, and minimum $z = 2 - 2\theta$.

   4. Modify the computational procedure of the revised simplex method to handle a parametric objective function.

   5. By means of the simplex method and a two-dimensional graph of the parameter space, solve the following two-parameter problems (Gass and Saaty [292]):

   a. Minimize

$$x_1 + \lambda_1 x_2 + \lambda_2 x_3$$

subject to

$$
\begin{aligned}
x_1 \quad\quad\quad\quad - x_4 \quad\quad - 2x_6 &= 5 \\
x_2 \quad + 2x_4 - 3x_5 + x_6 &= 3 \\
x_3 + 2x_4 - 5x_5 + 6x_6 &= 5 \\
x_j &\geq 0
\end{aligned}
$$

HINT: Set up the usual simplex tableau with separate cost rows for λ_1 and λ_2. Show that the initial solution $x_1 = 5$, $x_2 = 3$, $x_3 = 5$, with $z = 5 + 3\lambda_1 + 5\lambda_2$ is optimal (efficient) for the area defined by the inequalities $-1 + 2\lambda_1 + 2\lambda_2 \leq 0$, $-3\lambda_1 - 5\lambda_2 \leq 0$ and $-2 + \lambda_1 + 6\lambda_2 \leq 0$ by graphing the constraints in the (λ_1, λ_2)-plane. A new optimal basic feasible (efficient) solution is obtained by introducing (if possible) a nonbasic variable that corresponds to one of the defining inequalities. These inequalities are the usual $z_j - c_j \leq 0$ optimality conditions augmented to include the parameters. Not all values of the parameters yield finite optimal solutions, e.g., there are no such solutions when both parameters are negative. There are four optimal bases to this problem with each basis being optimal for (λ_1, λ_2) that lie in the corresponding regions of the (λ_1, λ_2)-plane of Fig. E5:

$\mathbf{B}_1$: (x_1, x_2, x_3)

$\mathbf{B}_2$: (x_1, x_3, x_4)

$\mathbf{B}_3$: (x_1, x_4, x_6)

$\mathbf{B}_4$: (x_1, x_2, x_6)

{For higher dimensions, Matheiss and Rubin [500] describe procedures for finding all vertices of convex polyhedral sets. How could you use such a technique in solving multiparameter problems (Gal and Nedoma [276])?}

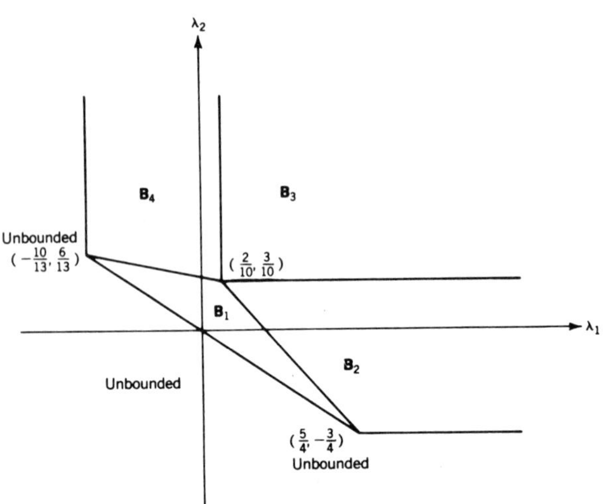

FIGURE E5

    b. Minimize

$$
x_1 + \lambda_1 x_2 + \lambda_2 x_3 + x_4 + x_5 + x_6 + x_7 + \lambda_1 x_8 + \lambda_2 x_9
$$

subject to

$$
\begin{array}{llllllll}
x_1 & & -x_4 & -2x_6 + 4x_7 + & 2x_8 + & x_9 = 4 \\
& x_2 & +2x_4 - 2x_5 + & x_6 + 2x_7 + & 4x_8 & = 2 \\
& x_3 & +x_5 + & x_6 + x_7 & + 2x_9 = 1
\end{array}
$$

$$x_j \geq 0$$

For this problem a feasible solution exists for all values and combinations of λ_1 and λ_2.

6. Describe procedures that will allow for the addition or deletion of a constraint.

7. Generalize the formulas for varying a single b_i to include variations in all b_i.

8. Generalize the formulas for varying a single c_j to include variations in all c_j for the set of basis variables.

9. In the corresponding optimal solutions, determine the range of the variation for the elements in the following problems:

a. In Exercise 1*a* of Chap. 4:
1. The cost coefficient c_1
2. The cost coefficient c_3
3. The right-hand-side coefficient b_1

b. In Exercise 1*c* of Chap. 4:
1. The cost coefficient c_1
2. The coefficient a_{24}
3. The coefficient a_{32}
4. The right-hand-side coefficient b_3
5. The cost coefficient c_4

10. Let the optimal basis to a linear-programming problem be denoted by $\mathbf{B} = (\mathbf{P}_1 \cdots \mathbf{P}_s \cdots \mathbf{P}_m)$. Let $\mathbf{P}_s$ be changed (perturbed) to $\mathbf{P}'_s$. Determine conditions on $\mathbf{P}'_s$ such that the $\mathbf{B}' = (\mathbf{P}_1 \cdots \mathbf{P}'_s \cdots \mathbf{P}_m)$ is the optimal basis (Barnett [39]). Similarly, for a vector $\mathbf{P}_k$ not in the optimal basis, let it be perturbed to $\mathbf{P}'_k$. What are the conditions on $\mathbf{P}'_k$ such that $\mathbf{B}$ remains the optimal basis?

11. For the linear-programming problem minimize $\mathbf{cX}$ subject to $\mathbf{AX} = \mathbf{b}$, $\mathbf{X} \geq \mathbf{0}$, we have determined the optimal solution with corresponding basis $\mathbf{B}$. Consider the new problem $\mathbf{AX} = \bar{\mathbf{b}}$ with $\bar{b}_i = b_i + \theta$ for all i and all values of θ. Develop a formula which tells for what range of θ the basis $\mathbf{B}$ will be an optimal basis for the new problem. Note that the basis $\mathbf{B}$ is optimal for $\theta = 0$ and $\bar{\mathbf{b}} = \mathbf{b} + \boldsymbol{\theta}$ where $\boldsymbol{\theta}$ is a column vector with elements all equal to θ. Show that if $\theta = 1$, then the value of the objective function is $z_0 + \sum w_i^0$, where the w_i^0 are the optimal dual variables (the final prices π_i). Use the results of this exercise to demonstrate that the optimal dual values can be interpreted as marginal prices. Note that the marginal prices hold for right-hand-side changes that allow the current optimal basis to stay feasible. Thus, a full additional unit change to a right-hand-side element may not be possible.

12. For the parametric dual problem of maximizing $\displaystyle\sum_{j=1}^{n} c_j x_j$ subject to

$$
\sum_{j=1}^{n} a_{ij}x_j = b_i + \theta b'_i \qquad i = 1, 2, \ldots, m
$$
$$
x_j \geq 0 \qquad \text{all } j
$$

prove that the objective function is a concave function of θ. Also, show for the parametric primal problem with the objective of maximizing $\displaystyle\sum_{j=1}^{n} (d_j + \lambda d'_j)x_j$ that the objective function is a convex function of λ. HINT FOR DUAL: Let $z_0(\theta) = \max \displaystyle\sum_{j=1}^{n} c_j x_j$ and show that $\alpha z_0(\theta_1) + (1 - \alpha)z_0(\theta_2) \leq z_0(\theta_3)$, where $0 \leq \alpha \leq 1$, $\theta_1 < \theta_2$, and $\theta_3 = \alpha\theta_1 + (1 - \alpha)\theta_2$. (See Chap. 12 for definitions of concave and convex functions.)

13. Discuss a computational procedure for solving the linear-programming problem when both the objective function and right-hand side are functions of the same parameter; are functions of different parameters (Orchard-Hays [532]).

14. Discuss the geometric aspects of the parametric right-hand side and illustrate using the constraints of Exercise 3 as inequalities; i.e., let x_1, x_2, x_3 be slack variables.

15. Discuss parametrization of a column vector $\mathbf{P}_j$, Kim [430].

16. Solve Example 1.2 by the parametric objective-function algorithm.

17. Given an optimal basic feasible solution to a linear-programming problem, discuss, for the following situations, the conditions under which the given solution remains optimal: (*a*) all basic variables have their cost coefficients changed to λc_j; (*b*) all variables change their costs from c_j to $c_j + \lambda$; (*c*) a nonbasic variable x_j changes its cost to $c_j + \lambda$; (*d*) a basic variable x_j changes its cost to $c_j + \lambda$.

18. Solve the furniture-manufacturer parametric objective-function problem of Sec. 1 and plot z as a function of the parameter.   Start with the solution given in Tableau 8.1.

19. Solve the furniture-manufacturer parametric right-hand-side problem of Sec. 2 and plot z as a function of the parameter.   Start with the solution given in Tableau 8.1.

20. In Theorem 2, explain why there are no feasible solutions if all $x_{ij} \geq 0$.   (HINT: Can the corresponding equation be satisfied?)

21. For the furniture-manufacturer problem, describe what happens geometrically as each b_i is varied.   When and why would a basis change occur?

22. Develop sensitivity analysis formulas to be used in conjunction with the revised simplex method.

23. For the symmetric primal and dual linear-programming problems min $f = \mathbf{cX}$, $\mathbf{AX} \geq \mathbf{b}$, $\mathbf{X} \geq \mathbf{0}$ and max $g = \mathbf{bW}$, $\mathbf{WA} \leq \mathbf{c}$, $\mathbf{W} \geq \mathbf{0}$, let $\mathbf{X}^0$ and $\mathbf{W}^0$ be corresponding optimal solutions.   Show that

$$\frac{\partial f}{\partial \mathbf{c}} = \mathbf{X}^0 \qquad \frac{\partial f}{\partial \mathbf{b}} = \mathbf{W}^0 \qquad \frac{\partial f}{\partial a_{ij}} = -x_j^0 w_i^0 \qquad \frac{\partial x_j}{\partial a_{ij}}\bigg|_{x_j^0} = \frac{-x_j^0 w_i^0}{c_j} \qquad \frac{\partial w_i}{\partial a_{ij}}\bigg|_{w_i^0} = \frac{-x_j^0 w_i^0}{b_i}$$

(Saaty [584] and Webb [691]).   Also see Williams [699] concerning the marginal value of a linear program.

24. Discuss the computational aspects of solving a linear-programming problem with the objective function that minimizes $\sum (c_j + \lambda c_j)x_j$; the parameter can be interpreted as an inflation factor for the costs.

25. For the furniture-manufacturer problem, solve and interpret the following parametric-programming problems for all values of the parameter:

   a.   Let the profit of a chair be $45 + \lambda$.

   b.   Let the number of labor-hours be $450 + \theta$.

   c.   Let the profit of a chair be $45 + \lambda$ and the profit of a desk be $80 - 2\lambda$.

26. Develop a parametric algorithm for the bounded-variable problem in which the objective function, the right-hand side of the constraints, and the upper-bound vector are functions of the same single parameter (Panwalkar [540]).   (See Chap. 9 for a definition of the bounded-variable problem.)

27. In finding efficient solutions by solving a multiparameter problem, why can we assume that the sum of the parameters is equal to 1?

28. Perform a complete sensitivity study of the furniture-manufacturer problem, i.e., determine how much each c_j, b_i, and a_{ij} can vary before the optimal basic solution changes.

29. Solve Example 1.3 graphically.

30. Solve the following quadratic parametric problem (Graves [327]): Maximize

$$z = -4(\lambda - 4)(\lambda - 9)x_3$$

subject to

$$x_1 \quad + x_3 = 7(\lambda - 5)(\lambda - 6)$$
$$x_2 + x_3 = -3(\lambda - 1)(\lambda - 10)$$

and $x_j \geq 0$.

Answer: The basic solution with x_1 and x_2 is optimal if the parameter lies in the intervals [1, 4] and [9, 10]; the solution with x_1 and x_3 has the interval [8, 9]; and the solution x_2 and x_3 has the intervals [4, 5] and [6, 8].   Note that the intervals are disconnected.

31. Most mathematical programming systems solve a parametric problem for nonnegative values of the parameter. Describe what you would need to do to find solutions for the full range of the parameter.

32. Describe the revised simplex algorithm for solving the parametric right-hand-side problem.

33. Solve the following parametric problem (Zlobec [733]): Minimize $z = x$, subject to $\theta x \leq 0$, $-\theta x \leq 0$, $-x \leq 1$. Note that the value of z is not continuous.

34. Solve Example 1.2 by the simplex algorithm.

35. For a multiobjective linear-programming problem, describe how you can determine if any of the objective functions are nonessential, i.e., if any can be removed without changing the set of efficient solutions (Gal [274]).

36. Solve the two-parameter furniture-manufacturer problem given in Sec. 4, by the technique of Exercise 5.

37. Find all extreme-point efficient solutions for the following problem (Zionts and Wallenius [732]): "Maximize"

$$\begin{pmatrix} c_1X \\ c_2X \\ c_3X \end{pmatrix} = \begin{pmatrix} 3x_1 + x_2 + 2x_3 + x_4 \\ x_1 - x_2 + 2x_3 + 4x_4 \\ -x_1 + 5x_2 + x_3 + 2x_4 \end{pmatrix}$$

subject to

$$2x_1 + x_2 + 4x_3 + 3x_4 \leq 60$$
$$3x_1 + 4x_2 + x_3 + 2x_4 \leq 60$$
$$(x_j \geq 0)$$

There are nine feasible extreme points and six are efficient.

38. Solve the following problem by graphical methods and determine the set of efficient solutions (Frank and Nygard [261]): "Maximize"

$$\begin{pmatrix} c_1X \\ c_2X \end{pmatrix} = \begin{pmatrix} x_1 + 2x_2 \\ 3x_1 - x_2 \end{pmatrix}$$

subject to

$$3x_1 + x_2 \leq 9$$
$$x_1 + 3x_2 \leq 19$$
$$x_1 + x_2 \leq 9$$
$$x_1 - x_2 \leq 3$$
$$(x_j \geq 0)$$

39. For multicriteria problems, it is usually computationally difficult to compute all efficient solutions and, even if the complete set is known, the ability of a decision maker to interpret the solutions and select one is open to question. A number of interactive computer procedures have been proposed that involve the decision maker in the selection of new adjacent efficient points; see for example Zionts and Wallenius [731, 732]. Describe your version of how such an interactive procedure should work.

40. For the linear vector maximum problem (4.1), consider a linear-programming tableau that has been augmented by p rows, with the rows corresponding to the objective functions. Assume the tableau has been transformed using the simplex transformations for maximization to a new tableau that contains an efficient, nondegenerate basic feasible solution X^0. Let x_j be a nonbasic variable that forms a new basic feasible solution X^1 when brought into the basis. Consider the following linear-programming problem: Maximize $y_0 = \sum_{k=1}^{p} y_k$ subject to $C_kU - y_k = \bar{c}_{kj}$ ($k = 1, \ldots, p$), $U \geq 0$ and all $y_k \geq 0$. Show that the edge (see Exercise 9, Chap. 3) joining X^0 and X^1 is in the efficient set of solutions if maximum $y_0 = 0$. Here the $\bar{c}_{kj}$ are the transformed elements in the objective function rows corresponding to x_j (Ecker and Kouada [232]). Use this result and the other theorems of Sec. 4 to develop an algorithm that generates efficient points and edges. Will you be able to find all the efficient points? Note that the set of efficient points is connected. Try your algorithm on the problems in Exercises 37 and 38, and the furniture-manufacturer problem. Start the algorithm with a simplex tableau that represents an initial efficient extreme-point solution.

41. Prove Theorems 6 and 7.

42. A *self-dual parametric algorithm* (Dantzig [152]). Consider the linear-programming problem to minimize

$$\sum_{j=1}^{n} c_j x_j$$

subject to

$$\sum_{j=1}^{n} a_{ij} x_j + x_{n+i} = b_i \qquad i = 1, \ldots, m$$

$$x_j \geq 0 \qquad j = 1, \ldots, n+m$$

If all $c_j \geq 0$ and all $b_i \geq 0$, then the solution $x_{n+i} = b_i$ is optimum. Assume some $c_j < 0$ and some $b_i < 0$. For negative c_j, define a new cost $c_j' = c_j + \theta$, and for negative b_i define a new right-hand side $b_i' = b_i + \theta$. Note that for $\theta \geq \max [-c_j, -b_i]$ for those $c_j < 0$ and $b_i < 0$, the solution $x_{n+i} = b_i$ for $b_i \geq 0$ and $x_{n+i} = b_i + \theta$ for $b_i < 0$ is optimum. Describe how you can apply a sequence of primal and dual simplex algorithm transformations so that you always have an optimal solution for some value(s) of θ and you find the true optimal solution that corresponds to $\theta = 0$.

43. A *tolerance approach to sensitivity analysis* (Wendell [692, 693]). Consider the following multiparameter problem: Maximize

$$\sum_{j=1}^{n} (c_j + \gamma_j c_j') x_j$$

subject to

$$\sum_{j=1}^{n} a_{ij} x_j = b_i + \delta_i b_i' \qquad i = 1, \ldots, m$$

$$x_j \geq 0 \qquad j = 1, \ldots, n$$

for given scalars b_i' and c_j' and where the δ_i and γ_j are considered to be multiplicative perturbations of the corresponding b_i' and c_j'. In particular, let $b_i' = b_i$ and $c_j' = c_j$. We can then interpret the δ_i and γ_j as percentage errors from the estimated values of b_i and c_j, respectively. For an optimal basis, define a number ρ as an allowable tolerance error for the right-hand-side perturbations if, whenever $-\rho \leq \delta_i \leq \rho$, for all i, the basis remains optimal. Similarly, define a number τ as an allowable tolerance error for the objective-function perturbations. Determine expressions for the maximum allowable tolerance errors for ρ and τ. Discuss how this approach enables you to investigate simultaneous changes in the c_j and b_i. Contrast this with regular sensitivity analysis, i.e., can you make simultaneous changes?

44. Solve the following problem for all efficient solutions (Kornbluth [440]): "Maximize"

$$\begin{pmatrix} 3x_1 + x_2 + x_3 \\ x_1 - x_2 + 2x_3 \\ x_1 + 2x_2 \end{pmatrix}$$

subject to

$$4x_1 + 2x_2 + 3x_3 \leq 10$$

$$x_1 + 3x_2 + 2x_3 \leq 8 \qquad (x_j \geq 0)$$

$$x_3 \leq 5$$

Solution: $(x_1 = \frac{7}{5}, x_2 = \frac{11}{5}, x_3 = 0)$, $(x_1 = 0, x_2 = \frac{4}{5}, x_3 = \frac{14}{5})$, $(x_1 = 0, x_2 = 0, x_3 = \frac{10}{3})$.

45. Perform a sensitivity analysis on the right-hand sides of the problem given in Exercise 30,

Chap. 6. This problem has multiple marginal values. Discuss which marginal values go with increases and decreases to the right-hand sides (Williams [702]).

46. Using the concepts and formulas of parametric programming, show how the usual sensitivity formulas can be derived.

47. Justify the results of Theorem 3 in terms of the requirements of Theorem 5.

48. For Figs. 8.6 and 8.9, explain why a point X in the feasible region of Fig. 8.6 yields a point $(c_1 X, c_2 X)$ that lies in the convex space of Fig. 8.7. (HINT: Any feasible X can be expressed as a convex combination of the extreme points of the feasible region.)

CHAPTER 9

Additional Computational Techniques

In this chapter we shall consider the question faced by investigators in the field who have formulated a linear-programming problem and determined the necessary coefficients and are ready to solve the associated numerical example. *What preliminary analysis and computational devices should they use in order to solve the problem with the minimum amount of computing?*

One reason that the above query is an open question is that the mathematical model of a specific application may lead to a simplified computational scheme. This is the case, for example, for the transportation and the production-scheduling problems discussed in Chaps. 10 and 11, respectively. In the former, a computationally simple variation of the simplex method is devised, while in the latter we could either reduce the number of equations in the system and solve it by the general simplex procedure or solve it by even simpler techniques. However, in the final analysis, it may prove to be more efficient not to worry about such schemes, since investigations of possible variations, although interesting, are in themselves time-consuming and may be unfruitful. Hence, we preface our discussion by emphasizing that under reasonable assumptions as to size and numerical characteristics *any* linear-programming problem can be solved by the tried and proven techniques of the standard and revised simplex procedures described in Chaps. 4 and 5. This is further reinforced by noting that computer-based mathematical programming systems, with their many refinements of the simplex method, are regularly used to solve large-scale problems from a wide variety of application areas.

Since the development of reduction and new computational methods is dependent upon the specific mathematical equations under investigation, we are here necessarily limited to discussing techniques for reducing the amount of computation that are applicable to most linear-programming problems. These schemes are all based on overcoming certain peculiarities of the simplex procedure and are designed to reduce the total number of iterations required to reach the minimum solution. Some of these are described in Sec. 1 below, but let us first review a few pertinent points of the simplex method.

The general simplex method is divided into two distinct computational phases: Phase I is concerned with determining a first basic feasible solution, while Phase II, which starts with this first solution, is concerned with obtaining a minimum feasible solution. Computational experience has shown that it takes approximately m to $2m$ iterations (where m is the number of equations in the final model) to solve the problem if the computation starts with an explicit basis of m unit vectors; approximately $2m$ to $3m$ iterations are required if a full artificial basis is used. Hence it appears to be advantageous to eliminate Phase I altogether, or at least to start with as few artificial vectors as possible. Further, if we are in Phase I, the selection of a new vector to be introduced into the basis depends only on a criterion that consistently reduces the contribution of the artificial variables to the value of the objective function. This criterion is used until the "artificial" part of the objective function is made zero. The criterion has no control over that part of the objective function that must eventually be reduced to a minimum by Phase II. This "real" part is allowed to fluctuate, and usually the value of the objective function for the first feasible solution is far from the minimum value. It seems reasonable to expect that the number of iterations required in Phase II depends on how close the first feasible solution is to the minimum. That is, the closer the first solution, the fewer the iterations required. What is more important is that the artificial-basis technique does not allow us to use certain information inherent in many programming problems. For example, from our knowledge of a particular problem, we may expect a specific set of m vectors to yield a solution that is close to the minimum. However, since there is no assurance that this set of vectors forms a feasible basis, it cannot, in general, be efficiently employed in the standard computational procedures. Again, it is felt that the total number of iterations will be reduced if we utilize some of our knowledge about the "expected solution."[1] The majority of the schemes developed for

---

[1] The cautious wording of the above sentences is due to the existence of the ever-present "counter-examples." Problems have been constructed that start with a full artificial basis and reach the minimum in exactly m iterations; problems also exist that start with a high value of the objective function and rapidly reach the minimum or, conversely, start close to the minimum and "creep" to the optimum.

speeding the computation are designed to offset the above points and to vary the initial simplex computational scheme in order to obtain a "good" first feasible solution. Mathematical programming systems employ special CRASH procedures that are designed to "crash through" to a good first feasible basis or to improve upon an initial basis or a user-specified solution. (See Gass [284], Orchard-Hays [532], IBM [387], Sperry Univac [619], and Benichou et al. [70].) For a discussion on the efficacy of Phase I methods, alternative rules for selecting vectors to enter the basis, and alternative forms for carrying out the simplex algorithm, see Sec. 3, Chap. 5, and Cutler and Wolfe [149], Quandt and Kuhn [553], and Goldfarb and Reid [315].

1. DETERMINING A FIRST FEASIBLE SOLUTION

A set of $m \times n$ constraints to a linear-programming problem can take on three basic configurations. Let us first assume that the final set of constraints of the model is given by

$$\mathbf{AX} = \mathbf{b} \qquad \text{with } \mathbf{b} \geq 0 \qquad m < n \tag{1.1}$$

We assume at least one $b_i > 0$.

For the set (1.1) we have three situations:

1. The set contains no unit vectors, and the computation starts with a set of m artificial vectors.
2. The set contains k distinct unit vectors, and the Phase I procedure starts with $m - k$ artificial vectors.
3. The set contains a basis of m unit vectors, and the computation begins with Phase II. (A column $\mathbf{P}_j$ of $\mathbf{A}$ which has only a single nonzero positive element, say a_{ij}, can be converted to a unit vector by simply dividing row i by a_{ij}.)

Next, if the model is of the form

$$\mathbf{AX} \leq \mathbf{b} \qquad \text{with } \mathbf{b} \geq 0 \tag{1.2}$$

we rewrite (1.2) as equalities in terms of nonnegative variables (as described in Sec. 4 of Chap. 2 and Sec. 4 of Chap. 4). The resulting set of equations contains a basis of m unit vectors, and the computation begins with Phase II.

A third possibility is that the constraints are of the form

$$\mathbf{AX} \geq \mathbf{b} \qquad \text{with } \mathbf{b} \geq 0 \tag{1.3}$$

Here we also rewrite (1.3) as equations in nonnegative variables, but the resulting system contains a set of m negative unit vectors. Let this system be given by

$$\mathbf{AX} - \bar{\mathbf{X}} = \mathbf{b} \tag{1.4}$$

where $\bar{\mathbf{X}} = (x_{n+1}, x_{n+2}, \ldots, x_{n+m})$ is a nonnegative column vector. By applying a simple transformation to the coefficients of $\mathbf{X}$, $\bar{\mathbf{X}}$, and the column vector $\mathbf{b}$, we can start the computation with only one artificial vector. The scheme calls for determining $\max\limits_{i} b_i = b_s$ and adding the sth row to the negative of every other row of (1.4). The resulting set of m equations will contain $m-1$ distinct positive unit vectors and will require one artificial vector, which corresponds to the sth variable. The transformation that is applied to (1.4) is given by

$$
\begin{aligned}
a'_{ij} &= -a_{ij} + a_{sj} &&\text{for } i \neq s &&j = 1, 2, \ldots, n+m \\
a'_{sj} &= a_{sj} &&&&j = 1, 2, \ldots, n+m \\
b'_i &= -b_i + b_s &&\text{for } i \neq s \\
b'_s &= b_s
\end{aligned}
\tag{1.5}
$$

To illustrate this scheme, let us consider the inequalities

$$
\begin{aligned}
x_1 - x_2 + 3x_3 &\geq 1 \\
2x_1 - x_2 + x_3 &\geq 0 \\
x_1 + x_2 - x_3 &\geq 2
\end{aligned}
\tag{1.6}
$$

We rewrite (1.6) as equalities to obtain[1]

$$
\begin{aligned}
x_1 - x_2 + 3x_3 - x_4 \qquad\qquad &= 1 \\
2x_1 - x_2 + x_3 \qquad - x_5 \qquad &= 0 \\
x_1 + x_2 - x_3 \qquad\qquad - x_6 &= 2
\end{aligned}
\tag{1.7}
$$

Here $\max\limits_{i} b_i = b_3 = 2$, and applying formulas (1.5) to (1.7) gives us

$$
\begin{aligned}
2x_2 - 4x_3 + x_4 \qquad - x_6 &= 1 \\
- x_1 + 2x_2 - 2x_3 \qquad + x_5 - x_6 &= 2 \\
x_1 + x_2 - x_3 \qquad\qquad - x_6 &= 2
\end{aligned}
$$

We can, of course, formulate a model that contains a mixture of inequalities and equalities, that is, $\mathbf{AX} \gtreqless \mathbf{b}$, where $\mathbf{b}$ is unrestricted. Here, by suitable manipulation of the appropriate constraints, we can, as above, introduce as many unit vectors as possible.

Up to now we have been assuming that the computation must begin with an explicit unit matrix for the first feasible basis. In Sec. 2 of Chap. 4 we briefly mentioned the possibility of selecting an arbitrary set of m vectors for the first admissible basis. Here the plan calls for the selection of m vectors to form an initial "expected solution." This selection should be based on a careful, but not time-consuming, analysis of the program objectives. We next attempt to compute the inverse of the corresponding $m \times m$ matrix. The following possibilities may occur:

---

[1] The transformation does not have to be applied to those rows which have their corresponding $b_i = 0$, as in the second equation, but here we illustrate the general application of the transformation.

1. The set of m vectors is linearly independent (the inverse exists).

   a. The corresponding solution is feasible.

   For this case the simplex procedure starts Phase II with a *preferred* solution.

   b. The corresponding solution is not feasible.

   Here, as will be described below, we can start Phase I with only one artificial vector.

2. The set of m vectors is linearly dependent and is of rank $k < m$. For this situation we can apply a transformation to the tableau to initiate Phase I with either $m - k$ or $m - k + 1$ artificial vectors (see Gass [284] for further discussion). We could also continue to choose other vectors for our initial basis from a secondary set of preferred vectors until we have chosen a linearly independent set of m vectors (see Orchard-Hays [529] for additional comments along these lines).

For Case 1*b* above, the following simple transformation, which is similar to the one applied to the set of inequalities $\mathbf{AX} \geq \mathbf{b}$, will enable us to begin Phase I with one artificial vector. For ease of discussion we shall describe the transformation in terms of the original simplex tableau of Chap. 4 (see Tableau 4.1). Let us assume that the first m vectors $\mathbf{P}_1, \mathbf{P}_2, \ldots, \mathbf{P}_m$ have been selected as the preferred set and are linearly independent but do not form an admissible basis. Let x_{ij} for $i = 1, 2, \ldots, m + 1$ and $j = 0, 1, \ldots, n$ be the transformed element in the ith row and jth column after the matrix $(\mathbf{P}_1\mathbf{P}_2 \cdots \mathbf{P}_m)$ has been inverted, and let $x_{ii} = 1$ for $i = 1, 2, \ldots, m$. Here we assume some $x_{i0} < 0$. Let $x_{s0} = \min_i x_{i0}$. Next, transform all the elements of the tableau by the formulas

$$x'_{ij} = x_{ij} - x_{sj} \qquad \text{for } i \neq s, m + 1$$
$$x'_{sj} = -x_{sj} \tag{1.8}$$

Formulas (1.8) hold for $j = 0, 1, \ldots, n$ and represent the subtraction of the sth row from all the other rows and the multiplication of the sth row by (-1). This transformation yields a basis consisting of $m - 1$ vectors of the selected m and one artificial vector whose corresponding variable $x'_{s0} > 0$. The new elements of the $(m + 1)$st row are

$$x'_{m+1,j} = x_{m+1,j} - x_{sj} \sum_{1}^{m} c_i \qquad \text{for } j \neq s$$
$$x'_{m+1,s} = -\sum_{1}^{m} c_i \tag{1.8'}$$

The elements in row s of the tableau guide the succeeding iterations until the artificial variable is reduced to zero or until it is determined that the problem is not feasible. Row s corresponds to the $(m + 2)$nd row of the artificial-basis tableau with one artificial vector. Since we wish to control the "real" part of the value of the objective function, we can modify the criterion that selects the new vector for the basis as follows: For the tableau given by (1.8) and (1.8'), let vector $\mathbf{P}_k$ correspond to

$$x'_{m+1,k} = \max_{x'_{sj}>0} x'_{m+1,j} \tag{1.9}$$

Formula (1.9) states that the vector selected has the "artificial" part of its $z_j - c_j$ equal to $x'_{sj} > 0$ and the "real" part equal to $x'_{m+1,j}$ as algebraically large as possible. This criterion will consistently reduce the artificial part and tend to reduce the real part of the objective function. If

$$\min_i \frac{x'_{i0}}{x'_{ik}} = \frac{x'_{s0}}{x'_{sk}} \qquad \text{for } x'_{ik} > 0$$

then the artificial vector will be eliminated, and Phase II begins. If x'_{sk} is not the pivot element, then the next iteration introduces a vector $\mathbf{P}_q$ with $x''_{m+1,q} = \max_{x''_{sj}>0} x''_{m+1,j}$, etc., until the artificial vector is eliminated or until it is determined that no feasible solutions exist.

For some linear-programming models the problem might call for the minimization of different objectives over the same set of linear constraints. This problem of multiple objective functions can be conveniently handled by setting up the basic constraints and solving the problem for the first objective function. Once the basic minimum feasible solution for this problem has been computed, it is a simple matter to determine whether this solution is also a minimum for any of the other objective functions. If we were using the original simplex procedure, we would include in the tableau an extra row for each objective function which transforms under the usual elimination formulas. By looking at the corresponding $z_j - c_j$ elements, we could determine the additional optimized objective functions. If we were using the revised procedure, then the optimality of the objective functions could be determined by means of the inverse of the current basis and the corresponding new cost coefficients. If the optimal solution for the first objective function were not the minimum for one of the other objective functions, then this first optimal solution would be used as the first feasible solution for the new objective function.

The dual of this problem, the problem of multiple constant vectors, may also arise, e.g., solutions to economic models for a number of different bills of goods might be desired (see Chap. 11, Sec. 2). This problem can be resolved by the following techniques: Let the original problem be that of minimizing

$$\mathbf{cX} \tag{1.10}$$

subject to

$$\mathbf{AX} = \mathbf{P}_0 \tag{1.11}$$

$$\mathbf{X} \geq \mathbf{0} \tag{1.12}$$

In addition to (1.10) to (1.12), we also wish to solve the following set of problems: To minimize

$$\mathbf{cX}$$

subject to

$$\mathbf{AX} = \mathbf{P}_{0q} \qquad q = 1, 2, \ldots, r$$

$$\mathbf{X} \geq 0$$

The problem is first solved for (1.10) to (1.12). In the original simplex method the additional $\mathbf{P}_{0q}$ are attached to the tableau and transformed in the usual manner. If, after the minimum feasible solution for $\mathbf{P}_0$ is obtained, all components of all the $\mathbf{P}_{0q}$ are nonnegative, then this solution is a minimum for all the $\mathbf{P}_{0q}$. Suppose that, for some $q = r$, not all the elements of the transformed vector $\mathbf{P}_{0r}$ are greater than or equal to zero. To develop a starting solution for this problem, we find the minimum element of the transformed vector $\mathbf{P}_{0r}$ and apply (1.8) and (1.8'). The resulting solution will consist of one artificial vector and $m - 1$ vectors of the previous minimum feasible solution. The problem is then solved in the usual manner for $\mathbf{P}_{0r}$, and similarly for any of the remaining $\mathbf{P}_{0q}$. However, instead of applying the procedure associated with Eqs. (1.8), we might look at the situation in the light of its dual interpretation.

Here we have selected a basis—let us say $\mathbf{B} = (\mathbf{P}_1\mathbf{P}_2\cdots\mathbf{P}_m)$—computed its inverse, and determined that $\mathbf{B}^{-1}\mathbf{P}_0 \geq 0$ and that

$$\mathbf{c}^0\mathbf{B}^{-1}\mathbf{P}_j - c_j = z_j - c_j \leq 0 \tag{1.13}$$

for all j, where $\mathbf{c}^0 = (c_1, c_2, \ldots, c_m)$ is a row vector. Hence $\mathbf{B}$ is an admissible basis for (1.10) to (1.12). We have also determined that $\mathbf{B}$ is not an admissible basis for the constraints

$$\mathbf{AX} = \mathbf{P}_{0r} \tag{1.14}$$

in that

$$\mathbf{B}^{-1}\mathbf{P}_{0r} \tag{1.15}$$

is not a nonnegative vector. We say, for the problem determined by (1.14), that the solution (1.15) is optimal but not feasible; i.e., all $z_j - c_j \leq 0$, but not all elements of $\mathbf{B}^{-1}\mathbf{P}_{0r}$ are nonnegative. The dual problem to (1.10), (1.14), and (1.12) is to maximize

$$\mathbf{WP}_{0r} \tag{1.16}$$

subject to

$$\mathbf{WA} \leq \mathbf{c}^\dagger \tag{1.17}$$

For the dual problem we see from (1.13) that the basis $\mathbf{B}$ yields a solution to (1.17), in that

$$\mathbf{W}^0\mathbf{P}_j - c_j \leq 0$$

for $\mathbf{W}^0 = \mathbf{c}^0\mathbf{B}^{-1}$. We also have that, for the dual, this solution is feasible but not optimal. A computational technique which enables us to employ efficiently the information concerning feasibility and optimality of the primal

---

$\dagger$ The dual problem is described in Chap. 6.

and dual contained in the above problem was first proposed by Lemke [467] and is called the *dual simplex method.*

2. THE DUAL SIMPLEX METHOD

To consider the general procedure, let us write our problem as follows: To minimize

$$\mathbf{cX}$$

subject to

$$\mathbf{AX} = \mathbf{b}$$
$$\mathbf{X} \geq \mathbf{0}$$

with its dual problem: To maximize

$$\mathbf{Wb}$$

subject to

$$\mathbf{WA} \leq \mathbf{c}$$

As we did in the case of multiple-constant vectors, let us assume we have selected a basis $\mathbf{B} = (\mathbf{P}_1\mathbf{P}_2\cdots\mathbf{P}_m)$ such that at least one element of $\mathbf{B}^{-1}\mathbf{b}$ is negative and $\mathbf{c}^0\mathbf{B}^{-1}\mathbf{P}_j \leq c_j$ for all j. A solution to the dual constraints is given by $\mathbf{W}^0 = \mathbf{c}^0\mathbf{B}^{-1}$ with its corresponding value of the objective function being

$$\mathbf{c}^0\mathbf{B}^{-1}\mathbf{b} \tag{2.1}$$

We wish to develop a computational procedure for the dual which will yield a maximizing solution, and hence, by the duality theorems of Chap. 6, a minimizing solution to the primal. This procedure must then determine a new basis for which

1. The dual inequalities will still be satisfied.
2. The value of the dual objective function will increase (or remain the same)[1] until the maximum or unbounded solution is reached.

In this fashion we will always preserve optimality of the primal and, in a finite number of steps, determine a feasible and optimal solution to the primal.

Let us denote the rows of $\mathbf{B}^{-1}$ by $\mathbf{B}_i$. Hence the m components of the *nonfeasible solution* of the primal are given by $x_{i0} = \mathbf{B}_i b$ for $i = 1, 2, \ldots,$ m. Let $x_{l0} = \mathbf{B}_l b = \min_i \mathbf{B}_i b < 0$. The vector $\mathbf{P}_l$, as we shall see, will be the one removed from the basis. For those vectors not in the basis, i.e., for those having $\mathbf{W}^0\mathbf{P}_j < c_j$, compute $x_{ij} = \mathbf{B}_i\mathbf{P}_j$. Let us assume that at least one $x_{lj} < 0$. For the set of $x_{lj} < 0$ form the ratios $(z_j - c_j)/x_{lj}$. Let

---

[1] The value of the dual objective for the new basis may not increase if the dual solution is degenerate. This is manifested by more than m of the quantities $\mathbf{c}^0\mathbf{B}^{-1}\mathbf{P}_j = c_j$.

$$\theta = \min_{x_{lj}<0} \frac{z_j - c_j}{x_{lj}} = \frac{z_k - c_k}{x_{lk}}^\dagger > 0 \tag{2.2}$$

The vector $\mathbf{P}_k$ is selected to replace $\mathbf{P}_l$, and the new basis will yield a solution to the dual constraints. The new basis is

$$\bar{\mathbf{B}} = (\mathbf{P}_1 \cdots \mathbf{P}_{l-1} \mathbf{P}_k \mathbf{P}_{l+1} \cdots \mathbf{P}_m)$$

and $\bar{\mathbf{B}}^{-1}$ is obtained by application of the elimination formulas on $\mathbf{B}^{-1}$ as described in Sec. 1 of Chap. 5. The reader can readily verify (Exercise 8) that the new solution to the dual constraints is

$$\bar{\mathbf{W}} = \mathbf{W}^0 - \theta \mathbf{B}_l \tag{2.3}$$

and the corresponding value of the objective function is

$$\bar{\mathbf{W}}\mathbf{b} = \mathbf{W}^0\mathbf{b} - \theta x_{l0} \tag{2.4}$$

The new solution to the primal constraints can be computed by the usual elimination formulas or directly from $\bar{\mathbf{X}} = \bar{\mathbf{B}}^{-1}\mathbf{b}$. If all $\bar{\mathbf{X}} \geq 0$, then we have determined an optimal feasible solution to the primal. If not, then we know we have made at least $\bar{x}_{l0} = \bar{\mathbf{B}}_l\mathbf{b} > 0$. For this situation, we repeat the above dual simplex process of selecting the vector to be removed and then the vector to be introduced into the basis, until we find a basis that solves the dual and is also an admissible basis for the primal or until we have determined that the dual has an unbounded solution and hence that there are no feasible solutions to the primal. The latter case arises when, in computing the $x_{lj} = \mathbf{B}_l\mathbf{P}_j$, all $x_{lj} \geq 0$. If this is true, we have from (2.2) and (2.3) that we can construct a solution to the dual constraints for any $\theta > 0$, since

$$(\mathbf{W}^0 - \theta \mathbf{B}_l)\mathbf{P}_j = \mathbf{W}^0\mathbf{P}_j - \theta \mathbf{B}_l\mathbf{P}_j = \mathbf{W}^0\mathbf{P}_j - \theta x_{lj} \leq \mathbf{W}^0\mathbf{P}_j < c_j$$

From (2.4) the corresponding value of the objective function can be made as large as possible, since $x_{l0} < 0$. This situation is revealed in the primal tableau, in that we would have all $x_{lj} \geq 0$, $x_{l0} < 0$, with the implication that the lth equation, which has been transformed to a nonnegative sum of nonnegative variables, is equal to a negative number. The dual simplex procedure can be employed as a variation of the original simplex procedure or of the revised procedure, since it uses exactly the information contained in their respective tableaus.

To handle degeneracy in the dual problem, Lemke [467] sets up the perturbed dual

$$\mathbf{W}\mathbf{P}_j \leq c_j + \epsilon^j \qquad j = 1, 2, \ldots, n$$

and shows that a scheme similar to the degeneracy method for the primal may

---

† This criterion for selecting a vector to be introduced into the basis is equivalent to the one used in the parametric dual problem (Theorem 2 of Sec. 2, Chap. 8). The proof of that theorem also applies to this situation.

be established.   In the dual, we must have a procedure for selecting a unique vector $\mathbf{P}_k$ to be introduced into the basis.   The situation is not unique when the value of θ in Eq. (2.2) assumes the minimum for more than one value of j.[†]   For these j, we first compute the set of ratios x_{1j}/x_{lj}.   If there is a unique minimum for these ratios, the index j of the minimum corresponds to the vector to be introduced.   If not, we next compute for the tied columns the ratios x_{2j}/x_{lj} and repeat the analysis, etc.   In this manner we select a unique $\mathbf{P}_k$, and the problem will not cycle.

In the above presentation of the dual procedure we assumed knowledge of an optimal but not feasible solution to the primal, i.e., a solution to the dual constraints.   This is equivalent to starting the simplex procedure with a known feasible but not optimal solution and hence eliminating the artificial Phase I computations.   The main advantage to using the dual simplex method is that the dual statements of many problems have explicit solutions.   These problems are characterized by having a set of nonnegative cost coefficients, e.g., the diet problem.   The dual constraints are immediately solved by the vector $\mathbf{W} = \mathbf{0}$.   Here we take the identity matrix as the corresponding artificial basis.   Then, with only minor changes, the dual method can be applied to determine the optimal solution to the primal (Dantzig [158]).

For those problems which do not have explicit solutions, i.e., which do not have all $c_j \geq 0$, a number of schemes have been proposed (Dantzig [158] and Vajda [662]).   The first technique sets all negative costs equal to zero and starts with the identity matrix as an initial basis.   Here the dual method, with the appropriate changes, is applied until the optimum is reached.   The corresponding basis will be an admissible one for the primal, and the usual simplex procedure can proceed to optimize the original objective function.   The second scheme constructs the dual analogue of the artificial-variable technique and solves the associated problem in two phases by the dual simplex method (see Exercise 6, Wagner [680], and Steinberg [624]).

As an example of the dual method, consider the following problem: Minimize

$$x_3 + x_4 + x_5$$

subject to

$$x_1 \quad - x_3 + x_4 - x_5 = -2$$
$$x_2 - x_3 - x_4 + x_5 = \quad 1$$
$$x_j \geq \quad 0$$

The dual objective is to maximize

$$-2w_1 + w_2$$

and the constraints are

[†] In practice, if there are ties for the minimum, then the ratio that corresponds to the smallest index is selected for θ.

$$w_1 \qquad \leq 0$$
$$w_2 \leq 0$$
$$-w_1 - w_2 \leq 1$$
$$w_1 - w_2 \leq 1$$
$$-w_1 + w_2 \leq 1$$

An initial basis is given by

$$\mathbf{B} = (\mathbf{P}_1 \mathbf{P}_2) = \begin{pmatrix} 1 & 0 \\ 0 & 1 \end{pmatrix}$$

Setting up the primal in the usual simplex tableau, we have the following:

i	Basis	$\mathbf{c}$		0	0	1	1	1
			$\mathbf{P}_0$	$\mathbf{P}_1$	$\mathbf{P}_2$	$\mathbf{P}_3$	$\mathbf{P}_4$	$\mathbf{P}_5$
1	$\mathbf{P}_1$	0	-2	1	0	$\left(-1\right)$	1	-1
2	$\mathbf{P}_2$	0	1	0	1	-1	-1	1
3			0	0	0	-1	-1	-1

Since all the $z_j - c_j$ elements are nonpositive, the basis $\mathbf{B}$ is a feasible basis for the dual, i.e., optimal but not feasible for the primal. The dual solution is

$$\mathbf{W}^0 = \mathbf{c}^0 \mathbf{B}^{-1} = (0 \quad 0)\begin{pmatrix} 1 & 0 \\ 0 & 1 \end{pmatrix} = (0 \quad 0)$$

Applying the dual algorithm, we see that $\mathbf{P}_1$ is to be eliminated

$$(x_l = x_1 = -2)$$

and vector $\mathbf{P}_3$ is to be introduced into the basis because

$$\theta = \frac{z_3 - c_3}{x_{13}} = \frac{z_5 - c_5}{x_{15}} = 1$$

Here we had a tie for θ and selected the first vector. The new tableau is as follows:

Basis	$\mathbf{c}$	$\mathbf{P}_0$	$\mathbf{P}_1$	$\mathbf{P}_2$	$\mathbf{P}_3$	$\mathbf{P}_4$	$\mathbf{P}_5$
$\mathbf{P}_3$	1	2	-1	0	1	-1	1
$\mathbf{P}_2$	0	3	-1	1	0	-2	2
		2	-1	0	0	-2	0

Since all $x_i \geq 0$ and since $z_j - c_j \leq 0$, we have determined an optimal admissible basis for the primal and dual consisting of vectors $\mathbf{P}_2$ and $\mathbf{P}_3$. The primal optimal solution is $x_2 = 3$ and $x_3 = 2$, and that for the dual is $w_1 = -1$ and $w_2 = 0$, with a common optimal value of the objective function of 2. The reader will note that, for the dual constraints that correspond to the vectors in the basis (the second and third constraints), equality holds, and since we have an alternate optimal solution with vector $\mathbf{P}_5$, we also have equality in the fifth constraint.

Elaborate and efficient procedures have been devised which enable one to combine the original simplex algorithm and the dual method to solve the general linear-programming problem. Two of these are the *composite simplex algorithm* (Dantzig [159] and Orchard-Hays [530]) and the *primal-dual algorithm*[1] (Dantzig, Ford, and Fulkerson [183]). These techniques place no restrictions on the signs of the c_j and enable one to obtain an initial basic optimal or feasible solution. The reader is also referred to another technique for solving the general linear-programming problem called the *method of leading variables* (Beale [53]). This procedure also offers a method for initiating the computation for any problem.

All the above techniques are intimately related to the standard simplex method. Only such methods have proved effective for solving the general linear-programming problem (Hoffman, Mannos, Sokolowsky, and Wiegmann [377] and Hoffman [374]). Other iterative and convergent techniques that solve the problem but are not as efficient as the simplex method include the double-description method (Raiffa, Thompson, and Thrall [556]), numerical methods for zero-sum two-person games by Brown [98] and von Neumann [524], relaxation techniques for solving linear inequalities (Agmon [5] and Motzkin and Schoenberg [509]), and a projection method for solving linear constraints (Tompkins [651]). A more recent procedure, the *ellipsoid method* of Khachiyan [429], solves the joint primal-dual constraint problem (2.15) of Chap. 6. Although the ellipsoid method exhibits a slow convergence rate and does not appear to be a practical alternative to the simplex method, it is of great theoretical importance (see Sec. 6 and Bland et al. [82]).

3. INTEGER PROGRAMMING

Since its introduction as a tool of applied mathematics, the outstanding computational problem of linear programming has been that of finding the optimal solution to a linear program in which all or some of the variables are restricted to integer values. The former case is termed a *pure-integer-programming problem*, while the latter is called a *mixed-integer-programming problem*. The need for computational procedures for solving such problems is emphasized by the great number of problems from the realm of combinatorial

---

[1] The reader is referred to Mueller and Cooper [510] for a discussion comparing the computational efficiency of the standard and the primal-dual simplex algorithms.

analysis and the areas of scheduling and production that have been formulated as programming problems with integer conditions. Many such problems are described in Dantzig [163, 164] and Garfinkel and Nemhauser [281]. Further references include Hirsch and Dantzig [369] on the fixed-charge problem, Dantzig et al. [184, 185] and Miller et al. [502] on the traveling-salesman problem, and Wagner [681] and Manne [493] on machine scheduling. Other discussions are found in Dantzig [165, 166], Gross [338, 339], and Wagner and Whitin [685]. Work of Gomory [320, 321, 322, 323], Benders, Catchpole, and Kuiken [67], Land and Doig [455], Little et al. [475], Balas [19], and others has led to the development of practical computational procedures for solving a number of such problems. Applications and algorithms are discussed in Garfinkel and Nemhauser [281], Taha [636], Salkin [591], Wagner [684], and Zionts [730].

The procedures for solving integer programs can be classified into two main computational categories: *cutting plane* and *enumeration.* The basic idea of a cutting-plane method is to alter the convex set of solutions to the related *continuous linear-programming problem* (i.e., the linear-programming problem that results by dropping the integer constraints) so that the optimal extreme point to the changed continuous problem is integer-valued. This is accomplished by systematically adding additional constraints (cutting planes) that cut off parts of the convex set that do not contain any feasible integer points and solving the resultant problems by the simplex algorithm. We shall discuss below a cutting-plane algorithm for solving the integer-programming problem. Cutting-plane methods have also been developed for solving mixed-integer problems (Benders [68], Hu [383]).

Enumerative methods are designed to systematically investigate a restricted subset of the possible large set of integer solutions. These methods can be used to solve both integer and mixed-integer problems and include the *tree-search* methods of *branch and bound* (Land and Doig [455], Little et al. [475], Dakin [150], Forrest et al. [256]), and *implicit enumeration* or *additive algorithms* (Balas [19]). Branch and bound procedures use information obtained from successive solutions to related continuous problems to generate new problems with bounds on selected variables and the objective function in such a manner as to restrict the total number of continuous problems which need to be solved. Implicit enumeration was designed basically to solve problems with *n binary variables*, i.e., variables restricted to the values of zero or one. As there are 2^n possible solutions, we require "a procedure for systematically enumerating part of the solutions to the binary problem and examining them in such a way as to ensure that by enumerating a (relatively small) number of solutions, we have implicitly examined all elements of the solution set" (Balas [19]). The procedure does not use any linear-programming (simplex) techniques but substitutes new 0-1 solutions based on systematic rules for improving a solution which rely on information implied by the original problem constraints. It is an additive procedure in that only additions and subtractions are used in the computation.

We shall describe only the all-integer procedure due to Gomory.[1] Surveys describing a wide variety of integer-programming methods and problems include Balinski [26, 28], Beale [54], Geoffrion and Marsten [300], and Salkin and Balinsky [593]; the surveys by Land and Powell [456, 457] describe the status of computer codes for solving pure and mixed-integer problems. Books include Garfinkel and Nemhauser [281], Greenberg [329], Hu [383], Saaty [585], Taha [636], and Zionts [730]. Fisher [249] describes the lagrangian relaxation method as applied to solving integer-programming problems; also see Geoffrion and Nauss [301] and Nauss [522] on parametric and postoptimality analysis for integer programs.

We should note that the successful application of the transportation model with integer availabilities and demands is partially due to the fact that each basic solution corresponds to an extreme point that has integer or zero values for the coordinates. This is due to the triangular matrix of 0's and 1's associated with the basis. We might ask what similar condition is sufficient for the general linear-programming model to have an integer solution. For this answer we need to refer to the basic transformation of the simplex procedure: the elimination transformations.

If x_{ij} represents the element in the ith row and jth column of the simplex tableau and x_{lk} is the pivot element, the transformed elements are given by

$$x'_{ij} = x_{ij} - \frac{x_{lj}}{x_{lk}} x_{ik} \qquad \text{for } i \neq l$$

$$x'_{lj} = \frac{x_{lj}}{x_{lk}}$$

$$(3.1)$$

For the purpose of this discussion we shall assume that the starting tableau is all-integer. We see that a very strong condition for the successive solutions to be integer is for each fraction in (3.1) to reduce to an integer. One unusual way for this to happen is for each pivot element to equal unity. This is, of course, what happens for the transportation problem. This property is equivalent to the selection of a basis whose associated matrix has a determinant equal to 1. As we shall see below, this is one factor in the computational scheme for solving the all-integer programming problem.

The integer-programming problem is to find a vector $\mathbf{X}$ which minimizes $\mathbf{cX}$ subject to $\mathbf{AX} = \mathbf{b}$, $\mathbf{X} \geq \mathbf{0}$, with the added (nonlinear) condition that the optimal-solution vector have integer coefficients. The geometry of the situation can be pictured by considering the solution to the following problem: Maximize

$$3x_1 + x_2$$

subject to

---

[1] See Exercise 41 for a discussion of the branch-and-bound procedure.

$$x_1 + 2x_2 \leq 8$$
$$3x_1 - 4x_2 \leq 12$$
$$x_1 \qquad \geq 0 \tag{3.2}$$
$$x_2 \geq 0$$

The optimal noninteger solution is given by $x_1 = \frac{28}{5}$ and $x_2 = \frac{6}{5}$ with the maximum value equal to 18.

This problem is pictured in Fig. 9.1. The convex set of solutions is bounded by the heavy lines, and, as indicated with heavy dots, this region also includes a number of integer points. The problem is to determine which of these points maximizes the objective function.

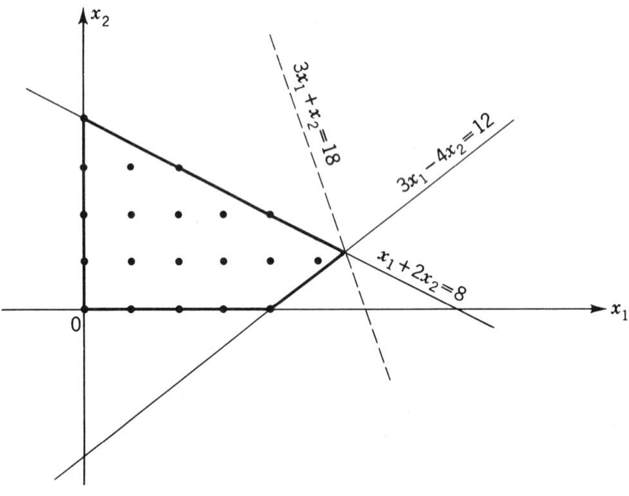

FIGURE 9.1

Since the basic computational scheme to be employed is the simplex method, we should review the geometry of this technique. The simplex method starts at an extreme point of the convex set of solutions. Each iteration, i.e., the selection of a new basis, determines a new extreme point which is a neighbor to the old point (has a boundary segment in common). This process continues and, after a finite number of steps, stops at the global optimum. We see that, in order to have an optimal integer solution, we must also have an associated extreme point that is integer-valued. For the problem in Fig. 9.1, the simplex process would probably start at the extreme point $(0, 0)$, move to $(4, 0)$, and finally to $\left(\frac{28}{5}, \frac{6}{5}\right)$. It is impossible to indicate what integer point in the solution space optimizes the objective function.

The question that now arises considers the possibility of changing the convex set of solutions in such a manner as to make the appropriate feasible integer point an extreme point of the new convex region. For example, in Fig. 9.2 we have introduced two arbitrary constraints to the problem which do the trick. This technique of putting in cutting hyperplanes was used in the past for specific formulations.   This concept can be motivated as follows.   We note for problem (3.2), as shown in Fig. 9.1, that the optimal basic feasible solution is $x_1 = \frac{28}{5}$ and $x_2 = \frac{6}{5}$, with both slack variables x_3 and x_4 equal to zero.   Thus, any integer solution (if one exists) must have at least the nonbasic variables x_3 or x_4 greater than or equal to 1.   We can then add the additional constraint $x_3 + x_4 \geq 1$ to the original problem, resolve, and hope that the new optimal is an integer.   From Fig. 9.2 we see that the optimal integer solution is $x_1 = 5$, $x_2 = 1$, with $x_3 = 1$, $x_4 = 1$.   Dantzig [172] proposed the use of such simple constraints involving the current set of nonbasic variables as cutting planes, but it was shown by Gomory and Hoffman [326] that the procedure will not, in general, converge to the integer optimal solution.   How to introduce these cuts systematically for any problem is the essence of the procedures developed by Gomory.

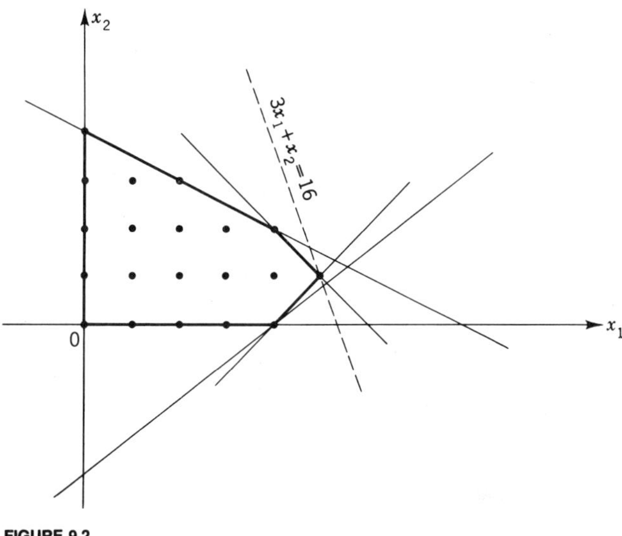

FIGURE 9.2

An early approach to the application of the cutting-plane method required solving the problem by the basic simplex algorithm, ignoring the integer constraints.[1]   If the optimal solution is not integer, then a cutting plane (i.e., a linear inequality) is introduced which preserves the optimality state of the primal problem but not the feasibility conditions.   A single application of the dual algorithm restores feasibility and yields an optimal answer to the reduced problem.   If this new solution is integer-valued, the process stops.   If not, a new constraint is introduced and the algorithm is repeated until an integer solution is found or until an indication is given that no integer solution exists for the original problem (Gomory [321]).   This method has been superseded

---

[1] Note that the optimal value of the objective function of the continuous (noninteger) problem is a lower bound for the integer problem's objective function.

by a more efficient procedure, in which the problem, assumed to be given in integers, is transformed by a modified simplex algorithm which preserves the integer-value characteristic of the complete tableau for all iterations (Gomory [323]). This algorithm is discussed next.

We assume for starting conditions that the simplex tableau is all-integer-valued and that we start with a feasible solution for the dual simplex algorithm but not for the primal. Letting $P_1, \ldots, P_m$ form a feasible basis for the dual problem, the initial Tableau 9.1 is given for a minimization problem.

In Tableau 9.1 x_{00} is the value of the objective function; some $x_{i0} < 0$ for $i \geq 1$ (solution is not primal feasible); $x_{0j} = z_j - c_j \leq 0$ for $j \geq 1$ (solution is dual feasible); and an additional set of $n - m$ constraints, $x'_j - x_j = 0$ for $j = m + 1, \ldots, n$, with $x'_j \geq 0$ and $c'_j = c_j$, has been added to the tableau. With this set of constraints, we are now forced to determine a basic feasible solution that contains n nonnegative variables instead of the usual m; hence we are given the means by which an interior point of the original convex set can be expressed. However, we still lack the means to force us into the interior and toward the optimal integer-valued point. This is accomplished by a proper selection of cutting planes.

TABLEAU 9.1

i	Basis	c	P_0	c_1 c_2 $\cdots$ c_l $\cdots$ c_m P_1 P_2 $\cdots$ P_l $\cdots$ P_m	c_{m+1} $\cdots$ c_n P_{m+1} $\cdots$ P_n	c_{m+1} $\cdots$ c_n P'_{m+1} $\cdots$ P'_n	P_{n+s}
0			x_{00}	0 0 $\cdots$ 0	$x_{0,m+1}$ $\cdots$ $x_{0,n}$	0 $\cdots$ 0	
1	P_1	c_1	x_{10}	1 0 $\cdots$ 0 $\cdots$ 0	$x_{1,m+1}$ $\cdots$ x_{1n}	0 $\cdots$ 0	
.	.	.	.	. . . .	. .	. .	
l	P_l	c_l	x_{l0}	0 0 $\cdots$ 1 $\cdots$ 0	$x_{l,m+1}$ $\cdots$ x_{ln}	0 $\cdots$ 0	
.	.	.	.	. . . .	.	. .	
m	P_m	c_m	x_{m0}	0 0 0 $\cdots$ 1	$x_{m,m+1}$ $\cdots$ x_{mn}	0 $\cdots$ 0	
$m+1$	P'_{m+1}	c_{m+1}	0	0 0 $\cdots$ 0 $\cdots$ 0	-1 $\cdots$ 0	1 $\cdots$ 0	
.	.	.	.	.	.	. .	
n	P'_n	c_n	0	0 0 $\cdots$ 0 $\cdots$ 0	0 $\cdots$ -1	0 $\cdots$ 1	
$n+1$							

We now wish to introduce new variables into the solution in a manner that will enable us to preserve the integer characteristic of the tableau and to move toward the optimal integer solution that is contained in the convex set defined by the equations of Tableau 9.1. This will be accomplished by determining cutting hyperplanes defined as equalities in terms of the nonbasic variables and a new slack vector. Eliminating the new slack vector from the basis with a pivot element[1] of -1 will preserve the integer tableau and move the solution point closer to the optimal integer point. A finite number of applications of these new constraints leads us to the optimal integer solution.

To determine the form of the cutting constraint, let us consider any equation from Tableau 9.1 that has a corresponding $x_{i0} < 0$, for example, the lth equation:

$$x_{l0} = x_l + x_{l,m+1}x_{m+1} + x_{l,m+2}x_{m+2} + \cdots + x_{ln}x_n \tag{3.3}$$

where $x_{m+1}, \ldots, x_n$ are nonbasic variables.

We next rewrite each coefficient of (3.3) as a multiple of an integer and a remainder, i.e., in the form $b_{lj}\lambda + r_{lj}$, where b_{lj} is an integer, r_{lj} is a remainder, and λ is an unspecified positive number to be determined. The coefficients can then be expressed as

$$x_{lj} = b_{lj}\lambda + r_{lj} = \left[\frac{x_{lj}}{\lambda}\right]\lambda + r_{lj} \qquad \text{for all } j$$

$$1 = \left[\frac{1}{\lambda}\right]\lambda + r \tag{3.4}$$

$$0 \le r_{lj} < \lambda \qquad 0 \le r < \lambda \qquad 0 < \lambda$$

where the brackets indicate "integer part of." If $x_{lj}/\lambda < 0$, then $[x_{lj}/\lambda] = b_{lj} < 0$ such that $b_{lj}\lambda + r_{ij} = x_{ij}$. Note that, for $\lambda > 1$, we have $[1/\lambda] = 0.$[†]

Substituting (3.4) into (3.3) and gathering appropriate terms, we have

---

[1] We select -1 as a pivot element instead of $+1$ as we are using the dual algorithm to change solutions.

[†] We illustrate expression (3.4) by the following examples. For $x_{lj} = 5$, $\lambda = \frac{2}{3}$, we have

$$5 = \left[\frac{5}{\frac{2}{3}}\right]\frac{2}{3} + \frac{1}{3} = 7\left(\frac{2}{3}\right) + \frac{1}{3}$$

For $x_{lj} = 5$, $\lambda = \frac{11}{2}$, we have

$$5 = \left[\frac{5}{\frac{11}{2}}\right]\frac{11}{2} + 5 = 0 + 5$$

For $x_{lj} = -\frac{3}{4}$, $\lambda = \frac{2}{3}$, we have

$$-\frac{3}{4} = \left[\frac{-\frac{3}{4}}{\frac{2}{3}}\right]\frac{2}{3} + \frac{7}{12} = \left[-\frac{9}{8}\right]\frac{2}{3} + \frac{7}{12} = -2\left(\frac{2}{3}\right) + \frac{7}{12}$$

$$\left[\frac{x_{l0}}{\lambda}\right]\lambda + r_{l0} = \left(\left[\frac{1}{\lambda}\right]\lambda + r\right)x_l + \left(\left[\frac{x_{l,m+1}}{\lambda}\right]\lambda + r_{l,m+1}\right)x_{m+1}$$

$$+ \left(\left[\frac{x_{l,m+2}}{\lambda}\right]\lambda + r_{l,m+2}\right)x_{m+2} + \cdots + \left(\left[\frac{x_{ln}}{\lambda}\right]\lambda + r_{ln}\right)x_n$$

or

$$r_{l0} + \lambda\left\{\left[\frac{x_{l0}}{\lambda}\right] - \left[\frac{x_{l,m+1}}{\lambda}\right]x_{m+1} - \left[\frac{x_{l,m+2}}{\lambda}\right]x_{m+2} - \cdots \right.$$

$$\left. - \left[\frac{x_{ln}}{\lambda}\right]x_n - \left[\frac{1}{\lambda}\right]x_l\right\} = r_{l,m+1}x_{m+1} + r_{l,m+2}x_{m+2} + \cdots + r_{ln}x_n + rx_l \qquad (3.5)$$

We note that any nonnegative integer values of the variables that satisfy Eq. (3.3) will also satisfy (3.5). Such a substitution will make the right-hand side of (3.5) a nonnegative number.

The term in braces can be rewritten as

$$\left[\frac{x_{l0}}{\lambda}\right] - \sum_{j \notin \mathbf{B}}\left[\frac{x_{lj}}{\lambda}\right]x_j - \left[\frac{1}{\lambda}\right]x_l^\dagger \qquad (3.6)$$

We note that the expression (3.6) not only must be an integer but also must be nonnegative. The first part is clear, since all variables and the bracketed quantities are integers. The second part is established by noting that $r_{l0} < \lambda$ in (3.5) and if (3.6) were a negative integer, the left-hand side of (3.5) would be negative, which contradicts the statement that the right-hand side of (3.5) is nonnegative.

For $\lambda > 1$, the last term of (3.6) is zero and by (3.4) and the above discussion we can rewrite (3.6) as

$$b_{l0} - \sum_{j \notin \mathbf{B}} b_{lj}x_j \geq 0 \qquad (3.6a)$$

Thus, (3.6a) is a cutting plane in terms of the current nonbasic variables, and the left-hand side of (3.6a) is a nonnegative integer. We transform (3.6a) into an equation by subtracting out the nonnegative slack variable x_{n+s}, i.e.,

$$b_{l0} - \sum_{j \notin \mathbf{B}} b_{lj}x_j - x_{n+s} = 0$$

or

$$b_{l0} = b_{l,m+1}x_{m+1} + b_{l,m+2}x_{m+2} + \cdots + b_{ln}x_n + x_{n+s} \qquad (3.7)$$

Variable x_{n+s} is also an integer variable, as it is equal to the left-hand side of

---

† The notation $\displaystyle\sum_{j \notin \mathbf{B}}$ means that the sum is taken for the set of indices j of those vectors not in the basis $\mathbf{B}$.

(3.6a).   For any $\lambda > 1$, Eq. (3.7)[1] is a constraint that must be satisfied by any integer solution to the original linear-programming problem.   The non-negative integer variable x_{n+s} is introduced as a new variable to the problem.   For the present solution, since all the nonbasic variables $x_{m+1} = \cdots = x_n = 0$, Eq. (3.7) is infeasible as $b_{l0} < 0$.   Thus, after a suitable selection of $\lambda > 1$, i.e., a value of λ which makes some element b_{lj} ($j \notin \mathbf{B}$) of (3.7) equal to -1, Eq. (3.7) can be used as a cutting constraint.

We note that $b_{l0} < 0$ since it was assumed that $x_{l0} < 0$.   Also, since we are applying the dual simplex method, some x_{lj} for j not in the basis must be negative or the problem is not feasible.   By selecting λ large enough, all $[x_{lj}/\lambda]$ for $x_{lj} < 0$ yields a $b_{lj} = -1$; hence, a pivot element of -1 is available and the integer tableau is preserved.   However, on the basis of the dual simplex transformations [Eqs. (2.2) to (2.4)] we note that a small λ will cause a larger improvement in the objective function than a larger λ since the change is a direct function of $b_{l0} = [x_{l0}/\lambda] < 0$; that is, the new value of the objective function is given by

$$x_{00} - \frac{x_{0k}}{-1}\, b_{l0} = x_{00} + x_{0k}b_{l0} = x_{00} + (z_k - c_k)\left[\frac{x_{l0}}{\lambda}\right]$$

where $\mathbf{P}_k$, the pivot column, must be the one such that

$$\theta = \min_{x_{lj} < 0} \frac{x_{0j}}{-1} = \frac{z_k - c_k}{-1}$$

(A $z_k - c_k = 0$ indicates degeneracy in the dual problem.)

In order to select λ in a precise fashion, we note the following: Since λ must be chosen such that $b_{lk} = [x_{lk}/\lambda] = -1$, we have from Eqs. (2.2)

$$\frac{z_k - c_k}{[x_{lk}/\lambda]} = \frac{z_k - c_k}{-1} \leq \frac{z_j - c_j}{[x_{lj}/\lambda]} = \frac{z_j - c_j}{b_{lj}} \qquad \text{for } x_{lj} < 0$$

---

[1] For $\lambda = 1$, and by substituting the value of x_l given by (3.3) into (3.6), a cutting constraint can be obtained of the form

$$x_{n+s} = -f_0 + \sum_{j \notin \mathbf{B}} f_j x_j$$

where the f are the fractional parts of the corresponding x_{ij}.   This constraint is used in a similar fashion to (3.7) to solve the general integer-programming problem, i.e., when not all the coefficients are integers but all the variables are restricted to integers (Gomory [321]).   This algorithm calls for the solution of the problem using the regular simplex algorithm; if the optimal solution is not all-integer, then a row with a fractional value of a variable is selected to generate the cutting plane.   It can be added to the set of original constraints and the new variable x_{n+s} put into the solution by an application of the dual simplex method (Sec. 2).   However, λ must be specified such that it yields a pivot element of -1 in Eq. (3.7) and at the same time causes the value of the objective function to be reduced as much as possible.

The reader will note that since $b_{l0} < 0$ and $x_{n+s} \geq 0$, (3.7) "cuts" away the current basic feasible solution with $x_{m+1} = \cdots = x_n = 0$, that is, forces some x_j not in the current basis into the new basis.

or

$$\frac{z_k - c_k}{-1} \leq \frac{z_j - c_j}{b_{lj}}$$

Let m_j be the largest integer for which

$$\frac{z_k - c_k}{-1} \leq \frac{z_j - c_j}{-m_j}$$

Then

$$-b_{lj} = -\left[\frac{x_{lj}}{\lambda}\right] \leq m_j \qquad \text{for } x_{lj} < 0 \qquad (3.8)$$

Hence, for each j, the smallest λ that satisfies (3.8) and allows $\mathbf{P}_k$ to be the pivot column, i.e., yields a pivot element $b_{lk} = -1$, is given by $\lambda_j = -(x_{lj}/m_j)$. As the minimum permissible λ must be at least as great as the largest λ_j, we have $\lambda_{\min} = \max_{x_{lj}<0} \lambda_j$. [If $\lambda_{\min} = \lambda_k = -x_{lk} = 1$, that is, the pivot element happens to be a -1, then we construct a suitable cutting constraint whose variables include only the nonbasic variables of the equation, as this implies a selection of $\lambda > 1$; see Eq. (3.6) and step 6 below.]

λ is not restricted to be an integer, and for $\mathbf{P}_k$ we have $m_k = 1$, $\lambda_k = -x_{lk}$, $\lambda_{\min} \geq \lambda_k$; hence $b_{lk} = [x_{lk}/\lambda_{\min}] = -1$.

The above procedure is summarized by the following steps:

1. Assume that a dual feasible solution with an all-integer tableau has been determined.
2. From the rows containing a negative constant term, select a row to be used to generate the cutting constraint (the lth row).[1]   If no such rows exist, an optimal all-integer solution has been determined.
3. The pivot column corresponds to the $\mathbf{P}_k$ whose

$$-(z_k - c_k) = \min_{x_{lj}<0} -(z_j - c_j) \qquad \text{for } j \geq 1$$

If no $x_{lj} < 0$ exists, then the problem is not feasible.

---

[1] The selection of the row that generates the cutting constraint is tied intimately to the proof that the all-integer algorithm converges. Gomory [323] cites the following rules which are consistent with the finiteness proof for selection of a row to generate the new constraint:

1. Always select the first row from the top having a negative element.
2. Select the rows by a cyclic process; i.e., on the first step look at the first row and if it does not have a negative constant, look at its successors; on the second step look at the second row and then its successors, etc.
3. If the rows are chosen at random, a finite process will result with probability 1.

The usual simplex rule of selecting the one with the largest negative term is not covered by the finiteness rule. In developing suitable computer codes for the all-integer algorithm, there has been much experimentation in the quest for a rule that is efficient for large classes of problems. Combinations of the simplex rule and other rules have worked rather well. The reader interested in developing a code would be well advised to consult the latest literature in the field.

4. For each j having a $x_{lj} < 0$, determine the largest integer

$$-(z_k - c_k) \leq \frac{-(z_j - c_j)}{m_j}$$

5. Define $\lambda_j = -(x_{lj}/m_j)$ for $x_{lj} < 0$.
6. Determine $\lambda_{min} = \max \lambda_j$.   If $\max \lambda_j = 1$, set $\lambda_{min} > 1$.
7. Develop the corresponding cutting constraint (3.7) and add it to Tableau 9.1, with the new equation being the $(n + 1)$st and the new basic variable x_{n+s} corresponding to column $\mathbf{P}_{n+s}$ (where $s = 1, 2, \ldots$, that is, the index of the iteration).
8. Apply the simplex transformation to the new tableau using $b_{lk} = -1$ as the pivot element. This causes x_{n+s} to be a nonbasic variable, and because the pivot element was -1, the tableau remains in integers. The new row is then dropped from further consideration and the process repeated.

We shall illustrate the procedure with an example due to Gomory [323]. Minimize

$$10x_1 + 14x_2 + 21x_3$$

subject to

$$8x_1 + 11x_2 + 9x_3 - x_4 \qquad\qquad = 12$$
$$2x_1 + 2x_2 + 7x_3 \qquad - x_5 \quad\;\; = 14$$
$$9x_1 + 6x_2 + 3x_3 \qquad\qquad - x_6 = 10$$
$$x_j \geq 0$$

The set of equations $x_1' = x_1$, $x_2' = x_2$, $x_3' = x_3$ and the conditions $c_j' = c_j$ have been added in Tableau 9.2, and row 7 has been added for the cutting constraint and column $\mathbf{P}_7$ for the new variable.

TABLEAU 9.2

i	Basis	c	$\mathbf{P}_0$	10	14	21	0	0	0	10	14	21	
				$\mathbf{P}_1$	$\mathbf{P}_2$	$\mathbf{P}_3$	$\mathbf{P}_4$	$\mathbf{P}_5$	$\mathbf{P}_6$	$\mathbf{P}_1'$	$\mathbf{P}_2'$	$\mathbf{P}_3'$	$\mathbf{P}_7$
0			0	-10	-14	-21	0	0	0	0	0	0	0
1	$\mathbf{P}_4$	0	-12	-8	-11	-9	1	0	0	0	0	0	0
2	$\mathbf{P}_5$	0	-14	-2	-2	-7	0	1	0	0	0	0	0
3	$\mathbf{P}_6$	0	-10	-9	-6	-3	0	0	1	0	0	0	0
4	$\mathbf{P}_1'$	10	0	-1	0	0	0	0	0	1	0	0	0
5	$\mathbf{P}_2'$	14	0	0	-1	0	0	0	0	0	1	0	0
6	$\mathbf{P}_3'$	21	0	0	0	-1	0	0	0	0	0	1	0
7	$\mathbf{P}_7$		-4	$\text{\textcircled{$-1$}}$	-1	-2	0	0	0	0	0	0	1

The second equation of the system is used to generate the cutting constraint. We have

$$\frac{z_k - c_k}{-1} = \min_{x_{2j}<0} \frac{z_j - c_j}{-1} = \frac{z_1 - c_1}{-1} = 10$$

Hence column $\mathbf{P}_1$ is to be the pivot column. To determine the m_j, we have

$$10 \leq \frac{10}{m_1} \qquad 10 \leq \frac{14}{m_2} \qquad 10 \leq \frac{21}{m_3}$$

or $m_1 = 1$, $m_2 = 1$, $m_3 = 2$. The $\lambda_j = -x_{1j}/m_j$ are then

$$\lambda_1 = \tfrac{2}{1} \qquad \lambda_2 = \tfrac{2}{1} \qquad \lambda_3 = \tfrac{7}{2}$$

and

$$\lambda = \max \lambda_j = \tfrac{7}{2}$$

The new constraint is then given by

$$\begin{bmatrix} -14 \\ \tfrac{7}{2} \end{bmatrix} = \begin{bmatrix} -2 \\ \tfrac{7}{2} \end{bmatrix} x_1 + \begin{bmatrix} -2 \\ \tfrac{7}{2} \end{bmatrix} x_2 + \begin{bmatrix} -7 \\ \tfrac{7}{2} \end{bmatrix} x_3 + x_7$$

$$-4 = \qquad -x_1 \qquad -x_2 \qquad -2x_3 + x_7$$

This equation is shown added to Tableau 9.2. By applying the elimination transformation with the pivot element as shown, Tableau 9.3 is obtained. Since all the elements of column $\mathbf{P}_1$ have been reduced to zeros, it can be eliminated from the tableau and column $\mathbf{P}_7$ substituted in its place. In fact, since vectors $\mathbf{P}_4$, $\mathbf{P}_5$, $\mathbf{P}_6$, $\mathbf{P}_1'$, $\mathbf{P}_2'$, $\mathbf{P}_3'$ will not change under the all-integer algorithm transformation, the tableau can be made quite concise as Tableau 9.4. Note that $x_1' = x_1$ is still in the basis and now has a value of 4. Here there is only one row with a negative constant term. The pivot column is $\mathbf{P}_3$ since

$$\frac{-1}{-1} = \min \left(\frac{-10}{-1}, \frac{-1}{-1} \right) \qquad 1 \leq \frac{10}{m_7} \qquad 1 \leq \frac{1}{m_3}$$

or $m_7 \leq 10$, $m_3 \leq 1$. Hence $\lambda_7 = \tfrac{2}{10}$, $\lambda_3 = \tfrac{3}{1}$, $\lambda = 3$. The new constraint is given by

$$-2 = -x_7 - x_3 + x_8$$

as shown in Tableau 9.4. The new solution is shown in Tableau 9.5.

Selecting row 3 as the row to generate the cutting constraint, we have $\mathbf{P}_7$ as the pivot column and $\lambda = 24$. The new constraint is $-1 = -x_7 + x_9$. The next solution is then given by Tableau 9.6. With row 2 as the generating row, we have $\mathbf{P}_8$ as the pivot column and $\lambda = 3$. The new constraint is $-1 = -x_8 + x_{10}$. This yields the final Tableau 9.7 and the optimal solution. The final solution is given by

$$x_1 = x_1' = 1 \qquad x_2 = x_2' = 0 \qquad x_3 = x_3' = 2$$

$$x_4 = 14 \qquad x_5 = 2 \qquad x_6 = 5 \qquad x_{00} = 52$$

TABLEAU 9.3

i	Basis	c	P_0	P_1	P_2	P_3	P_4	P_5	P_6	P'_1	P'_2	P'_3	P_7
0			40	0	-4	-1	0	0	0	0	0	0	-10
1	P_4	0	20	0	-3	7	1	0	0	0	0	0	-8
2	P_5	0	-6	0	0	-3	0	1	0	0	0	0	-2
3	P_6	0	26	0	3	15	0	0	1	0	0	0	-9
4	P'_1	10	4	0	1	2	0	0	0	1	0	0	-1
5	P'_2	14	0	0	-1	0	0	0	0	0	1	0	0
6	P'_3	21	0	0	0	-1	0	0	0	0	0	1	0
7													

TABLEAU 9.4

i	Basis	P_0	P_7	P_2	P_3	P_8
0		40	-10	-4	-1	
1	P_4	20	-8	-3	7	
2	P_5	-6	-2	0	-3	
3	P_6	26	-9	3	15	
4	P'_1	4	-1	1	2	
5	P'_2	0	0	-1	0	
6	P'_3	0	0	0	-1	
7	P_8	-2	-1	0	$\boxed{-1}$	1

TABLEAU 9.5

i	Basis	P_0	P_7	P_2	P_8	P_9
0		42	-9	-4	-1	
1	P_4	6	-15	-3	7	
2	P_5	0	1	0	-3	
3	P_6	-4	-24	3	15	
4	P'_1	0	-3	1	2	
5	P'_2	0	0	-1	0	
6	P'_3	2	1	0	-1	
7	P_9	-1	$\boxed{-1}$	0	0	1

TABLEAU 9.6

i	Basis	P_0	P_9	P_2	P_8	P_{10}
0		51	-9	-4	-1	
1	P_4	21	-15	-3	7	
2	P_5	-1	1	0	-3	
3	P_6	20	-24	3	15	
4	P'_1	3	-3	1	2	
5	P'_2	0	0	-1	0	
6	P'_3	1	1	0	-1	
7	P_{10}	-1	0	0	$\boxed{-1}$	1

TABLEAU 9.7

i	Basis	P_0	P_9	P_2	P_{10}
0		52	-9	-4	-1
1	P_4	14	-15	-3	7
2	P_5	2	1	0	-3
3	P_6	5	-24	3	15
4	P'_1	1	-3	1	2
5	P'_2	0	0	-1	0
6	P'_3	2	1	0	-1

From a theoretical point of view, the cutting-plane techniques of Gomory and others should enable us to solve any integer-programming problem. The digital-computer codes based on these algorithms have, in most instances, proved unpredictable in their ability to guarantee convergence to the optimum. The success of such algorithms varies according to the problem and the rules used to develop the cutting planes. This is in sharp contrast to the highly successful use of the simplex method to solve almost any standard (continuous) linear problem. Computer codes combining branch and bound procedures with an efficient simplex algorithm for solving the related continuous problems appear to be the most effective for solving special problems with up to 1,000 or so integer variables (see Exercise 41 for a description of the general branch and bound algorithm). Computational procedures and test results of commercially available integer-programming algorithms (all branch and bound and part of an integrated mathematical programming system) are reported in Land and Powell [457], Forrest et al. [256], Benichou et al. [70], McCall [482], and Peeters [543]. Although such systems can process mixed-variable problems with many thousands of integer variables, their ability to find and prove out optimal solutions in a reasonable amount of time may not be consistent. Problems with a few hundred integer variables can usually be solved optimally without difficulty. For larger problems, a good if not optimal integer solution may be found in an acceptable amount of time (10 to 60 minutes on an advanced computer). The user of an integer-mathematical programming system should become familiar with the system options and the experiences of other users before trying to solve a new problem, especially a large-scale one. (For example, see the discussion of the formulation and solution of a production allocation and distribution problem in Mairs et al., Sec. 13 of the Bibliography.) The theoretical and computational aspects of integer programming are undergoing intensive investigation and experimentation. Advanced work here includes the combining of branch and bound and cutting-plane techniques to solve rather large $(0, 1)$ pure integer problems, e.g., a 756 row and 2,756 variable problem solved in 54.42 minutes on an IBM 370/168 (Crowder et al. [144]). Also, special procedures have been developed to solve important classes of problems such as the $(0, 1)$ set-partitioning and set-covering problems (see definitions and discussion below) (Balas [20], Balas and Ho [21], Marsten [497], and Spielberg [620]). The reader is referred to Balinski [26], Salkin and Balinsky [593], and Geoffrion [298] for discussions comparing the efficacy of different cutting-plane, branch and bound, and implicit enumeration algorithms and related computational experience.

The ability to solve integer-programming problems opens up a wide range of important application areas, especially those of a combinatorial nature (see Exercise 42 and the papers by Ibaraki [386], Geoffrion and Graves [299], and Mairs et al. in Sec. 13 of the Bibliography). A combinatorial problem that can be formulated in terms of integer variables is the traveling-salesman problem as developed by Miller, Tucker, and Zemlin [502].[1]

> A salesman is required to visit each of n cities, indexed by $1, \ldots, n$. He leaves from a "base city" indexed by 0, visits each of the n other cities exactly once, and returns to city 0. During his travels he must return to 0 exactly t times, including his final return (here t may be allowed to vary), and he must visit no more than p cities in one tour. (By a tour we mean a succession of visits to cities without stopping at city 0.) It is required to find such an itinerary which minimizes the total distance traveled by the salesman.

Note that if t is fixed, then for the problem to have a solution we must have $tp \geq n$. For $t = 1$, $p \geq n$, we have the standard traveling-salesman problem.

Let d_{ij} $(i \neq j = 0, 1, \ldots, n)$ be the distance covered in traveling from city i

---

[1] See Bellmore and Nemhauser, Parker and Rardin, Bibliography, Sec. 18, and Gomory [325] for reviews of the traveling-salesman problem. Also see Dantzig, Fulkerson, and Johnson [185] for another integer formulation of the traveling-salesman problem.

to city j. The following integer-programming problem is shown to be a model of the problem:

Minimize the linear form

$$\sum_{0 \le i \ne j \le n} \sum d_{ij} x_{ij}$$

over the set determined by the relations

$$\sum_{\substack{i=0 \\ i \ne j}}^{n} x_{ij} = 1 \qquad j = 1, \ldots, n$$

$$\sum_{\substack{j=0 \\ j \ne i}}^{n} x_{ij} = 1 \qquad i = 1, \ldots, n$$

$$u_i - u_j + p x_{ij} \le p - 1 \qquad 1 \le i \ne j \le n$$

where the x_{ij} are nonnegative integers and the u_i ($i = 1, \ldots, n$) are arbitrary real numbers and can be restricted to nonnegative integers. If t is fixed, we must add the restriction $\sum_{i=1}^{n} x_{i0} = t$. Solutions to reasonable-size (50 to 325 cities) traveling-salesman problems have been found by applying branch and bound methods (Little et al. [475] and Balas and Toth [25]) or by using a combination of branch and bound and cutting-plane procedures (Crowder and Padberg [145]). The latter approach was used to solve a 318-city problem that dealt with the routing of a numerically controlled drilling machine, with the drilling being done by a laser. The data are given as coordinates of the 318 points in the plane in milli-inches. (The problem is due to Lin and Kernigham [474].) See Parker and Rardin [542] for a review of algorithms for solving the traveling-salesman problem.

An important special class of integer-programming problems is that of the *set-covering problems*, Balinski [26], Salkin and Saha [595], and Garfinkel and Nemhauser [281]. We illustrate the nature of these problems by the following example (other applications are given in Exercise 31 of Chap. 11).

Space-agency scientists want to design an instrumented pay load for an orbiting laboratory that will perform three experiments. Packages of instruments can be obtained which can carry out each one of the three experiments; or multipurpose, weight-saving packages can be purchased which can perform more than one experiment. Specifically, we have six packages: packages 1, 2, and 3 can perform only the corresponding experiments; package 4 can do experiments 1 and 3, package 5 can perform experiments 1 and 2, and package 6 can carry out experiments 2 and 3. For each package we can define an integer variable x_j equal to 1 if package j is chosen for the pay load and equal to 0 if it is not selected. Also, associated with each x_j is a weight (or cost) c_j of the corresponding package. We wish to select the set of packages so that each experiment can be accomplished (i.e., covered) at the minimum pay-load weight.

For this problem we can display the data in column-vector form as in Tableau 9.8. We see that if $x_2 = 1$ and $x_4 = 1$, the 1's in the corresponding columns add up to the 1's in the experiments column; i.e., the experiments are covered by a pay load with weight $(c_2 + c_4)$. However, if $x_5 = 1$ and $x_6 = 1$, experiments 1 and 3 are covered but experiment 2 is covered twice; i.e., both packages can do project 2 at a weight of $(c_5 + c_6)$, thus providing the added benefit of having backup for experiment 2. Although most set-covering problems do not require any such backup or redundancy considerations, they are so defined that, in order to ensure the existence of a feasible solution, we need to allow for multiple coverage. (This would be the case for the above problem if there were no single-purpose packages, i.e., $x_1 = x_2 = x_3 = 0$.)

TABLEAU 9.8

x_1	x_2	x_3	x_4	x_5	x_6	Experiments	
1	0	0	1	1	0	1	Experiment 1
0	1	0	0	1	1	1	Experiment 2
0	0	1	1	0	1	1	Experiment 3

In general, let $\mathbf{e} = (1, \ldots, 1)$ be an m-dimensional column vector whose components are all 1's. Define $\mathbf{E}_j$ as an m-dimensional column vector of 1's and 0's with a 1 appearing if the definition of the column is such that it covers a corresponding 1 of $\mathbf{e}$, and 0 otherwise. Let c_j be the cost associated with vector $\mathbf{E}_j$ and let x_j be the corresponding integer variable restricted to either 0 or 1. We can then define the set-covering problem which allows for multiple coverage to select the set of columns $\mathbf{E}_j$ such that we minimize

$$\sum_{j=1}^{n} c_j x_j$$

subject to

$$\sum_{j=1}^{n} \mathbf{E}_j x_j \geq \mathbf{e} \qquad x_j = 0 \text{ or } 1 \qquad j = 1, 2, \ldots, n \tag{3.9}$$

The selection process is accomplished by the optimal setting of the x_j at either 0 or 1. If equations are given in (3.9), then the problem is termed a *set-partitioning problem;* i.e., no multiple coverage is allowed. For a maximizing objective function and for less than or equal to constraints similar to (3.9), the problem is termed a *packing problem.*

Because of the 0-1 structure of the matrix of coefficients, special computational approaches have been developed to solve set-covering and partitioning problems. Some of these methods rely on demonstrating that certain vectors and rows can be eliminated from the problem, i.e., reducing the size of the problem, without changing the optimal solution, Garfinkel and Nemhauser

[281]. Experience in solving set-covering problems using Gomory's all-integer method is discussed in Salkin and Koncal [594], and the use of branch and bound procedures in Marsten [497]. Padberg [539] surveys the connections between covering, packing, and knapsack problems.

4. THE DECOMPOSITION OF LARGE-SCALE SYSTEMS

Although it is theoretically possible to solve any given linear-programming model, analysts quickly become aware of certain limitations which restrict their endeavors. Chief among these limitations is the problem of dimensionality. Almost all difficulties that arise in the development of a programming problem can be related to its size. This is certainly true for such restrictive items as the cost of data gathering, matrix preparation, computing costs, and validity of the linear model.

The development of procedures for the solution of large-scale systems is reviewed in Dantzig [167] and Gomory [324], and a number of procedures are given in Graves and Wolfe [328]. Dantzig discusses techniques to reduce the computational requirements of large systems from two points of view: (1) by decreasing the number of iterations, and (2) by finding a compact form for the inverse and/or by taking advantage of any special structure of the system of equations.[1] To cut down the number of iterations, variants of the simplex method have been proposed to replace the usual Phase I of the simplex method: *method of leading variables*, Beale [53]; *composite simplex algorithm*, Orchard-Hays [530] and Wolfe [718]; and *primal-dual algorithm*, Dantzig, Ford, and Fulkerson [183].

Proposals for finding a compact form for the inverse or for taking advantage of special structures include the sparse-basis technique, Markowitz [495]; block-triangular basis, Dantzig [156, 169]; the solution of a dynamic Leontief model with substitution, Dantzig [168]; large-scale economic systems, Rech [562]; and for the important class of block-angular systems and multistage systems of the staircase type we have the *decomposition algorithm*, Dantzig and Wolfe [196]; *partition programming*, Benders [68], Rosen [576], Winkler [708], and Ritter [568]; *pseudo-basic variable procedure*, Beale [55]; and the *dualplex method*, Gass [286]. Expositions and summaries of these and other procedures for solving large-scale problems are given in Dantzig [175], Geoffrion [298], and Lasdon [460]; some applications are discussed in Himmelblau [368] and Beale [59]. In this section we shall describe only the decomposition algorithm of Dantzig and Wolfe which, when applied to block-angular systems, has an interesting and important economic interpretation in terms of decentralized planning; see Dantzig [152] and Baumol and Fabian [49].

For many problems the constraints consist of rather large independent subsets of equations which refer to the same time period or same production

---

[1] See Sec. 3 of Chap. 5 for a general discussion of iteration-reducing procedures and compact inverse representations.

facility. These subsets are usually tied together by a small set of equations. These "tie-in equations" might represent restrictions on the availability of basic resources, total demand of a product, or total budgetary constraint. In problems of this sort we have, in a sense, a number of separate linear-programming problems whose joint solution must satisfy a set of additional restrictions. If we boxed in the sets of constraints and corresponding part of the objective function, Fig. 9.3 would result.

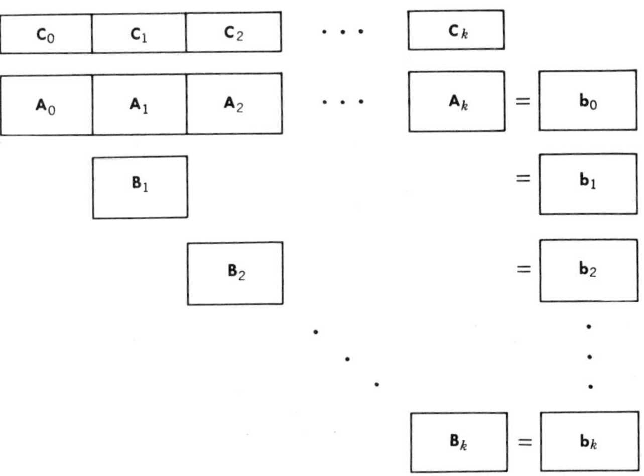

FIGURE 9.3

In Fig. 9.3 we have partitioned the original problem of $\mathbf{AX} = \mathbf{b}$, $\mathbf{X} \geq \mathbf{0}$, $\mathbf{cX}$ a minimum, into the *decomposed program*[1] of finding the vectors $\mathbf{X}_p \geq \mathbf{0}$ ($p = 0, 1, \ldots, k$) which minimize

$$\sum_{p=0}^{k} \mathbf{C}_p \mathbf{X}_p \tag{4.1}$$

subject to

$$\sum_{p=0}^{k} \mathbf{A}_p \mathbf{X}_p = \mathbf{b}_0 \tag{4.2}$$

$$\mathbf{B}_p \mathbf{X}_p = \mathbf{b}_p \qquad p = 1, 2, \ldots, k \tag{4.3}$$

where $\mathbf{A}_p$ is an $m_0 \times n_p$ matrix, $\mathbf{B}_p$ is an $m_p \times n_p$ matrix, $\mathbf{C}_p$ is an n_p-component row vector, $\mathbf{b}_p$ is an m_p-component column vector, and $\mathbf{X}_p$ is an n_p-variable column vector, i.e., $\mathbf{X}_p = (x_{p1}, \ldots, x_{pn_p})$. Hence, the above problem has $m = \sum_{p=0}^{k} m_p$ constraints and $n = \sum_{p=0}^{k} n_p$ variables. In explicit matrix notation, (4.2) and (4.3) are given by

[1] Here we are adhering to the notation of Harvey [357].

$$\begin{pmatrix} \mathbf{A}_0 & \mathbf{A}_1 & \mathbf{A}_2 & \cdots & \mathbf{A}_k \\ & \mathbf{B}_1 & & & \\ & & \mathbf{B}_2 & & \\ & & & \ddots & \\ & & & & \mathbf{B}_k \end{pmatrix} \begin{pmatrix} \mathbf{X}_0 \\ \mathbf{X}_1 \\ \mathbf{X}_2 \\ \vdots \\ \mathbf{X}_k \end{pmatrix} = \begin{pmatrix} \mathbf{b}_0 \\ \mathbf{b}_1 \\ \mathbf{b}_2 \\ \vdots \\ \mathbf{b}_k \end{pmatrix}$$

To better understand what follows, let us look at the geometry of a very simple decomposed problem: Minimize $c_{11}x_{11} + c_{12}x_{12}$ subject to $\mathbf{X}_1 = (x_{11}, x_{12}) \geq 0$ and

$$\left. \begin{array}{l} x_{11} + 4x_{12} \leq 12 \quad \Big\}\, \mathbf{A}_1\mathbf{X}_1 \leq \mathbf{b}_0 \\[4pt] \begin{array}{l} -x_{11} + 2x_{12} \leq 2 \\ 3x_{11} + 4x_{12} \leq 24 \end{array} \Big\}\, \mathbf{B}_1\mathbf{X}_1 \leq \mathbf{b}_1 \end{array} \right\} \tag{4.4}$$

The nonnegative solution set to (4.4) is shown in Fig. 9.4. The hatched area is the solution set to $\mathbf{B}_1\mathbf{X}_1 \leq \mathbf{b}_1$ and the shaded area is the solution space to the complete problem (4.4). The extreme points of $\mathbf{B}_1\mathbf{X}_1 \leq \mathbf{b}_1$ are indicated by squares, while the extreme points of the complete problem are indicated by circles. Any solution to the complete problem must, of course, satisfy $\mathbf{B}_1\mathbf{X}_1 \leq \mathbf{b}_1$. But since the solution set of linear constraints is a convex polyhedron, any solution to $\mathbf{B}_1\mathbf{X}_1 \leq \mathbf{b}_1$ can be expressed as a convex combination of the extreme points of its solution set (we assume this convex set is bounded). Hence, if we knew all the extreme points of $\mathbf{B}_1\mathbf{X}_1 \leq \mathbf{b}_1$ or could find the right ones, all we would need to do is to find the convex combination of these points which satisfies $\mathbf{A}_1\mathbf{X}_1 \leq \mathbf{b}_0$ and optimizes the objective function. This is the essence of the Dantzig-Wolfe decomposition algorithm. In Fig. 9.4 we note that any circled point is also a convex combination of the squared points and that any point in the shaded area can be represented as a convex combination of the squared points.

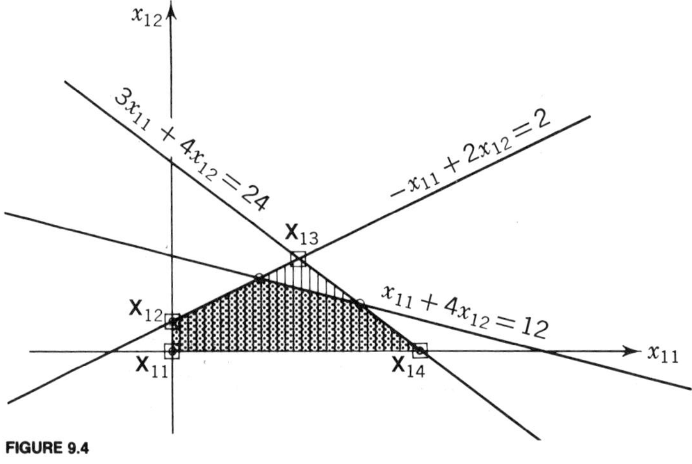

FIGURE 9.4

Extending this idea to the general problem of (4.1) to (4.3), we assume that for each p there is available the corresponding solution set $\mathbf{S}_p$ to the constraints $\mathbf{B}_p\mathbf{X}_p = \mathbf{b}_p$, $\mathbf{X}_p \geq 0$. Then the solution of the original problem could be thought of as the selection of a convex combination of solution points (i.e., extreme points) from each $\mathbf{S}_p$ so as to satisfy the tie-in restrictions $\sum_{p=0}^{k} \mathbf{A}_p\mathbf{X}_p = \mathbf{b}_0$ and make $\sum_{p=0}^{k} \mathbf{C}_p\mathbf{X}_p$ a minimum. If our original problem is looked at in the above manner, we shall see that we would then be required to optimize a new problem with $m_0 + k$ constraints subject to the solution of k individual $m_p \times n_p$ subproblems instead of one large problem with $\sum_{p=0}^{k} m_p$ constraints.[1]

The new problem to be considered is called the *extremal problem* or *master problem* and arises in the following fashion: Let the extreme points of each $\mathbf{S}_p$ be denoted by $\mathbf{X}_{pj}$ for $j = 1, 2, \ldots, N_p$, where $\mathbf{X}_{pj}$ are solution vectors for the pth subproblem and N_p is the total number of extreme points for $\mathbf{S}_p$. Any feasible solution $\mathbf{X}_p$ to the pth subproblem can be expressed as a convex combination of extreme points of $\mathbf{S}_p$, that is,

$$\mathbf{X}_p = \sum_{j=1}^{N_p} \lambda_{pj}\mathbf{X}_{pj} \qquad\qquad (4.4a)$$

where

$$\sum_{j=1}^{N_p} \lambda_{pj} = 1 \qquad \text{and} \qquad \lambda_{pj} \geq 0 \qquad\qquad (4.4b)$$

We are temporarily assuming that each $\mathbf{S}_p$ is a bounded polyhedron.

From (4.4a), substituting the expressions for $\mathbf{X}_p$ into (4.1) and (4.2) we have, respectively,

$$\mathbf{C}_0\mathbf{X}_0 + \sum_{p=1}^{k} \mathbf{C}_p\left(\sum_{j=1}^{N_p} \lambda_{pj}\mathbf{X}_{pj}\right)$$

and

$$\mathbf{A}_0\mathbf{X}_0 + \sum_{p=1}^{k} \mathbf{A}_p\left(\sum_{j=1}^{N_p} \lambda_{pj}\mathbf{X}_{pj}\right) = \mathbf{b}_0$$

or

$$\mathbf{C}_0\mathbf{X}_0 + \sum_{p=1}^{k} \sum_{j=1}^{N_p} \mathbf{C}_p\mathbf{X}_{pj}\lambda_{pj} \qquad\qquad (4.4c)$$

and

$$\mathbf{A}_0\mathbf{X}_0 + \sum_{p=1}^{k} \sum_{j=1}^{N_p} \mathbf{A}_p\mathbf{X}_{pj}\lambda_{pj} = \mathbf{b}_0 \qquad\qquad (4.4d)$$

---

[1] From the design of present computer procedures for solving linear-programming problems, we note that the number of constraints is usually the restrictive element.

Define the transformations

$$\mathbf{P}_{pj} = \mathbf{A}_p \mathbf{X}_{pj} \quad \text{and} \quad f_{pj} = \mathbf{C}_p \mathbf{X}_{pj} \qquad p = 1, 2, \ldots, k \tag{4.5}$$

and substituting in (4.4c) and (4.4d) we have that problem (4.1) to (4.3) can now be written in the equivalent form of finding $\lambda_{pj} \geq 0$, $\mathbf{X}_0 \geq \mathbf{0}$, which minimizes

$$\mathbf{C}_0 \mathbf{X}_0 + \sum_{p=1}^{k} \sum_{j=1}^{N_p} f_{pj} \lambda_{pj} \tag{4.6}$$

subject to

$$\mathbf{A}_0 \mathbf{X}_0 + \sum_{p=1}^{k} \sum_{j=1}^{N_p} \mathbf{P}_{pj} \lambda_{pj} = \mathbf{b}_0 \tag{4.7}$$

$$\sum_{j=1}^{N_p} \lambda_{pj} = 1 \qquad p = 1, 2, \ldots, k \tag{4.8}$$

This problem is the extremal problem. The first m_0 constraint rows are termed the *transfer rows* and the last k rows are the *convexity rows*. In (4.7) and (4.8) the columns $\begin{pmatrix} \mathbf{A}_{0j} \\ \mathbf{0} \end{pmatrix}$, that is, the columns of $\mathbf{A}_0$ with k zeros attached, are *natural columns*; and the columns $\begin{pmatrix} \mathbf{P}_{pj} \\ \mathbf{e}_p \end{pmatrix}$ are *extremal columns*, where $\mathbf{e}_p$ is the pth column of a unit matrix of order k. We should emphasize that (4.6) to (4.8) are just the representation of (4.1) to (4.3) in terms of all the extreme points of all the k subproblems. The transformations $\mathbf{P}_{pj}$ and f_{pj} are just the representation of these extreme points in terms of their contribution to the tie-in equations and objective function, respectively. Given the optimal λ_{pj} to the extremal problem (4.6) to (4.8), then the corresponding optimal solution in terms of each $\mathbf{X}_p$ is given by $\mathbf{X}_p = \sum_j \lambda_{pj} \mathbf{X}_{pj}$.

To illustrate the above let us develop the extremal problem for problem (4.4). The original problem in equation form, after the addition of slack variables x_{01}, x_{13}, and x_{14}, is to minimize

$$c_{11} x_{11} + c_{12} x_{12}$$

subject to

x_{01}	$+\, x_{11} + 4x_{12}$	$= 12$
	$-\, x_{11} + 2x_{12} + x_{13} \qquad = 2$ $3x_{11} + 4x_{12} \qquad\quad + x_{14} = 24$	

with all $x_{ij} \geq 0$. Here $\mathbf{A}_0 = (1)$, $\mathbf{A}_1 = (1, 4, 0, 0)$. $\mathbf{B}_1 = \begin{pmatrix} -1 & 2 & 1 & 0 \\ 3 & 4 & 0 & 1 \end{pmatrix}$, $\mathbf{b}_0 = (12)$, $\mathbf{b}_1 = \begin{pmatrix} 2 \\ 24 \end{pmatrix}$, $\mathbf{C}_0 = (0)$, $\mathbf{C}_1 = (c_{11}, c_{12}, 0, 0)$. In Fig. 9.4 the extreme points of the solution set to $\mathbf{B}_1\mathbf{X}_1 \leq \mathbf{b}_1$ are denoted by $\mathbf{X}_{11}$, $\mathbf{X}_{12}$, $\mathbf{X}_{13}$, $\mathbf{X}_{14}$.   We note that we are now dealing with equations and the $\mathbf{X}_{ij}$'s have four components corresponding to x_{11}, x_{12}, and the slack variables x_{13}, x_{14}.[†]   Following Eq. (4.5), we have

$$\mathbf{P}_{11} = \mathbf{A}_1\mathbf{X}_{11} = (1, 4, 0, 0)\mathbf{X}_{11} = (1, 4, 0, 0) \cdot (0, 0, 2, 24) = (0)$$
$$\mathbf{P}_{12} = \mathbf{A}_1\mathbf{X}_{12} = (1, 4, 0, 0)\mathbf{X}_{12} = (1, 4, 0, 0) \cdot (0, 1, 0, 20) = (4)$$
$$\mathbf{P}_{13} = \mathbf{A}_1\mathbf{X}_{13} = (1, 4, 0, 0)\mathbf{X}_{13} = (1, 4, 0, 0) \cdot (4, 3, 0, 0) \; = (16)$$
$$\mathbf{P}_{14} = \mathbf{A}_1\mathbf{X}_{14} = (1, 4, 0, 0)\mathbf{X}_{14} = (1, 4, 0, 0) \cdot (8, 0, 10, 0) = (8)$$

and

$$f_{11} = \mathbf{C}_1\mathbf{X}_{11} = (c_{11}, c_{12}, 0, 0)\mathbf{X}_{11} = (0)$$
$$f_{12} = \mathbf{C}_1\mathbf{X}_{12} = (c_{11}, c_{12}, 0, 0)\mathbf{X}_{12} = (c_{12})$$
$$f_{13} = \mathbf{C}_1\mathbf{X}_{13} = (c_{11}, c_{12}, 0, 0)\mathbf{X}_{13} = (4c_{11} + 3c_{12})$$
$$f_{14} = \mathbf{C}_1\mathbf{X}_{14} = (c_{11}, c_{12}, 0, 0)\mathbf{X}_{14} = (8c_{11})$$

The extremal problem, that is, the original problem expressed in terms of all the extreme points of the subproblem $\mathbf{B}_1\mathbf{X}_1 \leq \mathbf{b}_1$, is to find $\lambda_{1j} \geq 0$ which minimizes

$$f_{11}\lambda_{11} + f_{12}\lambda_{12} + f_{13}\lambda_{13} + f_{14}\lambda_{14} \qquad\qquad (4.6a)$$

subject to

$$x_{01} + \mathbf{P}_{11}\lambda_{11} + \mathbf{P}_{12}\lambda_{12} + \mathbf{P}_{13}\lambda_{13} + \mathbf{P}_{14}\lambda_{14} = 12 \qquad (4.7a)$$
$$\lambda_{11} + \lambda_{12} + \lambda_{13} + \lambda_{14} = 1 \qquad\qquad\qquad\quad (4.8a)$$

Using the numerical equivalents of the $\mathbf{P}_{pj}$ and the f_{pj} given above, the extremal problem is to minimize

$$c_{12}\lambda_{12} + (4c_{11} + 3c_{12})\lambda_{13} + 8c_{11}\lambda_{14}$$

subject to

$$x_{01} + 4\lambda_{12} + 16\lambda_{13} + 8\lambda_{14} = 12$$
$$\lambda_{12} + \lambda_{13} + \lambda_{14} = 1 \qquad\qquad\qquad (4.4e)$$

As mentioned above, this transformation yields a problem with only $m_0 + k$ constraints, the m_0 transfer rows (4.7), and the set of k convexity constraints (4.8).   However, the number of variables has been increased to the total of all extreme points of the convex polyhedra $\mathbf{S}_p$.   The saving element of the decomposition principle is that we need consider only a small number of this usually rather large total, and we need only the explicit representation of those to be considered.   We shall see that the decomposition algorithm embodies the basic elements of the pricing vector and basis transformation of

---

[†] To avoid notational confusion, the reader should bear in mind that $\mathbf{X}_1$ is the variable vector with variables (components) x_{11}, x_{12}, x_{13}, x_{14}, and the extreme points $\mathbf{X}_{1p}$ are particular solution vectors with specified values of the variables.

the revised simplex method. We next outline this algorithm and give a numerical example.

We treat the extremal problem (4.6) to (4.8) as we would any other linear-programming problem except that we have to allow for the fact that we are unable to write out the matrix of the system corresponding to (4.7). However, let us assume that we know enough about the system so that we have an initial basic feasible solution to the extremal problem. This first solution could have been obtained using an artificial basis or, based on the structure of the subproblems, we were able to find extreme-point solutions to the subproblems which combined into a feasible solution to the extremal problem (see example). This basis is of order $m_0 + k$ and must contain at least one extremal column from each subproblem solution space S_p, that is, contain in total at least k extremal columns and m_0 additional extremal or natural columns or both. Denote the associated simplex multiplier vector[1] by $\mathbf{\Pi} = (\boldsymbol{\pi}, \bar{\boldsymbol{\pi}})$, where the m_0-component row vector $\boldsymbol{\pi}$ is associated with the m_0 transfer rows of (4.7) and the k-component row vector $\bar{\boldsymbol{\pi}}$ is associated with the convexity rows (4.8). The components of $\boldsymbol{\pi}$ are called the *transfer prices* and will be denoted by π_i and those of $\bar{\boldsymbol{\pi}}$ by $\bar{\pi}_p$. For the vectors in the basis we must have $f_{pj} = \boldsymbol{\pi}\mathbf{P}_{pj} + \bar{\pi}_p$; that is, the corresponding $c_j - z_j = 0$. As in the regular simplex method, we next have to determine whether or not the initial feasible solution can be improved by introducing some new natural or extremal column. Pricing out an extremal column from some subproblem p, we have from (4.5) relative costs (that is, $c_j - z_j$)

$$f_{pj} - \mathbf{\Pi}\begin{pmatrix} \mathbf{P}_{pj} \\ \mathbf{e}_p \end{pmatrix} = f_{pj} - \boldsymbol{\pi}\mathbf{P}_{pj} - \bar{\boldsymbol{\pi}}\mathbf{e}_p = \mathbf{C}_p\mathbf{X}_{pj} - \boldsymbol{\pi}\mathbf{A}_p\mathbf{X}_{pj} - \bar{\pi}_p$$

$$= (\mathbf{C}_p - \boldsymbol{\pi}\mathbf{A}_p)\mathbf{X}_{pj} - \bar{\pi}_p \qquad p = 1, 2, \ldots, k \qquad (4.9)$$

Pricing out the natural columns, we have, for each j, the relative costs

$$c_{0j} - \boldsymbol{\pi}\mathbf{A}_{0j} \qquad (4.10)$$

where c_{0j} are the elements of $\mathbf{C}_0$. Since the extremal problem is a minimization problem, if all the relative costs (4.9) and (4.10) are nonnegative for all p and j, then the simplex criterion tells us that we have an optimum, since (4.9) and (4.10) correspond to the set of $c_j - z_j$ of the simplex algorithm, and for a minimization problem we are optimal if all $z_j - c_j \leq 0$ or if all $c_j - z_j \geq 0$. The costs (4.10) are readily calculated, as we know all the $\mathbf{A}_{0j}$. However, in order to determine if any of the relative costs (4.9) are negative, we need to solve p optimization problems which arise as follows. For any p we need to know whether or not for the set of extreme points $\mathbf{X}_{pj}$ we have

$$\underset{\substack{\mathbf{B}_p\mathbf{X}_{pj} = \mathbf{b}_p \\ \mathbf{X}_{pj} \geq 0}}{\text{minimum}} [(\mathbf{C}_p - \boldsymbol{\pi}\mathbf{A}_p)\mathbf{X}_{pj} - \bar{\pi}_p] < 0 \qquad p = 1, 2, \ldots, k \qquad (4.11)$$

[1] We assume we are employing the revised simplex method to perform the optimization of the extremal problem; see Chap. 5.

But (4.11) states that for each p the quantity in brackets is minimized over those extreme points X_{pj} which satisfy the constraints of the pth subproblem. If the minimum is nonnegative, then all extreme points of the pth subproblem price out optimally.

Since the condition (4.11) is independent of the scalar $\bar{\pi}_p$, (4.11) reduces to solving for each p the linear-programming problem, which is to minimize

$$(\mathbf{C}_p - \pi \mathbf{A}_p)\mathbf{X}_p \tag{4.12a}$$

subject to

$$\mathbf{B}_p\mathbf{X}_p = \mathbf{b}_p \tag{4.12b}$$
$$\mathbf{X}_p \geq \mathbf{0}$$

We note that since the solution to (4.12b) by the simplex procedure will generate only extreme-point solutions, then our solution to subproblem (4.12a and b) will also satisfy (4.11). We also note that each subproblem (4.12a and b) is equivalent to the given subproblem except that the corresponding cost coefficients have been reduced by $\pi \mathbf{A}_p$; the $(\mathbf{C}_p - \pi \mathbf{A}_p)$ are called *adjusted costs*. We see that $\pi \mathbf{A}_p$ is found by taking n_p inner products and can be readily calculated. We let $\mathbf{X}_{pq}$ be an optimal solution to (4.12a and b) for a given p ($\mathbf{X}_{pq}$ will be an extreme point if $\mathbf{S}_p$ is bounded) and let the corresponding optimal value of the objective function be denoted by

$$Z_{pq} = (\mathbf{C}_p - \pi \mathbf{A}_p)\mathbf{X}_{pq}$$

If $Z_{pq} - \bar{\pi}_p < 0$, then the new extremal column formed by the transformation (4.5), that is, $\mathbf{P}_{pq} = \mathbf{A}_p\mathbf{X}_{pq}$ with its associated cost $f_{pq} = \mathbf{C}_p\mathbf{X}_{pq}$, is a candidate to enter the basis. The new extreme point $\mathbf{X}_{pq}$ is called a *proposal vector*, the transformation vector $\mathbf{P}_{pq}$ its *transfer vector*, and f_{pq} its *transfer cost*.

Given the set of minimum costs (4.9) and the costs (4.10), the simplex method tells us to take as the next column to enter the extremal-problem basis the one corresponding to the min $(c_j - z_j)$, that is,

$$\min_{p} \left[\min_{p} (Z_{pq} - \bar{\pi}_p), \ \min_{j} (c_{0j} - \pi \mathbf{A}_{0j})\right] \tag{4.13}$$

where p ranges over the set of subproblems and j ranges over the nonbasic natural columns. If (4.13) is nonnegative, then the current basic feasible solution of the extremal problem yields an optimum to the original problem. If (4.13) is negative, we select the corresponding extremal or natural column to enter the basis. For this latter case, we determine the pivot row as usual, determine the new $\mathbf{\Pi}$, and repeat the procedure until an optimum is reached. The optimal solution to the original problem is given by

$$\left[\mathbf{X}_0, \mathbf{X}_p = \sum_{j} \lambda_{pj}\mathbf{X}_{pj} \quad (p = 1, 2, \ldots, k)\right] \tag{4.14}$$

where the summation is taken over those extreme points of the pth subproblem which are in the optimal solution to the extremal problem.

Up to this point we have assumed that the polyhedra S_p are bounded. For the general situation let us assume that this is not the case. Then in optimizing one of the subproblems (4.12a and b), we could obtain an optimal solution with an unbounded value Z_{pq} of the objective function; see Sec. 1, Chap. 4. For this case the optimal solution to the subproblem will yield a corresponding transfer vector $\mathbf{P}_{pq}$ as a candidate for the extremal-problem basis, but we cannot restrict the associated extremal-problem variable λ_{pq} to be bounded by the convexity equation; that is, λ_{pq} is not bounded above. The reason for this is as follows: In the simplex method an unbounded solution is represented by a basic feasible solution plus a nonnegative sum of vectors (see Exercise 16).[1] For the standard linear-programming problem this latter sum is a solution to the homogeneous equations $\mathbf{AX} = \mathbf{0}$ and the constraints $\mathbf{X} \geq \mathbf{0}$. For the decomposition problem, we would have a nonnegative solution to the homogeneous equations $\mathbf{B}_p\mathbf{X}_p = \mathbf{0}$. We let $\mathbf{X}_{pq}$ denote this solution, $\mathbf{B}_q\mathbf{X}_{pq} = \mathbf{0}$, and treat it as any other proposal vector (note that $\mathbf{X}_{pq}$ is not an extreme-point solution to the subproblem). As $\mathbf{B}_p(\lambda_{pq}\mathbf{X}_{pq}) = \mathbf{0}$ for all $\lambda_{pq} \geq 0$, we have that a nonnegative sum of $\mathbf{X}_{pq}$ with a convex combination of extreme points of S_p can be found which yields a vector satisfying the constraints $\mathbf{B}_p\mathbf{X}_p = \mathbf{b}_p$. Whenever we encounter an unbounded solution $\mathbf{X}_{pq}$ in any subproblem p, the corresponding convexity row is rewritten so as not to include the appropriate λ_{pq}, that is,

$$\sum_{\substack{j=1 \\ j \neq q}}^{N_p} \lambda_{pj} = 1$$

From a computational point of view we should note the following. Each time it is necessary to reoptimize a subproblem for a new objective function, the previous optimal feasible solution should be used as the first solution to the new problem. In the selection of the new nonbasic variable to enter the basis, we need not obtain the true minimum (4.13), but can select any vector which would yield an improvement. Computational experience has shown that it appears to be best if a number of vectors which would individually yield an improvement, possibly some from each nonoptimal subproblem, are simultaneously considered as transfer vectors in the optimization of the extremal problem. That is, we do not restrict an iteration of the extremal problem to just a preselected basis change, but allow the old basis vectors and a number of new transfer vectors to be candidates in the formation of the new basis. We should note that the application of the decomposition algorithm is not always successful in that convergence to the optimum might require an inordinate number of iterations; see Orchard-Hays [532], Beale [58], Müller-Merbach [517], and Ho and Loute [372] for computational considerations of the decomposition algorithm. For an application to decentralized planning, see Chris-

---

[1] See Sec. 3 of Chap. 2 for a discussion of extreme directions and the representation theorem of convex polyhedra.

tensen and Obel, Sec. 13 of the Bibliography; Ho and Loute [372] describe a number of test problems.   The field of large-scale linear-programming systems will always be a challenging one and still requires much experimentation and new developments.   For general discussions and reviews of approaches to solving large-scale systems, see Gomory [324], Geoffrion [298], Dantzig, Dempster, and Kallio [181], Himmelblau [368], Ten Kate [638], Molina [506], and Lasdon [460].

It should be stressed that although the decomposition algorithm was developed for structured problems, any linear-programming problem can be decomposed.   For example, if a computer code is limited to the handling of a problem with 500 constraints, an appropriately designed decomposition code could subdivide a larger problem into two parts which could be readily solved.   Also problems which involve transportation or network problems as the subproblems, e.g., the multicommodity problem, can use the special algorithms to solve the subproblems and the revised simplex method to solve the extremal problem.   The fact that any linear-programming problem can be decomposed in a manner that best suits its special form is the basis of an application of the decomposition principle to the transportation problem (Williams [700]).

Extensions of the decomposition principle include (1) the generalized-programming problem of Wolfe (Dantzig [152]), where each column vector of a linear-programming problem is allowed to be selected from a convex set (see Gilmore and Gomory [304] and Chap. 11 for applications of this technique), and (2) the dual solution of the extremal problem and parametric decomposition procedures of Abadie and Williams [4].   As the solution to the extremal problem can yield a nonbasic optimal solution to the original problem, Abadie [2] modifies the computational procedure so that the current solution to the extremal problem is always a basic one for the original problem.

Example 4.1.   Minimize

$$-x_{11} + x_{12} - 3x_{21} - 2x_{22}$$

subject to

<table>
<tr><td>$x_{11} - 3x_{12}$
x_{12}</td><td>$-2x_{21} + x_{22}$
$+3x_{21} - x_{22}$</td><td>≤ 6
≤ 4</td></tr>
<tr><td>$3x_{11} + 2x_{12}$
$-x_{11} + 3x_{12}$</td><td></td><td>≤ 12
≤ 6</td></tr>
<tr><td></td><td>$3x_{21} + x_{22}$
$x_{21} + 2x_{22}$</td><td>≤ 12
≤ 8</td></tr>
</table>

$$(4.15)$$

and

$$x_{ij} \geq 0$$

Rewriting (4.15) in terms of equations with slack variables x_{01}, x_{02}, x_{13}, x_{14}, x_{23}, x_{24}, we have

$$
\begin{array}{|c|c|c|c|}
\hline
x_{01} & +x_{11} - 3x_{12} & -2x_{21} + x_{22} & = 6 \\
\quad x_{02} & + x_{12} & +3x_{21} - x_{22} & = 4 \\
\hline
 & 3x_{11} + 2x_{12} + x_{13} & & = 12 \\
 & -x_{11} + 3x_{12} \quad\quad + x_{14} & & = 6 \\
\hline
 & & 3x_{21} + x_{22} + x_{23} & = 12 \\
 & & x_{21} + 2x_{22} \quad\quad + x_{24} & = 8 \\
\hline
\end{array}
\tag{4.16}
$$

$$
\mathbf{A}_0 = \begin{pmatrix} 1 & 0 \\ 0 & 1 \end{pmatrix} \quad
\mathbf{A}_1 = \begin{pmatrix} 1 & -3 & 0 & 0 \\ 0 & 1 & 0 & 0 \end{pmatrix} \quad
\mathbf{A}_2 = \begin{pmatrix} -2 & 1 & 0 & 0 \\ 3 & -1 & 0 & 0 \end{pmatrix}
$$

$$
\mathbf{B}_1 = \begin{pmatrix} 3 & 2 & 1 & 0 \\ -1 & 3 & 0 & 1 \end{pmatrix} \quad
\mathbf{B}_2 = \begin{pmatrix} 3 & 1 & 1 & 0 \\ 1 & 2 & 0 & 1 \end{pmatrix}
$$

$$
\mathbf{b}_0 = \begin{pmatrix} 6 \\ 4 \end{pmatrix} \quad
\mathbf{b}_1 = \begin{pmatrix} 12 \\ 6 \end{pmatrix} \quad
\mathbf{b}_2 = \begin{pmatrix} 12 \\ 8 \end{pmatrix}
$$

$$
\mathbf{C}_0 = (0, 0) \quad \mathbf{C}_1 = (-1, 1, 0, 0) \quad \mathbf{C}_2 = (-3, -2, 0, 0)
$$

Treating the subproblems as inequalities with nonnegative slack vectors, the solution spaces $\mathbf{S}_1$ and $\mathbf{S}_2$ to the subproblems are given in Fig. 9.5a and b. We have indicated all the feasible extreme-point solutions.

For $\mathbf{B}_1\mathbf{X}_1 = \mathbf{b}_1$

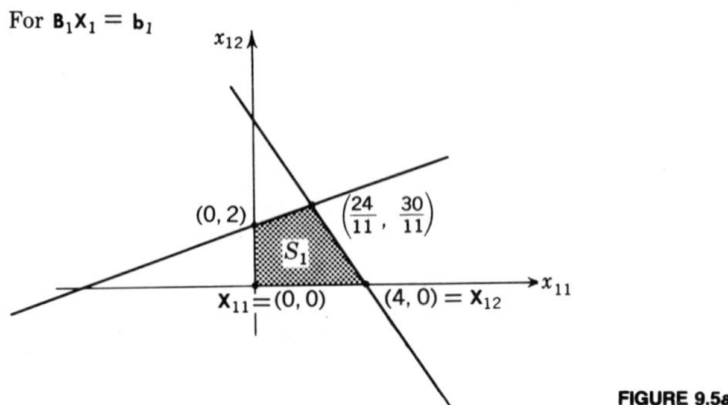

FIGURE 9.5a

For $\mathbf{B}_2\mathbf{X}_2 = \mathbf{b}_2$

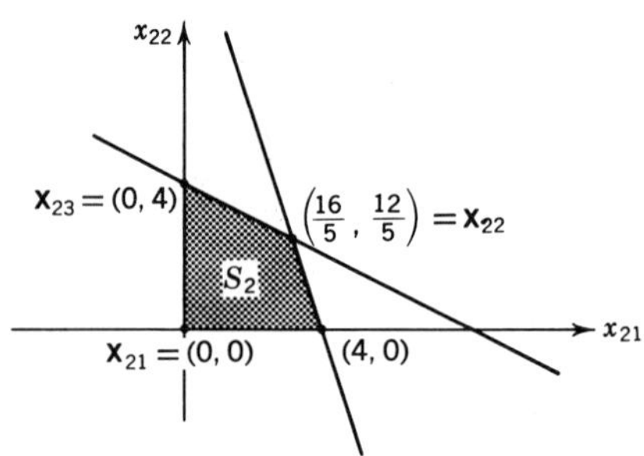

FIGURE 9.5b

To find a first feasible solution to the extremal problem,[1] we consider the extreme-point solution to the first subproblem $\mathbf{X}_{11} = (0, 0, 12, 6)$ and the extreme-point solution of the second subproblem $\mathbf{X}_{21} = (0, 0, 12, 8)$, where the last two components of $\mathbf{X}_{11}$ are the slack variables x_{13} and x_{14}, and the last two components of $\mathbf{X}_{21}$ are the slack variables x_{23} and x_{24}. Then $\mathbf{P}_{11} = \begin{pmatrix} 0 \\ 0 \end{pmatrix}$, $\mathbf{P}_{21} = \begin{pmatrix} 0 \\ 0 \end{pmatrix}$, $f_{11} = 0$, $f_{21} = 0$, and the extremal problem is given by

$$
\begin{pmatrix} 1 \\ 0 \\ 0 \\ 0 \end{pmatrix} x_{01} + \begin{pmatrix} 0 \\ 1 \\ 0 \\ 0 \end{pmatrix} x_{02} + \begin{pmatrix} 0 \\ 0 \\ 1 \\ 0 \end{pmatrix} \lambda_{11} + \begin{pmatrix} 0 \\ 0 \\ 0 \\ 1 \end{pmatrix} \lambda_{21} = \begin{pmatrix} 6 \\ 4 \\ 1 \\ 1 \end{pmatrix}
$$

with the first feasible solution of $x_{01} = 6$, $x_{02} = 4$, $\lambda_{11} = 1$, $\lambda_{21} = 1$, and as all c_{0j} and f_{pj} are zero, the value of the objective function is zero. Here the starting basis $\mathbf{B}$ of the extremal problem is

$$
\mathbf{B} = \begin{pmatrix} \mathbf{A}_{01} & \mathbf{A}_{02} & \mathbf{P}_{11} & \mathbf{P}_{21} \\ \mathbf{0} & \mathbf{0} & \mathbf{e}_1 & \mathbf{e}_2 \end{pmatrix} = \begin{pmatrix} 1 & 0 & 0 & 0 \\ 0 & 1 & 0 & 0 \\ 0 & 0 & 1 & 0 \\ 0 & 0 & 0 & 1 \end{pmatrix}
$$

Thus $\mathbf{B}^{-1} = \mathbf{I}$. As all the cost coefficients of the initial basis vectors are zero, the pricing vector $\mathbf{\Pi} = (0, 0, 0, 0)$. For this pricing vector $\mathbf{\Pi}$, we need to see if any other extreme points from the subproblems can improve the value of the objective function. Since both $\pi = \mathbf{0}$ and $\bar{\pi} = \mathbf{0}$, the optimization of the subproblems just involves the original data of the subproblems; i.e., the objective functions of the subproblems are not changed. We then have to find the extreme-point solutions of the following problems: Minimize

$$-x_{11} + x_{12}$$

subject to

$$
\begin{aligned}
3x_{11} + 2x_{12} + x_{13} \quad\quad &= 12 \\
-x_{11} + 3x_{12} \quad\quad + x_{14} &= 6 \\
x_{ij} &\geq 0
\end{aligned}
$$

$$(4.17)$$

and minimize

$$-3x_{21} - 2x_{22}$$

subject to

$$
\begin{aligned}
3x_{21} + x_{22} + x_{23} \quad\quad &= 12 \\
x_{21} + 2x_{22} \quad\quad + x_{24} &= 8 \\
x_{ij} &\geq 0
\end{aligned}
$$

$$(4.18)$$

[1] See Exercise 19 for a general procedure for finding a first feasible solution.

We would ordinarily use the simplex method to solve (4.17) and (4.18), but since we have the complete set of extreme-point solutions in Fig. 9.5a and b, respectively, we shall just evaluate each extreme point in the corresponding objective function and select the minimum. For the first subproblem we have $\mathbf{X}_{12} = (4, 0, 0, 10)$, $Z_{12} = -4$, and for the second subproblem we have $\mathbf{X}_{22} = (\frac{16}{5}, \frac{12}{5}, 0, 0)$, $Z_{22} = -\frac{72}{5}$. We note that all the vectors of $\mathbf{A}_0$ are in the current extremal basis and that min $[Z_{12}, Z_{22}] = -\frac{72}{5}$. Therefore, by (4.13) we select $\mathbf{X}_{22}$ to enter the basis of the extremal problem. Its transfer vector $\mathbf{P}_{22} = \begin{pmatrix} -4 \\ \frac{36}{5} \end{pmatrix}$ and its transfer cost $f_{22} = -\frac{72}{5}$. As we are using the revised simplex method, the vector to be introduced must be transformed by the current $\mathbf{B}^{-1}$. Here we have

$$\mathbf{B}^{-1}\begin{pmatrix} \mathbf{P}_{22} \\ \mathbf{e}_2 \end{pmatrix} = \mathbf{I} \begin{pmatrix} -4 \\ \frac{36}{5} \\ 0 \\ 1 \end{pmatrix} = \begin{pmatrix} -4 \\ \frac{36}{5} \\ 0 \\ 1 \end{pmatrix}$$

The extremal problem in tableau form is now

$\mathbf{b}$	x_{01}	x_{02}	λ_{11}	λ_{21}	λ_{22}
6	1	0	0	0	-4
4	0	1	0	0	$\left(\frac{36}{5}\right)$
1	0	0	1	0	0
1	0	0	0	1	1
0	0	0	0	0	$-\frac{72}{5}$

To improve the solution, we introduce λ_{22} into the basis, and to preserve feasibility, we eliminate x_{02}. This is shown in the next tableau.

	x_{01}		λ_{11}	λ_{21}	λ_{22}
$\frac{74}{9}$	1		0	0	0
$\frac{5}{9}$	0		0	0	1
1	0		1	0	0
$\frac{4}{9}$	0		0	1	0
8	0		0	0	0

We drop the column associated with x_{02}; if it is needed for a future basis, it will be brought back to the extremal problem at that time.[1] Our extremal solution is $x_{01} = \frac{74}{9}$, $\lambda_{11} = 1$, $\lambda_{21} = \frac{4}{9}$, $\lambda_{22} = \frac{5}{9}$, $Z = -8$. The corresponding

---

[1] For more involved problems, it appears as if it pays to keep the past extreme points as candidates to reenter the extremal basis as long as it is computationally convenient.

feasible solution to the original problem (4.16) is $\mathbf{X}_0 = (\frac{74}{9}, 0)$, $\mathbf{X}_1 = \lambda_{11}\mathbf{X}_{11} = (0, 0, 12, 6)$, and

$$\mathbf{X}_2 = \lambda_{21}\mathbf{X}_{21} + \lambda_{22}\mathbf{X}_{22} = \tfrac{4}{9}(0, 0, 12, 8) + \tfrac{5}{9}(\tfrac{16}{5}, \tfrac{12}{5}, 0, 0) = \tfrac{1}{9}(16, 12, 48, 32)$$

with $Z = -8$.   The new extremal problem basis is

$$\mathbf{B} = \begin{pmatrix} \mathbf{A}_{01} & \mathbf{P}_{22} & \mathbf{P}_{11} & \mathbf{P}_{21} \\ \mathbf{0} & \mathbf{e}_2 & \mathbf{e}_1 & \mathbf{e}_2 \end{pmatrix} = \begin{pmatrix} 1 & -4 & 0 & 0 \\ 0 & \tfrac{36}{5} & 0 & 0 \\ 0 & 0 & 1 & 0 \\ 0 & 1 & 0 & 1 \end{pmatrix}$$

with

$$\mathbf{B}^{-1} = \begin{pmatrix} 1 & \tfrac{5}{9} & 0 & 0 \\ 0 & \tfrac{5}{36} & 0 & 0 \\ 0 & 0 & 1 & 0 \\ 0 & -\tfrac{5}{36} & 0 & 1 \end{pmatrix}$$

The cost coefficients for these vectors are, respectively $(0, -\frac{72}{5}, 0, 0)$, yielding $\boldsymbol{\pi} = (0, -2, 0, 0)$; that is, $\pi = (0, -2)$, $\bar{\pi} = (0, 0)$.   The reader should recall that $\boldsymbol{\pi} = \mathbf{f}\mathbf{B}^{-1}$, where $\mathbf{f}$ is the vector of cost coefficients of the vectors in the current basis $\mathbf{B}$.   Using this pricing vector, we repeat the process to see if a new extreme-point solution from the subproblems can improve the solution.   From (4.12a) the subproblems now have the respective objective functions of

$$\text{minimize } -x_{11} + 3x_{12} \qquad \text{and} \qquad \text{minimize } 3x_{21} - 4x_{22}$$

For subproblem 1, $\mathbf{X}_{12}$ yields the optimum with $Z_{12} = -4$; for subproblem 2, $\mathbf{X}_{23}$ yields the optimum with $Z_{23} = -16$.   Since $\min_{j} (c_{0j} - \boldsymbol{\pi}\mathbf{A}_{0j}) = 0$, we then select $\mathbf{X}_{23}$ to be the proposal vector with the transfer vector of $\mathbf{P}_{23} = \begin{pmatrix} 4 \\ -4 \end{pmatrix}$ and transfer cost of $f_{23} = -8$.   The *complete* new tableau of the extremal problem is now

b		x_{01}	x_{02}	λ_{11}	λ_{21}	λ_{22}	λ_{23}
6		1	0	0	0	-4	4
4		0	1	0	0	$\left(\tfrac{36}{5}\right)$	-4
1		0	0	1	0	0	0
1		0	0	0	1	1	1
0		0	0	0	0	$-\tfrac{72}{5}$	-8

Introducing λ_{22} as before, we have

b		x_{01}	x_{02}	λ_{11}	λ_{21}	λ_{22}	λ_{23}
$\tfrac{74}{9}$		1		0	0	0	$\tfrac{16}{9}$
$\tfrac{5}{9}$		0		0	0	1	$-\tfrac{5}{9}$
1		0		1	0	0	0
$\tfrac{4}{9}$		0		0	1	0	$\left(\tfrac{14}{9}\right)$
8		0		0	0	0	-16

The reader will note that the extremal column associated with P_{23} has now been transformed, as would be the case in the revised simplex method; i.e., it has been multiplied by the inverse of the current basis of the extremal problem. In practice, we would not set up the first complete tableau, but just multiply the new extremal column, here $\begin{pmatrix} P_{23} \\ e_2 \end{pmatrix}$, by the inverse of the current extremal problem basis, and introduce the corresponding variable into the solution by the usual elimination transformation. We next introduce λ_{23} and remove λ_{21} to obtain

	x_{01}	λ_{11}	λ_{22}	λ_{23}
$\frac{54}{7}$	1	0	0	0
$\frac{5}{7}$	0	0	1	0
1	0	1	0	0
$\frac{2}{7}$	0	0	0	(1)
$\frac{88}{7}$	0	0	0	0

The current extremal problem basis is now

$$\mathbf{B} = \begin{pmatrix} \mathbf{A}_{01} & \mathbf{P}_{22} & \mathbf{P}_{11} & \mathbf{P}_{23} \\ 0 & \mathbf{e}_2 & \mathbf{e}_1 & \mathbf{e}_2 \end{pmatrix} = \begin{pmatrix} 1 & -4 & 0 & 4 \\ 0 & \frac{36}{5} & 0 & -4 \\ 0 & 0 & 1 & 0 \\ 0 & 1 & 0 & 1 \end{pmatrix}$$

with

$$\mathbf{B}^{-1} = \begin{pmatrix} 1 & \frac{5}{7} & 0 & -\frac{8}{7} \\ 0 & \frac{5}{56} & 0 & \frac{5}{14} \\ 0 & 0 & 1 & 0 \\ 0 & -\frac{5}{56} & 0 & \frac{9}{14} \end{pmatrix}$$

The cost coefficients are $(0, -\frac{72}{5}, 0, -8)$, which yields for this solution the pricing vector $\mathbf{\Pi} = (0, -\frac{4}{7}, 0, -\frac{72}{7})$; that is, $\boldsymbol{\pi} = (0, -\frac{4}{7})$, $\bar{\boldsymbol{\pi}} = (0, -\frac{72}{7})$, and the new objective functions for the subproblems are

$$\text{minimize } -x_{11} + \tfrac{11}{7}x_{12} \quad \text{and} \quad \text{minimize } -\tfrac{9}{7}x_{21} - \tfrac{18}{7}x_{22}$$

For subproblem 1 the minimum is assumed for $\mathbf{X}_{12} = (4, 0, 0, 10)$ with $Z_{12} = -4$ and $f_{12} = -4$; for subproblem 2 the minimum is assumed for both $\mathbf{X}_{22} = (\frac{16}{5}, \frac{12}{5}, 0, 0)$ and $\mathbf{X}_{23} = (0, 4, 0, 0)$ with $Z_{22} = Z_{23} = -\frac{72}{7}$. Also $c_{02} - \boldsymbol{\pi}\mathbf{A}_{02} = \frac{4}{7} > 0$. We then select $\mathbf{X}_{12}$ as the proposal vector since $Z_{12} - \bar{\pi}_1 = -4 - 0 = -4 < 0$ and $Z_{22} - \bar{\pi}_2 = Z_{23} - \bar{\pi}_2 = -\frac{72}{7} + \frac{72}{7} = 0$. (Note that the extremal column corresponding to $\mathbf{X}_{22}$ is a current basis vector of the extremal problem; hence $Z_{22} - \bar{\pi}_2 = 0$.) The transfer vector is $\mathbf{P}_{12} = \begin{pmatrix} 4 \\ 0 \end{pmatrix}$, with its transfer cost $f_{12} = -4$. In applying the revised simplex method to the extremal problem, we need to determine for the current basis the product of the basis inverse and the new vector to be introduced into the basis, i.e., here we calculate

$$\mathbf{B}^{-1}\begin{pmatrix}\mathbf{P}_{12}\\ \mathbf{e}_1\end{pmatrix} = \mathbf{B}^{-1}\begin{pmatrix}4\\0\\1\\0\end{pmatrix} = \begin{pmatrix}4\\0\\1\\0\end{pmatrix}$$

In this step, the vector remains the same.   The new tableau is then given by

	x_{01}	λ_{11}	λ_{22}	λ_{23}	λ_{12}
$\frac{54}{7}$	1	0	0	0	4
$\frac{5}{7}$	0	0	1	0	0
1	0	1	0	0	(1)
$\frac{2}{7}$	0	0	0	1	0
$\frac{88}{7}$	0	0	0	0	-4

We introduce λ_{12} and remove λ_{11} to obtain

	x_{01}	λ_{22}	λ_{23}	λ_{12}
$\frac{26}{7}$	1	0	0	0
$\frac{5}{7}$	0	1	0	0
1	0	0	0	1
$\frac{2}{7}$	0	0	1	0
$\frac{116}{7}$	0	0	0	0

The new $\mathbf{\Pi} = (0, -\frac{4}{7}, -4, -\frac{72}{7})$, and the reader can verify that for this pricing vector no improvement can be obtained.   The optimal solution to the original problem is

$$\mathbf{X}_0 = (\tfrac{26}{7}, 0) \qquad \mathbf{X}_1 = (4, 0, 0, 10)$$
$$\mathbf{X}_2 = \tfrac{5}{7}(\tfrac{16}{5}, \tfrac{12}{5}, 0, 0) + \tfrac{2}{7}(0, 4, 8, 0) = \tfrac{1}{7}(16, 20, 16, 0)$$

with $-\frac{116}{7}$ the minimum value of the objective function.

Special computational procedures have been developed for solving large-scale multistage systems with structures related to the one solved by the decomposition method.   The dual problem to equations (4.1) to (4.3) can be interpreted as consisting of independent linear programs tied together by a small set of variables, i.e., weakly coupled systems.   This type of problem arises in dynamic economic systems and storage and production problems (see

Chap. 10) (Gass [286] and Rosen [576]). More generally, some multistage systems require both tie-in equations and variables (Ritter [568] and Winkler [707]).

5. BOUNDED-VARIABLE PROBLEMS

In this section we shall discuss two important classes of linear-programming problems which impose certain restrictions on the values of the variables. Many applications require each variable not to exceed a specified limit (e.g., production levels or storage facilities), while other applications restrict the total value of certain linear sums of the variables (e.g., total shipments from many locations to a retail store). The former case is termed the *simple upper-bounded* (SUB) *problem*, while the latter, assuming that each variable appears in only one sum, is called the *generalized upper-bounded* (GUB) *problem*. We next discuss in turn the structure and special computational procedures for solving these problems.

　　　a. The simple upper-bounded problem. For the simple upper-bounded (SUB) problem, we have the following mathematical statement: Minimize

$$z = \mathbf{c}\mathbf{X} \tag{5.1}$$

subject to

$$\mathbf{A}\mathbf{X} = \mathbf{b} \tag{5.2}$$

$$\mathbf{X} \geq \mathbf{0} \tag{5.3}$$

$$x_j \leq u_j \qquad \text{for all or some } j \tag{5.4}$$

We have, of course, each $u_j > 0$. To generalize the discussion, we assume that all x_j are bounded above as we can always bound a variable by a suitably large number. The given upper-bound restrictions (or secondary constraints) can represent capacity limitations, production requirements, or resource limitations, and arise naturally in many applications.

　　　One way of solving this type of problem would be to replace the set of $x_j \leq u_j$ by an equivalent set of equalities in nonnegative variables (using slack variables). This, however, greatly increases the number of constraints to be handled. A more efficient technique has been developed (Charnes and Lemke [122] and Dantzig [160]) which entails only a few modifications of the standard simplex procedure and enables one to solve the problem without explicit representation of the upper-bound constraints. (The reader will note that lower-bound conditions offer no difficulty.[1])

---

[1] We stress that such lower-bound inequalities do not increase the size of the linear-programming model. In general, if our lower bounds are l_j, then the set of inequalities is of the form $x_j \geq l_j$. Introducing nonnegative slack variables, we have $x_j - y_j = l_j$ or $x_j = y_j + l_j$. In our original set of constraints we need only substitute $y_j + l_j$ for the corresponding x_j, and hence the number of constraints and variables remains the same as in the original formulation. The substitution takes into account the fact that the lower bounds must be met by all constraints; e.g., there are available resources to produce at least the lower-bound production.

The justification of the special procedure for solving the SUB problem can be seen from the following (Dantzig [152] and Simonnard [611]). In the general theory of the simplex method we deal with basic feasible solutions which are obtained by arbitrarily setting $n - m$ nonbasic variables equal to zero and solving the resulting $m \times m$ square system, assuming we have an admissible basis. For the bounded-variable problem (5.1) to (5.4), in order to employ a similar technique, we must allow the nonbasic variables to take on values other than zero. However, the situation is simplified, as we can restrict the value of a nonbasic variable to be either zero or the corresponding upper bound. For this problem, we then define an *extended basic feasible solution* to be one for which $n - m$ variables have been set equal to either their lower or upper bounds (zero or u_j), the matrix $\mathbf{B}$ of the associated square system is a basis, and the solution to the square system yields values of the basic variables which satisfy the upper- and lower-bound constraints (5.3) and (5.4).[†] We then have the following theorem.

Theorem 1.   *An extended basic feasible solution is optimal if the $z_j - c_j$ of the nonbasic variables satisfy the conditions*

$$z_j - c_j \leq 0 \qquad \text{if } x_j = 0 \tag{5.5}$$

and

$$z_j - c_j \geq 0 \qquad \text{if } x_j = u_j \tag{5.6}$$

Proof: Theorem 1 can be proved by noting the following. For $\mathbf{B}$, the basis of the extended basic feasible solution, let $\boldsymbol{\pi} = (\pi_1, \ldots, \pi_m)$ be the associated simplex multipliers (see discussion in Chap. 5). Then multiplying (5.2) by $\boldsymbol{\pi}$ and (5.1) by -1 and adding, we have

$$\boldsymbol{\pi}\mathbf{A}\mathbf{X} - \mathbf{c}\mathbf{X} = \boldsymbol{\pi}\mathbf{b} - z$$

or

$$z = \boldsymbol{\pi}\mathbf{b} - (\boldsymbol{\pi}\mathbf{A} - \mathbf{c})\mathbf{X}$$

For $\mathbf{B}$, we have $z = \bar{z} - \sum_{j \notin \mathbf{B}}(z_j - c_j)x_j$, $\bar{z} = \boldsymbol{\pi}\mathbf{b}$ is the value of the objective function for the basic variables.[1] Thus

$$\min z = \min \left[\bar{z} - \sum_{j \neq \mathbf{B}} (z_j - c_j)x_j \right] \tag{5.7}$$

If conditions (5.5) and (5.6) hold, then any changes in the nonbasic variables will increase the total value of the objective function; thus (5.7) is at its minimum.

[†] The extended basic feasible solution is also an extreme-point solution to the system (5.2) to (5.4). See Exercise 31.

[1] The notation $\sum_{j \notin \mathbf{B}}$ means that the sum is taken for the set of indices j of those vectors not in the basis $\mathbf{B}$.

If the conditions of Theorem 1 do not hold, we can change the current solution and determine a new, improved one. An analysis of the computational considerations enables us to apply the simplex algorithm to the primary constraints (5.2) and (5.3), but modified to ensure that the secondary constraints (5.4) are not violated. Thus, the size of the basis will be only of dimensions $m \times m$. Compared with solving the complete bounded-variable problem by the simplex method, the modified procedure represents quite a computational saving. By using the bounded-variable algorithm, we do not reduce the number of iterations, but decrease the time per iteration and are able to solve rather large bounded-variable problems. A discussion of the bounded-variable algorithm follows (Garvin [282] and Simonnard [611]).[1]

Assume that we have an extended basic feasible solution with basis **B** and that conditions (5.5) or (5.6) do not hold for some nonbasic variable x_k. If we want to increase x_k, then either some basic variable will reach its upper or lower bound or x_k will attain its upper bound. Similarly, if we want to decrease x_k, then either some basic variable will reach its upper or lower bound or x_k will attain its lower bound. If x_k reaches one of its bounds first, then we will not have to change the current basis **B**. If some basic variable, say x_l, attains one of its bounds first, then x_l will be replaced by x_k in the basis. We next have to determine the appropriate value of x_k for which all the constraints of the problem will remain satisfied. An improved solution will be obtained as we selected an x_k for which (5.5) or (5.6) is not satisfied.

Case 1. Determination of a new basis if $x_k = 0$.

For the given solution we have $x_k = 0$ and $z_k - c_k > 0$. In terms of the simplex transformation for the current basis (see Tableau 4.1, Chap. 4), each variable x_i in the basis is given by

$$x_i = x_{i0} - \sum_{j \in U} x_{ij} u_j - x_{ik} x_k \tag{5.8}$$

where **U** is the set of indices for those nonbasic variables which equal their upper bounds.[2] Letting $\bar{b}_i = x_{i0} - \sum_{j \in U} x_{ij} u_j$, (5.8) becomes

$$x_i = \bar{b}_i - x_{ik} x_k \qquad \text{for all } i \in \mathbf{B}$$

For those i for which $x_{ik} > 0$, we see that an increase in x_k from its current value of zero will decrease the corresponding x_i. Thus we must restrict any increase to x_k to a value which does not allow any of the current basic variables

---

[1] Other approaches to the bounded-variable problem employing dual relationships are given in Eisemann [236], Gass [288], and Wagner [680]. A SUB procedure is incorporated in most mathematical programming systems; see IBM [387], Sperry Univac [619], and Ketron [428].

[2] The notation $\sum\limits_{j \in U}$ means that the sum is taken for all j in the set of indices **U**; $i \in \mathbf{B}$ represents the set of indices of the basis vectors.

to become negative.   We must have for $x_{ik} > 0$,

$$\bar{b}_i - x_{ik}x_k \geq 0$$

or

$$x_k \leq \frac{\bar{b}_i}{x_{ik}}$$

or

$$x_k \leq \frac{\bar{b}_p}{x_{pk}} = \min_{x_{ik}>0} \frac{\bar{b}_i}{x_{ik}} \tag{5.9}$$

Similarly, for $x_{ik} < 0$, an increase in x_k will also increase the corresponding x_i.   but each x_i is bounded above.   Thus we must have for $x_{ik} < 0$,

$$\bar{b}_i - x_{ik}x_k \leq u_i$$

$$x_k \leq \frac{u_i - \bar{b}_i}{-x_{ik}}$$

$$x_k \leq \frac{u_q - \bar{b}_q}{-x_{qk}} = \min_{x_{ik}<0} \frac{u_i - \bar{b}_i}{-x_{ik}} \tag{5.10}$$

Finally, we also must have that

$$x_k \leq u_k \tag{5.11}$$

Taking (5.9) to (5.11) together, the maximum increase in x_k is given by

$$\max x_k = \min_{i \in \mathbf{B}} \left[\frac{\bar{b}_i}{x_{ik}} \text{ for } x_{ik} > 0, \frac{u_i - \bar{b}_i}{-x_{ik}} \text{ for } x_{ik} < 0, u_k \right]$$

$$\max x_k = \min \left[\frac{\bar{b}_p}{x_{pk}}, \frac{u_q - \bar{b}_q}{-x_{qk}}, u_k \right] \tag{5.11a}$$

If max x_k is limited by (5.9), then x_k replaces x_p in the basis; if it is limited by (5.10), then x_k replaces x_q in the basis; if it is limited by (5.11), then x_k stays nonbasic but its value is now u_k.   In the first two situations the basis is changed using x_{pk} or x_{qk} as the corresponding pivot.   We note that the pivot could be negative.

Case 2.   Determination of a new basis if $x_k = u_k$ and $z_k - c_k < 0$. For this case we have

$$x_i = x_{i0} - \sum_{\substack{j \in \mathbf{U} \\ j \neq k}} x_{ij}u_j - x_{ik}x_k \tag{5.12}$$

Letting

$$\bar{b}'_i = x_{i0} - \sum_{\substack{j \in \mathbf{U} \\ j \neq k}} x_{ij}u_j$$

(5.12) becomes $x_i = \bar{b}'_i - x_{ik}x_k$ for all $i \in \mathbf{B}$.   For $x_{ik} > 0$, a decrease in x_k will increase x_i, while for $x_{ik} < 0$, a decrease in x_k will decrease x_i. We must ensure that a decrease in x_k will not allow an x_i in the basis to go below zero or above its upper bound. Thus we must have

$$x_i = \bar{b}'_i - x_{ik}x_k \leq u_i \qquad \text{for } x_{ik} > 0$$

or

$$x_k \geq \frac{\bar{b}'_p - u_p}{x_{pk}} = \max_{x_{ik} > 0} \frac{\bar{b}'_i - u_i}{x_{ik}} \tag{5.13}$$

and

$$x_i = \bar{b}'_i - x_{ik}x_k \geq 0 \qquad \text{for } x_{ik} < 0$$

or

$$x_k \geq \frac{\bar{b}'_q}{x_{qk}} = \max_{x_{ik} < 0} \frac{\bar{b}'_i}{x_{ik}} \tag{5.14}$$

and finally,

$$x_k \geq 0 \tag{5.15}$$

Combining (5.13) to (5.15), we have that the minimum value of x_k is given by

$$\min x_k = \max \left[\frac{\bar{b}'_p - u_p}{x_{pk}}, \frac{\bar{b}'_q}{x_{qk}}, 0 \right] \tag{5.15a}$$

If $\min x_k$ is limited by (5.13), then x_k replaces x_p in the basis; if it is limited by (5.14), then x_k replaces x_q in the basis; if it is limited by (5.15), the old basis stays the same and x_k is a nonbasic variable equal to zero.

In either case, we obtain the largest decrease in the value of the objective function by making the variable x_k equal to the value which constrains it; that is, we increase x_k as much as possible in Case 1 and decrease it as much as possible in Case 2. After an appropriate change in the variable x_k, the above procedure is repeated until the conditions of Theorem 1 hold. Since all $x_j \leq u_j$, the given problem has a finite minimum.

The process of changing bases described above can be illustrated by examining the following problem and its graphical representation, Fig. 9.6. Consider the constraints

$$
\begin{aligned}
2x_1 + x_2 &\leq 6 \\
8x_1 + 9x_2 &\leq 36 \\
x_1 &\leq 2 \\
x_2 &\leq 3 \\
x_j &\geq 0
\end{aligned}
$$

Adding slack variables to the first inequalities, we then have

$$2x_1 + x_2 + x_3 \qquad\quad = 6$$
$$8x_1 + 9x_2 \qquad\ + x_4 = 36$$
$$x_1 \qquad\qquad\qquad \leq 2 \qquad\qquad\qquad (5.16)$$
$$x_2 \qquad\qquad\quad \leq 3$$
$$x_j \geq 0$$

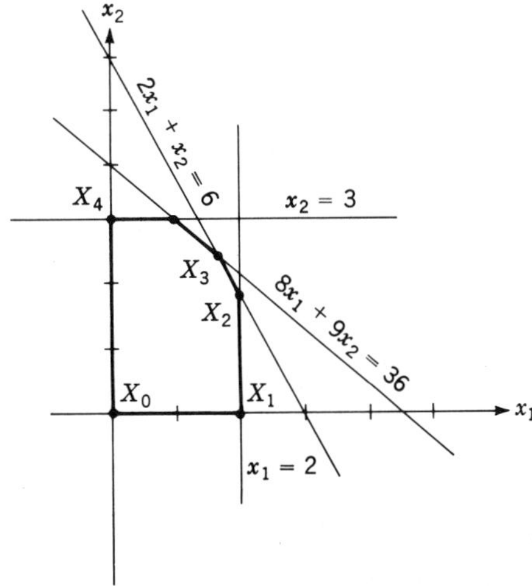

FIGURE 9.6

A first extended basic feasible solution would be $x_1 = x_2 = 0$, $x_3 = 6$, $x_4 = 36$, which corresponds to extreme point $\mathbf{X}_0$ in Fig. 9.6. The initial basis $\mathbf{B}$ consists of the unit vectors $\mathbf{P}_3$ and $\mathbf{P}_4$. Rewriting the constraints in the form (5.8), i.e., expressing the current basic variables in terms of the nonbasic variables, we have

$$x_3 = 6 - 2x_1 - x_2$$
$$x_4 = 36 - 8x_1 - 9x_2$$

As none of the nonbasic variables x_1 and x_2 is at its upper bound, the set $\mathbf{U}$ is empty. For some hypothesized objective function to be minimized, let us assume that the simplex criteria called for the introduction of x_1 into the solution (Case 1), i.e., $z_1 - c_1 > 0$. Then, as $\mathbf{U}$ is empty and $x_2 = 0$, the form of (5.8) to be considered is

$$x_3 = 6 - 2x_1$$
$$x_4 = 36 - 8x_1$$

Note that the $x_{i1} = (2, 8)$ and are positive. As we want to increase x_1, we apply (5.9) and (5.11) to determine the largest possible increase; i.e., by (5.11a) we have max $x_1 = \min\ [\frac{6}{2}, \frac{36}{8};\ -;\ 2] = 2$. If we did not have the upper bound of $x_1 \leq 2$, the normal simplex process would have introduced x_1 into the basis at a value of $\frac{6}{2} = 3$; but here $x_1 = 2$ and stays nonbasic. The basis variables are still the slack variables and they take on the new values of $x_3 = 2$ and $x_4 = 20$, with $x_2 = 0$. This corresponds to extreme point $\mathbf{X}_1$ in Fig. 9.6.

If now the simplex criteria called for the introduction of x_2 into the solution (Case 1), i.e., $z_2 - c_2 \geq 0$, we see, since $\mathbf{U} = \{1\}$, we can rewrite the expressions (5.8) for the basis variables as

$$x_3 = (6 - 2 \cdot 2) - x_2 = 2 - x_2$$
$$x_4 = (36 - 8 \cdot 2) - 9x_2 = 20 - 9x_2$$

The terms in parentheses yield the $\bar{b}_i = (2, 20)$. As the $x_{i2} = (1, 9)$, we have from (5.11a) that max $x_2 = \min \left[\frac{2}{1}, \frac{20}{9}; -; 3\right] = 2$. Here variable $x_2 = 2$ and becomes a basic variable replacing the slack variable x_3; this solution corresponds to extreme point $\mathbf{X}_2$ in Fig. 9.6. The new basis consists of vectors $\mathbf{P}_2$ and $\mathbf{P}_4$, and the new extended basic feasible solution is obtained by the application of the elimination transformation to Eqs. (5.16) to obtain

$$\begin{aligned} 2x_1 + x_2 + x_3 \qquad &= \quad 6 \\ -10x_1 \qquad - 9x_3 + x_4 &= -18 \end{aligned} \tag{5.17}$$

with $x_1 = 2$, $x_2 = 2$, $x_3 = 0$, $x_4 = 2$. The reader should note that although a negative right-hand-side value of -18 occurs in (5.17), we have $x_4 = 2$ as x_1 is at its upper bound and equals 2.

Continuing the process once more, let us assume that we can improve the value of the objective function by reducing x_1 as $z_1 - c_1 < 0$ (Case 2). Rewriting the constraints of the current transformed solution (5.17) in terms of the basis variables x_2 and x_4, we have by (5.12)

$$x_2 = \quad 6 - 2x_1$$
$$x_4 = -18 + 10x_1$$

Here $\mathbf{U} = \{1\}$ and $k = 1$; thus the $\bar{b}'_i = (6, -18)$. As the $x_{i1} = (2, -10)$, we apply (5.13), (5.14), and (5.15), i.e., (5.15a), to obtain

$$\min x_1 = \max \left[\frac{6-3}{2}; \frac{-18}{-10}; 0\right] = \frac{18}{10}$$

Variable $x_1 = \frac{18}{10}$ and replaces x_4 in the basis; $x_2 = \frac{24}{10}$ and is in the basis. Neither variable is at an upper bound. This solution corresponds to extreme point $\mathbf{X}_3$ in Fig. 9.6.

For a given objective function, the above process can be systematized by developing a computational tableau similar to the regular simplex tableau; see Exercise 21. With this in mind, the reader should reexamine the above example for the specific objective function of minimizing $-2x_1 - x_2$.

b. The generalized upper-bounded problem. The generalized upper-bounded (GUB) problem is a linear-programming problem with additional restrictions which limit positive, linear sums of the variables below specified bounds. Such problems are found in many areas including transportation, production-distribution, and resource allocation. As the number of this type of additional restriction can be quite large (in the tens of thousands), it is appropriate to investigate the resultant structure of the problem to determine if a simplifying computational procedure can be developed.

The added constraints can assume the following forms:

$$\sum_{\mathbf{J}} x_j = 1 \tag{5.18a}$$

or

$$\sum_{\mathbf{J}} x_j = u_{\mathbf{J}} \tag{5.18b}$$

where $\mathbf{J}$ refers to a set of indices from $j = 1, 2, \ldots, n$ and each j can appear at most once in some $\mathbf{J}$. The $x_j \geq 0$ and an equation's upper bound $u_{\mathbf{J}} > 0$. The constraints (5.18) could have been originally stated as less than or equal to inequalities, but here we assume the necessary slack variables have been added. For discussion purposes, as we can divide (5.18b) by $u_{\mathbf{J}}$ and redefine new variables so that their sum equals 1, we shall consider only the form (5.18a).

A simple numerical example is to minimize

$$-3x_3 + x_4 + 2x_5 + x_6$$

subject to

$$\begin{aligned}
x_1 \quad + \ x_3 - \ x_4 + \ x_5 - 2x_6 &= 1 \\
x_2 + 2x_3 + 3x_4 - 2x_5 + \ x_6 &= 4 \\
x_3 + \ x_4 \qquad\qquad\quad &= 1 \\
x_5 + \ x_6 &= 1 \\
x_j &\geq 0
\end{aligned}$$

This is a regular linear-programming problem, but we can consider the last two equations as special GUB constraints.

As the structure of a GUB problem is just a simplified form of the decomposition problem of Sec. 4 in which each subproblem consists of a single constraint of the form (5.18a), it is convenient to state a GUB problem using the notation of that section. Thus the above problem can be written as: Minimize

$$-3x_{11} + x_{12} + 2x_{21} + x_{22}$$

subject to

$$\begin{aligned}
x_{01} \quad + \ x_{11} - \ x_{12} + \ x_{21} - 2x_{22} &= 1 \\
x_{02} + 2x_{11} + 3x_{12} - 2x_{21} + \ x_{22} &= 4 \\
x_{11} + \ x_{12} \qquad\qquad\quad &= 1 \\
x_{21} + \ x_{22} &= 1 \\
x_{ij} &\geq 0
\end{aligned}$$

Each GUB constraint corresponds to a subproblem.

Using the decomposition notation, the general statement of the GUB problem is to find vectors $\mathbf{X}_p \geq 0$ for $p = 0, 1, \ldots, k$ which minimize

$$\mathbf{C}_0\mathbf{X}_0 + \mathbf{C}_1\mathbf{X}_1 + \cdots + \mathbf{C}_k\mathbf{X}_k$$

subject to

$$\mathbf{A}_0\mathbf{X}_0 + \mathbf{A}_1\mathbf{X}_1 + \cdots + \mathbf{A}_k\mathbf{X}_k = \mathbf{b}_0$$
$$\mathbf{B}_1\mathbf{X}_1 \qquad\qquad = b_1$$
$$\vdots$$
$$\mathbf{B}_k\mathbf{X}_k = b_k$$

where each $\mathbf{A}_p$ is an $m \times n_p$ matrix, $\mathbf{b}_0$ is an $m \times 1$ column vector, $\mathbf{B}_p = (1, 1, \ldots, 1)$ for $p = 1, 2, \ldots, k$ is a row vector with n_p components all equal to 1, and each $b_p = 1$. The total number of equations is $m + k$, of which k are GUB constraints, and the total number of variables is $n = \sum_{p=0}^{k} n_p$.

More explicitly, letting $\mathbf{A}_{pj}$ be the jth column of matrix $\mathbf{A}_p$ for $j = 1, \ldots, n_p$, $\mathbf{C}_p = (c_{p1}, \ldots, c_{pn_p})$, and $\mathbf{X}_p = (x_{p1}, \ldots, x_{pn_p})$, we can rewrite the above decomposition form of the GUB problem in column-vector terms as: Minimize

$$c_{01}x_{01} + \cdots + c_{0n_0}x_{0n_0} + c_{11}x_{11} + \cdots + c_{1n_1}x_{1n_1} + \cdots + c_{k1}x_{k1} + \cdots + c_{kn_k}x_{kn_k}$$

$$(5.19)$$

subject to

S_0	S_1	$\cdots$	S_k		
$\mathbf{A}_{01}x_{01} + \cdots + \mathbf{A}_{0n_0}x_{0n_0}$	$+ \mathbf{A}_{11}x_{11} + \cdots + \mathbf{A}_{1n_1}x_{1n_1}$	$+ \cdots$	$+ \mathbf{A}_{k1}x_{k1} + \cdots + \mathbf{A}_{kn_k}x_{kn_k}$	$= \mathbf{b}_0$	(5.20)
	$x_{11} + \cdots + x_{1n_1}$			$= 1$	
		$\vdots$		$\vdots$	(5.21)
			$x_{k1} + \cdots + x_{kn_k}$	$= 1$	

with $x_{pj} \geq 0$ for $p = 0, 1, \ldots, k$ and $j = 1, 2, \ldots, n_p$.

The problem (5.19), (5.20), and (5.21) can be solved using a standard simplex-method approach which would employ a basis of order $m + k$ or by decomposition, but as the set of k GUB constraints (5.21) is usually rather large in comparison with the m constraints of (5.20) and has a rather simple structure, we seek out a computational simplification that would enable us to employ a basis of less than $(m + k)$ vectors. To this end, the approach due to Dantzig and Van Slyke [192] has proved effective and has been successfully employed on computer systems to solve very large problems. Other approaches to the GUB problem and its generalizations are given in Gass [288],

Grigoriadis [336], Hartman and Lasdon [356], Hax [360], Kaul [420], Lasdon [460], and Sakarovitch and Saigal [590].   The Dantzig and Van Slyke method, which we discuss next, requires a modification of the revised simplex algorithm and employs a "working" basis of order m.   In that sense, it is similar to and a generalization of the simplex procedure applied to the bounded-variable problem discussed above.

The problem as stated in (5.19) to (5.21) has $k + 1$ subsets of columns (vectors) denoted by S_p for $p = 0, 1, \ldots, k$.   The columns of each S_p have $m + k$ components.   For S_0 the columns are denoted by

$$\begin{pmatrix} \mathbf{A}_{0j} \\ \mathbf{0}_k \end{pmatrix}$$

where $\mathbf{0}_k$ is a k-dimensional column vector whose components are all zero; and for each S_p, $p = 1, 2, \ldots, k$, the columns are denoted by

$$\begin{pmatrix} \mathbf{A}_{pj} \\ \mathbf{e}_p \end{pmatrix}$$

where $\mathbf{e}_p$ is a k-dimensional unit column vector with unity in the pth position.   For each $p = 0, 1, \ldots, k$, we denote these $m + k$-dimensional column vectors by $\{\bar{\mathbf{A}}_{pj}\}$, where $j = 1, \ldots, n_p$.   For the numerical example above we have $m = 2$, $k = 2$, with subsets of columns S_0, S_1, and S_2.   Typical columns are

$$\bar{\mathbf{A}}_{02} = \begin{pmatrix} \mathbf{A}_{02} \\ \mathbf{0}_2 \end{pmatrix} = \begin{pmatrix} 0 \\ 1 \\ 0 \\ 0 \end{pmatrix} \quad \text{and} \quad \bar{\mathbf{A}}_{21} = \begin{pmatrix} \mathbf{A}_{21} \\ \mathbf{e}_2 \end{pmatrix} = \begin{pmatrix} 1 \\ -2 \\ 0 \\ 1 \end{pmatrix}$$

For the system of Eqs. (5.20) and (5.21) in nonnegative variables we note that any feasible basis is of dimension $m + k$ (assuming full rank for the matrix of the system of equations).   Any such basis requires at least one vector from each subset S_p, $p = 1, 2, \ldots, k$; i.e., to satisfy Eqs. (5.21) which are nonnegative sums of nonnegative variables equal to one, we must have for each $p = 1, 2, \ldots, k$ at least one $x_{pj} > 0$.   Thus, k of any set of $m + k$ basic variables can be assigned to a corresponding subset S_p for $p = 1, 2, \ldots, k$.   The remaining m variables in a basis can, of course, fall into any of the subsets S_p for $p = 0, 1, \ldots, k$.   From this we can conclude that the number of subsets S_p which can have two or more basic variables is at most m, and it would be m if all S_p ($p \neq 0$) had two or fewer basic variables.   The significance of the above remarks will become clearer as we next describe the rationale of the GUB algorithm.

Assume a first basic feasible solution to Eqs. (5.20) and (5.21) is known.   Denote the corresponding basis of order $(m + k)$ by $\bar{\mathbf{B}}$.   Matrix $\bar{\mathbf{B}}$ consists of $(m + k)$ vectors from the sets of vectors $\{\bar{\mathbf{A}}_{pj}\}$, $p = 0, 1, \ldots, k$ and $j = 1, 2, \ldots, n_p$.

For each set of vectors S_p, $p = 1, 2, \ldots, k$, we denote one of the basic variables as the set's *key variable* for the current basis (from the discussion above we know that there is at least one positive variable in each set). Except for set S_0, all sets will then have a corresponding key variable. For discussion purposes, we assume here that each key variable is the first one of each set, i.e., the key variables $x_{p1} > 0$ and key columns $\bar{\mathbf{A}}_{p1}$ are in $\bar{\mathbf{B}}$ for $p = 1, 2, \ldots, k$.

From (5.21) we see that for $p = 1, 2, \ldots, k$

$$x_{p1} = 1 - x_{p2} - x_{p3} - \cdots - x_{pn_p} = 1 - \sum_{j=2}^{n_p} x_{pj}$$

Substituting these expressions into (5.20) is equivalent to subtracting the $\mathbf{A}_{p1}$ from each column of the corresponding subset S_p and the right-hand-side vector of Eqs. (5.20). We illustrate this for a problem with the following set of $m + 2$ equations with nonnegative variables:

$$\mathbf{A}_{01}x_{01} + \mathbf{A}_{02}x_{02} + \mathbf{A}_{11}x_{11} + \mathbf{A}_{12}x_{12} + \mathbf{A}_{13}x_{13} + \mathbf{A}_{21}x_{21} + \mathbf{A}_{22}x_{22} + \mathbf{A}_{23}x_{23} = \mathbf{b}_0 \qquad (5.20a)$$

$$x_{11} + x_{12} + x_{13} \qquad\qquad = 1$$
$$x_{21} + x_{22} + x_{23} = 1 \qquad (5.21a)$$

Let x_{11} and x_{21} be the key variables in the first and second sets of columns, respectively. Then from (5.21a), we have

$$x_{11} = 1 - x_{12} - x_{13} \qquad \text{and} \qquad x_{21} = 1 - x_{22} - x_{23}$$

Substituting for x_{11} and x_{21} in (5.20a) yields

$$\mathbf{A}_{01}x_{01} + \mathbf{A}_{02}x_{02} + \mathbf{A}_{11}(1 - x_{12} - x_{13}) + \mathbf{A}_{12}x_{12} + \mathbf{A}_{13}x_{13}$$
$$+ \mathbf{A}_{21}(1 - x_{22} - x_{23}) + \mathbf{A}_{22}x_{22} + \mathbf{A}_{23}x_{23} = \mathbf{b}_0$$

Rearranging the terms in the above expression, we then have

$$\mathbf{A}_{01}x_{01} + \mathbf{A}_{02}x_{02} + (\mathbf{A}_{12} - \mathbf{A}_{11})x_{12} + (\mathbf{A}_{13} - \mathbf{A}_{11})x_{13}$$
$$+ (\mathbf{A}_{22} - \mathbf{A}_{21})x_{22} + (\mathbf{A}_{23} - \mathbf{A}_{21})x_{23} = \mathbf{b}_0 - \mathbf{A}_{11} - \mathbf{A}_{21}$$

We see from this last expression that we have eliminated the key variables from the m major constraints (5.20a).

For the general problem (5.20) and (5.21), letting each key variable correspond to the first column $\bar{\mathbf{A}}_{p1}$ for $p = 1, 2, \ldots, k$, we substitute as above for each of the k key variables x_{p1} in the system (5.20) and (5.21) to obtain the m equations

$$\sum_{j=1}^{n_0} \mathbf{A}_{0j}x_{0j} + \sum_{j=1}^{n_1} (\mathbf{A}_{1j} - \mathbf{A}_{11})x_{1j} + \cdots + \sum_{j=1}^{n_k} (\mathbf{A}_{kj} - \mathbf{A}_{k1})x_{kj} = \mathbf{b}_0 - \sum_{p=1}^{k} \mathbf{A}_{p1}$$

We now define an $m \times m$ "working" basis to be the set of m column vectors which form the matrix

$$\mathbf{B} = (\{\mathbf{A}_{0j}\}\{\mathbf{A}_{1j} - \mathbf{A}_{11}\} \cdots \{\mathbf{A}_{kj} - \mathbf{A}_{k1}\})$$

for those j in the given basic feasible solution which are *not key*; i.e., $\{\mathbf{A}_{0j}\}$ is a set of vectors corresponding to nonkey basic variables from S_0 and $\{\mathbf{A}_{pj} - \mathbf{A}_{p1}\}$ is a set of vectors corresponding to nonkey basic variables from S_p $(p \neq 0)$, where the total number of such vectors is m. We next show that the set of vectors which form $\mathbf{B}$ is linearly independent; i.e., the vectors form a basis.

We consider the original $m + k$ basis $\bar{\mathbf{B}}$ and assume that its columns have been grouped in terms of key columns and nonkey columns into a partitioned matrix as follows:

$$
\begin{array}{cc}
k \text{ key} & m \text{ nonkey} \\
\text{columns} & \text{columns}
\end{array}
$$

$$
\bar{\mathbf{B}} = \left(\left\{ \begin{matrix} \mathbf{A}_{p1} \\ \mathbf{e}_p \end{matrix} \right\} \ \middle| \ \left\{ \begin{matrix} \mathbf{A}_{0j} \\ \mathbf{0}_k \end{matrix} \right\} \ \middle| \ \left\{ \begin{matrix} \mathbf{A}_{pj} \\ \mathbf{e}_p \end{matrix} \right\} \right)
$$

for j in $\bar{\mathbf{B}}$ and nonkey and $p = 1, 2, \ldots, k$. For each $1 \leq p \leq k$, we subtract the corresponding key column $\begin{pmatrix} \mathbf{A}_{p1} \\ \mathbf{e}_p \end{pmatrix}$ from the corresponding columns $\begin{pmatrix} \mathbf{A}_{pj} \\ \mathbf{e}_p \end{pmatrix}$ to obtain the transformed matrix

$$
\begin{array}{cc}
k \text{ key} & \\
\text{columns} & m \text{ nonkey columns}
\end{array}
$$

$$
\bar{\mathscr{B}} = \left(\left\{ \begin{matrix} \mathbf{A}_{p1} \\ \mathbf{e}_p \end{matrix} \right\} \ \middle| \ \left\{ \begin{matrix} \mathbf{A}_{0j} \\ \mathbf{0}_k \end{matrix} \right\} \ \middle| \ \left\{ \begin{matrix} \mathbf{A}_{pj} - \mathbf{A}_{p1} \\ \mathbf{e}_p - \mathbf{e}_p \end{matrix} \right\} \right)
$$

We note that the first m rows of the nonkey columns of $\bar{\mathscr{B}}$ form the working basis $\mathbf{B}$, while the last k rows of these columns form a $k \times m$ matrix $\mathbf{0}$ with all components equal to zero, i.e.,

$$
\bar{\mathscr{B}} = \left(\left\{ \begin{matrix} \mathbf{A}_{p1} \\ \mathbf{e}_p \end{matrix} \right\} \ \middle| \ \begin{matrix} \mathbf{B} \\ \mathbf{0} \end{matrix} \right)
$$

for $p = 1, 2, \ldots, k$. Since we applied elementary column transformations to $\bar{\mathbf{B}}$ to obtain $\bar{\mathscr{B}}$, then $\bar{\mathscr{B}}$ retains the same $m + k$ rank as $\bar{\mathbf{B}}$. The column vectors of $\bar{\mathscr{B}}$ form a linearly independent set of $m + k$ vectors and hence the subset of these vectors given by the last m columns, i.e., by the submatrix $\begin{pmatrix} \mathbf{B} \\ \mathbf{0} \end{pmatrix}$, is also linearly independent. By applying the definition of linear independence to this $(m + k) \times m$ set of vectors, we see that since the last k elements of the vectors of this set are equal to zero, then the m-dimensional vectors formed by the first m components of these vectors must be a linearly independent set; i.e., $\mathbf{B}$ forms a basis in m space.

The essence of the GUB procedure is that by using the working basis $\mathbf{B}$ of order $(m \times m)$ the information necessary to carry out the revised simplex algorithm to solve the given $(m + k) \times n$ problem can be determined more effectively in terms of computational requirements. In the discussion which follows, it will aid the reader to bear in mind the comparable process as applied to the SUB problem and to apply the GUB procedure to that problem; i.e., by adding a nonnegative slack variable s_j to each upper-bound inequality $x_j + s_j = u_j$ and setting each $u_j = 1$, the SUB problem is then of the more general form under investigation here. The $\mathbf{A}_{pj}$ vectors corresponding to the s_j are, of course, all zero vectors (see Exercise 26). We next show how the computations of the revised simplex procedure can be carried out by using the working basis $\mathbf{B}$ to determine if the present solution is optimal and, if not, how to improve the solution and change the basis.

In applying the revised simplex method of Chap. 5 to the GUB problem (5.19) to (5.21), we require the inverse of the current basis in order to determine the following elements: the pricing vector $\mathbf{\Pi}$, the vector $\bar{\mathbf{A}}_{qs}$ to be introduced into the basis, and the vector $\bar{\mathbf{A}}_{rt}$ to be removed from the basis. We discuss, in turn, how this information is developed assuming that we start with a given basic feasible solution, a corresponding basis $\bar{\mathbf{B}}$, and working basis inverse $\mathbf{B}^{-1}$.

To generalize the discussion, let the set of indices $\{\mathbf{I}\}$ denote the indices of the key columns of the basis $\bar{\mathbf{B}}$. There are k members in the set of indices, each corresponding to a basis vector from a set S_p, $p = 1, 2, \ldots, k$, with exactly one vector from each S_p. Where in our discussion above we let the key columns be the first in each set, here the selection of the k key columns is arbitrary. We let $\{\mathbf{J}\}$ denote the set of indices of the nonkey columns of $\bar{\mathbf{B}}$, with $\{\mathbf{J}_p\}$ the subset of indices of the nonkey columns in set S_p, $p = 0, 1, \ldots, k$. There are m members in the set $\{\mathbf{J}\}$. We shall assume that the basis vectors in $\bar{\mathbf{B}}$ are ordered in the manner in which they appear in Eqs. (5.20) and (5.21) and the indices in $\{\mathbf{I}\}$ and $\{\mathbf{J}\}$ are similarly ordered.

We redefine

$$
k \text{ key columns} \quad m \text{ nonkey columns}
$$

$$
\bar{\mathbf{B}} = \left(\quad \begin{Bmatrix} \mathbf{A}_{pi} \\ \mathbf{e}_p \end{Bmatrix} \quad \Bigg| \quad \begin{Bmatrix} \mathbf{A}_{0j} \\ \mathbf{0}_k \end{Bmatrix} \quad \Bigg| \quad \begin{Bmatrix} \mathbf{A}_{pj} \\ \mathbf{e}_p \end{Bmatrix} \quad \right) \tag{5.22}
$$

for i in $\{\mathbf{I}\}$, j in $\{\mathbf{J}\}$, and $p = 1, 2, \ldots, k$. Similarly, we define

$$
\bar{\mathscr{B}} = \left(\begin{Bmatrix} \mathbf{A}_{pi} \\ \mathbf{e}_p \end{Bmatrix} \quad \Bigg| \quad \begin{Bmatrix} \mathbf{A}_{0j} \\ \mathbf{0}_k \end{Bmatrix} \quad \Bigg| \quad \begin{Bmatrix} \mathbf{A}_{pj} - \mathbf{A}_{pi} \\ \mathbf{0} \end{Bmatrix} \right) \tag{5.23}
$$

and

$$
\mathbf{B} = \left(\{\mathbf{A}_{0j}\} \,\big|\, \{\mathbf{A}_{pj} - \mathbf{A}_{pi}\} \right)
$$

We note that the ordering assumption above causes the k key columns in $\bar{\mathbf{B}}$ and $\bar{\mathscr{B}}$ to be arranged so that the associated $\mathbf{e}_p$ form a $k \times k$ identity matrix. Define the $m + k$-dimensional pricing row vector $\mathbf{\Pi} = (\boldsymbol{\pi}_1, \boldsymbol{\pi}_2)$, where $\boldsymbol{\pi}_1 = (\pi_{11}, \ldots, \pi_{1m})$ and $\boldsymbol{\pi}_2 = (\pi_{21}, \ldots, \pi_{2k})$ are the prices corresponding to the m equations of (5.20) and k equations of (5.21), respectively. The vector $\mathbf{\Pi}$ is by definition [see Eq. (1.4) of Chap. 5]

$$
\mathbf{\Pi} = \bar{\mathbf{C}}\bar{\mathbf{B}}^{-1}
$$

where $\bar{\mathbf{C}} = (\{c_{pi}\}|\{c_{0j}\}|\{c_{pj}\})$ for i in $\{\mathbf{I}\}$ and j in $\{\mathbf{J}\}$, $p = 1, 2, \ldots, k$, is a $k + m$ row vector whose components are the ordered costs of the vectors in the given basis $\bar{\mathbf{B}}$. To determine $\mathbf{\Pi}$, we need to solve the $(m + k) \times (m + k)$ set of equations

$$\mathbf{\Pi}\bar{\mathbf{B}} = \bar{\mathbf{C}} \tag{5.24}$$

We shall show that this can be accomplished by knowledge of the inverse of the corresponding $m \times m$ working basis $\mathbf{B}$.

We wish to apply the key-nonkey column-subtraction transformation described above to both sides of (5.24) so as to obtain an equivalent form in terms of $\bar{\mathscr{B}}$. To formalize this transformation, we rewrite $\bar{\mathbf{B}}$ as given in (5.22) in the partitioned form

$$\bar{\mathbf{B}} = \left(\left\{ \begin{matrix} \mathbf{A}_{pi} \\ \mathbf{e}_p \end{matrix} \right\} \quad \middle| \quad \frac{\{\{\mathbf{A}_{0j}\}|\{\mathbf{A}_{pj}\}\}}{\mathbf{E}} \right)$$

where $\mathbf{E}$ is a $k \times m$ submatrix whose first $\mathbf{J}_0$ columns are all zero and whose last $\{\mathbf{J} - \mathbf{J}_0\}$ columns have unity in the pth position for *each* j in $\mathbf{J}_p$, $p = 1, 2, \ldots, k$. We define the nonsingular $(m + k) \times (m + k)$ partitioned transformation matrix

$$\mathbf{T} = \begin{pmatrix} \mathbf{I}_k & -\mathbf{E} \\ \mathbf{0} & \mathbf{I}_m \end{pmatrix}$$

where $\mathbf{0}$ is an $m \times k$ zero submatrix. The reader can verify that $\bar{\mathbf{B}}\mathbf{T} = \bar{\mathscr{B}}$. Applying $\mathbf{T}$ to (5.24), we have

$$\mathbf{\Pi}\bar{\mathbf{B}}\mathbf{T} = \bar{\mathbf{C}}\mathbf{T}$$

or

$$\mathbf{\Pi}\bar{\mathscr{B}} = \mathbf{\Pi}\left(\left\{ \begin{matrix} \mathbf{A}_{pi} \\ \mathbf{e}_p \end{matrix} \right\} \quad \middle| \quad \left\{ \begin{matrix} \mathbf{A}_{0j} \\ \mathbf{0} \end{matrix} \right\} \quad \middle| \quad \left\{ \begin{matrix} \mathbf{A}_{pj} - \mathbf{A}_{pi} \\ \mathbf{0} \end{matrix} \right\} \right) = \mathscr{C} \tag{5.25}$$

where $\mathscr{C} = (\{c_{pi}\}|\{c_{0j}\}|\{c_{pj} - c_{pi}\})$ for i in $\mathbf{I}$ and j in $\mathbf{J}$ and $p = 1, 2, \ldots, k$ is a $k + m$-dimensional row vector.[1] The first k elements of the row vector $\mathscr{C}$ are the corresponding c_{pi} for the key columns i in $\{\mathbf{I}\}$, and the last m elements are of the form c_{0j} for j in $\{\mathbf{J}_0\}$ or $c_{pj} - c_{pi}$ for j in $\{\mathbf{J}_p\}$ and $p = 1, 2, \ldots, k$. Multiplying the columns of $\bar{\mathscr{B}}$ by $\mathbf{\Pi} = (\boldsymbol{\pi}_1, \boldsymbol{\pi}_2)$, we obtain three sets of equations which relate the components of $\mathbf{\Pi}$:

$$\mathbf{\Pi} \left\{ \begin{matrix} \mathbf{A}_{pi} \\ \mathbf{e}_p \end{matrix} \right\} = \boldsymbol{\pi}_1 \mathbf{A}_{pi} + \boldsymbol{\pi}_{2p} \tag{5.26a}$$

$$\mathbf{\Pi} \left\{ \begin{matrix} \mathbf{A}_{0j} \\ \mathbf{0} \end{matrix} \right\} = \boldsymbol{\pi}_1 \mathbf{A}_{0j} \tag{5.26b}$$

$$\mathbf{\Pi} \left\{ \begin{matrix} \mathbf{A}_{pj} - \mathbf{A}_{pi} \\ \mathbf{0} \end{matrix} \right\} = \boldsymbol{\pi}_1 (\mathbf{A}_{pj} - \mathbf{A}_{pi}) \tag{5.26c}$$

---

[1] The reader can verify that this transformation is equivalent to subtracting appropriate equations from (5.24) to obtain (5.25).

for i in $\{\mathbf{I}\}$, j in $\{\mathbf{J}\}$, and $p = 1, 2, \ldots, k$. Equations (5.26b) and (5.26c) are the components of the product $\boldsymbol{\pi}_1\mathbf{B}$. From Eq. (5.25) we equate corresponding left- and right-hand-side members to solve for $\boldsymbol{\pi}_1$, i.e.,

$$\boldsymbol{\pi}_1\mathbf{B} = \boldsymbol{\pi}_1(\{\mathbf{A}_{0j}\}|\{\mathbf{A}_{pj} - \mathbf{A}_{pi}\}) = (\{c_{0j}\}|\{c_{pj} - c_{pi}\})$$

or, for $\mathbf{C} = (\{c_{0j}\}|\{c_{pj} - c_{pi}\})$ an m-dimensional row vector,

$$\boldsymbol{\pi}_1 = \mathbf{C}\mathbf{B}^{-1} \tag{5.27}$$

Knowing $\mathbf{B}^{-1}$, we can compute $\boldsymbol{\pi}_1$ and from (5.25) and (5.26a) we can determine $\boldsymbol{\pi}_2$ by the k equations

$$\boldsymbol{\pi}_1\mathbf{A}_{pi} + \pi_{2p} = c_{pi} \tag{5.28}$$

for i in $\{\mathbf{I}\}$ and $p = 1, 2, \ldots, k$. Thus, we see from (5.27) and (5.28) that for the current $(m + k) \times (m + k)$ basis $\bar{\mathbf{B}}$, the vector $\boldsymbol{\Pi}$, i.e., the set of simplex multipliers (prices) as defined by (5.24), can be found by having knowledge of the inverse of the corresponding $m \times m$ working basis $\mathbf{B}$.

The vector $\boldsymbol{\Pi}$ enables us to determine for those vectors not in the current basis $\bar{\mathbf{B}}$ the corresponding $z_{pj} - c_{pj}$; i.e., we can price out each nonbasic vector $\bar{\mathbf{A}}_{pj}$ by calculating $\boldsymbol{\Pi}\bar{\mathbf{A}}_{pj} - c_{pj}$. As our given problem is minimization, if all $\boldsymbol{\Pi}\bar{\mathbf{A}}_{pj} - c_{pj} \leq 0$, then the current solution is optimal. If some $\boldsymbol{\Pi}\bar{\mathbf{A}}_{pj} - c_{pj} > 0$, then the current solution is not optimal and we determine a new vector to enter the basis as given by the simplex method; i.e., vector $\bar{\mathbf{A}}_{qs}$ from set S_q enters the basis where

$$\boldsymbol{\Pi}\bar{\mathbf{A}}_{qs} - c_{qs} = \max_{\bar{\mathbf{A}}_{pj} \notin \bar{\mathbf{B}}} (\boldsymbol{\Pi}\bar{\mathbf{A}}_{pj} - c_{pj}) > 0 \tag{5.29}$$

for $p = 0, 1, \ldots, k$. When $p = 0$, $\boldsymbol{\Pi}\bar{\mathbf{A}}_{pj} = \boldsymbol{\pi}_1\mathbf{A}_{0j}$, and when $1 \leq p \leq k$, $\boldsymbol{\Pi}\bar{\mathbf{A}}_{pj} = \boldsymbol{\pi}_1\mathbf{A}_{pj} + \pi_{2p}$.

The vector to be removed from the current basis $\bar{\mathbf{B}}$ is selected based on the simplex-method prescription for preserving feasibility. Hence, we need to determine the corresponding θ_0 ratio which requires the expression of the new vector $\bar{\mathbf{A}}_{qs}$ in terms of the vectors of $\bar{\mathbf{B}}$, along with the nonnegative solution vector $\mathbf{X}_{00}$ associated with $\bar{\mathbf{B}}$. We next show how this information can be obtained with knowledge of $\mathbf{B}^{-1}$.

We require the solutions of the two sets of equations

$$\bar{\mathbf{B}}\mathbf{X}_{qs} = \bar{\mathbf{A}}_{qs} \tag{5.30}$$

and

$$\bar{\mathbf{B}}\mathbf{X}_{00} = \mathbf{b} \tag{5.31}$$

where $\mathbf{X}_{qs} = (x_{qs}^{(1)}, \ldots, x_{qs}^{(k)}, x_{qs}^{(k+1)}, \ldots, x_{qs}^{(k+m)})$ is a $k + m$-dimensional column vector which expresses the new vector in terms of the basis vectors, $\mathbf{X}_{00} = (x_{00}^{(1)}, \ldots, x_{00}^{(k)}, x_{00}^{(k+1)}, \ldots, x_{00}^{(k+m)})$ is the current $k + m$-dimensional nonnegative-solution column vector, and vector $\mathbf{b}$ is the $k + m$ right-hand-side column vector of the GUB equations (5.20) and (5.21).

For (5.30) we wish to transform $\bar{\mathbf{B}}$ into the form $\bar{\mathscr{B}}$ to obtain a set of equations which can be solved readily in terms of $\mathbf{B}^{-1}$. Where this was done to obtain (5.25) by a column transformation matrix $\mathbf{T}$ applied to both sides of the equation (5.24), here we apply the same $\mathbf{T}$ as a means of transforming the $\mathbf{X}_{qs}$ to another set of variables $\mathbf{Y}_{qs}$ which can be determined from equations involving $\bar{\mathscr{B}}$ and hence $\mathbf{B}$. To accomplish this we define

$$\mathbf{Y}_{qs} = \mathbf{T}^{-1}\mathbf{X}_{qs}$$

or

$$\mathbf{X}_{qs} = \mathbf{T}\mathbf{Y}_{qs} \tag{5.32}$$

Substituting for $\mathbf{X}_{qs}$ in (5.30), we have

$$\bar{\mathbf{B}}(\mathbf{T}\mathbf{Y}_{qs}) = \bar{\mathbf{A}}_{qs}$$

or

$$\bar{\mathscr{B}}\,\mathbf{Y}_{qs} = \bar{\mathbf{A}}_{qs} \tag{5.33}$$

For discussion purposes let us assume that the vector to be introduced into the basis $\bar{\mathbf{A}}_{qs}$ is from S_q, where $1 \le q \le p$. (We will note the changes which follow if $\bar{\mathbf{A}}_{qs}$ is from the set S_0.) Letting

$$\mathbf{Y}_{qs} = \begin{pmatrix} y_{qs}^{(1)} \\ \vdots \\ y_{qs}^{(k)} \\ y_{qs}^{(k+1)} \\ \vdots \\ y_{qs}^{(k+m)} \end{pmatrix}$$

and noting that

$$\bar{\mathscr{B}} = \left(\overbrace{\begin{Bmatrix} \mathbf{A}_{pi} \\ \mathbf{e}_p \end{Bmatrix}}^{k \text{ key columns}} \quad \middle| \quad \begin{matrix} \mathbf{B} \\ \mathbf{0} \end{matrix} \right)$$

and that the last k components of $\bar{\mathbf{A}}_{qs}$ are formed by the unit vector $\mathbf{e}_q$ which has unity in the qth position, the equation (5.33) can be written as

$$\left(\begin{Bmatrix} \mathbf{A}_{pi} \\ \mathbf{e}_p \end{Bmatrix} \;\middle|\; \begin{matrix} \mathbf{B} \\ \mathbf{0} \end{matrix} \right) \begin{pmatrix} y_{qs}^{(1)} \\ \vdots \\ y_{qs}^{(k)} \\ y_{qs}^{(k+1)} \\ \vdots \\ y_{qs}^{(k+m)} \end{pmatrix} = \begin{pmatrix} \mathbf{A}_{qs} \\ 0 \\ \vdots \\ 1 \\ \vdots \\ 0 \end{pmatrix} \leftarrow \text{row } m+q \tag{5.34}$$

Recalling that the last k rows of the key columns of $\bar{\mathscr{B}}$ form a unit matrix, we carry out the multiplications of (5.34) and set corresponding left- and right-hand sides equal to obtain

$$y_{qs}^{(i)} = 0 \qquad \text{for } i = 1, 2, \ldots, k \text{ and } i \neq q \tag{5.35}$$

$$y_{qs}^{(q)} = 1$$

and by substituting (5.35) into (5.34), we have

$$\mathbf{A}_{qi} + \mathbf{B} \begin{pmatrix} y_{qs}^{(k+1)} \\ \vdots \\ y_{qs}^{(k+m)} \end{pmatrix} = \mathbf{A}_{qs} \tag{5.36}$$

We note that $\mathbf{A}_{qi}$ corresponds to the key column from set S_q and $\mathbf{A}_{qs}$ corresponds to the incoming column, which is also from S_q. [If $q = 0$, then $y_{qs}^{(q)} = 0$ and no $\mathbf{A}_{qi}$ appears in (5.36).]   Letting

$$\bar{\mathbf{Y}}_{qs} = \begin{pmatrix} y_{qs}^{(k+1)} \\ \vdots \\ y_{qs}^{(k+m)} \end{pmatrix}$$

Eq. (5.36) can be written as

$$\mathbf{B}\bar{\mathbf{Y}}_{qs} = (\mathbf{A}_{qs} - \mathbf{A}_{qi})$$

or

$$\bar{\mathbf{Y}}_{qs} = \mathbf{B}^{-1}(\mathbf{A}_{qs} - \mathbf{A}_{qi}) \tag{5.37}$$

(or $\bar{\mathbf{Y}}_{qs} = \mathbf{B}^{-1}\mathbf{A}_{qs}$ if $q = 0$).

The transformation vector $\mathbf{Y}_{qs}$ can be obtained from (5.35) and (5.37) and the desired $\mathbf{X}_{qs}$ from (5.32).   The relationships between the components of $\mathbf{X}_{qs}$ and $\mathbf{Y}_{qs}$ are then given by (recalling the definition of $\mathbf{E}$)

$$\begin{pmatrix} x_{qs}^{(1)} \\ \vdots \\ x_{qs}^{(k)} \\ x_{qs}^{(k+1)} \\ \vdots \\ x_{qs}^{(k+m)} \end{pmatrix} = \begin{pmatrix} \mathbf{I}_k & -\mathbf{E} \\ \mathbf{0} & \mathbf{I}_m \end{pmatrix} \begin{pmatrix} 0 \\ \vdots \\ 1 \\ \vdots \\ 0 \\ y_{qs}^{(k+1)} \\ \vdots \\ y_{qs}^{(k+m)} \end{pmatrix} \leftarrow \text{row } q$$

which yields

$$x_{qs}^{(q)} = 1 - \sum_{j \in \{\mathbf{J}_q\}} y_{qs}^{(k+j)} \tag{5.32a}$$

$$x_{qs}^{(p)} = - \sum_{j \in \{\mathbf{J}_p\}} y_{qs}^{(k+j)} \qquad p = 1, 2, \ldots, k \text{ and } p \neq q \tag{5.32b}$$

and

$$x_{qs}^{(k+j)} = y_{qs}^{(k+j)} \qquad j = 1, 2, \ldots, m \tag{5.32c}$$

From (5.37) we see that $y_{qs}^{(k+j)} = \mathbf{B}_j(\mathbf{A}_{qs} - \mathbf{A}_{qi})$, where $\mathbf{B}_j$ is the jth row of $\mathbf{B}^{-1}$. (If $q = 0$, then $y_{qs}^{(k+j)} = \mathbf{B}_j\mathbf{A}_{qs}$ and $x_{qs}^{(p)} = \sum_{j \in \{J_p\}} y_{qs}^{(k+j)}$ for $p = 1, 2, \ldots, k$.)

To solve (5.31) for the solution vector $\mathbf{X}_{00}$, we let $\mathbf{X}_{00} = \mathbf{TW}_0$ to obtain

$$
\left(\begin{Bmatrix} \mathbf{A}_{pi} \\ \mathbf{e}_p \end{Bmatrix} \;\middle|\; \begin{matrix} \mathbf{B} \\ \mathbf{0} \end{matrix} \right)
\begin{pmatrix} w_{01} \\ \vdots \\ w_{0k} \\ w_{0,k+1} \\ \vdots \\ w_{0,k+m} \end{pmatrix}
= \begin{pmatrix} \mathbf{b}_0 \\ 1 \\ 1 \\ \vdots \\ 1 \end{pmatrix}
\tag{5.31a}
$$

which yields

$$
w_{0i} = 1 \qquad i = 1, 2, \ldots, k
$$

Substituting these values into (5.31a), we have

$$
\mathbf{B} \begin{pmatrix} w_{0,k+1} \\ \vdots \\ w_{0,k+m} \end{pmatrix} = \mathbf{b}_0 - \sum_{i \in \{I\}} \mathbf{A}_{pi}
$$

which yields the partial solution vector

$$
\bar{\mathbf{W}}_0 = \begin{pmatrix} w_{0,k+1} \\ \vdots \\ w_{0,k+m} \end{pmatrix} = \mathbf{B}^{-1}\left(\mathbf{b}_0 - \sum_{i \in \{I\}} \mathbf{A}_{pi} \right)
\tag{5.38}
$$

or

$$
w_{0,k+j} = \mathbf{B}_j\left(\mathbf{b}_0 - \sum_{i \in \{I\}} \mathbf{A}_{pi} \right) \qquad \text{for } j = 1, 2, \ldots, m
$$

The related $\mathbf{X}_{00}$ consists of the key variables $x_{00}^{(p)}$ for $p = 1, 2, \ldots, k$ and the nonkey variables $x_{00}^{(k+j)}$ for $j = 1, 2, \ldots, m$. The

$$
x_{00}^{(k+j)} = w_{0,k+j} \geq 0 \qquad j = 1, 2, \ldots, m
\tag{5.38a}
$$

and

$$
x_{00}^{(p)} = 1 - \sum_{j \in \{J_p\}} x_{00}^{(k+j)} \geq 0 \qquad p = 1, 2, \ldots, k
\tag{5.38b}
$$

The above relationships can be established by expanding the transformation $\mathbf{X}_{00} = \mathbf{TW}_0$ or, equivalently, by assuming that we solved the original equations (5.20) and (5.21) as if they were stated in terms of only the basic key and nonkey variables; i.e., we substituted for each of the k key variables

$$
x_{00}^{(p)} = 1 - \sum_{j \in \{J_p\}} x_{00}^{(k+j)} \qquad p = 1, 2, \ldots, k
$$

in (5.20) and solved for the m nonkey variables $x_{00}^{(k+j)}$ using the resultant $m \times m$ transformed system of Eqs. (5.20) which has $\mathbf{B}$ as its matrix of coefficients.

As we now have the vectors $\mathbf{X}_{qs}$ and $\mathbf{X}_{00}$, we can determine which vector is to leave the basis. First, if all the components $x_{qs}^{(j)}$ of $\mathbf{X}_{qs}$ are less than or equal to zero, then the problem is unbounded. [If the set S_0 is empty, then the GUB problem is bounded because of Eqs. (5.21).] Assume at least one $x_{qs}^{(j)} > 0$; then

$$\theta_0 = \frac{x_{00}^{(r)}}{x_{qs}^{(r)}} = \min_{x_{qs}^{(j)}>0} \frac{x_{00}^{(j)}}{x_{qs}^{(j)}} \qquad j = 1, 2, \ldots, k, k+1, \ldots, m+k \qquad (5.39)$$

where $\bar{\mathbf{A}}_{rt}$ is the vector to be eliminated and belongs to the set S_r. As a result of the way in which we ordered the basis vectors, if the ratio occurs for $1 \le j \le k$, then the vector to be replaced will be a key vector, and nonkey otherwise.

We have now developed revised-simplex-procedure formulas for determining whether a current basis $\bar{\mathbf{B}}$ of order $m+k$ for the GUB problem is optimal and if it is not, how to select a vector to enter the basis and a vector to leave the basis, using an associated reduced matrix $\mathbf{B}$ of order m. We next show how to update the formulas given the change vectors.

The incoming vector $\bar{\mathbf{A}}_{qs}$ is from the set S_q and the outgoing vector $\bar{\mathbf{A}}_{rt}$ is from the set S_r. We need to be concerned with the possibility that $\bar{\mathbf{A}}_{rt}$ is a key vector, and also whether the sets S_r and S_q are the same. If $\bar{\mathbf{A}}_{rt}$ is not key, then it can be dropped and $\bar{\mathbf{A}}_{qs}$ introduced without concern; if $\bar{\mathbf{A}}_{rt}$ is key and $r \neq q$, then some current nonkey basic vector from S_r must be made key; and finally, if $\bar{\mathbf{A}}_{rt}$ is key and $r = q$, then either some current nonkey basic vector from S_r is made key or $\bar{\mathbf{A}}_{qs}$ is made key for the set S_r. In any event, the major computational concern is the updating of the new working basis $\mathbf{B}$ and $\mathbf{B}^{-1}$ and the computation of the new solution vector. We next consider the three cases described above.

Case I. If $\bar{\mathbf{A}}_{rt}$ is not key, then in $\mathbf{B}$ we replace $(\mathbf{A}_{rt} - \mathbf{A}_{ri})$ by $(\mathbf{A}_{qs} - \mathbf{A}_{qi})$ and update $\mathbf{B}^{-1}$ by performing an elimination transformation using column $\mathbf{B}^{-1}(\mathbf{A}_{qs} - \mathbf{A}_{qi})$ as pivot column. $\mathbf{A}_{qi}$ is the previously designated key column of set S_q. As $\bar{\mathbf{A}}_{rt}$ is not key, then the minimum ratio (5.39) must have occurred for some $k+1 \le j \le m+k$ as the basis is ordered by the k key vectors followed by the m nonkey vectors. Thus in $\mathbf{B}^{-1}(\mathbf{A}_{qs} - \mathbf{A}_{qi})$ we pivot on the element in the same position as $x_{qs}^{(r)}$, i.e., row $r-k$. We note that the vector $\bar{\mathbf{Y}}_{qs}$ given by (5.37) is the new pivot column and has already been determined. [If $\bar{\mathbf{A}}_{rt}$ is from set S_0, then $(\mathbf{A}_{qs} - \mathbf{A}_{qi})$ replaces $\mathbf{A}_{rt}$ in $\mathbf{B}$ and $\mathbf{B}^{-1}$ is updated as above. If $\bar{\mathbf{A}}_{qs}$ is from S_0, then $\mathbf{A}_{rt}$ is replaced by $\mathbf{A}_{qs}$ if $\bar{\mathbf{A}}_{rt}$ is also from S_0, or otherwise $(\mathbf{A}_{rt} - \mathbf{A}_{ri})$ is replaced by $\mathbf{A}_{qs}$. In either event, $\mathbf{B}^{-1}$ is updated by the pivot column $\mathbf{B}^{-1}\mathbf{A}_{qs}$.]

The new solution vector is updated by applying the above pivot operation to vector $\bar{\mathbf{W}}_0$ of (5.38) and substituting the results in (5.38a) and (5.38b). Alternatively, if $\bar{\mathbf{X}}_{00} = (\bar{x}_{00}^{(1)}, \ldots, \bar{x}_{00}^{(k+m)})$ is the new solution vector after $\bar{\mathbf{A}}_{qs}$ replaces $\bar{\mathbf{A}}_{rt}$ in the basis, then $\bar{x}_{00}^{(j)} = x_{00}^{(j)} - \theta_0 x_{qs}^{(j)}$ for $j = 1, 2, \ldots, k+m$ and $j \neq r$, and $\bar{x}_{00}^{(r)} = \theta_0$.

Case II.   When $\bar{A}_{rt}$ is key and $r \neq q$, then by our previous discussions at least one other vector, say $\bar{A}_{rl}$, from S_r must be in the current basis $\bar{B}$.   To handle this case, we first rearrange the vectors in the current basis $\bar{B}$ and corresponding working basis B by making $\bar{A}_{rl}$ the key column of set S_r instead of $\bar{A}_{rt}$.   Once this is done, then the incoming vector $\bar{A}_{qs}$ replaces $\bar{A}_{rt}$, which is now nonkey, which is just Case I above.   We next show how the key column exchange between $\bar{A}_{rl}$ and $\bar{A}_{rt}$ can be readily accomplished by applying a simple transformation on B.

The current working basis B is of the form

$$B = (\{A_{0j}\} | \{A_{pj} - A_{pi}\})$$

Vector A_{rt} is one of the key A_{pi} and A_{rl} is one of the nonkey A_{pj}.   We wish to multiply B by an $m \times m$ column transformation matrix $\bar{T}$ which will cause the columns of B to appear as if $\bar{A}_{rl}$ was originally selected to be the key vector of S_r.   This can be accomplished by changing each column of B of the form $(A_{rj} - A_{rt})$ as follows:

$$(A_{rj} - A_{rl}) = (A_{rj} - A_{rt}) - (A_{rl} - A_{rt}) \qquad \text{if } j \neq l \tag{5.40}$$

$$(A_{rt} - A_{rl}) = -(A_{rl} - A_{rt}) \qquad\qquad \text{if } j = l$$

The above relationships are just the multiplying of column $(A_{rl} - A_{rt})$ of B by -1 and then adding of $-(A_{rl} - A_{rt})$ to every column $(A_{rj} - A_{rt})$ except for $j = l$.   The transformation $\bar{T}$ is of the form

$$\bar{T} = \begin{pmatrix} 1 & & & & & & & & & \\ & \ddots & & & & & & & & \\ & & 1 & & & & & & & \\ 0 \cdots 0 & -1 & \cdots & -1 & -1 & -1 & \cdots & -1 & 0 \cdots 0 \\ & & & & 1 & & & & & \\ & & & & & \ddots & & & & \\ & & & & & & & & & 1 \end{pmatrix}$$

and the rearranged working basis is $B\bar{T}$. $\bar{T}$ is a unit matrix except that the row corresponding to the column position of $(A_{rl} - A_{rt})$ in B has -1's in the positions corresponding to the vectors $(A_{rj} - A_{rt})$ and 0's elsewhere.   The reader can verify that the product $B\bar{T}$ performs the desired operations of formulas (5.40); i.e., $B\bar{T}$ is the new working basis after the rearrangement of the key variables for the basis $\bar{B}$.

We see that $(B\bar{T})^{-1} = \bar{T}^{-1}B^{-1}$.   By applying the elimination procedure for obtaining an inverse of a matrix (Chap. 2), the reader can demonstrate that $\bar{T}^{-1} = \bar{T}$.   [A reapplication of $\bar{T}$ to $B\bar{T}$, i.e., $(B\bar{T})\bar{T}$, will produce B.]   For Case II, once the transformation $\bar{T}$ is determined and applied to B^{-1}, we have the inverse $\bar{\mathbf{T}}\mathbf{B}^{-1}$ of the rearranged working basis. As the original current basis $\bar{\mathbf{B}}$ has also been rearranged, the terms of the solution vector $\mathbf{X}_{00}$ are permuted according to the ordering of the basis vectors; i.e., the values of $x_{00}{}^{(r)}$ and $x_{00}{}^{(k+l)}$ are interchanged in the positions where $\bar{\mathbf{A}}_{rl}$ replaces $\bar{\mathbf{A}}_{rt}$ and $(\bar{\mathbf{A}}_{rt} - \bar{\mathbf{A}}_{rl})$ replaces $(\bar{\mathbf{A}}_{rl} - \bar{\mathbf{A}}_{rt})$ in $\bar{\mathbf{B}}$. Also, in $\bar{\mathbf{W}}_0$, the lth term $w_{0,k+l}$ is replaced by the value of $x_{00}{}^{(r)} = 1 - \sum_{j \in \{\mathbf{J}_r\}} x_{00}{}^{(k+j)}.^{\dagger}$ With the above changes, the incoming vector $\bar{\mathbf{A}}_{qs}$ is now replacing a nonkey vector $\bar{\mathbf{A}}_{rt}$, and we have Case I.

Case III. If $\bar{\mathbf{A}}_{rt}$ is key and $r = q$, then two situations can occur. If there are other $\bar{\mathbf{A}}_{rj}$ from S_r in the current basis, replace $\bar{\mathbf{A}}_{rt}$ by one of these, and we proceed as in Case II. If $\bar{\mathbf{A}}_{rt}$ is the only vector from S_r in the basis, then the incoming vector $\bar{\mathbf{A}}_{qs} = \bar{\mathbf{A}}_{rs}$ must replace $\bar{\mathbf{A}}_{rt}$ as the key column for S_r. Here, however, $\mathbf{B}$ does not have any columns which involve $\mathbf{A}_{rt}$ or any $\mathbf{A}_{rj}$ from S_r and thus $\mathbf{B}$ and $\mathbf{B}^{-1}$ do not change as $\bar{\mathbf{A}}_{qs}$ is brought into the new basis. The new solution can be obtained from (5.38), as the right-hand side of that equation must be changed to reflect the new key vector, i.e.,

$$\mathbf{B}^{-1}\left(\mathbf{b}_0 - \sum_{i \in \{\mathbf{I}\}} \mathbf{A}_{pi} + \mathbf{A}_{rt} - \mathbf{A}_{qs}\right) = \bar{\mathbf{W}}_0 - \mathbf{B}^{-1}(\mathbf{A}_{qs} - \mathbf{A}_{rt})$$

and these values are substituted in (5.38a) and (5.38b). Note that the last term is given by (5.37).

Given the updated $\mathbf{B}$, $\mathbf{B}^{-1}$ and solution vector, the process can be reapplied until an optimum is found or it is determined that the problem is unbounded. We emphasize that in applying the GUB procedure care must be taken in keeping track of the key and nonkey variables and vectors, the ordering and position of the vectors in $\bar{\mathbf{B}}$, $\mathbf{B}$ and the solution vector. The GUB algorithm in its reduction in computational requirements places a concomitant burden on the bookkeeping aspects, as the reader can verify by solving Exercise 25.

All the above procedures can be looked at as special approaches that enable us to solve large-scale linear-programming problems. These algorithms, along with the decomposition and related techniques for handling the distinctive structures which usually arise in large-scale programming models, have added immeasurably to the utility of such models. However, as we now see the need to solve extremely large problems (of the order of many hundred

---

† This can be shown explicitly by solving for the new partial solution vector

$$\tilde{\mathbf{W}}_0 = (\mathbf{B}\bar{\mathbf{T}})^{-1}\left[\mathbf{b}_0 - \sum_{i \in \{\mathbf{I}\}} \mathbf{A}_{pi} - (\mathbf{A}_{rl} - \mathbf{A}_{rt})\right]$$

where the right-hand side is just the subtraction from $\mathbf{b}_0$ of the new set of key vectors which includes $\mathbf{A}_{rl}$ and not $\mathbf{A}_{rt}$. [This is equivalent to Eq. (5.38) and its development.] Then $\tilde{\mathbf{W}}_0 = \bar{\mathbf{T}}^{-1}\bar{\mathbf{W}}_0 - \bar{\mathbf{T}}^{-1}\mathbf{e}_l$, where $\mathbf{e}_l$ is a unit vector with unity in the lth position. Thus, $\tilde{\mathbf{W}}_0 = \bar{\mathbf{W}}_0$ except for the lth position, which equals $1 - \sum_{j \in \{\mathbf{J}_r\}} w_{0,k+j} = x_{00}{}^{(r)}$.

[See Eqs. (5.38a) and (5.38b).]

thousands of constraints and millions of variables) and as this need is now coupled with the incredible computing power of today's electronic computers, the development of further advances in this area should be a prime field of research. It should be noted that the mechanics of the SUB algorithm were discovered by the application of the standard simplex method to the enlarged problem. This fact, plus similar experiences, leads us to suggest that, whenever one is faced with a new linear-programming model, one should construct numerical examples, arrange them in the simplex tableau, and solve the problems. An investigation of the computational tableaus might reveal hidden characteristics of the equations which could possibly lead to a special and more efficient computational method. Todd [645] discusses a general approach to the handling of GUB and other side conditions, and the development and use of reduced working bases for classes of large-scale problems. The Dantzig and Van Slyke GUB procedure is incorporated in many mathematical programming systems (Sperry Univac [619] and Ketron [428]).

6. THE SIMPLEX ALGORITHM AND COMPUTATIONAL EFFICIENCY

The first successful solution of a linear-programming problem on a high-speed electronic digital computer occurred in January 1952, on the National Bureau of Standards computer, the SEAC. The computational method used was the original simplex procedure, and the application was an Air Force programming problem dealing with the deployment and support of aircraft to meet stipulated requirements. Since that time, the simplex algorithm, or variations of this procedure, has been coded for practically all general-purpose electronic computers.

The need for efficient means of solving linear-programming problems, especially large-scale problems, and the important variations described in Sec. 5 and the other chapters (e.g., parametric programming, transportation, decomposition, integer, SUB, GUB) has led to the design and implementation of highly complex computerized procedures, termed *mathematical programming systems* (MPS). The simplex method, i.e., its numerical/logical representation, is the central computational element of MPS. (See Sec. 3 of Chap. 5 for details of how the simplex method is implemented in a MPS.) Problems with up to 16,000 constraints (not including any SUB, GUB, and nonnegativity conditions) and many hundreds of thousands of variables can be solved in a reasonable amount of computer time. For example, a 4,426 row $\times$ 10,918 variable problem was solved on an IBM 3033 computer in 27.48 seconds using MPS-III and its WHIZARD in-core option (Ketron [428]).

There has been a great amount of interest in determining why the simplex algorithm works so well in practice. It has proved very efficient, with the measure of efficiency being the number of iterations (hence, the associated computer time) required to solve a particular problem. Most $m \times n$ problems are solved in m to $3m$ iterations, with the actual number depending on the algorithmic characteristics incorporated in the MPS and whether a first feasible basis (or an advanced starting basis) is known. Why it takes so few iterations is unclear.[1]

The simplex algorithm moves from one extreme point to an adjacent extreme point (along an edge of the defining convex polyhedron), improving the objective function value at each iteration (assuming nondegeneracy). At worst, the algorithm could generate all extreme points, with the last one reached being the optimal one. As the number of feasible extreme points can be rather large, the computational process could break down for large values of m and n, and even for reasonably sized problems. This has not proven to be the case in practice, especially when a MPS and its refinements are used.

In an attempt to answer the question, "How good is the simplex algorithm?" Klee and Minty [432] constructed a class of problems for which the simplex algorithm could be made to pass through all extreme points (see Exercise 40). These "worst-case" problems require $2^d - 1$ iterations, where d is the number of variables in an $m = 2d$ inequality problem (the equation problem has $m = 2d$ constraints and $n = 3d$ variables).

An algorithm has a *time-complexity function* that expresses its time requirements by giving, for each possible input length (size), the largest amount of time needed by the algorithm to solve a problem instance of that size. A *polynomial-time (polynomial-bounded) algorithm* is defined as one whose time complexity function is a polynomial function of its input length. An algorithm that cannot be bounded in this manner is called an *exponential-time algorithm* (Garey and Johnson [280]). An example of a polynomial-time algorithm is that of finding the shortest route in an acyclic network (see Sec. 4, Chap. 11) as the number of computations is bounded by a function of order n^2 (Lawler [462]). The Klee-Minty problems described above show that the simplex method is an exponential-time algorithm. Other examples of exponential-time algorithms are the procedures for the complete enumeration of all traveling-salesman tours or all solutions to an integer-programming problem. Note that exponential rates of growth include $n!$, k^n for $k > 1$, and n^n, as well as the basic one of 2^n. For an n-city traveling-salesman problem, there are $(n - 1)!/2$ tours. For detailed discussions on computational efficiency and its relationship to linear programming and other problems, see Papadimitriou and Steiglitz [541], Lenstra et al. [472], Karp [417], Dobkin and Reiss [210], Telgen [637], Traub and Woźniakowski [653], and Chandrasekaran et al. [108].

As noted above, the simplex algorithm does not have a polynomial bound but works well in practice. The question of whether a polynomial-bounded algorithm for solving linear-programming problems could be discovered was resolved with the development of the procedure termed the *ellipsoid algorithm*

---

[1] Smale [616] shows that for a fixed number of constraints m, the average number of pivot steps (iterations) required to solve an "average" linear programming problem using the self-dual parametric simplex algorithm (see Dantzig [152] and Exercise 42, Chap. 8) "grows slower than any prescribed root of the number of variables"; also see Borgwardt [84].

(Khachiyan [429]). This method embeds the solution space of the joint primal-dual inequality problem [see Eqs. (2.15), Chap. 6] in progressively smaller (in volume) ellipsoids, each of which is guaranteed to contain the optimal solution, if one exists. An iterative step is the determination of the new ellipsoid. Khachiyan showed that the optimal solution (here a solution to the primal-dual conditions) can be accomplished within a polynomial-bounded number of iterations. Although of theoretical importance, the ellipsoid algorithm has exhibited poor computational properties in terms of the time and the number of iterations required to solve problems of reasonable size (see, for example, McCall [480] and Dantzig [177] for a discussion of the computational characteristics of the ellipsoid algorithm versus the simplex algorithm).[1] In contrast, the simplex algorithm, as represented in MPS, has demonstrated an amazing ability to solve a diverse set of practical problems. It has proven readily adaptable to take advantage of data sparsity and of special structures like SUB, GUB, and networks (see Chaps. 10 and 11); and it provides a full range of sensitivity and parametric procedures. As noted in Bland et al. [82], "It is important to point out that because of the limitations of finite precision arithmetic it is unlikely that any reasonable implementation of the (ellipsoid) method would be polynomial."

REMARKS

The more advanced MPS have a full array of data input, output, and formatting procedures, as well as analytical tools which aid in understanding the full meaning of an optimal solution, e.g., sensitivity analysis and dual-variable marginal analysis of resource constraints. The ability to design these systems calls for an understanding of all facets of linear programming—theoretical, computational, and applied. This knowledge must then be integrated with the computer system so as to take advantage, in a computational sense, of the interrelationships between the computer hardware and language and the mathematical requirements. The reader interested in this important aspect of the field should consult Orchard-Hays [532], Driebeek [221], and Beale [58], as well as the manuals supplied by the computer manufacturers and computer service companies. Also, analysts, who are concerned with the formulation of problems for solution using a mathematical programming system, are well advised to become familiar with the system's characteristics and data handling procedures as they can affect greatly the computational burden.

A status report on computing algorithms for mathematical programming systems is given in White [696]; see Land and Powell [457] for a review of MPS mixed-integer capabilities. Periodically, updated descriptive lists of these systems appear in the newsletters of the Special Interest

---

[1] In applying the ellipsoid algorithm, it is usually assumed that the problem data are given in integers and a certain computational precision is maintained. With these assumptions, the ellipsoid algorithm applied to the finding of a solution to a set of $m \times n$ linear inequalities will, after $32n(n + 1)L$ iterations, find a solution or indicate that none exists. Here L is a measure of the problem size, i.e., the number of bits required to store the problem's data. For the linear-programming problem, we can define $L = mn + [\log |P|]$, where P is the product of the nonzero coefficients a_{ij}, b_i, and c_j, and $[\log |P|]$ is the smallest integer greater than or equal to $\log |P|$ (any logarithmic base will do) (Papadimitriou and Steiglitz [541]). As applied to the primal-dual inequality system, the ellipsoid algorithm will take on the order of $(m + n)mn^4L$ arithmetic operations (Aspvall and Stone [13]).

Group for Mathematical Programming (SIGMAP) of the Association for Computing Machinery. Recent advances of all phases of mathematical programming appear in the technical journals cited in the References, especially *Mathematical Programming*, the journal of the Mathematical Programming Society. The Bibliography and References vividly demonstrate that linear programming is an ever-advancing scientific field, one that has established itself as a major element of decision analysis as applied to a most varied set of important industrial, economic, and governmental problems.

EXERCISES

1. Selecting vectors P_3, P_4, P_5 as an initial basis to the following equations, determine the corresponding solution and, if necessary, transform the system to a form suitable for Phase I computation:

$$x_1 - 3x_2 + 4x_3 + x_4 + x_5 = 9$$
$$-x_1 + 2x_2 - x_3 - 7x_4 + x_5 = 2$$
$$-x_1 + x_2 - 3x_3 - 2x_4 + 6x_5 = 5$$
$$x_j \geq 0$$

2. Solve the following problem by means of the dual simplex method: Maximize

$$x_0 = 7x_1 - 7x_2 + 2x_3 + x_4 + 6x_5$$

subject to

$$3x_1 - x_2 + x_3 - 2x_4 \qquad = -3$$
$$2x_1 + x_2 \qquad + x_4 + x_5 \qquad = 4$$
$$-x_1 + 3x_2 \qquad - 3x_4 \qquad + x_6 = 12$$
$$x_j \geq 0$$

Answer: $x_4 = 1.5$, $x_5 = 2.5$, $x_6 = 16.5$; $x_0 = 16.5$.

3. Solve the following problems in terms of integers, using the all-integer algorithm.

a. Minimize

$$x_1 + 8x_2$$

subject to

$$2x_1 + 3x_2 \geq 6$$
$$-3x_1 + x_2 \leq 1$$
$$x_1 + x_2 \leq 7$$
$$x_1 - 3x_2 \leq 1$$
$$x_j \geq 0$$

b. Minimize

$$x_1$$

subject to

$$3x_1 + 2x_2 \geq 12$$
$$-2x_1 + 6x_2 \geq 12$$
$$x_2 \leq 5$$
$$x_j \geq 0$$

c. Minimize

$$x_1 + 3x_2 + x_3$$

subject to

$$x_1 + 2x_2 + 3x_3 \geq 7$$
$$-2x_1 + x_2 + x_3 \leq 13$$
$$x_1 - 2x_2 + 2x_3 \leq 5$$
$$3x_1 + x_2 + 7x_3 \geq 14$$
$$x_j \geq 0$$

4. Solve a and b of Exercise 3 by graphing the constraints.

5. Solve the following problem using the decomposition algorithm: Minimize

$$2x_1 - 3x_2 + x_3 - 2x_4 + 6x_5 - 4x_6$$

subject to

$$x_1 + 2x_2 + 3x_3 + 4x_4 - 5x_5 - 2x_6 = 13$$
$$x_1 + 2x_2 + x_3 = 8$$
$$3x_1 - 2x_2 + 2x_3 = 7$$
$$x_4 + x_5 - x_6 = 3$$
$$2x_4 - x_5 + 3x_6 = 21$$
$$x_j \geq 0$$

6. Phase I procedure for the dual-simplex algorithm (Vajda [662] and Beale [53]). For the dual simplex algorithm let $\mathbf{B}$ be a basis such that $\mathbf{B}^{-1}\mathbf{b}$ has some negative components (not a feasible solution) with some associated $z_i - c_i > 0$ (the optimality conditions are not all satisfied). Let $\{\mathbf{J}\}$ be the set of indices for the nonbasic vectors for which $z_j - c_j > 0$. Add the equation $\sum_{j \in \mathbf{J}} x_j + x_0 = M$, where M is a very large positive number and $x_0 \geq 0$. Let $z_k - c_k = \max_{j \in \mathbf{J}} z_j - c_j$ and in the new equation, pivot x_k into the basis, which is now an $(m+1) \times (m+1)$ matrix. This procedure makes all the transformed $z_j - c_j \leq 0$, and the values of the basic variables are functions of M. From this point on, show how the problem can be solved in a manner similar to the Phase I procedure of the standard simplex method. What happens when x_0 goes into the basis? What if x_0 is not in the optimal feasible basis of this new problem? This procedure is the analogous dual-problem approach to the one artificial-vector technique of Sec. 1 in this chapter.

7. Transformation of problems into integer problems (Dantzig [152]).

a. The *fixed-charge problem* has as constraints $\mathbf{AX} = \mathbf{b}$, $\mathbf{X} \geq \mathbf{0}$ and the objective function $\mathbf{cX}$ to be minimized, where $c_j = d_j x_j + k_j$ if $x_j > 0$, and $c_j = 0$ if $x_j = 0$; that is, the k_j are fixed, one-time charges applied to the cost if the activity x_j is used at a positive level. Transform this problem into a mixed-integer problem, where the x_j are not restricted to be integers, but special integer variables are needed to express the fixed-charge requirements. HINT: Assume each x_j has a known, finite upper bound.

b. Show how any integer problem can be transformed to one in which all variables are restricted to be either 0 or 1.

c. Show how to express the condition that x_j can equal b_1, b_2, or b_3, using integer variables.

d. Using a variable restricted to be either 0 or 1, express the nonlinear condition $x_1 x_2 = 0$ by two constraints which involve the nonnegative variables x_1 and x_2 by themselves. HINT: Assume that x_1 and x_2 have known, finite upper bounds.

e. Using a variable restricted to be either 0 or 1, show how the following conditions can be expressed within a mixed-integer problem: either $l_i \leq \sum_j a_{ij} x_j \leq u_i$ or $\sum_j a_{ij} x_j = 0$ (the x_j are continuous variables).

8. Verify Eqs. (2.3) and (2.4).

9. Solve the following continuous linear-programming problem geometrically (Trauth and Woolsey [654]). Is the optimal solution rounded to the nearest integer a feasible solution to the problem? What is the integer optimal solution?

Maximize

$$x_1 + x_2$$

subject to

$$-9x_1 + 10x_2 \le 0$$
$$x_1 - 3x_2 \le 0$$
$$23x_1 - 22x_2 \le 24$$
$$x_j \ge 0$$

10. *The knapsack problem* (Dantzig [152]). Formulate the following problem as an integer linear-programming problem. A hiker has a choice of carrying a combination of n objects subject to a total weight restriction. Each object has a certain value c_j to the hiker and a weight a_j. If b is the total weight that can be carried, how should the hiker determine the combination of objects which maximizes the total value? How would you solve the continuous-variable version of this problem? See Chap. 11 for an application of the knapsack problem. {It can be shown that any bounded-integer linear-programming problem can be transformed into a bounded-variable knapsack problem (Bradley [87, 88], Garfinkel and Nemhauser [281]). The coefficients of the resulting knapsack problem tend to be rather large, thus decreasing the efficacy of available knapsack-problem algorithms (Hu [383], Gilmore and Gomory [305], and Dreyfus and Prather [220]}.

11. What assumptions are required to prove the finiteness of the decomposition algorithm?

12. Lower bound for the objective function in the decomposition algorithm (Dantzig [152]). Denote the objective function (4.6) of the extremal problem by z and let z_0 be the current value of the objective function. Show that for $\mathbf{A}_0 = (0)$ a lower bound for z is given by

$$\min z \ge z_0 + \sum_{p=1}^{k} (z_{pq} - \bar{\pi}_p)$$

HINT: See Exercise 3 in Chap. 4. Multiply Eqs. (4.7) and (4.8) by π and $\bar{\pi}$, respectively, and subtract from z.

13. Develop a decomposition algorithm for the transportation problem.

14. Discuss the decomposition of a SUB problem, where the full set of bounding constraints represents one subproblem (Hadley [344]).

15. Formulate the multicommodity network problem as a block-angular system and discuss how it should be decomposed.

16. Prove that if an unbounded feasible solution exists for

$$\mathbf{BX} = \mathbf{b}$$

$$\mathbf{X} \ge \mathbf{0}$$

then we can find a family of feasible solutions which satisfies the homogeneous equations

$$\mathbf{BX} = \mathbf{0}$$

$$\mathbf{X} \ge \mathbf{0}$$

HINT: See Sec. 1, Chap. 4.

17. Solve the following problem using the decomposition algorithm (Harvey [357]). Minimize

$$x_{00} = -6x_{01} - 4x_{02} \quad - 5x_{11} - 5x_{12} \qquad\qquad - 7x_{21} - 7x_{22}$$

subject to

x_{01}	x_{11}	$+\ x_{21}$	$= 10$
$\quad x_{02}$	$+\ x_{12}$	$+\ x_{22}$	$= 20$
	$x_{11} + x_{12} - x_{13}$		$= 6$
	$-2x_{11} + x_{12} \qquad + x_{14}$		$= 0$
		$x_{21} + x_{22} + x_{23}$	$= 25$
		$-2x_{21} + x_{22} \qquad + x_{24} = 0$	

$$x_{ij} \geq 0$$

Answer: $x_{00} = -198,$

$$\mathbf{X}_0 = (0, 0) \qquad \mathbf{X}_1 = 0 \begin{pmatrix} 6 \\ 0 \\ 0 \\ 12 \end{pmatrix} + 1 \begin{pmatrix} 2 \\ 4 \\ 0 \\ 0 \end{pmatrix} = \begin{pmatrix} 2 \\ 4 \\ 0 \\ 0 \end{pmatrix}$$

$$\mathbf{X}_2 = \tfrac{1}{25} \begin{pmatrix} 0 \\ 0 \\ 25 \\ 0 \end{pmatrix} + \tfrac{24}{25} \begin{pmatrix} \tfrac{25}{3} \\ \tfrac{50}{3} \\ 0 \\ 0 \end{pmatrix} = \begin{pmatrix} 8 \\ 16 \\ 1 \\ 0 \end{pmatrix}$$

18. Solve the following problem by the direct simplex method and the decomposition algorithm (Müller-Merbach [511]). Minimize

$$x_{00} = -8x_{11} - 3x_{12} - 8x_{21} - 6x_{22}$$

subject to

$$4x_{11} + 3x_{12} + x_{21} + 3x_{22} \leq 16$$
$$4x_{11} - x_{12} + 3x_{21} \qquad\quad \leq 12$$
$$x_{11} + 2x_{12} \qquad\qquad\quad \leq 8$$
$$3x_{11} + x_{12} \qquad\qquad\quad \leq 10$$
$$2x_{21} + 3x_{22} \leq 9$$
$$4x_{21} + x_{22} \leq 12$$
$$x_{ij} \geq 0$$

Answer: $x_{00} = -\tfrac{877}{20},$

$$x_{11} = \tfrac{107}{80} \qquad x_{12} = \tfrac{29}{20} \qquad x_{21} = \tfrac{27}{10} \qquad x_{22} = \tfrac{6}{5}$$

19. Develop an artificial-basis procedure for determining a first feasible solution to the extremal problem.   HINT: Use positive and negative unit vectors which correspond, respectively, to the positive and negative elements of $\mathbf{b}_0$. Each vector has a nonnegative artificial variable. Employ an additional constraint which forces the sum of the artificial variables to zero if a feasible solution exists; i.e., Phase I corresponds to minimizing a variable which represents the sum of the artificial variables (Dantzig [152]).

20. Solve the following problem by the decomposition algorithm using the first constraint for the extremal problem and the last two constraints as a single subproblem.   Maximize

$$x_1 + 2x_2$$

subject to

$$x_1 + x_2 \geq 3$$
$$-2x_1 + x_2 \leq 2$$
$$x_1 - x_2 \leq 1$$
$$x_j \geq 0$$

21. For the SUB problem, develop a rule for selection of the variable x_k to enter the basis and a computational tableau.

22. Solve the following problems by the SUB algorithm.

a. Maximize

$$x_1 + x_2$$

subject to

$$2x_1 + x_2 \le 18$$
$$x_1 - 2x_2 \le 6$$
$$-3x_1 + 6x_2 \le 9$$
$$x_1 \le 8$$
$$x_2 \le 3$$
$$x_j \ge 0$$

b. Maximize $x_1 - x_2$ subject to the constraints in *a*.

c. Gass [286].   Maximize

$$x_0 = 2x_1 + x_2 + x_3$$

subject to

$$x_1 + x_2 + x_3 \le 6$$
$$x_1 - x_2 + x_3 \le 1$$
$$x_1 \le 3$$
$$x_2 \le 10$$
$$x_3 \le 4$$
$$x_j \ge 0$$

Answer: $x_1 = 3$, $x_2 = 3$, $x_3 = 0$; $x_0 = 9$ and $x_1 = 3$, $x_2 = \frac{5}{2}$, $x_3 = \frac{1}{2}$; $x_0 = 9$.

23. Make the necessary changes to the SUB algorithm if some x_j are not bounded above.

24. Write out the maximizing dual problem of the SUB problem. Determine the complementary slackness expression $f(\mathbf{X}) - g(\mathbf{W})$, where $g(\mathbf{W})$ is the dual objective function. If $\mathbf{X}^0$ and $\mathbf{W}^0$ are feasible solutions to the primal and dual problems, respectively, then by the duality theorem $f(\mathbf{X}^0) - g(\mathbf{W}^0) = 0$ implies that these solutions are also optimal. Show how the conditions of Theorem 1 can be obtained by analysis of the expression $f(\mathbf{X}^0) - g(\mathbf{W}^0)$.

25. Solve the following problems using the GUB algorithm.

a. Minimize

$$x_{00} = x_{12} + x_{21} + x_{22}$$

subject to

$$x_{01} + 3x_{11} + 2x_{12} + x_{13} + 2x_{21} - x_{22} + 2x_{23} = 4$$
$$x_{02} - x_{11} + x_{12} + 3x_{13} - 4x_{21} + 3x_{22} + 4x_{23} = 5$$
$$x_{11} + x_{12} + x_{13} = 1$$
$$x_{21} + x_{22} + x_{23} = 1$$
$$x_{ij} \ge 0$$

A first feasible solution is $x_{01} = 0$, $x_{02} = 0$, $x_{12} = 1$, and $x_{23} = 8$. The optimal solution is $x_{11} = \frac{1}{2}$, $x_{13} = \frac{1}{2}$, and $x_{23} = 1$, with $x_{00} = 0$. The initial $\bar{\mathbf{B}} = \bar{\mathscr{B}} = (\bar{\mathbf{A}}_{12}\bar{\mathbf{A}}_{23}|\bar{\mathbf{A}}_{01}\bar{\mathbf{A}}_{02})$, $\mathbf{B} = \begin{pmatrix} 1 & 0 \\ 0 & 1 \end{pmatrix} = \mathbf{B}^{-1}$ and $\mathbf{\Pi} = (0, 0, 1, 0)$.

b. Minimize

$$x_{00} = -3x_{11} + x_{12} + 2x_{21} + x_{22}$$

subject to

$$
\begin{aligned}
x_{01} \quad + \quad x_{11} - \quad x_{12} + \quad x_{21} - 2x_{22} &= 1 \\
x_{02} + 2x_{11} + 3x_{12} - 2x_{21} + \quad x_{22} &= 4 \\
x_{11} + \quad x_{12} \quad\quad\quad\quad\quad\quad &= 1 \\
x_{21} + \quad x_{22} &= 1 \\
x_{ij} &\geq 0
\end{aligned}
$$

An initial basic feasible solution is $x_{01} = 1$, $x_{02} = 3$, $x_{12} = 1$, $x_{21} = 1$. The optimal solution is $x_{01} = 2$, $x_{02} = 1$, $x_{11} = 1$, $x_{22} = 1$; $x_{00} = -2$.

26. Solve Eqs. (5.16) as a GUB problem with the objective function of minimizing $-2x_1 - x_2$.

27. Discuss how you would apply the GUB procedure to the transportation problem.

28. Show how the GUB procedure can be used if the GUB constraints sum to a positive number greater than one; and if the GUB constraints have coefficients of plus or minus one.

29. Develop the argument for why the decomposition algorithm will converge in a finite number of steps.

30. For the SUB problem with constraints $\mathbf{AX} = \mathbf{b}$ and $\mathbf{0} \leq \mathbf{X} \leq \mathbf{U}$, where each component of the upper-bound vector $\mathbf{U}$ is an integer, show that the continuous solution cannot have more than m variables noninteger (m is the number of equations in $\mathbf{AX} = \mathbf{b}$). Is this also true for a GUB problem?

31. Show that a solution to Eqs. (5.2) to (5.4) is an extreme point if $n - m$ components x_j are equal to either zero or the corresponding u_j, while the remaining m components are between zero and the corresponding u_j and are associated with a basis of (5.2).

32. Describe a revised simplex procedure for solving the SUB problem.

33. Develop a SUB algorithm for solving the linear-programming problem with constraints $\mathbf{AX} = \mathbf{b}$ and $\mathbf{0} \leq \mathbf{X} \leq \mathbf{U} + \theta \bar{\mathbf{U}}$, where $\mathbf{U}$ and $\bar{\mathbf{U}}$ have integer components and θ is a parameter.

34. Show that Theorem 1 of Sec. 5 on the optimal value of a SUB problem holds if $0 \leq l_j \leq x_j \leq u_j$.

35. Solve the following problem by using the decomposition algorithm (Krekó [443]). Minimize

$$-2x_{11} - x_{12} - 3x_{13} - 2x_{21} - x_{22}$$

subject to

$$
\begin{aligned}
x_{01} + \quad x_{11} + 2x_{12} + 4x_{13} + 2x_{21} + 2x_{22} &= 100 \\
x_{11} + \quad x_{12} \quad\quad\quad\quad\quad\quad &\leq 40 \\
2x_{12} + \quad x_{13} \quad\quad\quad\quad &\leq 50 \\
2x_{21} + \quad x_{22} &\leq 65 \\
x_{21} + \quad x_{22} &\leq 50 \\
x_{ij} &\geq 0
\end{aligned}
$$

36. Develop the GUB algorithm for the linear-programming problem that has only a single GUB constraint that involves all the variables.

37. Solve the following problem by the standard simplex method and the GUB algorithm (Müller-Merbach [517]). Maximize

$$x_0 = 7x_1 + 6x_2 + 8x_3 + 5x_4$$

subject to

$$x_1 + 5x_2 + 4x_3 + 3x_4 = 18$$
$$4x_1 + 2x_2 + 4x_3 + 2x_4 = 12$$
$$x_1 + x_2 \qquad\qquad = 1 \qquad (x_j \geq 0)$$
$$x_3 + x_4 = \tfrac{9}{2}$$

Answer: $x_1 = \tfrac{1}{5}$, $x_2 = \tfrac{4}{5}$, $x_3 = \tfrac{3}{10}$, $x_4 = \tfrac{21}{5}$; $x_0 = \tfrac{148}{5}$.

38. Cycling using the dual simplex algorithm (Beale [52]).

a. Solve the following problem using the dual simplex algorithm (when there are ties for the variable to enter the basis the one with the lowest index is selected). Minimize

$$x_7$$

subject to

$$x_1 \qquad\qquad -\tfrac{1}{4}x_5 - \tfrac{1}{2}x_6 \qquad\qquad = -\tfrac{3}{4}$$
$$x_2 \qquad + 8x_5 + 12x_6 \qquad\qquad = 20$$
$$x_3 + x_5 + \tfrac{1}{2}x_6 - x_7 = -\tfrac{1}{2} \qquad (x_j \geq 0)$$
$$x_4 - 9x_5 - 3x_6 \qquad\qquad = 6$$

b. Apply the simplex algorithm to the dual of the problem given in part *a* and show that it also cycles.

c. What is the dual simplex algorithm counterpart of the lowest-index rule described in Chap. 7? Apply the rule to the problem given in part *a*.

39. An upper bound on the number of feasible bases of a linear-programming problem is $\dfrac{n!}{m!(n-m)!}$. For a 3×5 problem, this bound is 10. Show that a better bound is at most 5. Also show that the number of extreme points is bounded by a function of order n^m, and that for m fixed, the simplex method is polynomially bounded; the bound for $m = 2$ is of the order $n \log n$ (Telgen [637]).

40. *Worst-case linear-programming problems.*

a. Show that the following problem has 2^d extreme points and describe a set of simplex iterations (a path) that would pass through all such points. Note that the problem is a "perturbed" cube (Klee and Minty [432], Klee [431], Papadimitriou and Steiglitz [541]). Minimize

$$-x_d$$

subject to

$$x_1 \geq 0$$
$$x_1 \leq 1$$
$$-\epsilon x_1 + x_2 \geq 0$$
$$\epsilon x_1 + x_2 \leq 1$$
$$-\epsilon x_2 + x_3 \geq 0$$
$$\epsilon x_2 + x_3 \leq 1$$
$$\vdots$$
$$-\epsilon x_{d-1} + x_d \geq 0$$
$$\epsilon x_{d-1} + x_d \leq 1$$

with $0 < \epsilon < \tfrac{1}{2}$.

b. For $d = 2$, the extreme points (in ascending value of x_d) are (ϵ, ϵ^2), $(1, \epsilon)$, $(1, 1-\epsilon)$, $(\epsilon, 1-\epsilon^2)$. Determine the extreme points for $d = 3$ and state the corresponding problem in equation form. Show that all bases are nondegenerate and that there is a simplex algorithm set of iterations that will pass through all extreme points, starting with the perturbed origin as the first basic feasible solution. What decision rule do you have to use in selecting the next variable x_k to enter the basis? What happens when you try the usual simplex rule of selecting x_k that corresponds to max $(z_j - c_j) > 0$? (Based on the work of Klee and Minty [432] and Jeroslow [400], it is conjectured that no matter what entering variable selection rule is used, problems can be constructed that will require an exponential number of pivots. Recall that the simplex algorithm's selection rule for the variable to enter the basis is arbitrary, while the rule that determines the variable to leave the basis is exact, except for degenerate situations.)

    c. Avis and Chvátal [15] give the following version of the Klee and Minty [432] problem stated in part *a*: Minimize

$$-\sum_{j=1}^{n} \epsilon^{n-j} x_j$$

subject to

$$2\sum_{j=1}^{i-1} \epsilon^{i-j} x_j + x_i \le 1 \qquad (i = 1, 2, \ldots, n)$$

$$x_j \ge 0 \qquad (j = 1, 2, \ldots, n)$$

with $0 < \epsilon < \frac{1}{2}$.

    Determine the extreme points of the solution space for $n = 2$ and $n = 3$. Solve the problem for $n = 4$ using the usual simplex rule for selecting the entering variable and then using the lowest-index rules described in Chap. 7. Start with the basic solution that has all $x_j = 0$. Can you transform and/or adjust the Klee and Minty or Avis and Chvátal problems so that the standard selection rule will cause the simplex algorithm to pass through all extreme points? (See Chvátal [128].)

    41. *Branch and bound algorithm for mixed-integer programming problems.* In their paper [455], Land and Doig describe a simplex-based procedure (later termed "branch and bound" because of the work by Little et al. [475]) for solving mixed-integer-programming problems. This scheme, modified by Dakin [150], has proved to be quite successful and it has been incorporated into most of the major mathematical programming systems. The basic branch and bound procedure is as follows. First solve the associated (minimizing) continuous problem (the integer conditions are *relaxed*) using the simplex method. If the solution satisfies the integer conditions, the process stops. If not, select some integer variable x_k whose value x_{k0} in the continuous solution is not integer. For x_k to assume an integer value, we must have either $x_k \le [x_{k0}]$ or $x_k \ge [x_{k0}] + 1$, where $[x_{k0}]$ indicates "integer part of." We next construct two continuous problems as before, with the first (branch) being augmented by the condition $x_k \le [x_{k0}]$ and the second (branch) by $x_k \ge [x_{k0}] + 1$. These problems are then solved by the simplex method. If either problem is infeasible, it (i.e., the branch) is excluded from any further consideration. If the new problems yield feasible solutions with objective function values less than the value of any currently known integer solution, we add them to an active consideration list. If not, we can eliminate (bound) them from further consideration (why?). (We usually can select a reference integer solution to start the process or, at worst, assume that the best known value is $+\infty$.) Select from the active problem list the one that is associated with the lowest value of the objective function. If its (continuous) solution satisfies the integer constraints, then it is optimal (why?). If not, select some integer variable that has a fractional value (the one with the largest fraction is usually selected) and branch as above and solve the resulting two continuous problems. Repeat the steps, adding any feasible problems to the active list depending on how their objective function values compare to the best known integer solution. This process will exhaust and/or eliminate all possibilities (can you prove why?). Branch and bound procedures can be viewed best in terms of a connected graph or tree in which the original continuous problem is the root node, with each new problem another node connected to a succeeding node by a branch defined by one of the bounding constraints on the selected integer variable x_k. The original Land and Doig procedure branched on the two conditions $x_k = [x_{k0}]$ and $x_k = [x_{k0}] + 1$. Describe why this would tend to be inefficient when compared with the inequality bounds used above and proposed by Dakin.

    a. Carry out the branch and bound calculations that solve the following problem and produce the accompanying branch and bound tree.   (From Müller-Merbach [516].)   Minimize

$$x_0 = -2x_1 - 5x_2$$

subject to

$$2x_1 - x_2 \le 9$$
$$2x_1 + 8x_2 \le 31$$

$(x_j \ge 0 \text{ and integer})$

Optimal solution $x_1 = 3$, $x_2 = 3$; $x_0 = -21$.   The branch and bound tree is shown in Fig. E41a.

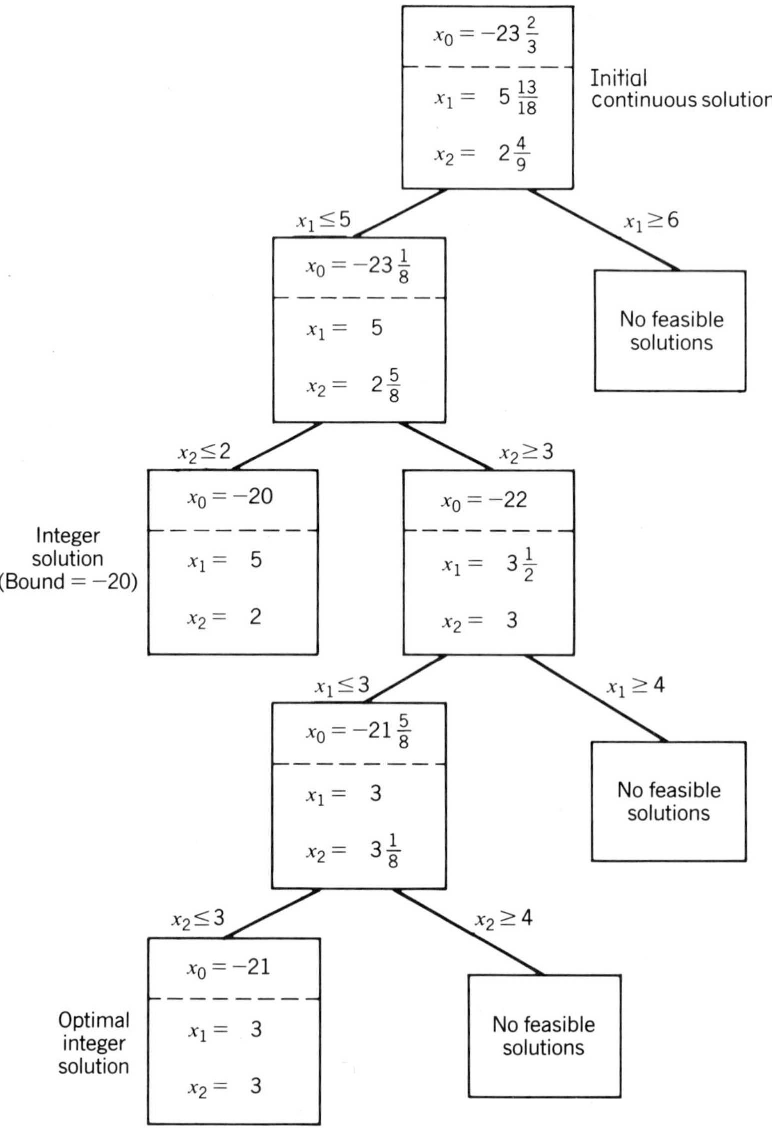

FIGURE E41a

b. Solve the integer problem of Part *a* by graphical procedures and Gomory's cutting-plane method. Also, for the branch and bound tree, graph the solution set for each problem defined at a node.

c. Solve the following integer problem by the branch and bound algorithm (Müller-Merbach [516]). Minimize

$$x_0 = -38x_1 - 20x_2 - 41x_3 - 35x_4$$

subject to

$$2x_1 + 2x_2 + 2x_3 + x_4 \leq 32$$
$$x_1 \quad\quad - 3x_3 + 5x_4 \leq 2 \quad\quad (x_j \geq 0 \text{ and integer})$$
$$2x_1 - 2x_2 + 5x_3 + 3x_4 \leq 17$$

Optimal solution: $x_1 = 7$, $x_2 = 6$, $x_3 = 3$, $x_4 = 0$; $x_0 = 509$.

d. In developing a procedure that implements the steps of the branch and bound algorithm, describe how the dual simplex algorithm and/or the simple upper-bounded procedure may be used to speed up the process.

e. Solve the following problem by the branch and bound procedure (Garfinkel and Nemhauser [281]): Minimize

$$x_0 = 7x_1 + 3x_2 + 4x_3$$

subject to

$$x_1 + 2x_2 + 3x_3 \geq 8$$
$$3x_1 + x_2 + x_3 \geq 5 \quad\quad (x_j \geq 0 \text{ and integer})$$

Optimal solution: $x_1 = 0$, $x_2 = 5$, $x_3 = 0$; $x_0 = 15$.

f. Many mixed-integer-programming problems contain constraints that require exactly one out of a set of (0, 1) variables to be positive, with each such variable being in at most one such set. For example, let x_j equal to 0 or 1 be a *binary choice variable* that represents the selection of location *j* at which a warehouse is to be located (0 = not selected, 1 = selected; see Exercise 42*b*). Then the constraints $x_1 + x_2 + x_3 = 1$ and $x_4 + x_5 + x_6 = 1$ state that exactly one of the locations 1, 2, 3 and one of the locations 4, 5, 6 must be selected. These types of (convexity) constraints are termed *special ordered sets of Type 1*. (See Exercise 8, Chap. 12, on Type 2 constraints.) Discuss in the context of a branch and bound algorithm how a branch can be accomplished on a set instead of the variables that are included in the set (Forrest et al. [256]). HINT: The set can be dichotomized to yield two problems, one with variables $x_1 = \cdots = x_k = 0$ and the other with $x_{k+1} = \cdots = x_p = 0$, where the (0, 1) variables $x_1, \ldots, x_k, x_{k+1}, \ldots, x_p$ are in the set. Type 1 constraints are also termed *multiple-choice constraints*.

g. For the branch and bound algorithm, discuss Dakin's bounds in terms of cutting planes and illustrate their impact by solving Exercises 3*a* and 3*b* by the branch and bound algorithm, accompanied by graphical solutions of the generated continuous problems.

42. *Integer-programming problem formulations* (Salkin [591], Wagner [684], Bradley et al. [92]).

a. Capital budgeting (Lorie and Savage, Sec. 4 of Bibliography). Given *n* competing and independent investment opportunities (or projects) and a total investment budget of *b* dollars. Let a_j be the cost of investment *j* and c_j the expected return. No partial investments can be made. We wish to maximize the return. Formulate this as an integer-programming problem in which the variables are no/yes (no-go/go) 0 or 1 variables. How would you generalize the formulation to include more conditions (projects that use limited resources such as labor and space) or investments that are made over many time periods (project *j* requires an investment of a_{ij} dollars in time period *i*, with b_i dollars available in period *i*)? (Also see the knapsack problem, Exercise 10.)

b. Location problems. A company has to decide which of *m* warehouses to build to better serve its *n* customers. The *j*th customer has a known yearly demand d_j for the company's product.   During a year, the ith warehouse will be able to serve a total demand of D_i.   The warehouse at location i will have a fixed overhead cost of F_i if built.   Let c_{ij} be the operating cost plus transportation cost of the ith warehouse shipping one unit of the product to the jth customer.   Formulate this as a mixed-integer variable problem, assuming that all the customer demands can be met and we want to minimize the total cost of the company's distribution system.   (HINT: Let x_{ij} be the amount shipped from warehouse i to customer j.   Define subsidiary variables y_i equal to 0 or 1 depending on whether the ith location is not built or built, respectively.   One set of constraints stipulates that the customer's demands are satisfied, while a second set states that the shipments out of a warehouse cannot exceed its capacity.   The trick here is to turn the ith warehouse's capacity off or on by using the y_i and, in a similar fashion, add the fixed costs into the objective function.)   How would you change the formulation if each customer had to be served by only one warehouse?   (Also see Exercises 28 and 31 of Chap. 11.)

    c.   The following is a problem in *project selection.*   A company is planning to select its research program from projects being proposed by its three research departments.   Each department is given a budget limit and is told that at least one of its projects will be selected.   The total budget of all projects selected must not exceed the company's stated limit.   Also, as research labor-hours and laboratory space are premium items, the total hours and space are restricted.   The data given in the following tableau have been submitted to the company's research director.   Discuss how this problem can be analyzed by a 0-1 integer-programming problem and how the director can use the model in selecting a total research program.   What objective function should the company use (possibilities include: Minimize the cost of the total program, maximize the expected worth of the research results, maximize the number of projects selected, minimize labor or space, maximize labor or space, maximize some weighted sum of the projects)?

	P_1	P_2	P_3	P_4	P_5	P_6	P_7	P_8	P_9	P_{10}	P_{11}	P_{12}	Limit
Hours	50	60	30	50	90	20	70	5	80	40	30	65	≤ 200
Space	10	80	70	100	30	40	15	10	120	60	75	40	≤ 250
Budget-1	100	80	90	100									≤ 300
Budget-2					20	75	50	15					≤ 200
Budget-3									200	60	70	80	≤ 350
Total budget	100	80	90	100	20	75	50	15	200	60	70	80	≤ 700

    43.   Consider a linear-programming problem in which the same linear sum of the variables is bounded from below and above as follows:

$$l_i \leq \sum_{j=1}^{n} a_{ij} x_j \leq u_i$$

Show that these conditions can be transformed to a single upper-bounded constraint and a single upper-bounded variable.   HINT: For the upper-bound condition, form an equality using a slack variable.

PART 3

Applications

CHAPTER 10

The Transportation Problem

One of the earliest and most fruitful applications of linear-programming techniques has been the formulation and solution of the transportation problem as a linear-programming problem. The basic transportation problem was originally stated by Hitchcock [370] and later discussed in detail by Koopmans [436]. An earlier approach was given by Kantorovich [409]. The linear-programming formulation and the associated systematic method for solution were first given in Dantzig [154]. The computational procedure is an adaptation of the simplex method applied to the system of equations of the associated linear-programming problem. In this chapter we shall first discuss the general problem and related theorems which describe important mathematical characteristics, and then develop in terms of the revised simplex method a computational procedure for solving the problem. In addition, a number of applications which can be viewed as transportation problems are also described. As preliminary reading it is suggested that the reader refer to the applications described in Sec. 2 of Chap. 1.

1. THE GENERAL TRANSPORTATION PROBLEM

A homogeneous product is to be shipped in the amounts $a_1, a_2, \ldots, a_m$, respectively, from each of *m shipping origins* and received in amounts $b_1, b_2,$

..., b_n, respectively, by each of *n shipping destinations*. The cost of shipping a unit amount from the ith origin to the jth destination is c_{ij} and is known for all combinations (i, j). The problem is to determine the amounts x_{ij} to be shipped over all routes (i, j) so as to minimize the total cost of transportation.

To develop the constraints of the problem, we set up Tableau 10.1. The amount shipped from origin i to destination j is x_{ij}; the total shipped from origin i is $a_i \geq 0$, and the total received by destination j is $b_j \geq 0$. Here we temporarily impose the restriction that the total amount shipped is equal to the total amount received; that is, $\sum_i a_i = \sum_j b_j = A$. The total cost of shipping x_{ij} units is $c_{ij}x_{ij}$.[†] Since a negative shipment has no valid interpretation for the problem as stated, we restrict each $x_{ij} \geq 0$. From the tableau we have the mathematical statement of the transportation problem: Find values for the variables x_{ij} which minimize the total cost

$$x_{00} = \sum_{i=1}^{m} \sum_{j=1}^{n} c_{ij}x_{ij} \tag{1.1}$$

subject to the constraints

$$\sum_{j=1}^{n} x_{ij} = a_i \qquad i = 1, 2, \ldots, m \tag{1.2}$$

$$\sum_{i=1}^{m} x_{ij} = b_j \qquad j = 1, 2, \ldots, n \tag{1.3}$$

and

$$x_{ij} \geq 0 \tag{1.4}$$

TABLEAU 10.1

Origins \ Destinations	(1)	(2)	...	(j)	...	(n)	
(1)	x_{11}	x_{12}	...	x_{1j}	...	x_{1n}	a_1
(2)	x_{21}	x_{22}	...	x_{2j}	...	x_{2n}	a_2
.	.	.	...	.	...	.	.
.	.	.	...	.	...	.	.
(i)	x_{i1}	x_{i2}	...	x_{ij}	...	x_{in}	a_i
.	.	.	...	.	...	.	.
.	.	.	...	.	...	.	.
(m)	x_{m1}	x_{m2}	...	x_{mj}	...	x_{mn}	a_m
	b_1	b_2	...	b_j	...	b_n	

[†] We are assuming linear cost relationships. Recognizing that the cost function for many transportation problems is not linear, e.g., parcel post rates, the reader should reconcile how this assumption can be invoked for a given application. The c_{ij} could be considered as representing average costs, distances, time, etc.

Equations (1.2) represent the row sums of Tableau 10.1 and (1.3) the column sums.   In order for Eqs. (1.2) and (1.3) to be consistent, we must have the sum of Eqs. (1.2) equal to the sum of Eqs. (1.3); that is,

$$\sum_{i=1}^{m} \sum_{j=1}^{n} x_{ij} = \sum_{j=1}^{n} \sum_{i=1}^{m} x_{ij} = \sum_{i} a_i = \sum_{j} b_j = A$$

We note that the system of Eqs. (1.1) to (1.4) is a linear-programming problem with $m + n$ equations in mn variables.

Theorem 1.   *The transportation problem has a feasible solution.*

Proof: Since $\sum_{i=1}^{m} a_i = \sum_{j=1}^{n} b_j = A$, we have the feasible solution $x_{ij} = a_i b_j / A$ for all (i, j).   Each $x_{ij} \geq 0$, and (1.2) is satisfied, since

$$\sum_{j=1}^{n} x_{ij} = \sum_{j=1}^{n} \frac{a_i b_j}{A} = \frac{a_i \sum_{j=1}^{n} b_j}{A} = a_i$$

and (1.3) is satisfied, since

$$\sum_{i=1}^{m} x_{ij} = \sum_{i=1}^{m} \frac{a_i b_j}{A} = \frac{b_j \sum_{i=1}^{m} a_i}{A} = b_j$$

For $m = 3$ and $n = 5$, let us write out the equations that correspond to (1.2) and (1.3).   We then obtain the following 8 (that is, $m + n$) equations in 15 (that is, mn) unknowns.

$$
\begin{array}{llll}
(a) & x_{11} + x_{12} + x_{13} + x_{14} + x_{15} & & = a_1 \\
(b) & \qquad x_{21} + x_{22} + x_{23} + x_{24} + x_{25} & & = a_2 \\
(c) & \qquad\qquad x_{31} + x_{32} + x_{33} + x_{34} + x_{35} & = a_3 \\
(d) & x_{11} \qquad + x_{21} \qquad + x_{31} & & = b_1 \\
(e) & \quad x_{12} \qquad + x_{22} \qquad + x_{32} & & = b_2 \\
(f) & \quad\ x_{13} \qquad + x_{23} \qquad + x_{33} & & = b_3 \\
(g) & \quad\ x_{14} \qquad + x_{24} \qquad + x_{34} & = b_4 \\
(h) & \quad\ x_{15} \qquad + x_{25} \qquad + x_{35} & = b_5 \\
\end{array}
$$

If we sum Eqs. (d) to (h) and subtract Eqs. (b) and (c) from this sum, the result is Eq. (a).   Hence Eq. (a) is redundant and does not need to be included in the system.   Generalizing, we note that one equation from the system (1.2) or (1.3) can be eliminated, and the transportation problem reduces to $m + n - 1$ independent equations in mn variables.[1]   For the preceding equations we have the following matrix for the reduced system [here we have

---

[1] The reader should be able to prove that the rank of the reduced system is $m + n - 1$ by finding a nonzero determinant of that order; see Eqs. (1.5).   In general, a matrix of order $m + n - 1$ can be formed from the reduced system (eliminating the first equation) by the first n columns P_{11} to P_{1n} and the $m - 1$ columns $P_{21}, \ldots, P_{m1}$.

eliminated the first equation of (1.2)]:

$$\mathbf{A} = \begin{array}{ccccccccccccccc} \mathbf{P}_{11} & \mathbf{P}_{12} & \mathbf{P}_{13} & \mathbf{P}_{14} & \mathbf{P}_{15} & \mathbf{P}_{21} & \mathbf{P}_{22} & \mathbf{P}_{23} & \mathbf{P}_{24} & \mathbf{P}_{25} & \mathbf{P}_{31} & \mathbf{P}_{32} & \mathbf{P}_{33} & \mathbf{P}_{34} & \mathbf{P}_{35} \end{array}$$
$$\mathbf{A} = \begin{pmatrix} 0 & 0 & 0 & 0 & 0 & 1 & 1 & 1 & 1 & 1 & 0 & 0 & 0 & 0 & 0 \\ 0 & 0 & 0 & 0 & 0 & 0 & 0 & 0 & 0 & 0 & 1 & 1 & 1 & 1 & 1 \\ 1 & 0 & 0 & 0 & 0 & 1 & 0 & 0 & 0 & 0 & 1 & 0 & 0 & 0 & 0 \\ 0 & 1 & 0 & 0 & 0 & 0 & 1 & 0 & 0 & 0 & 0 & 1 & 0 & 0 & 0 \\ 0 & 0 & 1 & 0 & 0 & 0 & 0 & 1 & 0 & 0 & 0 & 0 & 1 & 0 & 0 \\ 0 & 0 & 0 & 1 & 0 & 0 & 0 & 0 & 1 & 0 & 0 & 0 & 0 & 1 & 0 \\ 0 & 0 & 0 & 0 & 1 & 0 & 0 & 0 & 0 & 1 & 0 & 0 & 0 & 0 & 1 \end{pmatrix} \qquad (1.5)$$

and

$$\mathbf{P}_0 = \begin{pmatrix} a_2 \\ a_3 \\ b_1 \\ b_2 \\ b_3 \\ b_4 \\ b_5 \end{pmatrix}$$

The vector $\mathbf{P}_{ij}$ corresponds to the variable x_{ij}.   Letting

$$\mathbf{X} = (x_{11}, \ldots, x_{ij}, \ldots, x_{mn})$$

be a column vector with mn variables and $\mathbf{c} = (c_{11}, \ldots, c_{ij}, \ldots, c_{mn})$ be a row vector of cost coefficients, the reduced transportation problem can be represented as follows: Minimize

$$\mathbf{cX}$$

subject to

$$\begin{aligned} \mathbf{AX} &= \mathbf{P}_0 \\ \mathbf{X} &\geq \mathbf{0} \end{aligned} \qquad (1.6)$$

From the discussion in Chaps. 3 and 4, we have that the minimum feasible solution to (1.6) requires at most $m + n - 1$ positive x_{ij}'s.

Theorem 2.   *Construction of a basic feasible solution.*[1]   *A solution of at most $m + n - 1$ positive x_{ij}'s exists.*

---

[1] In the paper by Demuth [198] (translated by Doig [211]), it is shown that if N is the minimum number of basic feasible solutions to a transportation problem with m origins and n destinations ($m \leq n$), then N is bounded as follows:

$$n^{m-1} \leq N \leq n^{m-1}m^{n-1} \qquad \text{(nondegenerate problem)}$$

$$\frac{n!}{(n - m + 1)!} \leq N \leq n^{m-1}m^{n-1} \qquad \text{(degenerate problem)}$$

The upper bounds are due to Simonnard and Hadley [612].

Proof: To demonstrate this theorem, we shall use the procedure for obtaining a first basic feasible solution proposed by Dantzig [154] and termed the "northwest-corner rule" by Charnes and Cooper [111]. We shall apply this scheme to the following 3×4 tableau:

$$
\begin{array}{cccc|c}
x_{11} & x_{12} & x_{13} & x_{14} & a_1 \\
x_{21} & x_{22} & x_{23} & x_{24} & a_2 \\
x_{31} & x_{32} & x_{33} & x_{34} & a_3 \\
\hline
b_1 & b_2 & b_3 & b_4 &
\end{array}
$$

We first determine a value for the northwest-corner variable x_{11}. We let $x_{11} = \min (a_1, b_1)$; and if $a_1 \leq b_1$, then $x_{11} = a_1$, and all $x_{1j} = 0$ for $j = 2, 3, 4$. If $a_1 \geq b_1$, then $x_{11} = b_1$, and all $x_{i1} = 0$ for $i = 2, 3$. For discussion purposes, assume that the former is true; then this initial step transforms the tableau as shown in Step 1 below. Here the total left to be shipped from origin 1 has been reduced to 0, and the total left to be shipped to destination 1 is $b_1 - a_1$.

STEP 1   Assume $b_1 > a_1$.

$$
\begin{array}{cccc|c}
x_{11} = a_1 & 0 & 0 & 0 & 0 \\
x_{21} & x_{22} & x_{23} & x_{24} & a_2 \\
x_{31} & x_{32} & x_{33} & x_{34} & a_3 \\
\hline
b_1 - a_1 & b_2 & b_3 & b_4 &
\end{array}
$$

We next determine a value for the first variable in row 2. We let $x_{21} = \min (a_2, b_1 - a_1)$. If we assume $a_2 > b_1 - a_1$, we have $x_{21} = b_1 - a_1$ and $x_{31} = 0$. This is shown in Step 2. The total left to be shipped from origin 2 is now $a_2 - b_1 + a_1$, and the total to be shipped to destination 1 is 0.

STEP 2   Assume $a_2 > b_1 - a_1$.

$$
\begin{array}{cccc|c}
x_{11} = a_1 & 0 & 0 & 0 & 0 \\
x_{21} = b_1 - a_1 & x_{22} & x_{23} & x_{24} & a_2 - b_1 + a_1 \\
0 & x_{32} & x_{33} & x_{34} & a_3 \\
\hline
0 & b_2 & b_3 & b_4 &
\end{array}
$$

Similarly, for the following steps, we determine a value of a variable x_{ij} and reduce to zero either the amount to be shipped from i or the amount to be shipped to j, or both.

STEP 3   Assume $a_2 - b_1 + a_1 > b_2$.

$$
\begin{array}{cccc|c}
x_{11} = a_1 & 0 & 0 & 0 & 0 \\
x_{21} = b_1 - a_1 & x_{22} = b_2 & x_{23} & x_{24} & a_2 - b_1 + a_1 - b_2 \\
0 & 0 & x_{33} & x_{34} & a_3 \\
\hline
0 & 0 & b_3 & b_4 &
\end{array}
$$

STEP 4   Assume $a_2 - b_1 + a_1 - b_2 < b_3$.

$x_{11} = a_1$	0	0	0	0
$x_{21} = b_1 - a_1$	$x_{22} = b_2$	$x_{23} = a_2 - b_1 + a_1 - b_2$	0	0
0	0	x_{33}	x_{34}	a_3
0	0	$b_3 - a_2 + b_1 - a_1 + b_2$	b_4	

From Step 4, we see that $x_{33} = b_3 - a_2 + b_1 - a_1 + b_2$ and $x_{34} = b_4$. It should be noted that the values for the x_{ij} were all obtained by adding and subtracting various combinations of the a_i and b_j. Hence, if the a_i and b_j were originally nonnegative integers, then the feasible solution obtained by the above procedure would also consist of nonnegative integers. Since the northwest-corner rule of determining each x_{ij} eliminates either a row or a column from further consideration, while the last allocation eliminates both a row and a column, this feasible solution can have at most $m + n - 1$ positive x_{ij}'s. Depending on the values of the a_i and b_j and on the assumptions made in obtaining the feasible solution to the above example, we have as the six variables with possible positive values

$$x_{11} = a_1 \qquad x_{21} = b_1 - a_1$$
$$x_{22} = b_2 \qquad x_{23} = a_2 - b_1 + a_1 - b_2$$
$$x_{33} = b_3 - a_2 + b_1 - a_1 + b_2 \qquad x_{34} = b_4$$

The following argument shows that the feasible solution obtained by the northwest-corner rule is also a basic solution to the reduced transportation problem. We consider the system in explicit equation form and set all variables not in the northwest-corner solution equal to zero. Also, for discussion purposes, we form the reduced system by dropping the last $n + m$ equation from the system, i.e., the equation corresponding to the requirement b_n. We thus have formed an $(m + n - 1) \times (m - 1)$ square system of linear equations. Using the method of Theorem 2, the northwest element x_{11} is uniquely determined in that it is equal to the min (a_1, b_1). From this we note that one of the equations of the square system must contain only x_{11}. We then apply the elimination procedure to x_{11}, eliminating it from the rest of the equations of the system. This step produces a new set of uniquely determined right-hand-side elements. The northwest-corner rule now forces the selection of another variable, either some x_{1i} or some x_{1j} $(i, j \neq 1)$, equal to the corresponding new right-hand-side element. We then eliminate this variable from further consideration and repeat the process, each time being forced to select a unique value for the variable under consideration. Recalling that $\sum_i a_i = \sum_j b_j$, the final northwest-corner step yields a unique value for the variable x_{mn} in equation m. Thus, the complete application of the northwest-corner rule to the $(m + n - 1) \times (m + n - 1)$ system of equations yields a unique solution to the transportation problem with exactly $m +$

$n-1$ positive x_{ij}'s (assuming nondegeneracy).   The columns of the associated $(m+n-1) \times (m+n-1)$ matrix must form a basis as the elimination procedure is equivalent to multiplying the right-hand side by the inverse of the matrix, and a square system of equations has a unique solution if the matrix has an inverse.

We note the northwest-corner rule and the above discussion can be generalized to the selection of any x_{ij} as the initial variable to be evaluated; see Secs. 2*a*, 2*b*, and 2*c*.   We next exhibit two numerical examples and their northwest-corner-rule basic feasible solutions.

Example 1.1

$$
\begin{array}{ccccc|c}
2 & 0 & 0 & 0 & 0 & 2 \\
1 & 2 & 1 & 0 & 0 & 4 \\
0 & 0 & 3 & 2 & 2 & 7 \\
\hline
3 & 2 & 4 & 2 & 2 &
\end{array}
$$

Example 1.2

$$
\begin{array}{ccccc|c}
1 & 2 & 0 & 0 & 0 & 3 \\
0 & 1 & 3 & 0 & 0 & 4 \\
0 & 0 & 0 & 2 & 5 & 7 \\
\hline
1 & 3 & 3 & 2 & 5 &
\end{array}
$$

In Example 1.2, only $m+n-2=6$ of the x_{ij}'s are positive, and we have determined a degenerate basic feasible solution. This will happen whenever the amount left to be shipped from origin i is exactly equal to the amount to be shipped to destination j, except when both $i=m$ and $j=n$.   In Example 1.2 this is true for $x_{23} = \min(a_2 - x_{22}, b_3) = (3, 3) = 3$. (See discussion of degeneracy at the end of this section.)

Associated with each such solution is a set of $m+n-1$ linearly independent vectors.   For Example 1.1, the vectors associated with the positive x_{ij} are $\mathbf{P}_{11}, \mathbf{P}_{21}, \mathbf{P}_{22}, \mathbf{P}_{23}, \mathbf{P}_{33}, \mathbf{P}_{34}, \mathbf{P}_{35}$.   Let the basis formed by these vectors be denoted by $\mathbf{B}$.   Then the solution to the system $\mathbf{BX} = \mathbf{P}_0$ must yield the solution obtained by the northwest-corner rule.   Here $\mathbf{X} = (x_{11}, x_{21}, x_{22}, x_{23}, x_{33}, x_{34}, x_{35})$ is a column vector.   We can similarly select an associated set of $m+n-1$ linearly independent vectors for the degenerate solution of Example 1.2.   This will be discussed below.   The solutions obtained by the northwest-corner rule (and similar schemes) are extreme-point solutions, and only such solutions need be considered as candidates for the minimum feasible solution.

Since virtually all applications of the transportation problem require the shipping of only whole units of the item being considered, it is useful to establish the following important property of the transportation problem:

Theorem 3. *Assuming the a_i and b_j are nonnegative integers, then every basic feasible solution (i.e., extreme-point solution) has integral values.*

We prove Theorem 3 with the aid of the following statement:

Lemma 1. *Every set of $m + n - 1$ linearly independent vectors of the reduced-transportation-problem system of equations can be arranged into a triangular matrix.*[1]

To illustrate this point, let us rearrange the basis of Example 1.1 into a triangular matrix.   The original system given by $\mathbf{BX} = \mathbf{P}_0$ is

$$
\begin{array}{c}
(a)\\(b)\\(c)\\(d)\\(e)\\(f)\\(g)
\end{array}
\begin{pmatrix}
0 & 1 & 1 & 1 & 0 & 0 & 0\\
0 & 0 & 0 & 0 & 1 & 1 & 1\\
1 & 1 & 0 & 0 & 0 & 0 & 0\\
0 & 0 & 1 & 0 & 0 & 0 & 0\\
0 & 0 & 0 & 1 & 1 & 0 & 0\\
0 & 0 & 0 & 0 & 0 & 1 & 0\\
0 & 0 & 0 & 0 & 0 & 0 & 1
\end{pmatrix}
\begin{pmatrix}
x_{11}\\x_{21}\\x_{22}\\x_{23}\\x_{33}\\x_{34}\\x_{35}
\end{pmatrix}
=
\begin{pmatrix}
4\\7\\3\\2\\4\\2\\2
\end{pmatrix}
$$

where the column headers are $\mathbf{P}_{11}\ \mathbf{P}_{21}\ \mathbf{P}_{22}\ \mathbf{P}_{23}\ \mathbf{P}_{33}\ \mathbf{P}_{34}\ \mathbf{P}_{35}$.

To rearrange this $\mathbf{B}$, we need only interchange the rows of $\mathbf{B}$ and corresponding elements of $\mathbf{P}_0$, as shown below.   Since the rearrangement of $\mathbf{B}$ does not require any column interchanges, the elements of vector $\mathbf{X}$ are not permuted.   The transformed system is

$$
\begin{array}{c}
(c)\\(a)\\(d)\\(e)\\(b)\\(f)\\(g)
\end{array}
\begin{pmatrix}
1 & 1 & 0 & 0 & 0 & 0 & 0\\
0 & 1 & 1 & 1 & 0 & 0 & 0\\
0 & 0 & 1 & 0 & 0 & 0 & 0\\
0 & 0 & 0 & 1 & 1 & 0 & 0\\
0 & 0 & 0 & 0 & 1 & 1 & 1\\
0 & 0 & 0 & 0 & 0 & 1 & 0\\
0 & 0 & 0 & 0 & 0 & 0 & 1
\end{pmatrix}
\begin{pmatrix}
x_{11}\\x_{21}\\x_{22}\\x_{23}\\x_{33}\\x_{34}\\x_{35}
\end{pmatrix}
=
\begin{pmatrix}
3\\4\\2\\4\\7\\2\\2
\end{pmatrix}
$$

where the column headers are $\mathbf{P}_{11}\ \mathbf{P}_{21}\ \mathbf{P}_{22}\ \mathbf{P}_{23}\ \mathbf{P}_{33}\ \mathbf{P}_{34}\ \mathbf{P}_{35}$.

From the system of equations determined by the above triangular system, we see immediately that $x_{35} = 2$ and $x_{34} = 2$.   Then by substituting these values in Eq. (b), we have $x_{33} = 3$.   In a similar manner we solve in order Eqs. (e), (d), (a), and (c) to obtain $x_{23} = 1$, $x_{22} = 2$, $x_{21} = 1$, and $x_{11} = 2$. Since the matrix associated with a basic feasible solution is triangular and consists of elements that are either 1 or 0 and since the solution of a set of equations determined by this type of triangular matrix requires only additions and subtractions, the values of the variables will be nonnegative integers.

[1] See Exercise 6.

Theorem 4. *A finite minimum feasible solution always exists.*

Proof: By Theorem 1 the problem is feasible. Since the coefficients of Eqs. (1.2) and (1.3) are nonnegative and since we assume that all a_i and b_j are nonnegative and finite, no x_{ij} can be made arbitrarily large. In fact, the x_{ij} cannot be greater than the corresponding a_i or b_j. (We may note that, for any basic feasible solution, at least one x_{ij} equals its corresponding a_i or b_j.) If the a_i and b_j are integral, then a basic minimum feasible solution has finite integral values.

The reduced system of $m + n - 1$ equations in mn variables can, of course, be solved by the general simplex procedure. However, for even small values of m and n, the resulting system of equations becomes unwieldy for manual computation. Such systems could possibly be too large to be solved efficiently on computing machinery. This dilemma is resolved by the special adaptation of the simplex algorithm to the transportation problem. This iterative procedure requires the knowledge of exactly $m + n - 1$ nonnegative variables that are associated with $m + n - 1$ linearly independent vectors. We know that the vectors that are associated with the variables of a basic feasible solution (i.e., an extreme-point solution) are linearly independent. If the basic solution is nondegenerate, it satisfies the requirement of the procedure. This is not true if the solution is degenerate, as is the solution for Example 1.2. If a degenerate solution has $k < m + n - 1$ positive variables, then we have to select $m + n - 1 - k$ zero variables to be in the solution. The $m + n - 1 - k$ vectors associated with these zero variables plus the k vectors associated with the positive variables must be linearly independent. This selection is accomplished by the following "ϵ perturbation" method for the transportation problem (Dantzig [154]).

Let us refer to the 3×4 tableaus used to obtain a first feasible solution by the northwest-corner rule. In Step 1, if $x_{21} = b_1 - a_1 = a_2$, then neither x_{31} nor x_{22} could be positive. Or in Step 2, if

$$x_{22} = b_2 = a_2 - b_1 + a_1$$

then neither x_{32} nor x_{23} could be positive. Whenever situations of this sort arise, the number of positive variables in the basic solution is reduced by one. These degenerate cases arise when, in evaluating some x_{ij}, its two possible values are equal. We have then encountered a partial sum of the a_i and b_j which is equal to some a_i or b_j. In Example 1.2 we have a degenerate solution occurring because $x_{23} = \min\ (b_3,\ a_2 - b_2 + a_1 - b_1)$ and $b_3 = a_2 - b_2 + a_1 - b_1 = 3$.

To avoid these degenerate situations, we have to make sure that no partial sums of the a_i and b_j can be equal to some a_i or b_j. This is done by "perturbing" (i.e., modifying) the values of the a_i and b_j. We set up a new problem where

$$\bar{a}_i = a_i + \epsilon \qquad i = 1, 2, \ldots, m$$

and

$$\bar{b}_j = b_j \qquad\qquad j = 1, 2, \ldots, n-1$$
$$\bar{b}_n = b_n + m\epsilon$$

for $\epsilon > 0$.

We select an ϵ small enough that the final solution rounded to the same number of significant digits as the original a_i and b_j will yield the correct solution. This is possible since the computational procedure involves only additions and subtractions. Orden [536] proves that any $\epsilon < \delta/2m$ will do, where δ is a 1 in the last significant place in the quantities a_i and b_j. In practice, ϵ is taken to be 1×10^{-d}, where d is determined by $\delta/2m$. (For machine computation, it is easier always to add a 1 to each a_i and to add m to b_n in the last significant place that can be carried in the computer.)

Applying the ϵ procedure to Example 1.2, we have the solution

$$
\begin{array}{ccccc|c}
1 & 2+\epsilon & 0 & 0 & 0 & 3+\epsilon \\
0 & 1-\epsilon & 3 & 2\epsilon & 0 & 4+\epsilon \\
0 & 0 & 0 & 2-2\epsilon & 5+3\epsilon & 7+\epsilon \\
\hline
1 & 3 & 3 & 2 & 5+3\epsilon &
\end{array}
$$

The basic solution now contains $x_{24} = 2\epsilon > 0$. We see that the ϵ determines which of two variables with real values of zero should be brought into the solution. Here we had a choice of x_{24} or x_{33}. The ϵ procedure does away with any such choice and enables the computation to proceed without any degenerate solutions. Theoretically, the selection of either x_{24} or x_{33} would have yielded a basis of $m+n-1$ vectors. As will be seen, the computational procedure requires that we need only note which of these zero variables is selected to be in the basic solution. We can then dispense with the ϵ procedure and select either one of the two zero variables. It seems best to select the one with the smallest c_{ij}. No transportation problem has ever been known to cycle when solved using the variable selection rules of the simplex algorithm; see Sec. 2 and Exercise 20. (For further discussion of basic feasible solutions and degeneracy in the transportation problem, see Exercises 54 and 55, Chap. 11.)

2. COMPUTATIONAL PROCEDURE FOR SOLVING THE TRANSPORTATION PROBLEM

As noted earlier, we can solve the transportation problem by direct application of the simplex algorithm. However, for reasonable size values of m origins and n destinations the dimensions of the resultant problem of $m+n-1$ equations and mn variables become rather large, and it pays to investigate how one can simplify the computational load. As was the case for large-scale decomposition-type problems, this suggests looking at the special structure of the transportation-problem matrix to see if it can be used to any advantage.   We have already shown that a basic solution has a triangular matrix structure.   We use this fact, combined with the procedure of the revised simplex algorithm and duality considerations, to develop a computationally efficient version of the simplex algorithm as applied to the transportation problem.

We first consider the $m + n$ equation transportation problem as given by Eqs. (1.1) and (1.4) as the primal problem of an unsymmetric primal-dual problem set (see Sec. 1 of Chap. 6).   For this problem it is convenient to consider the set of $m + n$ dual variables $\mathbf{W}$ as consisting of two subsets of variables $(u_1, \ldots, u_m)$ and $(v_1, \ldots, v_n)$, i.e., $\mathbf{W} = (u_1, \ldots, u_m, v_1, \ldots, v_n)$ and is a row vector.   The maximizing dual problem of the transportation problem is then obtained by premultiplying the matrix of the primal problem by $\mathbf{W}$.   For the (3×5) problem used to illustrate Theorem 1, the corresponding dual problem is to: Maximize

$$a_1 u_1 + a_2 u_2 + a_3 u_3 + b_1 v_1 + b_2 v_2 + b_3 v_3 + b_4 v_4 + b_5 v_5$$

subject to

$$
\begin{aligned}
u_1 & & + v_1 & & & & & & & \leq c_{11} \\
u_1 & & & + v_2 & & & & & & \leq c_{12} \\
u_1 & & & & + v_3 & & & & & \leq c_{13} \\
u_1 & & & & & + v_4 & & & & \leq c_{14} \\
u_1 & & & & & & + v_5 & & & \leq c_{15} \\
u_2 & & + v_1 & & & & & & & \leq c_{21} \\
u_2 & & & + v_2 & & & & & & \leq c_{22} \\
u_2 & & & & + v_3 & & & & & \leq c_{23} \\
u_2 & & & & & + v_4 & & & & \leq c_{24} \\
u_2 & & & & & & + v_5 & & & \leq c_{25} \\
u_3 & + v_1 & & & & & & & & \leq c_{31} \\
u_3 & & + v_2 & & & & & & & \leq c_{32} \\
u_3 & & & + v_3 & & & & & & \leq c_{33} \\
u_3 & & & & + v_4 & & & & & \leq c_{34} \\
u_3 & & & & & + v_5 & & & & \leq c_{35}
\end{aligned}
$$

with all u_i and v_j unrestricted.   In general, these constraints can be written as

$$u_i + v_j \leq c_{ij} \qquad \text{for all } (i, j) \tag{2.1}$$

Letting $f(\mathbf{X})$ be the value of the primal objective function and $g(\mathbf{W})$ the value of the dual objective function and applying complementary-slackness-like arguments to the primal-dual problems (see Sec. 2 of Chap. 6), we have for the transportation problem the relationship

$$\sum_i \sum_j (c_{ij} - u_i - v_j) x_{ij} = f(\mathbf{X}) - g(\mathbf{W}) \geq 0 \tag{2.2}$$

for any $\mathbf{X}$ and $\mathbf{W}$ which satisfy their corresponding constraints.[1] We note that $x_{ij} \geq 0$, that all $c_{ij} - u_i - v_j \geq 0$, and that (2.2) is the sum of products of nonnegative numbers. Since $\min f(\mathbf{X}) \geq \max g(\mathbf{W})$, and as we know an optimal bounded solution to the transportation problem exists, solutions $\mathbf{X}$ and $\mathbf{W}$ which cause (2.2) to be equal to zero will be optimal solutions to their corresponding problems. We shall be using this property to prove the optimality of the solution obtained by the simplex algorithm, as adapted to the transportation problem. Since we wish to work with an $m + n - 1$ feasible basis to the transportation problem, i.e., an extreme-point solution, we shall drop the first equation of the primal from further consideration, which corresponds to setting $u_1 = 0$ in the dual. The dual then has $m + n - 1$ unrestricted variables in mn constraints.

Assume we are given a first basic feasible solution $\mathbf{X}_0$ to the transportation problem, e.g., the northwest-corner solution. We wish to determine if $\mathbf{X}_0$ is an optimal solution and if not, which variable (vector) to introduce into the new solution. Let $\mathbf{B}$ be the corresponding basis which consists of $m + n - 1$ vectors $\mathbf{P}_{ij}$ of the reduced system, and let $\mathbf{c}_0$ be a row vector of cost coefficients corresponding to those $\mathbf{P}_{ij}$ in the basis. The revised simplex algorithm requires the computation of the pricing vector defined as $\mathbf{\Pi} = \mathbf{c}_0 \mathbf{B}^{-1}$. To solve for the vector $\mathbf{\Pi}$, we investigate the corresponding $(m + n - 1) \times (m + n - 1)$ system of equations

$$\mathbf{\Pi B} = \mathbf{c}_0 \tag{2.3}$$

Let $\mathbf{\Pi} = (u_2, \ldots, u_m, v_1, \ldots, v_n)$, where here $\mathbf{\Pi}$ replaces the general dual-variable vector $\mathbf{W}$ which has $u_1 = 0$. Then the $m + n - 1$ equations represented by (2.3) can be written as

$$u_i + v_j = c_{ij} \qquad \text{for all } (i, j) \in \mathbf{B} \tag{2.4}$$

This is a set of $m + n - 1$ equations in $m + n - 1$ unrestricted variables. Since $\mathbf{B}$ is a triangular matrix, we can readily solve for all u_i and v_j without explicit knowledge of $\mathbf{B}^{-1}$. The computation involves only additions and subtractions and is described below. Given the u_i and v_j which satisfy (2.4), i.e., the $\mathbf{\Pi}$ which satisfies (2.3), we can next determine if the given $\mathbf{X}_0$ is optimal. The revised simplex algorithm calls for the computation of the $z_j - c_j$ elements by premultiplying a vector not in the basis by $\mathbf{\Pi}$ and subtracting the corresponding cost coefficient. For the transportation problem, this product term, the *indirect cost*, is $\bar{c}_{ij} = \mathbf{\Pi P}_{ij} = (u_2, \ldots, u_m, v_1, \ldots, v_n)\mathbf{P}_{ij}$. In general, as $\mathbf{P}_{ij}$ is a vector with unity in the ith and jth positions and zero elsewhere, we have

---

[1] The right-hand side of (2.2) holds by the duality theorem and $f(\mathbf{X}) = \sum_i \sum_j c_{ij} x_{ij}$. By multiplying Eqs. (1.2) by the corresponding u_i and (1.3) by the corresponding v_j and summing over all i and j, we have $\sum_i \sum_j u_i x_{ij} + \sum_i \sum_j v_j x_{ij} = \sum_i u_i a_i + \sum_j v_j b_j = g(\mathbf{W})$. The left-hand side of (2.2) is obtained by subtracting the expression of $g(\mathbf{W})$ in terms of the x_{ij} from that of $f(\mathbf{X})$ and factoring.

$\bar{c}_{ij} = u_i + v_j$ for all (i, j), with $u_1 = 0$. This important relationship can thus be readily calculated for $\mathbf{P}_{ij}$ once Π is determined. For those vectors not in the basis, the simplex algorithm requires the calculation of the indirect cost minus the direct cost, i.e., the set of $\bar{c}_{ij} - c_{ij} = u_i + v_j - c_{ij}$. If some $\bar{c}_{ij} - c_{ij} > 0$, then the current solution $\mathbf{X}_0$ is not optimal and we select the variable x_{ij} to enter the solution which corresponds to the max $(\bar{c}_{ij} - c_{ij}) > 0$. If all $\bar{c}_{ij} - c_{ij} \leq 0$, then $\mathbf{X}_0$ is optimal by the theory of Chaps. 4 and 5. Also, if all $\bar{c}_{ij} - c_{ij} = u_i + v_j - c_{ij} \leq 0$, i.e., $c_{ij} - u_i - v_j \geq 0$ and thus the dual constraints (2.1) are satisfied, and if the only $x_{ij} > 0$ are those which are in the current basis and hence must have their corresponding $c_{ij} - u_i - v_j = 0$, then (2.2) will sum to zero. By the duality theory of Chap. 6, the x_{ij} forms an optimal feasible solution to the primal transportation problem and the corresponding u_i and v_j will form an optimal feasible solution to its dual.

Assuming that the given solution $\mathbf{X}_0$ is not optimal, we next have to effect a change in the basis as dictated by the simplex algorithm. Let max $(\bar{c}_{ij} - c_{ij}) = \bar{c}_{pq} - c_{pq} > 0$. Variable x_{pq}, which is currently equal to zero, should be introduced into the solution and made as large as possible. This will cause a decrease in the value of the objective function equal to $x_{pq}(\bar{c}_{pq} - c_{pq})$. To make the corresponding basis change, the simplex algorithm requires the expression of the new vector entering the basis, here $\mathbf{P}_{pq}$, in terms of the current basis vectors. We denote this expression by $\mathbf{X}_{pq}$, and it is given by the $(m + n - 1) \times (m + n - 1)$ set of equations

$$\mathbf{B}\mathbf{X}_{pq} = \mathbf{P}_{pq} \tag{2.5}$$

As noted in Lemma 1, $\mathbf{B}$ can be arranged into a triangular matrix with positive ones along the main diagonal, and thus $|\mathbf{B}| = 1$. Solving (2.5) for the $m + n - 1$ components of $\mathbf{X}_{pq}$ by Cramer's rule of determinants[1] requires the evaluation of $m + n - 1$ determinants obtained by successively substituting $\mathbf{P}_{pq}$ for each of the columns of $\mathbf{B}$ and dividing each value by $|\mathbf{B}| = 1$. Each of the determinants will have a value of $+1$, -1, or 0. This is seen as follows. If all the columns of a transportation problem determinant have two ones, then the value of the determinant is zero as the first $(m - 1)$ rows and the last n rows sum to rows containing all ones (see the properties of determinants, Sec. 1, Chap. 2). This situation can arise when the vector $\mathbf{P}_{pq}$ is substituted for a column of $\mathbf{B}$ which has only a single one. (Lemma 1 shows that there must be at least one such column in $\mathbf{B}$.) If the determinant has at least one column with exactly a single one, we can expand the determinant along this column in terms of cofactors of order $m + n - 2$. There will be only one term in the expansion. Following the discussion above, this determinant of lower order has a value of zero or can be expanded along a column which has only a single one. Continuing in this fashion, the final expansion leads to a value of the original determinant being equated to a (1×1) determinant with a value of $+1$, -1, or 0, with the sign

---

[1] See Sec. 5, Chap. 2.

depending on the row-column location of the nonzero coefficients of each expansion. Thus, we see that the components of $\mathbf{X}_{pq}$ are either $+1$, -1, or 0.

To introduce $\mathbf{P}_{pq}$ into the basis in a manner which preserves the non-negativity of the basis variables, we have to consider the following. The current solution is given by

$$\mathbf{B}\mathbf{X}_0 = \mathbf{P}_0 \tag{2.6}$$

where $\mathbf{P}_0$ is the reduced $m + n - 1$ right-hand-side column vector. Multiplying (2.5) by $\theta \geq 0$ and subtracting from (2.6), we obtain

$$\mathbf{B}(\mathbf{X}_0 - \theta\mathbf{X}_{pq}) + \theta\mathbf{P}_{pq} = \mathbf{P}_0 \tag{2.7}$$

Equation (2.7) is equivalent to Eq. (3.3) of Chap. 3, and it is used to determine a value of the new variable $x_{pq} = \theta$ so that a column of $\mathbf{B}$ is replaced by $\mathbf{P}_{pq}$ and the new solution with x_{pq} is feasible.

Since the components of $\mathbf{X}_{pq}$ are $+1$, -1, or 0, and we must determine θ such that all components of

$$(\mathbf{X}_0 - \theta\mathbf{X}_{pq}) \geq 0 \tag{2.8}$$

we see that θ is limited by those components of (2.8) for which the corresponding component of $\mathbf{X}_{pq}$ is $+1$. To make θ as large as possible and preserve feasibility, θ is then equal to the minimum component of $\mathbf{X}_0$ which corresponds to a component of $\mathbf{X}_{pq}$ that is $+1$.

In order to determine the components of $\mathbf{X}_{pq}$ in an efficient manner without recourse to the calculation of $\mathbf{B}^{-1}$ and the explicit solution of (2.5), we rely on the structure of the transportation equations. Thus given that variable $x_{pq} = \theta$ is to enter the solution, we need only consider the set of restricted transportation equations given by the $(m + n - 1) \times (m + n)$ system

$$\mathbf{B}\mathbf{X} + \theta\mathbf{P}_{pq} = \mathbf{P}_0$$

or

$$\mathbf{B}\mathbf{X} = \mathbf{P}_0 - \theta\mathbf{P}_{pq} \tag{2.9}$$

As in the ordinary simplex method, the incoming variable θ replaces a variable of $\mathbf{X}$ by assuming a value which causes the outgoing variable to become zero. Recall that $\mathbf{P}_{pq}$ has, in general, a $+1$ in the pth and qth positions and zero everywhere else. Treating θ as a parameter, we subtract it from the pth and qth positions of $\mathbf{P}_0$ and solve the resulting square transportation problem (2.9) in terms of θ. Because of the triangular nature of $\mathbf{B}$, the solution of (2.9) is straightforward, and it yields the terms $(\mathbf{X}_0 - \theta\mathbf{X}_{pq})$ of Eq. (2.7); i.e., the solution to (2.9), which is equivalent to multiplication by $\mathbf{B}^{-1}$, gives a new solution vector $\bar{\mathbf{X}}_0 = \mathbf{X}_0 - \theta\mathbf{X}_{pq} \geq 0$. Selecting θ as described above makes one component of this $\bar{\mathbf{X}}_0$ equal to zero (assuming nondegeneracy) and yields a value of $\theta = x_{pq}$ equal to the smallest value of those components of $\mathbf{X}_0$ which correspond to positive components of $\mathbf{X}_{pq}$. The systemization of this computation is accomplished readily, and we next illustrate the routinization of the above discussion by solving a sample 3×4 problem.

In the problem we have three origins, with availabilities of $a_1 = 6$, $a_2 = 8$, and $a_3 = 10$, and four destinations, with requirements of $b_1 = 4$, $b_2 = 6$, $b_3 = 8$, and $b_4 = 6$. We note that $\sum_i a_i = \sum_j b_j = 24$. The costs between each origin and destination are given in the following cost matrix:

$$
\text{Origins}\;\begin{array}{c}\\1\\2\\3\end{array}
\begin{array}{|c|c|c|c|}
\hline
1 & 2 & 3 & 4 \\
\hline
4 & 3 & 2 & 0 \\
\hline
0 & 2 & 2 & 1 \\
\hline
\end{array} = (c_{ij})
$$

(Destinations 1 2 3 4 across the top.)

For example, the cost for shipping one unit between origin 3 and destination 2 is $c_{32} = 2$. In general, the c_{ij} can be any positive or negative numbers. Using the northwest-corner rule, we obtain the first feasible solution:

$$
(x_{ij}) =
\begin{array}{|c|c|c|c|}
\hline
4 & 2 & & \\
\hline
& 4 & 4 & \\
\hline
& & 4 & 6 \\
\hline
\end{array}
\begin{array}{c}6\\8\\10\end{array}
$$

with column totals $4\quad 6\quad 8\quad 6$.

where $x_{11} = 4$, $x_{12} = 2$, $x_{22} = 4$, $x_{23} = 4$, $x_{33} = 4$, $x_{34} = 6$, and all the other $x_{ij} = 0$. The value of the objective function $x_{00} = 42$. Since this is a nondegenerate basic feasible solution, the first solution does not require our perturbing the problem. As we are doing a simple hand computation, it is easier to resolve degenerate situations by not employing the technique at all. Instead, when a degenerate solution occurs, we shall select, from the variables that can be in the solution with a value of zero, the ones having the smallest c_{ij}. The example will illustrate this point. We shall always keep track of a solution with exactly $m + n - 1$ nonnegative variables. If the problem were to be solved on a computer, it would be more efficient to start the computation with the perturbed problem.

We are next required to determine, for those variables in the basic solution, m numbers u_i and n numbers v_j such that

$$
\begin{aligned}
u_1 + v_1 &= c_{11} = 1 \\
u_1 + v_2 &= c_{12} = 2 \\
u_2 + v_2 &= c_{22} = 3 \\
u_2 + v_3 &= c_{23} = 2 \\
u_3 + v_3 &= c_{33} = 2 \\
u_3 + v_4 &= c_{34} = 1
\end{aligned}
\qquad (2.10)
$$

Here we have seven variables in six (that is, $m + n - 1$) equations. Since (2.10) is an underdetermined set of linear equations (i.e., the number of unknowns exceeds the number of equations), this system has an infinite number of solutions. We determine a solution by arbitrarily letting any one of the variables equal its corresponding c_{ij}. This reduces the number of unknowns by one and forces a unique solution of the $m + n - 1$ equations in the remaining $m + n - 1$ variables.

Following the discussion above, in (2.10) we let $u_1 = 0$; then it is an easy matter to solve for the remaining u_i and v_j. We see that, with $u_1 = 0$, then $v_1 = 1$ and $v_2 = 2$; with $v_2 = 2$ then $u_2 = 1$; with $u_2 = 1$ then $v_3 = 1$; with $v_3 = 1$ then $u_3 = 1$; and finally, with $u_3 = 1$ then $v_4 = 0$. This computation can be readily accomplished by setting up the following table containing the cost coefficients (in bold type) of the variables in the basic solution:

u \ v				
	1	**2**		
		3	**2**	
			2	**1**

By letting $u_1 = 0$, we can compute, as was done above, the resulting u_i and v_j and enter them in the corresponding positions as follows:

u \ v	1	2	1	0
0	**1**	**2**		
1		**3**	**2**	
1			**2**	**1**

Since all the equations of (2.10) are satisfied, we have $u_i + v_j = c_{ij}$ for those x_{ij} in the basic feasible solution. We then compute $\bar{c}_{ij} = u_i + v_j$ for all combinations (i, j) and place these figures in their corresponding cells of the indirect-cost table. ($\bar{c}_{ij} = c_{ij}$ for all x_{ij} in the solution.) The indirect-cost table ($\bar{c}_{ij}$) for the first solution, along with a copy of the direct-cost table (c_{ij}), follows:[1]

$$(\bar{c}_{ij}) = \begin{array}{|c|c|c|c|}
\hline
1 & 2 & 1 & 0 \\
\hline
2 & 3 & 2 & 1 \\
\hline
2 & 3 & 2 & 1 \\
\hline
\end{array}
\qquad
\begin{array}{|c|c|c|c|}
\hline
1 & 2 & 3 & 4 \\
\hline
4 & 3 & 2 & 0 \\
\hline
0 & 2 & 2 & 1 \\
\hline
\end{array} = (c_{ij})$$

For example, $\bar{c}_{14} = u_1 + v_4 = 0$. As will be shown below, the above three steps can be combined into one efficient computational step. We next compute the differences $\bar{c}_{ij} - c_{ij}$. If all $\bar{c}_{ij} - c_{ij} \leq 0$, then the solution that yielded the indirect-cost table is a minimum feasible solution. If at least one $\bar{c}_{ij} - c_{ij} > 0$, we have not found a minimum solution. We can, as will be described below, readily obtain a new basic feasible solution which contains a variable associated with a $\bar{c}_{ij} - c_{ij} > 0$. As in the general simplex procedure, we select for entry into the new basis the variable corresponding to $\bar{c}_{pq} - c_{pq} = \max\ (\bar{c}_{ij} - c_{ij} > 0)$. This scheme will yield a new basic solution whose value of the objective function will be less than the value for the preceding solution (or possibly equal to it, if the preceding solution is degenerate).

For our first solution, we find that $\bar{c}_{pq} - c_{pq} = \bar{c}_{31} - c_{31} = 2$. Hence we select $x_{pq} = x_{31}$ to be introduced into the solution. If there were ties, we would select the variable with the smaller c_{ij}. We now wish to determine a new basic feasible solution which contains x_{31}. We first illustrate the process of obtaining the new solution by direct reference to Eq. (2.9) and then show how this process is further simplified. From the first solution matrix (x_{ij}) and following the discussion for solving Eq. (2.9) with $x_{pq} = x_{31} = \theta \geq 0$, we now want to solve the transportation problem with origin and destination quantities given by

[1] A concise way of combining the allocation table and the indirect-cost table is to form the following tableau (shown for a 3×4 problem):

u_i \ v_j	v_1	v_2	v_3	v_4	
u_1	c_{11} (x_{11})	c_{12} (x_{12})	c_{13} $\bar{c}_{13} - c_{13}$	c_{14} $\bar{c}_{14} - c_{14}$	a_1
u_2	c_{21} $\bar{c}_{21} - c_{21}$	c_{22} (x_{22})	c_{23} (x_{23})	c_{24} $\bar{c}_{24} - c_{24}$	a_2
u_3	c_{31} $\bar{c}_{31} - c_{31}$	c_{32} $\bar{c}_{32} - c_{32}$	c_{33} (x_{33})	c_{34} (x_{34})	a_3
	b_1	b_2	b_3	b_4	x_{00}

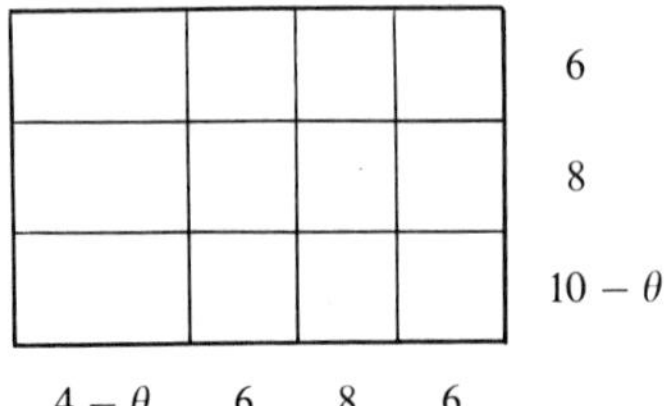

with the values of the nonzero x_{ij}'s restricted to those in the first solution. This is equivalent to solving an $m + n - 1$ nonsingular triangular system whose right-hand side is that of Eq. (2.9). We do this in a manner similar to the northwest-corner rule, treating θ as a nonnegative number. We start the process by first determining the largest possible value of a variable which appears by itself in either a row or a column, here x_{11} or x_{34}. The values of the remaining variables are obtained by selecting values which eliminate a row or column from further consideration, each time preserving the row or column balance. For the first restricted problem we have the new values of the variables in terms of θ given by

$4 - \theta$	$2 + \theta$			6
	$4 - \theta$	$4 + \theta$		8
		$4 - \theta$	6	$10 - \theta$
$4 - \theta$	6	8	6	

We see that the size of θ is restricted by those x_{ij} from which it is subtracted, i.e., we have determined the conditions of inequality (2.8). θ cannot be larger than the smallest x_{ij} from which it is subtracted. Here θ must be less than or equal to 4 and must be greater than zero in order to preserve feasibility. Since we wish to eliminate one of the variables from the old solution and to introduce x_{31}, we let $\theta = 4$. However, since $x_{11} - \theta = x_{22} - \theta = x_{33} - \theta = 4 - \theta$, we would eliminate three variables and obtain a degenerate solution with four positive variables. To keep a solution with exactly $m + n - 1$ nonnegative variables, we retain two of these three variables with values of zero. We select x_{11} and x_{33}, because they correspond to the smaller c_{ij}. The new solution with $\theta = x_{31} = 4$ is

$$(x_{ij}) = \begin{array}{|c|c|c|c|c|}
\hline
0 & 6 & & & 6 \\
\hline
 & & 8 & & 8 \\
\hline
4 & & 0 & 6 & 10 \\
\hline
4 & 6 & 8 & 6 & \\
\end{array}$$

The objective function for this solution[1] is equal to

$$42 - [\max\,(\bar{c}_{ij} - c_{ij} > 0)]\theta = 42 - (2)(4) = 34$$

The corresponding combined $(u_i,\ v_j)$ table and $\bar{c}_{ij}$ matrix for this solution is given in the following table:

$(\bar{c}_{ij}) =$

$u \backslash v$	1	2	3	2
0	**1**	**2**	**3**	2
-1	0	1	**2**	**1**
-1	**0**	1	**2**	**1**

1	2	3	4
4	3	2	0
0	2	2	1

$= (c_{ij})$

where the bold numbers correspond to the c_{ij} for the basis vectors. We see that $\max\,(\bar{c}_{ij} - c_{ij} > 0) = \bar{c}_{24} - c_{24} = 1$ and we next introduce $x_{24} = \theta \geq 0$ into cell (2, 4). Instead of resorting to the analysis that required the solution of Eq. (2.9), as was done in the first solution change above, we simplify the process by using the procedure described in the footnote below. Thus we use the old solution tableau to introduce θ into the appropriate cell positions, and for the current step we add and subtract $\theta = x_{24}$ from the previous x_{ij} as follows:

0	6			6
		$8 - \theta$	θ	8
4		$0 + \theta$	$6 - \theta$	10
4	6	8	6	

[1] The above analysis is equivalent to adding θ in position (3, 1) and then adding and subtracting θ from the current solution values x_{ij} in order to maintain the necessary equation balances, as shown in the following tableau:

$4 - \theta$	$2 + \theta$			6
	$4 - \theta$	$4 + \theta$		8
θ		$4 - \theta$	6	10
4	6	8	6	

Note that the tableau expresses the feasible solution given by Eq. (2.7) as $(4 - \theta)\mathbf{P}_{11} + (2 + \theta)\mathbf{P}_{12} + (4 - \theta)\mathbf{P}_{22} + (4 + \theta)\mathbf{P}_{23} + (4 - \theta)\mathbf{P}_{33} + 6\mathbf{P}_{34} + \theta\mathbf{P}_{31} = \mathbf{P}_0$. From this we see that $\mathbf{X}_{pq} = \mathbf{X}_{31} = \mathbf{P}_{11} - \mathbf{P}_{12} + \mathbf{P}_{22} - \mathbf{P}_{23} + \mathbf{P}_{33}.$

We have $\theta = 6$; x_{34} is eliminated, and x_{24} is introduced with a value of $\theta = 6$. The new value of the objective function is

$$34 - (\bar{c}_{24} - c_{24})\theta = 34 - (1)(6) = 28$$

The new solution is

$$(x_{ij}) = \begin{array}{|c|c|c|c|c}
\hline
0 & 6 & & & 6 \\
\hline
 & & 2 & 6 & 8 \\
\hline
4 & & 6 & & 10 \\
\hline
\end{array}$$

$$\begin{array}{cccc} 4 & 6 & 8 & 6 \end{array}$$

The corresponding (u_i, v_j) table and $(\bar{c}_{ij})$ matrix are given by

$(\bar{c}_{ij}) =$

u \ v	1	2	3	1
0	**1**	**2**	3	1
-1	0	1	2	**0**
-1	**0**	1	2	0

1	2	3	4
4	3	2	0
0	2	2	1

$= (c_{ij})$

Here, all $\bar{c}_{ij} - c_{ij} \leq 0$, and this last solution ($x_{11} = 0$, $x_{12} = 6$, $x_{23} = 2$, $x_{24} = 6$, $x_{31} = 4$, $x_{33} = 6$, and all other $x_{ij} = 0$) is a degenerate minimum feasible solution. The value of the objective function is 28. We note that $\bar{c}_{13} - c_{13} = 0$ and that x_{13} is not in the solution. We can then introduce x_{13} into this last solution and obtain an alternative minimum solution. We put $\theta \geq 0$ into cell $(1, 3)$ and obtain

$$\begin{array}{|c|c|c|c|c}
\hline
0 - \theta & 6 & \theta & & 6 \\
\hline
 & & 2 & 6 & 8 \\
\hline
4 + \theta & & 6 - \theta & & 10 \\
\hline
\end{array}$$

$$\begin{array}{cccc} 4 & 6 & 8 & 6 \end{array}$$

We then have $x_{13} = \theta = 0$ and a new degenerate basic minimum solution as follows:

$$(x_{ij}) = \begin{array}{|c|c|c|c|c}
\hline
 & 6 & 0 & & 6 \\
\hline
 & & 2 & 6 & 8 \\
\hline
4 & & 6 & & 10 \\
\hline
\end{array}$$

$$\begin{array}{cccc} 4 & 6 & 8 & 6 \end{array}$$

Of course, the value of the objective function is 28. For this multiple solution the value of the objective function did not change because $\bar{c}_{13} - c_{13} = 0$, while the values of the variables did not change because $\theta = 0$.

The solution with all $\bar{c}_{ij} - c_{ij} \leq 0$ is optimal by the following discussion. We now have $\bar{c}_{ij} - c_{ij} \geq 0$ or $c_{ij} - u_i - v_j \geq 0$ for all (i, j). A $c_{ij} - u_i - v_j > 0$ has an $x_{ij} = 0$, as it is not a variable in the basic feasible solution. Also, an $x_{ij} > 0$ has its corresponding $c_{ij} - u_i - v_j = 0$. Thus, we see that the final u_i and v_j satisfy the dual-problem constraints, the final x_{ij} satisfy the primal-problem constraints, and they cause the complementary-slackness condition [Eq. (2.2)] to be equal to zero. The duality theorem states that the u_i and v_j are optimal values to the dual variables, while the x_{ij} are the optimal values of the primal variables.

The optimality of these solutions can also be seen from the following. As a constant can be added or subtracted from each element of a row or column in the cost matrix without changing the optimal allocation (see Exercise 4), the final set of u_i and v_j elements can be subtracted from the corresponding rows or columns of the original cost table. This results in a cost table whose elements are either zero or positive, that is,

$$c_{ij} - u_i - v_j = c_{ij} - \bar{c}_{ij} \geq 0$$

When this transformed cost table is used for our transportation problem, an optimal solution would, of course, be one that allocated shipments only to those routes with zero costs. This is just the situation encountered in the final allocation determined by the above variation of the simplex method; hence the solution is optimum.

We next summarize the steps for solving a transportation problem.

1. Find a first feasible solution by the northwest-corner rule or an equivalent procedure, e.g., row minimum (see below), and develop the solution table (x_{ij}). Let x_{00} be the current value of the objective function.
2. Develop the combined (u_i, v_j) and indirect-cost table $(\bar{c}_{ij})$; i.e., solve for the u_i and v_j by requiring $u_1 = 0$ and $c_{ij} = u_i + v_j$ for x_{ij} in the basis and $\bar{c}_{ij} = u_i + v_j$ for all (i, j).
3. If all $\bar{c}_{ij} - c_{ij} \leq 0$, then the current solution is optimal. If some $\bar{c}_{ij} - c_{ij} > 0$, select the nonbasic variable x_{pq} to enter the basis corresponding to $\bar{c}_{pq} - c_{pq} = \max (\bar{c}_{ij} - c_{ij} > 0)$. (As any x_{ij} corresponding to a $\bar{c}_{ij} - c_{ij} > 0$ may be selected to enter the basis, a modified selection rule as noted below can be used.)
4. Determine the value of x_{pq} and the new basic feasible solution by entering the surrogate parameter $\theta = x_{pq}$ into position (p, q) of the current solution table (x_{ij}) and add and subtract θ from the x_{ij} to maintain row and column balances. Then for those positions in which θ is subtracted from x_{ij}, determine the minimum x_{ij}, and this value is equal to $\theta = x_{pq}$; i.e., we solve the set of constraints $x_{ij} - \theta \geq 0$ and select the maximum value of θ possible. If there are ties, we obtain a degenerate solution in that more than one of the current basic variables will become zero. We drop only one of these zero variables (any one) and retain the others to form the new basic

solution in order to construct the corresponding $(\bar{c}_{ij})$ table. The new value of the objective function $\bar{x}_{00} = x_{00} - x_{pq}(\bar{c}_{pq} - c_{pq})$.

5. Then apply Steps 2 to 4 to the new solution. If in the optimal solution some $\bar{c}_{ij} - c_{ij} = 0$ for a nonbasic variable, then a multiple optimal solution can be found by introducing the corresponding variable into the solution by applying Step 4 and treating the variable as x_{pq}.

In solving transportation problems by hand, the reader can save time and reduce errors by organizing work sheets so as to afford a ready access to the data, especially when calculating the $\bar{c}_{ij} - c_{ij}$ terms. The tableau given in the footnote on page 335 is a concise form for keeping the data and calculations and can be used for determining θ and the u_i's and v_j's. (It is quite effective on a blackboard where the old data can be erased readily.) Sensitivity analysis of the transportation problem's data (the c_{ij}, a_i, b_j) by procedures similar to those described in Chap. 8 can also be applied; see Exercise 14.

Some transportation problems might have the situation where the total of availabilities, $\sum_i a_i$, is less than the total of requirements, $\sum_j b_j$. Even though we cannot satisfy all the requirements, we can still allocate the items at the origins to the destinations in a manner that minimizes the total shipping cost. For this case, we assume that we have a "fictitious" origin that has on hand a total of $\sum_j b_j - \sum_i a_i > 0$ units. The costs of shipping a unit between this fictitious or $(m+1)$st origin and the destinations are assumed to be zero. If our problem was originally a 3×4 problem, we would set up the following 4×4 problem and solve this problem with the fictitious origin like any other transportation problem:

Destinations

	1	2	3	4	
1	c_{11}	c_{12}	c_{13}	c_{14}	a_1
2	c_{21}	c_{22}	c_{23}	c_{24}	a_2
3	c_{31}	c_{32}	c_{33}	c_{34}	a_3
4	0	0	0	0	$\sum_j b_j - \sum_i a_i$
	b_1	b_2	b_3	b_4	

Origins

By letting all the $c_{m+1,j} = c_{4j} = 0$, we make the minimum value of the objective function for the fictitious problem equal to the minimum value of the original problem.

If the problem is given with $\sum_i a_i > \sum_j b_j$, we set up a similar fictitious problem. Here we assume a fictitious destination that requires $\sum_i a_i - \sum_j b_j > 0$ units. The fictitious shipping costs are again zero. For a 3×4 problem we would then solve the following 3×5 problem:

Destinations

	1	2	3	4	5	
1	c_{11}	c_{12}	c_{13}	c_{14}	0	a_1
Origins 2	c_{21}	c_{22}	c_{23}	c_{24}	0	a_2
3	c_{31}	c_{32}	c_{33}	c_{34}	0	a_3
	b_1	b_2	b_3	b_4	$\sum_i a_i - \sum_j b_j$	

In solving any linear-programming problem, we should, in general, expect the total number of iterations required to depend on how close the value of the objective function for the first feasible solution is to the actual minimum. Since the northwest-corner rule does not consider the size of the c_{ij}, we cannot expect the corresponding value of the objective function to be close to the minimum. A number of alternative methods have been suggested that can be adapted to machine computation. We shall illustrate three of them for the following example taken from Dantzig [154]:

3	2	1	2	3	1
5	4	3	-1	1	5
0	2	3	4	5	7
3	3	3	2	2	

The northwest-corner solution yields a value of the objective function of 52.

a. Row minimum. Let the minimum element in the first row be c_{1k}. (If there is more than one minimum element, select the one with the smallest index j.) We let $x_{1k} = a_1$ if $a_1 \leq b_k$ or $x_{1k} = b_k$ if $a_1 > b_k$. In the first case, we have shipped all the a_1 units and go to the second row after changing b_k to $b_k - a_1$. We next find the minimum element in the second row and repeat the process. In the second case, we have allocated only b_k units of the a_1; hence we change a_1 to $a_1 - b_k$ and b_k to zero and find the next smallest c_{ij} in the first row and repeat the process. Using this scheme, we have a first feasible solution as shown below:

$$(x_{ij}) =$$

			1			1
			1	2	2	5
	3	3	1			7
	3	3	3	2	2	

This is a minimum feasible solution with an objective-function value of 13.

b. Column minimum. We do a calculation similar to that using the row minimum, except that we start with the first column and proceed to the last column. Using this scheme, the first feasible solution is

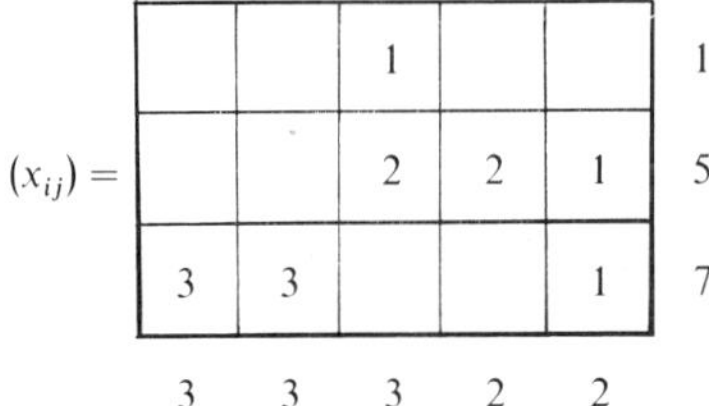

with an objective-function value of 17.

c. Matrix minimum. Here we search the whole matrix for the smallest element and allocate accordingly. We repeat this procedure until all the units are shipped. Using this scheme on the preceding example, we obtain the minimum solution obtained by the row-minimum procedure. Even though these schemes produce good results for the example in question, they are not foolproof. Examples have been constructed whose northwest-corner solution is much better than solutions obtained by any of the above procedures. However, experience has shown that, in general, the extra computational time required to obtain a first feasible solution by one of the above methods will be more than made up because the total iterations will be fewer. This is especially true for large problems.

For high-speed digital computers, where searching for elements can be time-consuming, a number of procedures have been employed for determining which variable should enter the basis, i.e., a rule that does not select the x_{ij} associated with the max $(\bar{c}_{ij} - c_{ij} > 0)$ for all (i, j). Dennis [199] describes rules for searching the cost data which appear to be very effective. In particular, a complete row of the indirect-cost matrix is examined and the x_{ij} in this row with the greatest $\bar{c}_{ij} - c_{ij} > 0$ is chosen. For the next iteration, the following row is examined in the same way, and the procedure continues in this cyclic fashion. As reported in Glover et al. [308], the row-minimum initial solution coupled with the variable selection process which scans the rows of the $(\bar{c}_{ij} - c_{ij})$ table until it finds a row with a positive element and then selects the variable in that row with the largest positive $\bar{c}_{ij} - c_{ij}$ to enter the basis appears to be quite efficient when used with the above primal simplex approach (here the problem should be structured so that $m \le n$). As the transportation problem is a special case of the minimum-cost network problem (see Chap. 11), the efficient computational algorithms for solving network problems can be applied to the transportation problem.

3. VARIATIONS OF THE TRANSPORTATION PROBLEM

a. The personnel-assignment problem.[1]   We shall illustrate the basic problem by the following example: A business concern needs to fill three jobs, which demand different abilities and training.   Three applicants, who can be hired for identical salaries, are available.   However, because of their different abilities, training, and experience, their value to the company depends upon the job in which each is placed.   The estimate of the value of each applicant to the company each year if assigned to any one of the three jobs is given below.

	Jobs		
	1	2	3
1	5	4	7
Applicants 2	6	7	3
3	8	11	2

It is desired to assign the applicants to the jobs in such a way that the total value to the company is a maximum.   In this problem, there are $3! = 6$ possible assignments.   Here we wish to determine the *value of the assignment* x_{ij} of the ith applicant to the jth job.   We can then restrict x_{ij} to be either 0 or 1.   An assignment value of 0 means that the ith person does not get the jth job, and a value of 1 means that the ith person does get the jth job.   Since each applicant can be assigned only one job and since, conversely, each job requires only one person, the total assignment value of the ith applicant, $\sum_{j} x_{ij}$, must equal 1.

Similarly, the total assignment value associated with each job, $\sum_{i} x_{ij}$, must also equal 1.   If we let the contribution of the ith person to the common effort if assigned to job j be c_{ij}, then the above example has the following linear-programming interpretation: Maximize

$$\sum_{i=1}^{3} \sum_{j=1}^{3} c_{ij} x_{ij}$$

subject to

$$\sum_{j=1}^{3} x_{ij} = 1 \qquad i = 1, 2, 3$$

$$\sum_{i=1}^{3} x_{ij} = 1 \qquad j = 1, 2, 3$$

$$x_{ij} \geq 0$$

The form of this problem is that of the transportation problem if we minimize

[1] The material in this section is from Dwyer [227].

the negative of the objective function, and hence it can be solved by means of the same computational procedure. The integer-solution property of the transportation problem assures our being able to make the required 0 or 1 assignment of an individual. Basic solutions to the assignment problem are highly degenerate; an $m \times m$ problem has $2m - 1$ basic variables, with m of them equal to 1 and the remaining $m - 1$ equal to 0. When applying the simplex algorithm in such situations, we would expect many degenerate pivots to occur, in effect stalling at the same extreme point and not changing the value of the objective function. Barr et al. [41] found that for large assignment problems, say $m = 500$, approximately 90 per cent of the pivots were degenerate. Special primal-simplex-based algorithms have proved quite effective. They take advantage of the assignment problem's network structure (see Chap. 11) and are thus able to reduce both the total number of pivots and the number of degenerate ones (Barr et al. [41, 42] and Cunningham [146, 147]).[1]

Frequently there are many identical jobs which demand the same basic qualifications. Such jobs can be combined into a job category. We shall assume there are n such categories and let b_j denote the number of such jobs grouped in the jth job category. If different individuals have identical or approximately the same c_{ij} values, then these individuals can be grouped into personnel categories. We shall assume there are m such categories and let a_i denote the number of persons in the ith personnel category. The linear-programming formulation of this general personnel-assignment problem is then: Maximize

$$\sum_{i=1}^{m} \sum_{j=1}^{n} c_{ij} x_{ij}$$

subject to

$$\sum_{j=1}^{n} x_{ij} = a_i \qquad i = 1, 2, \ldots, m$$

$$\sum_{i=1}^{m} x_{ij} = b_j \qquad j = 1, 2, \ldots, n$$

$$x_{ij} \geq 0$$

Here, x_{ij} is either zero or a positive integer indicating the number of persons in personnel category i assigned to job category j. We imply that the total number of jobs available is equal to the total number of people to be assigned, that is, that $\sum_i a_i = \sum_j b_j$. If $\sum_i a_i < \sum_j b_j$, then a fictitious personnel category containing $\sum_j b_j - \sum_i a_i$ people is added to the problem. When $\sum_j b_j < \sum_i a_i$, then a fictitious job category containing $\sum_i a_i - \sum_j b_j$ jobs is used.

[1] Glover and Klingman [311] report on other assignment problem algorithms with fast computer implementations; also see McGinnis [484] and Bertsekas [75].

Both Dwyer [227] and Votaw and Orden [676] discuss methods for obtaining "near-optimal" solutions to the personnel-assignment problem and, of course, to the transportation problem. The special problem of assigning n individuals to n jobs, i.e., all $a_i = 1$ and all $b_j = 1$, has led to the development of a number of computationally efficient algorithms (Balinski and Gomory [30], Kuhn [446], and Ford and Fulkerson [253]). Another problem which takes on this formulation is the *machine-assignment problem* in which n tasks are to be assigned n different machines with the cost or time of setting up the jth machine for the ith job being c_{ij}. The objective is to assign the jobs so as to minimize the total setup cost or time. For additional techniques for handling the assignment problem, the reader is referred to von Neumann [524], Egerváry [234], and Munkres [520]. See also Exercise 13.

b. The contract-award problem.[1] Although the contract-award problem is peculiar to government operations, its formulation and solution as a linear-programming problem represent an outstanding application of linear-programming techniques. Whenever a government agency wishes to procure items from civilian sources, producers of the items must be invited to participate in the bidding for contracts. The individual manufacturing company submits bids in which it states:

1. The price per unit of article or articles.
2. The maximum and minimum quantity of each item that can be produced at the stated price.
3. *Any other conditions it wishes to impose*

The bid reflects the company's desire for profit, its guess about the other companies' bids, and its own peculiar limitations. The agency must award contracts in such a way that the total dollar cost to the government is at a minimum.

In evaluating the bids, the procurement office must add shipping and other related costs to each of the bidders' quoted prices. Similarly, any savings that could be effected by agreeing to certain conditions are subtracted; e.g., a discount may be allowed for payments made within a certain time. Once the contracts have been awarded, the procurement office must be ready to demonstrate that the total cost to the government was the least possible. The application of linear-programming techniques to this problem guarantees that the solution is minimum.

Let us first consider the *simple* contract-award problem. Here there are m separate bidders willing to furnish the needs of n supply depots. The ith bidder wishes to produce an amount not exceeding a_i, for $i = 1, 2, \ldots, m$, and the jth depot requires the amount b_j, for $j = 1, 2, \ldots, n$. It costs c_{ij} dollars to purchase and deliver a unit from the ith bidder to the jth depot. If x_{ij} denotes the quantity purchased from the ith manufacturer for shipment to the jth

---

[1] For a more detailed discussion, see Stanley, Honig, and Gainen [623].

depot, then our problem is: Minimize

$$\sum_{i=1}^{m} \sum_{j=1}^{n} c_{ij} x_{ij}$$

subject to

$$\sum_{j=1}^{n} x_{ij} \le a_i \qquad i = 1, 2, \ldots, m$$

$$\sum_{i=1}^{m} x_{ij} = b_j \qquad j = 1, 2, \ldots, n$$

$$x_{ij} \ge 0$$

This is a problem of the same type as the transportation problem. Note that $\sum_i a_i \ne \sum_j b_j$. In general, we would expect more to be offered for sale than is required in the total award; that is, $\sum_i a_i \ge \sum_j b_j$. If we establish a ficti-tious depot, the $(n + 1)$st depot, which absorbs the amount $\sum_i a_i - \sum_j b_j = b_{n+1}$, assigning zero costs $c_{i,n+1}$ for all i, we have then formulated the simple contract-award problem as a general transportation problem. The problem is called "simple" because it does not consider the many conditions that can be stipulated by the manufacturers. For example, bidder one may wish to make a_1 units but is willing to accept an award for any part of a_1. Bidder two bids for a_2 units and will not accept a contract for less than a_2. Bidder three bids for a_3 units but will accept as little as $a_3' < a_3$ units. Bidder four is bidding for a total of a_4 units but has different costs on the sublots that total to the final award. By using various computational techniques, these and other such conditions can be handled while the problem still retains the basic form of the transportation problem. Some of these devices are discussed in Stanley, Honig, and Gainen [623]. A typical contract-award problem and its solution is given in Gass [287]. Other applications are cited in the Bibliography.

The transportation-problem formulation with its remarkable mathematical properties and simplified computational procedure makes it an important model for industry and government. This is further emphasized by the fact that a number of other problems can be cast into the transportation for-mat. These problems include those cited above and the transshipment prob-lem described in Orden [537] (see Exercise 5), the caterer problem of Jacobs [397] (see Sec. 5, Chap. 11), production-scheduling problems (see Bowman [85] and Sec. 1, Chap. 11), and the relationship between the transportation problem and network problems (see Sec. 4, Chap. 11). In addition, problems that consider the integration of the purchase of raw materials, the manufacturing of products, and their distribution have major blocks of constraints which derive from transportation considerations. Many applications of transportation theory are given in the Bibliography.

REMARKS

Numerous other methods exist for solving the transportation and assignment problems. These include the dual method of Ford and Fulkerson [253], the *stepping-stone method* of Charnes and Cooper [111], network procedures (see Sec. 4, Chap. 11), *primal partitioning programming* of Rosen as adapted by Grigoriadis and Walker [337], the *dualplex method* of Gass [286], the *Hungarian algorithm* of Kuhn [446] and its transportation-problem extension by Munkres [520] and Egerváry [234] based on a theorem of König (see Exercise 13), the shortest-path approach of Hoffman and Markowitz [378] and its modification and generalization by Lageman [454], the decomposition approach of Williams [700], and a primal Hungarian method by Balinski and Gomory [30]. It is difficult to state which of all the available procedures for solving transportation problems is most efficient in terms of computational time. Glover et al. [308] tested simplex-like methods and a network approach and reached the conclusions noted above. In some instances, one method may be superior because of the structure of the problem, e.g., where m is much less than n (Brandt and Intrator [94]). For assignment problems, although they can be solved by transportation methods, the Hungarian method and related variations are quite effective (McGinnis [484]); the method of Barr et al. [41, 42] and other algorithms discussed in Glover and Klingman [311] have proven to have faster computer implementations. Some analysts prefer to solve transportation problems using computerized mathematical programming systems that employ the revised simplex method instead of specially designed transportation codes, as the latter are not, in general, supported by the computer manufacturers and do not have the data-handling procedures of the mathematical programming systems. Such use is found for those who must solve transportation problems repetitively using inputs from the same but updated extensive data base.

Procedures also exist for handling such variations of the transportation problem as parametric availabilities and requirements (Gal [274]), the capacitated transportation problem where each $x_{ij} < u_{ij}$ (Garvin [282] and Hadley [343]; see Exercise 25, Chap. 9, and Exercise 12), the generalized transportation problem (machine-loading problem of Exercise 6, Chap. 11) (Eisemann and Lourie [237] and Balas and Ivănescu [22]). Methods for solving the transportation problem with additional constraints are discussed in Klingman and Russel [434], and for the fixed-charge transportation problem in Bellmore and Frank [63] and Balinski [27]. A nontechnical discussion of the transportation computational process is given in Henderson and Schlaifer [365].

Some additional mathematical aspects of the transportation problem and the corresponding simplex method of solution are given in Dantzig [154] and Klee and Witzgall [433].

EXERCISES

1. For the transportation problem of Exercise 1, Chap. 1, obtain a first feasible solution by means of the northwest-corner rule and the column-minimum and matrix-minimum procedures. Answer: Minimums of 4,925, 3,600, 3,650.

2. Using the northwest-corner solution as the first feasible solution, determine a minimum feasible solution to Exercise 1, Chap. 1. Are there multiple minimum feasible solutions? Answer: Minimum of 3,600. No.

3. Write the dual problem to the transportation problem of Exercise 2 and show that the u_i and v_j associated with any minimum feasible solution are also a solution to the dual.

4. Show that the optimal set of x_{ij} to any transportation problem is equal to the optimal set of x_{ij} of the corresponding transportation problem for which a constant k has been added to or subtracted from all c_{ij} elements belonging to either the same row or the same column of the cost-coefficient matrix.

5. The transshipment problem is a transportation problem in which each origin and destination can act as an intermediate point through which goods can be temporarily received and then transshipped to other points or to the final destination. Show how this problem can be formulated and solved as a transportation problem (Orden [537]). HINT: Set up an enlarged tableau with $m + n$ origins and $m + n$ destinations, assuming costs c_{ij} between all combinations of the origins and destinations are known. Make sure enough units are available to allow for transshipment.

6. Prove Lemma 1: Every set of $m + n - 1$ linearly independent vectors of the reduced-transportation-problem system can be arranged into a triangular matrix. HINT: At least one column must have a single 1.

7. Solve the following transportation problems.

a. Gass [287].

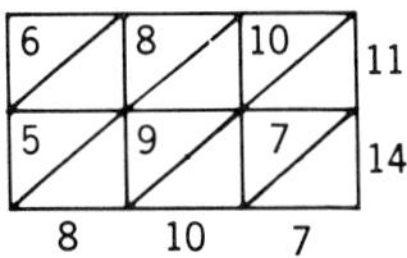

Answer: $x_{11} = 1$, $x_{12} = 10$, $x_{13} = 0$, $x_{21} = 7$, $x_{22} = 0$, $x_{23} = 7$, $x_{00} = 170$.

b.

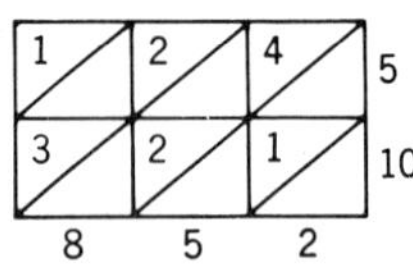

Answer: $x_{00} = 26$.

c.

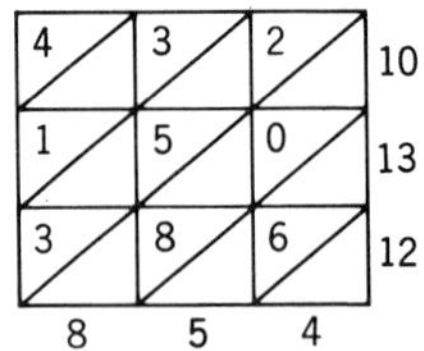

d. Balinski and Gomory [30].

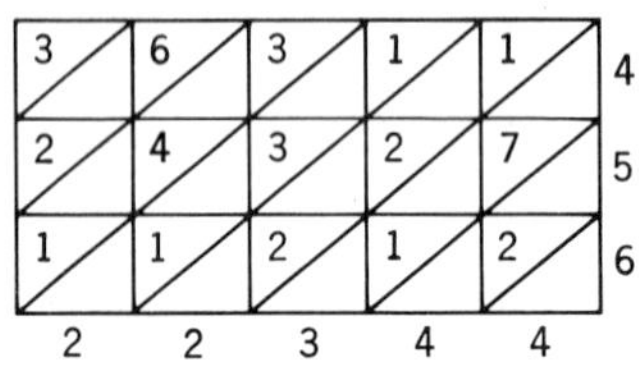

Answer: $x_{15} = 4$, $x_{21} = 1$, $x_{24} = 4$; $x_{31} = 1$, $x_{32} = 2$, $x_{33} = 3$; $x_{00} = 23$.

8. Solve the following assignment problem for the minimum assignment, Balinski and Gomory [30].

2	3	1	1
5	8	3	2
4	9	5	1
8	7	8	4

Answer: $x_{00} = 13$.

9.   Show explicitly that every minor determinant of the full transportation problem is equal to either 0, +1, or −1; i.e., the matrix of the transportation problem is unimodular.

10.   Explain why the multipliers u_i and v_j are integers if the c_{ij} are all integers.

11.   Why do all transportation problems (1.1) to (1.4) have a finite minimum?

12.   Modify the bounded-variable algorithm and the transportation-problem procedure to solve the capacitated-transportation problem, i.e., the transportation problem in which each shipment $x_{ij} < u_{ij}$, where the u_{ij} are given integer-upper bounds, Ford and Fulkerson [253], Garvin [282], and Krekó [443].   HINT: Develop a complementary-slackness relationship between the variables of the primal capacitated problem and its dual.   Show that for solutions to the primal and dual to be optimum we must have $u_i + v_j = c_{ij}$ if $0 < x_{ij} < u_{ij}$, $u_i + v_j \geq c_{ij}$ if $x_{ij} = u_{ij}$, and $u_i + v_j \leq c_{ij}$ if $x_{ij} = 0$.

13.   König's theorem: If **A** is a matrix and m is the maximum number of independent zero elements of **A**, then there are m lines which contain all the zero elements of **A**.   (A set of elements of a matrix is said to be independent if no two elements lie in the same line, i.e., a row or column.)   Using König's theorem and the results of Exercise 4, develop a computational approach for solving an assignment problem, Kuhn [446], Munkres [520], and Krekó [443].   HINT: We need to subtract a sequence of positive quantities from the cost table so that resulting elements in the cost table stay nonnegative and we eventually obtain m independent zeros.   These zeros represent the locations of the assignments.   The procedures can be generalized for the transportation problem.

14.   Develop a sensitivity-analysis approach for the transportation problem (Müller-Merbach [513] and Gal [274]).   HINT: Use the optimal (u, v) table for changes to the costs and the optimal solution table for changes to the a_i and b_j.   What assumption must you make with respect to the changes in the a_i and b_j?

15.   Show that any equation of the transportation problem (1.2) and (1.3) with $\sum_i a_i = \sum_j b_j$ can be expressed as a linear combination of the remaining $m + n - 1$ equations.

16.   Apply the results of Exercise 3, Chap. 4, to the transportation problem to obtain a test for a near-optimal solution.

17.   Develop a parametric analysis algorithm for the transportation problem (Gal [274]).

18.   *The fixed-charge transportation problem* (Hirsch and Dantzig [369] and Balinski [27]; also see Exercise 7, Chap. 9).   Consider a transportation problem in which the costs are defined to be $c_{ij} = d_{ij}x_{ij} + k_{ij}$ if $x_{ij} > 0$ and $c_{ij} = 0$ if $x_{ij} = 0$.   Transform this problem into an integer-programming problem and show that an optimal solution to the integer problem corresponds to an extreme point of the relaxed (continuous) problem.   (Procedures for finding approximate solutions to the fixed-charge transportation problem are described in Balinski [27], Cooper and Drebes [133], and Robers and Cooper [570].   Exact procedures based on branch and bound are described in Barr et al. [44], Kennington and Unger [427], and Soland [617].   Hirsch and Dantzig [369] and Cooper and Debres [133] show that for a general fixed-charge linear-programming problem, an optimal solution occurs at an extreme point of the defining solution space.   See Exercise 22, Chap. 12.)

19.   Develop an interpretation of the dual to the transportation problem (1.1) to (1.4).   Here the primal and dual problems fit the unsymmetric case in which the dual variables are unrestricted (free).   Note that the dual variables are in units of money.   HINT: It may be easier to first consider the symmetric case in which the origin constraints are less than or equal to inequalities and the destination constraints are greater than or equal to inequalities.   You will need to make all the primal inequalities greater than or equal to; introduce an entrepreneur who will purchase the goods at the origins and deliver them to the destinations (Gale [277] and Hsiao [381]).

20.   An assignment problem that cycles (Gassner [294]).   Given a 4×4 assignment problem (i.e., all $a_i = b_j = 1$) with the cost matrix

$$(c_{ij}) = \begin{array}{|c|c|c|c|}
\hline
3 & 5 & 5 & 11 \\
\hline
9 & 7 & 9 & 15 \\
\hline
7 & 7 & 11 & 13 \\
\hline
13 & 13 & 13 & 17 \\
\hline
\end{array}$$

show that the northwest-corner solution of

$$(x_{ij}) = \begin{array}{|c|c|c|c|}
\hline
1 & 0 & & \\
\hline
 & 1 & 0 & \\
\hline
 & & 1 & 0 \\
\hline
 & & & 1 \\
\hline
\end{array}$$

will repeat if the following sequence of new variables is introduced into the basis:

$$x_{13},\ x_{42},\ x_{32},\ x_{41},\ x_{43},\ x_{21},\ x_{31},\ x_{24},\ x_{23},\ x_{14},\ x_{34},\ x_{21}$$

In the first iteration let x_{13} eliminate x_{12} and do not allow the variable coming into the basis in iteration k to replace the variable that came into the basis in iteration $k-1$. For this set of iterations, $u_i + v_j - c_{ij} = \Delta = 2$ for all basis changes. For the northwest-corner solution, the initial u_i's are given by $u = (1, 3, 5, 9)$ and v_j's by $v = (2, 4, 6, 8)$. Letting u' and v' be the set of u_i's and v_j's for the final iteration, note that $u' = (u_1 + 6\Delta,\ u_2 + 6\Delta,\ u_3 + 6\Delta,\ u_4 + 6\Delta)$ and $v' = (v_1 - 6\Delta,\ v_2 - 6\Delta,\ v_3 - 6\Delta,\ v_4 - 6\Delta)$.

21. Solve Exercise 20 for the maximum of the objective function by (*a*) the degeneracy procedure of the transportation problem, and (*b*) the lowest-index rules (see Chap. 7).

22. Given the transportation problem defined by the following table:

3	6	6	5	9
8	1	3	6	8
2	7	3	5	

The problem has the additional constraint $3x_{13} + 2x_{22} = 10$. Solve the total problem using the decomposition algorithm with the only subproblem being the transportation problem (Dantzig and Wolfe [196]).

Answer: $x_{00} = 57$.

23. Solve the following assignment problem for the maximum assignment (Kuhn [446]).

$$(c_{ij}) = \begin{array}{|c|c|c|c|}
\hline
8 & 7 & 9 & 9 \\
\hline
5 & 2 & 7 & 8 \\
\hline
6 & 1 & 4 & 9 \\
\hline
2 & 3 & 2 & 6 \\
\hline
\end{array}$$

Answer: $x_{00} = 27$; $x_{11} = x_{23} = x_{34} = x_{42} = 1$.

CHAPTER 11

General Linear-Programming Applications

In Secs. 1 to 4 of this chapter we shall discuss and formulate examples of the more important (and classical) applications of linear-programming techniques. In Sec. 5 we present a selected sample of other applications which extend the formulative concepts of the earlier discussions and which serve to illustrate the adaptability of the linear-programming framework, and in Sec. 6 we consider linear programming in relation to the theory of games. For purposes of exposition we have grouped the linear-programming applications of Secs. 1 to 4 into four general categories: production-scheduling and inventory-control problems, interindustry problems, diet problems, and network problems.[1] In each section we shall give a verbal statement of the problem and then formulate the problem as a linear-programming problem. For some problems this is followed by a solution of a numerical example. Although these applications will give the reader some idea of the adaptability of linear-programming methods, it is not our purpose here to describe the

[1] The second and third categories are sometimes consolidated into the category of product mix, i.e., the combination of raw materials and resources to yield finished products. The important category of transportation problems is discussed in Chap. 10. See the Bibliography for a more detailed listing of applications.

complete range of linear-programming applications. Instead, these applications were selected mainly for their ability to illustrate the basic techniques of formulating linear-programming models. It is hoped that these discussions and the exercises will enable the reader to investigate and develop new applications.

At this point we should like to stress that there are no set procedures and techniques one can apply to determine and test the linear-programming formulation of a given problem. The casting of a new and complex problem into a linear-programming form appears to be a process of evolution. One's first impulse is to define what appear to be the variables of the problem and to determine the interrelationships between these variables, the resulting constraints, and appropriate objective function. Problems based on either actual or test data are then solved, using the initial model, and the solutions compared with expected results. These studies will in all probability suggest that new variables must be defined and introduced into the system and that previously defined variables can be eliminated; changes in the objective function and constraints, a reevaluation of the test data, etc., may also result. Additional problems are then solved, and this process for evolving a correct model, i.e., *model validation*, continues until the investigator is satisfied that the resulting model approximates the real situation to an acceptable degree (or, possibly, is satisfied that the problem cannot be handled by linear-programming techniques). If, in developing the constraints of the problem, it is determined that some of the relationships are not linear, we are faced with three alternatives:

1. Approximate the troublesome expressions by appropriate linear functions
2. Redefine the problem to fit the linear-programming format
3. Use other techniques to solve the problem

In selecting the first two alternatives, the investigator should avoid a drastic transformation of the problem, for this would tend to dilute any valid interpretation of the solution. For the third alternative, we might possibly use some of the techniques of nonlinear programming (see Chap. 12).

Validation of linear-programming models that describe industrial processes and operations [e.g., refinery operations (Aronofsky et al., Sec. 8g of Bibliography)] can be rather straightforward. The results from such models can be evaluated first by operational personnel and then tested with historical data and under real-world conditions. More difficult to evaluate are models that attempt to replicate decision structures in corporate long-range planning or public policy planning [e.g., corporate problems based on assumed future economic conditions, with conflicting multiple criteria to be optimized; and energy policy models (Manne et al., Sec. 5 of Bibliography)]. General approaches to model validation are described in Gass [291]; validation of linear-programming models is discussed in Williams [701].

When formulating a linear-programming model, it is quite natural to investigate its structure by seeing how the variables combine to form the constraints (rows), e.g., the transportation problem. Often, however, it is more appropriate to study the problem form by defining the activities (columns) and determining how they interact. This is especially true for models that describe integrated process and industrial operations and/or include many time periods.[1] Experience has shown that a judicious interplay between row and column formulation techniques will aid the problem formulator to develop a more accurate model, one that can be interpreted and understood by users and, thus, one that is more susceptible to validation and use. For further discussion see Murtagh [521], Kahan [407], Williams [701], Aronofsky et al., Sec. 8*g* of Bibliography, and Driebeek [221].

To simplify the formulation of the problems described below, we shall discuss all but the diet problem from the ideal point of view; i.e., we shall start with a complete and valid statement of the problem being considered.

1. PRODUCTION-SCHEDULING AND INVENTORY-CONTROL PROBLEMS

Let us consider the plight of a company that manufactures a seasonal item and must determine its monthly production schedule of the item for the next 12 months. The demand for its product fluctuates, as is illustrated in the sales-forecast chart for the coming year (Fig. 11.1). The company must always meet the monthly requirements given on the chart. It can fulfill the individual demands either by producing the desired amount during the month or by producing part of the desired amount and making up the difference by using the overproduction from previous months.

In general, any such scheduling problem has many different schedules that will satisfy the requirements. For example, the manufacturer could produce

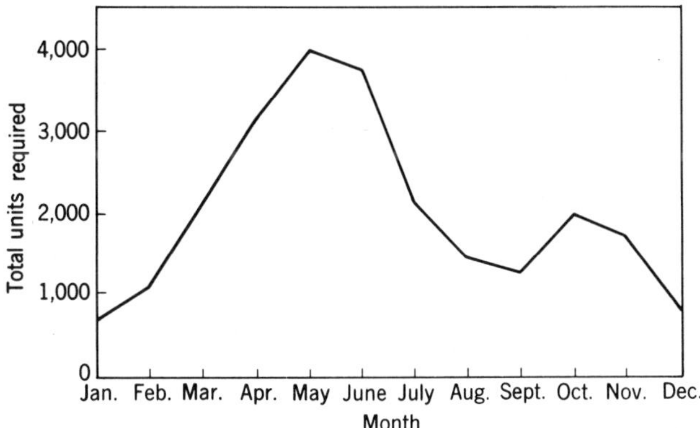

FIGURE 11.1

[1] Set-covering and personnel-scheduling problems are stated more readily in terms of columns; see Exercise 31. Most mathematical programming systems require the problem matrix to be entered by columns.

each month the exact number of units required by the sales forecast. However, since a fluctuating production schedule is costly to maintain, because of overtime costs in high-production months and because of the costs associated with the release of personnel and machinery in low-production months, this type of production schedule is not an efficient one. On the other hand, the manufacturer faced with fluctuating requirements could overproduce in periods of low requirements, store the surplus, and use the excess in periods of high requirements. The production pattern can thus be made quite stable. However, because of the cost of keeping a manufactured item in storage, such a solution may be undesirable if it yields comparatively large monthly surpluses.

Problems of this nature illustrate the difficulties that arise whenever conflicting objectives are inherent in a problem. Here the desire is to determine a production schedule that minimizes the sum of the costs due to output fluctuations and to inventories. For these problems, efficient scheduling means the determination and acceptance of a middle-ground solution lying between the two extreme solutions—the one that minimizes surpluses and the one that minimizes fluctuations. The optimal production schedule will depend on the relative costs assigned to the conflicting objectives.

Let us next develop a mathematical model of this scheduling problem. At the beginning of the first month the manufacturer has in storage a certain amount, say s_0, of the item available from previous production. If the item to be produced is of a new design or style, then $s_0 = 0$. Let

x_t = the number of units produced in month t, that is, the production
r_t = the number of finished units that must be available in month t, that is, the requirement
s_t = the number of finished units that are not required in month t, that is, the storage

Owing to the nature of the problem, we have $x_t \geq 0$, $r_t \geq 0$, and $s_t \geq 0$ for all values of t. For the first month, the production x_1 and the previously stored items s_0 must be such that their sum is greater than or equal to the requirement r_1. This yields the relationship

$$x_1 + s_0 \leq r_1 \tag{1.1}$$

If the equality in (1.1) holds, then the storage in the first month, s_1, must equal zero. If the inequality holds, then $s_1 > 0$. In either case, we have

$$x_1 + s_0 - r_1 = s_1$$

or

$$x_1 + s_0 - s_1 = r_1$$

For the second month, the production x_2 plus the previously stored items s_1 must be greater than or equal to the second month's requirement r_2. We then have

$$x_2 + s_1 \geq r_2$$

or

$$x_2 + s_1 - s_2 = r_2$$

In general, the production x_t, the storage s_t, and the requirement r_t are related by

$$x_t + s_{t-1} - s_t = r_t \tag{1.2}$$

The manufacturer desires to minimize the fluctuations in the production schedule and obtain a smooth pattern of production. The difference between any two successive months' production, say $x_t - x_{t-1}$, will represent the corresponding increase or decrease in production. Since any number has an equivalent representation in terms of two nonnegative numbers, we set

$$x_t - x_{t-1} = y_t - z_t \tag{1.3}$$

where $y_t \geq 0$ represents an increase in production and $z_t \geq 0$ represents a decrease in production. From (1.2) and (1.3) our basic equations for this production model are

$$\begin{aligned} x_t + s_{t-1} - s_t &= r_t \\ x_t - x_{t-1} - y_t + z_t &= 0 \end{aligned} \tag{1.4}$$

where $x_t \geq 0$, $s_t \geq 0$, $y_t \geq 0$, $z_t \geq 0$, and $t = 1, 2, \ldots, n$. If at the end of the production year we desire the surplus of finished items to be zero, then we set $s_n = 0$. Depending on the conditions of the model, $x_0 \geq 0$ and $s_0 \geq 0$.

The manufacturer is, of course, interested in maximizing profit. As we have seen, profit will depend on the fluctuations in the production schedule and the associated monthly storage of manufactured items. From the cost records in previous years, how much it costs to increase production by one unit from month $t - 1$ to month t is known and also how much it costs to store one unit for 1 month. Let these costs be a dollars and b dollars, respectively, where $a > 0$ and $b > 0$. The manufacturer then wants to minimize

$$a \sum_{t=1}^{n} y_t + b \sum_{t=1}^{n} s_t$$

If we let $a/b = \lambda$, where λ measures the cost of a unit increase in output relative to that of storing a unit for 1 month, we then wish to minimize

$$\lambda \sum_{t=1}^{n} y_t + \sum_{t=1}^{n} s_t \tag{1.5}$$

If a value of λ is known, it can be substituted in (1.5), and the corresponding linear-programming problem can be solved by means of the simplex method. It should be noted that a minimum solution cannot, for the same month, t, have both a $y_t > 0$ and a $z_t > 0$.[†]   Solving the problem in this manner will yield a unique optimal production schedule or possibly additional production schedules having the same value of the objective function.   In many instances a value of λ cannot be determined, or the manufacturer wishes to study various production schedules associated with a range of values for λ.   For these situations an explicit value of λ need not be given, and by means of a modification of the simplex method, optimal production schedules for ranges of λ extending from $\lambda = 0$ to $\lambda < +\infty$ can be computed.[1]

If a is very small, then the corresponding optimal solution will be the schedule that requires the most fluctuation and the least storage.   If b is very small, then the associated minimum solution will be the schedule that calls for the most storage and least fluctuation.   In the first case we are minimizing only storage, and in the second case we are concerned with minimizing only fluctuations.

The cost function (1.5) ignores the costs of the monthly decreases z_t and is directed toward smoothing out the increases in production.   Letting $c > 0$ be the cost of decreasing production by one unit from month $t - 1$ to month t, the more general cost function would be to minimize

$$a \sum_{t=1}^{n} y_t + b \sum_{t=1}^{n} s_t + c \sum_{t=1}^{n} z_t$$

We can convert this problem into a two-parameter linear-programming problem by letting $b/a = \lambda_1$ and $c/a = \lambda_2$.

The following example from Directorate of Management Analysis [206] illustrates the preceding discussion:

The number of monthly graduates required from an aircraft mechanics' training school (for a 12-month period) is shown in Fig. 11.2.   The total

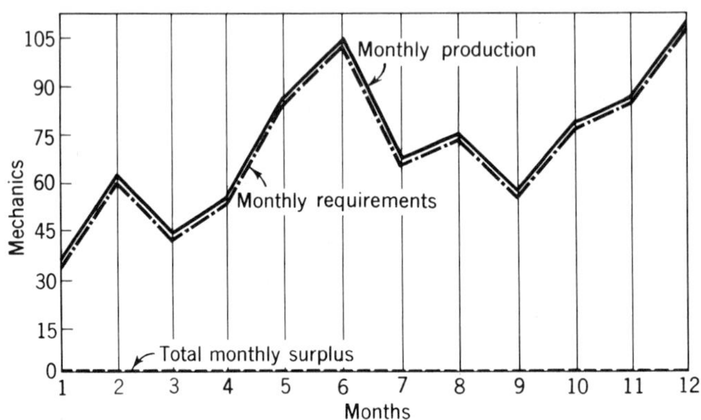

FIGURE 11.2   Minimum surplus solution.
(From Directorate of Management Analysis [206].)

[†] See Exercise 10, Chap. 4.
[1] See Chap. 8 for a discussion of parametric and multiobjective methods.

production of mechanics over the 12-month period must equal the total required; i.e., there cannot be a surplus or shortage of mechanics at the end of the program.

For this problem, the possible schedules range from the one where all individuals are assigned to duty on the day they graduate (minimum surplus, Fig. 11.2) to one where a fixed training load is provided (minimum fluctuation, Fig. 11.3).   The former schedule, by producing each month the graduates as needed, requires no replacement pool but creates the heaviest problems in operating the training school.   The second schedule eliminates fluctuations in training output but puts the greatest burden on the replacement pool.   An intermediate solution (Fig. 11.4) results in a more manageable training program, which at the same time places a lighter load on the replacement pool.

The reader will note that for only 12 time periods, that is, for $n = 12$, the linear-programming model involves 24 equations in 48 variables.   Problems of this size are difficult to solve by manual computation.   It is advantageous to reduce the size of the problem whenever possible, not only for manual

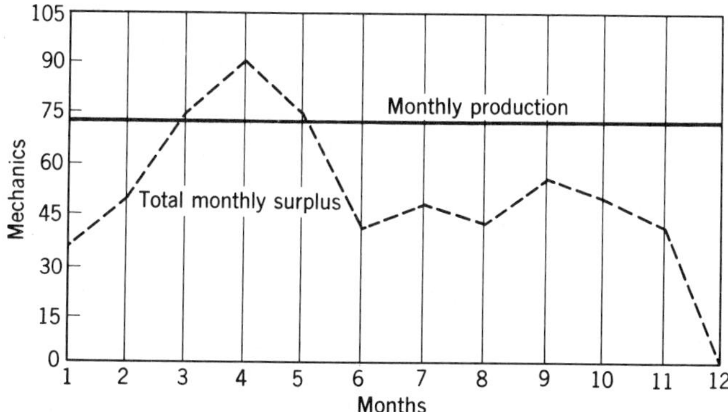

FIGURE 11.3   Minimum fluctuation solution.
(*From Directorate of Management Analysis* [206].)

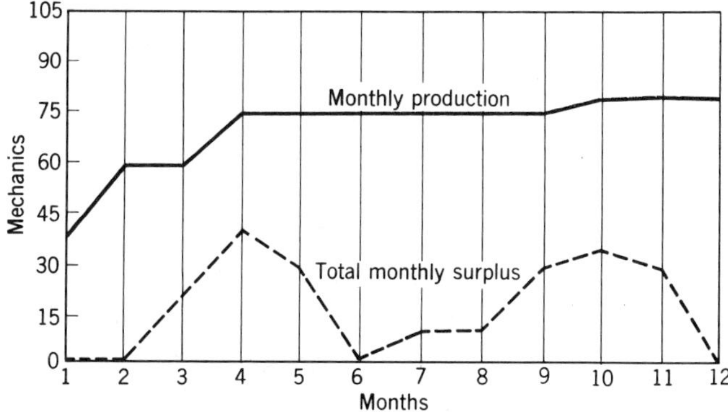

FIGURE 11.4   An intermediate solution.
(*From Directorate of Management Analysis* [206].)

computations, but also for computations being done on electronic computers. In some instances, a careful analysis of the problem will reveal equations and variables that can be eliminated from the computation. The above production problem is such a problem.

From Eq. (1.2) we have

$$x_t = r_t + s_t - s_{t-1}$$

and

$$x_{t-1} = r_{t-1} + s_{t-1} - s_{t-2}$$

By substituting the above expressions for x_t and x_{t-1} in (1.3), we eliminate all the x_t from the original model and obtain the following set of equations:

$$y_t - z_t + 2s_{t-1} - s_t - s_{t-2} = r_t - r_{t-1} \tag{1.6}$$

and

$$t = 1, 2, \ldots, n \qquad \text{and} \qquad s_{-1} = r_0 = 0$$

Equations (1.6) and the objective function (1.5) now represent the linear-programming model of the production problem. We have eliminated one equation from each time period and the 12 variables x_t. When we solve this reduced system, it is an easy matter to go back to the original equations and obtain the desired production schedule consisting of the x_t's. However, since the reduced set of equations does not restrict the $x_t \geq 0$, we have no assurance that the nonnegative values of y_t, z_t, and s_t associated with the optimal solution to the reduced problem will yield nonnegative values of x_t. Before any such elimination is accepted, the validity of the reduction must be demonstrated. To do this, we first assume that an optimal solution to the reduced system is given. We then need to show that the optimal values of y_t, z_t, and s_t cannot yield a negative x_t and at the same time be optimal. The proof is left to the reader (Exercise 60).

In addition to the above reduction in the dimensions of the problem, this formulation of the production-scheduling problem has also been subjected to further interpretations which enable one to solve it using much simpler computational devices than the general simplex method. Bowman [85] gives a variation of the problem that is written as a transportation problem and solved by the corresponding algorithm. Dantzig and Johnson [186] have developed a systematic and rapid graphical method for determining the optimal solutions.

The production-smoothing problem considers the cost of producing each unit in an indirect way by relating the cost of production to monthly increases and decreases. In the following production-scheduling problem (Magee [487]), the cost function attempts to balance the direct costs of monthly production against monthly inventory charges. Limited production smoothing is obtained by setting upper bounds on monthly production. The problem is as follows.

Consider a company that has one production line upon which it produces a single homogeneous commodity. We suppose that the commodity sells for a fixed unit price; that the costs for regular-time production, overtime production, and storage are known and vary between time periods; that the rate of production per unit time is known; and that an accurate sales forecast in the form of demand during each of a number of successive time periods is known. It is desired to formulate a production schedule that will meet the sales forecast and minimize the combined costs of production and storage.

n = number of time periods

r_t = number of units of finished product to be sold during tth time period (the sales requirements)

s_0 = initial inventory

p_t = maximum number of units that can be produced each time period on regular time

q_t = maximum number of units that can be produced each time period on overtime

a_t = cost of production of one unit on regular time during time period t

b_t = cost of production of one unit on overtime during time period t

c_t = cost of storage of one unit of product during time period t

x_t = regular-time production during tth time period

y_t = overtime production during tth time period

s_t = inventory at the end of the tth time period

For concreteness, it will be assumed that each time period is 1 month in length and that the inventory for each month is taken on the last day of that month. This is equivalent to adding the month's production to inventory and withdrawing the month's sales from inventory at the end of each month.

We first note that all $x_t \geq 0$, $y_t \geq 0$, and $s_t \geq 0$. Investigating the required constraints for each time period, we have for $t = 1$ that

$$x_1 + y_1 + s_0 \geq r_1$$
$$x_1 \leq p_1$$
$$y_1 \leq q_1$$

Since s_1 is the inventory at the end of time period 1, we have for $t = 1$

$$x_1 + y_1 + s_0 - s_1 = r_1$$
$$x_1 \leq p_1$$
$$y_1 \leq q_1$$

For $t = 2$, we have

$$x_2 + y_2 + s_1 - s_2 = r_2$$
$$x_2 \leq p_2$$
$$y_2 \leq q_2$$

and in general for time period t we have

$$x_t + y_t - s_t + s_{t-1} = r_t$$
$$x_t \le p_t$$
$$y_t \le q_t$$

We illustrate the formulation of the production-scheduling problem as a linear program for a three-period time horizon: Minimize

$a_1 x_1 + b_1 y_1 + c_1 s_1$	$+ a_2 x_2 + b_2 y_2 + c_2 s_2$	$+ a_3 x_3 + b_3 y_3 + c_3 s_3$

subject to

$x_1 + y_1 - s_1 + s_0$			$= r_1$
x_1			$\le p_1$
y_1			$\le q_1$
s_1	$+\ x_2 + y_2 - s_2$		$= r_2$
	x_2		$\le p_2$
	y_2		$\le q_2$
	s_2	$+\ x_3 + y_3 - s_3$	$= r_3$
		x_3	$\le p_3$
		y_3	$\le q_3$

$$(1.7)$$

with all $(x_t,\ y_t,\ s_t) \ge 0$. The form of this system illustrates the staircase structure and the sparse nature of the nonzero elements inherent in most time-dependent problems. Also, it is a loosely connected system in that only a few variables (the s_t's) join the time periods together; i.e., we would have independent problems if there were no storage.

For a given set of r_t, p_t, and q_t, the linear-programming formulation could lead to a set of constraints which are infeasible; i.e., the regular and overtime production bounds might not allow for enough goods to be made and stored to meet the given sales forecasts. Although the dimensions of the problem for $n = 12$ months and $n = 52$ weeks are not large and can be solved readily, we note that the problem can be cast into the bounded-variable format. We are given that all x_t and y_t are bounded by the p_t and q_t, respectively. We can impose a bound on each s_t by defining

$$\bar{s} = s_0 + n \sum_{t=1}^{n} (p_t + q_t)$$

and then letting $s_t \le \bar{s}$ for all i. ($\bar{s}$ is the maximum production possible plus the initial inventory.) To guard against production going to zero in any month, we can add lower-bound constraints of the form $0 \le l_t \le x_t$. These imply that at least l_t must be produced each month. By making the change in variables $x_t = \bar{x}_t + l_t$ in the other constraints of the problem, we both account for and eliminate the lower-bound constraints.[1]

The above production-scheduling problem of Magee [487] can be interpreted and transformed into a transportation problem and solved by the transportation algorithm (Bowman [85] and Bishop [76]). Here we consider each month's regular and overtime production as origins of goods and each month's sales requirement as a destination. The shipments represent initial inventory plus production manufactured and shipped in the same time period or stored and shipped in a later time period; see Exercises 29 and 48. Further analysis of this problem shows that it can be solved by inspection of the transportation-cost matrix without recourse to the transportation algorithm (Johnson [404]).

We next offer some comments with respect to the above problems which also apply to linear-programming formulations in general. Central to the realistic solution of production and similar operationally based problems is our ability to obtain accurate cost data on the production and inventory activities, as well as accurate sales forecasts. As noted earlier, most of the problems discussed in this chapter are placed in an ideal setting where we assume knowledge of the necessary data. We cannot emphasize too strongly that the collection, analysis, and interpretation of the data necessary to "fill out" the linear-programming model is in many ways the most difficult task in problem formulation.

Assuming accurate data, a related question is "How does the manufacturing company actually use the optimal solution generated by the linear-programming formulation of its problem?" We must recognize that an optimal solution which calls for 152 units of regular time production and 48 units of overtime production during a given month, with 36 units in inventory at the end of the month, might be difficult to accomplish in the real-life production environment. A machine breaks down, a shift is short a crew, physical resources do not arrive on schedule, etc. Also, the formulation of the problem—the manner in which we have stated it in mathematical terms—might have required certain simplifications or approximations, e.g., the use of average cost per unit of production instead of costs which vary by the number produced. If the model is a "reasonable" mathematical representation of the real problem, the plans based on the predicted production will not lead us too far astray and, in fact, will allow us to plan the operation in a more efficient and a more profitable manner. If the linear-programming model and its solution represent the untarnished ideal, the differences between the optimal solution and what actually comes off the production floor give production supervisors (and their managers) some goal to be measured against and indications where improvements can be made. The optimal solution is thus a plan which is reshaped and refined by the operational real world.

It should now be clear that our ability to formulate production, trans-

---

[1] See footnote page 282.

portation, and many industrial problems which fit the linear-programming format is simply based on the development of constraints which represent what we may term to be "accounting" or "bookkeeping" information.[1]   Except for the use of mathematical shorthand, the production clerk has to develop similar statements which record how production was generated and distributed between sales and inventory.   The transportation clerk records where shipments originated and how much each destination received.   Their books capture the full set of constraints and in most cases represent linear and additive relationships.   Thus, we submit that a good place to start in developing a statement of a problem and its constraints is by studying the associated accounting function.   It also brings the investigator close to the data sources.

Other production-type problems include the activity-analysis problem described in Chap. 1 and those in the references that follow.   These different formulative approaches can be combined into multiperiod, multiproduct activity-analysis problems with production-smoothing or production-scheduling considerations.   For additional reading concerning production and scheduling problems, the reader is referred to Hoffman and Jacobs [375], Antosiewicz and Hoffman [9], Jacobs [397], Magee [487], Directorate of Management Analysis [206], Charnes, Cooper, and Mellon [119], Cottle and Karup (Sec. 20 of Bibliography), and Sec. 13 of Bibliography.[2]

2. INTERINDUSTRY PROBLEMS

The first application of linear-programming techniques in the field of economics was to the area of interindustry or input-output analysis (Koopmans [437], Leontief [473], and Morgenstern [507]).   By way of introduction to the general formulation of input-output models, let us consider an economy that consists of only the three basic industries—railroad, steel, and coal—and a fourth category of all other industries (Glaser [306]).   We wish to analyze the interrelationships of these industries in terms of sales to each other and to other elements of the economy, as described below, during a base time period, e.g., 1 year.   This analysis can be readily accomplished by reference to the Input-Output Table.

Each element in the table represents a total sales activity of each industry that occurred during the base time period.   For example, the first row of the table describes the sales of the railroad industry (in terms of dollars) to each of the other industries.   The first element of row 1 represents the total sale of the railroad industry to the railroad industry (intra-industry sale); the second element describes the total sale of the railroad industry to the steel industry; the third element, the total sale to the coal industry; the fourth element, the

---

[1] We stress that the proper formulation of some problems requires an understanding of the physical processes involved, e.g., blending of aviation gasolines (Charnes, Cooper, and Mellon [119]), and blast-furnace burdening [Fabian (Sec. 8*e* of Bibliography)].

[2] Also see the production and distribution network model described in Sec. 4.

INPUT-OUTPUT TABLE

	Sales to RR		Sales to steel		Sales to coal		Sales to other		Sales to FD		Total sales
Sales by RR	RR to RR x_{11}	+	RR to steel x_{12}	+	RR to coal x_{13}	+	RR to other x_{14}	+	RR to FD y_1	=	Total sales by RR x_1
Sales by steel	Steel to RR x_{21}	+	Steel to steel x_{22}	+	Steel to coal x_{23}	+	Steel to other x_{24}	+	Steel to FD y_2	=	Total sales by steel x_2
Sales by coal	Coal to RR x_{31}	+	Coal to steel x_{32}	+	Coal to coal x_{33}	+	Coal to other x_{34}	+	Coal to FD y_3	=	Total sales by coal x_3
Sales by other	Other to RR x_{41}	+	Other to steel x_{42}	+	Other to coal x_{43}	+	Other to other x_{44}	+	Other to FD y_4	=	Total sales by other x_4

total sale to the other industries; and the fifth element, the total sale of the railroad industry to what is termed the final-demand consumers. In general, the final demand consists of those elements of the economy which consume the various commodities produced but do not contribute, or feed back, to the economy a product of their own. In the final-demand category we usually include foreign trade, government operations, and households. The total production required by the final-demand segments of the economy is called the bill of goods. The sum of the sales included in these five elements represents the total sales of the railroad industry during the base period. A similar discussion holds for the other three rows of the table. By columns, we see that each element represents the total purchase of an industry from the other industries to support its operations during the base period.

Let us define

$0 < x_i$ = the total output, in dollars, of industry i during the base period (only active industries are considered)

$0 \leq x_{ij}$ = total sales, in dollars, of industry i to industry j during the base period

$0 \leq y_i$ = amount of final demand, in dollars, for industry i

where $i, j = 1, 2, \ldots, m$.

With these definitions we can construct the following set of four linear equations, which define the interrelationships of our reduced economy, as described in the table:

$$x_1 - x_{11} - x_{12} - x_{13} - x_{14} = y_1$$
$$x_2 - x_{21} - x_{22} - x_{23} - x_{24} = y_2 \tag{2.1}$$
$$x_3 - x_{31} - x_{32} - x_{33} - x_{34} = y_3$$
$$x_4 - x_{41} - x_{42} - x_{43} - x_{44} = y_4$$

In the form (2.1), our input-output information explicitly defines the relationship between total outputs, total sales, and final demands for a given time period.

In the 1930s Leontief was able to gather the necessary data and write sets of equations similar to (2.1) for a 45-industry classification for the years 1919 and 1929. He assumed that this linear model of the economy for a base time period could be used to analyze the structure of the economy in future time periods. Let us denote the x_i and x_{ij} that are known for the given time period by $\bar{x}_i$ and $\bar{x}_{ij}$. For the railroad industry we see that the ratios

$$\frac{\bar{x}_{11}}{\bar{x}_1} \qquad \frac{\bar{x}_{21}}{\bar{x}_1} \qquad \frac{\bar{x}_{31}}{\bar{x}_1} \qquad \frac{\bar{x}_{41}}{\bar{x}_1}$$

represent the percentages of input by each of the other industries required to produce one unit of output by the railroad industry; for example, $\bar{x}_{21}/\bar{x}_1$ is the input of the steel industry necessary to produce one unit of output of the railroad industry.   Let

$$0 \leq a_{ij} = \frac{\bar{x}_{ij}}{\bar{x}_j} = \text{amount of industry } i \text{ which is necessary to produce one unit}$$
$$\text{of commodity } j$$

The a_{ij} are called *input-output coefficients.*
For the base time period we have

$$\bar{x}_1 - a_{11}\bar{x}_1 - a_{12}\bar{x}_2 - a_{13}\bar{x}_3 - a_{14}\bar{x}_4 = \bar{y}_1$$
$$\bar{x}_2 - a_{21}\bar{x}_1 - a_{22}\bar{x}_2 - a_{23}\bar{x}_3 - a_{24}\bar{x}_4 = \bar{y}_2$$
$$\bar{x}_3 - a_{31}\bar{x}_1 - a_{32}\bar{x}_2 - a_{33}\bar{x}_3 - a_{34}\bar{x}_4 = \bar{y}_3 \tag{2.2}$$
$$\bar{x}_4 - a_{41}\bar{x}_1 - a_{42}\bar{x}_2 - a_{43}\bar{x}_3 - a_{44}\bar{x}_4 = \bar{y}_4$$

We can then write (2.2) in the form

$$(\mathbf{I} - \mathbf{A})\bar{\mathbf{X}} = \bar{\mathbf{Y}}$$

where

$$\mathbf{A} = (a_{ij}) \qquad \bar{\mathbf{X}} = \begin{pmatrix} \bar{x}_1 \\ \bar{x}_2 \\ \bar{x}_3 \\ \bar{x}_4 \end{pmatrix} \qquad \bar{\mathbf{Y}} = \begin{pmatrix} \bar{y}_1 \\ \bar{y}_2 \\ \bar{y}_3 \\ \bar{y}_4 \end{pmatrix}$$

The matrix $(\mathbf{I} - \mathbf{A})$ is known as a *Leontief matrix.*   By assuming that this linear structure of the economy, i.e., the input-output coefficients, describes the activity of the economy not only for the base time period but for future time periods as well, we can determine a production vector $\mathbf{X}$ which satisfies a predicted final-demand vector $\mathbf{Y}$.   From this we have that the general problem for input-output economics is to find a vector $\mathbf{X}$ which satisfies the constraints

$$(\mathbf{I} - \mathbf{A})\mathbf{X} = \mathbf{Y}$$
$$\mathbf{X} \geq \mathbf{0} \tag{2.3}$$

where $\mathbf{Y}$ is a given nonnegative and nonzero final-demand vector and $\mathbf{A}$ is a given matrix of input-output coefficients.   For the problem being considered, the one with the nonzero final demand, it can be shown not only that $a_{ij} \geq 0$ but also that $\sum_{i=1}^{m} a_{ij} < 1$ for $j = 1, 2, \ldots, m$ (Morgenstern [507]).   This structure is called an "open model."   It is also shown in Morgenstern [507] that, if a matrix $\mathbf{A}$ satisfies the condition

$$\sum_{i=1}^{m} |a_{ij}| < 1 \qquad \text{for } j = 1, 2, \ldots, m$$

then $(\mathbf{I} - \mathbf{A})$ is nonsingular.   We then have that Eqs. (2.3) have the solution

$$\mathbf{X} = (\mathbf{I} - \mathbf{A})^{-1}\mathbf{Y}$$

The initial treatment of the interindustry model as a linear-programming problem interpreted the equalities of (2.3) as a system of inequalities

$$(\mathbf{I} - \mathbf{A})\mathbf{X} \leq \mathbf{Y} \tag{2.4}$$

Here we are interested in solutions that do not have to satisfy the final-demand requirements. An obvious but impractical solution to (2.4) is $\mathbf{X} = \mathbf{0}$; hence the problem is feasible. An alternative formulation of the problem might require the system (2.4) to be a mixture of equalities and inequalities. However, for the inequalities (2.4), the corresponding system of equalities is

$$(\mathbf{I} - \mathbf{A})\mathbf{X} + \mathbf{W} = \mathbf{Y} \tag{2.5}$$

where $\mathbf{W}$ is an m-dimensional column vector whose components w_i are nonnegative slack vectors. The system (2.5) consists of m equations in $2m$ variables. The objective function can take on various interpretations. For example, if c_j is the profit per unit of commodity j produced, then an appropriate objective function would be to maximize the total profit $\mathbf{cX}$, where $\mathbf{c} = (c_1, c_2, \ldots, c_m)$ is a row vector. The cost coefficients of the slack variables are taken to be zero. Another objective function could call for the solution that maximizes the output of a particular industry or combination of industries, e.g., to maximize $x_1 + x_3 + x_7$.

In addition to the constraints (2.3) or (2.4), the interindustry model might stipulate that the production (activity) level x_i of industry i must not exceed a known available capacity for production by industry i. Let us denote these capacity levels by l_i and let the capacity-level vector $\mathbf{L} = (l_1, l_2, \ldots, l_m)$ be a nonnegative column vector. Then $\mathbf{X} \leq \mathbf{L}$. As we are here looking at our economy for a particular time period, e.g., a month, we can assume that for some industries a portion of the production in the previous time period was used to stockpile finished units. These units will be made available for distribution in the following time period and can be applied to satisfy the new final-demand requirements. In our model let s_{0i} represent the stock of production i available from previous production, and let the vector of stocks, $\mathbf{S}_0 = (s_{01}, s_{02}, \ldots, s_{0m})$, be a nonnegative column vector. We can then expand the system (2.5) into the following linear-programming problem: Maximize

$$\mathbf{cX}$$

subject to

$$\begin{aligned}
(\mathbf{I} - \mathbf{A})\mathbf{X} + \mathbf{W} &= \mathbf{Y} - \mathbf{S}_0 \\
\mathbf{X} + \mathbf{U} &= \mathbf{L} \\
\mathbf{X} &\geq \mathbf{0}
\end{aligned} \tag{2.6}$$

where $\mathbf{U} = (u_1, u_2, \ldots, u_m)$ is a nonnegative column vector and where the u_i represent the unused capacity of industry i. If, for some i, $y_i - s_{0i} < 0$, then the final demand for industry i will be satisfied from stocks, and the effective final demand for industry i during this time period is set equal to zero. However, we assume that at least one $y_i - s_{0i} > 0$.

The system (2.6) is called a static Leontief model because it considers the economy over only a single time period.   A more challenging application of the interindustry model is its use in interpreting the behavior of an economy over several time periods.   We shall next discuss a particular linear-programming formulation of a similar but dynamic model, as given in Wagner [677].   To do this, we must introduce the following changes in our notation: Let n equal the total number of time periods considered, and let $t = 1, 2, \ldots, n$ be any particular time period.   Then for any t we have the nonnegative column vectors

$$\mathbf{X}_t = (x_{t1}, x_{t2}, \ldots, x_{tm}) = \text{the production vector}$$
$$\mathbf{Y}_t = (y_{t1}, y_{t2}, \ldots, y_{tm}) = \text{the final-demand vector}$$
$$\mathbf{S}_t = (s_{t1}, s_{t2}, \ldots, s_{tm}) = \text{the storage vector due to unused production up to}$$
$$\text{and including month } t; \text{ these stocks are available}$$
$$\text{in month } t + 1$$
$$\mathbf{U}_t = (u_{t1}, u_{t2}, \ldots, u_{tm}) = \text{the unused-capacity vector}$$

We assume that we are given the initial storage vector $\mathbf{S}_0$ and that the basic capacity-level vector $\mathbf{L}$ is the same for all time periods.

The main difference between our static and dynamic models is that in the dynamic model we shall make provision for the expansion of the capacity levels of each industry to meet the requirements of future final demands.   From time period to time period we shall be adding to $\mathbf{L}$ a nonnegative column vector $\mathbf{V}_t$, which represents additional available capacities.   To do this, we need, in addition to the appropriate Leontief matrix, $(\mathbf{I} - \mathbf{A})$, the knowledge of the corresponding m-dimensional square matrix $\mathbf{B}$ of capital coefficients. $\mathbf{B}$ is a matrix of nonnegative numbers in which the jth column represents the inputs from each industry needed to build an additional unit of capacity for the jth industry.   Let

$$\mathbf{V}_t = (v_{t1}, v_{t2}, \ldots, v_{tm}) = \text{the nonnegative capacity-expansion vector}$$

where v_{ti} is the additional capacity for industry i in period t.   Then the ith row of the product $\mathbf{BV}_t$, that is, $b_{i1}v_{t1} + b_{i2}v_{t2} + \cdots + b_{im}v_{tm}$, represents the amount of the ith industry's production that is used to build additional capacity in time period t for all the industries in our economy.   This production is not available to meet the final-demand requirements.   In this model we assume that the additional production capacity is available in the next time period $(t + 1)$.   We also assume that the matrices $(\mathbf{I} - \mathbf{A})$ and $\mathbf{B}$ are applicable to all the n time periods and that the final-demand requirements will be satisfied.   We can, for each time period, summarize the above conditions as follows:

$$(\mathbf{I} - \mathbf{A})\mathbf{X}_t + \mathbf{S}_{t-1} = \mathbf{Y}_t + \mathbf{BV}_t + \mathbf{S}_t \tag{2.7}$$

$$\mathbf{X}_t + \mathbf{U}_t = \mathbf{L} + \sum_{q=1}^{t-1} \mathbf{V}_q \tag{2.8}$$

for $t = 1, 2, \ldots, n$.   Equation (2.7) states that, for any t, the total output plus the previous stocks is equal to the final-demand and capacity-expansion requirements for output plus the current period's unused stocks.   Equation

(2.8) equates the total used and unused production to the sum of the initial production capacity and the previous additional increases in production capacity. We can rearrange these equations to read

$$(\mathbf{I} - \mathbf{A})\mathbf{X}_t - \mathbf{B}\mathbf{V}_t - \mathbf{S}_t + \mathbf{S}_{t-1} = \mathbf{Y}_t$$

$$\mathbf{X}_t - \sum_{q=1}^{t-1} \mathbf{V}_q + \mathbf{U}_t = \mathbf{L} \tag{2.9}$$

For any given set of $\mathbf{Y}_t$, $\mathbf{L}$, and $\mathbf{S}_0$ we can set up the tableau of coefficients, as shown in Tableau 11.1. Systems of this sort are termed *block-triangular*. Given an appropriate objective function, which could involve production, expansion, and storage costs, the problem could be solved by the standard simplex procedure. However, even for just a few time periods, the number of equations and variables may become large. Some work has been done that enables one to reduce the amount of computation by taking advantage of the block-triangular configuration (with its low density of nonzero elements) or to reduce the number of equations by transforming the system. The former is discussed in Dantzig [156] and the latter in Wagner [677].

TABLEAU 11.1

	$\mathbf{X}_1$	$\mathbf{V}_1$	$\mathbf{S}_1$	$\mathbf{U}_1$	$\mathbf{X}_2$	$\mathbf{V}_2$	$\mathbf{S}_2$	$\mathbf{U}_2$	$\cdots$	$\mathbf{X}_n$	$\mathbf{V}_n$	$\mathbf{S}_n$	$\mathbf{U}_n$
$\mathbf{Y}_1 - \mathbf{S}_0$	$(\mathbf{I} - \mathbf{A})$	$-\mathbf{B}$	$-\mathbf{I}$										
$\mathbf{L}$	$\mathbf{I}$			$\mathbf{I}$									
$\mathbf{Y}_2$			$\mathbf{I}$		$(\mathbf{I} - \mathbf{A})$	$-\mathbf{B}$	$-\mathbf{I}$						
$\mathbf{L}$		$-\mathbf{I}$			$\mathbf{I}$			$\mathbf{I}$					
$\vdots$									$\cdots$				
$\mathbf{Y}_n$										$(\mathbf{I} - \mathbf{A})$	$-\mathbf{B}$	$-\mathbf{I}$	
$\mathbf{L}$		$-\mathbf{I}$				$-\mathbf{I}$				$\mathbf{I}$			$\mathbf{I}$

The concepts of interindustry analysis and the use of the input-output table can be applied to many other situations which involve interdependent processes or organizational entities. In particular, Isard, Sec. 5 of Bibliography, has considered the ecologic system as a large set of such processes which are intimately connected to economic activities.

3. DIET PROBLEMS

In Sec. 2 of Chap. 1 we defined and formulated the basic diet problem in terms of a linear-programming problem. We next discuss a particular linear-programming model of a diet problem and review its corresponding solution to determine whether the formulation of Chap. 1 does apply.   Such an analysis is typical for linear-programming problems in general, and it is recommended when one attempts to describe a complex situation by an elementary linear model.   The models of such situations usually start as simple ones, and using the solutions to these modest models for a base, the investigator is able to evolve more realistic models.   To illustrate this process, we shall briefly describe various adjustments to a diet-problem model that tend to make it a truer representation of the real-life situation.

Historically, the diet problem of Stigler [630] was the first rather long and complicated linear-programming problem to be solved by the simplex method.   Here the problem was to determine what quantities of 77 foods should be bought in order not only to yield the minimum cost, but also to satisfy the minimum requirements of nine nutritive elements, e.g., vitamin A, niacin, and thiamine.   The resulting purchases were to form the complete diet to sustain a person for a year.   The basic linear-programming formulation of this problem was in terms of 9 equations and 86 variables, including 9 slack variables.   The final solution obtained by the simplex method was, of course, a true minimum solution.   However, since the simplex procedure deals only in terms of basic solutions, only nine of the possible foods were represented at a positive level in the minimum solution.   Thus the diet obtained by linear-programming methods called for the purchase of varying amounts of wheat flour, corn meal, evaporated milk, peanut butter, lard, beef liver, cabbage, potatoes, and spinach, and cost \$39.67 (for the year 1939).   The solution obtained by Stigler using a systematic trial-and-error procedure required only five foods: wheat flour, evaporated milk, cabbage, spinach, and dried navy beans, and cost \$39.93.   Such diets, although quite inexpensive, are certainly unpalatable over any period of time, and the selected foods would do justice to the chief dietitian of a slave-labor camp.   As Stigler points out, "No one recommends these diets (*i.e., true minimum-cost diets*) to anyone, let alone everyone."   He also cites a low-cost diet for 1939 that was constructed by a dietitian and cost \$115.   The difference in cost was attributed to the dietitian's concern with the requirements of palatability, variety of diet, and prestige value of certain foods.   Starting with our basic formulation, how can we modify our procedure or model to meet these additional dietary requirements?

Here we have a correct solution to the problem as stated, but this correct solution turns out to be unacceptable to execute, i.e., it is not a valid solution.   To overcome this defect, we could look for alternative optimal solutions and form various convex combinations of these solutions.   In this manner we would be able to select a diet that included more than nine foods.   Or we could look at the diets that just preceded the minimum-cost diet in the hope that one of them would be acceptable.   Or, finally, we could reevaluate the linear-programming formulation of the problem to see if it really described the problem under investigation.

To correct the deficiencies inherent in the solution to the original problem, a reformulation of this linear-programming problem is called for.   The new model should enable more than nine foods to be in the minimum solution and also take into consideration human taste preferences for certain foods. These elements can be introduced into the model by inequalities that force some foods to be into the final solution in at least a minimum amount. We could also determine preference weights and add them to the corresponding cost coefficients. Another approach to the general diet problem that would introduce more variety would be to subdivide the problem into smaller diet problems, each of which would involve only a single class of foods. In this process of suboptimizing, we might have the problem of selecting the minimum-cost diets for vegetables, or fruits, or meats, and the composite diet would be a solution to the general problem. In these attempts to introduce more realism into the problem, we suffer a corresponding increase in the cost of the resulting diet. If the problem is further constrained, the cost, in general, will increase, and it is up to the investigator to determine the relative value of dollar cost versus taste and variety.[1]

An advance over the standard diet problem is to interpret the problem setting in terms of selecting a daily set of menus which collectively meet the desired minimum daily requirements (Balintfy [34]). Menu planning is designed to overcome the inability of the diet-problem formulation to meet the human requirements of palatability and variety, while still achieving the nutritional and economic objectives. A menu item is a serving of food and may be an individual item like orange juice or combination of foods as found in beef stew. These items are then classified as to menu component (breakfast, appetizer, dinner entree, dessert, etc.). Given the nutrient content of each menu item, the problem is to select components which form a breakfast menu, lunch menu, and dinner menu which meet the minimum daily requirements at least cost. Here, for example, we have to ensure against the selecting of more than one lunch entree or more than one dessert at dinner.

A single menu problem is contained in the data of Tableau 11.2 (Balintfy

TABLEAU 11.2

	Menu items					
	Entrees (serving)			Desserts (serving)		
	Stew x_1	Ham x_2	Veal x_3	Gelatin x_4	Cobbler x_5	Cake x_6
Protein (g)	16	20	25	15	21	25
Vitamin C (mg)	20	15	12	10	8	5
Cost (cents)	23	30	38	20	25	35

[1] The reader is referred to work by Wolfe [710] where he solves a modified Stigler diet problem with a quadratic preference (objective) function. This treatment introduces more realism and variety into the optimal diet.

[35]). This problem is to select an entree and dessert to meet the stipulated requirements of 38 g of protein and 20 mg of vitamin C.

The *integer-programming problem* is to minimize

$$23x_1 + 30x_2 + 38x_3 + 20x_4 + 25x_5 + 35x_6$$

subject to

$$16x_1 + 20x_2 + 25x_3 + 15x_4 + 21x_5 + 25x_6 \geq 38$$
$$20x_1 + 15x_2 + 12x_3 + 10x_4 + 8x_5 + 5x_6 \geq 20$$
$$x_1 + x_2 + x_3 \qquad\qquad = 1$$
$$x_4 + x_5 + x_6 = 1$$

and

$$x_j = 0 \text{ or } 1$$

The two GUB equations ensure that only one entree and only one dessert will be picked. Similar generalized upper-bound constraints would be required for each menu component, e.g., green vegetables. Of course, we need to guard against formulating a problem which is infeasible in that such a restricted selection process may not allow for a set of menu items to be selected which meet the requirements. However, since the real-life problem is one in which the dietitians do resolve (although not necessarily at least cost), infeasibility here would probably indicate that the constraints do not reflect the problem statement. The optimal integer solution for the above problem is $x_2 = 1$ and $x_5 = 1$, with a cost of 55 cents. The optimal continuous linear-programming solution is $x_1 = \frac{8}{9}$, $x_3 = \frac{1}{9}$, and $x_5 = 1$, with a cost of 49.67 cents. The rounded answers of $x_1 = 1$ and $x_5 = 1$ do not form a feasible solution.

Other types of considerations can be handled using the integer formulation; e.g., certain items with starches or which are of the same color should not be selected on the same menu. An inequality of the form $x_2 + x_4 + x_6 \leq 1$, with the variables either 0 or 1, accomplishes this restricted selection process. If the menus are to be planned over time, e.g., the selection of a set of menus for a hospital over the next 30 days, then we must also guard against menu items appearing too often during the planning horizon. This can be accomplished by changing the right-hand side of the equalities from 1 to 30, adding appropriate integer upper bounds on each menu item, and multiplying the minimum daily requirements by 30. This would allow for a solution which selected the correct number of entrees, desserts, etc., which then have to be grouped into menus and meals. The daily menus would then meet the minimum daily requirements on the average (see Gue [340] for a discussion on related formulations designed to overcome this concern). Another approach would be the successive daily solution of a restricted problem, i.e., a menu-planning problem in which items are not allowed to enter the current day's solution if their past usage exceeds dietary and variety specifications. The use of noninteger procedures to solve large menu-planning problems appears to be effective in that the optimal or near-optimal integer solutions are found or the continuous solution can be rounded to an acceptable, low-cost feasible solution (Balintfy [35]).

There are many situations where the linear-programming formulation of the basic diet problem is directly applicable. These problems are concerned with minimum-cost feed mixtures for farm animals or with the mixing of various elements, e.g., chemicals or fertilizers, to meet requirements at least cost. The application of linear-programming techniques to these problems is fairly straightforward; see, for example, Waugh [690], Goldstein [319], Brigham [95], Jewell [401], and Sec. 1 of Bibliography.

4. NETWORK-FLOW PROBLEMS[1]

Let us first consider the mathematical formulation of the following problem. We are given a *transportation network* (pipeline system, railroad system, communication links) through which we wish to send a homogeneous commodity (oil, freight cars, message units) from a particular point of the network called the *source node* to a designated destination called the *sink node.* In addition to the source and sink nodes, the network consists of a set of *intermediate nodes* which are connected to each other or to the source and sink nodes by *arcs* or *links* of the network. These intermediate nodes can be interpreted as switching or *transshipment points.* We shall label the source node 0 and the sink node m and refer to the intermediate nodes by numbers or as node $i, j, k, \ldots$. We designate the arc connecting nodes i and j by the ordered pair (i, j) and assume that the flow of the commodity is directed from i to j; that is, an arc is a one-way street. If the flow can also go from j to i, the network would include both *directed arcs* (i, j) and (j, i). Each arc can accommodate a nonnegative flow, and we shall assume that each arc has a finite upper bound on its capacity which we designate u_{ij}. If we let x_{ij} be the unknown flow from node i to node j, then we have $0 \le x_{ij} \le u_{ij}$.

The flow of the commodity which originates at the source 0 is sent along the arcs to the intermediate nodes and then transshipped along additional arcs to other intermediate nodes or to the sink until all the commodity which began the trip at node 0 finally arrives at node m. That is, we impose upon the network the condition of *conservation of flow* at the nodes; what is shipped into a node is shipped out. The problem is to determine the maximum amount of flow f which can be sent from the source to the sink. We also, of course, wish to determine which arcs are used and to what extent of their capacity. A typical network is shown in Fig. 11.5. The number of each arc represents the capacity u_{ij} on arc (i, j).

---

[1] This section is based on lectures by Prof. Robert Oliver, University of California, Berkeley, CA, and Ford and Fulkerson [253].

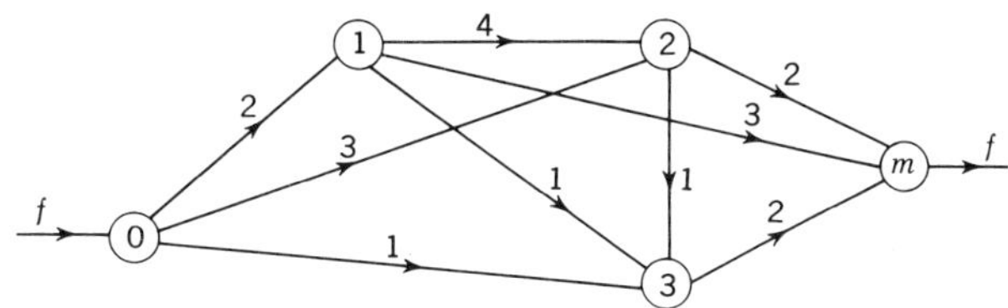

FIGURE 11.5

We shall use the convention that if an arc (i, j) does not exist, we let $u_{ij} = 0$; that is, all arcs are possible, but flow is restricted to that set for which the corresponding $u_{ij} > 0$.   Since we require the total flow into a node to equal the total flow out of it, the linear-programming model of the *maximal-flow network problem* is given by the following: Maximize

$$f$$

subject to

$$\sum_j x_{0j} - \sum_j x_{j0} - f = 0 \tag{4.1}$$

$$\sum_j x_{ij} - \sum_j x_{ji} = 0 \qquad i \neq 0, m \tag{4.2}$$

$$\sum_j x_{mj} - \sum_j x_{jm} + f = 0 \tag{4.3}$$

$$0 \leq x_{ij} \leq u_{ij} \tag{4.4}$$

Equation (4.1) represents the conservation of flow at the source; i.e., the total amount shipped from node 0 minus the total amount shipped back to node 0 must equal the total flow in the network; Eqs. (4.2) represent the conservation at the intermediate nodes, while Eq. (4.3) represents the conservation equation at the sink.   We note that the x_{ij} are nonnegative variables with finite upper bounds and that $x_{ij} = 0$, $f = 0$ is a feasible solution.   Hence, the maximal-flow problem is always feasible; i.e., the problem is a feasible linear-programming problem.

Let us write in explicit form the matrix of the network of Fig. 11.5.   If we let the arcs correspond to the columns of the matrix and the nodes correspond to the rows, we obtain the following tableau (ignoring the capacity constraints):

nodes \\ arcs	x_{01}	x_{02}	x_{03}	x_{12}	x_{13}	x_{1m}	x_{23}	x_{2m}	x_{3m}	f	
0	1	1	1							-1	$= 0$
1	-1			1	1	1					$= 0$
2		-1		-1			1	1			$= 0$
3			-1		-1		-1		1		$= 0$
m						-1		-1	-1	1	$= 0$

$$\tag{4.5}$$

The matrix of the equations is called the *node-arc incidence matrix.* For such matrices we note that one of the equations is redundant. We can, for example, drop the last equation which is given by the negative sum of the remaining equations. If we were to solve this problem by the simplex method, we would couple m equations from (4.5) with the corresponding upper-bound constraints and apply the simplex algorithm for bounded variables. The objective function would be to maximize f. However, because of its simple structure, more efficient procedures, to be discussed below, are available. The reader will note that any problem which takes on the form of a node-arc incidence matrix can be interpreted as a network problem; i.e., each column has one $+1$ and no more than one -1. Before describing the procedure for solving the maximal-flow problem, we shall introduce some appropriate concepts from the theory of graphs and networks (Lawler [462], Tucker [656], Wilson [706], Ford and Fulkerson [253], Busacker and Saaty [102], and Simonnard [611]).

A *graph* denoted by $G = (N, A)$ is a finite set N of elements $i, j, k, \ldots$ called *nodes* and a finite set A of unordered pairs of nodes (i, j) called *arcs.* The arc (i, j) joins the nodes i and j. A *directed graph* (or *digraph*) is a graph in which each arc (i, j) is an ordered pair that stipulates that the flow or direction between nodes i and j is restricted to be from i to j. The digraph can also contain arc (j, i). A *network* is a graph all of whose arcs (i, j) have been assigned a nonnegative number u_{ij} termed the *capacity.* We shall discuss only *directed networks,* i.e., networks that correspond to a directed graph. We indicate the direction of an arc pictorially by an arrow, as shown in Fig. 11.5. We assume that our networks have $m + 1$ nodes that are indexed from $0, 1, 2, \ldots, m$, with the source node designated by 0 and the sink node by m.

A sequence of arcs (i, j), (j, k), $\ldots$, (p, q) is called a *chain* which connects node i to node q. If a graph has such a chain, then nodes i and q are said to be *connected.* If $i = q$, then the chain is termed a *cycle* or *loop.* If we ignore the directions of the arcs, then an *undirected chain* or *path* from i to q is any sequence of arcs which form a connected set of arcs from i to q. A *connected graph* is a graph in which all pairs of nodes are connected. A *tree* is a connected graph that contains no cycles. A tree has the property that there is a unique chain joining each pair of nodes in the tree; also, a tree on n nodes has exactly $n - 1$ arcs. For Fig. 11.5, we have the tree formed by the arcs $(0, 1)$, $(1, 2)$, $(2, 3)$, $(3, m)$. Directed trees are also called *rooted trees.* A *root of a tree* is a designated node such that there is a unique chain from the root to any other node in the tree. A rooted tree has a unique root. Node 0 is the root for the tree in Fig. 11.5. An undirected tree can be rooted by selecting one node as the root and directing all arcs away from the root. A tree that connects all nodes of a graph G is termed a *spanning tree.*

Let all the nodes of a given network be divided into two disjoint subsets S and $\bar{S}$; i.e., a node will be either in S or $\bar{S}$, but not in both. In addition, let $(S, \bar{S})$ denote the set of arcs that leads from those $i \in S$ to those $j \in \bar{S}$, where the symbol $\in$ means "contained in." For example, in Fig. 11.5, if $S = (0, 1, 2)$ and $\bar{S} = (3, m)$, then $(S, \bar{S}) = (0, 3), (1, 3), (2, 3), (1, m), (2, m)$. A *cut* is the set of arcs $(S, \bar{S})$ with $0 \in S$ and $m \in \bar{S}$.[†] For Fig. 11.5, $S = (0, 1, 2)$, and $\bar{S} = (3, m)$, the cut is shown by dotted lines in Fig. 11.6. A network has a finite number of cuts. Another cut for the given network is $(0, 1), (0, 2), (0, 3)$, as shown in Fig. 11.7; here $S = (0)$ and $\bar{S} = (1, 2, 3, m)$. We define the *capacity of a cut* $f(S, \bar{S})$ to be the sum of the individual capacities of all the arcs in the cut; that is,

$$f(S, \bar{S}) = \sum_{\substack{i \in S \\ j \in \bar{S}}} u_{ij}$$

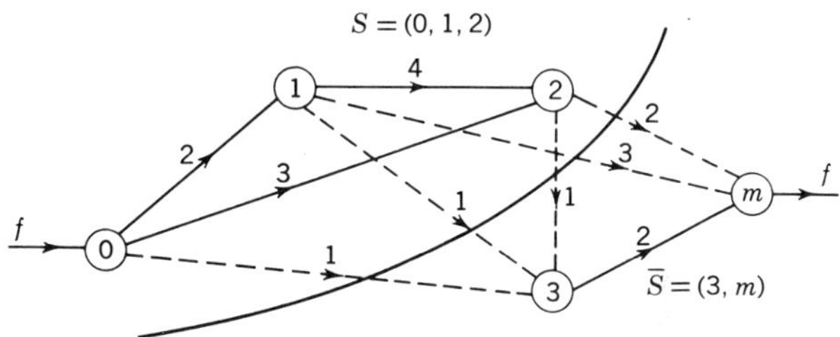

FIGURE 11.6

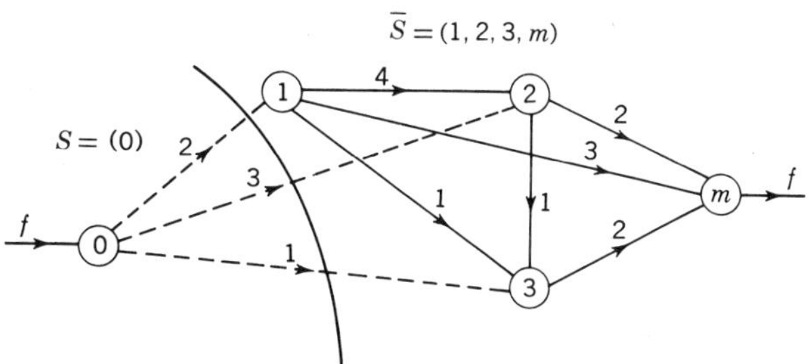

FIGURE 11.7

The cut in Fig. 11.6 has a capacity of 8, while the cut in Fig. 11.7 has a capacity of 6. Since if we remove a cut from a network there would be no chains from source to sink, i.e., the maximal flow would be zero, the maximal flow cannot exceed the capacity of the minimal cut capacity. The cut of Fig. 11.6 limits the flow to no more than 8, while the cut of Fig. 11.7 limits the flow to no more than 6. The following theorem, the *max-flow min-cut theorem*, establishes the existence of the maximum flow whose capacity is exactly equal to the minimal cut capacity. For our example, this is a flow of 6.

[†] An alternative definition of a cut is that it is a set of directed arcs such that every chain from the source node to the sink node contains at least one arc of the cut. Hence, if all the arcs of a cut were deleted from the network, there would be no chain from 0 to m.

Max-flow min-cut theorem. *For any network the maximal-flow value from node 0 to node m is equal to the minimal cut capacity.*

Proof: The proof is given in two parts. We first show that the maximum flow is less than or equal to the minimum cut and then show that a flow exists which is equal to the minimal cut capacity.

For any cut $(S, \bar{S})$ we sum Eqs. (4.1) and (4.2) for all $i \in S$ to obtain

$$f = \sum_{i \in S} \left(\sum_j x_{ij} - \sum_j x_{ji} \right)$$

This can be rewritten as

$$f = \sum_{\substack{i \in S \\ j \in S}} x_{ij} + \sum_{\substack{i \in S \\ j \in \bar{S}}} x_{ij} - \sum_{\substack{i \in S \\ j \in S}} x_{ji} - \sum_{\substack{i \in S \\ j \in \bar{S}}} x_{ji}$$

or

$$f = \sum_{\substack{i \in S \\ j \in \bar{S}}} x_{ij} - \sum_{\substack{i \in S \\ j \in \bar{S}}} x_{ji} + \left(\sum_{\substack{i \in S \\ j \in S}} x_{ij} - \sum_{\substack{i \in S \\ j \in S}} x_{ji} \right) \tag{4.6}$$

As both indices i and j of the term in parentheses range over the same set of values, this term will be equal to zero. We then have

$$f = \sum_{\substack{i \in S \\ j \in \bar{S}}} x_{ij} - \sum_{\substack{i \in S \\ j \in \bar{S}}} x_{ji} \tag{4.7}$$

Since each flow $x_{ji} \geq 0$ and each $x_{ij} \leq u_{ij}$, (4.7) becomes

$$f \leq \sum_{\substack{i \in S \\ j \in \bar{S}}} x_{ij} \leq \sum_{\substack{i \in S \\ j \in \bar{S}}} u_{ij} = f(S, \bar{S}) \tag{4.8}$$

and hence, the maximal flow is bounded above by the capacity of the arbitrary cut $(S, \bar{S})$, and thus f must be bounded above by the minimal cut capacity; that is, $f \leq \min f(S, \bar{S})$ for all possible cuts.

We next establish that for some flow and corresponding cut (to be defined), the equality of (4.8) holds. Since all arc capacities are finite, a maximum flow exists, although its value could be zero. Let the value of the maximal flow be f' and for this f' and corresponding arc flows x_{ij} we define a cut $(S', \bar{S}')$ as follows.

Let S' include the source node 0. For any $i \in S'$ if $x_{ij} < u_{ij}$, then $j \in S'$. In addition, if $i \in S'$ and $x_{ji} > 0$, then $j \in S'$. Let $\bar{S}'$ be all nodes not in S'. We first note that the set of nodes S' includes the sink node m. If $m \notin \bar{S}'$, that is, both nodes 0 and m are in S', then we can find a path from 0 to m such that for any arc in the path either $u_{ij} - x_{ij} > 0$ or $x_{ji} > 0$. We could then increase the flow from 0 to m by adding a small amount to those arcs for which $u_{ij} - x_{ij} > 0$ and subtracting the same amount from those arcs having $x_{ji} > 0$. Hence, we would have a flow from 0 to m which would be greater than f', contradicting the assumption that f' is the maximal flow. Therefore $m \in \bar{S}'$, and $(S', \bar{S}')$ is a cut. We next note that $x_{ij} = u_{ij}$ when $i \in S'$ and $j \in \bar{S}'$, and $x_{ji} = 0$ for $i \in S'$ and $j \in \bar{S}'$. This results from the definition of S' and $\bar{S}'$. Then for the cut $(S', \bar{S}')$, Eq. (4.7) becomes

$$ f = \sum_{\substack{i \in S' \\ j \in \bar{S}'}} x_{ij} - \sum_{\substack{j \in S' \\ j \in \bar{S}'}} x_{ji} = \sum_{\substack{i \in S' \\ j \in \bar{S}'}} x_{ij} = \sum_{\substack{i \in S' \\ j \in \bar{S}'}} u_{ij} = f(S', \bar{S}') $$

But from (4.8) $f \le f(S, \bar{S})$ for all cuts $(S, \bar{S})$ and since $f = f(S', \bar{S}')$, then $(S', \bar{S}')$ must yield the minimal cut capacity and max $f = f' = \min f(S, \bar{S}) = f(S', \bar{S}')$.

At this point it is instructive to look at the dual problem of the maximal-flow problem. Recalling that f is a nonnegative variable and letting π_i be the dual variables associated with the ith node constraint given by (4.1), (4.2), and (4.3), and w_{ij} be the dual variables associated with the arc-capacity constraints (4.4), the dual problem as given by the node-arc incidence matrix is as follows: Minimize

$$ \sum_i \sum_j u_{ij} w_{ij} $$

subject to

$$
\begin{aligned}
-\pi_0 + \pi_m &\ge 1 \\
\pi_i - \pi_j + w_{ij} &\ge 0 \qquad \text{all } (i, j) \\
w_{ij} &\ge 0 \qquad \text{all } (i, j)
\end{aligned}
$$

The π_i are unrestricted as to sign as they correspond to primal equations. If we know a minimal cut $(S, \bar{S})$, we can construct the optimal solution to the dual problem as follows:

$$
\begin{aligned}
\pi_i &= 0 \qquad \text{for } i \in S \\
\pi_i &= 1 \qquad \text{for } i \in \bar{S} \\
w_{ij} &= 1 \qquad \text{for } (i, j) \in (S, \bar{S}) \\
w_{ij} &= 0 \qquad \text{for } (i, j) \notin (S, \bar{S})
\end{aligned}
\tag{4.9}
$$

This is a feasible solution, and the minimum value of the dual objective function is equal to the maximum value of the primal objective function. As pointed out in Ford and Fulkerson [253], if we drop the redundant source equation, i.e., take $\pi_0 = 0$, it can be shown that all extreme-point solutions of the dual are of the form given by (4.9) for some S with $0 \in S$. Hence, the max-flow min-cut theorem would result from the duality theorem of linear programming.

We next describe an algorithm termed the *labeling method* for solving maximal-flow problems (Ford and Fulkerson [253, 255]). The process converges in a finite number of iterations if the arc capacities u_{ij} are all rational numbers.[1]   Its validity is based on the max-flow min-cut theorem.   The steps of the algorithm are given below and are designed to find a path over which a positive flow can be sent from source to sink.   The steps are repeated until no such path can be found.

Maximal-flow algorithm

STEP 1   Find an initial feasible solution for the network.   We may, for example, begin with all $x_{ij} = 0$.

STEP 2   Start with the source node 0 and give it a label $[-, \infty]$.   The general label for any nodes i and j is indicated by $[i\pm, v_j]$, with v_j being a positive number representing a change in flow between i and j, $i+$ representing an increase of flow by the amount v_j from i to j, and $i-$ representing a decrease of flow by the amount v_j from j to i.   The source label $[-, \infty]$ indicates that an unlimited amount of commodity is available at the source for shipment to the sink.

STEP 3   Select any unlabeled node i.   Initially only node 0 is labeled.

a.   For any unlabeled node j for which $x_{ij} < u_{ij}$, assign the label $[i+, v_j]$ to node j, where $v_j = \min (v_i, u_{ij} - x_{ij})$.   This limits the amount sent from i to j to either the amount already sent to i or the remaining capacity of the arc connecting i to j, whichever is smaller.

b. To all nodes j that are unlabeled and such that $x_{ji} > 0$, assign the label $[i-, v_j]$, where $v_j = \min (v_i, x_{ji})$.   This enables us to reroute a flow going into i away from i.   (For example, in Fig. 11.8 if we started with the initial solution $x_{01} = 2$, $x_{12} = 1$, $x_{20} = 1$, $x_{1m} = 1$, $x_{2m} = 0$, we have, starting with the origin label $[-, \infty]$, that the label for node 2 should be $[0-, 1]$ since $v_2 = \min (\infty, x_{20} = 1) = 1$.)

As node j receives a label, it is processed through Step 3 until all labeled nodes have been looked at with respect to the unlabeled nodes connecting them.   Either this process will bring us along a path to the sink node which then gets a label (Step 4), or we cannot find such a path which leads to the sink (Step 5).

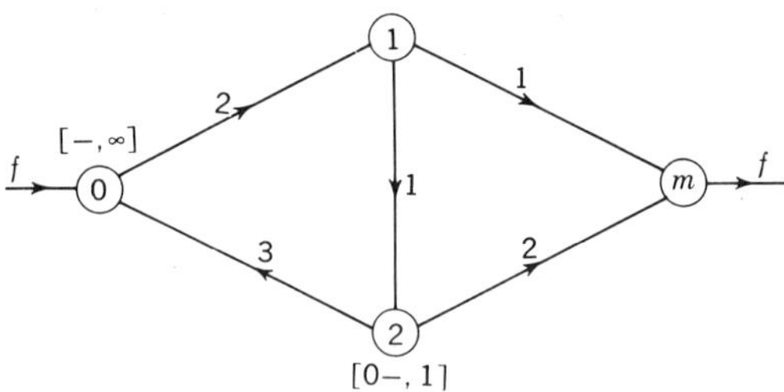

FIGURE 11.8

[1] See Edmonds and Karp [233] and Lawler [462] on how the algorithm can be modified to converge for irrational arc capacities.

STEP 4 The sink node has received a label $[i\pm, v_m]$ with $i \neq m$ and $v_m > 0$ by the selection rule for v_m. Hence we have found what is termed a flow-augmenting path and can add or subtract v_m to the arcs of the path leading from 0 to m. The new feasible solution is

$$x'_{ij} = x_{ij} + v_m \qquad \text{for } (i, j) \text{ in path and } j \text{ has label } [i+, v_j]$$

$$x'_{ji} = x_{ji} - v_m \qquad \text{for } (j, i) \text{ in path and } j \text{ has label } [i-, v_j]$$

$$x'_{ij} = x_{ij} \qquad \text{for } (i, j) \text{ not in path}$$

The labels are all erased, and we repeat the process starting with Step 2 and the new feasible flow x'_{ij}.

STEP 5 The process terminates in that we cannot find a path from source to sink. The maximal flow f is equal to the sum of all the v_m generated in the applications of Step 3, assuming the initial solution all $x_{ij} = 0$. To show that this flow is maximal, we let the set of labeled nodes in the final iteration correspond to the set of nodes S' in the proof of the max-flow min-cut theorem.

We illustrate the algorithm by finding the value of the maximal flow for the network of Fig. 11.5. We let the initial feasible solution be $x_{ij} = 0$ for all i, j. We indicate the current feasible solution and the arc capacities by the ordered pair of numbers (x_{ij}, u_{ij}) attached to each arc of the network as shown in Fig. 11.9. The node labels are indicated by $[i\pm, v_j]$. Along the flow-augmenting path $(0, 1, 2, m)$ of Fig. 11.10 we can send an additional two units of flow $(v_m = 2)$ which yields a new feasible solution as shown in Fig. 11.11. We again find an unsaturated path as shown by the labels of Fig. 11.11. Here $v_m = 1$. The new feasible solution which sends a total of three units from source to sink is shown in Fig. 11.12, along with a new path over which an additional unit can be sent. The resulting feasible solution is shown

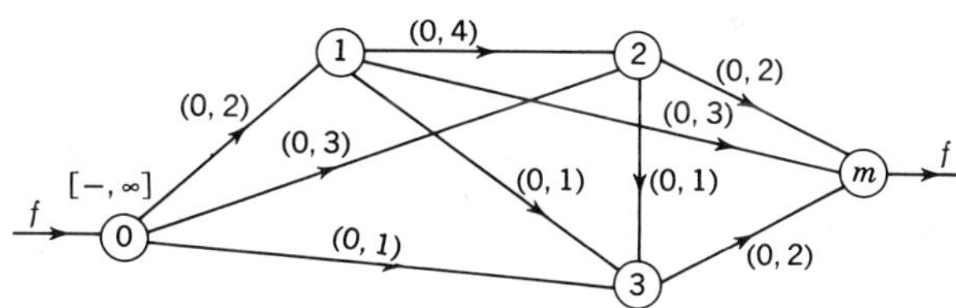

FIGURE 11.9

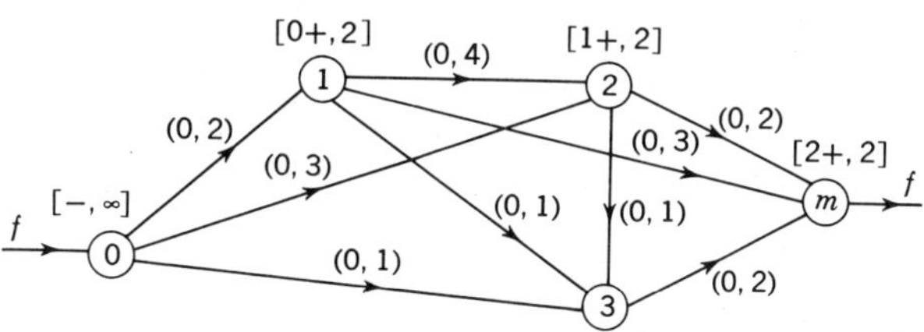

FIGURE 11.10

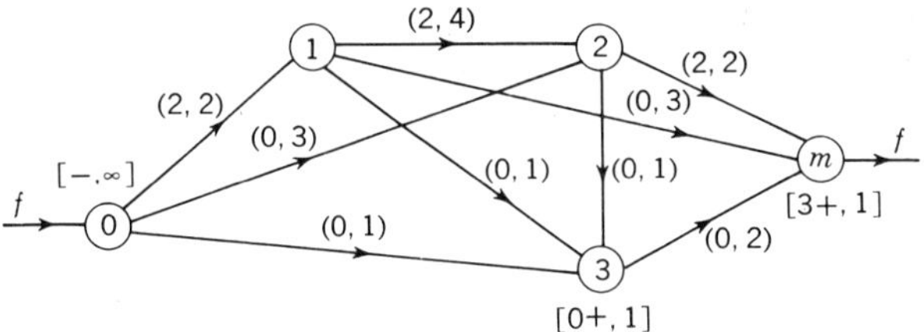

FIGURE 11.11

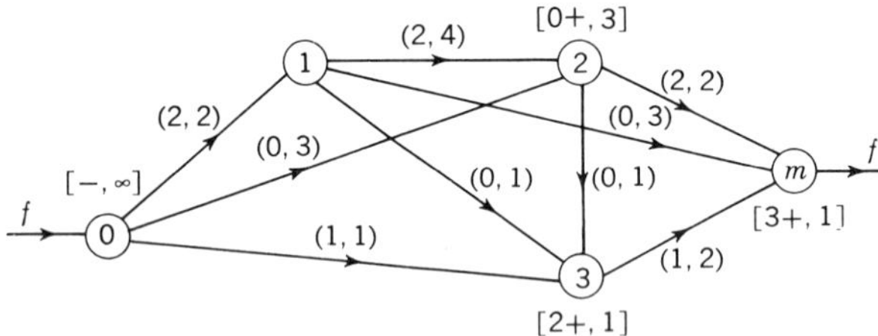

FIGURE 11.12

in Fig. 11.13, along with a new unsaturated path for which $v_m = 2$. The reader will note that the finding of the labeled path of Fig. 11.13 required an application of the rerouting Step 3*b*. The corresponding feasible solution is shown in Fig. 11.14. This solution has a flow of six units and is optimum as we cannot label any node except the origin. The corresponding cut is given by arcs (0, 1), (0, 2), (0, 3) and has a cut capacity of 6.

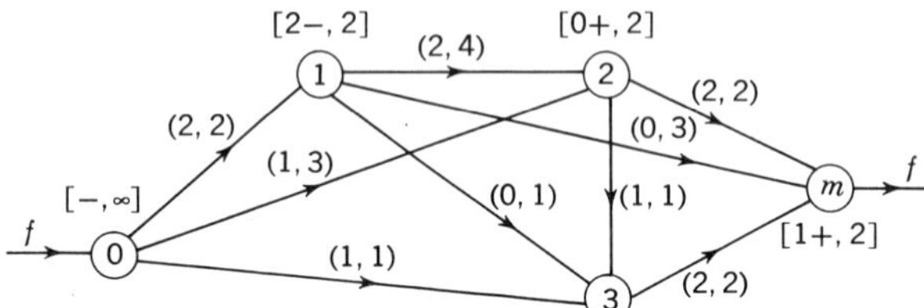

FIGURE 11.13

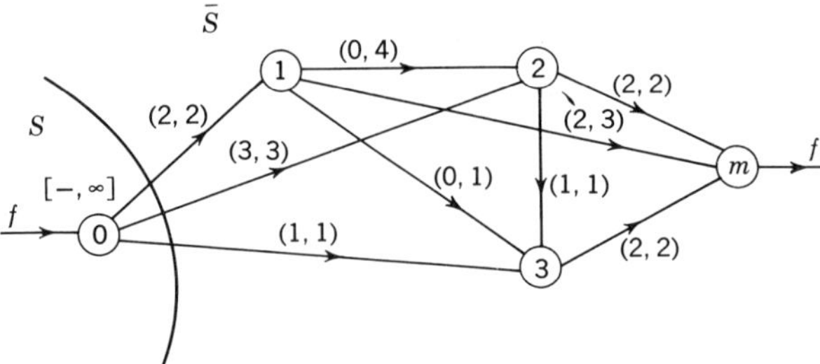

FIGURE 11.14

Many node-labeling flow-augmenting algorithms have been developed to solve the maximal-flow problem (see Cheung [124]). However, studies comparing maximal-flow algorithms have shown that adaptations of the primal-simplex method to the structure of the maximal-flow network, when combined with advanced computer-based data organization techniques, are more efficient (Glover et al. [312], Glover and Klingman [311]).

A number of modified maximal-flow network problems can be formulated in terms of the basic network model described above. For example, if a network has two (or more) sources, it can be recast into a single-source network by joining the given sources to a third source by artificial arcs which have infinite capacity, and similarly if we have multiple sinks. The added source or sink is identified by 0 or m, respectively. Another variation is found when in addition to the capacities on the arcs we also have node capacities. These latter capacities could, for example, represent the maximum amount of the commodity that can be processed by the facilities at the node. We transform this problem into a maximal-flow problem by replacing node i with capacity k_i by two nodes i' and i'', where node i' is connected to the network by the same arcs going into i, nodes i' and i'' are connected by a directed arc from i' to i'' with capacity k_i (that is, $u_{i'i''} = k_i$), and node i'' is connected to the network by the arcs going out of i. In Fulkerson [269] it is shown how to solve the maximal-flow problem in which the capacities of the arcs of the network can be increased within the limits of a given budget.

If the arc capacities u_{ij} are integers, then the maximal-flow problem defined by Eqs. (4.1) to (4.4) has an optimal integer solution. This can be seen by noting that the problem matrix [after converting (4.4) to equations] is *unimodular*, i.e., every minor determinant is equal to 0, $+1$, or -1.[1] Thus, any basis **B** to the maximal-flow problem has an inverse with integral components. The matrix formed by the $m + 1$ conservation of flow equations (4.1) to (4.3) also has the unimodular property. In fact, if we drop one equation, e.g., Eq. (4.3), then we can show that the rank of the remaining equations is m and that any basis is triangular. As was done for the transportation problem, these properties can be exploited when applying the revised simplex algorithm to solve network-structured problems (see Exercise 64).

Another important class of network problems is the set of *minimal-cost network-flow problems*. As we shall see, this class of problems includes, among others, the transportation problem and the shortest-route problem. We shall first develop the general formulation and then specialize it to particular problems.

For the minimal-cost flow problem we are given, as in the maximal-flow problem, a general network over which units of a homogeneous commodity are to be shipped from the source to the sink. Associated with each arc (i, j) is a cost c_{ij} of shipping one unit of the commodity from node i to node j.[2] We

---

[1] See Hoffman and Kruskal [376] for a discussion on unimodular matrices and linear programming.

[2] We assume a linear cost relationship.

must ship a given quantity of F units from the source to the sink so as to minimize the total cost of shipping the F units to the sink. Assuming conservation of flow at the nodes and that the flow x_{ij} along an arc is bounded by both lower- and upper-capacity restrictions, $0 \leq l_{ij} \leq x_{ij} \leq u_{ij}$, the linear-programming model of the *minimal-cost network-flow problem* is given by the following: Minimize

$$\sum_i \sum_j c_{ij} x_{ij} \tag{4.10a}$$

subject to

$$\sum_j x_{0j} - \sum_j x_{j0} = F \tag{4.10b}$$

$$\sum_j x_{ij} - \sum_j x_{ji} = 0 \qquad i \neq 0, m \tag{4.10c}$$

$$\sum_j x_{mj} - \sum_j x_{jm} = -F \tag{4.10d}$$

$$0 \leq l_{ij} \leq x_{ij} \leq u_{ij} \tag{4.10e}$$

A typical arc of a minimal-cost network is identified as shown in Fig. 11.15.   It is usually assumed that all coefficients are integers.   An example of a minimal-cost network with $F = 5$ is shown in Fig. 11.16.   Note that $F \leq f$ (the network's minimal cut capacity) for the problem to be feasible.   Given integer values for F and the u_{ij}, the unimodular property of the system (4.10b) to (4.10e) guarantees that an optimal integer solution exists (Exercise 64).

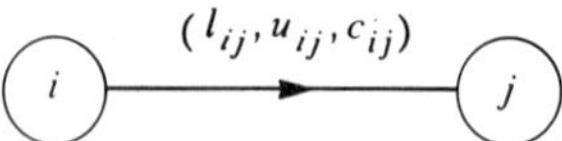

FIGURE 11.15

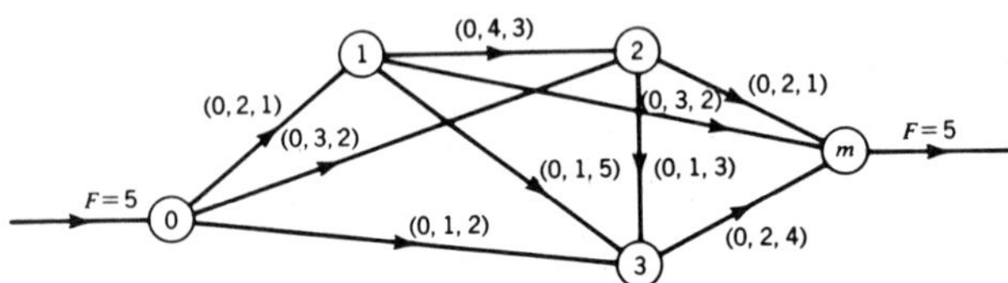

FIGURE 11.16

We next develop the dual problem to the primal minimal-cost network problem (4.10a) to (4.10e).   The dual variables are the π_i corresponding to the node i conservation of flow equations and the v_{ij} and w_{ij} associated with the lower and upper bounds of each arc x_{ij}, respectively.   The dual problem is then the following: Maximize

$$(\pi_0 - \pi_m)F + \sum_i \sum_j l_{ij} v_{ij} - \sum_i \sum_j u_{ij} w_{ij} \tag{4.11a}$$

subject to

$$\pi_i - \pi_j + v_{ij} - w_{ij} \leq c_{ij} \tag{4.11b}$$

$$(v_{ij}, w_{ij}) \geq 0 \quad \text{and} \quad \pi_i \text{ unrestricted} \tag{4.11c}$$

Complementary slackness considerations (Chap. 6) for this primal-dual pair yields the optimality conditions of

$$x_{ij} > 0 \qquad \text{implies } \pi_i - \pi_j + v_{ij} - w_{ij} = c_{ij}$$
$$v_{ij} > 0 \qquad \text{implies } x_{ij} = l_{ij}$$
$$w_{ij} > 0 \qquad \text{implies } x_{ij} = u_{ij}$$

or

$$x_{ij} = l_{ij} \qquad \text{implies } \pi_i - \pi_j \leq c_{ij}$$
$$l_{ij} < x_{ij} < u_{ij} \qquad \text{implies } \pi_i - \pi_j = c_{ij} \tag{4.12}$$
$$x_{ij} = u_{ij} \qquad \text{implies } \pi_i - \pi_j \geq c_{ij}$$

Conditions (4.12) are referred to as *kilter conditions* and the finding of a feasible flow that satisfies them is the essence of the *out-of-kilter algorithm* for solving minimum-cost network problems (Ford and Fulkerson [253], Fulkerson [268], Simonnard [612], Durbin and Kroenke [225], Barr et al. [40], Minty [503], Potts and Oliver [550], and Lawler [462]).   (The kilter conditions are the optimality conditions for a simple upper-bounded problem, here augmented with lower-bound conditions; see Theorem 1 of Sec. 5, Chap. 9, and let $\pi_i - \pi_j = z_j$.)   Although computational experience with the out-of-kilter algorithm has proven satisfactory, specializations of the revised simplex method to the minimal-cost network problem have greatly advanced our ability to solve very large problems efficiently.   Such primal *network-simplex algorithms* take advantage of the tree structure of basic solutions and employ advanced computational data structures and pivot strategies in their computer implementation (see Exercise 41) (Cunningham [146, 147], Barr, Glover, and Klingman [41, 42, 43], and Bradley, Brown, and Graves [90]).   As Glover and Klingman [311] point out, current primal network-simplex algorithms are substantially superior to out-of-kilter methods.[1]

The minimal-cost network problem can be generalized to include a number of sources and sinks, with each source, sink, and intermediate node able to serve as a transshipment point for the goods flowing through the network.   Let $a_i > 0$ be the amount of goods to be shipped from source node i and $b_i > 0$ be the amount of goods to be received by sink node i.   We assume that $\sum_{i \text{ a source}} a_i = \sum_{i \text{ a sink}} b_i = F$.   Applying the conservation of flow assumption, we then have the following *minimal-cost transshipment problem*: Minimize

$$\sum_i \sum_j c_{ij} x_{ij} \tag{4.13a}$$

---

[1] Computational approaches for solving the minimal-cost problem that are not based on linear-programming concepts are given in Hu [383].

subject to

$$\sum_j x_{ij} - \sum_j x_{ji} = a_i \qquad i \text{ a source node} \qquad\qquad (4.13b)$$

$$\sum_j x_{ij} - \sum_j x_{ji} = 0 \qquad i \text{ an intermediate node} \qquad (4.13c)$$

$$\sum_j x_{ij} - \sum_j x_{ji} = -b_i \qquad i \text{ a sink node} \qquad\qquad (4.13d)$$

$$0 \le l_{ij} \le x_{ij} \le u_{ij} \qquad\qquad\qquad\qquad\qquad\qquad\qquad (4.13e)$$

An example of a two-source and two-sink transshipment network is shown in Fig. 11.17. The transshipment problem can be transformed into the single-source and single-sink format of problem (4.10a) to (4.10e) by adding a fictitious source with flow in of F and a fictitious sink with flow out of F. Each original source is connected to the fictitious source by an arc having lower and upper bounds both equal to the corresponding a_i; similarly, each original sink is connected to the fictitious sink by an arc having lower and upper bounds equal to b_i. In both cases, the costs of the new arcs are zero. The transshipment problem with arc capacities can also be transformed into an uncapacitated standard transportation problem (Orden [537], Wagner [683], Lawler [462], and Papadimitriou and Steiglitz [541]). In turn, as described below, the transportation problem can be viewed as a network problem. If all a_i, b_i, l_{ij}, and u_{ij} are integers, then the system (4.13a) to (4.13e) will have an optimal integer solution.

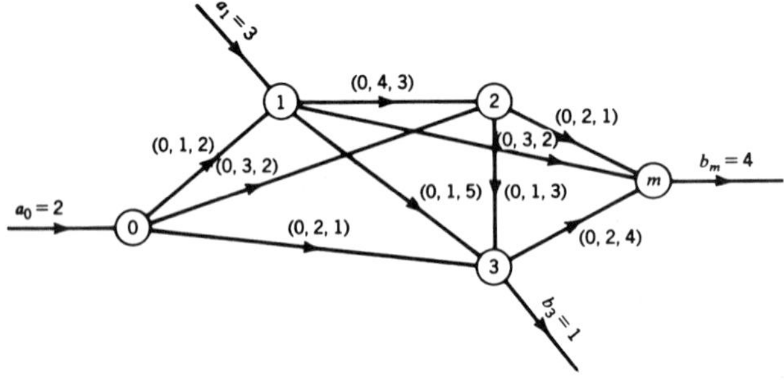

FIGURE 11.17

    The structure of many problems in industry and government can be formulated as assignment, transportation, or transshipment problems (Glover and Klingman, Sec. 13 of Bibliography). In particular, consider the manufacturer who has two factories that must meet the demands of two distribution centers. The product (e.g., automobiles) can be made in three models. The problem is to determine how many of each model should be produced at the factories and how many of each model should be shipped from the factories to meet the model demands at the centers so that the resulting production-distribution plan minimizes the total cost of production and distribution. The key to the proper network formulation of this problem is the designation of factory/model nodes and center/model nodes that enables the minimal-cost network to differentiate between model types.[1] The network for this problem (as given in Glover and Klingman) is shown in Fig. 11.18. Note that only models 1 and 2 are made in factory 2, and each arc can be given lower and upper bounds and costs depending on the requirements of the problem (see Exercise 44 for further discussion).

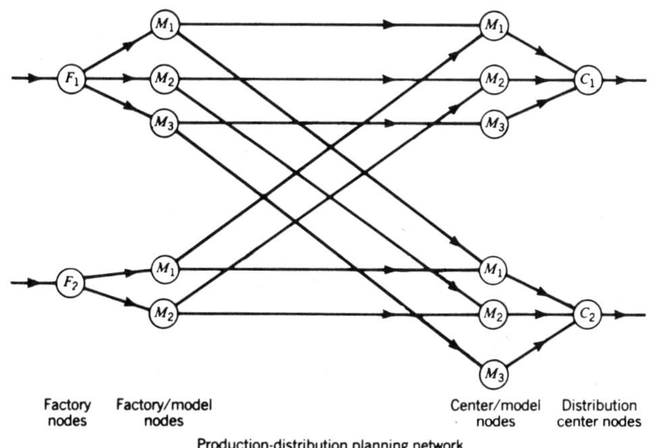

FIGURE 11.18

If we let the network be a roadmap with the source being the origin city, the sink the destination city, and the c_{ij} being distances between cities (intermediate nodes), we can convert the minimal-cost network problem $(4.10a)$ to $(4.10e)$ to that of finding the shortest distance between the origin and the destination by letting $F = 1$ and all $l_{ij} = 0$ and $u_{ij} = 1$. This problem can, of course, be solved by the regular techniques of linear programming, but more efficient, specialized algorithms exist, for example, the simple combinatorial labeling scheme of Ford and Fulkerson [253]; see also Dantzig [152] and Dijkstra [203]. This procedure, described below, also yields the shortest routes from the source to all other nodes.

Shortest-route algorithm (general network)

STEP 1 Assign all nodes a label of the form $[-, \pi_i]$, where the first component indicates the preceding node in the shortest route and π_i indicates the shortest distance from node 0 to node i. Node 0 starts with a label of $\pi_0 = 0$, and its index is always $-$; all other nodes start with $\pi_i = \infty$.

---

[1] These nodes allow the shipment of multicommodities, i.e., the three model types, in a single-commodity network format.

STEP 2   For any arc (i, j) for which $\pi_i + c_{ij} < \pi_j$, change the label of node j to $[i, \pi_i + c_{ij}]$ and continue the process until no such arc can be found.   In the latter situation the process is terminated, and the node labels indicate the shortest distance from node 0 to node j.

Ford and Fulkerson [253] prove that the algorithm will terminate under the assumption that the sum of the costs around any directed cycle is non-negative.   The simple example of Fig. 11.19 illustrates the algorithm.   The labels can be readily checked as being those which yield the minimum distances from the origin to the corresponding node.   The minimum distance is 11.

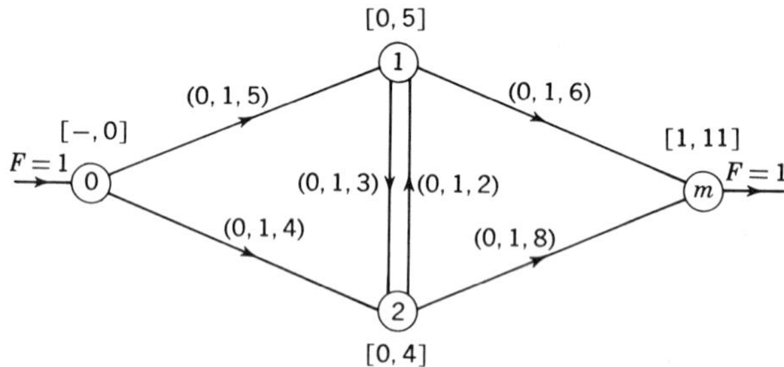

FIGURE 11.19

Networks which do not contain any cycles are termed *acyclic*.   The networks of Figs. 11.5 and 11.20 are acyclic, while the one of Fig. 11.19 is not.   Networks which are acyclic have the property that their nodes can be numbered sequentially from 0, 1, 2, ..., m, such that for any node i, the arcs leaving node i connect nodes j such that $j > i$.   This is illustrated in Fig. 11.20.   Also, if a shortest-distance network is acyclic, then the labeling algorithm is simplified; i.e., if we determine the shortest routes of each node from node 0 sequentially, then the node labels do not change.   Thus, for an acyclic network the shortest-route algorithm is as follows.

Shortest-route algorithm (acyclic network)

STEP 0   Number the nodes from 0 to m such that for any node i the arcs (i, j) connect nodes j such that $j > i$.   (The numbering can also be considered to order the nodes such that any arc ending at node j must come from a node i such that $j > i$.)

STEP 1   Assign all nodes a label as in Step 1 of the general shortest-route algorithm.

STEP 2   Determine $\pi_j = $ minimum $(\pi_i + c_{ij})$ sequentially for $j = 1, 2, \ldots,$ m.   (For $j = 1$, we have only to consider $\pi_1 = \pi_0 + c_{01} = c_{01}$.)

The node numbers in Fig. 11.20 illustrate the above steps and the shortest route.

There are a number of shortest-route (path) problems which we cite here for reference purposes.   Including the problem considered above, these prob-

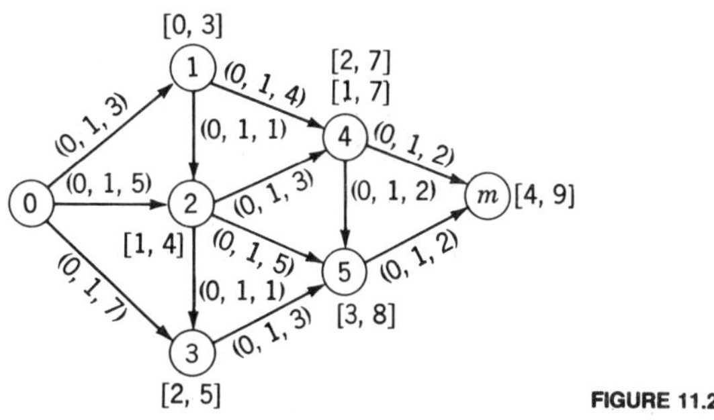

FIGURE 11.20

lems are (1) determining the shortest paths between two specified nodes of a network; (2) determining the shortest paths between all pairs of nodes of a network; (3) determining the second, third, etc., shortest path; (4) determining the fastest path through a network with travel times depending on departure time; and (5) finding the shortest path between specified intermediate points (Dreyfus [219]).[1]   The works of Dreyfus [219] and Witzgall [709] contain descriptions of associated algorithms which are discussed in terms of computational steps and computer-programming requirements.   As the problem of determining the shortest paths between all pairs of nodes is of importance, we note the references to the algorithms of Floyd [251], Hoffman and Winograd [379], and Yen [722].   The shortest-route problem can be transformed into an assignment problem (Lawler [462]); see Exercise 45.

The converse problem of finding the longest route through a network can be accomplished by appropriate changes in the above algorithms (see Exercise 18).   A particular application which considers the longest route is that of the management techniques termed PERT (program evaluation and review technique) or CPM (critical-path method), which are used to analyze the scheduling of construction, manufacturing, and other complex projects (Wiest and Levy [698] and Kelley [423]).   A project is divided into well-defined, independent activities (put up walls, lay shingles) and important events or milestones (painting finished, roof completed).   Each activity takes a certain length of time, and the activities are ordered in that, for example, the walls must be put up before they are painted.   If we picture the activities leading from one event to another and let activities be denoted by directed arcs and events by nodes, then, owing to the ordering requirements of the activities, the total project can be viewed as an acyclic network.   We can let Fig. 11.20 be a PERT network, with the distances interpreted as the time to complete an activity, node 0 the start of the project, node m the completion of the project, and the intermediate nodes specific events.   The minimum time required to complete the entire project is the largest total time for any path from node 0 to node m.   The corresponding path is termed the *critical path* and is the longest route

---

[1] As our networks are directed, the term "path" here and what follows coincides with the term "chain."

(time) from node 0 to node m.   The activities on the critical path represent the set of activities which need to be managed closely to ensure against slippage in the project-completion time.   Also, if these critical activities can be speeded up by a greater expenditure of resources, then the project time will be decreased accordingly until a new path becomes critical.   We can view the timing aspects of a PERT network in a linear-programming framework as follows.   Let variable x_i represent the occurrence time of event (node) i and let $t_{ij} \geq 0$ be the assumed duration of activity (arc) (i, j).   The objective is to determine the earliest start time for each activity which will then result in the shortest total project time; i.e., we want to minimize the time difference between event 0 and event m which is given by $x_m - x_0$.   The problem is then: Minimize

$$x_m - x_0$$

subject to

$$x_j - x_i \geq t_{ij} \qquad \text{for all } (i, j) \tag{4.14}$$

with the $x_j \geq 0$.[†]   As the start time of node 0 is arbitrary, it can be set equal to zero.   The dual of this problem is the finding of the longest path in the network from node 0 to node m, which corresponds to the critical path.   By the duality theorem the value of the longest path will be equal to the minimum value of x_m (see Exercise 26).   We illustrate the above formulation for the PERT network of Fig. 11.21, where the t on each arc represents the activity time.

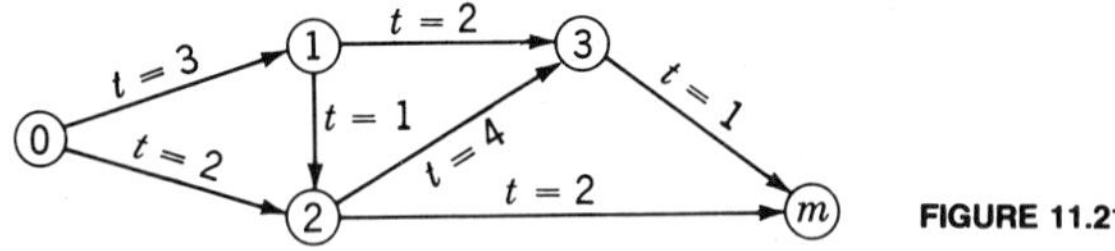

FIGURE 11.21

Minimize

$$x_m$$

subject to

$$
\begin{aligned}
x_1 &\geq 3 \\
x_2 &\geq 2 \\
x_2 - x_1 &\geq 1 \\
x_3 - x_1 &\geq 2 \\
x_3 - x_2 &\geq 4 \\
x_m - x_2 &\geq 2 \\
x_m - x_3 &\geq 1
\end{aligned}
$$

[†] See Exercise 4.

The optimal answer is $x_1 = 3$, $x_2 = 4$, $x_3 = 8$, $x_m = 9$, with the critical path of $(0, 1)$, $(1, 2)$, $(2, 3)$, $(3, m)$.

Other aspects of PERT/CPM can be handled by linear-programming formulations. In particular, under the assumption that the total project time can be decreased by the expenditure of resources on particular activities, we have the following minimum-cost problem for an acyclic PERT network. Let t_{ij} be the variable activity time for activity (i, j) to be determined, $l_{ij} \geq 0$ the minimum time in which the activity can be completed under priority conditions, and $u_{ij} \geq 0$ the normal completion time. We assume that the cost of activity (i, j) is given as a linear cost function of the form $c_{ij} = a_{ij} - b_{ij}t_{ij}$, with $a_{ij} \geq 0$ and $b_{ij} \geq 0$. As we wish to determine the minimum project cost under various completion schedules, we treat the project-completion time as a parameter $T \geq 0$. Then imposing the above constraints and considerations on the previous formulation (4.14), we have the parametric problem of minimizing

$$\sum_i \sum_j (a_{ij} - b_{ij}t_{ij})$$

subject to

$$x_j - x_i - t_{ij} \geq 0 \qquad \text{for all } (i, j)$$
$$t_{ij} \leq u_{ij} \qquad \text{for all } (i, j)$$
$$t_{ij} \geq l_{ij} \qquad \text{for all } (i, j)$$
$$x_m \leq T$$

and all $x_j \geq 0$ and $t_{ij} \geq 0$. The problem can be converted to a maximizing problem by maximizing $\sum_i \sum_j b_{ij}t_{ij}$. $\left(\text{The } \sum_i \sum_j a_{ij} \text{ is a constant and can be dropped.}\right)$

We next transform a transportation problem into a special network termed *bipartite network*.[1] For exposition purposes, let us consider a particular 2×3 transportation problem with costs c_{ij}, availabilities a_i, and requirements b_j, with $\Sigma a_i = \Sigma b_j$, as shown in Tableau 11.3 (see Chap. 10). Here we use triangles and squares to distinguish between the original origins and the destinations. The minimal-cost network of Fig. 11.22 represents an equivalent statement of the above transportation problem.

TABLEAU 11.3

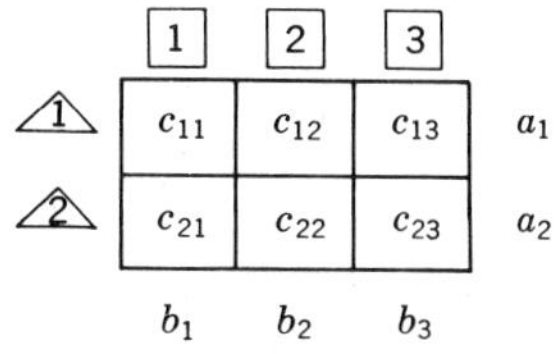

[1] A bipartite network is a network in which the nodes can be divided into two subsets such that the arcs of the network join the nodes of one subset to the other. In Fig. 11.22, we have a bipartite network if we omit nodes 0 and m.

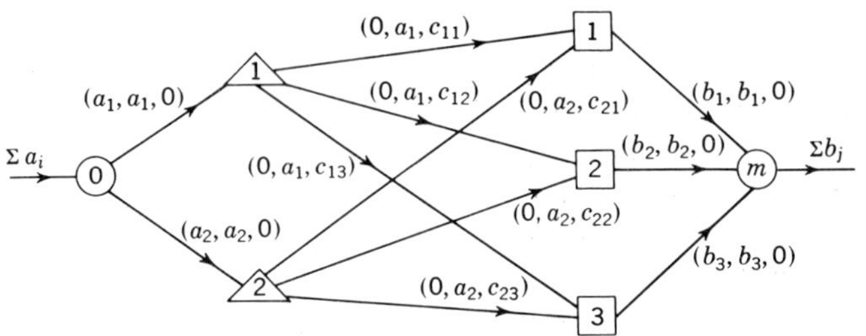

FIGURE 11.22

Other special network problems include the following: (1) The *capacitated transportation problem*, which is a standard transportation problem with the added set of upper-bound constraints $0 \le x_{ij} \le u_{ij}$ for all shipments between origin i and destination j (Ford and Fulkerson [253]).   (2) The *multicommodity network-flow problem* in which more than one commodity can flow through the same network.   Each commodity has a separate origin and destination, and the capacity of an arc cannot be exceeded by the sum of the individual flows of each commodity through that arc.   In addition, each arc has specified capacity restrictions for each type of commodity.   The object is to maximize the sum of the values of all the simultaneous flows (Ford and Fulkerson [253], Hu [382], Assad [14], Kennington [424], and Ali et al. [7]).   (3) A *network with gains* is a single commodity network in which the flow x_{ij} sent from node i to node j along arc (i, j) is multiplied by a factor k_{ij} before it reaches node j; that is, $k_{ij}x_{ij}$ is the amount which arrives at node j.   The multiplicative factor could represent the loss of a fluid due to evaporation or the gain of a product due to breeding[1] (Jewell [402], Glover and Klingman [310, 311], and Glover, Hultz, Klingman, and Stutz, Sec. 13 of Bibliography).

For a general discussion of efficient algorithmic approaches for solving the transportation problem, the maximum-flow problem and the minimum-cost network problem, see Edmonds and Karp [233].   A survey of the theory and applications of network analysis is given in Elmaghraby (Sec. 17 of the Bibliography); also see Moder and Elmaghraby [504].

5. SAMPLE OF APPLICATIONS

In this section we present problem statements and formulations of a number of applications which utilize and extend the formulative procedures discussed in the previous sections.   These problems and the exercises at the end of this chapter should aid the reader in developing approaches to and in becoming proficient at analyzing new problem situations.

a. The trim problem.   This problem is concerned with cutting a solid material into a desired number of subparts such that the amount of waste is

---

[1] The unimodular property does not hold for the multicommodity and network with gains problems; see Exercises 8, 9, 11, and 14.

minimized.[1]   It was first formulated by Paull (Sec. *8f* of Bibliography) and has led to some important formulative and theoretical considerations. Two of these to be discussed are the manner in which the variables are defined and the concept of delayed-data generation. A formal statement of the problem as given in Eisemann [235] is as follows.

A problem of primary significance to a variety of industries is the suppression of trim losses in cutting rolls of paper, textiles, cellophane, metallic foil, or other material, for the execution of business orders. We define the measurement taken between the two circular ends of a standard roll as its "width," the measurement perpendicular thereto when completely unwound as its "length." A number of cutting machines are at our disposal and their knives can be set for any combination of widths for which the combined total does not exceed the overall roll width. The knives then slice through the completely wound rolls much in the same way as a loaf of bread is sliced.[2] Rolls are marketed in several widths. Orders specify desired widths and may prescribe either the number of rolls or, alternatively, the total length to be supplied for each width. In fitting the list of orders to the available rolls and machines, it is generally found that trimming losses due to odd end pieces are unavoidable; this wasted material represents a total loss, which may be somewhat alleviated by selling it as scrap. The problem then consists in fitting orders to rolls and machines in such a way as to subdue trimming losses to an absolute minimum.

For our discussion we shall consider one paper-cutting machine available to cut standard newsprint rolls into the desired smaller widths (see Exercise 3 for a two-machine problem). Each standard roll is 60 in, and a number of them must be cut to produce the current orders for thirty 28-in-wide rolls, sixty 20-in-wide rolls, and forty-eight 15-in-wide rolls. It is assumed that there are enough of the large standard rolls available to satisfy the order and that any leftover less than 15 in wide is to be considered a trim loss. (We ignore the recycling or other uses for the trim.) The data for the orders are summarized:

Widths ordered	No. of rolls ordered
28 in.	30
20 in.	60
15 in.	48

Key to the formulation of the corresponding linear model is the definition of the appropriate variables. Each setting of the cutting blades yields a set of smaller rolls, and the problem then becomes the determination of how the

---

[1] The paper-cutting trim problem considered here is a one-dimensional problem. See Hahn (Sec. *8f* of Bibliography) for a discussion of the two-dimensional problem in which flat sheets must be cut to minimize loss.

[2] Some manufacturing processes slice the material as it is unwound from the standard roll.

blades should be positioned and, for a given setting of the blades, how many large rolls should be cut.   For example, to obtain rolls 28 in wide, two blades can be used to cut the 60-in-wide roll into three pieces—two rolls 28 in wide and one roll 4 in wide.   The former rolls can be used to fill the orders, while the latter roll is trim loss.   Thus, the first thing to be done is to determine which setting of the blades yields rolls that can be used.   Each distinct setting is a variable of the problem, and the value of each variable represents how many 60-in rolls should be cut at the corresponding setting of the blades.   We let x_1 be the number of times we cut a 60-in roll into two 28-in rolls, with a 4-in trim loss, and let other appropriate blade settings be variables as noted in the following table:

Widths ordered, in.	x_1	x_2	x_3	x_4	x_5	x_6	x_7	No. of rolls ordered
28	2	1	1	0	0	0	0	30
20	0	1	0	3	2	1	0	60
15	0	0	2	0	1	2	4	48
Trim loss	4	12	2	0	5	10	0	

Variable x_3 represents the number of times a 60-in roll is cut into one 28-in roll and two 15-in rolls, with a trim loss of 2 in.   The other variables are similarly defined.   Each variable is nonnegative—we cut rolls at that blade setting or we do not.

We would like to produce the exact number of rolls ordered, but because of the limited ways we can cut a standard roll, it is unclear whether a combination of allowable blade settings exists which will exactly yield orders of 30, 60, 48; i.e., we do not know if the problem has any feasible solutions.   Thus, in developing the constraints of the problem, we need to allow for more rolls than ordered to be produced.   The mathematical statement for the above numerical problem is then: Minimize

$$4x_1 + 12x_2 + 2x_3 \qquad + 5x_5 + 10x_6$$

subject to

$$2x_1 + \quad x_2 + \quad x_3 \qquad\qquad\qquad\qquad \geq 30$$
$$x_2 \quad + 3x_4 + 2x_5 + \quad x_6 \qquad \geq 60$$
$$2x_3 \qquad + \quad x_5 + \quad 2x_6 + 4x_7 \geq 48$$
$$x_j \geq 0$$

This problem has a basic feasible solution obtained by inspection of $x_1 = 15$, $x_4 = 20$, $x_7 = 12$ with a trim loss of 60.   This solution states that we should cut 15 large rolls using the blade positions of x_1 (cut two 28-in rolls and one 4-in roll), cut 20 large rolls with the cutting pattern of x_4 (cut three 20-in rolls), and cut 12 large rolls as given by x_7 (cut four 15-in rolls)—this yields a loss of 60 in.   Note that all the constraints are satisfied exactly; i.e., there is no overproduction.   Another solution having a 60-in trim loss is $x_1 = 3$, $x_3 = 24$, and $x_4 = 20$, with all slack variables equal to zero.   We leave it as an exercise to show that both of these solutions are optimal.   (In general, a solution with all the slack variables equal to zero would be more acceptable to one that forces some overproduction.   However, a solution which requires some over-production but fewer, different cutting patterns might be more acceptable as less setup costs for changing blades are incurred; e.g., the solution $x_3 = 30$ and $x_4 = 20$ also has a trim loss of 60 in but produces 12 extra units of 15-in rolls.)

Another objective function that is applied to the trim problem is to minimize the total sum of the standard rolls that have to be cut, i.e., minimize $\sum_j x_j$.   The two formulations would be equivalent if the trim-loss objective function included any excess rolls represented by the slack variables; see Exercise 23.

The trim problem as stated is one in which the variables x_j are continuous, and thus the optimal solution could call for fractional applications of cutting-blade settings, which, of course, have no physical interpretation.   Problems of this sort can usually have their solutions rounded without causing much concern with respect to production or storage facilities and related costs.

The numerical example above is quite restrictive, as we are considering only three widths—all other widths are unacceptable or are considered trim loss.   In most applications there are many allowable widths and the number of possible blade settings or variables can become quite large.   When this prob-lem was first solved on a computer, special data generators were developed which automatically calculated all possible combinations of widths and cor-responding trim loss, and then formatted the problem to be solved by a linear-programming computer code.   Thus, manual data preparation was kept to a minimum, which resulted in both a cost savings and a reduction in errors.   However, as some problems could have millions of cutting patterns, new approaches were required to make them computationally tractable.   In particular, the concept of *delayed data generation,* or in this case, *delayed column generation,* was developed and applied to the trim problem (Gilmore and Gomory [304]).

We have seen some instances of delayed data generation in other chap-ters.   The standard, full-tableau simplex method of Chap. 4 requires all the data to be on hand at each step of the process.   In contrast, the revised simplex method of Chap. 5, using the explicit inverse, calculates the represen-tation of the vectors not in the basis in terms of the basis vectors only when that information is required.   A stronger example is that of the decomposition algorithm in which the extreme points of the subproblems are calculated one at a time as needed for the master problem.

In the trim problem, as we change from one basis to the next, we would usually search out a new vector from the complete set of vectors which had already been generated; i.e., we look for a vector—a cutting pattern—associated with some nonbasic variable x_j which has the largest positive $z_j - c_j$ element. We next examine how this search can be accomplished without having an explicit list of all possible cutting patterns.

For the 3×7 numerical problem above, we define the generic, unknown cutting-pattern vector by $\mathbf{P} = \begin{pmatrix} a_1 \\ a_2 \\ a_3 \end{pmatrix}$, where a_1 represents the number of 28-in rolls, a_2 the number of 20-in rolls, and a_3 the number of 15-in rolls to be cut. The a_i's must satisfy the 60-in standard-roll constraint of $28a_1 + 20a_2 + 15a_3 \leq 60$, with each a_i a positive integer. We let c represent the corresponding nonnegative trim loss, so that the constraint is $28a_1 + 20a_2 + 15a_3 + c = 60$. Using the revised simplex method, assuming we have found a feasible basis, we need to apply the corresponding pricing vector $\mathbf{\Pi} = (\pi_1, \pi_2, \pi_3)$ to vector $\mathbf{P}$ to determine if there is any cutting pattern not in the basis, i.e., a set of a_i's which will improve the solution. For a given set of a_i's, if $\mathbf{\Pi P} - c > 0$, then the associated vector is a candidate to enter the basis. However, the simplex algorithm requires us to find the set of a_i's such that $\mathbf{\Pi P} - c$ is the maximum for all possible cutting patterns. This selection problem reduces to the integer-programming problem of maximizing

$$\pi_0 = \pi_1 a_1 + \pi_2 a_2 + \pi_3 a_3 - c$$

subject to

$$28a_1 + 20a_2 + 15a_3 + c = 60$$

Here the a_i and c are integer variables and the π_i are the known components of the pricing vector $\mathbf{\Pi}$ and are determined from the current basic feasible solution. The slack variable c can be eliminated from the objective function by substitution as $c = 60 - 28a_1 - 20a_2 - 15a_3$, yielding

$$\pi_0 = (\pi_1 + 28)a_1 + (\pi_2 + 20)a_2 + (\pi_3 + 15)a_3 - 60$$

If max $\pi_0 > 0$, then the corresponding solution is a cutting-pattern vector which can enter the basis and improve the solution. If max $\pi_0 \leq 0$ and the nonbasic negative slack vectors price out nonpositive, then the solution is optimal. The simple integer-maximizing problem is of the form of a knapsack problem (see Exercise 10, Chap. 9) and can be solved by special integer procedures (Gilmore and Gomory [305]). We leave as an exercise the development of the corresponding basis improvement criteria for the objective function which minimizes the total number of rolls to be cut—see Exercise 22.

In general, column generation is possible if a mathematical/logical statement can be given as to how the columns are developed. We are then required to solve a side optimization problem to determine the vector to be introduced into the basis. This can require the solution of integer-knapsack problems, as is the case for the trim problem, regular linear programs, or more complex side problems, e.g., the generalized linear program in which the columns of the problem can be selected from convex sets (Dantzig [152]). We hope the side calculations will require less effort than that of the regular revised simplex method. Thus, the decision to use a column-generating approach or explicitly calculate all possible columns is a function of the computational load and data-handling concerns, and appropriate trade-offs must be evaluated for each special formulation; see Balinski and Cobb [29] for other problems for which delayed-data concepts are applicable.

   b. Absolute value, numerical analysis, and goal-programming problems. In this section we describe how certain problems involving absolute-value terms can be transformed into a standard linear-programming formulation. The first problem is to minimize

$$\sum_j |x_j|$$

subject to

$$\mathbf{AX} = \mathbf{b}$$

$$\mathbf{X} \text{ unrestricted}$$

The nonlinear condition of the absolute values can be removed by letting each $x_j = x_j' - x_j''$, where $x_j' \geq 0$ and $x_j'' \geq 0$. The equivalent linear-programming problem is to minimize

$$\sum_j x_j' + x_j''$$

subject to

$$\mathbf{AX}' - \mathbf{AX}'' = \mathbf{B}$$
$$\mathbf{X}' \qquad \geq \mathbf{0}$$
$$\mathbf{X}'' \geq \mathbf{0}$$

where the vectors $\mathbf{X}'$ and $\mathbf{X}''$ are the variable vectors which contain the x_j' and x_j'', respectively. Since the coefficient column of x_j' is $\mathbf{P}_j$ and that of x_j'' is $-\mathbf{P}_j$, we note that both x_j' and x_j'' cannot appear at a positive level in a basic feasible solution.

   If the original objective function is to minimize $\sum_j c_j |x_j|$, then for the above transformation to work and for the simplex algorithm to always find the optimal solution, we must have all the $c_j \geq 0$; similarly, if the objective function is to maximize $\sum_j c_j |x_j|$, then we require all the $c_j \leq 0$ (Shanno and Weil [607]) (see Exercise 19).

   There have been a number of applications of linear programming in the area of numerical analysis and the numerical solution of linear equations (Rabinowitz [554], Rosen [578], and Rust and Burrus [582]). We next describe two applications dealing with statistical-estimation problems.

The problem of *minimizing the sum of absolute deviations* can be handled as follows (Wagner [682]). Let a_{ij}, $i = 1, 2, \ldots, k$ and $j = 1, 2, \ldots, p$ denote a set of k observational measurements on p independent variables and b_i, $i = 1, 2, \ldots, k$ denote the associated measurements on the dependent variable. The problem is to find the regression coefficients x_j such that we minimize $\sum_i \left| \sum_j a_{ij}x_j - b_i \right|$; that is, we wish to find values of the regression coefficients such that the sum of the absolute differences $\left| \sum_j a_{ij}x_j - b_i \right|$ is a minimum.

Let $z_i' - z_i'' = b_i - \sum_j a_{ij}x_j$, $z_i' \geq 0$, $z_i'' \geq 0$. Since for any set of x_j the expression $b_i - \sum_j a_{ij}x_j$ can be positive or negative, we represent this difference as the difference of two nonnegative numbers. We can now rewrite the problem as follows: Minimize

$$\sum_i (z_i' + z_i'')$$

subject to

$$\sum_j a_{ij}x_j + z_i' - z_i'' = b_i \qquad i = 1, \ldots, k$$
$$z_i' \geq 0$$
$$z_i'' \geq 0$$

with the variables x_j unrestricted as to sign. The x_j must, of course, be written as the difference of two nonnegative variables to put the problem into the standard linear-programming format. Since in a basic feasible solution z_i' and z_i'' cannot both be positive, the optimal basic solution will select a set of x_j which minimizes the sum of the absolute differences.

In a similar vein, the *Chebyshev problem* for a set of observations is to find a set of coefficients x_j such that we minimize $\{\max_{x_j} \, {}_i |\Sigma a_{ij}x_j - b_i|\}$; that is, we wish to find a set of x_j such that the maximum deviation between the corresponding observations is a minimum. Let $z \geq \left| \sum_j a_{ij}x_j - b_i \right|$ for all i. The variable z is nonnegative and we wish to have z a minimum. This inequality in absolute terms can be rewritten for each i as two inequalities; i.e., we have that the value of $\sum_j a_{ij}x_j - b_i$ can lie between z and $-z$, or $-z \leq \sum_j a_{ij}x_j - b_j \leq z$. The Chebyshev problem is now as follows: Minimize

$$z$$

subject to

$$\sum_j a_{ij}x_j - b_i - z \leq 0$$

$$i = 1, \ldots, k$$

$$\sum_j a_{ij}x_j - b_i + z \geq 0$$

$$z \geq 0$$

and x_j unrestricted as to sign.

Given an $m \times m$ square set of linear equations $\mathbf{AX} = \mathbf{b}$, a linear-programming procedure for finding the vector $\mathbf{X}$ would be the following. Let $\mathbf{X} = \mathbf{X}' - \mathbf{X}''$, $\mathbf{X}' \geq \mathbf{0}$, and $\mathbf{X}'' \geq \mathbf{0}$, introduce an artificial variable $y_i \geq 0$ for each equation, and solve the linear-programming problem:[1] Minimize

$$z = \sum_i y_i$$

subject to

$$\Sigma a_{ij}x_j' - \Sigma a_{ij}x_j'' + y_i = b_i \qquad i = 1, \ldots, m$$

with $x_j' \geq 0$, $x_j'' \geq 0$, $y_i \geq 0$. If min $z = 0$, then the set of equations has a solution given by the corresponding $\mathbf{X}'$ and $\mathbf{X}''$; if min $z > 0$, then the set of equations is inconsistent and has no solution. The procedure generalizes to the finding of a solution to rectangular sets of equations. (See Sec. 2, Chap. 3, on a discussion relative to determining the determinant of a basis.) As mathematical programming systems use very efficient and accurate gaussian elimination procedures (see Chaps. 2 and 5), they can be used for finding an inverse of a nonsingular matrix. MPS usually include automatic steps for scaling the problem data which help to reduce round-off errors and which can also aid in solving an ill-conditioned square set of equations, i.e., a system with a determinant close to zero.

Related to the concept of absolute value is the *goal-programming* formulation introduced by Charnes and Cooper [112, 116]; also see Ijiri (Sec. 4 of Bibliography) and Ignizio [391]. For discussion purposes, we consider a set of constraints which represents the scheduling of personnel to meet a forecasted set of hourly requirements (see Exercises 31*f* and 65). As the personnel must be assigned to and work a full shift, and assuming a number of different shift possibilities, we will find, in general, that it is impossible to schedule the exact number of people required each hour. As the requirements will change from hour to hour, we will have some time periods with too many people, some with too little, and some with exactly the right amount. The problem, i.e., the goal, is to attempt to assign people to shifts in a manner which minimizes the sum of

---

[1] We can also use the somewhat simpler transformation of $x_j' = x_j + x_0$ for the n unrestricted variables x_j, where $x_0 = -\min{(0, \min{x_j})}$ is an unknown translating variable and $x_j' \geq 0$, $x_0 \geq 0$. (See Deutsch [201], Exercise 19, Chap. 2, and Exercise 34, Chap. 4.)

the over and under assignments. We illustrate this for a problem with five time periods and six possible shift combinations as given in the following tableau:

Time period	Shifts						Personnel requirements
	x_1	x_2	x_3	x_4	x_5	x_6	
1	1	1	1	0	0	0	6
2	0	1	1	1	1	0	5
3	1	0	0	1	1	1	3
4	1	1	1	1	0	1	7
5	0	0	1	0	1	1	8

For example, the x_1 column states that persons assigned to the first shift work the first time period, take a break in the second time period, work the third and fourth time periods, and are off beginning the fifth time period. The variables x_j represent the number of people to be assigned to the corresponding jth shift. (We are ignoring the integer aspects of the variables for this discussion.) We denote the matrix of shifts by $\mathbf{A}$, the vector of variables by $\mathbf{X}$, and the personnel-requirements vector by $\mathbf{b}$. The ith element of the product $\mathbf{AX}$, i.e., the total people assigned to work the ith time period, can be greater than, less than, or equal to the corresponding b_i element of the vector $\mathbf{b}$. We then define for each row i an unrestricted variable y_i which measures the positive, negative, or zero deviation of the ith left-hand-side element from the corresponding b_i, and let $\mathbf{Y}$ be the column vector of such variables. The goal-programming problem is to minimize

$$\sum_i |y_i|$$

subject to

$$\mathbf{AX} + \mathbf{Y} = \mathbf{b}$$

and $\mathbf{X} \geq \mathbf{0}$. If we now let $y_i = y_i' - y_i''$, with $y_i' \geq 0$ and $y_i'' \geq 0$, or $\mathbf{Y} = \mathbf{Y}' - \mathbf{Y}''$, we have the linear-programming problem: Minimize

$$\sum_i (y_i' + y_i'')$$

subject to

$$\mathbf{AX} + \mathbf{Y}' - \mathbf{Y}'' = \mathbf{b}$$
$$\mathbf{X} \qquad\qquad \geq \mathbf{0}$$
$$\mathbf{Y}' \quad\; \geq \mathbf{0}$$
$$\mathbf{Y}'' \geq \mathbf{0}$$

If the goal is to never exceed the requirements vector $\mathbf{b}$ but to come as close as possible, then the formulation requires that $\mathbf{Y}'' = \mathbf{0}$; conversely, if the goal is to meet the requirements with the minimum amount of surplus people, then $\mathbf{Y}' = \mathbf{0}$.[†] The y_i' and y_i'' are called *deviation variables.*

The basic concept of goal programming, i.e., the treatment of constraints as goals that need not be satisfied exactly, can be used to develop the *linear goal-programming model* for solving problems with multiple objectives. In general, we have the usual system of constraints $\mathbf{AX} = \mathbf{b}$, $\mathbf{X} \geq \mathbf{0}$, and a set of p linear objective functions $\mathbf{c}_k \mathbf{X}$, $k = 1, \ldots, p$, where $\mathbf{c}_k = (c_{k1}, \ldots, c_{kn})$. We associate with the kth objective a corresponding goal b_k^*. For any feasible $\mathbf{X}$, the deviation from the kth goal is given by $|b_k^* - \mathbf{c}_k \mathbf{X}|$. This difference can be redefined by $y_k' - y_k'' = b_k^* - \mathbf{c}_k \mathbf{X}$, where the y_k' and y_k'' are nonnegative deviation variables. The multiobjective problem can then be interpreted as the following linear goal-program: Minimize

$$\sum_{k=1}^{p} y_k' + y_k''$$

subject to

$$\mathbf{c}_k \mathbf{X} + y_k' - y_k'' = b_k^* \qquad k = 1, \ldots, p$$
$$\mathbf{AX} = \mathbf{b} \tag{5.1}$$
$$(\mathbf{X}, \, y_k', \, y_k'') \geq \mathbf{0}$$

To illustrate, consider the furniture-manufacturer problem of Chap. 1 where we had the objective of maximizing the profit function $45x_1 + 80x_2$ subject to the resource constraints of

$$5x_1 + 20x_2 \leq 400$$
$$10x_1 + 15x_2 \leq 450$$
$$(x_1, x_2) \geq \mathbf{0}$$

For the same constraints, we now consider the profit objective to have a goal of \$2,100, and introduce a second objective that requires the total units of production $x_1 + x_2$ to have a goal of 45. The goal-programming model is then to minimize

$$y_1' + y_1'' + y_2' + y_2''$$

[†] Since in the general goal-programming problem we do not restrict the form of the matrix $\mathbf{A}$ and we assume that the equation set $\mathbf{AX} = \mathbf{b}$ has no nonnegative (feasible) solutions, we usually restrict both $\mathbf{Y}' \geq \mathbf{0}$ and $\mathbf{Y}'' \geq \mathbf{0}$. The individual y_i' and y_i'' can be given suitable weights other than unity in the objective function.

subject to

$$45x_1 + 80x_2 + y_1' - y_1'' \qquad\qquad = 2100$$
$$x_1 + \quad x_2 \qquad\quad + y_2' - y_2'' = \quad 45$$
$$5x_1 + 20x_2 \qquad\qquad\qquad \leq \quad 400 \qquad\qquad\qquad (5.2)$$
$$10x_1 + 15x_2 \qquad\qquad\qquad \leq \quad 450$$

$$(x_j,\ y_k',\ y_k'') \geq \mathbf{0}$$

We could assign positive or negative weights w_k' and w_k'' to the corresponding deviation variables and solve the above goal program with respect to the weighted deviation objective of minimizing $w_1'y_1' + w_1''y_1'' + w_2'y_2' + w_2''y_2''$. The weights represent an attempt to give priorities or preferences to the meeting of the goals.

Goal-programming procedures extend the treatment of multiobjectives by assuming that the objectives can be ranked or given *preemptive priorities*, i.e., the high-order priority objective must be satisfied as much as possible before the second objective, and so on. Let c_1X, c_2X, $\ldots$, c_kX be the objectives in priority order. Each objective is represented by $c_kX + y_k' - y_k'' = b_k^*$. Using the simplex algorithm, the preemptive procedure first solves problem (5.1) with only the first objective function, i.e., with $k = 1$. Let the optimal value of the first objective function be z_1^*. The problem is then solved for the second objective function with the added constraint that fixes the first objective function at z_1^*. The process continues in a similar fashion through all objectives. The simplex algorithm can be readily adapted for preemptive objectives (see Lee [466], Ignizio [389, 391], Chankong and Haimes [109], and Exercises 48 and 68). A more general goal-programming model in which all objectives and constraints can be treated as goals, as well as a more general approach to preemptive priorities, is given in Ignizio [391]. The relationship between efficient solutions and goal programming is discussed in Ignizio [390]. The reader should contrast the ideas and computational procedures of efficient solutions and goal programming with other multiobjective procedures and their applications (Chankong and Haimes [109], Zeleny [728], Raff and Zeleny [555], and Raff and Ignizio in Sec. 20 of Bibliography). The resolution of a multi-objective problem requires the close interaction between the problem analyst and the ultimate decision maker.

We next consider a scheduling problem whose structure includes elements of the covering, generalized-upper-bounds, and goal-programming problems. The problem is due to Dantzig [174] and is stated in terms of hospital-patient admissions based on bed availability.

At the beginning of the time period to be analyzed (e.g., a 30-day planning horizon), the hospital administrator estimates the number of beds required for those patients already admitted or scheduled to be admitted; we then let b_d be equal to the remaining number of beds available for assignment on day d, for $d = 1, 2, \ldots, D$. A request for admission of a patient is characterized by three numbers (e, f, l), where e is the earliest date a patient can be admitted, f

the latest date, and l the anticipated length of stay. We assume patients can be grouped by these numbers, and those patients having the same three (e_j, f_j, l_j) values are designated as type j patients. (At the extreme, each patient will be of a different type.) Let p_j be the number of patients of type j requesting admission during the days under consideration and let J be the number of different types. We define the variables x_{jt} to be the number of patients of type j to be admitted starting on date t, $t = 1, 2, \ldots, D$; and x_{j0} the number of patients of type j who are not admitted by day D, i.e., the corresponding slack variable. The patient-availability constraints are then given by

$$x_{j0} + \sum_{t=e_j}^{f_j} x_{jt} = p_j \qquad j = 1, 2, \ldots, J$$

We note that for a given patient type j the corresponding $x_{jt} = 0$ for all $t < e_j$ and $t > f_j$.

In order to express the bed-capacity constraints we need to define a set of coefficients for each x_{jt} as follows. Let $\delta_d(j, t)$ be defined as a coefficient for x_{jt} whose value is a 1 if a patient of type j admitted on day t will be using a bed on day d; otherwise the coefficient is 0. For example, if a patient of type one requires a 3-day stay and is admitted on day $t = 2$, then $\delta_2(1, 2) = 1$, $\delta_3(1, 2) = 1$, and $\delta_4(1, 2) = 1$ and $\delta_d(1, 2) = 0$ otherwise. The bed-capacity constraints can then be written as

$$\sum_j \sum_t \delta_d(j, t) x_{jt} \le b_d \qquad d = 1, 2, \ldots, D$$

The variables $x_{jt} \ge 0$ and are integer-valued.

For each patient type j a cost c_{jt} is assigned for those days for which $e_t \le t \le f_t$, where the cost represents the value to the medical system (patient, doctor, hospital) in having a patient of type j admitted on day t. A high cost represents an undesirable admission date, and we would expect the costs to be low for the days in the middle of the time span (e_j, f_j) and increase for the days at the end points. The cost c_{j0} corresponding to the nonadmitted patients x_{j0} would be set rather high to reflect the desire of not wanting to turn away a patient. The objective function is then to minimize $\sum_j \sum_t c_{jt} x_{jt}$, where $t = 0$, 1, $\ldots$, D ($t = 0$ represents the slack assignments) and $j = 1, 2, \ldots, J$.

The structure of the above problem is seen best by developing the explicit matrix for a typical but small problem, as shown in Tableau 11.4 (Dantzig [174]). Here our planning horizon is 4 days and there are three types of patients. Patients of type 1 require a 2-day stay and can be admitted on days 1, 2, 3; type 2 require a 3-day stay and can start on days 1, 2; while type 3 require a 3-day stay and can start on days 1, 2, 3, 4.

The patient-availability constraints (1), (2), and (3) form generalized upper bounds and, since J can be quite large, the GUB algorithm could be appropriate, although rather large problems of this type can be solved by standard computer codes. The bed constraints (4), (5), (6), and (7) form a covering problem of special structure in that if row (6) is subtracted from row (7), row (5) from (6), and row (4) from row (5), the resultant constraints form a network node-arc incidence matrix. The reader should give some thought to whether this transformation can be of computational value. Of course, the problem might be considered in decomposition terms where the bed constraints (there are only D of them) are the master problem and all the GUB constraints are considered as a single subproblem. If the problem is solved as a continuous linear program, some of the x_{jt} will be fractional, but since at most D x_{jt} out of a basic set of $D + J$ will be noninteger (see Exercise 30, Chap. 9), a rounded solution would probably be appropriate for planning purposes.

TABLEAU 11.4

	Type 1				Type 2			Type 3					Bed slack					
	x_{10}	x_{11}	x_{12}	x_{13}	x_{20}	x_{21}	x_{22}	x_{30}	x_{31}	x_{32}	x_{33}	x_{34}	s_1	s_2	s_3	s_4		
(1)	1	1	1	1													p_1	(Patients)
(2)					1	1	1										p_2	
(3)								1	1	1	1	1					p_3	
(4)		1			1				1				1				b_1	(Beds)
(5)		1	1		1	1			1	1				1			b_2	
(6)			1	1		1	1		1	1	1				1		b_3	
(7)				1			1			1	1	1				1	b_4	
	c_{10}	c_{11}	c_{12}	c_{13}	c_{20}	c_{21}	c_{22}	c_{30}	c_{31}	c_{32}	c_{33}	c_{34}	c_1	c_2	c_3	c_4		

We introduced slack variables s_d in the bed constraints, with costs of c_d which represent the implied value of having unassigned beds on day d. (If $c_d \neq 0$, then the proper terms must be added to the objective function.) The bed constraints could be treated as goal-programming constraints in that overadmissions as well as underadmissions, each with corresponding penalties, could be allowed.

   c. The caterer problem (Jacobs [397]). Although actual applications of the caterer problem appear to be limited, it is of interest because of the ease with which it can be stated as a regular linear-programming problem and then reinterpreted as either a transportation problem or a network problem. The formal statement of the problem is as follows.

   A caterer knows that, in connection with the meals to be served during the next n days, $r_t \geq 0$ fresh napkins will be needed on the tth day, with $t = 1, 2, \ldots, n$. Laundering normally takes p days; i.e., a soiled napkin sent for laundering immediately after use on the tth day is returned in time to be used again on the $(t + p)$th day. However, the laundry also has a higher-cost service which returns the napkins in $q < p$ days (p and q integers). Having no usable napkins on hand or in the laundry, the caterer will meet early needs by purchasing napkins at a cents each. Laundering expense is b and c cents per napkin for the normal slow and high-cost (fast) service, respectively. We assume $a > b$ and $a > c > b$. How does the caterer arrange matters to meet the needs and minimize the outlays for the n days?[1]

Define

$0 \leq x_t =$ the number of new napkins bought on day t
$0 \leq y_t =$ the number of soiled napkins sent to the slow laundry on day t
$0 \leq z_t =$ the number of soiled napkins to the fast laundry on day t
$0 \leq s_t =$ the number of soiled napkins not sent to any laundry on day t

We assume all purchases, uses, and shipments happen simultaneously at time t, $t = 1, 2, \ldots, n$, and all the variables are integer-valued. Also, as we want to minimize the total cost of maintaining the proper inventory of clean napkins, it does not pay to buy more napkins than needed for the current day, or to send out a napkin to be laundered unless it will be used at a future time. For discussion purposes we shall develop the equations for the set of requirements $r_t = (5, 6, 7, 8, 7, 9, 10)$, $p = 3$ days, $q = 2$ days, and $a = 25$ cents, $b = 10$ cents, $c = 15$ cents. We next develop the 14 equations, two for each day, corresponding to the daily requirements for clean napkins and disposition of dirty napkins:

Day	Clean napkins	Dirty napkins
1	$x_1 = 5$	$z_1 + y_1 + s_1 = 5$
2	$x_2 = 6$	$z_2 + y_2 + s_2 = 6 + s_1$
3	$x_3 + z_1 = 7$	$z_3 + y_3 + s_3 = 7 + s_2$
4	$x_4 + z_2 + y_1 = 8$	$z_4 + y_4 + s_4 = 8 + s_3$
5	$x_5 + z_3 + y_2 = 7$	$z_5 + s_5 = 7 + s_4$
6	$x_6 + z_4 + y_3 = 9$	$s_6 = 9 + s_5$
7	$x_7 + z_5 + y_4 = 10$	$s_7 = 10 + s_6$

$$(5.3)$$

For example, for day 1 we must purchase five new napkins and, once used, they are distributed to the fast or slow laundries, or left in a dirty inventory pile; for day 4, we must have eight clean napkins which can come from either new purchases, the shipment sent to the fast laundry on day 2, or the shipment sent to the slow laundry on day 1, with the resultant eight dirty napkins and the previous day's dirty pile available for laundering or storage; finally, we see that

---

[1] The original statement of the caterer problem had a military commander (the caterer) supplying aircraft engines (napkins) based on specific requirements (number of guests per day). The commander had a choice of buying new ones or scheduling the overhaul of repairable engines to make them available again to meet the requirements. The overhaul could be accomplished in an expedited manner (fast service) or in the normal, less costly fashion (slow service). Of course, a new engine costs more than either of the overhaul procedures.

dirty napkins are inventoried at the end of the time horizon in that the laundries would not be able to return them in time to be used.

The objective function is to minimize the total cost function of

$$25 \sum_t x_t + 10 \sum_t y_t + 15 \sum_t z_t.$$

The optimal solution has a total cost of \$8.80 with the nonzero variables being $x_1 = 5$, $x_2 = 6$, $x_3 = 7$, $x_4 = 3$, $z_3 = 1$, $z_4 = 3$, $z_5 = 5$, $y_1 = 5$, $y_2 = 6$, $y_3 = 6$, $y_4 = 5$, $s_5 = 2$, $s_6 = 9$, $s_7 = 10$.

The general statement of the caterer problem is given by: Minimize

$$a \sum_t x_t + b \sum_t y_t + c \sum_t z_t$$

subject to

$$y_t + z_t + s_t - s_{t-1} = r_t \qquad t = 1, 2, \ldots, n$$
$$x_t + y_{t-p} + z_{t-q} = r_t \qquad t = 1, 2, \ldots, n$$

with $x_t \geq 0$, $y_t \geq 0$, $z_t \geq 0$, $s_t \geq 0$; $y_{t-p} = 0$ if $t - p \leq 0$, $z_{t-q} = 0$ if $t - q \leq 0$, $y_t = 0$ if $t + p > n$, $z_t = 0$ if $t + q > n$; with all variables integer-valued.

To interpret the caterer problem as a transportation problem (Prager [551], Garvin [282]), we define x_{ij} to be the number of dirty napkins sent to a laundry (slow or fast) on day i to be returned as clean napkins for use on day j. Here we are allowed to launder napkins in advance and do not store dirty napkins unless they will not be laundered at all. Let x_{i0} be the number of dirty napkins retired from future service on day i. Thus, each day i represents an origin of dirty napkins with availability or r_i to be shipped (laundered or retired), and we must have

$$\sum_{j=1}^{n} x_{ij} + x_{i0} = r_i \qquad i = 1, 2, \ldots, n$$

Similarly, each day j represents a destination for clean napkins, and r_j napkins must be received from previous shipments or from new purchases; let x_{0j} be the number of new napkins purchased on day j. We then have

$$\sum_{i=1}^{n} x_{ij} + x_{0j} = r_j \qquad j = 1, 2, \ldots, n$$

As the process starts out with no napkins on hand and ends with all napkins retired, we have

$$\sum_{j=1}^{n} x_{0j} = \sum_{i=1}^{n} x_{i0}$$

The above equations represent a transportation problem with the origins being each day's dirty napkins and the supply of new napkins, and the destinations being each day's requirement of new napkins and the pile of retired dirty napkins. Of course, many of the x_{ij} are zero, as they represent impossible shipments and can be considered to have a very high cost, e.g., a shipment x_{31} of dirty napkins on day 3 which arrives on day 1. The transportation tableau (see Chap. 10) has $n + 1$ rows and $n + 1$ columns. To form the necessary balanced structure between the origin availabilities of napkins and the destination requirements, we let

$$\sum_{j=1}^{n} x_{0j} + x_{00} = \sum_{i=1}^{n} x_{i0} + x_{00} = \sum_{i=1}^{n} r_i$$

where x_{00} represents new napkins which would have had to be purchased if there were no laundries. The above equations, for the previous numerical problem, are represented in Tableau 11.5. The optimal solution is also shown.

To transform the caterer problem to a minimum-cost network-flow problem, we assume a source node of new napkins and a sink node for retired dirty napkins, with intermediate nodes representing the days in which clean napkins must be on hand and dirty napkins must be redistributed to the laundries and dirty-napkin storage. The network for the above 7-day example is shown in Fig. 11.23. We have labeled each arc with the corresponding variable or daily-napkin requirement. The reader will note how the network assumption of conservation of flow about a node forces the caterer-problem equations to be satisfied. Also, to obtain the proper balance of flow at the sink node, we

TABLEAU 11.5

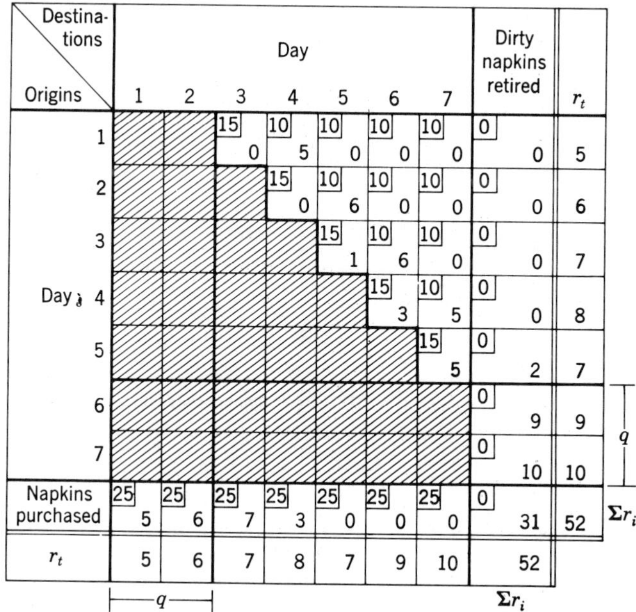

Each cell shows the cost in the small box (upper left) and the optimal shipment amount (lower right); blank hatched cells are impossible shipments.

Origins \ Destinations	Day 1	Day 2	Day 3	Day 4	Day 5	Day 6	Day 7	Dirty napkins retired	r_t
1			[15] 0	[10] 5	[10] 0	[10] 0	[10] 0	[0] 0	5
2				[15] 0	[10] 6	[10] 0	[10] 0	[0] 0	6
3					[15] 1	[10] 6	[10] 0	[0] 0	7
4						[15] 3	[10] 5	[0] 0	8
5							[15] 5	[0] 2	7
6								[0] 9	9
7								[0] 10	10
Napkins purchased	[25] 5	[25] 6	[25] 7	[25] 3	[25] 0	[25] 0	[25] 0	[0] 31	52 Σr_i
r_t	5	6	7	8	7	9	10	52	

(Rows 1–7 are labeled "Day i"; the columns 1–7 span "Day". The bracket q spans Days 1–7 across columns and the right; Σr_i is shown at the bottom right.)

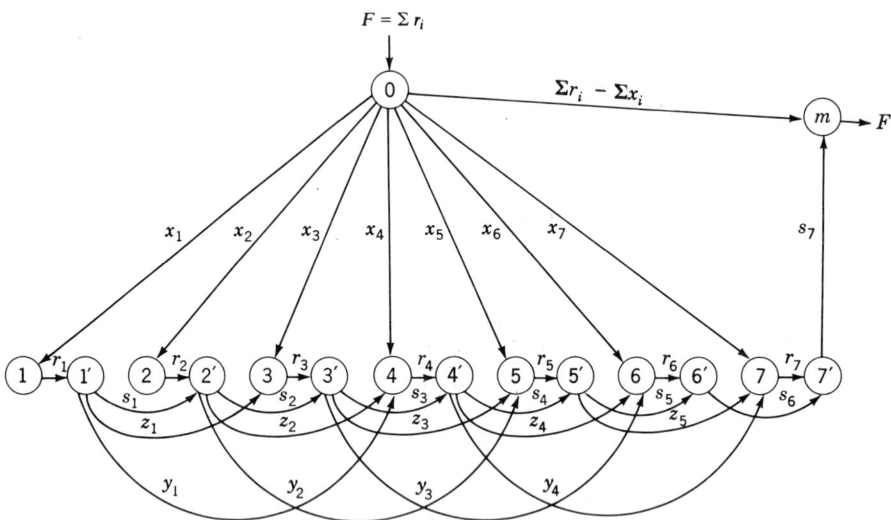

FIGURE 11.23

need to assume a total of $F = \sum_i r_i$ napkins at the source node and that any of the F not purchased are sent to the sink node.   We leave as an exercise the development of each arc's lower and upper bounds and associated cost.

6. LINEAR PROGRAMMING AND THE THEORY OF GAMES[1]

In this section we demonstrate the equivalence between the linear-programming model and the mathematical model of the important class of zero-sum two-person games.   The main concern of the general theory of games is the study of the following problem stated by von Neumann: "If n players P_1, P_2, ..., P_n play a given game Γ, how must the ith player, P_i, play to achieve the most favorable result?"

In interpreting this basic problem, we say that the term *game* refers to a set of rules and conventions for playing and a *play* refers to a particular possible realization of the rules, i.e., an individual contest.   We assume that, at the end of a play of Γ, each of the players P_i receives an amount of money v_i, called the *pay-off to player P_i*.   We also assume that, in order to receive the most favorable result, the only object of a player will be to maximize the amount of money received.   In most parlor games, e.g., poker, the total amount of money lost by the losing players is equal to the total amount of money won by the winning players.   Here we have $v_1 + v_2 + \cdots + v_n = 0$ for all

---

[1] For introductory and more detailed descriptions of the basic elements of the theory of games see previous editions of this book and Kuhn [445], McKinsey [485], von Neumann and Morgenstern [526], Luce and Raiffa [476], Owen [538], and Williams [703].

plays.   Note that v_i can be positive, negative, or zero, where $v_i > 0$ represents a gain, $v_i < 0$ represents a loss, and $v_i = 0$ represents no pay-off.   Games whose algebraic sum of the pay-offs is equal to zero are called *zero-sum games.*   In zero-sum games, wealth is neither created nor destroyed by the players.   An example of a non-zero-sum game would be poker in which a certain percentage of the pot is removed for the "house" before the final pay-off.

Games are also classified by the number of players and possible moves.   Chess is a two-person game with a finite number of possible moves, while poker is a many-person game, also with a finite number of possible moves.[1]   A duel in which the duelists may fire at any instant in a given time interval is a two-person game with an infinity of possible moves.   Games are further characterized by being cooperative or noncooperative.   In the former the players have the ability to form coalitions and work as teams, while in the latter players are concerned only with their individual results.   Two-person games are, of course, noncooperative.   As noted earlier, our main interest is in establishing the relationship between finite, zero-sum, two-person games and the linear-programming model.

As an example of a finite, zero-sum, two-person game, we have the familiar game of matching pennies.   To accomplish one move that for this game is equivalent to a play, the first player P_1 selects either heads or tails, and the second player P_2, not knowing P_1's choice, also selects either heads or tails.   After the choices have been made, P_2 pays P_1 1 unit if they match or -1 unit if they do not.   The pay-off of -1 represents the giving of 1 unit by P_1 to P_2.   The pay-off values of 1 and -1 are written in terms of a gain being a positive value to P_1 and a loss a negative value to P_1.   When this is the case, we say that P_1 is the *maximizing player* and P_2 the *minimizing player.*   We designate P_1 to be the maximizing player and write the pay-off values accordingly.

Definition 1.   A *zero-sum, two-person* or *matrix game* Γ is determined by any $m \times n$ matrix

$$\mathbf{A} = \begin{pmatrix} a_{11} & a_{12} & \cdots & a_{1n} \\ a_{21} & a_{22} & \cdots & a_{2n} \\ \hdotsfor{4} \\ a_{m1} & a_{m2} & \cdots & a_{mn} \end{pmatrix} = (a_{ij})$$

in which the elements a_{ij} are any real numbers.   The matrix $\mathbf{A}$ is termed the *pay-off matrix.*   The element a_{ij} represents the amount paid to P_1 by P_2 if P_1 selects the move associated with the ith row and P_2 selects the move associated with the jth column.

---

[1] Here we include appropriate "stop rules" in the conventions of the games.

Definition 2. A *mixed strategy for P_1* is a row vector $\mathbf{X} = (x_1, x_2, \ldots, x_m)$ of nonnegative numbers x_i such that

$$x_1 + x_2 + \cdots + x_m = 1$$

A *mixed strategy for P_2* is a column vector $\mathbf{Y} = (y_1, y_2, \ldots, y_n)$ of nonnegative numbers y_j such that

$$y_1 + y_2 + \cdots + y_n = 1$$

The elements x_i and y_j represent, respectively, the frequencies with which P_1 selects the ith move (row) and P_2 selects the jth move (column). For the matching-pennies game, a mixed strategy, i.e., the probabilities for selecting heads or tails, for P_1 could, for example, be any of the following: $(1, 0)\ (0, 1)\ (\frac{1}{4}, \frac{3}{4}),\ (\frac{1}{2}, \frac{1}{2})$. P_2 could employ similar mixed strategies.

Definition 3. For each $i = 1, 2, \ldots, m$, the mixed strategy that is 1 in the ith component and 0 elsewhere is called the *ith pure strategy for P_1* and will be denoted by i.

Similarly, the *jth pure strategy for P_2*, denoted by j, is a mixed strategy for P_2 which is 1 in the jth component and 0 elsewhere.

Let us next consider the matrix game determined by the pay-off matrix

$$\mathbf{A} = \begin{pmatrix} a_{11} & a_{12} & \cdots & a_{1n} \\ a_{21} & a_{22} & \cdots & a_{2n} \\ \cdots\cdots\cdots\cdots\cdots\cdots\cdots \\ a_{m1} & a_{m2} & \cdots & a_{mn} \end{pmatrix}$$

If P_1 chooses any pure strategy i, the player is sure of being paid at least $\min_j a_{ij}$. For example, if the matrix game has the following 3×4 pay-off matrix:

$$\begin{pmatrix} 1 & 5 & 0 & 4 \\ 2 & 1 & 3 & 3 \\ 4 & 2 & -1 & 0 \end{pmatrix}$$

we have

$$\min_j a_{1j} = a_{13} = 0 \qquad \min_j a_{2j} = a_{22} = 1$$

$$\tag{6.1}$$

$$\min_j a_{3j} = a_{33} = -1$$

Since P_1 can select any i, he or she, can, of course, select the i which *maximizes* the gain. That is, the player can select the pure strategy that will make $\min_j a_{ij}$ as large as possible. By using this pure strategy, P_1 can win at least $\max_i \min_j a_{ij}$.

For the 3×4 pay-off matrix we have from (6.1) that

$$\max_i \min_j a_{ij} = a_{22} = 1$$

If P_2 selects a j, the worst thing that can happen is to lose $\max_i a_{ij}$. P_2 can then select the pure strategy which *minimizes* his or her losses. By using this pure strategy, P_2 can keep P_1 from winning more than $\min_j \max_i a_{ij}$. For the 3×4 matrix we have

$$\max_i a_{i1} = a_{31} = 4 \qquad \max_i a_{i2} = a_{12} = 5$$

$$\max_i a_{i3} = a_{23} = 3 \qquad \max_i a_{i4} = a_{14} = 4$$

and

$$\min_j \max_i a_{ij} = a_{23} = 3$$

If

$$\max_i \min_j a_{ij} = \min_j \max_i a_{ij} = v \tag{6.2}$$

then P_1 can be sure of winning v, and he or she can be prevented from winning more than v by P_2. Any matrix game for which (6.2) holds can best be played by P_1 and P_2 playing the corresponding pure strategies. Since any deviation from this strategy by P_1 lowers P_1's opportunity for winning at least v and since any deviation from this strategy by P_2 raises P_1's opportunity for winning more than v, these pure strategies are called *optimal pure strategies.*

The following matrix game is one which has for its optimal pure strategies $\mathbf{X} = (1, 0, 0)$ and $\mathbf{Y} = (1, 0, 0)$:

$$\begin{pmatrix} 3 & 5 & 6 \\ 1 & 2 & 3 \\ 0 & 7 & 4 \end{pmatrix}$$

Here

$$\max_i \min_j a_{ij} = \min_j \max_i a_{ij} = a_{11} = 3$$

The reader will note that the element a_{11} is both the minimum element in its row and the maximum element in its column. Any such element is called a *saddle point.* If the element a_{kl} is a saddle point, then the pure strategies k and l for P_1 and P_2, respectively, are optimal pure strategies and $a_{kl} = v$.

Since not all matrix games can be optimally played by means of pure strategies (e.g., matching pennies), we need to consider the notion of optimal mixed strategies.

Definition 4.   The *pay-off function* for P_1, i.e., the mathematical expectation of P_1, is defined to be

$$E(\mathbf{X}, \mathbf{Y}) = \mathbf{XAY} = \sum_{i=1}^{m} \sum_{j=1}^{n} x_i a_{ij} y_j$$

where $\mathbf{X} = (x_1, x_2, \ldots, x_m)$ and $\mathbf{Y} = (y_1, y_2, \ldots, y_n)$ are any mixed strategies for P_1 and P_2, respectively.

Definition 5.   A *solution* to a matrix game Γ is a pair of mixed strategies

$$\bar{\mathbf{X}} = (\bar{x}_1, \bar{x}_2, \ldots, \bar{x}_m)$$
$$\bar{\mathbf{Y}} = (\bar{y}_1, \bar{y}_2, \ldots, \bar{y}_n)$$

and a real number v such that

$$E(\bar{\mathbf{X}}, j) \geq v \qquad \text{for the pure strategies } j = 1, 2, \ldots, n$$
$$E(i, \bar{\mathbf{Y}}) \leq v \qquad \text{for the pure strategies } i = 1, 2, \ldots, m$$

The $\bar{\mathbf{X}}$ and $\bar{\mathbf{Y}}$ are called *optimal strategies*, and the number v is called the *value* of the game.

From this definition we see that, if P_1 selects moves as determined by the probabilities of the corresponding optimal strategy, then no matter what moves P_2 selects, P_1 can expect to win at least v units.   A similar statement is true for P_2; i.e., P_2 can expect to lose no more than v units.   We should note that v can be positive, negative, or zero.   For the game of matching pennies, whose pay-off matrix is

$$\begin{pmatrix} 1 & -1 \\ -1 & 1 \end{pmatrix}$$

the value of the game is 0 and the optimal strategies are $\bar{\mathbf{X}} = (\tfrac{1}{2}, \tfrac{1}{2})$ and $\bar{\mathbf{Y}} = (\tfrac{1}{2}, \tfrac{1}{2})$.

In what follows we shall want to change the value v of a game by adding a fixed amount w to each element of the pay-off matrix (a_{ij}).   In this connection it can be shown that, for the new pay-off matrix $(a_{ij} + w)$, the optimal strategies are the same as for the original game and the value of the new game is $v + w$.

We next state, without proof, the following theorem:

The fundamental theorem of matrix games.[1]   *For every matrix game* $\max\limits_{\mathbf{X}} \min\limits_{\mathbf{Y}} E(\mathbf{X}, \mathbf{Y})$ *and* $\min\limits_{\mathbf{Y}} \max\limits_{\mathbf{X}} E(\mathbf{X}, \mathbf{Y})$ *exist and are equal to* v.   *That is, every matrix game has a solution* $[\bar{\mathbf{X}}, \bar{\mathbf{Y}}; v]$, *with* $E(\mathbf{X}, \bar{\mathbf{Y}}) \leq E(\bar{\mathbf{X}}, \bar{\mathbf{Y}}) = v \leq E(\bar{\mathbf{X}}, \mathbf{Y})$.

---

[1] The first statement is the min-max expression of von Neumann and Morgenstern, while the second is termed the saddle-point statement (Kuhn [445]).

We next show the equivalence of the matrix game and the problem of linear programming. We assume that we are given an arbitrary matrix game:

$$\mathbf{A} = \begin{pmatrix} a_{11} & a_{12} & \cdots & a_{1n} \\ a_{21} & a_{22} & \cdots & a_{2n} \\ \cdots\cdots\cdots\cdots\cdots\cdots \\ a_{m1} & a_{m2} & \cdots & a_{mn} \end{pmatrix}$$

From Definitions 4 and 5, the problem for P_1 is to find a vector

$$\mathbf{X} = (x_1, x_2, \ldots, x_m)$$

and a number v such that

$$\begin{aligned}
a_{11}x_1 + a_{21}x_2 + \cdots + a_{m1}x_m &\geq v \\
a_{12}x_1 + a_{22}x_2 + \cdots + a_{m2}x_m &\geq v \\
\cdots\cdots\cdots\cdots\cdots\cdots\cdots\cdots \\
a_{1n}x_1 + a_{2n}x_2 + \cdots + a_{mn}x_m &\geq v \\
x_1 + x_2 + \cdots + x_m &= 1 \\
x_1 \quad\quad\quad &\geq 0 \\
x_2 \quad\quad &\geq 0 \\
\vdots\quad\quad \\
x_m &\geq 0
\end{aligned} \tag{6.3}$$

Similarly, for P_2 we have

$$\begin{aligned}
a_{11}y_1 + a_{12}y_2 + \cdots + a_{1n}y_n &\leq v \\
a_{21}y_1 + a_{22}y_2 + \cdots + a_{2n}y_n &\leq v \\
\cdots\cdots\cdots\cdots\cdots\cdots\cdots\cdots \\
a_{m1}y_1 + a_{m2}y_2 + \cdots + a_{mn}y_n &\leq v \\
y_1 + y_2 + \cdots + y_n &= 1 \\
y_1 \quad\quad\quad &\geq 0 \\
y_2 \quad\quad &\geq 0 \\
\vdots\quad\quad \\
y_n &\geq 0
\end{aligned} \tag{6.4}$$

Since every element of $\mathbf{A}$ can be made positive by the addition of a suitable constant to all the a_{ij}, we can assume that $v > 0$. Let us divide each of the relationships in (6.3) and (6.4) by v and let

$$x'_i = \frac{x_i}{v} \quad \text{and} \quad y'_j = \frac{y_j}{v}$$

Note that

$$\sum_i x_i' = \frac{1}{v} \sum_i x_i = \frac{1}{v}$$

and

$$\sum_j y_j' = \frac{1}{v} \sum_j y_j = \frac{1}{v}$$

Hence, by minimizing $\sum_i x_i'$, P_1 will maximize the value of the game, and by maximizing $\sum_j y_j'$, P_2 will minimize the value of the game. We can then restate (6.3) and (6.4) in terms of equivalent linear-programming problems and obtain the following symmetric dual problems:

The primal problem. Find a vector $\mathbf{X}' = (x_1', x_2', \ldots, x_m')$ which minimizes

$$x_1' + x_2' + \cdots + x_m'$$

subject to

$$a_{11}x_1' + a_{21}x_2' + \cdots + a_{m1}x_m' \geq 1$$
$$a_{12}x_1' + a_{22}x_2' + \cdots + a_{m2}x_m' \geq 1$$
$$\cdots\cdots\cdots\cdots\cdots\cdots\cdots\cdots\cdots\cdots$$
$$a_{1n}x_1' + a_{2n}x_2' + \cdots + a_{mn}x_m' \geq 1$$
$$x_i' \geq 0$$

The dual problem. Find a vector $\mathbf{Y}' = (y_1', y_2', \ldots, y_n')$ which maximizes

$$y_1' + y_2' + \cdots + y_n'$$

subject to

$$a_{11}y_1' + a_{12}y_2' + \cdots + a_{1n}y_n' \leq 1$$
$$a_{21}y_1' + a_{22}y_2' + \cdots + a_{2n}y_n' \leq 1$$
$$\cdots\cdots\cdots\cdots\cdots\cdots\cdots\cdots\cdots\cdots$$
$$a_{m1}y_1' + a_{m2}y_2' + \cdots + a_{mn}y_n' \leq 1$$
$$y_j' \geq 0$$

Since every game has a solution, optimal solutions to the above problems exist and

$$\min_i \sum_i x_i' = \max_j \sum_j y_j' = \frac{1^\dagger}{v}$$

It should be noted that, if only the primal or only the dual is solved, the optimal strategy for the other problem is contained in the simplex tableau of

---

† The set of x_i' and y_j' which satisfies the linear-programming problems must, of course, be converted to the optimal x_i and y_j that solve the game problems.

the corresponding final solution. The optimal strategy corresponds to the $z_j - c_j$ elements for the slack vectors.

An alternative method for reducing the game problem to a linear-programming problem is the following: For P_1 the problem is again given by (6.3). Let us write the first n inequalities as equalities by subtracting a nonnegative variable from each to obtain

$$
\begin{aligned}
a_{11}x_1 + a_{21}x_2 + \cdots + a_{m1}x_m - x_{m+1} &&&= v \\
a_{12}x_1 + a_{22}x_2 + \cdots + a_{m2}x_m && - x_{m+2} &&= v \\
&\cdots\cdots\cdots\cdots \\
a_{1n}x_1 + a_{2n}x_2 + \cdots + a_{mn}x_m &&& - x_{m+n} = v \\
x_1 + x_2 + \cdots + x_m &&&= 1 \\
&&& x_i \geq 0
\end{aligned}
$$

By subtracting the first equation from the succeeding $n - 1$ equations and using the first equation for the objective function, since it is equal to v, we then have the equivalent linear-programming problem of maximizing

$$a_{11}x_1 + a_{21}x_2 + \cdots + a_{m1}x_m - x_{m+1} = v$$

subject to

$$
\begin{aligned}
(a_{12} - a_{11})x_1 + (a_{22} - a_{21})x_2 + \cdots + (a_{m2} - a_{m1})x_m + x_{m+1} - x_{m+2} &&&= 0 \\
&\cdots\cdots\cdots\cdots\cdots \\
(a_{1n} - a_{11})x_1 + (a_{2n} - a_{21})x_2 + \cdots + (a_{mn} - a_{m1})x_m + x_{m+1} && - x_{m+n} &&= 0 \\
x_1 + x_2 + \cdots + x_m &&&= 1 \\
&&& x_i \geq 0
\end{aligned}
$$

Example 6.1. Here we are given a matrix game with the following pay-off matrix:

$$
\mathbf{A} = \begin{pmatrix} 3 & -2 & -4 \\ -1 & 4 & 2 \\ 2 & 2 & 6 \end{pmatrix}
$$

The development of the corresponding linear-programming problem (using the above alternative method) is as follows: The initial system of constraints is

$$
\begin{aligned}
3x_1 - x_2 + 2x_3 &\geq v \\
-2x_1 + 4x_2 + 2x_3 &\geq v \\
-4x_1 + 2x_2 + 6x_3 &\geq v \\
x_1 + x_2 + x_3 &= 1 \\
x_1 &\geq 0 \\
x_2 &\geq 0 \\
x_3 &\geq 0
\end{aligned}
$$

Subtracting nonnegative slack variables, we have

$$3x_1 - x_2 + 2x_3 - x_4 \qquad\qquad = v$$
$$-2x_1 + 4x_2 + 2x_3 \qquad - x_5 \qquad = v$$
$$-4x_1 + 2x_2 + 6x_3 \qquad\qquad - x_6 = v$$
$$x_1 + x_2 + x_3 \qquad\qquad = 1$$
$$x_i \geq 0$$

By subtracting the first equation from the second and third, we find that the corresponding linear-programming problem is to maximize

$$3x_1 - x_2 + 2x_3 - x_4$$

subject to

$$-5x_1 + 5x_2 \qquad + x_4 - x_5 \qquad = 0$$
$$-7x_1 + 3x_2 + 4x_3 + x_4 \qquad - x_6 = 0$$
$$x_1 + x_2 + x_3 \qquad\qquad = 1$$
$$x_i \geq 0$$

As the computational techniques for the direct solution of large matrix games are rather cumbersome, the efficient computational techniques of linear programming are usually employed to solve such games. There are, however, rapid iterative methods for solving games that yield good approximations to the optimal solution (Williams [703] and Luce and Raiffa [476]).

Example 6.2. Matrix games in which a player has only two strategies can be solved by graphical procedures. We illustrate this approach for a game having the pay-off matrix of

$$\begin{pmatrix} 3 & -2 & 4 \\ -1 & 4 & 0 \\ 2 & 2 & 2 \\ -1 & 3 & -1 \end{pmatrix}$$

By applying the concepts of dominance, we see that row 4 is dominated by row 2 and strategy 4 will never be played by P_1; also column 3 is dominated by column 1 and strategy 3 will never be played by P_2. The reduced pay-off matrix is then

$$\begin{pmatrix} 3 & -2 \\ -1 & 4 \\ 2 & 2 \end{pmatrix}$$

For the reduced game, P_2 has only two strategies. Let y_1 and y_2, with $y_1 + y_2 = 1$, be the frequencies in which P_2 plays strategies 1 and 2, respectively, and v be the value of the game. From Eqs. (6.4) the problem of the minimizing player P_2 can be expressed as

$$3y_1 - 2y_2 \le v$$
$$-\,y_1 + 4y_2 \le v$$
$$2y_1 + 2y_2 \le v$$
$$y_1 + \;\; y_2 = 1$$
$$y_1 \qquad \ge 0$$
$$y_2 \ge 0$$

Substituting $y_2 = 1 - y_1$, the constraints become

$$5y_1 \;\; -v \;\le\; 2$$
$$-5y_1 \;\; -v \;\le -4$$
$$-v \;\le -2$$
$$y_1 \qquad \ge \;\; 0$$
$$y_1 \qquad \le \;\; 1$$

As the transformed constraints involve only two variables, they can be graphed as shown in Fig. 11.24. P_2's problem is to find a value of $0 \le y_1 \le 1$ which minimizes the objective function v. We see from Fig.

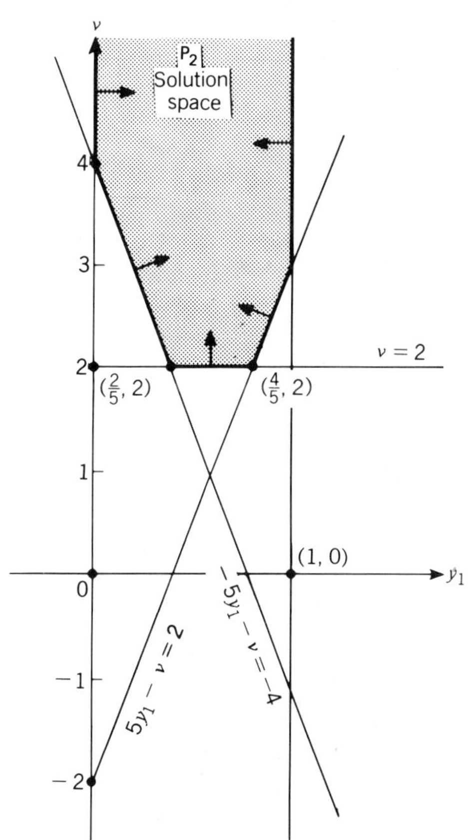

FIGURE 11.24

11.24 that $v = 2$ (the game is not fair for P_2) and that any $\frac{2}{5} \leq y_1 \leq \frac{4}{5}$ with a corresponding $y_2 = 1 - y_1$, $y_3 = 0$ yields an optimal strategy for P_2. (See Exercise 37 concerning the solution for P_1's optimal strategy.)

The converse problem of expressing a given linear-programming problem as a matrix game can be readily accomplished by considering the problem and its corresponding dual. Let the original problem be given in terms of the inequalities

$$a_{11} x_1 + \cdots + a_{1n} x_n \geq b_1$$
$$\cdots\cdots\cdots\cdots\cdots\cdots\cdots$$
$$a_{m1} x_1 + \cdots + a_{mn} x_n \geq b_m \tag{6.5}$$
$$c_1 x_1 + \cdots + c_n x_n \geq f \tag{6.6}$$

where $x_j \geq 0$ and f is the minimum of the objective function (6.6). The associated dual is

$$a_{11} w_1 + \cdots + a_{m1} w_m \leq c_1$$
$$\cdots\cdots\cdots\cdots\cdots\cdots\cdots$$
$$a_{1n} w_1 + \cdots + a_{mn} w_m \leq c_n \tag{6.7}$$
$$b_1 w_1 + \cdots + b_m w_m \leq g \tag{6.8}$$

where $w_i \geq 0$ and g is the maximum of the objective function (6.8). From the duality theorem of Chap. 6 we have that, if a finite optimal solution to either problem exists, then a finite optimal solution for the other also exists and $f = g$. Let us assume that such solutions do exist for the problems (6.5), (6.6) and (6.7), (6.8). Let $(x_1, \ldots, x_n)$ and $(w_1, \ldots, w_m)$ be solutions to the former and latter problems, respectively. Multiplying the first inequality of (6.5) by w_1, the second by w_2, etc., and summing the m inequalities, we have

$$b_1 w_1 + \cdots + b_m w_m \leq \left(\sum_{i=1}^{m} a_{i1} w_i \right) x_1 + \cdots + \left(\sum_{i=1}^{m} a_{in} w_i \right) x_n \tag{6.9}$$

From (6.7) we can replace the sums in (6.9) by their corresponding c_j to obtain the following inequality that is true for all solutions to the primal and dual:

$$b_1 w_1 + \cdots + b_m w_m \leq c_1 x_1 + \cdots + c_n w_n^{\dagger} \tag{6.10}$$

this implies that $g \leq f$.

The reverse inequality of (6.10) is given by

$$b_1 w_1 + \cdots + b_m w_m \geq c_1 x_1 + \cdots + c_n x_n$$

or

$$b_1 w_1 + \cdots + b_m w_m - (c_1 x_1 + \cdots + c_n x_n) \geq 0 \tag{6.11}$$

This inequality restricts $g \geq f$. Hence, a simultaneous solution to the complete set of inequalities (6.5), (6.7), and (6.11) will yield optimal solutions for the primal and dual with, of course, $f = g$.

Let us multiply each inequality of the set (6.5), (6.7), and (6.11) by an unknown positive variable z to obtain

---

† The inequality (6.10) is equivalent to formula (1.13) of Chap. 6.

$$a_{11}\bar{x}_1 + \cdots + a_{1n}\bar{x}_n - b_1 z \geq 0$$
$$\cdots\cdots\cdots\cdots\cdots\cdots\cdots\cdots\cdots\cdots\cdots\cdots$$
$$a_{m1}\bar{x}_1 + \cdots + a_{mn}\bar{x}_n - b_m z \geq 0$$

$$-(a_{11}\bar{w}_1 + \cdots + a_{m1}\bar{w}_m) \qquad\qquad + c_1 z \geq 0 \qquad\qquad (6.12)$$
$$\cdots\cdots\cdots\cdots\cdots\cdots\cdots\cdots\cdots\cdots\cdots\cdots$$
$$-(a_{1n}\bar{w}_1 + \cdots + a_{mn}\bar{w}_m) \qquad\qquad + c_n z \geq 0$$
$$(b_1\bar{w}_1 + \cdots + b_m\bar{w}_m) - (c_1\bar{x}_1 + \cdots + c_n\bar{x}_n) \qquad \geq 0$$

where $\bar{x}_j = z x_j$ and $\bar{w}_i = z w_i$. A solution to (6.12) with $z > 0$ will yield a solution to the original primal and dual problems. The above has the following representation in terms of matrices:

$$(\bar{\mathbf{X}}\bar{\mathbf{W}}z) \begin{pmatrix} \mathbf{0} & \mathbf{A}' & -\mathbf{c}' \\ -\mathbf{A} & \mathbf{0} & \mathbf{b}' \\ \mathbf{c} & -\mathbf{b} & \mathbf{0} \end{pmatrix} \geq (\mathbf{0}) \qquad\qquad (6.13)$$

Here $\mathbf{A}$ is an $m \times n$ matrix; $\mathbf{b}$, $\mathbf{c}$, $\bar{\mathbf{X}}$, and $\bar{\mathbf{W}}$ are row vectors; and the $\mathbf{0}$ matrices are of the right dimensions. We can consider the skew-symmetric matrix of (6.13) as defining a zero-sum two-person *symmetric game* and the row vector $(\bar{\mathbf{X}}\bar{\mathbf{W}}z)$ as the strategy vector for the maximizing player P_1. Since it can be shown that the value of a symmetric game is zero, we have from Definition 5 that a solution to inequalities (6.13) with

$$\sum_j \bar{x}_j + \sum_i \bar{w}_i + z = 1$$

is a solution to the corresponding game. By the fundamental theorem, a solution to this game always exists. However, the associated linear-programming problems may not have any solutions. This will be the case if the game solution yields an optimal strategy with $z = 0$. Hence a game solution with $z > 0$ will yield optimal solutions to the primal and dual linear-programming problems as defined by (6.5), (6.7), and (6.11), where $x_j = \bar{x}_j/z$ and $w_i = \bar{w}_i/z$ are the respective optimal solutions.

We next introduce some basic concepts of two-person non-zero-sum (noncooperative) games (Owen [538] and Luce and Raiffa [476]). As noted in Chap. 12, certain forms of such games are related to special nonlinear-programming problems.

A two-person non-zero-sum game is defined as follows. There are two players P_1 and P_2, with P_1 denoted as the row player and P_2 as the column player. Associated with P_1 is an $m \times n$ matrix $\mathbf{A} = (a_{ij})$ and with P_2 an $m \times n$ matrix $\mathbf{B} = (b_{ij})$. P_1 has m pure strategies $i = 1, 2, \ldots, m$ and P_2 has n pure strategies $j = 1, 2, \ldots, n$. If P_1 chooses the ith pure strategy and P_2 chooses the jth pure strategy, then the payoff vector is (a_{ij}, b_{ij}), where a_{ij} is the amount P_1 receives and b_{ij} is the amount P_2 receives. Games of this sort are termed *bimatrix games* and are denoted by $\Gamma(\mathbf{A}, \mathbf{B})$.

If $\mathbf{B} = -\mathbf{A}$, then the game reduces to a zero-sum game with payoff matrix $\mathbf{A}$. With the usual definition of mixed strategies (Definition 2), the payoff functions for P_1 and P_2 are $\mathbf{X}\mathbf{A}\mathbf{Y}$ and $\mathbf{X}\mathbf{B}\mathbf{Y}$, respectively.

A pair of mixed strategies $(\bar{\mathbf{X}}, \bar{\mathbf{Y}})$ for $\Gamma(\mathbf{A}, \mathbf{B})$ is an *equilibrium point* if, for any other mixed strategies $\mathbf{X}$ and $\mathbf{Y}$,

$$\mathbf{X}A\bar{\mathbf{Y}} \le \bar{\mathbf{X}}A\bar{\mathbf{Y}}$$

$$\bar{\mathbf{X}}B\mathbf{Y} \le \bar{\mathbf{X}}B\bar{\mathbf{Y}}$$

It can be shown that every bimatrix game has at least one equilibrium point. (See Sec. 4, Chap. 12 for references to nonlinear-programming procedures for finding equilibrium points.)

REMARKS

Most of the problems and exercises in this text are stated in rather simple terms and lead to correspondingly simple mathematical formulations. This approach is necessary in order to illustrate the mathematical and computational aspects of linear programming. Many real-world problems can be formulated in mathematical terms without difficulty, while others require complex and detailed notation to reflect the constraints and the relationships between the variables. A thorough mathematical statement of a complex problem is necessary as its development forces the analyst (and allows others) to understand completely the problem environment and the interactions of the variables and constraints. The mathematical statement also aids in the construction of the problem matrix, especially for processing by a computer, as the resultant mathematical and logical relations can be used to generate the constraints and data and/or check that the problem has been entered properly into the computer.

It is an open question as to what is the best way to develop one's formulative skills. Although problem formulation might be considered to be an art, it is felt that active participation and experience in analyzing problems, coupled with the studying of detailed case studies, as reported in the literature, is a preferred route. (See, for example, the applications described in Beale [59] and many of those given in the Bibliography.) The reader should bear in mind that every problem has related computational and data difficulties that are not usually discussed in any detail. Also, while studying a case the reader should consider how the problem formulation leads to implementable real-world solutions and how it can be used to evaluate constraints and requirements for planning and operational situations.

In most case studies, terminology, physical or organizational characteristics, and other elements peculiar to the specific areas of application are foreign to the reader. However, the studying of detailed cases is encouraged, as it is felt that through close review of the definitions, notation, and resultant models, the reader (with pencil in hand) will gain the necessary formulative insights.

EXERCISES

1. Solve the following production-scheduling problem, using the reduced set of Eqs. (1.6) with the initial conditions $s_{-1} = 0$, $s_n = 0$, $r_0 = 0$. The objective function to be minimized is

$$16s_0 + \sum_{t=1}^{4} s_t + 15 \sum_{t=1}^{5} y_t$$

and requirements are $r_1 = 5$, $r_2 = 12$, $r_3 = 6$, $r_4 = 6$, $r_5 = 8$.

2. Using the numerical data in Sec. 5, solve the caterer problem by the simplex algorithm.

3. Formulate and solve the following trim problem (Eisemann [235]). Assume that two machines A and B are available; machine A is capable of cutting a standard roll of 100 in, and machine B is capable of cutting a standard roll of 80 in. The following orders of various widths must be cut: 45-in roll, 862 rolls; 36-in roll, 341 rolls; 31-in roll, 87 rolls; and 14-in roll, 216 rolls. We assume that as many standard rolls as necessary are available and that, excluding the unavoidable trim loss, only the widths (that is, 45, 36, 31, 14 in) on order are cut.

4. Show that the constraints (4.14) force the PERT occurrence times x_j to be nonnegative.

5. *The multidimensional transportation problem.* Formulate the following example of a three-dimensional problem (Schell, Sec. 17 of Bibliography). Consider a soap manufacturer who has l factories in various parts of the country. Each of the l plants can manufacture m different types of soap. The soap is to be distributed from the factories to n different areas. Let

a_{ik} = required number of units to be shipped from factory i to area k
b_{jk} = required number of units of type j to be shipped to area k
d_{ij} = required number of units to be shipped from factory i of type j
x_{ijk} = amount of the jth type made in the ith plant shipped to the kth area
c_{ijk} = cost of shipping one unit of the jth type from the ith plant to the kth area
$x_{ijk} \geq 0$

Haley [346] gives a computational procedure for solving this problem. There is no guarantee that a solution will be integral if the a_{ik}, b_{jk}, and d_{ij} are integral (Motzkin [508]).

6. *The machine-loading problem.* Formulate the following problem. We are given m machines and n products and know the time a_{ij} required to process one unit of product j on machine i. Let x_{ij} be the number of units of product j produced on machine i, b_i the total time available on machine i, and d_j the number of units of product j which must be processed. We also know the cost (or time) c_{ij} of processing one unit of product j on machine i. (We assume each product needs to be processed on only one machine.) We wish to minimize the total cost. This problem can be looked at as a generalized transportation problem, and special algorithms exist for solving it (Eisemann and Lourie [237] and Jewell [402]).

7. Form the dual of the shortest-route problem of Fig. 11.19 and show that the final π_i of the Ford-Fulkerson algorithm represent an optimal solution to the dual.

8. Develop the equations for a two-commodity network with three nodes with a separate origin and destination for each commodity.

9. Construct an example to show that a multicommodity network with integer parameters can have a noninteger optimal solution.

10. Find the shortest-distance route from the origin to the destination in the network shown in Fig. E10:

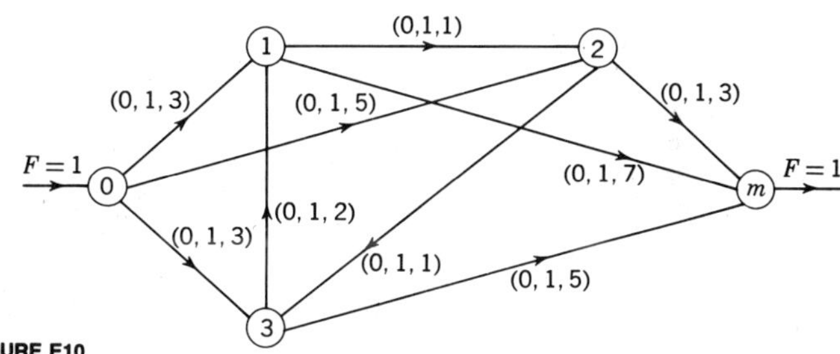

FIGURE E10

Answer: $(0, 1)(1, 2)(2, m)$; 7.

11. Show that the unimodular property holds for the systems (4.1) to (4.4), (4.10b) to (4.10e), and (4.13b) to (4.13e). For each system, if the nonzero bounding constraints on the x_{ij} are dropped, show that the remaining equations that form the node-arc incidence matrix have rank m, i.e., one equation is redundant, and any basis is triangular. If you include the bounding constraints, are bases for the full systems also triangular? Note that any problem that has each variable appearing at most twice with nonzero coefficients, with one coefficient positive and the other negative, has an equivalent network representation (bounding constraints are not included). HINT: See the discussion on the triangularity of the transportation problem bases, Chap. 10; also see Exercise 65 in this chapter.

12. For the transshipment problem of Fig. 11.17, restructure the network so it has only one source and one sink.

13. *a.* Write out the explicit equations of the maximal-flow network shown in Fig. E13.

  b. Solve the problem using the maximal-flow algorithm.

Answer: $f = 18$

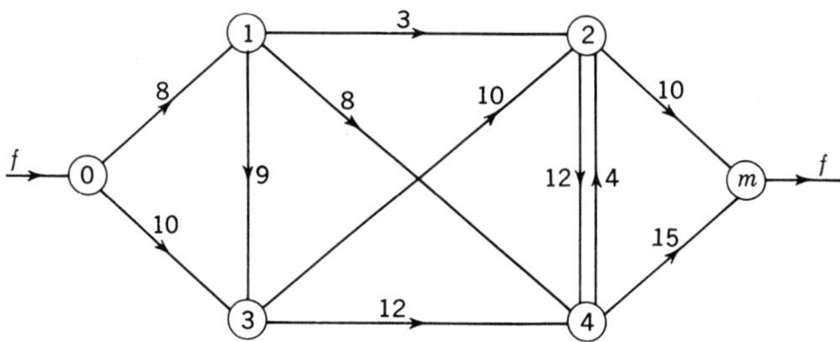

FIGURE E13

14. Develop the mathematical formulation of the network-with-gains problem (Jewell [402]).

15. Transform the following problem into a linear-constraint problem:

$$\mathbf{AX} = \mathbf{b}$$

$$|x_j| \le 1, \ x_j \text{ unrestricted}$$

16. Formulate a maximal-flow network problem in which capacities on each arc can be increased, but the total capacity is constrained by a given budget. Assume a linear cost function for each arc's increased cost. The problem is to allocate the budget to arcs so that the total flow is maximized (Fulkerson [269]).

17. Prove the max-flow min-cut theorem by use of the duality relationships and complementary slackness conditions (Frank and Frisch [259]).

18. If a distance network contains no cycles of positive length, show that the longest-route problem can be transformed into a standard shortest-path problem by reversing the signs of the distances of all arcs. For an acyclic network, what changes must be made in the corresponding shortest-route algorithm to solve the longest-route problem? Find the longest route in the network of Fig. 11.20.

19. Show that the simplex algorithm applied to the following problem will not find the correct answer if we transform the variable as the difference of two nonnegative variables (Shanno and Weil [607]). Maximize

$$|x_1|$$

subject to

$$x_1 \le \ \ 2$$
$$x_1 \ge -4$$

20. Find the Chebyshev approximation to the following problem (Zukhovitskiy and Avdeyeva [736]):

$$2x_1 - \ x_2 - 3 = 0$$
$$x_1 + \ x_2 + 2 = 0$$
$$x_1 - 3x_2 + 1 = 0$$
$$x_1 - 2x_2 + 2 = 0$$

Answer: $x_1 = \tfrac{1}{3}, \ x_2 = 0$; minimum $z = \tfrac{7}{3}$.

21. Formulate and solve the trim-loss problem which has a 215-in standard roll and allowable widths and orders as follows (Paull, Sec. 8*f* of Bibliography):

Widths ordered	No. of rolls ordered
64	782
60	624
48	142
45	118
33	144
12	826
16	188

Answer: Minimum total trim loss of 3,730 in.

22.   Develop the column-generating basis improvement criteria for the trim problem which has the objective function of minimizing Σx_j.

23.   For the trim problem, show that the problem which minimizes trim loss is equivalent to the one which minimizes the total number of rolls cut if the former objective function also includes the total material represented by the slack variables (Strum [633]).

24.   Solve the numerical trim problem of Sec. 5.

25.   Formulate the following "frankfurter" problem as a linear-programming problem. A frankfurter consists of a mix of bull meat, cow meat, beef trimmings, beef-head meat, regular pork trimmings, skinned jowls, and picnic meat. Each of these ingredients contains a certain percentage of fat which is, respectively: 10 per cent, 15 per cent, 30 per cent, 30 per cent, 65 per cent, 70 per cent, 30 per cent. The restrictions on the amounts of these ingredients for 100 lb of total mix are as follows: (1) combined cow meat and bull meat must be greater than or equal to 42 per cent of the batch weight; (2) combined cow meat, bull meat, beef trimmings, and beef-head meat must be greater than or equal to 30 per cent; (3) beef-head meat must be less than or equal to 10 per cent; (4) combined regular pork trimmings, skinned jowls, and picnic meat must be greater than or equal to 30 per cent; and (5) total fat must be equal to 42 per cent. Assume costs for each ingredient are known. Discuss how the manufacturer can treat the fat content as a parameter which depends on government specifications. As the prices of the ingredients change each week, what procedure should the manufacturer establish to routinize the computation of the optimal mix?

26.   Show that the dual of the PERT minimum-time problem (4.14) is a network-flow problem. What interpretation can you give to the dual variables and how are the dual variables related to the critical path (Charnes and Cooper [113] and Wiest and Levy [698])?

27.   Formulate the following as a linear program, Cahn [104] and Garvin [282]. A warehouse has a fixed capacity C and an initial stock s_0 of a certain product which is subject to known seasonal fluctuation in selling price and cost. We want to determine the optimal pattern of purchases, storage, and sales for the next n months. For each month i, let c_i be the cost per unit, p_i the selling price per unit, x_i the amount bought, and y_i the amount sold. (See Charnes and Cooper [112] for a special approach to solving the problem based on duality relationships.)

28.   Some government agencies (e.g., the National Science Foundation) receive grant requests to perform research and subject the requests to a review by a peer-group selection panel. The panel assigns a worth rating to each request from a high of 1,000 to a low of 100. Based on these ratings and a number of organizational and financial constraints, a set of grants is selected for funding. Formulate the selection process as an integer-programming problem which incorporates the following conditions: (a) At least a certain number of grants from each discipline (mathematics, physics, etc.) must be selected; (b) geographic dispersion must be attained in that a state must be represented at least once if a request has been submitted from that state; and (c) the total budget constraint must not be exceeded. What assumptions must be made concerning the ratings, and how would you use such a model in the actual administration of the selection process, e.g., a change in budget due to Congressional actions?

29.   Develop the transportation tableau which represents the production-scheduling problem (1.7) discussed in Sec. 1.

30.   Describe how the column-generating procedure could be applied to the goal-programming shift-scheduling problem of Sec. 5; also for Exercise 31*f*.

31.   Set-covering problems; see Chap. 9.

a.   Formulate the following *delivery problem* as an integer-programming problem (Balinski and Quandt [32]).   A delivery firm has three large boxes to deliver and six different trucks which can carry loads of varying sizes.   Truck 1 can carry only order 1, truck 2 must carry orders 1 and 3 together, truck 3 must carry orders 1 and 2 together, truck 4 must carry all the orders at the same time, truck 5 can carry only order 2, and truck 6 can carry only order 3.   Each truck can make one trip a day and all deliveries must be made on the same day.   The cost of each truckload combination is c_j.   Are there feasible solutions to this problem?

b.   Formulate the following targeting problem as an integer-programming problem.   We wish to target multiwarhead missiles so as to hit (cover) four targets with the minimum number of warheads.   Missile system 1 can knock out targets 1 and 3 (i.e., has two independently targeted warheads); missile system 2 can take care of targets 2 and 3; missile system 3 can hit targets 1, 3, and 4; missile system 4 can hit targets 2, 3, and 4; while missile number 5 can knock out targets 1, 2, and 5.   For this problem, will a target get hit more than once?   How does the formulation change if we wish to minimize the number of missiles?

c.   A city is divided into 100 crime-reporting blocks.   For a given time period, say 6 months, we forecast the level of crime for each block.   Defining a patrol unit's work load as the total amount of crime in the blocks covered by the patrol beat, develop an integer-programming problem which divides the 100 blocks into 10 patrol beats so as to equalize the work load on each beat.   Assume a block cannot be further subdivided.   (See Gass, Sec. 14 of Bibliography; also papers by Garfinkel and Nemhauser and Weaver for a similar problem dealing with political redistricting, Sec. 14 of Bibliography.)

d.   Interpret the data of problem 31*b* above in terms of a truck delivery problem in which the targets are stores and each missile represents a route which takes the truck to the corresponding targeted stores.   Each store must receive a delivery.   For this problem we can add four additional routes each of which goes only to one store.   What is an appropriate objective function?

e.   Interpret the data of problem 31*b* above in terms of an airline-crew-scheduling problem in which the targets are flight legs and a missile represents a possible routing of a crew.   What would be an appropriate objective function?   (See the survey on airline-crew-scheduling problems by Arabeyre et al. [10] and other references in Sec. 8*a* of Bibliography.)

f.   Formulate the following simplified telephone-operator-scheduling problem as a set-covering problem.   Daily, 24-h forecasts for each 15-min time period are made for the number of telephone operators required to supply the required level of service.   An operator can start an 8-h shift at the beginning of any 15-min time period, must take one 15-min break between the second and third hours, a 30-min lunch break between the 4th and 5th hours, and a 15-min break between the 6th and 7th hours.   Assume all shifts have this structure, although in practice there are split shifts, short shifts, and certain union regulations proscribing certain shifts.   The objective is to minimize the total number of operators which exceed or fall below each 15-min forecast, i.e., the sum of the absolute difference between what is scheduled and what is required.   The variables will be integer and represent the number of operators assigned to each of the 96 possible shifts which can start at the beginning of a 15-min period.   What other objective functions would be appropriate?   (See Segal, Sec. 12 of the Bibliography, and Bartholdi et al. [47].)

32.   Write the pay-off matrix and the equivalent linear-programming problem to the following game: Each player selects an integer from the set of integers (1, 2, 3).   The player with the lower number wins 2 points unless it is 1 lower, in which case 4 points are lost.   When the numbers are equal, there is no pay-off.

33.   Transform the following matrix games into their corresponding primal and dual linear-programming problems:

$$a.\ \begin{pmatrix} 2 & 1 & 0 & -2 \\ 1 & 0 & 3 & 2 \end{pmatrix} \qquad b.\ \begin{pmatrix} 0 & -1 & 1 \\ 1 & 1 & -1 \\ 1 & -1 & 0 \end{pmatrix}$$

34. Write out the associated game matrix for the following linear-programming problems:
Minimize

$$x_2 - x_3$$

subject to

$$x_1 + x_2 - x_3 \geq 0$$
$$-x_1 + x_2 + 3x_3 \geq 1$$
$$x_1 - 3x_2 + x_3 \leq 1$$
$$x_j \geq 0$$

35. Solve the following parametric game for all values of the parameter:

$$\mathbf{A} = \begin{pmatrix} \lambda & \lambda & \lambda \\ -1 & 4 & 0 \\ 3 & -1 & 2 \end{pmatrix}$$

Answer: $-\infty < \lambda \leq \frac{8}{7}$, $\mathbf{X} = (0, \frac{3}{7}, \frac{4}{7})$; $\frac{8}{7} \leq \lambda < +\infty$, $\mathbf{X} = (1, 0, 0)$

36. By letting $v = v_1 - v_2$, that is, the difference of two nonnegative variables, in Eqs. (6.1), determine a third linear-programming formulation of the matrix game.

37. For a $2 \times n$ or $m \times 2$ matrix game, describe how you can determine optimal strategies for both players. Apply your results to Example 6.2 and Exercise 33a. (HINT: For a matrix game, show that if $P_1(P_2)$ has only two strategies, then $P_2(P_1)$ has an optimal mixed strategy which has at most two positive elements. This result can be generalized as follows: For a $m \times n$ matrix game, each player has an optimal strategy that has at most k positive elements, where $k = $ minimum (m, n); see McKinsey [485].)

38. Solve the skin game by the simplex and graphical methods, Kuhn [445]. The two players are each provided with an ace of diamonds and an ace of clubs. P_1 is also given the two of diamonds and P_2 the two of clubs. In the first move, P_1 shows one card, and P_2, ignorant of P_1's choice, shows one card. P_1 wins if the suits match, and P_2 wins if they do not. The amount of the pay-off is the numerical value of the card shown by the winner. If the two deuces are shown, the pay-off is zero.

39. Show that the value of a symmetric game is zero. Note that $a_{ij} = -a_{ji}$.

40. Develop and solve the transportation tableau for the caterer problem of Sec. 5 when $q = 1$ and $p = 3$.

41. *The network-simplex algorithm* (Dantzig [152], Glover, Hultz, and Klingman [307], Karney and Klingman [416], Cunningham [146, 147], Bradley, Brown, and Graves [90], and Ali et al. [6]). Specialized adaptations of the revised simplex algorithm have been very successful in solving large-scale minimal-cost transshipment and other network problems. They are centered about the properties that any basis of the corresponding conservation of flow equations (with or without lower and upper bounds) is a triangular and unimodular matrix (Exercise 11), and that any basis can be represented as a (spanning) tree connecting all nodes of the defining network. Given a starting feasible basis $\mathbf{B}$, the corresponding solution vector can be readily found by solving the associated triangular set of equations; similarly, the pricing vector $\boldsymbol{\pi} = (\pi_i)$ can be found by solving the equations $\boldsymbol{\pi}\mathbf{B} = \mathbf{c}$, where $\mathbf{c}$ is the vector of costs for the flows x_{ij} in the basis. We take $\pi_0 = 0$ as we discard the conservation of flow equation determined by the source node 0. [For discussion purposes, we ignore the lower- and upper-bound conditions, but note that the lower bounds can be eliminated and made zero (footnote, page 282) and the upper bounds can be accommodated by the simple-upper-bound procedure, Chap. 9.] The nonbasic flow variables x_{ij} are priced out by calculating $z_{ij} - c_{ij} = \pi_i - \pi_j - c_{ij}$. If all $z_{ij} - c_{ij} \leq 0$, then the basis and its solution are optimal. If some $z_{ij} - c_{ij} > 0$, a corresponding arc (p, q) (i.e., variable x_{pq}) is selected to enter the basis. {The x_{pq} can correspond to the usual choice rule of $z_{pq} - c_{pq} = \max (\pi_p - \pi_q - c_{pq}) > 0$, but computational efficiency is improved if a modified rule is employed; see Glover, Karney, and Klingman [308].} The new arc is added to the current basis tree and, thus, a cycle is formed. As we now allow flow across arc (p, q) to the maximum extent possible, conservation-of-flow adjustments need to be made to the other arcs in the cycle until one of their flows goes to zero. A corresponding arc is dropped from the cycle and a new basis tree is obtained. (Adjustments are made in the optimality criteria and in the increase or decrease of the entering x_{pq} flow if upper bounds are present.) And the process continues until an optimal solution is determined. The effective computer-based implementation of this procedure requires the efficient storage and manipulation of each basis as a rooted spanning tree of the network (see Glover, Hultz, and Klingman [307], Barr, Glover, and Klingman [43], Jensen and Barnes [398], Bradley, Brown, and Graves [90], and Srinivasan and Thompson [621]).

The reader should prove the statements relating network spanning trees to a basis, i.e., the graph corresponding to a basic set of variables is a spanning tree of the network, and every spanning tree of a network corresponds to a basic set of variables (Dantzig [152]). HINT: Show that no basic set can have a graph with a cycle (uniqueness of basic solutions), and that the x_{ij} of a tree are uniquely determined (starting with the root and noting that the sum of the goods at the sources equals the sum of the goods at the sinks).

42. For the problem depicted in Fig. 11.17, develop the corresponding mathematical model of the two-source two-sink minimal-cost transshipment problem and the associated dual problem. Adapt the revised simplex algorithm, as described in Exercise 41, to solve this problem. How do you find a first basic feasible solution? (Glover, Karney, and Klingman [308].)

43. Show how to transform a transshipment problem into a transportation problem (Orden [537], Wagner [684], Dantzig [152], and Garvin [282]).

44. Using the following data, develop the mathematical model of the production-distribution planning network of Fig. 11.18. Assume that there are no lower and upper bounds on capacities

Factory production/cost	M_1	M_2	M_3	Total
F_1	300/\$4,000	200/\$5,000	100/\$6,000	600
F_2	400/\$4,025	150/\$4,800	150/\$5,900	700

Center demand	M_1	M_2	M_3	Total
C_1	400	175	130	705
C_2	300	180	115	595

Transportation cost (same for all models)	C_1	C_2
F_1	\$200	\$225
F_2	\$180	\$240

except those implied by the data. Solve the problem for the optimal distribution. The above data assumes that we know exactly how many automobiles by model type to produce at each factory and, thus, the different costs of production do not influence the previous formulation and its solution. To use this model as a production planning aid, assume the demands as given but production not known at the factories. Discuss how we can then let the minimal-cost network model determine the total production at each factory, as well as the production levels by model. Here you will need to assume lower- and upper-bound conditions for the production levels or else a factory may go out of business or not produce any models of a given type. How could you use this model to determine where you should open or close down a factory or distribution center, or to increase or decrease production capacity?

45.   Transform the shortest-route problem into an assignment problem (Lawler [462]).   What happens if in a shortest-route problem there is a cycle the sum of whose costs are negative?   (A negative cost or distance could represent a subsidy for using an arc.)

46.   For the production-distribution problem of Exercise 44, describe how you would change the network if the demands at the centers for each model and the totals were considered to be goals, i.e., show how you can handle goal-programming considerations within the minimal-cost network structure.   HINT: You will need to split the arcs into and out of the distribution center nodes to allow for under-and overachievement of the goals and assign appropriate costs (weights); see Price, Sec. 12 of Bibliography.

47.   Describe how you would formulate the diet problem and the trim problem as goal-programming models.

48.   Solve the goal-programming problem (5.2).   Also solve it by assuming preemptive priorities for the two objectives and develop the corresponding simplex tableau structure.   HINT: A $z_j - c_j$ row must be added to the tableau for each objective.

49.   Formulate the following as regular linear-programming problems, and as transportation and minimal-cost network models.

a.   A manufacturer schedules the production for the company's product 6 months in advance.   The factory has a production capacity of 1,500 units per month on its regular shift, with a regular-time cost of $900 per unit.   The factory can go into overtime and produce 600 more units a month at a unit cost of $1,000.   It costs the company $50 per month to keep a unit in storage if it is not sold in the month it is produced.   The following monthly sales forecasts have to be met: 1,200, 2,300, 2,900, 3,500, 2,600, 1,600.

b.   For the manufacturer in part *a*, assume no overtime is allowed and production is fixed at 1,500 per month.   At the beginning of the production period, the unit cost of the product is $900, but it is expected that this cost will increase $25 per month each month.   Formulate the problem to meet the sales forecasts of part *a*.

50.   Solve the problems of Exercise 49 using any appropriate formulation.

51.   Formulate and solve goal-programming models of the problems given in Exercise 49.   What are the goal-programming network models (see Exercise 46)?

52.   Reformulate the caterer's problem equations (5.3) for $p = 4$ and $q = 1$ and find an optimal solution.

53.   Given a transportation problem with the additional constraint(s) that require the sum of some of the variables in a row (column) to be bounded above, show how the total problem can be reformulated as a transportation problem (Dantzig [152] and Wagner [683]).

54.   If in determining a first feasible solution to a transportation problem we encounter a degenerate solution, we then need to make a choice of variable(s) that will form a basis.   Show that this selection is equivalent to constructing a spanning tree for the bipartite transportation network whose zero arc flows correspond to the $x_{ij} = 0$ in the degenerate basic feasible solution.   The degenerate solution to Example 1.2, Chap. 10, is given by the disconnected graph of Fig.

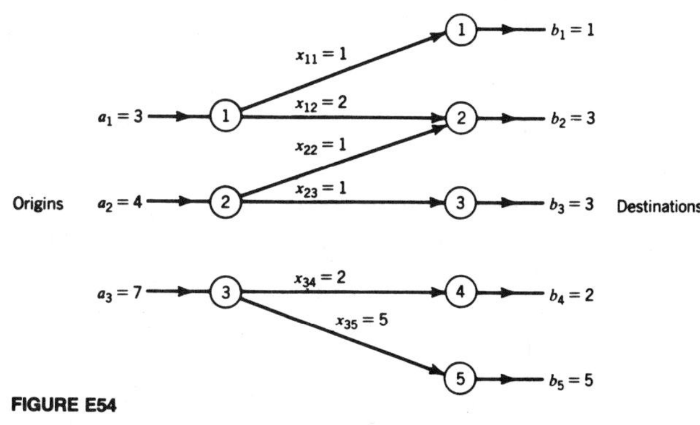

FIGURE E54

E54.   Which arcs (variables) would form the spanning tree and what decision rule would you use to select one?   (Collatz and Wetterling [131].)

55.   In Fig. E54, complete the spanning tree by selecting $x_{24} = 0$ to be in the first basic feasible solution.   Let x_{33} be the next variable to enter the basis.   Using your spanning tree and the cycle formed by adding the arc $(3, 3)$, show how to determine the value of x_{33}.   Note that your cycle is directed, but the added flow x_{33} can be added or subtracted from the arcs in the cycle to maintain conservation of flow and the origin and destination requirements.

56.   Discuss how the kilter conditions (4.12) are deduced.

57.   *The transportation problem paradox (Szwarc [635]).*   The following transportation problem has an optimal objective function value of 241.   If one unit is added to both a_3 and b_1, show that the new optimal value is 239, i.e., we can ship more for less.   Can you explain the paradox?

25	10	2	30	10
5	15	20	10	15
100	65	0	2	20
5	12	13	15	

58.   The U.S. Department of Agriculture suggests the following as recommended daily allowances (RDAs) for the listed important nutrients:

Protein—65 grams
Vitamin A—5000 International Units
Vitamin C—60 milligrams
Thiamin—1.5 milligrams
Riboflavin—1.7 milligrams
Niacin—20 milligrams
Calcium—1.0 gram
Iron—18 milligrams

The following table lists the calories and the percentage of the RDAs recommended for males and females between the ages of 19 and 22 years.

		Percentage of the RDAs							
	Calories	Protein	A	C	Thiamin	Riboflavin	Niacin	Calcium	Iron
Male	3000	85	100	75	100	110	60	80	60
Female	2100	75	80	75	75	85	35	80	100

Using these figures, construct and solve a diet problem that involves at least 15 foods, using actual cost and dietary data for each food.   You should attempt to use constraints that reflect your preferences for certain foods, e.g., steak or vegetables being forced into or out of the solution, and to ensure that a diversity of food types is selected, e.g., some meat, fish, greens, fruits, etc.   How do these preferences impact on the cost of your diet?   See Peterkin et al. [547] and the food containers for nutritional value data.   Discuss the use of the diet problem formulation to develop special diets for diabetics, runners, and other athletes, etc.   You might also try to develop a special food mix for your cat or dog.   Discuss how you determine the validity of these diets.

59.   Figure E59 represents a museum with nine rooms and ten connecting openings.   Determine the minimal number of guards required so that the guards are in positions

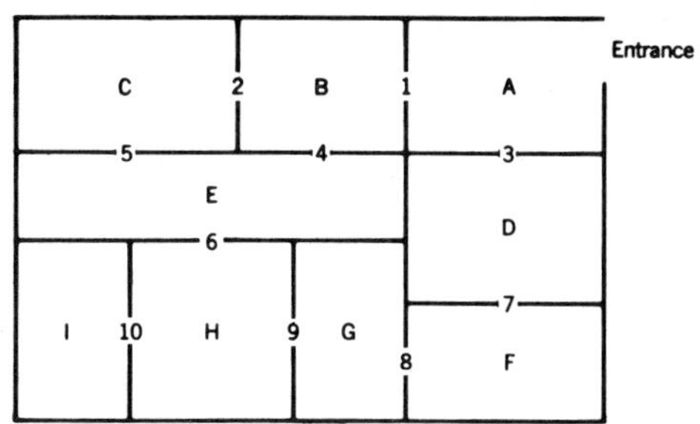

FIGURE E59

(openings) to observe all the rooms. A room can be considered guarded if a guard is situated in the opening between two rooms. No guard is required at the entrance. HINT: Formulate the problem as a set-covering problem. You should be able to determine the minimal number of guard positions required without resorting to detailed computations (Vajda [666]).

60. Determine a blend of coffee beans that suits your requirements for body, acidity, aroma, and caffeine. Collect data on eight to ten different types of beans. You will need to develop scales for the measurement of the qualitative attributes (see, for example, Scanlan [598]). You want to minimize cost. It takes about 19.1 oz of green beans to make 16 oz of roasted beans. (Problem due to Marta Quinn.)

61. Show that the elimination of the x_t in (1.6) enables us to replace (1.4) with (1.6).

62. Consider the following *block-triangular system* of equations and costs that defines a minimizing linear-programming problem:

	x_{11}	x_{12}	x_{13}	x_{21}	x_{22}	x_{23}	x_{31}	x_{32}	x_{33}		
	4	6	3							$= 4$	
	-2	-1	-4	2	1	2				$= 2$	$(x_{ij} \ge 0)$
	-3	-2	-1	-4	-1	-3	6	8	4	$= 3$	
Costs	8	9	4	3	2	6	4	3	1		

Develop a solution algorithm that does not require you to solve the total problem at once. HINT: Think revised simplex algorithm and note that you can begin your solution procedure with only the last block. Show that your optimal basis is independent of the right-hand side. The problem matrix is known as a *Leontief substitution model*; see Dantzig [168, 179], Veinott [674], and Koehler et al. [435] for the general form, its properties, and applications.
Solution: $x_{12} = \frac{2}{3}$, $x_{21} = \frac{4}{3}$, $x_{33} = \frac{29}{12}$; $x_0 = \frac{149}{12}$.

63. Show that a knapsack problem can be reformulated in terms of finding the longest distance in an acyclic network (Fulkerson [270]).

64. In Ref. 147, Cunningham illustrates cycling in a network that has three nodes and twelve arcs. All the a_i and b_i are zero. For each distinct pair of nodes i and j, there are two directed arcs from i to j, with one having a cost of 1 and the other a cost of -1. For example, there are two directed arcs connecting nodes 1 and 2, and two directed arcs connecting nodes 2 and 1. Starting with an initial spanning tree consisting of arc (1, 2) with cost 1 and arc (2, 3) with cost -1, can you determine a cycle of twelve different trees? Construct the graph of the above network and its explicit mathematical representation. Try to solve it as a standard linear-programming problem using anticycling procedures (Chap. 7) and by means of a computer-based simplex algorithm. Cunningham and Klincewicz [148] show that a network having basis size 2 must have a cycle length of at least 10.

65. Formulate and solve the following problem (Wagner [684]). A police department estimates that the following number of police officers should be on duty during the corresponding time periods:

Shift periods	Time periods	No. of officers required
1	0200–0600	22
2	0600–1000	55
3	1000–1400	88
4	1400–1800	110
5	1800–2200	44
6	2200–0200	33

Each officer works eight consecutive hours, i.e., two successive periods. A shift can start at the beginning of any of the six periods. The problem is to at least meet the requirements with the minimum number of officers on duty during the 24-hour day that runs from period 1 to period 6. Show that the constraint structure can be transformed into a network and, thus, that there is an optimal integer solution. Construct the network. If r_i is the requirement in period i, show that the optimal number of officers will be the larger of R_1 or R_2 where $R_1 = \max(r_1 + r_3 + r_5, r_2 + r_4 + r_6, r_1 + r_4, r_2 + r_5, r_3 + r_6)$ and $R_2 = \frac{1}{2}(r_1 + r_2 + r_3 + r_4 + r_5 + r_6)$ for R_2 rounded up. (See Baker, Sec. 12 of Bibliography, Bartholdi et al. [47], and Exercise 31f.) As there is no solution to the above problem that meets the requirements exactly, formulate the problem as a goal-programming problem with the requirements as goals, assign numerical weights to the deviation variables, and solve and compare the two solutions.
Answer: x_1 (number assigned shift period 1) $= 55$, $x_3 = 99$, $x_4 = 11$, $x_5 = 33$; minimum $= 198$.

66. Discuss how a company, a university, or the military can use network models as work-force planning tools. Given future requirements, the models should be able to hire, promote, or fire individuals (or similar groups) over the time horizon of the plan; see Secs. 12 and 19 of Bibliography.

67. *a.* For the minimal-cost transshipment problem, let $\Sigma a_i < \Sigma b_i$. What adjustments do you need to make to the network? HINT: Set up a dummy source that can ship to each sink. How can you set priorities on those sinks not satisfied?

b. For the minimal-cost transshipment problem, let the b_i be interpreted as goals. Describe how you can adjust the network so that shipments to a sink can be above or below its goal. Also, how can you treat an arc capacity as a goal? (See Exercise 46.)

68. Solve the goal-programming problem (5.2) with a profit goal of $2,300 and a total unit production goal of 45 [see problem (4.1a) of Chap. 8 and Fig. 8.6]. Also solve it by assuming preemptive priorities (Exercise 48).

PART 4

Nonlinear Programming

CHAPTER 12

Nonlinear Programming

The linear-programming model, along with the implied concept of viewing complex systems in terms of interrelated activities,[1] has proved to be a major contribution to the field of scientific decision-making. Applications of linear programming exist in every category of industrial, business, and government endeavors. The versatility and adaptability of this deceptively simple mathematical model appears to have no bounds. However, early workers in the field recognized certain implied limitations of the linear model and began to develop theoretical, as well as computational, foundations for special and general nonlinear-programming problems.[2] One can, for example, trace such developments by studying the growth of programming applications within the petroleum industry (Charnes, Cooper, and Mellon [119], Manne [491], Symonds [634], Garvin et al. [283], and Catchpole [106]). This industrial area yielded some of the first successful nonmilitary applications of

----

[1] See Dantzig [152], Chap. 3, for a detailed discussion along these lines.

[2] The reader is referred to the work of, among others, Arrow, Hurwicz, and Uzawa [11], Barankin and Dorfman [37], Beale [57], Charnes and Lemke [123], Dantzig [152], Dennis [200], John [403], Kuhn and Tucker [450], Markowitz [494], Slater [615], and Wolfe [713]. See Kuhn [444] for a history of the early developments in nonlinear programming.

linear programming. Extensions of these applications led to the need and development of many nonlinear-programming techniques. Nonlinearities in the programming model of an oil refinery arise in both the objective function and the system of constraints. As Catchpole [106] and others point out, the cost of crude oil may increase with the quantity required, while the use of lead as an octane improver has a decreasing effect as more lead is used; i.e., the mathematical relations which describe these statements are not linear.

The consideration of these and other nonlinearities has advanced our ability to develop mathematical models of complex systems. We can now describe and deal with many generalizations of the linear model that take us into the field of nonlinear programming—or into the more inclusive field of *mathematical programming*.[1]

The purpose of this chapter is to present selected basic material necessary to the understanding of the foundations of nonlinear programming.[2] The material requires a broader mathematical background than does that covered in previous chapters. However, except for basic notions from calculus, this chapter is essentially self-contained, assuming knowledge of the preceding discussions.

1. THE GENERAL PROBLEM OF MATHEMATICAL PROGRAMMING

As noted in Sec. 1 of Chap. 1, the mathematical model of the linear-programming problem can be described using relationships of the form

$$a_1 x_1 + a_2 x_2 + \cdots + a_j x_j + \cdots + a_n x_n = b$$

From a mathematical point of view this is a rather restrictive relationship in that it excludes a wide variety of possible interactions between the variables. For example, the output of an activity might be functionally related in a nonlinear fashion to the output of other activities or even to itself. Nonlinear expressions of the form

$$g(\mathbf{X}) = g(x_1, x_2, \ldots, x_n) = a_1 x_1^2 + a_2 x_2 x_5 + \cdots + a_n x_n^3 = b$$

and trigonometric, logarithmic, and other special functional forms can occur in the constraint set and objective function of practical, as well as theoretical, problems. Other nonlinearities include the restriction of a variable to integer values; for example, $x_1 = 0, 1, 2, 3, \ldots$, as discussed in Chap. 9.

---

[1] See the books of Abadie [1], Boot [83], Saaty and Bram [586], Hadley [344], Kunzi and Krelle [451], Vajda [662], Zoutendijk [735], Fiacco and McCormick [248], McCormick [483], Zangwill [727], Mangasarian [489], and the survey article by Dorn [215].

[2] The reader interested in specific applications should consult the references given above, Bracken and McCormick [86], and Lasdon and Waren [461]. Although apparently not as plentiful and pervasive as linear programming, nonlinear applications include chemical equilibrium, least-square and Chebyshev approximations (see Chap. 11), stock-portfolio selection, and many problems from mathematics and economics (see Bibliography).

For our purposes, we define the problem of mathematical programming to be:

Find a vector $\mathbf{X} = (x_1, x_2, \ldots, x_n)$ which minimizes (maximizes) the objective function $f(\mathbf{X})$ subject to

$$g_i(\mathbf{X}) \leq 0 \qquad \text{for } i = 1, 2, \ldots, m \tag{1.1}$$

Here $f(\mathbf{X})$ and the $g_i(\mathbf{X})$ can be considered to be general functions of the variables $x_1, x_2, \ldots, x_n$. If a problem contains equality constraints or inequality constraints going in the opposite direction, these constraints can be expressed in the form (1.1), so there is no loss of generality in assuming the constraints to be of the above form.

For a linear-programming problem, problem (1.1) becomes: Minimize

$$f(\mathbf{X}) = c_1 x_1 + \cdots + c_n x_n$$

subject to

$$g_1(\mathbf{X}) = a_{11}x_1 + \cdots + a_{1n} x_n - b_1 \leq 0$$

$$\cdots\cdots\cdots\cdots\cdots\cdots\cdots\cdots\cdots\cdots$$

$$g_m(\mathbf{X}) = a_{m1}x_1 + \cdots + a_{mn}x_n - b_m \leq 0$$
$$g_{m+1}(\mathbf{X}) = x_1 \qquad\qquad\qquad \geq 0$$

$$\cdots\cdots\cdots\cdots\cdots\cdots\cdots\cdots\cdots\cdots$$

$$g_{m+n}(\mathbf{X}) = \qquad\qquad x_n \qquad \geq 0$$

A simple example of a nonlinear problem involving logarithmic functions in the objective function (Dorn [217]) is as follows: Minimize

$$f(x_1, x_2) = -\log x_1 - \log x_2$$

subject to

$$x_1 + x_2 \leq 2$$
$$x_1 \qquad\quad \geq 0$$
$$x_2 \geq 0$$

The graph of the constraint set and the objective function, for various values of the objective function, is shown in Fig. 12.1. The solution space is the shaded triangle and the optimum occurs at the point (1, 1), not at an extreme point of the solution space. When the constraint set can contain nonlinear functions, we are not necessarily limited to the nice mathematical situation found in linear programming, i.e., the search for an optimizing extreme point of a convex solution space. The mathematical-programming problem could have a nonconvex solution space, an interior optimal solution, and other characteristics which compound both theoretical and computational developments.

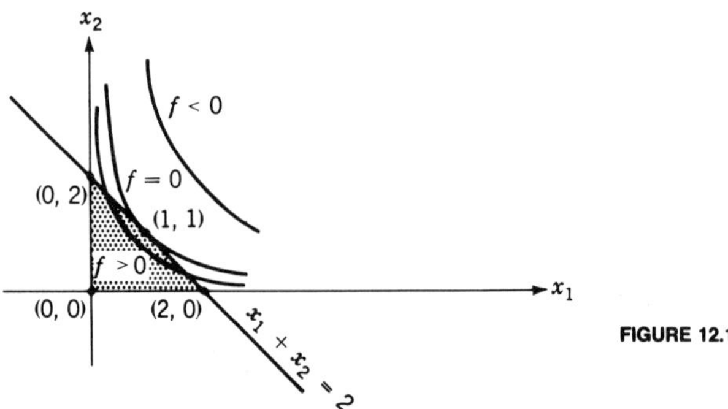

FIGURE 12.1

One of the main difficulties encountered in the realm of mathematical programming is the determination of a solution point which appears not only to optimize the objective function at that point, but to optimize the function over the complete range of the solution space. More precisely, we have:

Definition 1. A function $f(\mathbf{X})$ has a *global minimum* at a point $\mathbf{X}^0$ of a set of points $\mathbf{K}$ if and only if $f(\mathbf{X}^0) \leq f(\mathbf{X})$ for all $\mathbf{X}$ in $\mathbf{K}$.

Definition 2. A function $f(\mathbf{X})$ has a *local minimum* at a point $\mathbf{X}^0$ of a set of points $\mathbf{K}$ if and only if there exists a positive number ϵ such that $f(\mathbf{X}^0) \leq f(\mathbf{X})$ for all $\mathbf{X}$ in $\mathbf{K}$ at which $\|\mathbf{X}^0 - \mathbf{X}\| < \epsilon$.

In one dimension, it is easy to illustrate local and global minima for a function $f(\mathbf{X})$ as in Fig. 12.2. Here $\mathbf{K}$ is the set of all nonnegative x. The points $x = 0$ and $x = x_2$ are local minima, while $x = x_1$ is the global minimum. (A global minimum is also a local minimum.) The reader should note that global and local maxima can also be defined by obvious changes in Definitions 1 and 2.

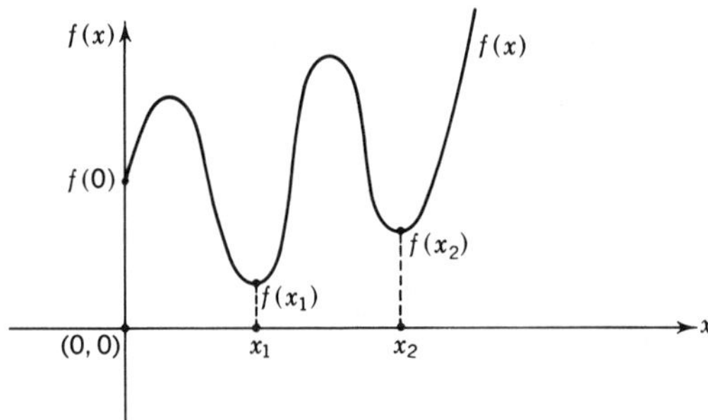

FIGURE 12.2

Although nonlinear optimization problems call for the determination of a global minimum, computational procedures will, in general, lead to a solution which is only a local minimum. Moreover, it is usually impossible to determine if a local minimum is really a global minimum. Even if this could be done, computational procedures have no way of proceeding from a local minimum to a global minimum. This is not the case, of course, for the linear-programming problem. For that problem, we can show that the simplex algorithm arrives at a solution point which is not only a local minimum, but also a global minimum. This is a key characteristic of most computationally tractable nonlinear-programming problems. Thus, our discussions below will be directed toward problems for which we can demonstrate mathematically the coincidence of the local and global minima. To accomplish this end, we next describe and develop some necessary mathematical notions.

2. MATHEMATICAL BACKGROUND

a. Quadratic forms. In some mathematical-programming problems, the objective function is restricted to a specialized expression, a positive-definite quadratic form.

Definition 3. A *quadratic form* $f(\mathbf{X})$ in n variables is an expression of second degree which can be written as $f(\mathbf{X}) = \sum\limits_{i=1}^{n} \sum\limits_{j=1}^{n} c_{ij}x_ix_j$, or

$$f(\mathbf{X}) = c_{11}x_1x_1 + c_{12}x_1x_2 + \cdots + c_{1n}x_1x_n + \cdots + c_{n1}x_nx_1 + \cdots + c_{nn}x_nx_n$$

For $n = 2$ we have

$$f(x_1, x_2) = c_{11}x_1^2 + c_{12}x_1x_2 + c_{21}x_1x_2 + c_{22}x_2^2$$

If $\mathbf{C} = (c_{ij})$, then the quadratic form is expressed as $f(\mathbf{X}) = \mathbf{X'CX}$, where $\mathbf{X} = (x_1, \ldots, x_n)$ is an $n \times 1$ column vector and $\mathbf{C}$ is an $n \times n$ matrix. For $n = 2$,

$$f(x) = \begin{pmatrix} x_1 \\ x_2 \end{pmatrix}' \mathbf{C} \begin{pmatrix} x_1 \\ x_2 \end{pmatrix} = (x_1 x_2) \begin{pmatrix} c_{11} & c_{12} \\ c_{21} & c_{22} \end{pmatrix} \begin{pmatrix} x_1 \\ x_2 \end{pmatrix}$$

$$= (c_{11}x_1 + c_{21}x_2 \quad c_{12}x_1 + c_{22}x_2) \begin{pmatrix} x_1 \\ x_2 \end{pmatrix}$$

$$= c_{11}x_1^2 + c_{12}x_1x_2 + c_{21}x_2x_1 + c_{22}x_2^2$$

$$= c_{11}x_1^2 + (c_{12} + c_{21})x_1x_2 + c_{22}x_2^2$$

Noting that for any $\mathbf{X}$ and $\mathbf{C}$ the product $\mathbf{X'CX}$ is a scalar and thus $\mathbf{X'CX} = (\mathbf{X'CX})' = (\mathbf{CX})'(\mathbf{X'})' = \mathbf{X'C'X}$, we can write

$$\mathbf{X'CX} = \tfrac{1}{2}[\mathbf{X'CX} + \mathbf{X'C'X}] = \frac{\mathbf{X'(C + C')X}}{2}$$

The matrix $\tfrac{1}{2}(\mathbf{C} + \mathbf{C'})$ is a symmetric matrix with the general element $\tfrac{1}{2}(c_{ij} + c_{ji})$. In this form, the coefficient of x_ix_j is $\tfrac{1}{2}(c_{ij} + c_{ji})$ and of x_jx_i is $\tfrac{1}{2}(c_{ji} + c_{ij})$. However, for the total product term of x_i and x_j, the coefficient is

$(c_{ij} + c_{ji})$, as it is in the original form $\mathbf{X'CX}$. For any $\mathbf{C}$ we can redefine the coefficients of $x_i x_j$ to be $\frac{1}{2}(c_{ij} + c_{ji})$ and of $x_j x_i$ to also be $\frac{1}{2}(c_{ij} + c_{ji})$ so that the matrix representation of the quadratic form is in terms of a symmetric matrix. In what follows, we shall always assume that the matrix $\mathbf{C}$ is symmetric.

Definition 4. A quadratic form is called *positive definite* if $\mathbf{X'CX} > 0$ for $\mathbf{X} \neq 0$. For example $f(\mathbf{X}) = x_1^2 + x_2^2$ is positive definite.

Definition 5. A quadratic form is called *positive semidefinite* if $\mathbf{X'CX} \geq 0$ for all $\mathbf{X}$ and there exists at least one vector $\mathbf{X} \neq 0$ such that $\mathbf{X'CX} = 0$. The quadratic form

$$f(\mathbf{X}) = x_1^2 - x_1 x_2 - x_2 x_1 + x_2^2 = x_1^2 - 2x_1 x_2 + x_2^2 = (x_1 - x_2)^2 \geq 0$$

for all x_1 and x_2 and equals zero for $x_1 = x_2 = 1$, and thus $f(\mathbf{X})$ is positive semidefinite.

Negative definite and negative semidefinite forms can also be defined by appropriate reversal of the inequality signs in the above definitions.

For quadratic forms of small dimension we can test for positive definiteness by the following theorem (Hadley [342]):

Theorem 1. *The quadratic form $f(\mathbf{X}) = \mathbf{X'CX}$ is positive definite if and only if the principal minors, i.e., the minors obtained by taking successively larger minors along the principal diagonal (see below), are all strictly positive. Thus for $n = 4$ we must have*

$$c_{11} > 0 \qquad \begin{vmatrix} c_{11} & c_{12} \\ c_{21} & c_{22} \end{vmatrix} > 0 \qquad \begin{vmatrix} c_{11} & c_{12} & c_{13} \\ c_{21} & c_{22} & c_{23} \\ c_{31} & c_{32} & c_{33} \end{vmatrix} > 0 \qquad |\mathbf{C}| > 0$$

b. Convex functions. The main mathematical concept required in our development of nonlinear programming is that of a convex function. Given that the constraints $g_i(\mathbf{X})$ and the objective function $f(\mathbf{X})$ are convex functions we can, as shall be shown below, prove that a local minimum is also a global minimum.

Definition 6. A function $f(\mathbf{X}) = f(x_1, x_2, \ldots, x_n)$ defined over a set of points which lie in a convex set $\mathbf{K}$ in E_n is called a *convex function* if for any two points $\mathbf{X}_1$ and $\mathbf{X}_2$ in $\mathbf{K}$ and any $0 \leq \lambda \leq 1$ we have

$$f[\lambda \mathbf{X}_1 + (1 - \lambda)\mathbf{X}_2] \leq \lambda f(\mathbf{X}_1) + (1 - \lambda)f(\mathbf{X}_2) \tag{2.1}$$

If strict inequality holds in (2.1), $f(\mathbf{X})$ is said to be *strictly convex*. Similarly, a function $f(\mathbf{X})$ is a *concave function* if (2.1) holds with the inequality reversed, that is, $-f(\mathbf{X})$ is convex.

A convex function $f(\mathbf{X})$ in one variable is pictured in Fig. 12.3. The reader will note that the expression $\lambda f(x_1) + (1-\lambda)f(x_2)$ is the ordinate corresponding to an abscissa of $\lambda x_1 + (1-\lambda)x_2$ for the line segment joining the points $[x_1, f(x_1)]$ and $[x_2, f(x_2)]$. Thus, a function is convex if every point on each such line segment overestimates the corresponding point on the function; i.e., the line segment lies above the function. A linear function is both convex and concave; the sum of convex functions is also a convex function; i.e., if $f_i(\mathbf{X})$, $i = 1, \ldots, k$ are convex functions, then $f(\mathbf{X}) = \sum_{i=1}^{k} f_i(\mathbf{X})$ is convex.

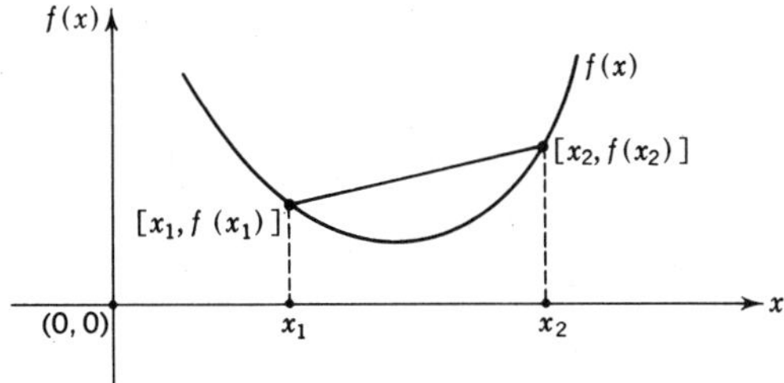

FIGURE 12.3

Theorem 2. *The positive semidefinite quadratic form $f(\mathbf{X}) = \mathbf{X'CX}$ is a convex function for all $\mathbf{X}$ in $\mathbf{E}_n$.*

Proof: We wish to show for $0 \le \lambda \le 1$ and all $\mathbf{X}_1, \mathbf{X}_2$ in $\mathbf{E}_n$ that

$$f[\lambda \mathbf{X}_1 + (1-\lambda)\mathbf{X}_2] \le \lambda f(\mathbf{X}_1) + (1-\lambda)f(\mathbf{X}_2)$$

or

$$f[\lambda \mathbf{X}_1 + (1-\lambda)\mathbf{X}_2] - \lambda f(\mathbf{X}_1) - (1-\lambda)f(\mathbf{X}_2) \le 0 \tag{2.2}$$

Noting that $\mathbf{X}_2'\mathbf{CX}_1 = \mathbf{X}_1'\mathbf{CX}_2$, the left-hand side of (2.2) can be rewritten as

$$[\lambda \mathbf{X}_1 + (1-\lambda)\mathbf{X}_2]'\mathbf{C}[\lambda \mathbf{X}_1 + (1-\lambda)\mathbf{X}_2] - \lambda \mathbf{X}_1'\mathbf{CX}_1 - (1-\lambda)\mathbf{X}_2'\mathbf{CX}_2$$

$$= \lambda^2 \mathbf{X}_1'\mathbf{CX}_1 + (1-\lambda)^2\mathbf{X}_2'\mathbf{CX}_2 + 2\lambda(1-\lambda)\mathbf{X}_1'\mathbf{CX}_2 - \lambda \mathbf{X}_1'\mathbf{CX}_1 - (1-\lambda)\mathbf{X}_2'\mathbf{CX}_2$$

$$= (\lambda^2 - \lambda)\mathbf{X}_1'\mathbf{CX}_1 + (1-\lambda)[(1-\lambda)\mathbf{X}_2'\mathbf{CX}_2 - \mathbf{X}_2'\mathbf{CX}_2] + 2\lambda(1-\lambda)\mathbf{X}_1'\mathbf{CX}_2$$

$$= \lambda(\lambda - 1)\mathbf{X}_1'\mathbf{CX}_1 + \lambda(\lambda - 1)\mathbf{X}_2'\mathbf{CX}_2 - 2\lambda(\lambda - 1)\mathbf{X}_1'\mathbf{CX}_2$$

$$= \lambda(\lambda - 1)[\mathbf{X}_1'\mathbf{CX}_1 + \mathbf{X}_2'\mathbf{CX}_2 - 2\mathbf{X}_1'\mathbf{CX}_2]$$

$$= \lambda(\lambda - 1)[(\mathbf{X}_1 - \mathbf{X}_2)'\mathbf{C}(\mathbf{X}_1 - \mathbf{X}_2)] \tag{2.3}$$

Since the form $\mathbf{X'CX} \ge 0$ for any $\mathbf{X}$, for (2.3) we have

$$[(\mathbf{X}_1 - \mathbf{X}_2)'\mathbf{C}(\mathbf{X}_1 - \mathbf{X}_2)] \ge 0$$

Since $\lambda(\lambda - 1) < 0$ for $0 < \lambda < 1$ and $\lambda(\lambda - 1) = 0$ for $\lambda = 0$ or 1, then (2.3) will always be less than or equal to zero for all vectors $\mathbf{X}_1$ and $\mathbf{X}_2$, which was to be shown. Thus $f(\mathbf{X}) = \mathbf{X'CX}$ is convex for all $\mathbf{X}$ in $\mathbf{E}_n$.

Theorem 3.   *If $f(\mathbf{X})$ is convex on a convex set $\mathbf{K}$, then $f(\mathbf{X})$ has at most one local minimum.   If there is such a minimum, it is a global minimum and is attained on a convex set.*

Proof: Suppose there is a local minimum at a point $\mathbf{X}^0$.   For *any* $\mathbf{X} = \bar{\mathbf{X}}$ in $\mathbf{K}$, by definition of a local minimum and convexity of $f(\mathbf{X})$, we have

$$f(\mathbf{X}^0) \leq f[(1-\lambda)\mathbf{X}^0 + \lambda\bar{\mathbf{X}}] \leq (1-\lambda)f(\mathbf{X}^0) + \lambda f(\bar{\mathbf{X}}) \tag{2.4}$$

for λ a sufficiently small positive number.   From the extremes of (2.4) we have

$$f(\mathbf{X}^0) \leq (1-\lambda)f(\mathbf{X}^0) + \lambda f(\bar{\mathbf{X}})$$

$$\lambda f(\mathbf{X}^0) \leq \lambda f(\bar{\mathbf{X}})$$

and since $\lambda > 0$, then $f(\mathbf{X}^0) \leq f(\bar{\mathbf{X}})$, which implies $f(\mathbf{X}^0)$ is a global minimum by definition, as $\bar{\mathbf{X}}$ is any point in $\mathbf{K}$.

If $\mathbf{X}^0$ and $\mathbf{X}^1$ are two points at which $f(\mathbf{X})$ attains its minimum value z_0, then for $0 \leq \lambda \leq 1$

$$z_0 \leq f[(1-\lambda)\mathbf{X}^0 + \lambda\mathbf{X}^1] \leq (1-\lambda)f(\mathbf{X}^0) + \lambda f(\mathbf{X}^1) = z_0$$

Hence, $f(\mathbf{X})$ also attains its minimum at $\mathbf{X} = (1-\lambda)\mathbf{X}^0 + \lambda\mathbf{X}^1$ and thus the set of solutions is convex.

Theorem 4.   *Let $\mathbf{K}$ be the set of points of $\mathbf{E}_n$ which satisfy the constraints*

$$g_i(\mathbf{X}) \leq 0 \qquad i = 1, 2, \ldots, m$$

$$\mathbf{X} \geq \mathbf{0}$$

If the $g_i(\mathbf{X})$ are convex functions, then $\mathbf{K}$ is a convex set.   (We assume $\mathbf{K}$ is not empty.)

Proof: Let $\mathbf{X}_1$ and $\mathbf{X}_2$ be any two points in $\mathbf{K}$ and define

$$\bar{\mathbf{X}} = \lambda\mathbf{X}_1 + (1-\lambda)\mathbf{X}_2$$

for $0 \leq \lambda \leq 1$.   We need to show that $\bar{\mathbf{X}}$ satisfies the constraints.

$\bar{\mathbf{X}} \geq \mathbf{0}$ as it is the nonnegative sum of nonnegative quantities.   For any i, we have

$$g_i(\bar{\mathbf{X}}) = g_i[\lambda\mathbf{X}_1 + (1-\lambda)\mathbf{X}_2] \leq \lambda g_i(\mathbf{X}_1) + (1-\lambda)g_i(\mathbf{X}_2)$$

But $g_i(\mathbf{X}_1) \leq 0$ and $g_i(\mathbf{X}_2) \leq 0$ and for $0 \leq \lambda \leq 1$ we must have

$$g_i(\bar{\mathbf{X}}) \leq \lambda g_i(\mathbf{X}_1) + (1-\lambda)g_i(\mathbf{X}_2) \leq 0$$

N.B.   The preceding two theorems state that for the general programming problem of minimizing $f(\mathbf{X})$ subject to $g_i(\mathbf{X}) \leq 0$ ($i = 1, \ldots, m$) and $\mathbf{X} \geq \mathbf{0}$, if $f(\mathbf{X})$ and all $g_i(\mathbf{X})$ are convex functions, then a local minimum of $f(\mathbf{X})$ subject to the constraint set is also a global minimum.

c. The gradient vector and saddle point. The following concepts will enable us to develop necessary and sufficient conditions for a point $\mathbf{X}^0$ to be a solution of certain restricted mathematical-programming problems.

Definition 7. If a function $f(\mathbf{X})$ and all its first derivatives are continuous over some subset of $\mathbf{E}_n$, then for each point $\mathbf{X}^0$ in this subset we define the n-component column vector $\nabla f(\mathbf{X}^0)$, termed the gradient vector of $f(\mathbf{X})$ at $\mathbf{X}^0$, to be

$$\nabla f(\mathbf{X}^0) = \begin{pmatrix} \dfrac{\partial f(\mathbf{X}^0)}{\partial x_1} \\ \cdot \\ \cdot \\ \dfrac{\partial f(\mathbf{X}^0)}{\partial x_n} \end{pmatrix}$$

$\nabla f(\mathbf{X}^0)$ is a vector perpendicular to the contour of $f(\mathbf{X})$ which passes through $\mathbf{X}^0$. Its direction is the direction of maximum increase of $f(\mathbf{X})$, and its length is the magnitude of that maximum rate of increase. This direction is often referred to as the direction of steepest ascent.

From the calculus we note that a necessary condition for $\mathbf{X}^0$ to be a minimum for the general differentiable function $f(x_1, \ldots, x_n) = f(\mathbf{X})$ is that

$$\left[\frac{\partial f(\mathbf{X})}{\partial x_j} \right]_{\mathbf{X}^0} = 0$$

for $\mathbf{X}^0$ an interior point of the region of definition (Buck [99] and Widder [697]). To find the minimum of $f(\mathbf{X})$ with $a_j \le x_j \le b_j$, we must investigate the boundary points as well as the interior points. For example, if $f(\mathbf{X})$ is a convex function of the single variable x, we could have the three cases illustrated in Fig. 12.4a to c.

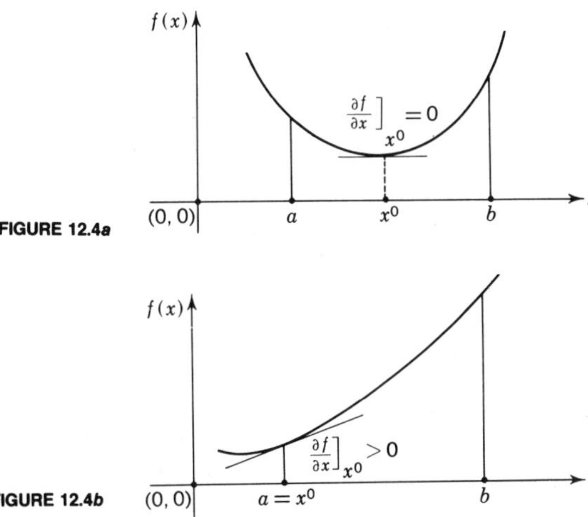

FIGURE 12.4a

FIGURE 12.4b

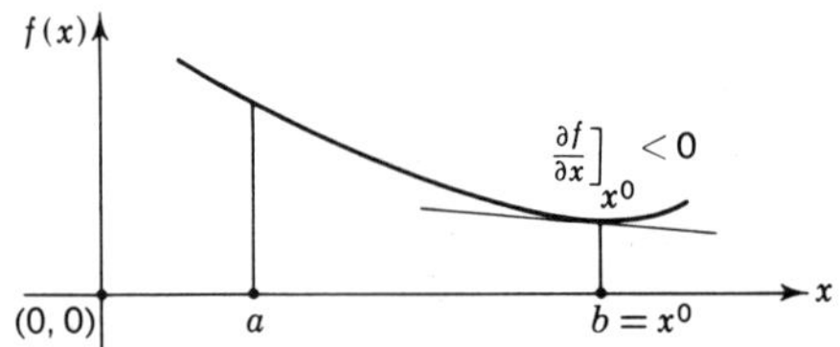

FIGURE 12.4c

The next two theorems relate the notions of convexity, the gradient and conditions for a point $\mathbf{X}^0$ to minimize a convex function over a convex set.

Theorem 5. *If $f(\mathbf{X})$ is defined on an open convex set $\mathbf{K}$ and $f(\mathbf{X})$ is differentiable in $\mathbf{X}$, then $f(\mathbf{X})$ is convex if and only if*

$$f(\mathbf{X}_1) - f(\mathbf{X}_2) \geq (\mathbf{X}_1 - \mathbf{X}_2)' \nabla f(\mathbf{X}_2)$$

for all $\mathbf{X}_1$ and $\mathbf{X}_2$ in $\mathbf{K}$.

Proof: (*a*) We first prove if $f(\mathbf{X}_1) - f(\mathbf{X}_2) \geq (\mathbf{X}_1 - \mathbf{X}_2)' \nabla f(\mathbf{X}_2)$, then $f(\mathbf{X})$ is convex (sufficiency).

For $\mathbf{X}_1$ and $\mathbf{X}_2$ in $\mathbf{K}$ and $0 \leq \lambda \leq 1$, let $\mathbf{X}_3 = \lambda \mathbf{X}_1 + (1 - \lambda)\mathbf{X}_2$. We note that $\mathbf{X}_3$ is also in $\mathbf{K}$. For $\mathbf{X}_1$ and $\mathbf{X}_3$ we have

$$f(\mathbf{X}_1) - f(\mathbf{X}_3) \geq (\mathbf{X}_1 - \mathbf{X}_3)' \nabla f(\mathbf{X}_3) \tag{2.5}$$

and for $\mathbf{X}_2$ and $\mathbf{X}_3$ we have

$$f(\mathbf{X}_2) - f(\mathbf{X}_3) \geq (\mathbf{X}_2 - \mathbf{X}_3)' \nabla f(\mathbf{X}_3) \tag{2.6}$$

Multiplying (2.5) by λ and (2.6) by $(1 - \lambda)$ and adding, we obtain

$$\lambda f(\mathbf{X}_1) - \lambda f(\mathbf{X}_3) + (1 - \lambda)f(\mathbf{X}_2) - (1 - \lambda)f(\mathbf{X}_3)$$
$$\geq \lambda (\mathbf{X}_1 - \mathbf{X}_3)' \nabla f(\mathbf{X}_3) + (1 - \lambda)(\mathbf{X}_2 - \mathbf{X}_3)' \nabla f(\mathbf{X}_3)$$

or

$$\lambda f(\mathbf{X}_1) + (1 - \lambda)f(\mathbf{X}_2)$$
$$\geq f(\mathbf{X}_3) + [\lambda \mathbf{X}_1' + (1 - \lambda)\mathbf{X}_2'] \nabla f(\mathbf{X}_3) - \mathbf{X}_3' \nabla f(\mathbf{X}_3) \tag{2.7}$$

Applying the definition of $\mathbf{X}_3$, (2.7) becomes

$$\lambda f(\mathbf{X}_1) + (1 - \lambda)f(\mathbf{X}_2) \geq f[\lambda \mathbf{X}_1 + (1 - \lambda)\mathbf{X}_2]$$

which implies that $f(\mathbf{X})$ is convex.

(*b*) We next prove that if $f(\mathbf{X})$ is convex we must also have that $f(\mathbf{X}_1) - f(\mathbf{X}_2) \geq (\mathbf{X}_1 - \mathbf{X}_2)' \nabla f(\mathbf{X}_2)$ (necessity).

For $0 < \lambda \leq 1$ we have for any $\mathbf{X}_1$ and $\mathbf{X}_2$ in $\mathbf{K}$ that

$$\lambda f(\mathbf{X}_1) + (1 - \lambda)f(\mathbf{X}_2) \geq f[\lambda \mathbf{X}_1 + (1 - \lambda)\mathbf{X}_2]$$

Regrouping terms,

$$\lambda f(\mathbf{X}_1) - \lambda f(\mathbf{X}_2) \geq f[\lambda \mathbf{X}_1 + (1 - \lambda)\mathbf{X}_2] - f(\mathbf{X}_2)$$

or

$$f(\mathbf{X}_1) - f(\mathbf{X}_2) \geq \frac{f[\mathbf{X}_2 + \lambda\,(\mathbf{X}_1 - \mathbf{X}_2)] - f(\mathbf{X}_2)}{\lambda}$$

Taking the limit of the right-hand side as λ approaches zero, we have from calculus the result

$$f(\mathbf{X}_1) - f(\mathbf{X}_2) \geq (\mathbf{X}_1 - \mathbf{X}_2)'\nabla f(\mathbf{X}_2)$$

Theorem 6. *Let $f(\mathbf{X})$ be a convex continuously differentiable function defined on a convex set* $\mathbf{K}$. *Let* $\mathbf{X}^0$ *be in* $\mathbf{K}$. *Then* $f(\mathbf{X}^0) \leq f(\mathbf{X})$ *for all* $\mathbf{X}$ *in* $\mathbf{K}$ [*that is,* $\mathbf{X}^0$ *minimizes* $f(\mathbf{X})$ *for all* $\mathbf{X}$ *in* $\mathbf{K}$] *if and only if* $(\mathbf{X} - \mathbf{X}^0)'\nabla f(\mathbf{X}^0) \geq 0$ *for all* $\mathbf{X}$ *in* $\mathbf{K}$.

Proof: (*a*) (Necessity.) For $f(\mathbf{X}^0) \leq f(\mathbf{X})$, the minimum point $\mathbf{X}^0$ is either an interior point of $\mathbf{K}$ or a boundary point of $\mathbf{K}$. If $\mathbf{X}^0$ is interior, by the calculus we must have $\nabla f(\mathbf{X}^0) = 0$ and, of course, $(\mathbf{X} - \mathbf{X}^0)'\nabla f(\mathbf{X}^0) = 0$. For $\mathbf{X}^0$ *any* minimum point, we have for $f(\mathbf{X})$ convex and $\mathbf{K}$ a convex set (see Theorem 3)

$$f(\mathbf{X}^0) \leq f[\lambda \mathbf{X} + (1 - \lambda)\mathbf{X}^0]$$

for all $\mathbf{X}$ in $\mathbf{K}$ and $0 \leq \lambda \leq 1$. For $\lambda > 0$, then

$$\frac{f[\mathbf{X}^0 + \lambda\,(\mathbf{X} - \mathbf{X}^0)] - f(\mathbf{X}^0)}{\lambda} \geq 0$$

Taking the limit as λ approaches zero, we have

$$(\mathbf{X} - \mathbf{X}^0)'\nabla f(\mathbf{X}^0) \geq 0$$

(*b*) (Sufficiency.) Given that $(\mathbf{X} - \mathbf{X}^0)'\nabla f(\mathbf{X}^0) \geq 0$ for all $\mathbf{X}$ in $\mathbf{K}$, by convexity of $f(\mathbf{X})$ we have from Theorem 5 that

$$f(\mathbf{X}) - f(\mathbf{X}^0) \geq (\mathbf{X} - \mathbf{X}^0)'\nabla f(\mathbf{X}^0) \geq 0$$

Hence $f(\mathbf{X}) \geq f(\mathbf{X}^0)$ for all $\mathbf{X}$ in $\mathbf{K}$ and $\mathbf{X}^0$ minimizes $f(\mathbf{X})$ over $\mathbf{K}$.

From a geometric point of view, the above theorem states that $\mathbf{X}^0$ is a minimum point if the angle between the vectors $\nabla f(\mathbf{X}^0)$ and $(\mathbf{X} - \mathbf{X}^0)'$ for all $\mathbf{X}$ in $\mathbf{K}$ is less than or equal to $90°$. See Fig. 12.5.

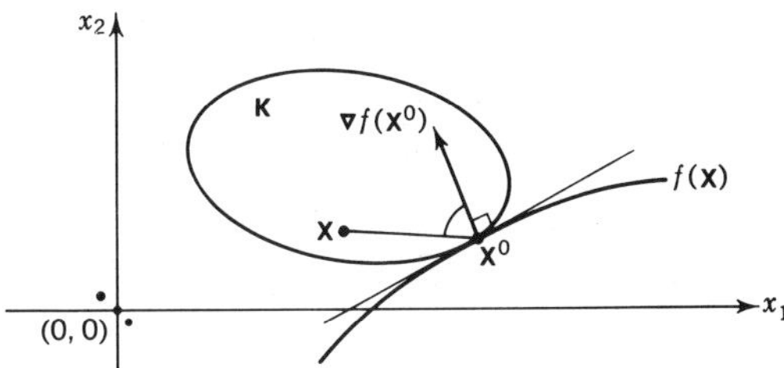

FIGURE 12.5

For a point $\mathbf{X}_1$ not the minimum and if for some $\mathbf{X}_2$ we have $f(\mathbf{X}_1) \geq f(\mathbf{X}_2)$, then from Theorem 5

$$0 \geq f(\mathbf{X}_2) - f(\mathbf{X}_1) \geq (\mathbf{X}_2 - \mathbf{X}_1)' \nabla f(\mathbf{X}_1)$$

i.e., the angle between the gradient at $\mathbf{X}_1$ and $(\mathbf{X}_2 - \mathbf{X}_1)'$ is no less than 90°.

In the following section, we shall need the concept of a saddle point of a function. The reader will recognize the extension of the saddle point of a matrix developed for the theory of games in Sec. 6, Chap. 11.

Definition 8. The point $(\mathbf{X}^0, \boldsymbol{\pi}^0)$ is said to be a *saddle point* of a function $F(\mathbf{X}, \boldsymbol{\pi})$ if there exists an $\epsilon > 0$ such that for all $\mathbf{X}$ and $\boldsymbol{\pi}$ in an ϵ neighborhood of $\mathbf{X}^0$ and $\boldsymbol{\pi}^0$, that is, with $\|\mathbf{X} - \mathbf{X}^0\| < \epsilon$ and $\|\boldsymbol{\pi} - \boldsymbol{\pi}^0\| < \epsilon$, we have

$$F(\mathbf{X}^0, \boldsymbol{\pi}) \leq F(\mathbf{X}^0, \boldsymbol{\pi}^0) \leq F(\mathbf{X}, \boldsymbol{\pi}^0) \tag{2.8}$$

If (2.8) holds for all $\mathbf{X}$ and $\boldsymbol{\pi}$ then $(\mathbf{X}^0, \boldsymbol{\pi}^0)$ is a global saddle point. $\boldsymbol{\pi}^0$ is an m-component vector.

The last theorem of this section is stated without proof (Valentine [667]).

Theorem 7. *Theorem of the separating hyperplane. If two convex sets $\mathbf{K}_1$ and $\mathbf{K}_2$ have, at most, common boundary points, there exists a hyperplane $\mathbf{aX} = b$, $\mathbf{a} \neq \mathbf{0}$ which separates $\mathbf{K}_1$ and $\mathbf{K}_2$ in the following sense:*

$$\mathbf{aX}_1 \leq b \quad \text{for all } \mathbf{X}_1 \text{ in } \mathbf{K}_1$$

$$\mathbf{aX}_2 \geq b \quad \text{for all } \mathbf{X}_2 \text{ in } \mathbf{K}_2$$

Here $\mathbf{a}$ is an n-component row vector and b is a scalar.

3. THE CONVEX-PROGRAMMING PROBLEM[1]

We can now specialize our discussion to the important class of nonlinear problems termed *convex programs* and develop necessary and sufficient conditions for solution to such problems, the *Kuhn-Tucker conditions*. In this section we consider:

The convex-programming problem. Minimize

$$f(\mathbf{X})$$

subject to

$$g_i(\mathbf{X}) \leq 0 \quad i = 1, 2, \ldots, m \tag{3.1}$$
$$\mathbf{X} \geq \mathbf{0}$$

where $f(\mathbf{X})$ and all the $g_i(\mathbf{X})$ are convex, continuously differentiable functions.

[1] This section follows the development of Kunzi and Krelle [451]. See also Karlin [415] and Vajda [662].

For these functions, define the *Lagrangian function*

$$F(\mathbf{X}, \boldsymbol{\pi}) = f(\mathbf{X}) + \sum_{i=1}^{m} \pi_i g_i(\mathbf{X})$$

where the m variables π_i are called *Lagrange multipliers*. We can now describe a new problem whose solution has a direct relationship to the convex-programming problem.

The saddle-point problem. Find nonnegative vectors $\mathbf{X}^0$, $\boldsymbol{\pi}^0$ for the Lagrangian function $F(\mathbf{X}, \boldsymbol{\pi})$ such that

$$F(\mathbf{X}^0, \boldsymbol{\pi}) \leq F(\mathbf{X}^0, \boldsymbol{\pi}^0) \leq F(\mathbf{X}, \boldsymbol{\pi}^0) \tag{3.2}$$

for all $\mathbf{X} \geq 0$ and $\boldsymbol{\pi} \geq 0$. The point $(\mathbf{X}^0, \boldsymbol{\pi}^0)$ is termed a nonnegative saddle point. We can interpret (3.2) to read

$$F(\mathbf{X}^0, \boldsymbol{\pi}^0) = \max_{\pi \geq 0} \left[\min_{\mathbf{X} \geq 0} F(\mathbf{X}, \boldsymbol{\pi}) \right] = \min_{\mathbf{X} \geq 0} \left[\max_{\pi \geq 0} F(\mathbf{X}, \boldsymbol{\pi}) \right]^\dagger$$

† This relationship can be developed as follows: From (3.2) we have

$$F(\mathbf{X}^0, \boldsymbol{\pi}) \leq F(\mathbf{X}^0, \boldsymbol{\pi}^0) \tag{3.2a}$$
$$F(\mathbf{X}^0, \boldsymbol{\pi}^0) \leq F(\mathbf{X}, \boldsymbol{\pi}^0) \tag{3.2b}$$

From (3.2a) we have

$$\max_{\pi \geq 0} F(\mathbf{X}^0, \boldsymbol{\pi}) \leq F(\mathbf{X}^0, \boldsymbol{\pi}^0)$$

and from (3.2b)

$$F(\mathbf{X}^0, \boldsymbol{\pi}^0) \leq \min_{\mathbf{X} \geq 0} F(\mathbf{X}, \boldsymbol{\pi}^0)$$

so that

$$\max_{\pi \geq 0} f(\mathbf{X}^0, \boldsymbol{\pi}) \leq F(\mathbf{X}^0, \boldsymbol{\pi}^0) \leq \min_{\mathbf{X} \geq 0} f(\mathbf{X}, \boldsymbol{\pi}^0) \tag{3.2c}$$

Since

$$\min_{\mathbf{X} \geq 0} \max_{\pi \geq 0} f(\mathbf{X}, \boldsymbol{\pi}) \leq \max_{\pi \geq 0} F(\mathbf{X}^0, \boldsymbol{\pi})$$

and

$$\min_{\mathbf{X} \geq 0} F(\mathbf{X}, \boldsymbol{\pi}^0) \leq \max_{\pi \geq 0} \min_{\mathbf{X} \geq 0} F(\mathbf{X}, \boldsymbol{\pi})$$

we conclude from (3.2c) that

$$\min_{\mathbf{X} \geq 0} \max_{\pi \geq 0} f(\mathbf{X}, \boldsymbol{\pi}) \leq F(\mathbf{X}^0, \boldsymbol{\pi}^0) \leq \max_{\pi \geq 0} \min_{\mathbf{X} \geq 0} F(\mathbf{X}, \boldsymbol{\pi}) \tag{3.2d}$$

But for $F(\mathbf{X}, \boldsymbol{\pi})$ we have for any $\mathbf{X}$ and $\boldsymbol{\pi}$

$$\min_{\mathbf{X} \geq 0} F(\mathbf{X}, \boldsymbol{\pi}) \leq F(\mathbf{X}, \boldsymbol{\pi})$$

and

$$F(\mathbf{X}, \boldsymbol{\pi}) \leq \max_{\pi \geq 0} F(\mathbf{X}, \boldsymbol{\pi})$$

The point $(\mathbf{X}^0, \boldsymbol{\pi}^0)$ of the saddle-point problem and the optimal solution to the convex-programming problem are related by

Theorem 8. *The Kuhn-Tucker theorem.* $\mathbf{X}^0$ *is a solution to the convex-programming problem* (3.1) *if and only if a vector* $\boldsymbol{\pi}^0$ *exists such that* $(\mathbf{X}^0, \boldsymbol{\pi}^0)$ *is a solution to the saddle-point problem.*

The proof of this theorem enables us to evolve the necessary and sufficient conditions which a vector $\mathbf{X}^0$ must satisfy in order to be a solution to (3.1). The knowledge of such conditions leads to the development of appropriate algorithms and justifies their validity.

Proof: We first show that the saddle-point conditions are sufficient to ensure that $\mathbf{X}^0$ is a solution to the convex-programming problem. For the saddle point $(\mathbf{X}^0, \boldsymbol{\pi}^0)$ and $F(\mathbf{X}, \boldsymbol{\pi}) = f(\mathbf{X}) + \sum_{i=1}^{m} \pi_i g_i(\mathbf{X})$ we have from (3.2)

$$f(\mathbf{X}^0) + \sum_{i=1}^{m} \pi_i g_i(\mathbf{X}^0) \le f(\mathbf{X}^0) + \sum_{i=1}^{m} \pi_i^0 g_i(\mathbf{X}^0) \le f(\mathbf{X}) + \sum_{i=1}^{m} \pi_i^0 g_i(\mathbf{X}) \qquad (3.3)$$

for all $\mathbf{X} \ge \mathbf{0}$ and $\boldsymbol{\pi} \ge \mathbf{0}$. For the left-hand inequality of (3.3) to hold for all $\boldsymbol{\pi} \ge \mathbf{0}$, we must have both $g_i(\mathbf{X}^0) \le 0$ for all i and $\sum_{i=1}^{m} \pi_i^0 g_i(\mathbf{X}^0) = 0$. We show this as follows. The left-hand inequality is

$$f(\mathbf{X}^0) + \sum_{i=1}^{m} \pi_i g_i(\mathbf{X}^0) \le f(\mathbf{X}^0) + \sum_{i=1}^{m} \pi_i^0 g_i(\mathbf{X}^0)$$

or

$$\sum_{i=1}^{m} \pi_i g_i(\mathbf{X}^0) \le \sum_{i=1}^{m} \pi_i^0 g_i(\mathbf{X}^0) \qquad (3.4)$$

If some $g_i(\mathbf{X}^0) > 0$, then a corresponding $\pi_i > 0$ can be chosen large enough that the inequality (3.4) does not hold. Therefore $g_i(\mathbf{X}^0) \le 0$ for all i. Since $\mathbf{X}^0 \ge \mathbf{0}$ we have shown that $\mathbf{X}^0$ satisfies the constraints (3.1) of the

---

and hence

$$\min_{\mathbf{X} \ge \mathbf{0}} F(\mathbf{X}, \boldsymbol{\pi}) \le \max_{\boldsymbol{\pi} \ge \mathbf{0}} F(\mathbf{X}, \boldsymbol{\pi}) \qquad (3.2e)$$

Since the left-hand side of (3.2e) is independent of $\mathbf{X}$, we have

$$\min_{\mathbf{X} \ge \mathbf{0}} F(\mathbf{X}, \boldsymbol{\pi}) \le \min_{\mathbf{X} \ge \mathbf{0}} \max_{\boldsymbol{\pi} \ge \mathbf{0}} F(\mathbf{X}, \boldsymbol{\pi}) \qquad (3.2f)$$

Also, since the right-hand side of (3.2f) is independent of $\boldsymbol{\pi}$, we have

$$\max_{\boldsymbol{\pi} \ge \mathbf{0}} \min_{\mathbf{X} \ge \mathbf{0}} F(\mathbf{X}, \boldsymbol{\pi}) \le \min_{\mathbf{X} \ge \mathbf{0}} \max_{\boldsymbol{\pi} \ge \mathbf{0}} F(\mathbf{X}, \boldsymbol{\pi}) \qquad (3.2g)$$

Relationships (3.2d) and (3.2g) yield the desired result.

convex-programming problem. Also as (3.4) must hold for $\pi = 0$, we have $0 \leq \sum_{i=1}^{m} \pi_i^0 g_i(\mathbf{X}^0)$. But as $\pi^0 \geq \mathbf{0}$, all $g_i(\mathbf{X}^0) \leq 0$; then $\sum_{i=1}^{m} \pi_i^0 g_i(\mathbf{X}^0) \leq 0$, which implies $\sum_{i=1}^{m} \pi_i^0 g_i(\mathbf{X}^0) = 0$. Thus either $\pi_i^0 = 0$, $g_i(\mathbf{X}^0) = 0$, or both equal zero.

From the right-hand inequality of (3.3) we now can write

$$f(\mathbf{X}^0) \leq f(\mathbf{X}) + \sum_{i=1}^{m} \pi_i^0 g_i(\mathbf{X}) \qquad \text{for all } \mathbf{X} \geq \mathbf{0}$$

Since $\pi^0 \geq \mathbf{0}$, $f(\mathbf{X}^0) \leq f(\mathbf{X})$ for all $\mathbf{X} \geq \mathbf{0}$ for which $g_i(\mathbf{X}) \leq \mathbf{0}$. Hence, as $g_i(\mathbf{X}^0) \leq 0$, $\mathbf{X}^0$ is a solution to the convex-programming problem.

To prove necessity, i.e., that a solution to the convex-programming problem (3.1) solves the saddle-point problem, is a more complex task. Here we have to introduce an apparently mild assumption concerning the convex region of feasible solutions to (3.1). This regularity or constraint qualification, which is due to Slater [615], takes the place of the assumption dealing with the differentiability of the functions required by Kuhn and Tucker in their original proof [450]. The Kuhn-Tucker qualification rules out certain peculiarities on the boundary of the feasible-solution space. For our purposes, then, we require the following assumption: For the constraints (3.1) there exists an $\mathbf{X} \geq \mathbf{0}$ such that all $g_i(\mathbf{X}) < 0$.[1]

To continue with the proof, let $\mathbf{X}^0$ be a solution to (3.1). We shall show that a $\pi^0 \geq \mathbf{0}$ exists such that $(\mathbf{X}^0, \pi^0)$ satisfies the saddle-point problem. For this purpose we construct in $\mathbf{E}_{m+1}$ two point sets $\mathbf{K}_1$ and $\mathbf{K}_2$ with vectors $\mathbf{Y} = (y_0, y_1, \ldots, y_m)$. $\mathbf{K}_1$ is the set of all points $\mathbf{Y}$ for which there exists at least one $\mathbf{X} \geq \mathbf{0}$ with $f(\mathbf{X}) \leq y_0$ and $g_i(\mathbf{X}) \leq y_i$ for $i = 1, \ldots, m$; that is,

$$\mathbf{K}_1 = \{\mathbf{Y} | f(\mathbf{X}) \leq y_0;\ g_i(\mathbf{X}) \leq y_i,\ i = 1, \ldots, m, \text{ for at least one } \mathbf{X} \geq \mathbf{0}\}$$

$\mathbf{K}_2$ is defined in terms of $\mathbf{X}^0$ to be

$$\mathbf{K}_2 = \{\mathbf{Y} | f(\mathbf{X}^0) > y_0;\ 0 > y_i,\ i = 1, \ldots, m\}$$

Since $f(\mathbf{X})$ and all $g_i(\mathbf{X})$ are convex, $\mathbf{K}_1$ is a convex set. Also, $\mathbf{K}_2$ is convex as it represents an open subset of $\mathbf{E}_{m+1}$ which is bounded by planes parallel to the coordinate axes of $\mathbf{E}_{m+1}$. It consists of the interior of an orthant with vertex at $[f(\mathbf{X}^0), \mathbf{0}]$. As $\mathbf{X}^0$ minimizes $f(\mathbf{X})$ for all $\mathbf{X} \geq \mathbf{0}$, $\mathbf{K}_1$ and $\mathbf{K}_2$ have no points in common; i.e., the intersection of $\mathbf{K}_1$ and $\mathbf{K}_2$ is empty.

From the above and the theorem of the existence of a hyperplane which separates two nonintersecting convex sets, there exists a vector

---

[1] This assumption apparently rules out linear-equality restrictions such as $\mathbf{AX} - \mathbf{b} = 0$. However, it can be shown that the assumption is not required for those $g_i(\mathbf{X})$ which are linear. The Kuhn-Tucker theorem is true without restriction if all constraints are linear (Karlin [415] and Arrow, Hurwicz, and Uzawa [11]).

$\mathbf{a} \neq \mathbf{0}$, a constant b, such that the hyperplane $\mathbf{aY} = b$ separates $\mathbf{K}_1$ from $\mathbf{K}_2$; that is,

$$\mathbf{aY}_1 \geq \mathbf{aY}_2 \tag{3.5}$$

for all $\mathbf{Y}_1$ in $\mathbf{K}_1$ and $\mathbf{Y}_2$ in $\mathbf{K}_2$. Since the components of $\mathbf{Y}_2$ are allowed to be arbitrarily small, we can conclude that $\mathbf{a} \geq \mathbf{0}$. For if some component $a_i < 0$, then as the corresponding component of $\mathbf{Y}_2$ can be made as small as desired, the separating hyperplane inequality (3.5) would not be satisfied.

If we choose $\mathbf{Y}_1 = [f(\mathbf{X}), g_1(\mathbf{X}), \ldots, g_m(\mathbf{X})]$ and $\mathbf{Y}_2 = [f(\mathbf{X}^0), 0, \ldots, 0]$, since (3.5) is satisfied for the boundary points of $\mathbf{K}_2$, we have

$$a_0 f(\mathbf{X}) + a_1 g_1(\mathbf{X}) + \cdots + a_m g_m(\mathbf{X}) \geq a_0 f(\mathbf{X}^0) \tag{3.6}$$

for all $\mathbf{X} \geq \mathbf{0}$. If $a_0 = 0$, then from (3.6) for all $\mathbf{X} \geq \mathbf{0}$,

$$a_1 g_1(\mathbf{X}) + \cdots + a_m g_m(\mathbf{X}) \geq 0 \tag{3.7}$$

and all $a_i \geq 0$ and some $a_i \neq 0$. But by the regularity assumption there exists some $\mathbf{X} \geq \mathbf{0}$, say $\bar{\mathbf{X}}$, for which all $g_i(\bar{\mathbf{X}}) < 0$. For $\bar{\mathbf{X}}$, (3.7) would be strictly less than zero. Thus, we can conclude that $a_0 > 0$. Dividing (3.6) by a_0 we obtain

$$f(\mathbf{X}) + \pi_1^0 g_1(\mathbf{X}) + \cdots + \pi_m^0 g_m(\mathbf{X}) \geq f(\mathbf{X}^0) \tag{3.8}$$

with $\pi_i^0 = a_i/a_0 \geq 0$. This can be rewritten as

$$F(\mathbf{X}, \boldsymbol{\pi}^0) \geq f(\mathbf{X}^0) \tag{3.9}$$

for all $\mathbf{X} \geq \mathbf{0}$. Letting $\mathbf{X} = \mathbf{X}^0$ in (3.8) we have

$$f(\mathbf{X}^0, \boldsymbol{\pi}^0) \geq f(\mathbf{X}^0)$$

or

$$F(\mathbf{X}^0, \boldsymbol{\pi}^0) = f(\mathbf{X}^0) + \pi_1^0 g_1(\mathbf{X}^0) + \cdots + \pi_m^0 g_m(\mathbf{X}^0) \geq f(\mathbf{X}^0)$$

or

$$\pi_1^0 g_1(\mathbf{X}^0) + \cdots + \pi_m^0 g_m(\mathbf{X}^0) \geq 0$$

But as all $g_i(\mathbf{X}^0) \leq 0$ and all $\pi_i^0 \geq 0$ we must have

$$\pi_1^0 g_1(\mathbf{X}^0) + \cdots + \pi_m^0 g_m(\mathbf{X}^0) = 0$$

and thus

$$F(\mathbf{X}^0, \boldsymbol{\pi}^0) = f(\mathbf{X}^0) \tag{3.10}$$

Also, for all $\boldsymbol{\pi} \geq \mathbf{0}$, as $g_i(\mathbf{X}^0) \leq 0$ for all i, it follows that

$$f(\mathbf{X}^0) \geq f(\mathbf{X}^0) + \sum_{i=1}^m \pi_i g_i(\mathbf{X}^0) = F(\mathbf{X}^0, \boldsymbol{\pi}) \tag{3.11}$$

Combining (3.9) to (3.11) we obtain

$$F(\mathbf{X}^0, \boldsymbol{\pi}) \leq f(\mathbf{X}^0) = F(\mathbf{X}^0, \boldsymbol{\pi}^0) = f(\mathbf{X}^0) \leq F(\mathbf{X}, \boldsymbol{\pi}^0)$$

or

$$F(\mathbf{X}^0, \boldsymbol{\pi}) \leq F(\mathbf{X}^0, \boldsymbol{\pi}^0) \leq F(\mathbf{X}, \boldsymbol{\pi}^0)$$

This completes the proof. For purposes of reference we summarize the above in the following:

Theorem 9. *The equivalence of the saddle-point and convex-programming problems. Let $f(\mathbf{X})$ and $g_i(\mathbf{X})$ be convex functions defined on the convex set $\mathbf{X} \geq 0$ of $\mathbf{E}_n$, with the property that there exists some $\bar{\mathbf{X}} \geq 0$ for which $g_i(\bar{\mathbf{X}}) < 0$ for all $i = 1, \ldots, m$. If $\mathbf{X}^0$ is a point at which $f(\mathbf{X})$ achieves its minimum subject to $g_i(\mathbf{X}) \leq 0$, $\mathbf{X} \geq 0$, then there exists some $\boldsymbol{\pi}^0 \geq 0$ such that for $F(\mathbf{X}, \boldsymbol{\pi}) = f(\mathbf{X}) + \sum_{i=1}^{m} \pi_i g_i(\mathbf{X})$, we have*

$$F(\mathbf{X}^0, \boldsymbol{\pi}) \leq F(\mathbf{X}^0, \boldsymbol{\pi}^0) \leq F(\mathbf{X}, \boldsymbol{\pi}^0) \tag{3.12}$$

for all $\mathbf{X} \geq 0$, $\boldsymbol{\pi} \geq 0$, and $\sum_{i=1}^{m} \pi_i^0 g_i(\mathbf{X}^0) = 0$. Conversely, if $(\mathbf{X}^0, \boldsymbol{\pi}^0)$ satisfies (3.12), then $\mathbf{X}^0$ minimizes $f(\mathbf{X})$ subject to all $\mathbf{X}$ satisfying $\mathbf{X} \geq 0$ and $g_i(\mathbf{X}) \leq 0$ for all i.

If $f(\mathbf{X})$ and all the $g_i(\mathbf{X})$ are continuously differentiable as well as convex functions, then the saddle-point inequalities are equivalent to certain conditions which are necessary and sufficient to provide a solution to the convex-programming problem. These are given by the following theorem:

Theorem 10. *For $\mathbf{X} \geq 0$ and $f(\mathbf{X})$ and all $g_i(\mathbf{X})$ continuously differentiable convex functions, (3.13) and (3.14) are necessary and sufficient conditions for a point $(\mathbf{X}^0, \boldsymbol{\pi}^0)$ to satisfy the constraints (3.2) of the saddle-point problem.*

The Kuhn-Tucker conditions:

$$\frac{\partial F(\mathbf{X}^0, \boldsymbol{\pi}^0)}{\partial x_j} = \frac{\partial f(\mathbf{X}^0)}{\partial x_j} + \sum_{i=1}^{m} \pi_i^0 \frac{\partial g_i(\mathbf{X}^0)}{\partial x_j} \geq 0 \tag{3.13a}$$

$$(\mathbf{X}^0)' \left[\frac{\partial F(\mathbf{X}^0, \boldsymbol{\pi}^0)}{\partial \mathbf{X}} \right] = \sum_{j=1}^{n} x_j^0 \left\{ \frac{\partial f(\mathbf{X}^0)}{\partial x_j} + \sum_{i=1}^{m} \pi_i^0 \frac{\partial g_i(\mathbf{X}^0)}{\partial x_j} \right\} = 0 \tag{3.13b}$$

$$\mathbf{X}^0 = \{x_j^0\} \geq 0 \tag{3.13c}$$

and

$$\frac{\partial F(\mathbf{X}^0, \boldsymbol{\pi}^0)}{\partial \pi_i} = g_i(\mathbf{X}^0) \leq 0 \tag{3.14a}$$

$$\boldsymbol{\pi}^0 \left[\frac{\partial F(\mathbf{X}^0, \boldsymbol{\pi}^0)}{\partial \boldsymbol{\pi}} \right] = \sum_{i=1}^{m} \pi_i^0 g_i(\mathbf{X}^0) = 0 \tag{3.14b}$$

$$\boldsymbol{\pi}^0 = \{\pi_i^0\} \geq 0 \tag{3.14c}$$

From the nonnegativity implications of (3.13*a* and *c*) we have that the individual terms of (3.13b)

$$x_j^0 \left\{ \frac{\partial f(\mathbf{X}^0)}{\partial x_j} + \sum_{i=1}^{m} \pi_i^0 \frac{\partial g_i(\mathbf{X}^0)}{\partial x_j} \right\} = 0$$

Similarly, from (3.14a and c) we have that the individual terms of (3.14b) $\pi_i^0 g_i(\mathbf{X}^0) = 0$.

Proof: For necessity, i.e., assuming $(\mathbf{X}^0, \boldsymbol{\pi}^0)$ is a saddle-point solution, (3.13) and (3.14) assert that $F(\mathbf{X}, \boldsymbol{\pi}^0)$ has a local minimum at $F(\mathbf{X}^0, \boldsymbol{\pi}^0)$ and $F(\mathbf{X}^0, \boldsymbol{\pi})$ has a local maximum at $F(\mathbf{X}^0, \boldsymbol{\pi}^0)$, respectively. This must be the case for $F(\mathbf{X}^0, \boldsymbol{\pi}) \leq F(\mathbf{X}^0, \boldsymbol{\pi}^0) \leq F(\mathbf{X}, \boldsymbol{\pi}^0)$. For first assume (3.13$a$ and b) do not hold with $\mathbf{X}^0 \geq \mathbf{0}$. We fix $\boldsymbol{\pi} = \boldsymbol{\pi}^0$ and note that $F(\mathbf{X}, \boldsymbol{\pi}^0)$ is a convex function of $\mathbf{X}$. If some $x_j^0 > 0$, that is, we have an interior point, and $\partial F(\mathbf{X}^0, \boldsymbol{\pi}^0)/\partial x_j \neq 0$, then we could decrease $F(\mathbf{X}^0, \boldsymbol{\pi}^0)$ by taking a different value of $\mathbf{X}$, say $\bar{\mathbf{X}}$, such that $F(\bar{\mathbf{X}}, \boldsymbol{\pi}^0) < F(\mathbf{X}^0, \boldsymbol{\pi}^0)$. See Fig. 12.6. If $x_j^0 = 0$, that is, a boundary point, and $\partial F(\mathbf{X}^0, \boldsymbol{\pi}^0)/\partial x_j < 0$, then again we could find $\bar{\mathbf{X}} > \mathbf{0}$ such that $F(\bar{\mathbf{X}}, \boldsymbol{\pi}^0) < F(\mathbf{X}^0, \boldsymbol{\pi}^0)$. See Fig. 12.7. These situations cannot arise as $F(\mathbf{X}, \boldsymbol{\pi})$ is minimized by $\mathbf{X}^0$ for all $\boldsymbol{\pi}$. Thus (3.13a and b) must hold, and, of course, $\mathbf{X}^0 \geq \mathbf{0}$, which is (3.13$c$).[1]

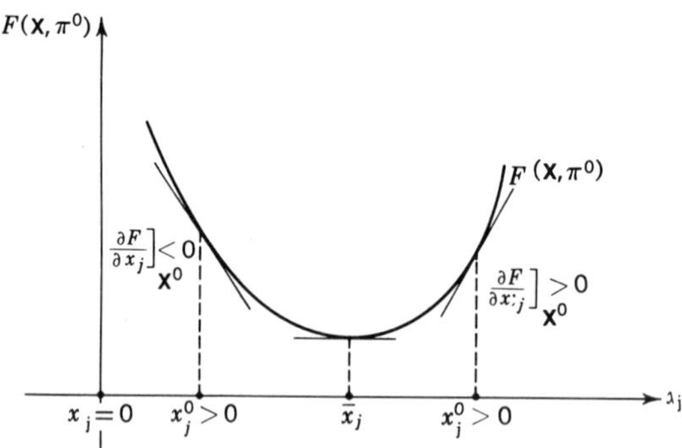

FIGURE 12.6

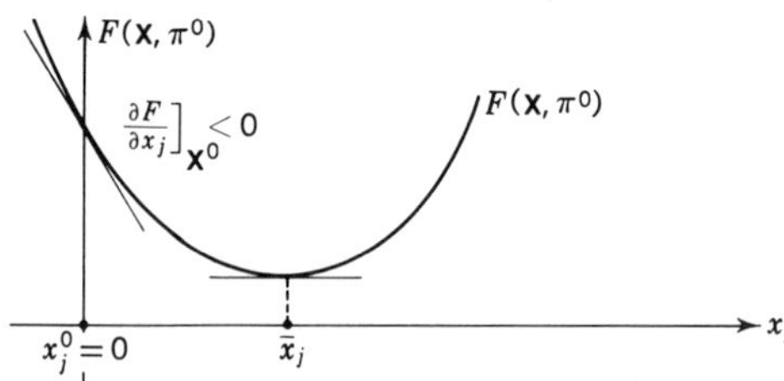

FIGURE 12.7

[1] From Theorem 6 it is possible to prove that $\partial f(\mathbf{X}^0, \boldsymbol{\pi}^0)/\partial x_j = 0$ when $x_j^0 > 0$ and $\partial F(\mathbf{X}^0, \boldsymbol{\pi}^0)/\partial x_j \geq 0$ when $x_j^0 = 0$.

Similarly, for $\mathbf{X} = \mathbf{X}^0$, we note that $F(\mathbf{X}^0, \boldsymbol{\pi})$ is a linear function of the π_i. From (3.4) we have for all $\boldsymbol{\pi} \geq \mathbf{0}$ and $(\mathbf{X}^0, \boldsymbol{\pi}^0)$,

$$\sum_{i=1}^{m} \pi_i g_i(\mathbf{X}^0) \leq \sum_{i=1}^{m} \pi_i^0 g_i(\mathbf{X}^0) \tag{3.15}$$

Since $\partial F(\mathbf{X}^0, \boldsymbol{\pi}^0)/\partial \pi_i = g_i(\mathbf{X}^0)$, if $\partial F(\mathbf{X}^0, \boldsymbol{\pi}^0)/\partial \pi_i > 0$, then $g_i(\mathbf{X}^0) > 0$. From (3.15), if a $g_i(\mathbf{X}^0) > 0$, we could select π_i very large such that this condition could not hold. Therefore, we must have all $\partial F(\mathbf{X}^0, \boldsymbol{\pi}^0)/\partial \pi_i = g_i(\mathbf{X}^0) \leq 0$. We have already shown that $\sum_{i=1}^{m} \pi_i^0 g_i(\mathbf{X}^0) = 0$ (Theorem 8) and, of course, $\boldsymbol{\pi}^0 \geq \mathbf{0}$. Thus, (3.14) holds.

For the sufficiency, i.e., given that the Kuhn-Tucker conditions hold, we have from Theorem 5 and the fact that $F(\mathbf{X}, \boldsymbol{\pi}^0)$ is convex in $\mathbf{X}$,

$$F(\mathbf{X}, \boldsymbol{\pi}^0) \geq F(\mathbf{X}^0, \boldsymbol{\pi}^0) + (\mathbf{X} - \mathbf{X}^0)' \left[\frac{\partial F(\mathbf{X}^0, \boldsymbol{\pi}^0)}{\partial \mathbf{X}} \right]$$

or

$$F(\mathbf{X}, \boldsymbol{\pi}^0) \geq F(\mathbf{X}^0, \boldsymbol{\pi}^0) + \mathbf{X}' \left[\frac{\partial F(\mathbf{X}^0, \boldsymbol{\pi}^0)}{\partial \mathbf{X}} \right] - (\mathbf{X}^0)' \left[\frac{\partial F(\mathbf{X}^0, \boldsymbol{\pi}^0)}{\partial \mathbf{X}} \right] \tag{3.16}$$

From (3.13b) the last term of (3.16) is zero. Also, from (3.13a) and $\mathbf{X} \geq \mathbf{0}$ the middle term on the right-hand side of (3.16) is nonnegative. We thus have $F(\mathbf{X}, \boldsymbol{\pi}^0) \geq F(\mathbf{X}^0, \boldsymbol{\pi}^0)$. Since $F(\mathbf{X}^0, \boldsymbol{\pi})$ is also convex, i.e., linear in $\boldsymbol{\pi}$, we have

$$F(\mathbf{X}^0, \boldsymbol{\pi}) = F(\mathbf{X}^0, \boldsymbol{\pi}^0) + (\boldsymbol{\pi} - \boldsymbol{\pi}^0) \left[\frac{\partial F(\mathbf{X}^0, \boldsymbol{\pi}^0)}{\partial \boldsymbol{\pi}} \right]$$

or

$$F(\mathbf{X}^0, \boldsymbol{\pi}) = F(\mathbf{X}^0, \boldsymbol{\pi}^0) + \boldsymbol{\pi} \left[\frac{\partial F(\mathbf{X}^0, \boldsymbol{\pi}^0)}{\partial \boldsymbol{\pi}} \right] - \boldsymbol{\pi}^0 \left[\frac{\partial F(\mathbf{X}^0, \boldsymbol{\pi}^0)}{\partial \boldsymbol{\pi}} \right] \tag{3.17}$$

From (3.14a) and (3.14b) we have for $\boldsymbol{\pi} \geq \mathbf{0}$ that the last term of (3.17) is zero and the middle term is nonpositive; thus $F(\mathbf{X}^0, \boldsymbol{\pi}) \leq F(\mathbf{X}^0, \boldsymbol{\pi}^0)$. Together the results yield that $F(\mathbf{X}^0, \boldsymbol{\pi}) \leq F(\mathbf{X}^0, \boldsymbol{\pi}^0) \leq F(\mathbf{X}, \boldsymbol{\pi}^0)$, that is, $\mathbf{X}^0 \geq \mathbf{0}$ and $\boldsymbol{\pi}^0 \geq \mathbf{0}$ satisfy the saddle-point constraints. This completes the proof.

To recapitulate, we have shown the equivalence of the optimal solution $\mathbf{X}^0$ to the convex-programming problem and the $\mathbf{X}^0$ of the point $(\mathbf{X}^0, \boldsymbol{\pi}^0)$ which solves the saddle-point problem under the assumption of Slater's constraint qualification applied to the solution space to the convex-programming problem. Then, assuming differentiability of the convex functions, necessary and sufficient conditions for a point $(\mathbf{X}^0, \boldsymbol{\pi}^0)$ to solve the saddle-point problem were developed using arguments from calculus and the properties of convex functions. A number of computational procedures exist based on the above developments which enable us to solve convex-programming problems. We shall discuss below a procedure for solving *quadratic-programming problems*, i.e., problems in which the objective function is a quadratic form and the constraints are linear; we shall also discuss a procedure for solving *separable-programming problems* in which the objective function is a convex function that is separable in the variables and the constraints are linear. The reader is referred to the following books for discussions concerning more general situations and algorithms: Kunzi and Krelle [451], McCormick [483], Hadley [344], Bazaraa and Shetty [51], Shapiro [608], Cooper [132], Luenberger [477], and Cooper and Steinberg [135]. The reader should recognize that the field of nonlinear programming is indeed more difficult in an algorithmic sense than linear programming. New approaches are constantly being reported in the literature, but much work still needs to be done in this area to make the finding of a solution to a nonlinear-programming problem as efficient and as sure as finding one to a linear problem.

A number of special convex-programming problems are of interest in terms of their associated Kuhn-Tucker conditions. The complete developments are left as exercises for the reader, but we shall next discuss a few points. Suppose our problem is the primal linear program: Minimize $\mathbf{cX}$, subject to $\mathbf{AX} \geq \mathbf{b}$, $\mathbf{X} \geq \mathbf{0}$. We let $f(\mathbf{X}) = \mathbf{cX}$, $[g_i(\mathbf{X})] = \mathbf{b} - \mathbf{AX} \leq \mathbf{0}$, with each $g_i(\mathbf{X})$ a linear function. Then $F(\mathbf{X}, \boldsymbol{\pi}) = \mathbf{cX} + \boldsymbol{\pi}[\mathbf{b} - \mathbf{AX}]$. The Kuhn-Tucker conditions for this problem lead to the complementary slackness conditions for the optimal solutions to the primal and dual problems, and the equality of the extreme values of the primal and dual objective function. Here the π_i are the dual variables.

For the linear program: minimize $\mathbf{cX}$, $\mathbf{AX} = \mathbf{b}$, $\mathbf{X} \geq \mathbf{0}$, we transform the equalities to $[\mathbf{AX} \leq \mathbf{b}, \mathbf{AX} \geq \mathbf{b}]$ and let $[g(\mathbf{X})] = \begin{bmatrix} \mathbf{AX} - \mathbf{b} \\ \mathbf{b} - \mathbf{AX} \end{bmatrix} \leq \mathbf{0}$. We now have $2m$ inequalities. For this problem the $\boldsymbol{\pi}$ are unrestricted. This will be true if the more general convex problem is stated in terms of $g_i(\mathbf{X}) = 0$ and we convert the equality to two inequalities. Thus, if the *i*th constraint is an equality, the corresponding π_i will be unrestricted as to sign; i.e., it can be represented by the difference of two nonnegative variables. If the $\mathbf{X}$ are unrestricted in the convex-programming problem, then the Kuhn-Tucker conditions (3.13) reduce to

$$\frac{\partial f(\mathbf{X}^0)}{\partial x_j} + \sum_{i=1}^{m} \pi_i^0 \frac{\partial g_i(\mathbf{X}^0)}{\partial x_j} = 0$$

This can be shown by writing $\mathbf{X}$ as the difference of two nonnegative variables and direct application of (3.13).

As an example of the above, consider the following problem (Dorn [217]): Minimize $f(\mathbf{X}) = -\log x_1 - \log x_2$ subject to $x_1 + x_2 - 2 \leq 0$ and $x_j \geq 0$. Here $F(\mathbf{X}, \boldsymbol{\pi}) = -\log x_1 - \log x_2 + \pi_1(x_1 + x_2 - 2)$ and the Kuhn-Tucker conditions for optimal $\mathbf{X}^0$ and $\boldsymbol{\pi}^0$ are

$$-\frac{1}{x_1^0} + \pi_1^0 \geq 0 \qquad -\frac{1}{x_2^0} + \pi_1^0 \geq 0$$

$$x_1^0\left[-\frac{1}{x_1^0} + \pi_1^0\right] + x_2^0\left[-\frac{1}{x_2^0} + \pi_1^0\right] = 0$$

$$x_j^0 \geq 0$$

$$x_1^0 + x_2^0 - 2 \leq 0$$

$$\pi_1^0(x_1^0 + x_2^0 - 2) = 0$$

$$\pi_1^0 \geq 0$$

An optimal solution is $\mathbf{X}^0 = (1, 1)$ and $\pi_1^0 = 1$.

As in the case for linear programming, primal-dual problem definitions exist for nonlinear programming along with theorems relating the solutions of the two problems. A duality theorem for convex programs with linear constraints is given by Dorn [217]. More general theorems exist, and the reader is referred to the papers by Dantzig and Cottle, Kuhn, and Whinston in Abadie [1], Moeske [505], Shapiro [608], Bazaraa and Shetty [51], and Sinha [613].

4. QUADRATIC PROGRAMMING

An important class of convex-programming problems arises when the usual linear-constraint set of $\mathbf{AX} = \mathbf{b}$, $\mathbf{X} \geq \mathbf{0}$ is coupled with a convex objective function. This class includes the separable convex-programming problem and the quadratic-programming problem. The solution of both these problems can be accomplished by adaptations of the simplex method.

For our purposes we shall state the quadratic-programming problem as follows[1]: Minimize

$$f(\mathbf{X}) = \mathbf{pX} + \mathbf{X'CX} = \sum_{j=1}^{n} p_j x_j + \sum_{i=1}^{n}\sum_{j=1}^{n} c_{ij} x_i x_j$$

subject to

$$\mathbf{AX} = \mathbf{b} \tag{4.1}$$

$$\mathbf{X} \geq \mathbf{0}$$

where $\mathbf{p}$ is an n-component row vector of given constants. We assume the matrix $\mathbf{C}$ is symmetric and the quadratic form $\mathbf{X'CX}$ is positive semidefinite. From Theorem 2 we note that $\mathbf{X'CX}$ is a convex function for all $\mathbf{X}$ in $\mathbf{E}_n$ and, since $\mathbf{pX}$ is linear, $f(\mathbf{X})$ is a convex function. We next develop the Kuhn-Tucker conditions for problem (4.1).

We rewrite the equations of (4.1) as inequalities in order to formulate the

[1] Applications which lead to quadratic-programming formulations include stock-portfolio selection, quadratic diet problem, chemical equilibrium, allocation of resources, electrical networks, structural mechanics, and regression analysis (Boot [83], Dorn [215], Saaty and Bram [586], and Lasdon and Waren [461]).

problem in the convex-programming form of (3.1); that is, we have the following: Minimize

$$f(\mathbf{X}) = \mathbf{pX} + \mathbf{X'CX}$$

subject to

$$\mathbf{AX} - \mathbf{b} \leq \mathbf{0}$$
$$-\mathbf{AX} + \mathbf{b} \leq \mathbf{0} \tag{4.2}$$
$$\mathbf{X} \geq \mathbf{0}$$

If we let $\boldsymbol{\pi}_1$ and $\boldsymbol{\pi}_2$ be m-component row vectors representing the set of Lagrange multipliers π_{1i} and π_{2i} for the first and second set of inequalities, respectively, we can write the Lagrangian function for (4.2) as

$$F(\mathbf{X}, \boldsymbol{\pi}_1, \boldsymbol{\pi}_2) = \mathbf{pX} + \mathbf{X'CX} + \boldsymbol{\pi}_1(\mathbf{AX} - \mathbf{b}) + \boldsymbol{\pi}_2(-\mathbf{AX} + \mathbf{b}) \tag{4.3}$$

Noting that $\partial \mathbf{X'CX}/\partial \mathbf{X} = 2\mathbf{CX}$, we have for (4.3) the following derivatives:

$$\frac{\partial F(\mathbf{X}, \boldsymbol{\pi}_1, \boldsymbol{\pi}_2)}{\partial x_j} = p_j + 2 \sum_{i=1}^{m} c_{ij} x_i + \sum_{i=1}^{m} \pi_{1i} a_{ij} - \sum_{i=1}^{m} \pi_{2i} a_{ij}$$

$$\frac{\partial F(\mathbf{X}, \boldsymbol{\pi}_1, \boldsymbol{\pi}_2)}{\partial \pi_{1i}} = \sum_{j=1}^{n} a_{ij} x_j - b_i$$

$$\frac{\partial F(\mathbf{X}, \boldsymbol{\pi}_1, \boldsymbol{\pi}_2)}{\partial \pi_{2i}} = - \sum_{j=1}^{n} a_{ij} x_j + b_i$$

or by grouping terms and writing the above as column vectors

$$\frac{\partial F(\mathbf{X}, \boldsymbol{\pi}_1, \boldsymbol{\pi}_2)}{\partial \mathbf{X}} = \mathbf{p'} + 2\mathbf{CX} + \mathbf{A'}\boldsymbol{\pi}_1' - \mathbf{A'}\boldsymbol{\pi}_2'$$

$$\frac{\partial F(\mathbf{X}, \boldsymbol{\pi}_1, \boldsymbol{\pi}_2)}{\partial \boldsymbol{\pi}_1} = \mathbf{AX} - \mathbf{b}$$

$$\frac{\partial F(\mathbf{X}, \boldsymbol{\pi}_1, \boldsymbol{\pi}_2)}{\partial \boldsymbol{\pi}_2} = -\mathbf{AX} + \mathbf{b}$$

From (3.13) and (3.14) for $\mathbf{X}^0$, $\boldsymbol{\pi}_1^0$, $\boldsymbol{\pi}_2^0$ to solve the saddle-point problem we have the following Kuhn-Tucker conditions for the solution to the quadratic-programming problem:

$$\mathbf{p'} + 2\mathbf{CX}^0 + \mathbf{A'}\boldsymbol{\pi}_1^{0'} - \mathbf{A'}\boldsymbol{\pi}_2^{0'} \geq \mathbf{0} \tag{4.4a}$$

$$(\mathbf{X}^0)'[\mathbf{p'} + 2\mathbf{CX}^0 + \mathbf{A'}\boldsymbol{\pi}_1^{0'} - \mathbf{A'}\boldsymbol{\pi}_2^{0'}] = 0 \tag{4.4b}$$

$$\mathbf{X}^0 \geq \mathbf{0} \tag{4.4c}$$

$$\mathbf{AX}^0 - \mathbf{b} \leq \mathbf{0}$$
$$-\mathbf{AX}^0 + \mathbf{b} \leq \mathbf{0} \tag{4.5a}$$

$$[\boldsymbol{\pi}_1^0, \boldsymbol{\pi}_2^0][\mathbf{AX}^0 - \mathbf{b}, -\mathbf{AX}^0 + \mathbf{b}] = 0 \tag{4.5b}$$

$$\boldsymbol{\pi}_1^0 \geq \mathbf{0}$$
$$\boldsymbol{\pi}_2^0 \geq \mathbf{0} \tag{4.5c}$$

Inequalities (4.5a) are just $\mathbf{AX}^0 = \mathbf{b}$; Eq. (4.5$b$) is always true since it can be rewritten as $(\boldsymbol{\pi}_1{}^0 - \boldsymbol{\pi}_2{}^0)(\mathbf{AX}^0 - \mathbf{b}) = \mathbf{0}$; and from inequalities (4.5c) we can let $\boldsymbol{\pi}^0 = (\boldsymbol{\pi}_1{}^0 - \boldsymbol{\pi}_2{}^0)'$, where the $m \times 1$ column vector $\boldsymbol{\pi}^0$ is a vector of multipliers π_i^0 which are unrestricted as to sign. The conditions (4.4a to c) and (4.5a to c) can now be rewritten as

$$\mathbf{p}' + 2\mathbf{CX}^0 + \mathbf{A}'\boldsymbol{\pi}^0 \geq \mathbf{0}$$

$$(\mathbf{X}^0)'[\mathbf{p}' + 2\mathbf{CX}^0 + \mathbf{A}'\boldsymbol{\pi}^0] = 0$$

$$\mathbf{X}^0 \geq \mathbf{0}$$

$$\mathbf{AX}^0 = \mathbf{b}$$

$$(\mathbf{AX}^0 - \mathbf{b})'\boldsymbol{\pi}^0 = 0$$

$$\boldsymbol{\pi}^0 \text{ unrestricted}$$

Let $\mathbf{V}^0$ be an $n \times 1$ nonnegative column vector of slack variables v_j^0, that is, $\mathbf{p}' + 2\mathbf{CX}^0 + \mathbf{A}'\boldsymbol{\pi}^0 - \mathbf{IV}^0 = \mathbf{0}$ or $\mathbf{V}^0 = \mathbf{p}' + 2\mathbf{CX}^0 + \mathbf{A}'\boldsymbol{\pi}^0$. The Kuhn-Tucker conditions are now

$$\mathbf{p}' + 2\mathbf{CX}^0 + \mathbf{A}'\boldsymbol{\pi}^0 - \mathbf{IV}^0 = \mathbf{0} \tag{4.6a}$$

$$(\mathbf{X}^0)'\mathbf{V}^0 = 0 \tag{4.6b}$$

$$\mathbf{X}^0 \geq \mathbf{0}$$

$$\mathbf{V}^0 \geq \mathbf{0} \tag{4.6c}$$

$$\mathbf{AX}^0 = \mathbf{b} \tag{4.7a}$$

$$(\mathbf{AX}^0 - \mathbf{b})'\boldsymbol{\pi}^0 = 0 \tag{4.7b}$$

$$\boldsymbol{\pi}^0 \text{ unrestricted} \tag{4.7c}$$

Based on Theorem 8, the Kuhn-Tucker theorem on the equivalence of the solutions to the saddle-point problem and the convex-programming problem, we can restate the above in the following theorem:

Theorem 11. *The vector* $\mathbf{X}^0 \geq \mathbf{0}$ *solves the quadratic-programming problem if and only if* $\mathbf{AX}^0 = \mathbf{b}$ *and vectors* $\mathbf{V}^0 \geq \mathbf{0}$ *and* $\boldsymbol{\pi}^0$ *unrestricted exist such that*

$$\mathbf{p}' + 2\mathbf{CX}^0 + \mathbf{A}'\boldsymbol{\pi}^0 - \mathbf{IV}^0 = \mathbf{0} \tag{4.8}$$

$$(\mathbf{X}^0)'\mathbf{V}^0 = 0 \tag{4.9}$$

The reader will note that the conditions of the theorem are now linear except for (4.9), which states that the vector sum $\sum_{j=1}^{n} x_j^0 v_j^0 = 0$; that is, since $x_j^0 \geq 0$, $v_j^0 \geq 0$, then each term $x_j^0 v_j^0 = 0$. If $x_j^0 > 0$, then $v_j^0 = 0$, and conversely. The total number of linear constraints is $m + n$, m from $\mathbf{AX}^0 = \mathbf{b}$ and n from (4.8). The total number of variables to be considered is $2n$ nonnegative variables, $\mathbf{X}^0$ and $\mathbf{V}^0$, and m unrestricted variables $\boldsymbol{\pi}^0$. If we look at the equations which must be satisfied by the optimal $\mathbf{X}^0$, $\mathbf{V}^0$, $\boldsymbol{\pi}^0$, we have (dropping the superscript notation)

$$AX = b \tag{4.10}$$

$$2CX - IV + A'\pi = -p' \tag{4.11}$$

with $X \geq 0$ and $V \geq 0$. The side condition $\sum_{j=1}^{n} x_j v_j$ states that no more than n of the $2n$ variables x_j and v_j can be positive. Thus the possible solutions to the equations are restricted to those with no more than $n + m$ of the variables X, V, π nonzero; i.e., the solution set is restricted to the set of basic solutions to (4.10) and (4.11), with the implied nonnegativity of X and V. Using these concepts, Wolfe [713] developed a finite computational procedure—based on the simplex method—to determine the optimal basic solution to these equations. Wolfe's scheme is given in two forms, the short form for $p = 0$ or $X'CX$ positive definite and the long form for $p \neq 0$ or $X'CX$ positive semidefinite. We shall next describe these procedures.[1]

Wolfe's quadratic-programming problem algorithm. SHORT FORM Here we assume $p = 0$ or $X'CX$ is positive definite. We wish to find an $X \geq 0$, $V \geq 0$, and π unrestricted which satisfies the constraints

$$AX = b$$

$$2CX - IV + A'\pi = -p'$$

and such that $X'V = 0$. We first find a starting basic solution to these equations by a variation of the artificial-basis procedure of the simplex method. Assuming $b \geq 0$, let $W = (w_1, w_2, \ldots, w_m) \geq 0$, $Z_1 = (z_{11}, z_{12}, \ldots, z_{1n}) \geq 0$, $Z_2 = (z_{21}, z_{22}, \ldots, z_{2n}) \geq 0$ be artificial nonnegative variables and rewrite the equations of the system in the form

$$AX + IW = b \tag{4.12}$$

$$2CX - IV + A'\pi + IZ_1 - IZ_2 = -p' \tag{4.13}$$

Here, we take $w_i = b_i$ and either $z_{j1} = -p_j$ or $z_{j2} = p_j$ depending on whether $-p_j \geq 0$ or $-p_j < 0$, respectively. This yields a set of nonnegative values for the components of W, Z_1, and Z_2 which correspond to an artificial basic feasible solution to (4.12) and (4.13). Keeping $V = 0$ and $\pi = 0$, we next solve (4.12) and (4.13) as a linear-programming problem with an objective function of minimizing $\sum_{i=1}^{m} w_i$. This process will yield a feasible solution to $AX = b$, if one exists, and vectors Z_1 and Z_2 satisfying (4.13). [Of course, if a basic feasible solution to (4.12) is known with $W = 0$, then it can be utilized immediately by introducing it into (4.12) and (4.13), with appropriate changes made in the right-hand side of (4.13).]

---

[1] There are a number of other procedures for solving the quadratic-programming problem, most of which are described in Kunzi and Krelle [451], and Bazaraa and Shetty [51].

At this point we define a new artificial nonnegative vector

$$\mathbf{Z} = (z_1, z_2, \ldots, z_n) \geq \mathbf{0}$$

such that the components of $\mathbf{Z}$ are equal to the n nonnegative components of $\mathbf{Z}_1$ and $\mathbf{Z}_2$ which form the basic feasible solution to (4.13). That is, we now have a basic feasible solution $(\mathbf{X}, \mathbf{Z})$ to the equations

$$\mathbf{AX} = \mathbf{b} \tag{4.14}$$

$$2\mathbf{CX} - \mathbf{IV} + \mathbf{A}'\boldsymbol{\pi} + \bar{\mathbf{I}}\mathbf{Z} = -\mathbf{p}' \tag{4.15}$$

where the diagonal elements of $\bar{\mathbf{I}}$ are positive or negative ones, according to the sign of the corresponding $-p_j$ and all other elements of $\bar{\mathbf{I}}$ are zero. Next, we need to find a solution to (4.14) and (4.15) such that the artificial variables $\mathbf{Z}$ are removed with $\mathbf{X} \geq \mathbf{0}$, $\mathbf{V} \geq \mathbf{0}$, $\boldsymbol{\pi}$ unrestricted and $\mathbf{X}'\mathbf{V} = 0$. We note that our present artificial solution $(\mathbf{X}, \mathbf{Z}, \mathbf{V} = \mathbf{0}, \boldsymbol{\pi} = \mathbf{0})$ satisfies $\mathbf{X}'\mathbf{V} = 0$.

To accomplish the above we apply the simplex method to (4.14) and (4.15) with the objective function of minimizing $\sum\limits_{j=1}^{n} z_j$ and modify the method to account for the nonlinear constraint of $\mathbf{X}'\mathbf{V} = 0$. This modification calls for the following rule: If for any j, x_j is a basic variable, then the corresponding v_j is not allowed to become basic; similarly if v_j is basic, x_j is not allowed to become basic. Under this modification, Wolfe has shown that for $\mathbf{X}'\mathbf{CX}$ positive definite or $\mathbf{p} = \mathbf{0}$, the simplex algorithm will find an optimal solution in a finite number of iterations with $\sum\limits_{j=1}^{n} z_j = 0$, and hence, each $z_j = 0$. The corresponding $\mathbf{X}$ is an optimal solution to the quadratic-programming problem, and the final $\mathbf{X}$, $\mathbf{V}$, and $\boldsymbol{\pi}$ satisfy the Kuhn-Tucker conditions as given in (4.6a to c) and (4.7a to c).

In order to apply the simplex method to the constraints (4.14) and (4.15), we must convert the unrestricted variables $\boldsymbol{\pi}$ to nonnegative variables. Thus we let $\boldsymbol{\pi} = \boldsymbol{\pi}_1 - \boldsymbol{\pi}_2$, with $\boldsymbol{\pi}_1 \geq \mathbf{0}$, $\boldsymbol{\pi}_2 \geq \mathbf{0}$ and rewrite the starting problem as follows: Minimize

$$\sum_{j=1}^{n} z_j$$

subject to

$$\mathbf{AX} = \mathbf{b} \tag{4.16}$$

$$2\mathbf{CX} - \mathbf{IV} + \mathbf{A}'\boldsymbol{\pi}_1 - \mathbf{A}'\boldsymbol{\pi}_2 + \bar{\mathbf{I}}\mathbf{Z} = -\mathbf{p}' \tag{4.17}$$

with $\mathbf{X}'\mathbf{V} = 0$ and $\mathbf{X} \geq \mathbf{0}$, $\mathbf{V} \geq \mathbf{0}$, $\boldsymbol{\pi}_1 \geq \mathbf{0}$, $\boldsymbol{\pi}_2 \geq \mathbf{0}$, $\mathbf{Z} \geq \mathbf{0}$. This problem is solved by direct application of the simplex method as modified above.

LONG FORM　Here we consider the general quadratic-programming problem with $\mathbf{p} \neq \mathbf{0}$ and $\mathbf{X}'\mathbf{CX}$ positive semidefinite. To accomplish the optimization we consider the parametric objective function $f(\mathbf{X}) = \lambda\mathbf{pX} + \mathbf{X}'\mathbf{CX}$ subject to $\mathbf{AX} = \mathbf{b}$, $\mathbf{X} \geq \mathbf{0}$, $\lambda \geq 0$. We desire, of course, the optimal solution for

$\lambda = 1$. The problem equivalent to (4.16) and (4.17) is as follows: Minimize

$$\sum_{j=1}^{n} z_j$$

subject to

$$\mathbf{AX} = \mathbf{b} \tag{4.18}$$

$$2\mathbf{CX} - \mathbf{IV} + \mathbf{A}'\boldsymbol{\pi}_1 - \mathbf{A}'\boldsymbol{\pi}_2 + \bar{\mathbf{I}}\mathbf{Z} = -\lambda\mathbf{p}' \tag{4.19}$$

with $\mathbf{X}'\mathbf{V} = 0$, $(\mathbf{X}, \mathbf{V}, \boldsymbol{\pi}_1, \boldsymbol{\pi}_2, \mathbf{Z}, \lambda) \geq \mathbf{0}$. This problem is first solved for $\lambda = 0$ by applying the short-form algorithm. Next, restricting all $z_j = 0$, we apply the parametric simplex algorithm (see Chap. 8) with the modification rule of the short form which restricts $\mathbf{X}'\mathbf{V} = 0$. The parametrization, to find solutions for $\lambda > 0$, is accomplished by using the new objective function minimize $-\lambda$, that is, maximize λ. Wolfe [713] proves the following:

1. If no change of basis can be made starting with the short-form solution and the objective function minimize $-\lambda$, then the original quadratic-programming objective function $f(\mathbf{X})$ is unbounded for $\lambda > 0$.
2. Otherwise a sequence of solutions $\mathbf{X}_0$, $\mathbf{X}_1$, $\ldots$, $\mathbf{X}_t$ corresponding to parameter values $0 = \lambda_0, \lambda_1, \ldots, \lambda_t$ can be found such that for $\lambda_k \leq \lambda \leq \lambda_{k+1}$, the quadratic-programming problem for this range of λ is solved by

$$\mathbf{X} = \frac{\lambda_{k+1} - \lambda}{\lambda_{k+1} - \lambda_k}\mathbf{X}_k + \frac{\lambda - \lambda_k}{\lambda_{k+1} - \lambda_k}\mathbf{X}_{k+1}$$

and for $\lambda \geq \lambda_t$

$$\mathbf{X} = \mathbf{X}_t + (\lambda - \lambda_t)\mathbf{X}_\infty$$

where $\mathbf{X}_\infty$ is the final solution of the iterative process.

The reader is referred to Dantzig [152], Beale [57], and Van de Panne and Whinston [671, 672, 673] for other variations of the simplex method for handling both the positive definite and positive semidefinite cases. A comparison of the computational performance of the Wolfe, Dantzig, and Beale algorithms is given in Braitsch [93].

To establish a primal-dual relationship for quadratic programs, we define the following problems:

Primal: Minimize

$$f(\mathbf{X}) = \mathbf{pX} + \mathbf{X}'\mathbf{CX}$$

subject to

$$\mathbf{AX} = \mathbf{b}$$

$$\mathbf{X} \geq \mathbf{0}$$

Dual: Maximize

$$g(\mathbf{X}, \boldsymbol{\pi}) = \mathbf{b}'\boldsymbol{\pi} - \mathbf{X}'\mathbf{CX}$$

subject to

$$-2\mathbf{CX} + \mathbf{A'}\boldsymbol{\pi} \le \mathbf{p'}$$

We then have the following duality theorem of quadratic programming (Dorn [216]).

Theorem 12. *If there exists a vector $\mathbf{X}^0$ which minimizes $f(\mathbf{X})$ in the primal, then there exists a $\boldsymbol{\pi}^0$ for which $(\mathbf{X}^0, \boldsymbol{\pi}^0)$ is an optimal solution to the dual. Conversely, if $(\mathbf{X}^0, \boldsymbol{\pi}^0)$ maximizes $g(\mathbf{X}, \boldsymbol{\pi})$ of the dual, then $\mathbf{X}^0$ is an optimal solution to the primal. In either case, $f(\mathbf{X}^0) = g(\mathbf{X}^0, \boldsymbol{\pi}^0)$; that is, the minimum value for the primal problem is equal to the maximum value for the dual problem.*

We shall not prove this theorem, but the reader should be able to verify its validity by the derivation and discussion of the Kuhn-Tucker conditions. If $\mathbf{C} = \mathbf{0}$, the above quadratic dual statements reduce to the linear-programming unsymmetric dual problems. For the symmetric case, that is, $\mathbf{AX} \ge \mathbf{b}$, $\mathbf{X} \ge \mathbf{0}$, the quadratic dual constraints would be $-2\mathbf{CX} + \mathbf{A'}\boldsymbol{\pi} \le \mathbf{p'}$, $\boldsymbol{\pi} \ge \mathbf{0}$, with the same objective functions.

Consideration of the form of the quadratic-programming problem (4.16) and (4.17) and similarly structured problems has led to the statement and solution of the *linear complementarity problem* (Cottle and Dantzig [138, 139], Lemke [468, 469], Eaves [228], and Karamardian [414]). This problem is to find vectors $\mathbf{X} \ge \mathbf{0}$, $\mathbf{Y} \ge \mathbf{0}$ which satisfy the constraints

$$\mathbf{Y} = \mathbf{MX} + \mathbf{b}$$

$$\mathbf{XY} = 0$$

where $\mathbf{M}$ is an $n \times n$ matrix and $\mathbf{b}$ an n-component column vector. The $2n$ variables which form the components of $\mathbf{X}$ and $\mathbf{Y}$ are said to form a complementary solution if each complementary pair (x_j, y_j) satisfies the conditions $x_j y_j = 0$ for $j = 1, 2, \ldots, n$.

The symmetric primal and dual problems, the quadratic-programming problem, and the bimatrix two-person non-zero-sum game can all be transformed into the linear complementarity problem. Under varying assumptions on the form of the matrix $\mathbf{M}$ (e.g., positive-semidefinite matrices and matrices with nonnegative elements), Dantzig and Cottle [180], Lemke and Howson [470], and Lemke [468, 469] have developed algorithms for solving the specialization of the linear complementarity problem to bimatrix and quadratic-programming problems. See also Eaves [229] (general quadratic-objective function) and Karamardian [413] (the nonlinear complementarity problem). A computer procedure using the algorithm of Lemke [468] to solve quadratic and linear programs is described in Ravindran [561]. Approximate solutions to some quadratic-programming problems can be found by the methods of separable programming, which are discussed next.

5. SEPARABLE PROGRAMMING[1]

A continuous function $f(\mathbf{X}) = f(x_1, x_2, \ldots, x_n)$ is said to be separable in its variables if $f(\mathbf{X})$ can be expressed as a linear sum of n single-variable continuous functions $f_1(x_1), f_2(x_2), \ldots, f_n(x_n)$, i.e., $f(\mathbf{X}) = \sum_{j=1}^{n} f_j(x_j)$. For example, the following functions are separable:

$$f(\mathbf{X}) = x_1^2 - 3x_2^3 + \log x_3 + e^{x_4}$$

$$f(\mathbf{X}) = \sum_{j=1}^{n} c_j x_j = c_1 x_1 + c_2 x_2 + \cdots + c_n x_n$$

Separable functions occur explicitly as constraints and/or objective functions in many nonlinear formulations of industrial, financial, and engineering applications (portfolio analysis, petrochemical and refinery operations), or we are able to transform the nonlinear expressions into separable form (see Exercise 23). We define the *separable-programming problem* to be the following non-linear-programming problem: Minimize

$$f(\mathbf{X}) = \sum_{j=1}^{n} f_j(x_j)$$

subject to

$$g_i(\mathbf{X}) = \sum_{j=1}^{n} g_{ij}(x_j) = b_i \qquad i = 1, \ldots, m$$
$$0 \le l_j \le x_j \le u_j \qquad j = 1, \ldots, n$$

$$(5.1)$$

The objective function $f(\mathbf{X})$ and all the constraints are assumed to be given in separable form. The main idea of separable programming is that for each $f_j(x_j)$ and $g_{ij}(x_j)$ we can develop a suitable linear approximation and then employ standard or modified linear-programming techniques to solve the approximate problem. The validity of such a process depends on the problem structure and the correctness of the approximation.

To illustrate the general approach, let us consider the function $f_j(x_j)$ shown in Fig. 12.8. The x_{kj} are ordered *grid* (*mesh, break*) *points*. We assume that there are $p_j + 1$ of them and thus we have p_j intervals on the x_j axis. Taken together, the straight-line segments joining successive points $[x_{kj}, f_j(x_{kj})]$ and $[x_{k+1,j}, f_j(x_{k+1,j})]$ represent a linear approximation of $f_j(x_j)$. For any point $x_{ki} \le x_j \le x_{k+1,j}$, its corresponding approximate functional value, denoted by $\hat{f}_j(x_j)$, can be expressed (see Theorem 1, Chap. 2) as

$$\hat{f}_j(x_j) = f_j(x_{kj})\lambda_{kj} + f_j(x_{k+1,j})\lambda_{k+1,j}$$

with

---

[1] See Charnes and Lemke [123], Dantzig [152], Hadley [344], and Miller [501] for further discussion.

$$x_j = x_{kj}\lambda_{kj} + x_{k+1,j}\lambda_{k+1,j}$$

and

$$\lambda_{kj} + \lambda_{k+1,j} = 1 \qquad \lambda_{kj} \geq 0 \qquad \lambda_{k+1,j} \geq 0$$

We extend and generalize this representation as follows.   For $l_j = x_{0j} \leq x_j \leq x_{p_j j} = u_j$, we have

$$x_j = \sum_{k=0}^{p_j} x_{kj}\lambda_{kj} \tag{5.2}$$

$$\sum_{k=0}^{p_j} \lambda_{kj} = 1 \qquad \lambda_{kj} \geq 0$$

and the functional approximation

$$\hat{f}_j(x_j) = \sum_{k=0}^{p_j} f_j(x_{kj})\lambda_{kj}$$

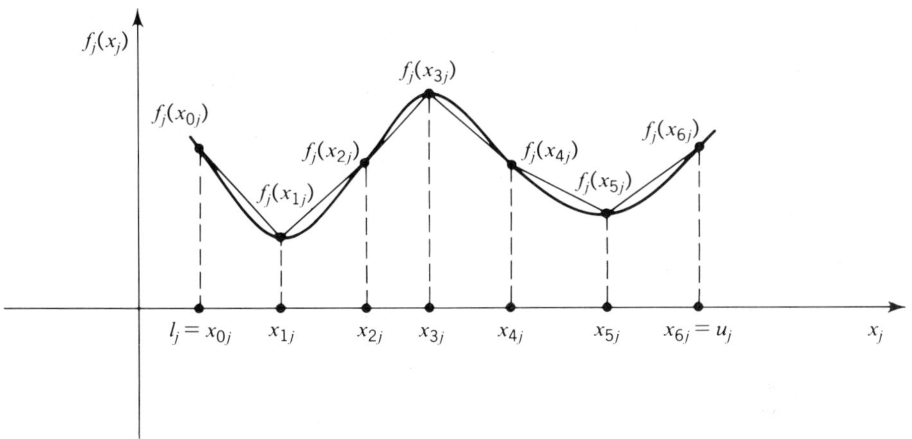

FIGURE 12.8

The approximation will be correct only if no more than two λ_{kj}'s are positive and they must be adjacent (approximation restriction).   This restriction forces x_j to be between two adjacent grid points (or equal to one of them) and the approximation $\hat{f}_j(x_j)$ to be on the corresponding line segment.   With similar notation for approximating the $g_{ij}(x_j)$, we can then state the *separable-programming approximating problem* (λ *method*): Minimize

$$\sum_{j=1}^{n}\sum_{k=0}^{p_j} f_j(x_{kj})\lambda_{kj} \tag{5.3}$$

subject to

$$\sum_{j=1}^{n}\sum_{k=0}^{p_j} g_{ij}(x_{kj})\lambda_{kj} = b_i \qquad i = 1, \ldots, m \tag{5.4}$$

$$\sum_{k=0}^{p_j} \lambda_{kj} = 1 \qquad j = 1, \ldots, n \tag{5.5}$$

$$\lambda_{kj} \geq 0 \qquad \begin{array}{l} j = 1, \ldots, n \\ k = 0, 1, \ldots, p_j \end{array} \tag{5.6}$$

and the approximation restriction applies. [The expression (5.2) for the x_j can be included or the x_j can be calculated after the solution in terms of the λ_{kj} has been obtained.]

We can solve the $(m + n) \times \left(n + \sum_{j=1}^{n} p_j \right)$ linear-approximating problem (5.3) to (5.6) by the simplex algorithm if we adjust for the approximating condition. This is accomplished by using a *restricted-basis entry rule* that allows no more than two λ_{kj} to be positive for a given j and, if two are positive, they are adjacent. This rule applied to the separable approximating problem will, in general, yield a local minimum (Miller [501]). It is possible for the approximating problem's solution to not even be a feasible solution to the original nonlinear problem (Hadley [344]). If the solution set defined by $\{g_{ij}(x_j) = b_i,\ \mathbf{X} \geq \mathbf{0}\}$ is convex, then a feasible solution to the approximating problem is feasible to the original problem. In addition, if the separable objective functions $f_j(x_j)$ are strictly convex, then the simplex algorithm, without the restricted-basis entry rule, will find the global minimum to the approximating problem (Bazaraa and Shetty [51]). This can be seen from Fig. 12.9. For the convex function $f_j(x_j)$, a solution that does not use two adjacent x_{kj}'s will overstate the approximate functional value for the corresponding $\bar{x}_j$ and the solution will not be minimal.[1]

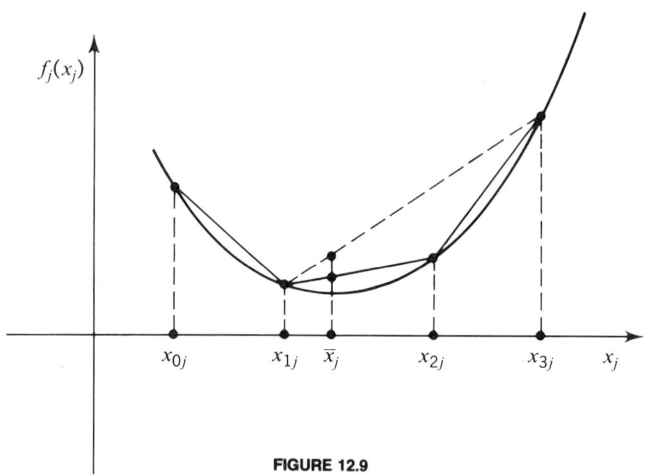

FIGURE 12.9

[1] The reader will note that the conditions (5.5) are GUB constraints. When solving the approximating problem (5.3) to (5.6), some mathematical programming systems consider such constraints to be *special ordered sets of type* 2 for which the computational procedure imposes the restricted-basis entry rule (see Ketron [428] and Beale and

An important application of separable procedures is when all of a problem's constraints are linear, but its objective function is separable in either convex functions or more general forms. We illustrate the former situation by developing the approximating problem for the following example.

Example 5.1.[†]   Minimize

$$x_1^2 + x_2^2$$

subject to

$$3x_1 + 2x_2 \geq 6$$
$$x_1 - x_2 \leq 1$$
$$-2x_1 + x_2 \leq 2$$
$$0 \leq x_1 \leq 2$$
$$0 \leq x_2 \leq 3$$

Approximating data: Figs. 12.9a and 12.9b

x_1	$f_1(x_1)$	x_2	$f_2(x_2)$
0.0	0.00	0.0	0.00
0.5	0.25	0.5	0.25
1.0	1.00	1.0	1.00
1.5	2.25	1.5	2.25
2.0	4.00	2.0	4.00
		2.5	6.25
		3.0	9.00

For $f_1(x_1) = x_1^2$ and $f_2(x_2) = x_2^2$, we have the above approximating data for the grid points shown in Figs. 12.9a and 12.9b:

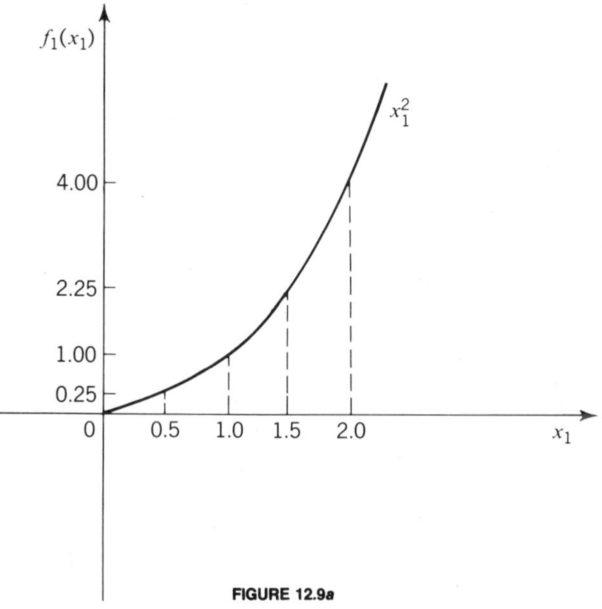

FIGURE 12.9a

Tomlin [62]). The number of grid points to use depends, of course, on the functions involved and the acceptable error. Beale [56, 58] argues for four to six points, with an upper limit of eleven. The disadvantage of using more points is the increase in computational time. Note that the grid points do not need to be evenly spaced and a finer grid can be used at the critical turning points. A solution to the approximating problem can be refined by redefining the grid about the solution point and resolving. Most MPS have procedures for solving separable problems using the λ method, or the δ method described at the end of this section.

[†] Adapted from Cooper [132].

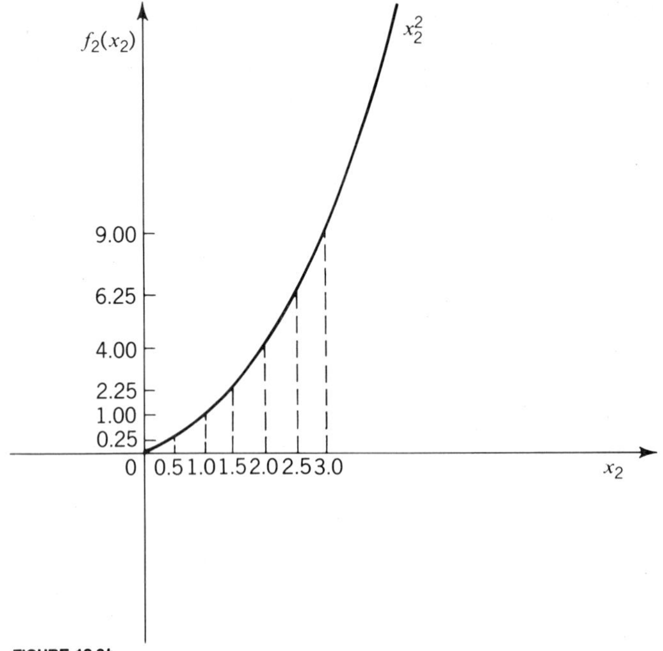

FIGURE 12.9b

The approximating problem can be expressed in terms of the λ_{1k}'s and λ_{2k}'s as follows: Minimize

$$x_0 = \sum_{k=0}^{4} f_1(x_{1k})\lambda_{1k} + \sum_{k=0}^{6} f_2(x_{2k})\lambda_{2k}$$

subject to

$$3 \sum_{k=0}^{4} x_{1k}\lambda_{1k} + 2 \sum_{k=0}^{6} x_{2k}\lambda_{2k} - s_1 = 6$$

$$\sum_{k=0}^{4} x_{1k}\lambda_{1k} - \sum_{k=0}^{6} x_{2k}\lambda_{2k} + s_2 = 1$$

$$-2 \sum_{k=0}^{4} x_{1k}\lambda_{1k} + \sum_{k=0}^{6} x_{2k}\lambda_{2k} + s_3 = 2$$

$$\sum_{k=0}^{4} \lambda_{1k} = 1$$

$$\sum_{k=0}^{6} \lambda_{2k} = 1$$

$$(\lambda_{1k},\ \lambda_{2k},\ s_1,\ s_2,\ s_3) \geq 0$$

where s_1, s_2, and s_3 are slack variables. The numerical data for the approximating problem are given in Tableau 12.1. Note that a column $\mathbf{A}_j$ of the original problem is multiplied by the value of each grid point of x_j. The optimal solution to the approximating problem is $\lambda_{12} = \frac{1}{3}$, $\lambda_{13} = \frac{2}{3}$, $\lambda_{22} = 1$, which corresponds to $x_1 = 1.3333$, $x_2 = 1$, and $x_0 = 2.8333$. The

TABLEAU 12.1

	λ_{10}	λ_{11}	λ_{12}	λ_{13}	λ_{14}	λ_{20}	λ_{21}	λ_{22}	λ_{23}	λ_{24}	λ_{25}	λ_{26}	s_1	s_2	s_3
$6 =$	0	1.5	3	4.5	6	0	1	2	3	4	5	6	-1	0	0
$1 =$	0	0.5	1	1.5	2	0	-0.5	-1	-1.5	-2	-2.5	-3	0	1	0
$2 =$	0	-1	-2	-3	-4	0	0.5	1	1.5	2	2.5	3	0	0	1
$1 =$	1	1	1	1	1	0	0	0	0	0	0	0	0	0	0
$1 =$	0	0	0	0	0	1	1	1	1	1	1	1	0	0	0
$x_0 =$	0	0.25	1	2.25	4	0	0.25	1	2.25	4	6.25	9	0	0	0

"true" optimal solution is $x_1 = 1.3846$, $x_2 = 0.9231$, and $x_0 = 2.7692$. The computation is left as an exercise.

An alternative separable procedure, the δ *method*, forms the approximations by expressing the x_j's as sums of bounded variables. This approach can be developed by reference to Fig. 12.10. There, for the convex function $f_j(x_j)$, we define

$$\Delta(x_{kj}) = x_{kj} - x_{k-1,j}$$

$$\Delta f_j(x_{kj}) = f_j(x_{kj}) - f_j(x_{k-1,j})$$

Then for any $x_{k-1,j} \le x_j \le x_{kj}$, we have $x_j = x_{k-1,j} + \Delta(x_{kj})\delta_{kj}$ with $0 \le \delta_{kj} \le 1$. In general, $l_j \le x_j \le u_j$ can be expressed as

$$x_j = \sum_{k=1}^{p_j} \Delta(x_{kj})\delta_{kj} + x_{0j}$$

with the corresponding approximate functional value given by

$$\hat{f}_j(x_j) = \sum_{k=1}^{p_j} \Delta f_j(x_{kj})\delta_{kj} + f_j(x_{0j})$$

For the convex-separable problem, the δ method yields the following approximating problem: Minimize

$$\sum_{j=1}^{n} \sum_{k=1}^{p_j} \Delta f_j(x_{kj})\delta_{kj} + \sum_{j=1}^{n} f_j(x_{0j})^\dagger \tag{5.7}$$

subject to

$$\sum_{j=1}^{n} \sum_{k=1}^{p_j} \Delta g_{ij}(x_{kj})\delta_{kj} = b_i - \sum_{j=1}^{n} g_{ij}(x_{0j}) \qquad i = 1, \ldots, m \tag{5.8}$$

$$0 \le \delta_{kj} \le 1 \qquad \begin{aligned} j &= 1, \ldots, n \\ k &= 1, \ldots, p_j \end{aligned} \tag{5.9}$$

The $\Delta g_{ij}(x_{kj}) = g_{ij}(x_{kj}) - g_{ij}(x_{k-1,j})$.

For the general separable problem, we must impose a restricted-basis entry rule which states that if $\delta_{kj} > 0$, then all preceding δ_{kj}'s must equal 1, i.e., we require $\delta_{1j} = \delta_{2j} = \cdots = \delta_{k-1,j} = 1$. For a convex programming problem, use of this restriction is not required (why?). The advantage in using the approximating problem (5.7) to (5.9) is that it involves only m constraints, as the restrictions on the δ_{kj} can be handled by simple-bounded variable procedures. Some mathematical programming systems use the δ method form and apply the simplex method modified to include the above restricted-basis entry rule (IBM [387], Sperry Univac [619]).

A general simplex method approach for solving the problem of minimizing a convex function subject to linear constraints is given in Zangwill [726, 727] and Bazaraa and Shetty [51]. The procedure, termed the *convex simplex method*, uses the tableau structure of the simplex method and maintains a basis

---

† The constant term in the objective function (5.7) can be dropped.

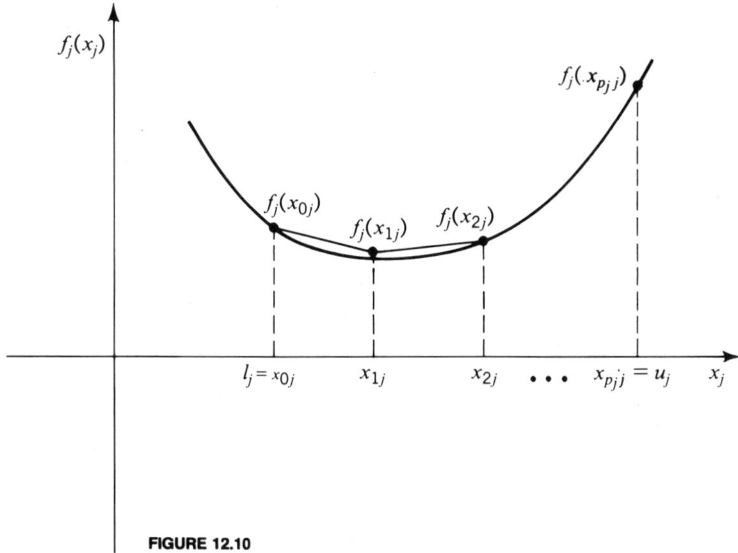

FIGURE 12.10

of m variables, although variables not in the basis may be at a nonzero level. It employs the gradient of the objective function to determine the new variable to be introduced into a basis, and the procedure reduces to the regular simplex method if the objective function is linear. Computational algorithms exist for solving many other variations of the convex-programming problem: *gradient methods* for linear and nonlinear constraints with convex objective function (Rosen [575], Zoutendijk [734], Wolfe [719], Hadley [344], and Kunzi and Krelle [451]); *cutting-plane methods* (Kelley [422]); *sequential uncon-strained minimization* (Fiacco and McCormick [248]); *partitioning procedures* (Rosen [576], Benders [68]; also see Lasdon [460] and Taha [636]); and *generalized programming* (Dantzig [152]). See Abadie [1], Luenberger [477], Mangasarian [489], Gill et al. [303], McCormick [483], Zangwill [727], and Saaty and Bram [586] for general discussions on nonlinear mathematics and optimization. Nonlinear computer codes are reviewed in Waren and Lasdon [689], and Land and Powell [456].

REMARKS

Many problems dealing with statistical and other approximation problems can be converted to linear programs (Charnes, Cooper, and Ferguson [117], Freed and Glover [263, 264], Kelley [421], Stiefel [628], Ward [688], Wagner [682], Sengupta et al. [605], Arthanari and Dodge [12], and Zukhovitskiy and Avdeyeva [736]). In addition, Dantzig [171], Rosen [577], Torng [652], Whalen [695], and Zadeh [724] have related linear-programming procedures to the area of optimal-control theory. In Sec. 5, Chap. 11, we discuss formulations of certain absolute-variable problems that have proved of value in these fields.

EXERCISES

1. Construct and graph two-dimensional programming problems with the following charac-teristics:

     a.   an interior point which is optimal

     b.   a nonconvex solution space

2. Prove that a linear function is both convex and concave.

3. Prove that the set of points $(\mathbf{X}, \mathbf{Y})$ for which $\mathbf{Y} \geq f(\mathbf{X})$, $f(\mathbf{X})$ a convex function, is a convex set, i.e., the region above and bounded by a convex function is convex.

4. Prove that any nonnegative combination of convex functions is convex.

5. Determine if the following functions are convex:

(a) $|x|$ for all x

(b) e^x for all x

(c) $(x-1)^3$ for $1 \leq x < \infty$

(d) $(x-1)^3$ for $0 \leq x < \infty$

(e) $\log x$ for $0 < x < \infty$

(f) e^{-x^2} for $0 \leq x < \infty$

6. Develop the Kuhn-Tucker conditions for the following problems:

a. Minimize $\mathbf{cX}$, subject to $\mathbf{AX} \geq \mathbf{b}$, $\mathbf{X} \geq \mathbf{0}$.

b. Minimize $\mathbf{cX}$, subject to $\mathbf{AX} = \mathbf{b}$, $\mathbf{X} \geq \mathbf{0}$.

c. Minimize $f(\mathbf{X})$, subject to $g_i(\mathbf{X}) \leq 0$, $i = 1, \ldots, m$ and $f(\mathbf{X})$ and all $g_i(\mathbf{X})$ are convex.

d. Minimize $f(\mathbf{X})$ subject to $\mathbf{AX} = \mathbf{b}$, $\mathbf{X} \geq \mathbf{0}$, $f(\mathbf{X})$ a convex function.

7. Develop the Kuhn-Tucker conditions for the following convex-programming problems:

a. Minimize

$$a_1 e^{-b_1 x_1} + a_2 e^{-b_2 x_2}$$

subject to

$$x_1 + x_2 - 1 \leq 0$$

$$x_j \leq 0$$

and $(a_1, a_2, b_1, b_2) > \mathbf{0}$ (Carr and Howe [105]).

b. Minimize

$$x_1^2 + 2x_2^2 + e^{x_3}$$

subject to

$$x_1 x_2 + x_2 \quad\;\; - 1 \leq 0$$

$$x_2 + x_3^2 - 2 \leq 0$$

and all $x_j \geq 0$.

c. Minimize

$$-x_1 - \; x_2 - x_3 + \tfrac{1}{2}(x_1^2 + x_2^2 + x_3^2)$$

subject to

$$x_1 + \; x_2 + x_3 \leq 1$$

$$4x_1 + 2x_2 \quad\;\; \leq \tfrac{7}{3}$$

all $x_j \geq 0$ (Moeske [505]). (Note that the objective function is the sum of a linear function and a convex function.)

8. Solve Example 5.1 by both the λ method and the δ method. Also solve it by graphical procedures.

9. Show that for $\mathbf{X'CX}$ positive definite the quadratic-programming problem cannot have an unbounded solution (Hadley [344]).

10. Develop the necessary changes in Wolfe's algorithm for $\mathbf{AX} \leq \mathbf{b}$, $\mathbf{X} \geq \mathbf{0}$.

11. Solve the following problems:

a. Wolfe [711]. Minimize

$$\lambda(x_1 - 2x_3) + \tfrac{1}{2}(x_1^2 + x_2^2 + x_3^2)$$

subject to

$$x_1 - x_2 + x_3 = 1$$

$$x_j \geq 0$$

Answer: $\lambda = 1$, $x_1 = 0$, $x_2 = \tfrac{1}{2}$, $x_3 = \tfrac{3}{2}$, $f(\mathbf{X}) = -\tfrac{7}{4}$.

b. Hadley [344]. Minimize

$$-2x_1 - x_2 + x_1{}^2$$

subject to

$$2x_1 + 3x_2 + x_3 \qquad = 6$$
$$2x_1 + x_2 \qquad + x_4 = 4$$
$$x_j \geq 0$$

Answer: $x_1 = \tfrac{2}{3}$, $x_2 = \tfrac{14}{9}$, $x_3 = 0$, $x_4 = \tfrac{10}{9}$, $f(\mathbf{X}) = -\tfrac{22}{9}$.

c. Carr and Howe [105]. Minimize

$$-10x_1 - 20x_2 - x_1x_2 - 2x_1{}^2 - 2x_2{}^2$$

subject to

$$x_2 + x_3 \qquad = 8$$
$$x_1 + x_2 \qquad + x_4 = 10$$
$$x_j \leq 0$$

Answer: $x_1 = 4$, $x_2 = 6$, $x_3 = 0$, $x_4 = 0$, $f(\mathbf{X}) = 288$.

d. Beale [57]. Minimize

$$-6x_1 + 2x_1{}^2 - 2x_1x_2 + 2x_2{}^2$$

subject to

$$x_1 + x_2 \leq 2$$
$$x_j \geq 0$$

Answer: $x_1 = \tfrac{3}{2}$, $x_2 = \tfrac{1}{2}$, $f(\mathbf{X}) = -\tfrac{11}{2}$.

12. Solve *b*, *c*, and *d* of Exercise 11 graphically.

13. Prove the sufficiency of Theorem 11; that is, if $\mathbf{X}^0$, $\mathbf{V}^0$, π^0 satisfy $\mathbf{AX}^0 = \mathbf{b}$, (4.8), and (4.9), then $f(\mathbf{X}^0)$ is the minimum solution. HINT: You need to show $f(\mathbf{X}^0) \leq f(\mathbf{X})$ for all other $\mathbf{X}$. Use the fact that $(\mathbf{X} - \mathbf{X}^0)'\mathbf{C}(\mathbf{X} - \mathbf{X}^0) \geq 0$ of Theorem 2.

14. Show for $\mathbf{X}'\mathbf{CX}$ positive semidefinite that $\mathbf{X}'\mathbf{CX} = 0$ implies that $\mathbf{CX} = \mathbf{0}$.

15. Form the complementarity problem for the symmetric-dual problem, the bimatrix game, and the quadratic-programming problem.

16. The *linear fractional-programming problem* has the form: Maximize

$$f(\mathbf{X}) = \frac{\mathbf{cX} + \alpha}{\mathbf{dX} + \beta}$$

subject to

$$\mathbf{AX} \leq \mathbf{b}$$

$$\mathbf{X} \geq \mathbf{0}$$

where α and β are scalars, and $\mathbf{c}$ and $\mathbf{d}$ are n-component row vectors of constants. Show that $f(\mathbf{X})$ is neither convex nor concave, that a local maximum is a global maximum, and if the maximum occurs at a finite point, then it occurs also at a basic feasible solution of the constraints. Show how this problem can be converted to equivalent linear-programming problems and solved using the simplex algorithm, Charnes and Cooper [112] and Dorn [218]. See also Martos [498], Schaible [599, 600], Thomas and Grunspan [640], and Wagner [684]. HINT: Define a new variable $t = \dfrac{1}{\mathbf{dX} + \beta}$.

17. Solve the following problems by both the λ method and the δ method:

a. Exercise 7*b*.

b. Exercise 7*c.*

c. Maximize

$$x_0 = 3x_1 - x_1^2 + 2x_2$$

subject to

$$2x_1^2 + 3x_2^2 \leq 35$$

and $(x_1, x_2) \geq 0$ (Cooper [132]).
Answer: $x_1 = 1.246$, $x_2 = 3.261$; $x_0 = 8.708$.

18. Solve Exercise 17*c* by graphical procedures.

19. Solve Example 5.1 by Wolfe's quadratic-programming algorithm.

20. Show how a general nonlinear-programming problem can be transformed into one that has a linear objective function subject to nonlinear constraints (Wolfe [719]).   HINT: Define the objective function as a variable.

21. Prove that the restricted-basis entry rule does not have to be used when solving the approximating problem of a convex-programming problem (Hadley [344]).

22. Prove that the maximum of a convex function $f(\mathbf{X})$ over a closed convex set $\mathbf{K}$ assumes its global maximum at one or more extreme points of $\mathbf{K}$ (Hadley [344]).   If $f(\mathbf{X})$ is concave, what are the analogous results to Theorem 3 and the above statement?   (See Hoffman [380] on minimizing concave functions over convex sets.)

23. Show how logarithms can be used to develop separable functions for expressions such as $x_1 x_2$ and $x_1 x_2 x_3$.   You will need to make a transformation that does not allow any of the variables to be zero.   Also, show how the term $x_1 x_2$ can be expressed as the difference of the squares of two other variables.   Note that when using any of the above transformations, additional defining constraints must be added to the approximating problem.

Bibliography of Linear-Programming Applications

The organization of this Bibliography is based on the "Bibliography of Linear Programming and Related Techniques" by Vera Riley and Saul I. Gass, Johns Hopkins Press, Baltimore, 1958. It is an annotated bibliography that includes over 1,000 items arranged by topic, e.g., introductory survey, computational techniques, game theory, convex sets and linear inequalities, applications, etc. The list of applications cited below is only a small part of the total references on linear-programming applications.[1]

1. Agricultural Applications

Bazaraa, M. S., and A. Bouzaher: A Linear Goal Programming Model for Developing Economies with an Illustration from the Agricultural Sector in Egypt, *Management Science*, vol. 27, no. 4, 1981.

Beneke, R. R., and R. Winterboer: "Linear Programming Applications to Agriculture," Iowa State University Press, Ames, IA, 1973.

Boles, James N.: Linear Programming and Farm Management Analysis, *Journal of Farm Economics*, vol. 37, no. 1, pp. 1–24, 1955.

Cabraal, R. A.: Production Planning in a Sri-Lanka Coconut Mill Using Parametric Linear Programming, *Management Science*, vol. 11, no. 3, 1981.

---

[1] Many of the early references that appeared in previous editions of this text have not been included in this revision.

Candler, Wilfred: A Modified Simplex Solution for Linear Programming with Variable Capital Restriction, *Journal of Farm Economics*, vol. 38, no. 4, pp. 940–955, 1956.

Chappell, A. E.: Linear Programming Cuts Costs in Production of Animal Feeds, *Operational Research Quarterly*, vol. 25, no. 1, 1974.

Fisher, Walter D., and Leonard W. Shruben: Linear Programming Applied to Feed-Mixing under Different Price Conditions, *Journal of Farm Economics*, vol. 35, no. 4, pp. 471–483, 1953.

Fox, Karl A., and Richard C. Taeuber: Spatial Equilibrium Models of the Livestock-Feed Economy, *The American Economic Review*, vol. 45, no. 4, pp. 584–608, 1955.

Glen, J. J.: A Mathematical Programming Approach to Beef Feedlot Optimization, *Management Science*, vol. 26, no. 5, 1980.

Harrison, H., and D. R. Wills: Product Assembly and Distribution Optimization in an Agribusiness Cooperative, *Interfaces*, vol. 13, no. 2, 1983.

Heady, E. O., and A. C. Egbert: Regional Programming of Efficient Agricultural Production Patterns, *Econometrica*, vol. 32, no. 3, 1964.

Katzman, Irwin: Solving Feed Problems through Linear Programming, *Journal of Farm Economics*, vol. 38, 1956.

Killen, L., and M. Keane: A Linear Programming Model of Seasonality in Milk Production, *Journal of the Operational Research Society*, vol. 29, no. 7, 1978.

Klingman, D., P. H. Randolph, and S. W. Fuller: A Cotton Ginning Problem, *Operations Research*, vol. 24, no. 4, 1976.

Markland, R. E., and R. J. Newett: An Application of Mathematical Programming to Soybean Processing, *Omega*, vol. 3, no. 3, 1975.

Pierson, D. R.: Farm Profits Up by 40 Percent, *Automatic Data Processing*, vols. 4, 5, 1962.

Swanson, Earl R.: Integrating Crop and Livestock Activities in Farm Management Activity Analysis, *Journal of Farm Economics*, vol. 37, no. 5, pp. 1249–1258, 1955.

Swanson, L. W., and J. G. Woodruff: A Sequential Approach to the Feed-Mix Problem, *Operations Research*, vol. 12, no. 1, 1964.

2. Contract Awards

Beged-Dov, A. G.: Contract Award Analysis by Mathematical Programming, *Naval Research Logistics Quarterly*, vol. 17, no. 3, 1970.

Percus, Jerome, and Leon Quinto: The Application of Linear Programming to Competitive Bond Bidding, *Econometrica*, vol. 24, no. 4, 1956.

Stark, R. M.: Unbalanced Highway Contract Tendering, *Operational Research Quarterly*, vol. 25, pp. 373–388, 1974.

Waggener, H. A., and G. Suzuki: Bid Evaluation for Procurement of Aviation Fuel at DFSC: A Case History, *Naval Research Logistics Quarterly*, vol. 14, no. 1, 1967.

3. Diet and Menu Planning

Armstrong, R. D., and P. Sinha: Application of Quasi-Integer Programming to the Solution of Menu Planning Problems with Variable Portion Size, *Management Science*, vol. 21, no. 4, 1974.

Balintfy, J. L.: Mathematical Models of Menus, *Naval Research Review*, July 1974.

Balintfy, J. L., W. J. Duffy, and P. Sinha: Modeling Food Preferences over Time, *Operations Research*, vol. 22, no. 4, 1974.

Balintfy, J. L.: The Cost of Decent Subsistence, *Management Science*, vol. 25, no. 10, 1979.

Balintfy, J. L., G. T. Ross, P. Sinha, and A. A. Zoltners: A Mathematical Programming System for Preference and Compatibility Maximized Menu Planning and Scheduling, *Mathematical Programming*, vol. 15, no. 1, 1978.

Smith, V. E.: Linear Programming Models for the Determination of Palatable Human Diets, *Journal of Farm Economics*, vol. 41, pp. 272–283, 1959.

Smith, V. E.: A Diet Model with Protein Quality Variable, *Management Science*, vol. 20, no. 8, 1974.

4. Economic Analysis, Banking, and Finance

Agarwala, R., and G. C. Goodson: A Linear Programming Approach to Designing an Optimum Tax Package, *Operational Research Quarterly*, vol. 21, no. 2, 1970.

Babad, Y. M., and P. D. Speer, Jr.: Optimization of the Arbitrage Process in Advance Refunding of Municipal Bonds, *Management Science*, vol. 24, no. 10, 1978.

Baker, K. R., and W. W. Damon: A Simultaneous Planning Model for Production and Working Capital, *Decision Sciences*, vol. 8, no. 1, 1977.

Balbirer, S. D., and D. Shaw: An Application of Linear Programming to Bank Financial Planning, *Interfaces*, vol. 11, no. 5, 1981.

Beckmann, Martin J., and Thomas Marschak: An Activity Analysis Approach to Location Theory, *Proceedings of the Second Symposium in Linear Programming*, vol. I, pp. 331–379, 1955.

Broyles, J. E.: Compact Formulations of Mathematical Programmes for Financial Planning Problems, *Operational Research Quarterly*, vol. 27, no. 4, 1976.

Carleton, W. T., G. Kendall, and S. Tandon: Application of the Decomposition Principle to the Capital Budgeting Problem in a Decentralized Firm, *Journal of Finance*, vol. 29, no. 3, 1974.

Chilcott, J. F., and T. W. Hendy: GAMBIT—A Linear-Programming Model to Balance Transmission Investment and Market Options, *Operational Research Quarterly*, vol. 21, no. 4, 1970.

Cohen, K. J., and F. S. Hammer: Optimal Coupon Schedules for Municipal Bonds, *Management Science*, vol. 12, no. 1, 1965.

Cohen, K. J., S. F. Maier, and J. H. Vander Weide: Recent Development in Management Science in Banking, *Management Science*, vol. 27, no. 10, 1981.

Cornuejos, G., M. L. Fisher, and G. L. Nemhauser: Location of Bank Accounts to Optimize Float: An Analytic Study of Exact and Approximate Algorithms, *Management Science*, vol. 23, no. 8, 1977.

Davis, S. G., and E. T. Reutzel: Point Determination of Machine Requirements and Shift Scheduling in Banking Operations, *Interfaces*, vol. 11, no. 1, 1981.

Faaland, B.: An Integer Programming Algorithm for Portfolio Selection, *Management Science*, vol. 20, no. 10, 1974.

Heroux, R. L., and W. A. Wallace: Linear Programming and Financial Analysis of the New Community Development Process, *Management Science*, vol. 19, no. 8, 1973.

Hobbs, J. A.: Linear Programming with Costs and Incomes in Two or More Separated Currencies, *Operational Research Quarterly*, vol. 17, no. 2, 1966.

Ijiri, Y.: "Management Goals and Accounting for Control," North-Holland Publishing Company, Amsterdam, 1965.

Itami, H. and R. S. Kaplan: An Activity Analysis Approach to Unit Costing with Multiple Interactive Products, *Management Science*, vol. 26, no. 8, 1980.

Jensen, R. E.: Optimal Allocation of Joint (Common) Costs: A Parametric Programming Approach, *Decision Sciences*, vol. 9, no. 3, 1978.

Krajewski, L., L. Ritzman and P. McKenzie: Shift Scheduling in Banking Operations: A Case Application, *Interfaces*, vol. 10, no. 2, 1980.

Lorie, J., and L. Savage: Three Problems in Capital Rationing, *Journal of Business*, vol. 28, pp. 229–239, 1955.

Maier, S. F., and J. H. Vander Weide: Capital Budgeting in the Decentralized Firm, *Management Science*, vol. 23, no. 14, 1976.

Markland, R. E., and R. M. Nauss: Improving Transit Check Clearing Operations at Maryland National Bank, *Interfaces*, vol. 13, no. 1, 1983.

Markowitz, Harry M.: Portfolio Selection, *Journal of Finance*, vol. 7, no. 1, pp. 77–91, 1952.

Martin, Alfred D., Jr.: Mathematical Programming of Portfolio Selections, *Management Science*, vol. 1, no. 2, pp. 152–156, 1955.

Modiano, E. M., and J. F. Shapiro: A Dynamic Optimization Model of Resources, *The Bell Journal of Economics*, vol. II, no. 1, 1980.

Morris, J. R.: An Application of the Decomposition Principle to Financial Decision Models, *Journal of Financial and Quantitative Analysis*, vol. 10, no. 1, 1975.

Nauss, R. M., and B. R. Keeler: Minimizing Net Interest Cost in Municipal Bond Bidding, *Management Science*, vol. 27, no. 4, 1981.

Nauss, R. M., and R. E. Markland: Theory and Application of an Optimizing Procedure for Lock Box Location Analysis, *Management Science*, vol. 27, no. 8, 1981.

Obersteiner, E.: Should the Foreign Affiliate Remit Dividends or Reinvest, *Financial Management*, vol. 2, no. 1, 1973.

Philippatos, G. C.: "Financial Management: Demand Techniques," Holden-Day, Inc., San Francisco, 1973.

Pogue, G. A., and R. N. Bussard: Linear Programming Model for Short Term Financial Planning under Uncertainty, *Sloan Management Review*, vol. 16, no. 2, 1972.

Samuelson, Paul A.: "Linear Programming and Economic Theory," *Proceedings of the Second Symposium in Linear Programming*, vol. I, pp. 251–272, 1955.

Sartoris, W. L., and M. L. Spruill: Goal Programming and Working Capital Management, *Journal of Financial Management*, pp. 67–73, 1974.

Sengupta, J. K., and K. A. Fox: "Optimization Techniques in Quantitative Economic Models," North-Holland Publishing Co., Amsterdam, 1969.

Sharp, J. F.: The Effects of Income Taxes on Linear Programming Models, *Decision Sciences*, vol. 6, no. 3, 1975.

Sharpe, W. F.: A Linear Programming Algorithm for Mutual Fund Portfolio Selection, *Management Science*, vol. 13, no. 7, 1967.

Sharpe, W. F.: "Portfolio Theory and Capital Markets," McGraw-Hill Book Company, New York, 1970.

Srinivasan, V.: A Transshipment Model for Cash Management Decisions, *Management Science*, vol. 20, no. 10, 1974.

Van Horne, J. C.: "Financial Management and Policy," Prentice-Hall, Inc., Englewood Cliffs, NJ, 1974.

Weingartner, H. M.: "Mathematical Programming and the Analysis of Capital Budgeting Problems," Prentice-Hall, Inc., Englewood Cliffs, N.J, 1967.

5. Energy and Environment

Cohon, J. L., C. S. ReVelle, J. Current, T. Eagles, and R. Eberhart: Application of a Multiobjective Facility Location Model to Power Plant Siting in a Six-State Region of the U.S., *Computers and Operations Research*, vol. 7, nos. 1 and 2, 1980.

Deonigi, D. E., and R. L. Engel: Linear Programming in Energy Modeling, pp. 144–176 of "Energy Modeling," Resources for the Future, Inc., Washington, DC, 1973.

Dutton, R., G. Hinman, and C. B. Millham: The Optimal Location of Nuclear-Power Facilities in the Pacific Northwest, *Operations Research*, vol. 22, no. 3, 1974.

Glassey, G. R.: Price Sensitive Consumer Demands in Energy Modeling, *Management Science*, vol. 24, no. 9, 1978.

Graves, J. S.: Developing Generation Construction Plans for the Potomac Electric Power Company: Management Science in Utility Regulation, *Interfaces*, vol. 12, no. 6, 1982.

Greenberg, M. R., "Applied Linear Programming for the Socioeconomic and Environmental Sciences," Academic Press, Inc., New York, 1978.

Isard, W.: "Some Notes on the Linkage of the Ecologic and Economic Systems," Regional Science Association: Papers XXII, Budapest Conference, 1968.

Knight, U. G. W.: The Logical Design of Electrical Networks Using Linear Programming Methods, *Institute of Electrical Engineers, Proceedings*, vol. 107, no. 33, 1960.

Kohn, Robert E.: "A Linear Programming Model for Air Pollution Control," The MIT Press, Cambridge, MA, 1978.

Lee, T. D., R. J. Graves, and L. F. McGinnis: A Procedure for Air Monitoring Instrumentation Location, *Management Science*, vol. 24, no. 14, 1978.

Liebman, J. C.: Systems Approaches to Solid Waste Collection Problems, chap. 15 in "Systems Approach to Environmental Pollution," G. K. Chacko (ed.), Operations Research Society of America, Arlington, VA, 1972.

Limaye, D. R. (ed.): "Energy Policy Evaluation," D. C. Heath and Company, Boston, 1974.

Loucks, D. P., C. S. ReVelle, and W. R. Lynn: Linear Programming Models for Water Pollution Control, *Management Science*, vol. 14, no. 4, 1967.

McMahan, W. L., and P. A. Roach: Site Energy Optimization, A Math Programming Approach, *Interfaces*, vol. 12, no. 6, 1982.

Manne, A. S., R. G. Richels, and J. P. Weyant: Energy Policy Modeling: A Survey, *Operations Research*, vol. 27, no. 1, 1979.

Moonan, F., and R. J. Giglio: Planning Electric Power Generation: A Nonlinear Mixed Integer Model Employing Benders Decomposition, *Management Science*, vol. 23, no. 9, 1977.

Neely, W. P., J. Sellers, and R. M. North: Goal Programming Priority Sensitivity Analysis: An Application in Natural Resource Decision Making Process, *Interfaces*, vol. 10, no. 5, 1980.

Nour, N. E., and H. H. Happ: The Application of Transportation Techniques in Modelling Power Pools, *Computers and Operations Research*, vol. 1, no. 1, 1974.

Pechan, E. H., E. S. Burton, and W. Sanjour: Computerized Regional Air Pollution Abatement and Fuel Use Modeling, *Computers and Operations Research*, vol. 1, no. 1, 1974.

Scherer, C. R., and L. Joe: Electric Power System Planning with Explicit Stochastic Reserves Constraint, *Management Science*, vol. 23, no. 4, 1977.

Sherali, H. D., A. L. Soyster, F. H. Murphy, and S. Sen: Linear Programming Based Analysis of Marginal Cost Pricing in Electric Utility Capacity Expansion, *European Journal of Operational Research*, vol. 11, no. 4, 1982.

Teller, A.: The Use of Linear Programming to Estimate the Cost of Some Alternative Air Pollution Abatement Policies, *Proceedings of the IBM Scientific Computing*

Symposium on Water and Air Resource Management, pp. 345–353, IBM, White Plains, NY, 1968.

6. Facility and Site Location

Akine U., and B. M. Khumawale: An Efficient Branch and Bound Algorithm for the Capacitated Warehouse Location Problem, *Management Science*, vol. 23, no. 6, 1977.

Drayer, E., and S. Seaburg: Facilities Expansion Model, *Interfaces*, vol. 5, no. 2, part 2, 1975.

Efroymson, M. A., and T. L. Ray: A Branch-Bound Algorithm for Plant Location, *Operations Research*, vol. 14, no. 3, 1966.

Elson, D. C.: Site Location via Mixed-Integer Programming, *Operational Research Quarterly*, vol. 23, no. 1, 1972.

Morris, J. C.: A Linear Programming Approach to the Solution of Constrained Multifacility Minimax Location Problems where Distances are Rectangular, *Operational Research Quarterly*, vol. 24, no. 3, 1973.

Proceedings of the Symposium on Operations Analysis of Education, *Socio-Economic Planning Sciences*, vol. 2, nos. 2, 3, 4, 1969.

ReVelle, C., D. Marks, and J. C. Liebman: An Analysis of Private and Public Sector Location Models, *Management Science*, vol. 16, no. 11, 1970.

Schniederjans, M. J., H. K. Kwak, and M. C. Helmer: An Application of Goal Programming to Resolve a Site Location Problem, *Interfaces*, vol. 12, no. 3, 1982.

Soland, R.: Optimal Facility Location with Concave Costs, *Operations Research*, vol. 22, no. 2, 1974.

Swain, R. W.: A Parametric Decomposition Approach for the Solution of Uncapacitated Location Problems, *Management Science*, vol. 21, no. 2, 1974.

Sweeney, D. J., and R. L. Tatham: An Improved Long-Run Model for Multiple Warehouse Location, *Management Science*, vol. 22, no. 7, 1976.

Trippi, R. R.: The Warehouse Location Formulation as a Special Type of Inspection Problem, *Management Science*, vol. 21, no. 9, 1975.

Van Roy, T. J., and D. Erlenkotter: A Dual-Based Procedure for Dynamic Facility Location, *Management Science*, vol. 28, no. 10, 1982.

Warszawaski, A., and S. Peer: Optimizing the Location of Facilities on a Building Site, *Operational Research Quarterly*, vol. 24, no. 1, 1973.

7. Health Care and Medical

Feldstein, M. S., M. A. Piot, and T. K. Sundaresan, Resource Allocation Model for Public Health Planning: A Case Study of Tuberculosis Control, supplement to vol. 48 of the *Bulletin of the World Health Organization*, WHO, Geneva, 1973.

Gould, F. J., and M. J. Magazine: A Mathematical Programming Model for Planning Contraceptive Deliveries, *Socio-Economic Planning Sciences*, vol. 5, no. 3, 1971.

Health Care Issue, *Operations Research*, vol. 24, no. 5, 1976.

Heiner, K., W. A. Wallace, and K. Young: A Resource Allocation and Evaluation Model for Providing Services to the Mentally Retarded, *Management Science*, vol. 27, no. 7, 1981.

Hilal, S. S., and W. Erikson: Matching Supplies to Save Lives: Linear Programming the Production of Heart Valves, *Interfaces*, vol. 11, no. 6, 1981.

Ittig, P. T.: A Model for Planning Ambulatory Health Services, *Management Science*, vol. 24, no. 10, 1978.

Jackman, H. W.: Financing Public Hospitals in Ontario: A Case Study in Rationing of Capital Budgets, *Management Science*, vol. 20, no. 4, 1973.

Jones, L., and N. K. Kwak: A Goal Programming Model for Allocating Human Resources for the Good Laboratory Practice Regulations, *Decision Sciences*, vol. 13, no. 1, 1982.

Kendall, K. E., and S. M. Lee: Formulating Blood Rotation Policies with Multiple Objectives, *Management Science*, vol. 26, no. 11, 1980.

Kolesar, P.: Testing for Vision Loss in Glaucoma Suspects, *Management Science*, vol. 26, no. 5, 1980.

Nackel, J. G., J. Goldman, and W. L. Fairman: A Group Decision Process for Resource Allocation in the Health Setting, *Management Science*, vol. 24, no. 12, 1978.

Or, I., and W. P. Pierskalla: A Transportation Location-Allocation Model for Regional Blood Banking, *AIIE Transactions*, vol. 11, no. 2, 1979.

Prastacos, G. P., and E. Brodheim: PBDS: A Decision Support System for Regional Blood Management, *Management Science*, vol. 26, no. 5, 1980.

Rasmasen, H. J.: Conflicting Strategic Objectives in Hospital Design, *OMEGA*, vol. 2, no. 2, 1974.

Ruth, R. J.: A Mixed Integer Programming Model for Regional Planning of a Hospital Inpatient Service, *Management Science*, vol. 27, no. 5, 1981.

Taylor, B. W., III, K. R. Davis, and L. J. Ryan: Compliance with the Occupational Safety and Health Act: A Mathematical Programming Framework, *Decision Sciences*, vol. 8, no. 4, 1977.

Vora, J. A.: Heuristics and Optimizing Techniques Applied to Long Range Hospital Planning for Ancillary Departments, *Management Science*, vol. 21, no. 4, 1974.

Warner, D. M., and J. Prawda: A Mathematical Programming Model for Scheduling Nursing Personnel in a Hospital, *Management Science*, vol. 19, no. 4, 1972.

8. Industrial Applications

a. Airlines Industry

Arabeyre, J. P., J. Fearnley, F. C. Steiger, and W. Teather: The Airline Crew Scheduling Problem: A Survey, *Transportation Science*, vol. 3, no. 2, 1969.

Austin, L. M., and W. W. Hogan: Optimizing the Procurement of Aviation Fuels, *Management Science*, vol. 22, no. 6, 1976.

Etschmaier, M. M., and M. Rothstein: Operations Research in the Management of the Airlines, *OMEGA*, vol. 2, no. 2, 1974.

Glover, F., R. Glover, J. Lorenzo, and C. McMillan: The Passenger-Mix Problem in the Scheduled Airlines, *Interfaces*, vol. 12, no. 3, 1982.

Hansen, P., and R. E. Wendell: A Note on Airline Commuting, *Interfaces*, vol. 12, no. 1, 1982.

Lee, R.: An Application of Integer Linear Programming to the Allocation of Itineraries between Aircraft, *Operational Research Quarterly*, vol. 23, no. 2, 1972.

Marsten, R. E., and M. R. Muller: A Mixed-Integer Programming Approach to Air Cargo Planning, *Management Science*, vol. 26, no. 11, 1980.

Marsten, R. E., M. R. Muller, and C. L. Killian: Crew Planning at Flying Tiger: A Successful Application of Integer Programming, *Management Science*, vol. 25, no. 12, 1979.

Nash, B.: A Simplified Alternative to Current Airline Fuel Allocation Models, *Interfaces*, vol. 11, no. 1, 1981.

Rubin, J.: A Technique for the Solution of Massive Set Covering Problems, with Application to Airline Crew Scheduling, *Transportation Science*, vol. 7, no. 1, 1973.

b. Chemical Industry

Dantzig, G. B., S. Johnson, and W. White: A Linear Programming Approach to the Chemical Equilibrium Problem, *Management Science*, vol. 5, no. 1, 1958.

Stadtherr, M. A., and D. F. Rudd: Response Management in the Petrochemical Industry, *Management Science*, vol. 24, no. 7, 1978.

c. Coal Industry

Henderson, James M.: A Short-Run Model for the Coal Industry, *Review of Economics and Statistics*, vol. 37, no. 4, pp. 336–346, 1955.

Meyer, M.: Applying Linear Programming to the Design of Ultimate Pit Limits, *Management Science*, vol. 16, no. 2, 1969.

Wagner, M. H.: Supply-Demand Decomposition of the National Coal Model, *Operations Research*, vol. 29, no. 6, 1981.

d. Communications and Computer Industries

Ben-Bassat, M., and J. Borovits: Computer Network Scheduling, *OMEGA*, vol. 3, no. 1, 1975.

Fisher, M., and D. S. Hochbaum: Database Location in Computer Networks, *Journal of the Association for Computing Machinery*, vol. 27, no. 4, 1980.

Kalaba, Robert E., and Mario L. Juncosa: Optical Design and Utilization of Communication Networks, *Management Science*, vol. 3, no. 1, pp. 33–44, 1956.

Kortanek, K. O., D. N. Lee, and G. G. Polak: A Linear Programming Model for Design of Communications Networks with Time Varying Probabilistic Demands, *Naval Research Logistics Quarterly*, vol. 28, no. 1, 1981.

Lopez, D. A., and P. Gray: The Substitution of Communication for Transportation—A Case Study, *Management Science*, vol. 23, no. 11, 1977.

Saaty, T. L., and G. Suzuki: A Nonlinear Programming Model in Optimum Communication Satellite Use, *SIAM Review*, vol. 7, no. 3, 1965.

e. Iron and Steel Industry

Fabian, Tibor: Blast Furnace Burdening and Production Planning, *Management Science*, vol. 14, no. 2, 1967.

Haessler, R. W., and M. A. Vonderembse: A Procedure for Solving the Master Slab Cutting Stock Problem in the Steel Industry, *AIIE Transactions*, vol. 11, no. 2, 1979.

Jain, S. K., K. L. Stott, and E. G. Vasold: Overbook Balancing Using a Combination of Linear Programming and Heuristic Techniques, *Interfaces*, vol. 9, no. 1, 1978.

Vonderembse, M. A., and R. W. Haessler: A Mathematical Programming Approach to Schedule Master Slab Casters in the Steel Industry, *Management Science*, vol. 28, no. 12, 1982.

f. Paper Industry

Beged-Dov, A. G.: The Paper Trim Problem: A Variable Demand Analysis, *AIIE Transactions*, vol. 6, no. 1, 1974.

Dyckhoff, H.: A New Linear Programming Approach to the Cutting Stock Problem, *Operations Research*, vol. 29, no. 6, 1981.

Gilmore, P. C., and R. E. Gomory: A Linear-Programming Approach to the Cutting-Stock Problem—Part I, *Operations Research*, vol. 9, no. 6, 1961.

Gilmore, P. C., and R. E. Gomory: A Linear-Programming Approach to the Cutting-Stock Problem—Part II, *Operations Research*, vol. 11, no. 6, 1963.

Glassey, C. R., and V. K. Gupta: A Linear Programming Analysis of Paper Recycling, *Management Science*, vol. 21, no. 4, 1974.

Haessler, R. W.: A Heuristic Programming Solution to a Nonlinear Cutting Stock Problem, *Management Science*, vol. 17, no. 12, 1971.

Haessler, R. W., and F. B. Talbot: A 0-1 Model for Solving the Corrugator Trim Problem, *Management Science*, vol. 29, no. 2, 1983.

Hahn, S. G.: On the Optimal Cutting of Defective Sheets, *Operations Research*, vol. 16, no. 6, 1968.

Morgan, J. I.: Survey of Operations Research, *Paper Mill News*, vol. 84, no. 11, 1961.

Paull, A. E.: Linear Programming: A Key to Optimum Newsprint Production, *Pulp and Paper Magazine of Canada*, vol. 57, no. 1, pp. 85–90, 1956, reissued, vol. 57, no. 4, pp. 145–150, 1956.

g. Petroleum Industry

Aronofsky, J. S., J. M. Dutton, and M. T. Tayyabkhan: "Managerial Planning with Linear Programming: In Process Industry Operations," John Wiley & Sons, New York, 1978.

Aronofsky, J. S., and A. C. Williams: The Use of Linear Programming and Mathematical Models in Underground Oil Production, *Management Science*, vol. 8, no. 4, 1962.

Brown, G. G., and G. W. Graves: Real-Time Dispatch of Petroleum Tank Trucks, *Management Science*, vol. 27, no. 1, 1971.

Catchpole, A. R.: The Application of Linear Programming to Integrated Supply Problems in the Oil Industry, *Operational Research Quarterly*, vol. 13, no. 2, 1962.

Charnes, Abraham, William W. Cooper, and Robert Mellon: Blending Aviation Gasolines—A Study in Programming Interdependent Activities in an Integrated Oil Company, *Econometrica*, vol. 20, no. 2, pp. 135–159, 1952.

Durrer, E. J., and G. E. Slater: Optimization of Petroleum and Natural Gas Production—A Survey, *Management Science*, vol. 24, no. 1, 1977.

Faur, P.: Elements for Selection of Investments in the Refining Industry, *Proceedings of the Second International Conference on Operational Research*, English Universities Press, Ltd., London, and John Wiley & Sons, Inc., New York, 1960.

Frair, L., and M. Devine: Economic Optimization of Offshore Petroleum Development, *Management Science*, vol. 21, no. 12, 1975.

Garvin, W. W., H. W. Crandell, J. B. John, and R. A. Spellman: Applications of Linear Programming in the Oil Industry, *Management Science*, vol. 3, no. 4, 1957.

Jackson, B. L.: Using LP for Grade Oil Sales at Elk Hills: A Case Study, *Interfaces*, vol. 10, no. 3, 1980.

Manne, Alan S.: "Concave Programming for Gasoline Blends," P-383, The RAND Corporation, Mar. 20, 1953.

————: "Scheduling of Petroleum Refinery Operations," 181 pp., Harvard Economic Studies, vol. 48, Harvard University Press, Cambridge, MA, 1956.

O'Neill, R. P., M. Willard, B. Pike, and R. Pike: A Mathematical Programming Model for Allocation of Natural Gas, *Operations Research*, vol. 27, no. 5, 1979.

Symonds, Gifford H.: "Linear Programming: The Solution of Refinery Problems," 74 pp., Esso Standard Oil Company, New York, 1955.

Zierev, T. K., W. A. Mitchell, and T. R. White: Practical Applications of Linear Programming to Shell's Distribution Problems, *Interfaces*, vol. 6, no. 4, 1976.

h. Railroad Industry

Brosk, L. C., R. J. Buck, W. H. Sparrow, and J. R. White: Boxcars, Linear Programming and the Sleeping Kitten, *Interfaces*, vol. 10, no. 6, 1980.

Charnes, Abraham, and M. H. Miller: A Model for Optimal Programming of Railway Freight Train Movements, *Management Science*, vol. 3, no. 1, pp. 74–93, 1956.

Crane, Roger R.: A New Tool: Operations Research, *Modern Railroads*, vol. 9, no. 1, pp. 146–152, 1954.

i. Other Industries

Advani, S.: A Linear-Programming Approach to Air-Cleaner Design, *Operations Research*, vol. 22, no. 2, 1974.

Androit, J., and J. Gaussens: Programme for Thermal Reactor and "Breeder Reactor" Power Stations—Fuel Economy Problem of Storage—Price of Plutonium, *Proceedings of the Second International Conference on Operational Research*, English Universities Press, Ltd., London, and John Wiley & Sons, Inc., New York, 1960.

Barr, B. M.: "The Soviet Wood-Processing Industry," University of Toronto Press, Toronto, Canada, 1970.

Barros, O., and A. Weintraub: Planning for a Vertically Integrated Forest Industry, *Operations Research*, vol. 30, no. 6, 1982.

Bender, F. E., A. Kramer, and G. Kahan: "Systems Analysis for the Food Industry," AVI Publishing Company, Inc., Westport, CN, 1976.

Bhatnagar, S. C.: Implementing Linear Programming in a Textile Unit: Some Problems and a Solution, *Interfaces*, vol. 11, no. 2, 1981.

Caswell, W. M., and A. Rao: A Practical Approach to the Large-Scale Forest Scheduling Problem, *Decisions Sciences*, vol. 5, no. 3, 1974.

Cavalieri, F., A. Roversi, and R. Ruggeri: Use of Mixed-Integer Programming to Investigate Optimum Running Conditions for a Thermal Power Station and Possible Extension to Capacity, *Operational Research Quarterly*, vol. 22, no. 3, 1971.

Chambers, M. L., and R. G. Dyson: The Cutting Stock Problem in the Flat Glass Industry—Selection of Stock Sizes, *Operational Research Quarterly*, vol. 27, no. 4, 1976.

Charkin, L. M., and J. L. Showers: Reducing Uncollectible Revenue from Residential Telephone Customers, *Interfaces*, vol. 11, no. 6, 1981.

Charnes, Abraham, William W. Cooper, and Robert O. Ferguson: Optimal Estimation of Executive Compensation by Linear Programming, *Management Science*, vol. 1, no. 2, 1955.

Danø, S.: "Linear Programming in Industry," Springer-Verlag OHG, Vienna, 1963.

Dwyer, F. R., and J. R. Evans: A Branch and Bound Algorithm for the List Selection Problem in Direct Mail Advertising, *Management Science*, vol. 27, no. 6, 1981.

Dyson, R. G., and A. S. Gregory: The Cutting Stock Problem in the Flat Glass Industry, *Operational Research Quarterly*, vol. 25, no. 1, 1974.

Gaitner, M.: A Stochastic Constrained Optimization Model for Determining Commercial Fishing Seasons, *Management Science*, vol. 26, no. 2, 1980.

Gleason, J. M., and C. C. Lilly: A Goal Programming Model for Insurance Agency Management, *Decision Sciences*, vol. 8, no. 1, 1977.

Horowitz, J., R. Lattes, and E. Parker: Some Operational Problems Connected with Power Reactor Discharges, *Proceedings of the Second International Conference on*

Operational Research, English Universities Press, Ltd., London, and John Wiley & Sons, Inc., New York, 1960.

Kalvaitis, R., and A. G. Posgay: An Application of Mixed Integer Programming in the Direct Mail Industry, *Management Science*, vol. 20, no. 5, 1974.

King, R. H., and R. R. Lowe, Jr.: Coordinating Decisions for Increased Profits, *Interfaces*, vol. 10, no. 6, 1980.

Kotak, D. B.: Application of Linear Programming to Plywood Manufacture, *Interfaces*, vol. 7, no. 1, part 2, 1976.

Lockett, G.: A Model of the Hide-Leather Sequence in a Tannery, *Operations Research Quarterly*, vol. 24, no. 4, 1973.

Lockett, A. G., and A. E. Gear: Linear Programming to Aid Resource Allocation in R & D, *The Journal of Management Studies*, vol. 7, no. 2, 1970.

Love, R. F., and L. Yerex: An Application of a Facilities Location Model in the Prestressed Concrete Industry, *Interfaces*, vol. 6, no. 4, 1976.

Perry, C., and M. Iliff: From the Shadows: Earthmoving on Construction Projects, *Interfaces*, vol. 13, no. 1, 1983.

Roy, A., E. E. DeFalomir, and L. Lasdon: An Optimization-Based Decision Support System for a Product Mix Problem, *Interfaces*, vol. 12, no. 2, 1982.

Schleaf, H. J.: Using Linear Programming for Planning Life Insurance Purchases, *Decision Sciences*, vol. 11, no. 3, 1980.

Smith, D.: "Linear Programming Models in Business," Polytech Publishers Ltd., Stockport, England, 1973.

Wardle, P. A.: Forest Management and Operational Research: A Linear Program Study, *Management Science*, vol. 11, no. 10, 1965.

Weintraub, A., and D. Navon: A Forest Management Planning Model Integrating Silvicultural and Transportation Activities, *Management Science*, vol. 22, no. 12, 1976.

Williams, H. P., and A. C. Redwood: A Structured Linear Programming Model in the Road Industry, *Operational Research Quarterly*, vol. 25, no. 1, 1974.

9. Land Use Planning

Bagby, G., and A. Thesen: A Network Flow Model for Allocation of Land Uses to Sectors, *Management Science*, vol. 22, no. 11, 1966.

Bammi, D., and D. Bammi: Land Use Planning: An Optimizing Model, *OMEGA*, vol. 3, no. 5, 1975.

Bammi, D., and D. Bammi: Development of a Comprehensive Land Use Plan by Means of a Multiple Objective Mathematical Programming Model, *Interfaces*, vol. 9, no. 2, part 2, 1979.

Brill, E. D., Jr., S.-Y. Chang, and L. D. Hopkins: Modeling To Generate Alternatives: The HSJ Approach and an Illustration Using a Problem in Land Use Planning, *Management Science*, vol. 28, no. 3, 1982.

Fong, C. O.: Planning for Industrial Estate Development in a Developing Country, *Management Science*, vol. 26, no. 10, 1980.

Nazareth, L.: A Land Management Model Using Dantzig–Wolfe Decomposition, *Management Science*, vol. 26, no. 5, 1980.

Penz, A. J.: Outdoor Recreation Areas: Capacity and the Formulation of Use Policy, *Management Science*, vol. 22, no. 2, 1975.

Winokur, H. S., Jr., J. B. Frick, and J. C. Bean: The Affair Between the Land Developer and the Management Scientist, *Interfaces*, vol. 11, no. 5, 1981.

10. Marketing and Advertising

Bass, F. M., and R. T. Lonsdale: An Exploration of Linear Programming in Media Selection, *Journal of Marketing Research*, vol. 3, pp. 179–188, 1966.

Charnes, A., W. W. Cooper, J. K. De Voe, D. B. Learner, and W. Reinecke: A Goal Programming Model for Media Planning, *Management Science*, vol. 14, no. 8, 1968.

Farquhar, P. H., and V. R. Rao: A Balance Model for Evaluating Subsets of Multi-attributed Items, *Management Science*, vol. 22, no. 5, 1976.

Hartung, P. H.: Brand Switching and Mathematical Programming in Market Expansion, *Management Science*, vol. 11, no. 10, 1965.

Lee, S. M., and R. E. Nicely: Goal Programming for Marketing Decisions: A Case Study, *Journal of Marketing*, vol. 38, pp. 24–32, 1974.

Lodish, L. M.: Sales Territory Alignment to Maximize Profit, *Journal of Marketing*, vol. XII, pp. 30–36, 1975.

Meidan, A.: "Marketing Applications of Operational Research Techniques," MCB Publications Limited, Bradford, England, 1981.

Pekelman, D., and S. K. Sen: Marketing Product Design Mathematical Programming Models for the Determination of Attribute Weights, *Management Science*, vol. 20, no. 8, 1974.

Shanker, R. J., R. E. Turner, and A. A. Zoltners: Sales Territory Design: An Integrated Approach, *Management Science*, vol. 22, no. 3, 1975.

Stasch, S. F.: Linear Programming and Space-Time Considerations in Media Selection *Journal of Advertising Research*, vol. 5, pp. 40–46, 1965.

Thomas, J.: Linear Programming Models for Production-Advertising Decisions, *Management Science*, vol. 17, no. 8, 1971.

Zoltners, A. A., and P. Sinhue: Integer Programming Models for Sales Resource Allocation, *Management Science*, vol. 26, no. 3, 1980.

11. Military

Bracken, J., and P. R. Burnham: A Linear Programming Model for Minimum-Cost Procurement and Operation of Marine Corps Training Aircraft, *Naval Research Logistics Quarterly*, vol. 15, no. 1, 1968.

Dellinger, D. C.: An Application of Linear Programming to Contingency Planning: A Tactical Airlift System Analysis, *Naval Research Logistics Quarterly*, vol. 18, no. 3, 1971.

Furman, G. G., and H. J. Greenberg: Optimal Weapon Allocation with Overlapping Area Defenses, *Operations Research*, vol. 21, no. 6, 1973.

Jacobs, Walter W.: The Caterer Problem, *Naval Research Logistics Quarterly*, vol. 1, no. 2, 1954.

———: Military Applications of Linear Programming, *Proceedings of the Second Symposium in Linear Programming*, vol. I, pp. 1–27, 1955.

Mellichamp, J. M., W. L. Dixon, Jr., and S. L. Mitchell: Ballistic Missile Defense Technology Management with Goal Programming, *Interfaces*, vol. 10, no. 5, 1980.

Miercort, F. A., and R. M. Soland: Optimal Allocation of Missiles against Area and Point Defenses, *Operations Research*, vol. 19, no. 3, 1971.

Saaty, T. L., and K. W. Webb: Sensitivity and Renewals in Scheduling Aircraft Overhaul, *Proceedings of the Second International Conference on Operational Research*, English Universities Press, Ltd., London, and John Wiley & Sons, Inc., New York, 1960.

Wollmer, R. D.: Algorithms for Targeting Strikes in a Lines-of-Communication Network, *Operations Research*, vol. 18, no. 3, 1970.

12. Personnel Assignment and Workforce Planning

Baker, K. A.: Workforce Allocation in Cyclical Scheduling Problems: A Survey, *Operational Research Quarterly*, vol. 27, no. 1, 1976.

Bandyopadhyay, J. K., S. K. Sengupta, and R. K. Bogh: A Resource Allocation Model for an Employability Planning Model, *Interfaces*, vol. 10, no. 5, 1980.

Bres, E. S., D. Burns, A. Charnes, and W. W. Cooper: A Goal Programming Model for Planning Accessions, *Management Science*, vol. 26, no. 8, 1980.

Byrne, J. L., and R. B. Potts: Scheduling of Toll Collectors, *Transportation Science*, vol. 7, no. 3, 1973.

Charnes, A., W. W. Cooper, and R. J. Niehaus: "Management Science Approaches to Manpower Planning and Organization Design," North-Holland Publishing Co., New York, 1978.

Collins, R. W., S. I. Gass, and E. E. Rosendahl: The ASCAR Model for Evaluating Military Manpower Policy, *Interfaces*, vol. 13, no. 4, 1983.

Glover, F., and D. Klingman: Network Applications in Industry and Government, *AIIE Transactions*, vol. 9, no. 4, 1977.

Goodman, D. A.: A Goal Programming Approach to Aggregate Planning of Production and Work Force, *Management Science,* vol. 20, no. 12, 1974.

Grinold, R.: Input Policies for a Longitudinal Manpower Flow Model, *Management Science*, vol. 22, no. 5, 1976.

Grinold, R. C., K. T. Marshall, and R. M. Oliver: Longitudinal Manpower Planning Models, *Naval Research Logistics Quarterly*, vol. 23, no. 2, 1970.

Hansemann, F., and S. W. Hess: A Linear Programming Approach to Production and Employment Scheduling, *Management Technology*, no. 1, 1960.

Harwood, G. B., and R. W. Lawless: Optimizing Organizational Goals in Assigning Faculty Teaching Schedules, *Decision Sciences*, vol. 6, no. 3, 1975.

Holz, B. W., and J. M. Wroth: Improving Strength Forecasts: Support for Army Manpower Management, *Interfaces*, vol. 10, no. 6, 1980.

Kahalas, H., and D. L. Groves: Public Manpower Planning for Senior Citizens, *Technological Forecasting and Social Change*, vol. 10, no. 3, 1977.

Koenigsberg, E., and P. Loya: Equality of Work Opportunity: A Linear Programming Application, *Journal of the Operational Research Society*, vol. 29, no. 11, 1978.

Lilien, G. L., and A. G. Rao: A Model for Manpower Management, *Management Science*, vol. 21, no. 2, 1975.

Martel, A.: Tactical Manpower Planning via Programming under Uncertainty, *Operational Research Quarterly*, vol. 24, no. 4, 1973.

Patz, A. L.: Linear Programming Applied to Manpower Management, *The Industrial Management Review*, vol. 11, no. 2, 1970.

Price, W. L.: Solving Goal-Programming Manpower Models Using Advanced Network Codes, *Journal of the Operational Research Society*, vol. 29, no. 12, 1978.

Ross, G. T., and A. A. Zoltners: Weighted Assignment Models and their Application, *Management Science*, vol. 25, no. 7, 1979.

Rothstein, M.: Scheduling Manpower by Mathematical Programming, *Industrial Engineering*, April 1972.

Schneider, D. P., and K. E. Kilpatrick: An Optimum Manpower Utilization Model for Health Maintenance Organizations, *Operations Research*, vol. 23, no. 5, 1975.

Segal, M.: The Operator-Scheduling Problem: A Network-Flow Approach, *Operations Research*, vol. 22, no. 4, 1974.

Segal, M., and D. B. Weinberger: Turfing, *Operations Research*, vol. 25, no. 3, 1977.

Shih, W., and J. A. Sullivan: Dynamic Course Scheduling for College Faculty via Zero-One Programming, *Decision Sciences*, vol. 8, no. 4, 1977.

Summers, E. L.: The Audit Staff Assignment Problem: A Linear Programming Analysis, *Accounting Review*, vol. 47, pp. 443–453, 1972.

Taylor, B. W., III, L. J. Moore, and E. R. Clayton: R & D Project Selector and Manpower Allocation with Integer Nonlinear Goal Programming, *Management Science*, vol. 28, no. 10, 1982.

Tien, J. M., and A. Kamiyama: On Manpower Scheduling Algorithms, *SIAM Review*, vol. 24, no. 3, 1982.

Tillett, P.: An Operations Research Approach to the Assignment of Teachers to Courses, *Socio-Economic Planning Sciences*, vol. 9, no. 3, 1975.

Trippi, R. R., A. W. Ash, and J. V. Ravenis, II: A Mathematical Approach to Large Scale Personnel Assignment, *Computers and Operations Research*, vol. 1, no. 1, 1974.

Trivedi, V. M.: A Mixed-Integer Goal Programming Model for Nursing Service Budgeting, *Operations Research*, vol. 29, no. 5, 1981.

13. Production Scheduling, Inventory Control, and Planning

Ahrsjoe, S., and S. Svedunger: A Production Planning System Based on Linear Programming, *OMEGA*, vol. 1, no. 4, 1973.

Barchi, R. H., F. T. Sparrow, and R. R. Vemuganti: Production, Inventory and Capacity Expansion Scheduling with Integer Variables, *Management Science*, vol. 21, no. 7, 1975.

Bellman, Richard E.: Mathematical Aspects of Scheduling Theory, *Journal of the Society for Industrial and Applied Mathematics*, vol. 4, no. 3, 1956.

Bitran, G. R., E. A. Haas, and A. C. Hax: Hierarchical Production Planning: A Two-Stage System, *Operations Research*, vol. 30, no. 2, 1982.

Bitran, G. R., and A. C. Hax: On the Design of Hierarchical Production Planning Systems, *Decision Sciences*, vol. 8, no. 1, 1977.

Byrd, J., Jr., and L. T. Moore: The Application of a Product Mix Linear Programming Model in Corporate Policy Making, *Management Science*, vol. 24, no. 15, 1978.

Charnes, Abraham, and William W. Cooper: Generalizations of the Warehousing Model, *Operational Research Quarterly*, vol. 6, no. 4, pp. 131–172, 1955.

————, William W. Cooper, and Donald Farr: Linear Programming and Profit Preference Scheduling for a Manufacturing Firm, *Journal of the Operations Research Society of America*, vol. 1, no. 3, 1953.

————, ————, and B. Mellon: A Model for Optimizing Production by Reference to Cost Surrogates, *Econometrica*, vol. 23, no. 3, 1955.

Christensen, J., and B. Obel: Simulation of Decentralized Planning in Two Danish Organizations Using Linear Programming Decomposition, *Management Science*, vol. 24, no. 15, 1978.

DeBoer, J., and J. VanderSloot: A Method of Cost-Controlled Production Planning, *Statistica Neerlandica* (Netherlands), vol. 16, no. 1, 1962.

Dorsey, R. C., T. J. Hodgson, and H. D. Ratliff: A Network Approach to a Multi-Facility, Multi-Product Production Scheduling Problem Without Backordering, *Management Science*, vol. 21, no. 7, 1975.

Fetter, R. B.: A Linear Programming Model for Long Range Capacity Planning, *Management Science*, vol. 7, no. 4, 1961.

Folie, M., and J. Tiffin: A Solution of a Multi-Product Manufacturing and Distribution Problem, *Management Science*, vol. 23, no. 3, 1976.

Geoffrion, A. M., and G. W. Graves: Scheduling Parallel Production Lines with Changeover Costs: Practical Application of a Quadratic Assignment/LP Approach, *Operations Research*, vol. 24, no. 4, 1976.

Glover, F., J. Hultz, and D. Klingman: Improved Computer-Based Planning Techniques, Parts I and II, *Interfaces*, vol. 8, no. 4, 1978, and vol. 9, no. 4, 1979.

Glover, F., J. Hultz, D. Klingman, and J. Stutz: Generalized Networks: A Fundamental Computer-Based Planning Tool, *Management Science*, vol. 24, no. 12, 1978.

Glover, F., G. Jones, D. Klingman, and J. Mote: An Integrated Production, Distribution, and Inventory Planning System, *Interfaces*, vol. 9, no. 5, 1979.

Glover, F., and D. Klingman: Network Application in Industry and Government, *AIIE Transactions*, vol. 9, no. 4, 1977.

Gomory, R. E., and B. P. Dzielinski: Optimal Programming of Lot Sizes, Inventory and Labor Allocation, *Management Science*, vol. 11, no. 9, 1965.

Hax, A. C., and D. Candea: "Production and Inventory Management," Prentice-Hall, Inc., Englewood Cliffs, N.J., 1984.

Horen, J. H.: Scheduling of Network Television Programs, *Management Science*, vol. 26, no. 4, 1980.

Johnson, Selmer M.: Optimal Two- and Three-Stage Production Schedules with Setup Times Included, *Naval Research Logistics Quarterly*, vol. 1, no. 1, 1954.

Kantorovich, L. V.: Mathematical Methods of Organizing and Planning Production, *Management Science*, vol. 6, no. 4, 1960.

Kelly, J. E., Jr.: Critical Path Planning and Scheduling: Mathematical Basis, *Operations Research*, vol. 9, no. 3, 1961.

King, J. R.: "Production Planning and Control," Pergamon Press, Oxford, England, 1975.

Koeningsberg, E: Some Industrial Applications of Linear Programming, *Operational Research Quarterly*, vol. 12, no. 2, 1961.

Lawler, E. L., and J. Labetoulle: Our Preemptive Scheduling of Unrelated Parallel Processors by Linear Programming, *Journal of the ACM*, vol. 25, no. 4, 1978.

Mairs, T. G., G. W. Wakefield, E. L. Johnson, and K. Spielberg: On a Production Allocation and Distribution Problem, *Management Science*, vol. 24, no. 15, 1978.

Markland, R. E., and R. J. Newett: Production Distribution Planning in a Large Scale Commodity Processing Network, *Decision Sciences*, vol. 7, no. 4, 1976.

May, J. G.: A Linear Program for Economic Lot Sizes Using Labor Priorities, *Management Science*, vol. 21, no. 3, 1974.

Miller, L. W.: Using Linear Programming to Derive Planning Horizons for a Production Smoothing Problem, *Management Science*, vol. 25, no. 12, 1979.

Miranda, C. F., E. J. Rhomberg, and V. M. R. Tummala: A Separable Convex Programming Model for Automobile Production, *Proceedings 1978 Northeast AIDS*.

Müller-Merbach, Heiner: The Optimum Allocation of Products to Machines by Linear Programming, *Fortschriftliche Betriebsfuhrung*, vol. 11, no. 1, 1962.

Murphy, F. H., and E. A. Stohn: A Mathematical Approach to the Scheduling of Sorting Operations, *Naval Research Logistics Quarterly*, vol. 25, no. 1, 1978.

Pinkus, C. E.: Optimal Design of Multi-Product Multi-Echelon Inventory Systems, *Decision Sciences*, vol. 6, no. 3, 1975.

Posner, M. E., and W. Szwarc: A Transportation Type Aggregate Production Model with Backordering, *Management Science*, vol. 29, no. 2, 1983.

Prabhakar, T.: A Production Scheduling Problem with Sequencing Considerations, *Management Science*, vol. 21, no. 1, 1974.

Rapoport, L. A., and W. P. Drews: Mathematical Approach to Long-Range Planning, *Harvard Business Review*, vol. 40, no. 3, 1962.

Ricketts, D. E., and R. K. Zimmer: A Dynamic Optimization Model for Planning in a Multi-Product Environment, *Decision Sciences*, vol. 6, no. 2, 1975.

Silver, E. A.: A Tutorial on Production Smoothing and Work Force Balancing, *Operations Research*, vol. 15, no. 6, 1967.

Smith, S. B.: Planning Transistor Production by Linear Programming, *Operations Research*, vol. 13, no. 1, 1965.

Von Lanzenauer, C. H.: A Production Scheduling Model by Bivalent Linear Programming, *Management Science*, vol. 17, no. 1, 1970.

Wright, F. K.: Dual Variables in Inventory Measurement, *The Accounting Review*, pp. 129–134, 1970.

14. Public Policy and Urban Affairs

Belford, P. C., and H. D. Ratliff: A Network Flow Model for Racially Balancing Schools, *Operations Research*, vol. 20, no. 3, 1972.

Beltrami, E., and L. Bodin: Network and Vehicle Routing for Municipal Waste Collection, *Journal of Networks*, vol. 4, no. 1, 1974.

Bessent, A., W. Bessent, J. Kennington, and B. Reagan: An Application of Mathematical Programming to Assess Productivity in the Houston Independent School District, *Management Science*, vol. 28, no. 12, 1983.

Bohigian, H.: "The Foundations and Mathematical Models of Operations Research with Extensions to the Criminal Justice System," Gazette Press, Yonkers, NY, 1971.

Bruno, J. E.: A Mathematical Programming Approach to School Finance, *Socio-Economic Planning Sciences*, vol. 3, no. 1, 1969.

————: Using Linear Programming Salary Evaluation Models in Collective Bargaining Negotiations with Teacher Unions, *Socio-Economic Planning Sciences*, vol. 3, no. 2, 1969.

————: A Mathematical Programming Approach to School System Finance, *Socio-Economic Planning Sciences*, vol. 3, no. 1, 1969.

Cheung, T.: An Interactive Graphic Display for Region Partitioning by Linear Programming, *Communications of the ACM*, vol. 17, no. 9, 1974.

Clark, R. M., and B. P. Helms: Decentralized Solid Waste Collection Facilities, Journal of the Sanitary Engineering Division, *Proceedings of the American Society of Civil Engineers*, 1970.

Courtney, J. F., Jr., T. D. Klastorin, and T. W. Ruefli: A Goal Programming Approach to Urban-Surburban Location Preferences, *Management Science*, vol., 18, no. 6, 1972.

Drake, A. W., R. L. Keeney, and P. M. Morse (eds.): "Analysis of Public Systems," The M. I. T. Press, Cambridge, MA, 1972.

Garfinkel, R. S., and G. L. Nemhauser: Optimal Political Districting by Implicit Enumeration Techniques, *Management Science*, vol. 16, no. 8, 1970.

Gass, S. I.: On the Division of Police Districts into Patrol Beats, *Proceedings, 1968 ACM National Conference*, Spartan Press, Washington, DC.

Gass, S. I., and R. L. Sisson (eds.): "A Guide to Models in Governmental Planning and Operations," Sauger Books, Potomac, MD, 1975.

Gleason, J.M.: A Set Covering Approach to Bus Stop Location, *OMEGA*, vol. 3, no. 5, 1975.

Golabi, K., R. M. Kulkarni, and G. B. Way: A Statewide Pavement Management System, *Interfaces*, vol. 12, no. 6, 1982.

Heckman, L. B., and H. M. Taylor: School Rezoning to Achieve Racial Balance: A Linear Programming Approach, *Socio-Economic Planning Science*, vol. 3, no. 2, 1969.

Holtzman, A. G.: Linear Programming and the Value of an Input to a Local Public School System, *Public Finance*, vol. 23, no. 3, 1968.

Ignall, E., P. Kolesar, and W. Walker: "Linear Programming Models of Crew Assignments for Refuse Collection," Report D-20497-NSF, New York City RAND Institute, New York, 1970.

Jenkins, L.: Parametric Mixed Integer Programming: An Application to Solid Waste Management, *Management Science*, vol. 28, no. 11, 1982.

Koenigsberg, E.: Mathematical Analysis Applied to School Attendance Areas, *Socio-Economic Planning Sciences*, vol. 1, no. 4, 1968.

Larson, R. C., and A. R. Odoni: "Urban Operations Research," Prentice-Hall, Inc., Englewood Cliffs, NJ, 1981.

Lee, S. M., and L. J. Moore: Multi-Criteria School Busing Models, *Management Science*, vol. 23, no. 7, 1977.

Liggett, R. S.: The Application of an Implicit Enumeration Algorithm to the School Desegregation Problem, *Management Science*, vol. 20, no. 2, 1973.

McGuire, C. B.: Some Team Models of a Sales Organization, *Management Science*, vol. 7, no. 2, 1961.

McKeown, P., and B. Workman: A Study in Using Linear Programming to Assign Students to Schools, *Interfaces*, vol. 6, no. 4, 1976.

McNamara, J.: Mathematical Programming Applications in Education, *Socio-Economic Planning Sciences*, vol. 7, no. 1, 1973.

Mills, G.: The Determination of Local Government Electoral Boundaries, *Operational Research Quarterly*, vol. 18, no. 3, 1967.

Nakamura, M.: Some Programming Problems in Population Projection, *Operations Research*, vol. 21, no. 5, 1973.

Newton, R. M., and W. H. Thomas: Design of School Bus Routes by Computer, *Socio-Economic Planning Sciences*, vol. 3, no. 1, 1969.

Plane, D. R., and T. E. Hendrick: Mathematical Programming and the Location of Fire Companies for the Denver Fire Department, *Operations Research*, vol. 25, no. 4, 1977.

Seward, S. M., D. R. Plane, and T. E. Hendrick: Municipal Resource Allocation: Minimizing the Cost of Fire Protection, *Management Science*, vol. 24, no. 16, 1978.

Urban Issues, *Management Science*, vol. 16, no. 12, 1970.

Urban Problems Issue, *Operations Research*, vol. 20, no. 3, 1972.

Walker, W. E.: Using the Set-Covering Problem to Assign Fire Companies to Fire Houses, *Operations Research*, vol. 22, no. 2, 1974.

Walker, W. E., J. M. Chaiken, and E. J. Ignall (eds.): "Fire Department Deployment Analysis: A Public Policy Analysis Case Study," Elsevier/North-Holland, New York, 1976.

Weaver, J. B.: "Fair and Equal Districts," National Municipal League, New York, 1967.

15. Structural Design

Charnes, Abraham, and Herbert J. Greenberg: Plastic Collapse and Linear Programming. Preliminary Report, abstracted in *Bulletin of the American Mathematical Society*, vol. 57, no. 6, p. 480, 1951.

Dorn, W. S., and Herbert J. Greenberg: "Linear Programming and Plastic Limit Analysis of Structures," 30 pp. (including tables), *Technical Report* 7, Carnegie Institute of Technology, Pittsburgh, PA, 1955.

Heyman, Jacques: Plastic Design of Beams and Plane Frames for Minimum Material Consumption, *Quarterly of Applied Mathematics*, vol. 8, no. 4, 1951.

Pang, J. S., I. Kaneko, and W. P. Hallman: On the Solution of Some (Parametric) Linear Complementary Problems with Applications to Portfolio Selection, Structural Engineering and Actuarial Graduation, *Mathematical Programming*, vol. 16, no. 3, 1979.

16. Traffic Analysis

Lansdowne, Z. F., and D. W. Robinson: Geographic Decomposition of the Shortest Path Problem, with an Application to the Traffic Assignment Problem, *Management Science*, vol. 28, no. 12, 1983.

Little, J. P. C.: The Synchronization of Traffic Signals by Mixed-Integer Linear Programming, *Operations Research*, vol. 14, no. 4, 1966.

Potts, R. B., and R. M. Oliver: "Flows in Transportation Networks," Academic Press, New York, 1972.

Tomlin, J. A.: A Mathematical Programming Model for the Combined Distribution-Assignment of Traffic, *Transportation Science*, vol. 5, no. 2, 1971.

17. Transportation and Networks

Appelgren, L. H.: A Column Generating Algorithm for a Ship Scheduling Problem, *Transportation Science*, vol. 3, no. 1, 1969.

————: Integer Programming Methods for a Vessel Scheduling Problem, *Transportation Science*, vol. 5, no. 1, 1971.

Assad, A. A.: Multicommodity Network Flows—A Survey, *Networks*, vol. 8, pp. 37–91, 1978.

Avramovich, D., T. M. Cook, G. D. Langston, and F. Sutherland: A Decision Support System for Fleet Management, *Interfaces*, vol. 12, no. 3, 1982.

Balinski, M. L., and R. E. Quandt: On an Integer Program for a Delivery Problem, *Operations Research*, vol. 12, no. 2, 1964.

Bennington, G., and S. Lubore: Resource Allocation for Transportation, *Naval Research Logistics Quarterly*, vol. 17, no. 4, 1970.

Bodin, L., B. Golden, A. Assad, and M. Ball: Routing and Scheduling of Vehicles and Crews: The State of the Art, *Computers and Operations Research*, vol. 10, no. 2, 1983.

Collins, M., L. Cooper, R. Helgason, J. Kennington, and L. Le Blanc: Solving the Pipe Network Analysis Problem Using Optimization Techniques, *Management Science*, vol. 24, no. 7, 1978.

Dantzig, G. B., and D. L. Johnson: Maximum Payloads per Unit Time Delivered through an Air Network, *Operations Research*, vol. 12, no. 2, 1964.

————, and J. H. Ramser: The Truck Dispatching Problem, *Management Science*, vol. 6, no. 1, 1959.

Elmaghraby, S. E.: The Theory of Networks and Management Science, Parts I and II, *Management Science*, vol. 17, nos. 1 and 2, 1970.

Evans, S. P.: A Relationship between the Gravity Model for Trip Distribution and the Transportation Problem in Linear Programming, *Transportation Science*, vol. 7, no. 1, 1973.

Flood, Merrill M.: Application of Transportation Theory to Scheduling a Military Tanker Fleet, *Journal of the Operations Research Society of America*, vol. 2, no. 2, 1954.

————: On the Hitchcock Distribution Problem, *Pacific Journal of Mathematics*, vol. 3, no. 2, 1953.

Fulkerson, D. R., and G. B. Dantzig: Computation of Maximal Flows in Networks, *Naval Research Logistics Quarterly*, vol. 2, no. 4, 1955.

Gavish, B., and P. Schweitzer: An Algorithm for Combining Truck Trips, *Transportation Science*, vol. 8, no. 1, 1974.

Glover, F., and D. Klingman: Network Application in Industry and Government, *AIIE Transactions*, vol. 9, no. 4, 1977.

Gould, F. J.: A Linear Programming Model for Cargo Movement Evaluation, *Transportation Science*, vol. 5, no. 4, 1971.

Haley, K. B.: A General Method of Solution for Special Structure Linear Programmes, *Operational Research Quarterly*, vol. 17, no. 1, 1966.

Harrison, H.: A Planning System for Facilities and Resources in Distribution Networks, *Interfaces*, vol. 9, no. 2, part 2, 1979.

Jones, G. A., and J. G. Wilson: Optimal Scheduling of Jobs on a Transmission Network, *Management Science*, vol. 25, no. 1, 1979.

Kennington, J. L.: A Survey of Linear Cost Multicommodity Network Flows, *Operations Research*, vol. 26, no. 2, 1978.

Klingman, D., and J. M. Mulvey (eds.): "Network Models and Associated Applications," Mathematical Programming Study 15, North-Holland Publishing Company, New York, 1981.

Lederman, J., L. Gleiberman, and J. F. Egan: Vessel Allocation by Linear Programming, *Naval Research Logistics Quarterly*, vol. 13, no. 3, 1966.

Mole, R. H.: Dynamic Optimization of Vehicle Size, *Operational Research Quarterly*, vol. 26, no. 1, 1975.

Nakagawa, Shizuaki: An Application of Operations Research to Optimization of Sailing Cost, *Operations Research as a Management Science* (Japan), vol. 6, no. 2, 1961.

Panagiotakopoulos, D.: Multi-Commodity Multi-Transformed Network Flows with an Application to Residuals Management, *Management Science*, vol. 22, no. 8, 1976.

Schell, E.: Distribution of a Product by Several Properties, pp. 615–642 of Directorate of Management Analysis [207].

Wright, J. W.: Reallocation of Housing by Use of Network Analysis, *Operational Research Quarterly*, vol. 26, no. 2, 1975.

18. Traveling-Salesman Problem

Bellmore, M., and G. L. Nemhauser: The Traveling Salesman Problem: A Survey, *Operations Research*, vol. 16, no. 3, 1968.

Christofides, N., and S. Eilon: Algorithms for Large-Scale Traveling Salesman Problems, *Operational Research Quarterly*, vol. 23, no. 4, 1972.

Dantzig, George B., Delbert R. Fulkerson, and Selmer Johnson: Solution of a Large-Scale Traveling-Salesman Problem, *Journal of the Operations Research Society of America*, vol. 2, no. 4, 1954.

Golden, B., L. Bodin, T. Doyle, and W. Stewart: Approximate Traveling Salesman Algorithms, *Operations Research*, vol. 28, no. 3, 1980.

Larson, R. C., and A. R. Odoni: "Urban Operations Research," Prentice-Hall, Inc., Englewood Cliffs, NJ, 1981.

Lenstra, J. K., and A. H. G. Rinnooy Kan: Some Simple Applications of the Traveling Salesman Problem, *Operational Research Quarterly*, vol. 26, no. 4, 1975.

Little, J. D. C., K. G. Murty, D. W. Sweeney, and C. Karel: An Algorithm for the Traveling-Salesman Problem, *Operations Research*, vol. 11, no. 6, 1963.

Parker, R. G., and R. L. Rardin: The Traveling Salesman Problem: An Update of Research, *Naval Research Logistics Quarterly*, vol. 30, no. 1, 1983.

19. University Management and Planning

Dyer, J. S., and J. M. Mulvey: An Integrated Optimization/Information System for Academic Departmental Planning, *Management Science*, vol. 22, no. 12, 1976.

Graves, R. J., and W. H. Thomas: A Classroom Location-Allocation Model for Campus Planning, *Socio-Economic Planning Sciences*, vol. 5, no. 3, 1971.

Hopkins, D. S., J.-C. Larréché, and W. F. Massy: Constrained Optimization of a University Administrator's Preference Function, *Management Science*, vol. 24, no. 4, 1977.

Hopkins, D. S., and W. F. Massy: "Planning Models for Colleges and Universities," Stanford University Press, Stanford, CA, 1981.

Kraft, D. H., and T. W. Hill, Jr.: The Journal Selection Problem in a University Library System, *Management Science*, vol. 19, no. 6, 1973.

Lee, S. M., and E. R. Clayton: A Goal Programming Model for Academic Resource Allocation, *Management Science*, vol. 18, no. 8, 1972.

Lee, S. M., and L. J. Moore: Optimizing University Admissions Planning, *Decision Sciences*, vol. 5, no. 3, 1974.

Ritzman, L., J. Bradford, and R. Jacobs: A Multiple Objective Approach to Space Planning for Academic Facilities, *Management Science*, vol. 25, no. 9, 1979.

Schroeder, R. G.: Resource Planning in University Management by Goal Programming, *Operations Research*, vol. 22, no. 4, 1974.

Walters, A., J. Mangold, and E. G. P. Haran: A Comprehensive Planning Model for Long-Range Academic Strategies, *Management Science*, vol. 22, no. 7, 1976.

Wehrung, D. A., D. S. P. Hopkins, and W. F. Massy: Interactive Preference Optimization for University Administrators, *Management Science*, vol. 24, no. 6, 1978.

20. Other Applications and General References

Arnold, L. R., and D. Botkin: "Portfolios to Satisfy Damage Judgements: A Linear Programming Approach, *Interfaces*, vol. 8, no. 2, 1978.

Aronofsky, J. S.: Growing Applications of Linear Programming, *Communications of the ACM*, vol. 7, no. 6, 1964.

Bajgier, S. M., and A. V. Hill: An Experimental Comparison of Statistical and Linear Programming Approaches to the Discriminant Problem, *Decision Sciences*, vol. 13, no. 4, 1982.

Balinski, M. L., and C. Lemarechal (eds.): "Mathematical Programming in Use," North-Holland Publishing Company, Amsterdam, 1978.

Beale, E. M. L. (ed.): "Applications of Mathematical Programming Techniques," American Elsevier Publishing Co., Inc., New York, 1970.

Charnes, A., M. J. L. Kirby, and A. S. Walters: Horizon Models for Social Development, *Management Science*, vol. 17, no. 4, 1970.

Christofides, N., A. Mingozzi, and P. Toth: Dynamic Loading and Unloading of Liquids into Tanks, *Operations Research*, vol. 28, no. 3, part II, 1980.

Cottle, R. W., and J. Karup (eds.): "Optimization Methods for Resource-Allocation," Proceedings of the NATO Conference, Elsinore, Denmark, English University Press, London, 1973.

Dantzig, George B., and Alan J. Hoffman: Dilworth's Theorem on Partially Ordered Sets, Paper 11 [pp. 207–214 (including references)] in "Linear Inequalities and Related Systems," *Annals of Mathematics Studies* 38, Princeton University Press, Princeton, NJ, 1956.

Economides, S. C., and W. T. Crawford: Radioisotope Production Scheduling in a Nuclear Reactor, *Decision Sciences*, vol. 13, No. 3, 1982.

Freeman, R. J., D. C. Gogerty, G. W. Graves, and R. B. S. Brooks: A Mathematical Model of Supply Support for Space Operations, *Operations Research*, vol. 14, no. 1, 1966.

Goffin, J.-L., and J.-M. Rousseau (eds.): "Mathematical Programming Study 20: Applications," North-Holland Publishing Company, Amsterdam, 1982.

Gray, P., and C. Cullinan-James: Applied Optimization—A Survey, *Interfaces*, vol. 6, no. 3, 1976.

Grinold, Richard C.: Mathematical Programming Methods of Pattern Classification, *Management Science*, vol. 19, no. 3, 1972.

Hoffman, Alan J., and Harold W. Kuhn: On Systems of Distinct Representatives, in "Linear Inequalities and Related Systems," *Annals of Mathematics Studies* 38, Princeton University Press, Princeton, NJ, 1956.

Johnson, M., A. A. Zoltners, and P. Sinha: An Allocation Model for Catalog Space Planning, *Management Science*, vol. 25, no. 2, 1979.

Kolesar, P. J.: Linear Programming and the Reliability of Multicomponent Systems, *Naval Research Logistics Quarterly*, vol. 14, no. 3, 1967.

Ladany, S. P.: Optimization of Pentathlon Training Plans, *Management Science*, vol. 21, no. 10, 1975.

Larson, R. C., and A. R. Odoni: "Urban Operations Research," Prentice-Hall, Inc., Englewood Cliffs, NJ, 1981.

Lasdon, L. S., and A. D. Waren: Survey of Nonlinear Programming Applications, *Operations Research*, vol. 28, no. 5, 1980.

Liittschwager, J. M., and C. Wang: Integer Programming Solution of a Classification Problem, *Management Science*, vol. 24, no. 14, 1978.

Moore, J. R.: A Linear Programming Approach to Scheduling Nuclear Isotope Production, *IEEE Transactions on Engineering Management*, vol. EM-20, no. 2, 1973.

Mulvey, J. M.: Reducing the U. S. Treasury Taxpayer Data Base by Optimization, *Interfaces*, vol. 10, no. 5, 1980.

Orden, Alex: Application of Linear Programming to Optical Filter Design (abstract), *Proceedings of the Second Symposium in Linear Programming*, vol. I, p. 185, 1955.

Raff, S., and J. P. Ignizio (eds.): Generalized Goal Programming, *Computers and Operations Research*, vol. 10, no. 4, 1983.

Salkin, H., and C. H. Lin: Aggregations of Subsidiary Firms for Minimal Unemployment Compensation Payments via Integer Programming, *Management Science*, vol. 25, no. 4, 1979.

Salkin, H. M., and J. Saha: "Studies in Linear Programming," North-Holland/American Elsevier, Amsterdam, 1975.

Waespy, C. M.: Linear Programming Solutions for Orbital Transfer Trajectories, *Operations Research*, vol. 18, no. 4, 1970.

Wagner, H. M., R. J. Giglio, and R. G. Glaser: Preventive Maintenance Scheduling by Mathematical Programming, *Management Science*, vol. 10, no. 2, 1964.

Wilson, R. C.: A Packaging Problem, *Management Science*, vol. 12, no. 4, 1965.

References

1. Abadie, J.: "Nonlinear Programming," North-Holland Publishing Company, Amsterdam, 1967.
2. Abadie, J.: "On the Decomposition Principle," ORC Report 63-20, Operations Research Center, University of California, Berkeley, CA, August 1963.
3. Abadie, J.: "Problèmes d'Optimisation," Tome II, Institut Blaise Pascal, Paris, June 1965.
4. Abadie, J., and A. C. Williams: Dual and Parametric Methods in Decomposition, pp. 149–158 of Graves and Wolfe [328].
5. Agmon, S.: The Relaxation Method for Linear Inequalities, *Canadian Journal of Mathematics*, vol. 6, 1954.
6. Ali, A., R. Helgason, J. Kennington, and H. Lall: "Primal Simplex Network Codes: State-of-the-Art Implementation Technology," Technical Report IEOR 76014, Department of Industrial Engineering and Operations Research, Southern Methodist University, Dallas, TX, 1977.
7. Ali, A., R. Helgason, J. Kennington, and H. Lall: Computational Comparison among Three Multicommodity Network Flow Algorithms, *Operations Research*, vol. 28, no. 4, 1980.
8. Almogy, Y., and O. Levin: The Fractional Fixed-Charge Problem, *Naval Research Logistics Quarterly*, vol. 18, no. 3, 1971.
9. Antosiewicz, H. A., and A. J. Hoffman: A Remark on the Smoothing Problem, *Management Science*, vol. 1, 1954–1955.
10. Arabeyre, J. P., J. Fearnley, F. C. Steiger, and W. Teather: The Airline Crew Scheduling Problem: A Survey, *Transportation Science*, vol. 3, no. 2, 1969.

11. Arrow, K. J., L. Hurwicz, and H. Uzawa (eds.): "Studies in Linear and Nonlinear Programming," Stanford University Press, Stanford, CA, 1958.

12. Arthanari, T. S., and Y. Dodge: "Mathematical Programming in Statistics," John Wiley & Sons, Inc., New York, 1981.

13. Aspvall, B., and R. E. Stone: Khachiyan's Linear Programming Algorithm, *Journal of Algorithms*, vol. 1, pp. 1–13, 1980.

14. Assad, A. A.: Multicommodity Network Flows—A Survey, *Networks*, vol. 8, pp. 37–91, 1978.

15. Avis, D., and V. Chvátal: "Notes on Bland's Pivoting Rule," Mathematical Programming Study 8, North-Holland Publishing Company, Amsterdam, 1978.

16. Azpeitia, A. G., and D. J. Dickinson: A Decision Rule in the Simplex Method that Avoids Cycling, *Numerische Mathematik*, vol. 6, pp. 324–331, 1964.

17. Bachem, A., M. Grötschel, and B. Korte (eds.): "Mathematical Programming: The State of the Art (Bonn 1982)," Springer-Verlag, Berlin, 1983.

18. Balachandran, V., and G. L. Thompson: An Operator Theory of Parametric Programming for the Generalized Transportation Problem, *Naval Research Logistics Quarterly*, vol. 22, nos. 1 and 2, 1975.

19. Balas, E.: An Additive Algorithm for Solving Linear Programs with Zero-One Variables, *Operations Research*, vol. 13, no. 4, 1965.

20. Balas, E.: Cutting Planes from Conditional Bounds: A New Approach to Set Covering, *Mathematical Programming*, vol. 12, pp. 19–36, 1980.

21. Balas, E., and A. Ho: Set Covering Algorithms Using Cutting Planes, Heuristics, and Subgradient Optimization: A Computational Study, *Mathematical Programming*, vol. 12, pp. 37–60, 1980.

22. Balas, E., and P. Ivǎnescu: On the Generalized Transportation Problem, *Management Science*, vol. 11, no. 2, 1964.

23. Balas, E., and C. H. Martin: Pivot and Complement—A Heuristic for 0–1 Programming, *Management Science*, vol. 26, no. 1, 1980.

24. Balas, E., and M. W. Padberg: Set Partitioning: A Survey, *SIAM Review*, vol. 18, no. 4, 1976.

25. Balas, E., and P. Toth: Branch and Bound Methods for the Traveling Salesman Problem, in Lawler et al. [463].

26. Balinski, M. L.: Integer Programming: Methods, Uses, Computation, *Management Science*, vol. 12, no. 3, 1965.

27. Balinski, M. L.: Fixed-Cost Transportation Problems, *Naval Research Logistics Quarterly*, vol. 8, pp. 41–54, 1961.

28. Balinski, M. L. (ed.): "Approaches to Integer Programming," Mathematical Programming Study 2, North-Holland/American Elsevier Publishing Company, Amsterdam, 1974.

29. Balinski, M. L., and R. H. Cobb: On Solving Large Structured Programs, *Proceedings of IFIP Congress 65*, vol. 2, Spartan Books, Washington, DC, 1965.

30. Balinski, M. L., and R. E. Gomory: A Primal Method for the Assignment and Transportation Problems, *Management Science*, vol. 10, no. 3, 1964.

31. Balinski, M. L., and E. Hellerman (eds.): "Computational Practice in Mathematical Programming," Mathematical Programming Study 4, North-Holland/American Elsevier Publishing Company, Amsterdam, 1975.

32. Balinski, M. L., and R. E. Quandt: On an Integer Program for a Delivery Problem, *Operations Research*, vol. 12, pp. 300–304, 1964.

33. Balinski, M. L., and A. W. Tucker: Duality Theory of Linear Programs: A Constructive Approach with Applications, *SIAM Review*, vol. 11, no. 3, 1969.

34. Balintfy, J. L.: Menu Planning by Computers, *Communications of The ACM*, vol. 7, 1964.

35. Balintfy, J. L.: Linear Programming Models for Menu Planning, in H. E. Smalley and J. R. Freeman (eds.), "Hospital Industrial Engineering," Reinhold Publishing Corporation, New York, 1966.

36. Bammi, D.: A Generalized-Indices Transportation Problem, *Naval Research Logistics Quarterly*, vol. 25, no. 4, 1978.

37. Barankin, E. W., and R. Dorfman: "On Quadratic Programming," University of California Publications in Statistics 2, University of California, Berkeley, CA, 1958.

38. Barker, V. A. (ed.): "Sparse Matrix Techniques," Springer-Verlag, New York, 1977.

39. Barnett, S.: Stability of the Solution to a Linear Programming Problem, *Operational Research Quarterly*, vol. 13, no. 3, 1962.

40. Barr, R., F. Glover, and D. Klingman: An Improved Version of the Out-of-Kilter Method and a Comparative Study of Computer Codes, *Mathematical Programming*, vol. 7, no. 1, 1974.

41. Barr, R., F. Glover, and D. Klingman: The Alternating Basis Algorithm for Assignment Problems, *Mathematical Programming*, vol. 13, no. 1, 1977.

42. Barr, R., F. Glover, and D. Klingman: The Generalized Alternating Path Algorithm for Transportation Problems, *European Journal of Operations Research*, vol. 2, pp. 137–144, 1978.

43. Barr, R., F. Glover, and D. Klingman: "Enhancements of Spanning Tree Labeling Procedures for Network Optimization," CCS 262, Center for Cybernetic Studies, The University of Texas, Austin, TX, 1979.

44. Barr, R., F. Glover, and D. Klingman: A New Optimization Method for Large-Scale Fixed Charge Transportation Problems, *Operations Research*, vol. 29, no. 3, 1981.

45. Barsov, A. S.: "What Is Linear Programming," D. C. Heath and Company, Boston, 1964.

46. Bartels, R. H., and G. H. Golub: The Simplex Method of Linear Programming Using LU Decomposition, *Communications of the ACM*, vol. 12, no. 5, 1969.

47. Bartholdi, J. J., III, J. B. Orlin, and H. D. Ratliff: Cyclic Scheduling via Integer Programs with Circular Ones, *Operations Research*, vol. 28, no. 5, 1980.

48. Baumol, W. J.: "Economic Theory and Operations Analysis," 2d ed., Prentice-Hall, Inc., Englewood Cliffs, NJ, 1965.

49. Baumol, W. J., and T. Fabian: Decomposition, Pricing for Decentralization and External Economies, *Management Science*, vol. 11, no. 1, 1964.

50. Bazaraa, M. S., and J. J. Jarvis: "Linear Programming and Network Flows," John Wiley & Sons, Inc., New York, 1977.

51. Bazaraa, M. S., and C. M. Shetty: "Nonlinear Programming," John Wiley & Sons, Inc., New York, 1979.

52. Beale, E. M. L.: Cycling in the Dual Simplex Algorithm, *Naval Research Logistics. Quarterly*, vol. 2, no. 4, 1955.

53. Beale, E. M. L.: An Alternative Method for Linear Programming, *Proceedings of the Cambridge Philosophical Society*, vol. 50, 1954.

54. Beale, E. M. L.: Survey of Integer Programming, *Operational Research Quarterly*, vol. 16, no. 2, 1965.

55. Beale, E. M. L.: The Simplex Method Using Pseudo-Basic Variables for Structured Linear Programming Problems, pp. 133–148 in Graves and Wolfe [328].

56. Beale, E. M. L.: Numerical Methods, chap. 7 in Abadie [1].

57. Beale, E. M. L.: On Quadratic Programming, *Naval Research Logistics Quarterly*, no. 6, 1959.

58. Beale, E. M. L.: "Mathematical Programming in Practice," John Wiley & Sons, Inc., New York, 1968.

59. Beale, E. M. L. (ed.): "Applications of Mathematical Programming Techniques," American Elsevier Publishing Company, Inc., New York, 1970.

60. Beale, E. M. L.: The Current Algorithmic Scope on Mathematical Programming, pp. 1–11 in Balinski and Hellerman [31].

61. Beale, E. M. L.: "Optimization Techniques Based on Linear Programming," pp. 447–466 in Dixon [209].

62. Beale, E. M. L., and J. A. Tomlin: Special Facilities in a General Mathematical Programming System for Non-Convex Problems Using Ordered Sets of Variables, pp. 447–454 in J. Lawrence (ed.), *Proceedings of the Fifth International Conference on Operational Research*, Tavistock, London, 1970.

63. Bellmore, M., and R. S. Frank: "On the Fixed-Charge Hitchcock Transportation Problem," Report HBS-73-17, Division of Research, Graduate School of Business Administration, Harvard University, Cambridge, MA, 1973.

64. Bellmore, M., and S. Hong: Transformation of Multisalesmen Problem to the Standard Traveling Salesman Problem, *Journal of the ACM*, vol. 21, no. 3, 1974.

65. Beltrami, E. J.: "Models for Public Systems Analysis," Academic Press, New York, 1977.

66. Benayoun, R., J. De Montgolfier, J. Tergny, and O. Laritchev: Linear Programming with Multiple Objective Functions: STEP Method (STEM), *Mathematical Programming*, vol. 1, no. 3, 1971.

67. Benders, F., Jr., A. R. Catchpole, and C. Kuiken: "Discrete Variables Optimization Problems," Koninklijke/Shell-Lab., Amsterdam, 1960.

68. Benders, J. F.: Partitioning Procedures for Solving Mixed Variable Programming Problems, *Numerische Mathematik*, 4, pp. 238–252, 1962.

69. Benichou, M., J. M. Gauthier, P. Girodet, G. Hentges, G. Ribière, and O. Vincent: Experiments in Mixed-Integer Linear Programming, *Mathematical Programming*, vol. 1, no. 1, 1971.

70. Benichou, M., J. M. Gauthier, G. Hentges, and G. Ribière: The Efficient Solution of Large-Scale Linear Programming Problems—Some Algorithmic Techniques and Computational Methods, *Mathematical Programming*, vol. 13, no. 3, 1977.

71. Ben-Israel, A., A. Ben-Tal, and S. Zlobec: Optimality Conditions in Convex Programming, pp. 153–169 in Prékopa [552].

72. Ben-Israel, A., A. Ben-Tal, and S. Zlobec: "Optimality in Nonlinear Programming," John Wiley & Sons, Inc., New York, 1981.

73. Bentz, W. F., L. A. Sherr, and R. E. Miller: Sensitivity Analysis with Interaction Effects, *Decision Sciences*, vol. 7, no. 3, 1976.

74. Berge, C., and A. Ghouila-Houri: "Programming, Games and Transportation Networks," John Wiley & Sons, Inc., New York, 1962.

75. Bertsekas, D. P.: A New Algorithm for the Assignment Problem, *Mathematical Programming*, vol. 21, no. 2, 1981.

76. Bishop, G. T.: On a Problem of Production Scheduling, *Operations Research*, vol. 5, no. 1, 1957.

77. Bisschop, J., and A. Meerans: Matrix Augmentation and Partitioning in the Updating of the Basis Inverse, *Mathematical Programming*, vol. 13, no. 3, 1977.

78. Bitran, G. R.: Linear Multiple Objective Problems with Interval Coefficients, *Management Science*, vol. 26, no. 7, 1980.

79. Bixby, R. E., and W. H. Cunningham: Converting Linear Programs to Network Problems, *Mathematics of Operations Research*, vol. 5, no. 3, 1980.

80. Bland, R. G.: New Finite Pivoting Rules for the Simplex Method, *Mathematics of Operations Research*, vol. 2, no. 2, 1977.

81. Bland, R. G.: The Allocation of Resources by Linear Programming, *Scientific American*, June 1981.

82. Bland, R. G., D. Goldfarb, and M. J. Todd: The Ellipsoid Method: A Survey, *Operations Research*, vol. 29, no. 6, 1981.

83. Boot, J. C.: "Quadratic Programming," Rand McNally & Company, Chicago, 1964.

84. Borgwardt, K.-H.: Some Distribution-Independent Results About the Asymptotic Order of the Average Number of Pivot Steps of the Simplex Method, *Mathematics of Operations Research*, vol. 7, no. 3, 1982.

85. Bowman, E. H.: Production Scheduling by the Transportation Method of Linear Programming, *Operations Research*, vol. 4, 1956.

86. Bracken, J., and G. P. McCormick: "Selected Applications of Nonlinear Programming," John Wiley & Sons, Inc., New York, 1968.

87. Bradley, G. H.: Equivalent Integer Programs and Canonical Problems, *Management Science*, vol. 17, no. 5, 1971.

88. Bradley, G. H.: Transformation of Integer Programs to Knapsack Problems, *Discrete Mathematics*, vol. 1, pp. 29–45, 1971.

89. Bradley, G. H.: Survey of Deterministic Networks, *AIIE Transactions*, September 1975.

90. Bradley, G. H., G. G. Brown, and G. W. Graves: Design and Implementation of Large Scale Primal Transshipment Algorithms, *Management Science*, vol. 24, no. 1, 1977.

91. Bradley, S. P., and S. C. Frey, Jr.: Fractional Programming with Homogeneous Functions, *Operations Research*, vol. 22, no. 2, 1974.

92. Bradley, S. P., A. C. Hax, and T. L. Magnanti: "Applied Mathematical Programming," Addison-Wesley Publishing Company, Reading MA, 1977.

93. Braitsch, R. J., Jr.: A Computer Comparison of Four Quadratic Programming Algorithms, *Management Science*, vol. 18, no. 11, 1972.

94. Brandt, A., and J. Intrator: Fast Algorithms for Long Transportation Problems, *Computers and Operations Research*, vol. 5, pp. 263–271, 1978.

95. Brigham, G.: A Classroom Example of Linear Programming, *Operations Research*, vol. 7, no. 4, 1959.

96. Bronson, R.: "Operations Research," Schaum's Outline Series, McGraw-Hill Book Company, New York, 1982.

97. Brown, G. G., and D. S. Thomen: Automatic Identification of Generalized Upper Bounds in Large-Scale Optimization Models, *Management Science*, vol. 26, no. 11, 1980.

98. Brown, G. W.: Iterative Solution of Games by Fictitious Play, chap. 24 of Koopmans [437].

99. Buck, R. C.: "Advanced Calculus," 2d ed., McGraw-Hill Book Company, New York, 1965.

100. Bulfin, R. L., R. G. Parker, and C. M. Shetty: Computational Results with a Branch-and-Bound Algorithm for the General Knapsack Problem, *Naval Research Logistics Quarterly*, vol. 26, no. 1, 1979.

101. Bunch, J. R., and D. J. Rose: "Sparse Matrix Computations," Academic Press, New York, 1976.

102. Busacker, R. G., and T. L. Saaty: "Finite Graphs and Networks," McGraw-Hill Book Company, New York, 1965.

103. Byrne, J. L., and R. B. Potts: Scheduling of Toll Collectors, *Transportation Science*, vol. 7, no. 3, 1973.

104. Cahn, A. J., Jr.: The Warehouse Problem, *Bulletin of the American Mathematical Society*, vol. 54, November 1948.

105. Carr, C. R., and C. W. Howe: "Quantitative Decision Procedures in Management and Economics," McGraw-Hill Book Company, New York, 1964.

106. Catchpole, A. R.: The Application of Linear Programming to Integrated Supply Problems in the Oil Industry, *Operational Research Quarterly*, vol. 13, no. 2, 1962.

107. Chadha, S. S.: A Decomposition Principle for Fractional Programming, *Opsearch*, vol. 4, no. 3, 1967.

108. Chandrasekaran, R., S. N. Kabadi, and K. G. Murty: Some NP-Complete Problems in Linear Programming, *Operations Research Letters*, vol. 1, no. 3, 1982.

109. Chankong, V., and Y. Y. Haimes: "Multiobjective Decision Making," North-Holland Publishing Company, New York, 1983.

110. Charnes, A.: Optimality and Degeneracy in Linear Programming, *Econometrica*, vol. 20, 1952.

111. Charnes, A., and W. W. Cooper: The Stepping Stone Method of Explaining Linear Programming Calculations in Transportation Problems, *Management Science*, vol. 1, 1954–1955.

112. Charnes, A., and W. W. Cooper: "Management Models and Industrial Application of Linear Programming," vols. I and II, John Wiley & Sons, Inc., New York, 1960.

113. Charnes, A., and W. W. Cooper: A Network Interpretation and a Directed Subdual Algorithm for Critical Path Scheduling, *Journal of Industrial Engineering*, vol. 13, no. 4, 1962.

114. Charnes, A., and W. W. Cooper: Deterministic Equivalents for Optimizing and Satisficing under Chance Constraints, *Operations Research*, vol. 11, no. 1, 1963.

115. Charnes, A., and W. W. Cooper: Programming with Linear Fractional Functions, *Naval Research Logistics Quarterly*, vol. 9, nos. 3, 4, 1962.

116. Charnes A., and W. W. Cooper: Goal Programming and Multiple Objective Optimizations, *European Journal of Operational Research*, vol. 1, pp. 39–54, 1977.

117. Charnes, A., W. W. Cooper, and R. O. Ferguson: Optimal Estimation of Executive Compensation by Linear Programming, *Management Science*, vol. 1, no. 2, 1955.

118. Charnes, A., W. W. Cooper, and A. Henderson: "Introduction to Linear Programming," John Wiley & Sons, Inc., New York, 1953.

119. Charnes, A., W. W. Cooper, and B. Mellon: Blending Aviation Gasolines—A Study in Programming Interdependent Activities in an Integrated Oil Company, *Econometrica*, vol. 20, 1952.

120. Charnes, A., W. W. Cooper, and G. L. Thompson: Some Properties of Redundant Constraints and Extraneous Variables in Direct and Dual Linear Programming Problems, *Operations Research*, vol. 10, no. 5, 1962.

121. Charnes, A., F. Glover, and D. Klingman: A Note on the Distribution Problem, *Operations Research*, vol. 18, no. 6, 1970.

122. Charnes, A., and C. E. Lemke: "The Bounded Variables Problem," *ONR Research Memorandum* 10, Graduate School of Industrial Administration, Carnegie Institute of Technology, Pittsburgh, PA, 1954.

123. Charnes, A., and C. E. Lemke: Minimization of Nonlinear Separable Functions, *Naval Research Logistics Quarterly*, vol. 1, no. 4, 1954.

124. Cheung, T.-Y.: Computational Comparison of Eight Methods for the Maximum Network Flow Problem, *ACM Transactions on Mathematical Software*, vol. 6, no. 1, 1980.

125. Childress, R. L.: "Sets, Matrices and Linear Programming," Prentice-Hall, Inc., Englewood Cliffs, NJ, 1974.

126. Christofides, N., and S. Korman: A Computational Survey of Methods for the Set Covering Problem, *Management Science*, vol. 21, no. 5, 1975.

127. Chung, An-min: "Linear Programming," Charles E. Merrill Books, Inc., Columbus, OH, 1963.

128. Chvátal, V.: "Linear Programming," W. H. Freeman and Company, New York, 1983.

129. Clark F. E.: Remark on the Constraint Sets in Linear Programming, *American Mathematical Monthly*, vol. 68, pp. 351–352, 1961.

130. Cohon, J. L: "Multiobjective Programming and Planning," Academic Press, New York, 1978.

131. Collatz, L., and W. Wetterling: "Optimization Problems," Springer-Verlag, New York, 1975.

132. Cooper, L.: "Applied Nonlinear Programming for Engineers and Scientists," Aloray Publishers, Inc., Englewood, NJ, 1974.

133. Cooper, L., and C. Drebes: An Approximate Solution Method for the Fixed Charge Problem, *Naval Research Logistics Quarterly*, vol. 14, no. 1, 1967.

134. Cooper, L., and J. Kennington: Nonextreme Point Solution Strategies for Linear Programs, *Naval Research Logistics Quarterly*, vol. 26, no. 3, 1979.

135. Cooper, L., and D. Steinberg: "Introduction to Methods of Optimization," W. B. Saunders Company, Philadelphia, PA, 1970.

136. Cooper, M. W.: A Survey of Methods for Pure Nonlinear Integer Programming, *Management Science*, vol. 27, no. 3, 1981.

137. Cooper, W. W., and A. Charnes: Linear Programming, *Scientific American*, August 1954.

138. Cottle, R. W., and G. B. Dantzig: Complementary Pivot Theory of Mathematical Programming, *Linear Algebra and Its Applications*, vol. 1, pp. 103–125, 1968.

139. Cottle, R. W., and G. B. Dantzig: A Generalization of the Linear Complementary Problem, *Journal of Combinatorial Theory*, 1969.

140. Cottle, R. W., and J. Krarup: "Optimization Methods for Resource Allocation," The English Universities Press, Ltd., London, 1974.

141. Cottle, R. W., and C. E. Lemke: "Nonlinear Programming," American Mathematical Society, Providence, RI, 1976.

142. Courtillot, M.: On Varying All the Parameters in a Linear-Programming Problem and Sequential Solution of a Linear-Programming Problem, *Operations Research*, vol. 10, no. 4, 1962.

143. Crowder, H., and J. M. Hattingh: Partially Normalized Pivot Selection in Linear Programming, pp. 12–25 in Balinski and Hellerman [31].

144. Crowder, H., E. L. Johnson, and M. W. Padberg: "Solving Large-Scale Zero-One Linear Programming Problems," IBM Research Report RC8888 (No. 38963), IBM, Yorktown Heights, NY, 1981.

145. Crowder, H., and M. W. Padberg: Solving Large-Scale Symmetric Travelling Salesman Problems to Optimality, *Management Science*, vol. 26, no. 5, 1980.

146. Cunningham, W. H.: A Network Simplex Method, *Mathematical Programming*, vol. 11, no. 2, 1976.

147. Cunningham, W. H.: Theoretical Properties of the Network Simplex Algorithm, *Mathematics of Operations Research*, vol. 4, no. 2, 1979.

148. Cunningham, W. H., and J. G. Klincewicz: On Cycling in the Network Simplex Method, *Mathematical Programming*, vol. 26, no. 2, 1983.

149. Cutler, Leola, and P. Wolfe: "Experiments in Linear Programming," pp. 177–200 in Graves and Wolfe [328].

150. Dakin, R. J.: A Tree Search Algorithm for Mixed Integer Programming Problems, *The Computer Journal*, vol. 8, pp. 250–255, 1965.

151. Danø, S.: "Linear Programming in Industry," Springer-Verlag OHG, Vienna, 1963.

152. Dantzig, G. B.: "Linear Programming and Extensions," Princeton University Press, Princeton, NJ, 1963.

153. Dantzig, G. B.: Maximization of a Linear Function of Variables Subject to Linear Inequalities, chap. 21 in Koopmans [437].

154. Dantzig, G. B.: Application of the Simplex Method to a Transportation Problem, chap. 23 in Koopmans [437].

155. Dantzig, G. B.: A Proof of the Equivalence of the Programming Problem and the Game Problem, chap. 20 in Koopmans [437].

156. Dantzig, G. B.: "Block Triangular Systems in Linear Programming," RAND Report RM-1273, The RAND Corporation, Santa Monica, CA, 1954.

157. Dantzig, G. B.: "Computational Algorithm of the Revised Simplex Method," RAND Report RM-1266, The RAND Corporation, Santa Monica, CA, 1953.

158. Dantzig, G. B.: "The Dual Simplex Algorithm," RAND Report RM-1270, The RAND Corporation, Santa Monica, CA, 1954.

159. Dantzig, G. B.: "Composite Simplex-Dual Simplex Algorithm—I," RAND Report RM-1274, The RAND Corporation, Santa Monica, CA, 1954.

160. Dantzig, G. B.: "Variables with Upper Bounds in Linear Programming," RAND Report RM-1271, The RAND Corporation, Santa Monica, CA, 1954.

161. Dantzig, G. B.: "Linear Programming under Uncertainty," RAND Report RM-1374, The RAND Corporation, Santa Monica, CA, 1954.

162. Dantzig, G. B.: Developments in Linear Programming, in Directorate of Management Analysis [207].

163. Dantzig, G. B.: Discrete-variable Extremum Problems, *Operations Research*, vol. 5, no. 2, 1957.

164. Dantzig, G. B.: On the Significance of Solving Linear Programming Problems with Some Integer Variables, *Econometrica*, vol. 28, no. 1, 1960.

165. Dantzig, G. B.: "On Integer and Partial Integer Linear Programming Problems," RAND Report P-1410, The RAND Corporation, Santa Monica, CA, 1958.

166. Dantzig, G. B.: Solving Linear Programs in Integers, "Notes on Linear Programming," Part XLVII, RM-2209, The RAND Corporation, Santa Monica, CA, 1958.

167. Dantzig, G. B.: On the Status of Multistage Linear Programming Problems, *Management Science*, vol. 6, no. 1, 1959.

168. Dantzig, G. B.: Optimal Solution of a Dynamic Leontief Model with Substitution, *Econometrica*, vol. 23, July 1955.

169. Dantzig, G. B.: Compact Basis Triangularization for the Simplex Method, pp. 125–132 in Graves and Wolfe [328].

170. Dantzig, G. B.: "Quadratic Programming: A Variant of the Wolfe-Markowitz

Algorithms," Research Report no. 2, Operations Research Center, University of California, Berkeley, CA, 1960.

171. Dantzig, G. B.: Application of Generalized Linear Programming to Control Theory, chap. 12 in Abadie [1].

172. Dantzig, G. B.: Note on Solving Linear Programs in Integers, *Naval Research Logistics Quarterly*, vol. 6, no. 1, 1959.

173. Dantzig, G. B.: Inductive Proof of the Simplex Method, *IBM Journal of Research and Development*, vol. 14, no. 5, 1960.

174. Dantzig, G. B.: "A Hospital Admission Problem," Technical Report 69-15, Operations Research House, Stanford University, Stanford, CA, December 1969.

175. Dantzig, G. B.: Large-Scale Systems and the Computer Revolution, pp. 51–72 in Kuhn [447].

176. Dantzig, G. B.: "Are Dual Variables Prices? If Not, How to Make them More So," Technical Report SOL 78-6, Department of Operations Research, Stanford University, Stanford, CA, 1978.

177. Dantzig, G. B.: "Comments on Khachian's Algorithm for Linear Programming," Technical Report SOL 79-22, Department of Operations Research, Stanford University, Stanford, CA, 1979.

178. Dantzig, G. B.: Reminiscences About the Origins of Linear Programming, *Operations Research Letter*, vol. 1, no. 2, 1982.

179. Dantzig, G. B.: Large Scale Linear Programming, pp. 77–92 in Dantzig and Veinott [193].

180. Dantzig, G. B., and R. W. Cottle: Positive (semi-) definite Programming, chap. 4 in Abadie [1].

181. Dantzig, G. B., M. A. H. Dempster, and M. Kallio (eds.): "Large-Scale Linear Programming," volumes 1 and 2, International Institute for Applied Systems Analysis, Laxenburg, Austria, 1981.

182. Dantzig, G. B., and B. C. Eaves (eds.): "Studies in Optimization," volume 10, The Mathematical Association of America, Washington, DC, 1974.

183. Dantzig, G. B., L. R. Ford, Jr., and D. R. Fulkerson: "A Primal-Dual Algorithm," RAND Report RM-1709, The RAND Corporation, Santa Monica, CA, 1956.

184. Dantzig, G. B., D. R. Fulkerson, and S. M. Johnson: Solution of a Large-Scale Traveling-Salesman Problem, *Operations Research*, vol. 2, no. 4, 1954.

185. Dantzig, G. B., D. R. Fulkerson, and S. M. Johnson: On a Linear Programming, Combinatorial Approach to the Traveling Salesman Problem, *Operations Research*, vol. 7, no. 1, 1959.

186. Dantzig, G. B., and S. Johnson: A Production Smoothing Problem, pp. 151–176 in Directorate of Management Analysis [207].

187. Dantzig, G. B., and W. Orchard-Hays: "Alternate Algorithm for the Revised Simplex Method," RAND Report RM-1268, The RAND Corporation, Santa Monica, CA, 1953.

188. Dantzig, G. B., W. Orchard-Hays, and G. Waters: "Product-Form Tableau for Revised Simplex Method," RAND Report RM-1268A, The RAND Corporation, Santa Monica, CA, 1954.

189. Dantzig, G. B., and A. Orden: A Duality Theorem Based on the Simplex Method, pp. 51–55 in Directorate of Management Analysis [205].

190. Dantzig, G. B., and A. Orden: "Duality Theorems," RAND Report RM-1265, The RAND Corporation, Santa Monica, CA, October 1953.

191. Dantzig, G. B., A. Orden, and P. Wolfe: "Generalized Simplex Method for Minimizing a Linear Form under Linear Inequality Restraints," RAND Report RM-1264, The RAND Corporation, Santa Monica, CA, 1954.

192. Dantzig, G. B., and R. M. Van Slyke: "Generalized Upper Bounded Techniques for Linear Programming," ORC Reports 64-17 and 64-18, Operations Research Center, University of California, Berkeley, CA, 1964; also see Generalized Upper Bounded Techniques, *Journal of Computer System Science*, vol. 1, pp. 213–226, 1967.

193. Dantzig, G. B., and A. F. Veinott, Jr.: "Mathematics of the Decision Sciences," Part 1, American Mathematical Society, Providence, RI, 1968.

194. Dantzig, G. B., and P. Wolfe: A Decomposition Principle for Linear Programs, *Operations Research*, vol. 8, no. 1, 1959.

195. Dantzig, G. B., and P. Wolfe: The Decomposition Algorithm for Linear Programs, *Econometrica*, vol. 29, no. 4, 1961.

196. Dantzig, G. B., and P. Wolfe: The Decomposition Algorithm for Linear Programs, pp. 160–174 in Dantzig and Eaves [182].

197. Dempster, M. A. H. (ed.): "Stochastic Programming," Academic Press, New York, 1980.

198. Demuth, O.: A Remark on the Transport Problem, *Čas. pěst. mat.*, vol. 86, pp. 103–110, 1961.

199. Dennis, J. B.: A High-Speed Computer Technique for the Transportation Problem, *Journal of the Association of Computing Machinery*, vol. 5, no. 2, 1958.

200. Dennis, J. B.: "Mathematical Programming and Electrical Networks," John Wiley & Sons, Inc., New York, 1959.

201. Deutsch, M. L.: Letter to the Editor, *Communications of the ACM*, vol. 2, no. 3, 1959.

202. Dickson, J. C., and F. P. Frederick: A Decision Rule for Improved Efficiency in Solving Linear Programming Problems with the Simplex Algorithm, *Communications of the ACM*, vol. 3, September 1960.

203. Dijkstra, E. W.: A Note on Two Problems in Connexion with Graphs, *Numerische Mathematik*, vol. 1, pp. 269–271, 1959.

204. Dinkelbach, W.: "Sensitivitatsanalysen und parametrische Programmierung," Springer-Verlag OHG, Berlin, 1969.

205. Directorate of Management Analysis: "Symposium on Linear Inequalities and Programming," A. Orden and L. Goldstein (eds.), DCS/Comptroller, Headquarters U.S. Air Force, Washington, DC, April 1952.

206. Directorate of Management Analysis: "The Application of Linear Programming Techniques to Air Force Problems," DCS/Comptroller, Headquarters U.S. Air Force, Washington, DC, December 1954.

207. Directorate of Management Analysis: "Proceedings of the Second Symposium in Linear Programming," H. Antosiewicz (ed.), vols. 1 and 2, DCS/Comptroller, Headquarters U.S. Air Force, Washington, DC, January 1955.

208. Dirickx, Y. M. I., and M. R. Rao: Networks with Gains in Discrete Dynamic Programming, *Management Science*, vol. 20, no. 11, 1974.

209. Dixon, L. C. W.: "Optimization in Action," Academic Press, New York, 1976.

210. Dobkin, D. P., and S. P. Reiss: The Complexity of Linear Programming, *Theoretical Computer Science*, vol. 11, pp. 1–18, 1980.

211. Doig, A. G.: "The Minimum Number of Basic Feasible Solutions to a Transport Problem," *Operational Research Quarterly*, vol. 14, no. 4, 1963.

212. Dorfman, R.: Mathematical, or "Linear," Programming, *American Economic Review*, vol. 43, 1953.

213. Dorfman, R.: "Application of Linear Programming to the Theory of the Firm," University of California Press, Berkeley, CA, 1951.

214. Dorfman, R., P. A. Samuelson, and R. Solow: "Linear Programming and Economic Analysis," McGraw-Hill Book Company, New York, 1958.

215. Dorn, W. S.: Nonlinear Programming: A Survey, *Management Science*, vol. 9, no. 2, 1963.

216. Dorn, W. S.: Duality in Quadratic Programming, *Quarterly of Applied Mathematics*, vol. 18, 1960.

217. Dorn, W. S.: A Duality Theorem for Convex Programs, *IBM Journal of Research and Development*, vol. 4, no. 4, 1960.

218. Dorn, W. S.: "Linear Fractional Programming," Research Report RC 830, IBM Research Center, Yorktown Heights, NY, November 1962.

219. Dreyfus, S. E.: An Appraisal of Some Shortest Path Algorithms, *Operations Research*, vol. 17, no. 3, 1969.

220. Dreyfus, S. E., and K. L. Prather: "Improved Algorithms for Knapsack Problems," Report ORC 70-30, Operations Research Center, University of California, Berkeley, CA, 1970.

221. Driebeek, N. J., "Applied Linear Programming," Addison-Wesley Publishing Company, Inc., Reading, MA, 1969.

222. Duff, I. S. (ed.): "Sparse Matrices and their Uses," Academic Press, New York, 1981.

223. Duff, I. S., and G. W. Stewart (eds.): "Sparse Matrix Proceedings 1978," SIAM Publications, Philadelphia, PA, 1979.

224. Duffin, R. T., E. L. Peterson, and C. Zener: "Geometric Programming," John Wiley & Sons, Inc., New York, 1967.

225. Durbin, E. P., and D. M. Kroenke: "The Out-of-Kilter Algorithm: A Primer," RAND Report RM-5472-PR, The RAND Corporation, Santa Monica, CA, 1967.

226. Dwyer, P. S.: "Linear Computations," John Wiley & Sons, Inc., New York, 1951.

227. Dwyer, P. S.: Solution to the Personnel Classification Problem with the Method of Optimal Regions, *Psychometrika*, vol. 19, 1954.

228. Eaves, B. C.: The Linear Complementary Problem, *Management Science*, vol. 17, no. 6, 1971.

229. Eaves, C. B.: On Quadratic Programming, *Management Science*, vol. 17, no. 11, 1971.

230. Ecker, J. G., N. S. Hegner, and I. A. Kouada: Generating All Maximal Efficient Faces for Multiple Objective Linear Programs, *Journal of Optimization Theory and Applications*, vol. 30, no. 3, 1980.

231. Ecker, J. G., and I. A. Kouada: Finding Efficient Points for Linear Multiple Objective Programs, *Mathematical Programming*, vol. 8, no. 3, 1975.

232. Ecker, J. G., and I. A. Kouada: Finding All Efficient Extreme Points for Multiple Objective Linear Programs, *Mathematical Programming*, vol. 14, no. 2, 1978.

233. Edmonds, J., and R. M. Karp: Theoretical Improvements in Algorithmic Efficiency for Network Flow Problems, *Journal of the ACM*, vol. 19, no. 2, 1972.

234. Egerváry, E.: On Combinational Properties of Matrices, *Matematikaés Fizikai Lapok*, vol. 38, 1931; translated as "On Combinatorial Properties of Matrices," by H. W. Kuhn, Office of Naval Research Logistics Project Report, Department of Mathematics, Princeton University, Princeton, NJ, 1953.

235. Eisemann, K.: The Trim Problem, *Management Science*, vol. 3, no. 3, 1957.
236. Eisemann, K.: The Primal-Dual Method for Bounded Variables, *Operations Research*, vol. 12, no. 1, 1964.
237. Eisemann, K., and J. R. Lourie: "The Machine Loading Problem," IBM Applications Library, New York, 1959.
238. Elam, J., F. Glover, and D. Klingman: A Strongly Convergent Primal Simplex Algorithm for Generalized Networks, *Mathematics of Operations Research*, vol. 4, no. 1, 1979.
239. Evans, J. P., and R. E. Steuer: A Revised Simplex Method for Linear Multiple Objective Programs, *Mathematical Programming*, vol. 5, no. 1, 1973.
240. Evans, J. R.: The Simplex Method for Integral Multicommodity Networks, *Naval Research Logistics Quarterly*, vol. 25, no. 1, 1978.
241. Evans, J. R.: On Equivalent Representatives of Certain Multicommodity Networks as Single Commodity Flow Problems, *Mathematical Programming*, vol. 15, no. 1, 1978.
242. Evans, J. R.: A Single-Commodity Transformation for Certain Multicommodity Networks, *Operations Research*, vol. 26, no. 4, 1978.
243. Evans, J. R., and N. R. Baker: Degeneracy and the (MIS) Interpretation of Sensitivity Analysis in Linear Programming, *Decision Science*, vol. 13, no. 2, 1982.
244. Faddeeva, V. N.: "Computational Methods of Linear Algebra," Dover Publications, Inc., New York, 1959.
245. Falk, J. E.: Exact Solution of Inexact Linear Programs, *Operations Research*, vol. 24, no. 4, 1976.
246. Fenchel, W.: "Convex Cones, Sets, and Functions," Princeton University Department of Mathematics, Princeton, NJ, 1953.
247. Fiacco, A. V., and G. P. McCormick: The Sequential Unconstrained Minimization Technique for Nonlinear Programming—A Primal-Dual Method, *Management Science*, vol. 10, no. 4, 1964.
248. Fiacco, A. V., and G. P. McCormick: "Nonlinear Programming," John Wiley & Sons, Inc., New York, 1968.
249. Fisher, Marshall L.: Lagrangian Relaxation Method for Solving Integer Programming Problems, *Management Science*, vol. 27, no. 1, 1981.
250. Flood, M. M.: A Transportation Algorithm and Code, *Naval Research Logistics Quarterly*, vol. 8, no. 3, 1961.
251. Floyd, R. W.: Algorithm 97, Shortest Path, *Communications of the ACM*, vol. 5, no. 6, 1962.
252. Ford, L. R., Jr., and D. R. Fulkerson: Solving the Transportation Problem, *Management Science*, vol. 3, no. 1, 1956.
253. Ford, L. R., Jr., and D. R. Fulkerson: "Flows in Networks," Princeton University Press, Princeton, NJ, 1962.
254. Ford, L. R., Jr., and D. R. Fulkerson: A Primal Dual Algorithm for the Capacitated Hitchcock Problem, *Naval Research Logistics Quarterly*, vol. 4, no. 1, 1957.
255. Ford, L. R., Jr., and D. R. Fulkerson: Maximal Flow through a Network, *Canadian Journal of Mathematics*, vol. 8, no. 3, 1956.
256. Forrest, J. J. H., J. P. H. Hirst, and J. A. Tomlin: Practical Solution of Large Mixed Integer Programming Problems with Umpire, *Management Science*, vol. 20, no. 5, 1974.
257. Forrest, J. J. H., and J. A. Tomlin: Updated Triangular Factors of the Basis to

Maintain Sparsity in the Product Form Simplex Method, *Mathematical Programming*, vol. 2, pp. 263–278, 1972.

258. Fourer, R.: Solving Staircase Linear Programs by the Simplex Method I: Inversion, *Mathematical Programming*, vol. 23, no. 3, 1982.

259. Frank, H., and I. T. Frisch: "Communication, Transmission, and Transportation Networks," Addison-Wesley Publishing Company, Inc., Reading, MA, 1971.

260. Frank, M., and P. Wolfe: An Algorithm for Quadratic Programming, *Naval Research Logistics Quarterly*, vol. 3, nos. 1, 2, 1956.

261. Frank, M. T., and K. E. Nygard: "A Fully Interactive Multiobjective Linear Programming System," Technical Report No. 10, College of Sciences and Mathematics, North Dakota State University, Fargo, ND, 1981.

262. Franklin, J.: Mathematical Methods of Economics, *The American Mathematical Monthly*, vol. 90, no. 4, 1983.

263. Freed, N., and F. Glover: "A Linear Programming Approach to the Discriminant Problem," *Decision Science*, vol. 12, no. 1, 1981.

264. Freed, N., and F. Glover: "Simple but Powerful Goal Programming Models for Discriminant Analysis," *European Journal of Operational Research*, vol. 7, no. 1, 1981.

265. French, S., R. Hartley, L. C. Thomas, and D. J. White: "Multi-Objective Decision Making," Academic Press, New York, 1983.

266. Frisch, R.: "The Multiplex Method for Linear Programming," Memorandum, Sosialökon. Inst., University of Oslo, Norway, 1958.

267. Fulkerson, D. R.: Flow Networks and Combinatorial Operations Research, *The American Mathematical Monthly*, vol. 73, no. 2, 1966.

268. Fulkerson, D. R.: An Out-of-Kilter Method for Minimal Cost Flow Problems, *Journal of the Society for Industrial and Applied Mathematics*, vol. 9, no. 1, 1961.

269. Fulkerson, D. R.: Increasing the Capacity of a Network: The Parametric Budget Problem, *Management Science*, vol. 5, no. 4, 1969.

270. Fulkerson, D. R.: Flow Networks and Combinatorial Operations Research, pp. 139–171 in part I of Fulkerson [271].

271. Fulkerson, D. R. (ed.): "Studies in Graph Theory—Part I and Part II," The Mathematical Association of America, Washington, DC, 1975.

272. Gács, P., and L. Lovász: Khachiyan's Algorithm for Linear Programming, *Mathematical Programming Study 14*, pp. 61–68, 1981.

273. Gal, T.: Rim Multiparametric Linear Programming, *Management Science*, vol. 21, no. 5, 1975.

274. Gal, T.: "Postoptimal Analyses, Parametric Programming, and Related Topics," McGraw-Hill Book Company, New York, 1979.

275. Gal, T.: A General Method for Determining the Set of All Efficient Solutions to a Linear Vectormaximum Problem, *European Journal of Operational Research*, vol. 1, pp. 307–322, 1977.

276. Gal, T., and J. Nedoma: Multiparametric Linear Programming, *Management Science*, vol. 18, no. 7, 1972.

277. Gale, D.: "The Theory of Linear Economic Models," McGraw-Hill Book Company, New York, 1960.

278. Gale, D., H. W. Kuhn, and A. W. Tucker: Linear Programming and the Theory of Games, chap. 19 in Koopmans [437].

279. Gantmacher, F. R.: "The Theory of Matrices," vols. I and II, Chelsea Publishing Co., New York, 1960.

280. Garey, M. R., and D. S. Johnson: "Computers and Intractability," W. H. Freeman and Company, San Francisco, 1979.

281. Garfinkel, R. S., and G. L. Nemhauser: "Integer Programming," John Wiley & Sons, Inc., New York, 1972.

282. Garvin, W. W.: "Introduction to Linear Programming," McGraw-Hill Book Company, New York, 1960.

283. Garvin, W. W., H. W. Crandell, J. B. John, and R. A. Spellman: Applications of Linear Programming in the Oil Industry, *Management Science*, vol. 3, no. 4, 1957.

284. Gass, S. I.: A First Feasible Solution to the Linear Programming Problem, pp. 495–508 in Directorate of Management Analysis [207].

285. Gass, S. I.: Recent Developments in Linear Programming, in F. Alt (ed.), "Advances in Computers," vol. II, Academic Press, New York, 1962.

286. Gass, S. I.: "The Dualplex Method for Large-Scale Linear Programs," ORC Report 66-15, Operations Research Center, University of California, Berkeley, CA, June, 1966.

287. Gass, S. I.: "An Illustrated Guide to Linear Programming," McGraw-Hill Book Company, New York, 1970.

288. Gass, S. I.: The Dualplex Method Applied to Special Linear Programs, "Proceedings of the International Federation of Information Processing Societies, 1971," pp. 1317–1323, North-Holland Publishing Company, Amsterdam, 1972.

289. Gass, S. I.: Comments on the Possibility of Cycling with the Simplex Method, *Operations Research*, vol. 27, no. 4, 1979.

290. Gass, S. I.: "On the Solution of Linear-Programming Problems with Free (Unrestricted) Variables," *Computers & Operations Research,* vol. 12, no. 3, 1985. Management, University of Maryland, College Park, MD, 1982.

291. Gass, S. I.: Decision-Aiding Models: Validation, Assessment, and Related Issues for Policy Analysis, *Operations Research*, vol. 31, no. 4, 1983.

292. Gass, S. I., and T. L. Saaty: Parametric Objective Function. Part II: Generalization, *Operations Research*, vol. 3, no. 4, 1955.

293. Gass, S. I., and T. L. Saaty: The Computational Algorithm for the Parametric Objective Function, *Naval Research Logistics Quarterly*, vol. 2, nos. 1 and 2, 1955.

294. Gassner, B. J.: Cycling in the Transportation Problem, *Naval Research Logistics Quarterly*, vol. 11, no. 1, 1964.

295. Geoffrion, A. M.: Solving Bicriterion Mathematical Programs, *Operations Research*, vol. 15, no. 1, 1967.

296. Geoffrion, A. M.: Proper Efficiency and the Theory of Vector Maximization, *Journal of Mathematical Analysis and Applications*, vol. 22, pp. 618–630, 1968.

297. Geoffrion, A. M.: Generalized Benders Decomposition, *Journal of Optimization Theory and Applications*, vol. 10, no. 4, 1972.

298. Geoffrion, A. M. (ed.): "Perspectives on Optimization," Addison-Wesley Publishing Company, Inc., Reading, MA, 1972.

299. Geoffrion, A. M., and G. W. Graves: Multicommodity Distribution System Design by Benders Decomposition, *Management Science*, vol. 20, no. 5, 1974.

300. Geoffrion, A. M., and R. E. Marsten: Integer Programming Algorithms: A Framework and State-of-the-Art Survey, *Management Science*, vol. 18, no. 7, 1972.

301. Geoffrion, A. M., and R. Nauss: Parametric and Postoptimality Analysis in Integer Linear Programming, *Management Science*, vol. 23, no. 5, 1977.

302. George, A., and J. W.-H. Liu: "Computer Solution of Large Sparse Positive Definite Systems," Prentice-Hall, Inc., Englewood Cliffs, NJ, 1981.

303. Gill, P. E., W. Murray, and M. H. Wright: "Practical Optimization," Academic Press, New York, 1981.

304. Gilmore, P. C., and R. E. Gomory: A Linear-Programming Approach to the Cutting-Stock Problem—Part I, *Operations Research*, vol. 9, no. 6, 1961; Part II, *Operations Research*, vol. 11, no. 6, 1963.

305. Gilmore, P. C., and R. E. Gomory: The Theory and Computation of Knapsack Functions, *Operations Research*, vol. 14, no. 6, 1966.

306. Glaser, E.: "Introductory Notes on Input-Output Analysis," mimeograph notes presented to the Washington section of the American Statistical Association, Washington, DC, 1953.

307. Glover, F., J. Hultz, and D. Klingman: "Network Versus Linear Programming Algorithms and Implementations," Research Report CCS 306, Center for Cybernetic Studies, University of Texas, Austin, TX, 1977.

308. Glover, F., D. Karney, and D. Klingman: Implementation and Computational Comparisons of Primal, Dual and Primal-Dual Computer Codes for Minimum Cost Network Flow Problems, *Networks*, vol. 4, no. 3, 1974.

309. Glover, F., D. Karney, D. Klingman, and A. Napier: A Computation Study on Start Procedures, Basis Change Criteria, and Solution Algorithms for Transportation Problems, *Management Science*, vol. 20, no. 5, 1974.

310. Glover, F., and D. Klingman: On the Equivalence of Some Generalized Network Problems to Pure Network Problems, *Mathematical Programming*, vol. 4, no. 3, 1973.

311. Glover, F., and D. Klingman: "Recent Developments in Computer Implementation Technology for Network Flow Algorithms," Research Report CCS 337, Center for Cybernetic Studies, University of Texas, Austin, TX, 1980.

312. Glover, F., D. Klingman, J. Mote, and D. Whitman: "Comprehensive Computer Evaluation and Enhancement of Maximum Flow Algorithms," Research Report CCS 356, Center for Cybernetic Studies, University of Texas, Austin, TX, 1979.

313. Glover, F., and J. Mulvey: Equivalence of the 0–1 Integer Programming Problem to Discrete Generalized and Pure Networks, *Operations Research*, vol. 28, no. 3, part II, 1980.

314. Goldfarb, D.: On the Bartels-Golub Decomposition for Linear Programming Bases, *Mathematical Programming*, vol. 13, no. 3, 1977.

315. Goldfarb, D., and J. K. Reid: A Practicable Steepest-Edge Simplex Algorithm, *Mathematical Programming*, vol. 12, no. 3, 1977.

316. Goldfarb, D., and M. J. Todd: Modification, and Implementation of the Ellipsoid Algorithm for Linear Programming, *Mathematical Programming*, vol. 23, no. 1, 1982.

317. Goldman, A. J., and A. W. Tucker: Polyhedral Conex Cones, paper 3 in Kuhn and Tucker [449].

318. Goldman, A. J., and A. W. Tucker: Theory of Linear Programming, paper 4 in Kuhn and Tucker [449].

319. Goldstein, L.: "The Simplex Solution of Waugh's Problem," mimeographed notes, Directorate of Management Analysis, Headquarters U.S. Air Force, Washington, DC, 1952.

320. Gomory, R. E.: Essentials of an Algorithm for Integer Solutions to Linear Programs, *Bulletin of the American Mathematical Society*, vol. 64, 1958.

321. Gomory, R. E.: An Algorithm for Integer Solution to Linear Programming, pp. 269–302 in Graves and Wolfe [328].

322. Gomory, R. E.: "An Algorithm for the Mixed Integer Problem," RAND Report RM-2597, The RAND Corporation, Santa Monica, CA, 1960.

323. Gomory, R. E.: All-Integer Programming Algorithm, pp. 193–206 in J. F. Muth and G. L. Thompson (eds.), "Industrial Scheduling," Prentice-Hall, Inc., Englewood Cliffs, NJ, 1963.

324. Gomory, R. E.: Large and Non-Convex Problems in Linear Programming, *Proceedings of Symposia in Applied Mathematics*, vol. 15, American Mathematical Society, Providence, RI, 1963.

325. Gomory, R. E.: The Traveling Salesman Problem, *Proceedings of the IBM Scientific Computing Symposium on Combinatorial Problems*, IBM, White Plains, NY, 1964.

326. Gomory, R. E., and A. J. Hoffman: On the Convergence of an Integer-Programming Process, *Naval Research Logistics Quarterly*, vol. 10, no. 2, 1963.

327. Graves, R. L.: Parametric Linear Programming, pp. 201–210 in Graves and Wolfe [328].

328. Graves, R. L., and P. Wolfe: "Recent Advances in Mathematical Programming," McGraw-Hill Book Company, New York, 1963.

329. Greenberg, H.: "Integer Programming," Academic Press, New York, 1971.

330. Greenberg, H. J. (ed.): "Design and Implementation of Optimization Software," Sijthoff & Noordhoff, Amsterdam, 1978.

331. Greenberg, H. J.: Pivot Selection Tactics, pp. 143–174 in Greenberg [330].

332. Greenberg, H. J.: The One-Dimensional Generalized Lagrange Multiplier Problem, *Operations Research*, vol. 25, no. 2, 1977.

333. Greenberg, H. J.: A Tutorial on Matricial Packing, pp. 109–142 in Greenberg [330].

334. Greenberg, H. J., and J. E. Kalan: An Exact Update of Harris' Tread, pp. 26–29 in Balinski and Hellerman [31].

335. Greig, D. M.: "Optimization," Longman Group Limited, London, 1980.

336. Grigoriadis, M. D.: A Dual Generalized Upper Bounding Technique, *Management Science*, vol. 17, no. 5, 1971.

337. Grigoriadis, M. D., and W. P. Walker: A Treatment of Transportation Problems by Primal Partitioning Programming, *Management Science*, vol. 14, no. 9, 1968.

338. Gross, O.: "A Class of Discrete-Type Minimization Problems," RAND Report RM-1644, The RAND Corporation, Santa Monica, CA, 1956.

339. Gross, O.: A Simple Linear Programming Problem Explicitly Solvable in Integers, "Notes on Linear Programming," Part XXVIII, RAND Report RM-1560, The RAND Corporation, Santa Monica, CA, 1955.

340. Gue, R. L.: "Mathematical Programming in Institutional Menu Planning," Computer Sciences Center, Southern Methodist University, Dallas, TX, 1968.

341. Guignard, M., K. Spielberg, and U. Suhl: "Survey of Enumerative Methods for Integer Programming," Technical Report No. 41, Department of Statistics, The Wharton School, University of Pennsylvania, Philadelphia, PA, 1981.

342. Hadley, G.: "Linear Algebra," Addison-Wesley Publishing Company, Inc., Reading, MA, 1961.

343. Hadley, G.: "Linear Programming," Addison-Wesley Publishing Company, Inc., Reading, MA, 1962.

344. Hadley, G.: "Nonlinear and Dynamic Programming," Addison-Wesley Publishing Company, Inc., Reading, MA, 1964.

345. Haessler, R. W.: A Note on Computational Modifications to the Gilmore-Gomory Cutting Stock Algorithm, *Operations Research*, vol. 28, no. 4, 1980.

346. Haley, K. B.: The Solid Transportation Problem, *Operations Research*, vol. 10, no. 4, 1962.

347. Hammer, P. L., E. L. Johnson, B. H. Korte, and G. L. Nemhauser: "Studies in Integer Programming," North-Holland, Amsterdam, 1977.

348. Hammer, P. L., and A. Zoutendijk (eds.): "Mathematical Programming in Theory and Practice," North-Holland, Amsterdam, 1975.

349. Haneveld, W. K. K., C. L. J. van der Meer, and R. J. Peters: A Construction Method in Parametric Programming, *Mathematical Programming*, vol. 16, no. 1, 1979.

350. Hannan, E. L.: Using Duality Theory for Identification of Primal Efficient Points and for Sensitivity Analysis in Multiple Linear Objective Linear Programming, *Journal of the Operational Research Society*, vol. 29, no. 7, 1978.

351. Hansen, P.: Methods of Nonlinear 0–1 Programming, *Annals of Discrete Mathematics*, vol. 5, pp. 53–70, 1979.

352. Hansotia, B. J.: Stochastic Linear Programming: A Tutorial, *Decision Sciences*, vol. 11, no. 1, 1980.

353. Harris, P. M. J.: Pivot Selection Methods of the Devex LP Code, pp. 30–57 in Balinski and Hellerman [31].

354. Harris, P. M. J.: Pivot Selection Methods of the Devex LP Code, *Mathematical Programming*, vol. 5, no. 1, 1973.

355. Harrison, J. O., Jr.: Linear Programming and Operations Research, in J. F. McCloskey and F. N. Trefethen (eds.), "Operations Research for Management," vol. I, pp. 217–237, The Johns Hopkins Press, Baltimore, MD, 1954.

356. Hartman, J. K., and L. S. Lasdon: A Generalized Upper Bounding Method for Doubly Coupled Linear Programs, *Naval Research Logistics Quarterly*, vol. 17, no. 4, 1970.

357. Harvey, R. P.: The Decomposition Principle for Linear Programs, *International Journal of Computer Mathematics*, vol. 1, no. 1, 1964.

358. Hassin, R.: The Minimum Cost Flow Problem: A Unifying Approach to Dual Algorithms and a New Tree-Search Algorithm, *Mathematical Programming*, vol. 25, no. 2, 1983.

359. Hatch, R. S.: Bench Marks Comparing Transportation Codes Based on Primal Simplex and Primal-Dual Algorithms, *Operations Research*, vol. 23, no. 6, 1975.

360. Hax, A.: "On Generalized Upper Bounded Techniques for Linear Programming," O.R. Center Report 67-17, *Notes on Operations Research-6*, University of California, Berkeley, CA, 1967.

361. Helgason, R. V., and J. L. Kennington: Spike Swapping in Basic Reinversion, *Naval Research Logistics Quarterly*, vol. 27, no. 4, 1980.

362. Heller, I., and C. B. Tompkins: An Extension of a Theorem of Dantzig, paper 14 in Kuhn and Tucker [449].

363. Hellerman, E., and D. Rarick: Reinversion with the Preassigned Pivot Procedure, *Mathematical Programming*, vol. 1, no. 2, 1972.

364. Hellerman, E., and D. Rarick: The Partitioned Preassigned Pivot Procedure, pp. 68–76 in Rose and Willoughby [574].

365. Henderson, A., and R. Schlaifer: Mathematical Programming, *Harvard Business Review*, vol. 32, 1954.

366. Hildebrand, F. B.: "Methods of Applied Mathematics," Prentice-Hall, Inc., Englewood Cliffs, NJ, 1958.

367. Himmelblau, D. M.: "Applied Nonlinear Programming," McGraw-Hill Book Company, New York, 1972.

368. Himmelblau, D. M. (ed.): "Decomposition of Large-Scale Problems," North-Holland Publishing Company, Amsterdam, 1973.

369. Hirsch, W. M., and G. B. Dantzig: The Fixed Charge Problem, *Naval Research Logistics Quarterly*, vol. 15, no. 3, 1968.

370. Hitchcock, F. L.: Distribution of a Product from Several Sources to Numerous Localities, *Journal of Mathematical Physics*, vol. 20, 1941.

371. Ho, J. K.: Pricing for Sparsity in the Revised Simplex Method, *R.A.I.R.O. Recherche Opérationnelle/Operations Research*, vol. 12, no. 3, 1978.

372. Ho, J. K., and E. Loute: An Advanced Implementation of the Dantzig-Wolfe Decomposition Algorithm for Linear Programming, *Mathematical Programming*, vol. 20, no. 3, 1981.

373. Hoffman, A.: "Cycling in the Simplex Algorithm," National Bureau of Standards Report, Washington, DC, 1953.

374. Hoffman, A. J.: How to Solve a Linear Programming Problem, pp. 397–424 in Directorate of Management Analysis [207].

375. Hoffman, A. J., and W. W. Jacobs: Smooth Patterns of Production, *Management Science*, vol. 1, no. 1, 1954.

376. Hoffman, A. J., and J. G. Kruskal: Integral Boundary Points of Convex Polyhedra, paper 13 in Kuhn and Tucker [449].

377. Hoffman, A., M. Mannos, D. Sokolowsky, and N. Wiegmann: Computational Experience in Solving Linear Programs, *Journal of the Society for Industrial and Applied Mathematics*, vol. 1, no. 1, 1953.

378. Hoffman, A. J., and H. M. Markowitz: A Note on Shortest Path, Assignment and Transportation Problems, *Naval Research Logistics Quarterly*, vol. 10, no. 4, 1963.

379. Hoffman, A. J., and S. Winograd: On Finding All Shortest Distances in a Directed Network, *IBM Journal of Research and Development*, vol. 16, no. 4, 1972.

380. Hoffman, K. L.: A Method for Globally Minimizing Concave Functions over Convex Sets, *Mathematical Programming*, vol. 20, no. 1, 1981.

381. Hsiao, J. C.: An Economic Interpretation of the Transportation Problem, *American Journal of Agricultural Economics*, vol. 53, no. 3, 1971.

382. Hu, T. C.: Multi-Commodity Network Flows, *Operations Research*, vol. 11, no. 3, 1964.

383. Hu, T. C.: "Integer Programming and Network Flows," Addison-Wesley Publishing Company, Inc., Reading, MA, 1969.

384. Hu, T. C.: "Combinatorial Algorithms," Addison-Wesley Publishing Company, Reading, MA, 1982.

385. Hu, T. C., and S. M. Robinson (eds.): "Mathematical Programming," Academic Press, New York, 1973.

386. Ibaraki, T.: Integer Formulation of Combinatorial Optimization Problems, *Discrete Mathematics*, vol. 16, pp. 39–52, 1976.

387. IBM: "Mathematical Programming System Extended/370 Program Reference Manual," IBM, White Plains, NY, 1979.

388. Ignizio, J. P.: "Goal Programming and Extensions," Lexington Books, Lexington, MA, 1976.

389. Ignizio, J. P.: A Review of Goal Programming: A Tool for Multiobjective Analysis, *Journal of the Operational Research Society*, vol. 29, no. 11, 1978.

390. Ignizio, J. P.: The Determination of a Subset of Efficient Solutions via Goal Programming, *Computers and Operations Research*, vol. 8, pp. 9–16, 1981.

391. Ignizio, J. P.: "Linear Programming in Single- and Multiple-Objective Systems," Prentice-Hall, Inc., Englewood Cliffs, NJ, 1982.

392. Ijiri, Y.: "Management Goals and Accounting for Control," North-Holland, Amsterdam, 1965.

393. Intriligator, M. D.: "Mathematical Optimization and Economic Theory," Prentice-Hall, Inc., Englewood Cliffs, NJ, 1971.

394. Isermann, H.: Proper Efficiency and the Linear Maximum Problem, *Operations Research*, vol. 22, no. 1, 1974.

395. Isermann, H.: The Enumeration of the Set of All Efficient Solutions for a Linear Multiple Objective Program, *Operational Research Quarterly*, vol. 28, no. 3, 1977.

396. Itai, A.: Two-Commodity Flow, *Journal of the ACM*, vol. 25, no. 4, 1978.

397. Jacobs, W. W.: The Caterer Problem, *Naval Research Logistics Quarterly*, vol. 1, no. 2, 1954.

398. Jensen, P. A., and J. W. Barnes: "Network Flow Programming," John Wiley & Sons, Inc., New York, 1980.

399. Jensen, P. A., and G. Bhaumik: A Flow Augmentation Approach to the Network with Gains Minimum Cost Flow Problem, *Management Science*, vol. 23, no. 6, 1977.

400. Jeroslow, R. G.: The Simplex Algorithm with the Pivot Rule of Maximizing Criterion Improvement, *Discrete Mathematics*, vol. 4, pp. 367–377, 1973.

401. Jewell, W. S.: A Classroom Example of Linear Programming, Lesson no. 2, *Operations Research*, vol. 8, no. 4, 1960.

402. Jewell, W. S.: Optimal Flow through Networks with Gains, *Operations Research*, vol. 10, no. 4, 1962.

403. John, F.: Extremum Problems with Inequalities as Subsidiary Conditions, "Studies and Essays," Courant Anniversary Volume, Interscience Publishers (Division of John Wiley & Sons, Inc.), New York, 1948.

404. Johnson, S. M.: Sequential Production Planning over Time at Minimum Cost, *Management Science*, vol. 3, no. 4, 1957.

405. Jones, P. C., and E. S. Marwil: Dimensional Reduction Variant of the Ellipsoid Algorithm for Linear Programming Problems, *Mathematics of Operations Research*, vol. 7, no. 2, 1982.

406. Joseph, J. A.: "The Application of Linear Programming to Weapon Selection and Target Analysis," *Technical Memorandum* 42, Operations Analysis Division, Headquarters U.S. Air Force, Washington, DC, 1954.

407. Kahan, G.: Walking through a Columnar Approach to Linear Programming of a Business, *Interfaces*, vol. 12, no. 3, 1982.

408. Kalan, J. E.: Aspects of Large-Scale In-Core Linear Programming, pp. 304–313 in *Proceedings of the ACM Conference*, Chicago, 1971.

409. Kantorovich, L. V.: On the Translocation of Masses, *Doklady Akad. Nauk SSR*, vol. 37, p. 227, 1942, translated in *Management Science*, vol. 5, no. 1, 1958.

410. Kantorovich, L. V.: "Mathematical Methods of Organization and Planning Production," Publication House of the Leningrad State University, 1939, translated in *Management Science*, vol. 6, no. 4, pp. 366–422, 1960.

411. Kaplan, E. L.: *Mathematical Programming*, John Wiley & Sons, Inc., New York, 1982.

412. Kaplan, S.: Comment on a Précis by Shanno and Weil, *Management Science*, vol. 17, no. 11, 1971.

413. Karamardian, S.: The Nonlinear Complementarity Problem with Applications, Parts I, II, *Journal of Optimization Theory and Applications*, vol. 4, nos. 2, 3, 1969.

414. Karamardian, S.: The Complementarity Problem, *Mathematical Programming*, vol. 2, no. 1, 1972.

415. Karlin, S.: "Mathematical Methods and Theory in Games, Programming, and Economics," vol. I, Addison-Wesley Publishing Company, Inc., Reading, MA, 1959.

416. Karney, D., and D. Klingman: Implementation and Computational Study on an In-Core, Out-of-Core Primal Network Code, *Operations Research*, vol. 24, no. 6, 1976.

417. Karp, R. M.: On the Computational Complexity of Combinatorial Problems. *Networks*, vol. 5, pp. 45–68, 1975.

418. Karwan, M. H., V. Lotfi, J. Telgen, and S. Zionts: "Redundancy in Mathematical Programming," Springer-Verlag New York, Inc., New York, 1982.

419. Katz, S., L. J. Risman, and M. Rodeh: A System for Constructing Linear Programming Models, *IBM Systems Journal*, vol. 19, no. 4, 1980.

420. Kaul, R. N.: "An Extension of Generalized Upper Bounded Techniques," O.R. Center Report 1965-27, University of California, Berkeley, CA, 1965.

421. Kelley, J. E., Jr.: An Application of Linear Programming to Curve Fitting, *Journal of the Society for Industrial and Applied Mathematics*, vol. 6, pp. 15–22, 1958.

422. Kelley, J. E., Jr.: The Cutting-Plane Method for Solving Convex Programs, *Journal of the Society for Industrial and Applied Mathematics*, vol. 8, pp. 703–712, 1960.

423. Kelley, J. E., Jr.: The Critical Path Planning and Scheduling: Mathematical Basis, *Operations Research*, vol. 9, no. 3, 1961.

424. Kennington, J.: A Survey of Linear Cost Multicommodity Network Flows, *Operations Research*, vol. 26, no. 2, 1978.

425. Kennington, J.: The Fixed-Charge Transportation Problem: A Computational Study with a Branch-and-Bound Code, *AIIE Transactions*, vol. 8, 1976.

426. Kennington, J., and R. V. Helgason: "Algorithms for Network Programming," John Wiley & Sons, Inc., New York, 1980.

427. Kennington, J., and E. Unger: A New Branch-and-Bound Algorithm for the Fixed Charge Transportation Problem, *Management Science*, vol. 22, no. 10, 1976.

428. Ketron, Inc.: "MPSIII—General Description," Ketron, Inc., Arlington, VA, 1982.

429. Khachiyan, L. G.: A Polynomial Algorithm for Linear Programming (Russian), *Doklady Akad. Nauk SSSR*, vol. 244, pp. 1093–1096, 1979; translated in *Soviet Mathematics Doklady*, vol. 20, pp. 191–194, 1979.

430. Kim, C.: Parametrizing an Activity Vector in Linear Programming, *Operations Research*, vol. 19, no. 7, 1971.

431. Klee, V.: A Class of Linear Programming Problems Requiring a Large Number of Iterations, *Numerische Mathematik*, vol. 7, pp. 313–321, 1965.

432. Klee, V., and G. J. Minty: How Good is the Simplex Method?, pp. 159–175 in O. Shisha (ed.), *Inequalities-III*, Academic Press, New York, 1972.

433. Klee, V., and C. Witzgall: Facets and Vertices of Transportation Polytypes, in Dantzig and Veinott [193].

434. Klingman, D., and R. Russell: A Streamlined Approach to the Singly Constrained Transportation Problem, *Naval Research Logistics Quarterly*, vol. 25, no. 4, 1978.

435. Koehler, G. J., A. B. Whinston, and G. P. Wright: The Solution of Leontief Substitution Systems Using Matrix Iterative Techniques, *Management Science*, vol. 21, no. 11, 1975.

436. Koopmans, T. C.: Optimum Utilization of the Transportation System, *Econometrica*, vol. 17, supplement, 1949.

437. Koopmans, T. C. (ed.): "Activity Analysis of Production and Allocation," *Cowles Commission Monograph* 13, John Wiley & Sons, Inc., New York, 1951.

438. Koopmans, T. C.: "Three Essays on the State of Economic Science," McGraw-Hill Book Company, New York, 1957.

439. Koopmans, T. C., and M. Beckman: Assignment Problems and the Location of Economic Activities, *Econometrica*, vol. 25, no. 1, 1957.

440. Kornbluth, J. S. H.: Duality, Indifference and Sensitivity Analysis in Multiple Objective Linear Programming, *Operational Research Quarterly*, vol. 25, no. 4, 1974.

441. Kornbluth, J. S. H., and G. R. Salkin: The Optimal Dual Solution in Linear Fractional Decomposition Problems, *Operations Research*, vol. 22, no. 1, 1974.

442. Kornbluth, J. S. H., and R. E. Steuer: Multiple Objective Linear Fractional Programming, *Management Science*, vol. 27, no. 9, 1981.

443. Krekó, B.: "Linear Programming," American Elsevier Publishing Company, Inc., New York, 1968.

444. Kuhn, H. W.: Nonlinear Programming: A Historical View, pp. 1–26 in Cottle and Lemke [141].

445. Kuhn, H. W.: "Lectures on the Theory of Games," *Annals of Mathematics Studies* 37, Princeton University Press, Princeton, NJ, 1957.

446. Kuhn, H. W.: The Hungarian Method for the Assignment Problem, *Naval Research Logistics Quarterly*, vol. 2, nos. 1, 2, 1955.

447. Kuhn, H. W. (ed.): "Proceedings of the Princeton Symposium on Mathematical Programming," Princeton University Press, Princeton, NJ, 1970.

448. Kuhn, H. W., and R. E. Quandt: An Experimental Study of the Simplex Method, *Proceedings Symposia in Applied Mathematics*, vol. 15, pp. 107–124, 1963.

449. Kuhn, H. W., and A. W. Tucker: "Linear Inequalities and Related Systems," *Annals of Mathematics Studies* 38, Princeton University Press, Princeton, NJ, 1956.

450. Kuhn, H. W., and A. W. Tucker: Nonlinear Programming, in "Proceedings of the Second Berkeley Symposium on Mathematical Statistics and Probability," University of California Press, Berkeley, CA, 1951.

451. Kunzi, H. P., and W. Krelle: "Nonlinear Programming," Blaisdell Publishing Co., Waltham, MA, 1966.

452. Kwak, N. K.: "Mathematical Programming with Business Applications," McGraw-Hill Book Co., New York, 1973.

453. Kydland, F.: Hierarchical Decomposition in Linear Economic Models, *Management Science*, vol. 21, no. 9, 1975.

454. Lageman, J. J.: A Method for Solving the Transportation Problem, *Naval Research Logistics Quarterly*, vol. 14, no. 1, 1967.

455. Land, A. H., and A. G. Doig: An Automatic Method for Solving Discrete Programming Problems, *Econometrica*, vol. 28, pp. 497–520, 1960.

456. Land. A. H., and S. Powell: "Fortran Codes for Mathematical Programming: Linear, Quadratic and Discrete," John Wiley & Sons, Inc., New York, 1973.

457. Land, A. H., and S. Powell: "A Survey of Available Computer Codes to Solve Integer Linear Programming Problems," Research Report 81-09, The London School of Economics and Political Science, London, England, 1981.

458. Larsen, L. J.: A Modified Inversion Procedure for Product Form of the Inverse-Linear Programming Codes, *Communications of the ACM*, vol. 5, no. 7, 1962.

459. Larson, R. E.: Normalizing Vogel's Approximation Method, *Mathematics Magazine*, vol. 45, no. 5, 1972.

460. Lasdon, L. S.: "Optimization Theory for Large Systems," The Macmillan Company, New York, 1970.

461. Lasdon, L. S., and A. D. Waren: Survey of Nonlinear Programming Applications, *Operations Research*, vol. 28, no. 5, 1980.

462. Lawler, E.: "Combinatorial Optimization: Networks and Matroids," Holt, Rinehart and Winston, New York, 1976.

463. Lawler E., J. K. Lenstra, and A. H. G. Rinnooy Kan (eds.): "The Traveling Salesman Problem," John Wiley & Sons, Inc., New York, 1983.

464. Lawler, E., and D. E. Wood: Branch and Bound Methods: A Survey, *Operations Research*, vol. 14, no. 4, 1966.

465. Lawrence, J. A., Jr.: Abstract Polytopes and The Hirsch Conjecture, *Mathematical Programming*, vol. 15, no. 1, 1978.

466. Lee, S. M.: "Goal Programming for Decision Analysis," Auerbach, Philadelphia, PA, 1972.

467. Lemke, C. E.: The Dual Method of Solving the Linear Programming Problem, *Naval Research Logistics Quarterly*, vol. 1, no. 1, 1954.

468. Lemke, C. E.: Bimatrix Equilibrium Points and Mathematical Programming, *Management Science*, vol. 11, no. 7, 1965.

469. Lemke, C. E.: On Complementary Pivot Theory, pp. 95–114 in Dantzig and Veinott [193].

470. Lemke, C. E., and J. T. Howson, Jr.: Equilibrium Points of Bimatrix Games, *SIAM Journal*, vol. 12, no. 2, 1964.

471. Lenstra, J. K., and A. H. G. Rinnooy Kan: Some Simple Applications of the Traveling Salesman Problem, *Operational Research Quarterly*, vol. 26, no. 4, 1975.

472. Lenstra, J. K., A. H. G. Rinnooy Kan, and P. Van Emde Boas: An Appraisal of Computational Complexity for Operations Researchers, *European Journal of Operations Research*, vol. 11, pp. 201–210, 1982.

473. Leontief, W. W.: "The Structure of the American Economy, 1919–1929," Oxford University Press, New York, 1951.

474. Lin, S., and B. W. Kernighan: An Effective Heuristic Algorithm for the Traveling-Salesman Problem, *Operations Research*, vol. 21, no. 2, 1973.

475. Little, J. D. C., K. G. Murty, D. W. Sweeney, and C. Karel: An Algorithm for the Traveling Salesman Problem, *Operations Research*, vol. 11, no. 6, 1963.

476. Luce, R. D., and H. Raiffa: "Games and Decisions," John Wiley & Sons, Inc., New York, 1957.

477. Luenberger, D. G.: "Introduction to Linear and Nonlinear Programming," Addison-Wesley Publishing Company, Inc., Reading, MA, 1973.

478. McBride, R. D.: A Spike Collection Dynamic Factorization Algorithm for the Simplex Method, *Management Science*, vol. 24, no. 10, 1978.

479. McBride, R. D., and J. S. Yormark: An Implicit Enumeration Algorithm for Quadratic Integer Programming, *Management Science*, vol. 26, no. 3, 1980.

480. McCall, E. H.: "A Study of the Khachian Algorithm for Real-World Linear Programming," Sperry Univac, St. Paul, MN, 1980.

481. McCall, E. H.: Performance Results of the Simplex Algorithm for a Set of Real-World Linear-Programming Models, *Communications of the ACM*, vol. 25, no. 3, 1982.

482. McCall, E. H.: "Large Scale Mixed Integer Programming Experiments with Sperry Univac's Series 1100 FMPS Product," Sperry Univac, Roseville, MN, 1973.

483. McCormick, G. P.: *Nonlinear Programming*, John Wiley & Sons, Inc., New York, 1983.

484. McGinnis, L. F.: Implementation and Testing of a Primal-Dual Algorithm for the Assignment Problem, *Operations Research*, vol. 31, no. 2, 1983.

485. McKinsey, J. C. C.: "Introduction to the Theory of Games," McGraw-Hill Book Company, New York, 1952.

486. Madansky, A.: Methods of Solution of Linear Programs under Uncertainty, *Operations Research*, vol. 10, no. 4, 1962.

487. Magee, John F.: "Studies in Operations Research. I. Application of Linear Programming to Production Scheduling," Arthur D. Little, Inc., Cambridge, MA, 1953.

488. Magnanti, T. L.: Optimization for Sparse Systems, pp. 147–176 in Bunch and Rose [101].

489. Mangasarian, O. L.: "Nonlinear Programming," McGraw-Hill Book Company, New York, 1969.

490. Mangasarian, O. L., R. R. Meyer, and S. M. Robinson (eds.): "Nonlinear Programming 3," Academic Press, New York, 1978.

491. Manne, A. S.: "Scheduling of Refinery Operations," Harvard University Press, Cambridge, MA, 1956.

492. Manne, A. S.: "Notes on Parametric Linear Programming," RAND Report P-468, The RAND Corporation, Santa Monica, CA, 1953.

493. Manne, A. S.: "On the Job Shop Scheduling Problem," *Cowles Commission for Research in Economics* Contract 358(01) NR 047-066, Office of Naval Research, Washington, DC, 1959.

494. Markowitz, H.: The Optimization of a Quadratic Function Subject to Linear Constraints, *Naval Research Logistics Quarterly*, vol. 3, nos. 1, 2, 1956.

495. Markowitz, H.: The Elimination Form of the Inverse and its Application to Linear Programming, *Management Science*, vol. 3, no. 3, 1957.

496. Marshall, K. T., and J. W. Suurballe: A Note on Cycling in the Simplex Method, *Naval Research Logistics Quarterly*, vol. 16, no. 1, 1969.

497. Marsten, R. E.: An Algorithm for Large Set Partitioning Problems, *Management Science*, vol. 20, no. 5, 1974.

498. Martos, B.: Hyperbolic Programming, *Naval Research Logistics Quarterly*, vol. 11, nos. 2, 3, 1964.

499. Martos, B.: "Nonlinear Programming," North-Holland, Amsterdam, 1975.

500. Matheiss, T. H., and D. S. Rubin: A Survey and Comparison of Methods for Finding All Vertices of Convex Polyhedral Sets, *Mathematics of Operations Research*, vol. 5, no. 2, 1980.

501. Miller, C. E.: "The Simplex Method for Local Separable Programming," pp. 89–100 in Graves and Wolfe [328].

502. Miller, C. E., A. W. Tucker, and R. A. Zemlin: Integer Programming Formulation of Traveling Salesman Problems, *Journal of the ACM*, vol. 7, no. 4, 1960.

503. Minty, G. J.: Monotone Networks, *Proceedings of the Royal Society of London*, Ser. A 257, pp. 194–212, 1960.

504. Moder, J. J., and S. E. Elmagraby (eds.): "Handbook of Operations Research," Van Nostrand Reinhold Company, New York, 1978.

505. Moeske, P. V.: A General Duality Theorem of Convex Programming, *Metroeconomica*, vol. 17, pp. 161–170, 1965.

506. Molina, F. W.: A Survey of Resource Directive Decomposition in Mathematical Programming," *Computing Surveys*, vol. 11, no. 2, 1979.

507. Morgenstern, O. (ed.): "Economic Activity Analysis," John Wiley & Sons, Inc., New York, 1954.

508. Motzkin, T. S.: The Multi-Index Transportation Problem, *Bulletin of the American Mathematical Society*, vol. 58, no. 4, 1952.

509. Motzkin, T. S., and I. J. Schoenberg: The Relaxation Method for Linear Inequalities, *Canadian Journal of Mathematics*, vol. 6, 1954.

510. Mueller, R. K., and L. Cooper: A Comparison of the Primal Simplex and Primal-Dual Algorithms in Linear Programming, *Communications of the ACM*, vol. 18, no. 11, 1965.

511. Müller-Merbach, H.: Das Verfahren der direkten Dekomposition in der linearen Planungsrechnung, *Ablauf- und Planungsforschung*, vol. 6, no. 2, 1965.

512. Müller-Merbach, H.: Lineare Planungsrechung mit parametrisch veränderten Koeffizienten der Bedingungsmatrix, *Ablauf- und Planungsforschung*, vol. 8, no. 3, 1967.

513. Müller-Merbach, H.: Sensibilitätsanalyse von Transportproblemen der linearen Planungsrechnung (mit ALGOL-Programm), *Electronische Datenverarbeitung*, vol. 10, no. 4, 1968.

514. Müller-Merbach, H.: "Operations Research," Verlag Franz Vahlen GmbH, Berlin, 1969.

515. Müller-Merbach, H.: On Round-Off Errors in Linear Programming, *Lecture Notes in Operations Research and Mathematical Systems*, vol. 37, Springer-Verlag OHG, Berlin, 1970.

516. Müller-Merbach, H.: "Tri-Branching in Integer Programming," Technical Report, Technische Hochschule, Darmstad, 1982.

517. Müller-Merbach, H.: Upper-Bounding-Technique, Generalized Upper-Bounding-Technique and Direct Decomposition in Linear Programming: A Survey on their General Principles Including a Report About Numerical Experience, pp. 167–180 in Himmelblau [368].

518. Mulvey, J. M.: Testing of a Large-Scale Network Optimization Program, *Mathematical Programming*, vol. 15, pp. 291–314, 1978.

519. Munakata, T.: "Matrices and Linear Programming," Holden-Day, Inc., San Francisco, 1979.

520. Munkres, J.: Algorithms for the Assignment and Transportation Problem, *Journal of the Society for Industrial and Applied Mathematics*, vol. 5, no. 1, 1957.

521. Murtagh, B. A.: "Advanced Linear Programming: Computation and Practice," McGraw-Hill Book Co., New York, 1981.

522. Nauss, R. M.: "Parametric Integer Programming," University of Missouri Press, Columbia, MO, 1977.

523. Nef, W.: "Linear Algebra," McGraw-Hill Book Company, New York, 1966.

524. Neumann, J. von: A Certain Zero-Sum Two-Person Game Equivalent to the Optimal Assignment Problem, pp. 5–12 in *Annals of Mathematics Studies* 28, Princeton University Press, Princeton, NJ, 1953.

525. Neumann, J. von: A Numerical Method to Determine Optimum Strategy, *Naval Research Logistics Quarterly*, vol. 1, no. 2, 1954.

526. Neumann, J. von, and O. Morgenstern: "Theory of Games and Economic Behavior," Princeton University Press, Princeton, NJ, 1947.

527. Ng, K. Y. K.: Solution of Navier-Stokes Equations by Goal-Programming, *Journal of Computational Physics*, vol. 39, 1981.

528. Orchard-Hays, W.: "Background, Development and Extensions of the Revised

Simplex Method," RAND Report RM-1433, The RAND Corporation, Santa Monica, CA, 1954.

529. Orchard-Hays, W.: "The RAND Code for the Simplex Method," RAND Report RM-1269, The RAND Corporation, Santa Monica, CA, 1954.

530. Orchard-Hays, W.: "A Composite Simplex Algorithm—II," RAND Report RM-1275, The RAND Corporation, Santa Monica, CA, 1954.

531. Orchard-Hays, W.: "Matrices, Elimination and the Simplex Method," CEIR, Inc., Bethesda, MD, 1961.

532. Orchard-Hays, W.: "Advanced Linear-Programming Computing Techniques," McGraw-Hill Book Company, New York, 1968.

533. Orchard-Hays, W.: History of Mathematical Programming Systems, pp. 1–26 in Greenberg [330].

534. Orchard-Hays, W.: "Some Additional Views on the Simplex Method and the Geometry of Constraint Space," Report RR-76-3, International Institute for Applied Systems Analysis, Laxenburg, Austria, 1976.

535. Orden, A.: Application of the Simplex Method to a Variety of Matrix Problems, pp. 28–50 of Directorate of Management Analysis [205].

536. Orden, A.: "A Procedure for Handling Degeneracy in the Transportation Problem," mimeograph, DCS/Comptroller, Headquarters U.S. Air Force, Washington, DC, 1951.

537. Orden, A.: The Transshipment Problem, *Management Science*, vol. 2, no. 3, 1956.

538. Owen, G.: "Game Theory," W. B. Saunders Company, Philadelphia, 1968.

539. Padberg, M. W.: Covering, Packing and Knapsack Problems, *Annals of Discrete Mathematics*, vol. 4, pp. 265–287, 1979.

540. Panwalkar, S. S.: Parametric Analysis of Linear Programs with Upper Bounded Variables, *Naval Research Logistics Quarterly*, vol. 20, no. 1, 1973.

541. Papadimitriou, C. H., and K. Steiglitz: "Combinatorial Optimization: Algorithms and Complexity," Prentice-Hall, Inc., Englewood Cliffs, NJ, 1982.

542. Parker, R. G., and R. L. Rardin: The Traveling Salesman Problem: An Update of Research, *Naval Research Logistics Quarterly*, vol. 30, no. 1, 1983.

543. Peeters, H.: "The User Interface and Performance of Commercial Mixed Integer Programming Software," *OR Spectrum*, vol. 2, pp. 233–249, 1981.

544. Perold, A. F.: A Degeneracy Exploiting LU Factorization for the Simplex Method, *Mathematical Programming*, vol. 19, no. 3, 1980.

545. Perold, A. F., and G. B. Dantzig: A Basis Factorization Method for Block Triangular Linear Programs, in Duff and Stewart [223].

546. Perry, C., and K. C. Crellin: The Precise Management Meaning of a Shadow Price, *Interfaces*, vol. 12, no. 2, 1982.

547. Peterkin, B., J. Nichols, and C. Cromwell: "Nutrition Labeling," U.S. Department of Agriculture Information Bulletin No. 382, U.S. Government Printing Office, Washington, DC, 1975.

548. Pfefferkorn, C. E., and J. A. Tomlin: "Design of a Linear Programming System for the ILLIAC IV," Systems Optimization Laboratory Report 76-8, Stanford University, Stanford, CA, 1976.

549. Philip, J.: Algorithms for the Vector Maximization Problem, *Mathematical Programming*, vol. 2, no. 2, 1972.

550. Potts, R. B., and R. M. Oliver: "Flows in Transportation Networks," Academic Press, New York, 1972.

551. Prager, W.: On the Caterer Problem, *Management Science*, vol. 3, pp. 15–23, 1956.

552. Prékopa, S. (ed.): "Survey of Mathematical Programming," vols. 1, 2, and 3, North-Holland, Amsterdam, 1979.

553. Quandt, R. E., and H. W. Kuhn: On Upper Bounds for the Number of Iterations in Solving Linear Programs, *Operations Research*, vol. 12, no. 1, 1964.

554. Rabinowitz, P.: Applications of Linear Programming to Numerical Analysis, *SIAM Review*, vol. 10, no. 2, 1968.

555. Raff, S. J., and M. Zeleny (eds.): Mathematical Programming with Multiple Objectives, *Computers and Operations Research*, vol. 7, nos. 1 and 2, 1980.

556. Raiffa, H., G. L. Thompson, and R. M. Thrall: An Algorithm for the Determination of All Solutions of a Two-Person Zero-Sum Game with a Finite Number of Strategies (Double Descriptive Method), pp. 100–114 of Directorate of Management Analysis [205].

557. Ralston, A., and P. Rabinowitz: "A First Course in Numerical Analysis," McGraw-Hill Book Co., New York, 1978.

558. Ramakrishnan, K. G.: Solving Two Commodity Transportation Problems with Coupling Constraints, *Journal of the ACM*, vol. 27, no. 4, 1980.

559. Randolph, P. H., and H. D. Meeks: "Applied Linear Optimization," Grid, Inc., Columbus, OH, 1978.

560. Randow, R. von (ed.): "Integer Programming and Related Areas: A Classified Bibliography 1978–1981," Lecture Notes in Economics and Mathematical Systems, vol. 197, Springer-Verlag, Berlin, 1982.

561. Ravindran, A.: A Computer Routine for Quadratic and Linear Programming Problems, *Communications of the ACM*, vol. 15, no. 9, 1972.

562. Rech, P.: "Decomposition and Interconnected Systems in Mathematical Programming," ORC Report 65-31, Operations Research Center, University of California, Berkeley, CA, 1965.

563. Reese, R. M., and A. C. Stedry: A Chance-Constrained Distribution Problem, *Naval Research Logistics Quarterly*, vol. 24, no. 1, 1977.

564. Reid, J. K.: Solution of Linear Systems of Equations: Direct Methods (General), pp. 102–129 in Barker [38].

565. Reid, J. K.: A Sparsity-Exploiting Variant of the Bartels-Golub Decomposition for Linear Programming Bases, *Mathematical Programming*, vol. 24, no. 1, 1982.

566. Richmond, T., and A. Ravindran: A Generalized Euclidean Procedure for Integer Linear Programs, *Naval Research Logistics Quarterly*, vol. 21, no. 1, 1974.

567. Riley, V., and S. I. Gass: "Bibliography on Linear Programming and Related Techniques," Johns Hopkins Press, Baltimore, MD, 1958.

568. Ritter, K.: "A Decomposition Method for Linear Programming Problems with Coupling Constraints and Variables," Mathematics Research Center, U.S. Army, MRC Report no. 739, University of Wisconsin, Madison, WI, 1967.

569. Roadman, G. M.: A Procedure for Optimal Stepwise MSAE Regression Analysis, *Operations Research*, vol. 22, no. 2, 1974.

570. Robers, P., and L. Cooper: A Study of the Fixed Charge Transportation Problem, *Computers and Mathematics with Applications*, vol. 2, pp. 125–135, 1976.

571. Rockafellar, R. T.: "Convex Analysis," Princeton University Press, Princeton, NJ, 1970.

572. Ronen, D.: On a Common Misinterpretation of Shadow Prices, *Management Science*, vol. 28, no. 12, 1982.

573. Roodman, G. M.: Post-Infeasibility Analysis in Linear Programming, *Management Science*, vol. 25, no. 9, 1979.

574. Rose, D. J., and R. A. Willoughby (eds.): "Sparse Matrices and their Applications," Plenum Press, New York, 1972.

575. Rosen, J. B.: The Gradient Projection Method for Nonlinear Programming, *SIAM Journal*, part I, vol. 8, no. 1, 1960; part II, vol. 9, no. 4, 1961.

576. Rosen, J. B.: Primal Partition Programming for Block Diagonal Matrices, *Numerische Mathematik*, vol. 6, pp. 250–260, 1964.

577. Rosen, J. B.: "Optimal Control and Convex Programming," chap. 13 in Abadie [1].

578. Rosen, J. B.: Chebyshev Solution of Large Linear Systems, *Journal of Computer and System Science*, vol. 1, pp. 29–43, 1967.

579. Ross, G. T., and R. M. Soland: A Branch and Bound Algorithm for the Generalized Assignment Problem, *Mathematical Programming*, vol. 8, no. 1, 1975.

580. Ross, G. T., D. Klingman, and A. Napier: A Computational Study of Effects of Problem Dimensions on Solution Times for Transportation Problems, *Journal of the ACM*, vol. 22, no. 3, 1975.

581. Russell, E. J.: Extension of Dantzig's Algorithm to Finding an Initial Near-Optimal Basis for the Transportation Problem, *Operations Research*, vol. 17, no. 1, 1969.

582. Rust, B. W., and W. R. Burrus: "Mathematical Programming and the Numerical Solution of Linear Equations," American Elsevier Publishing Company, Inc., New York, 1972.

583. Saaty, T. L.: The Number of Vertices of a Polyhedron, *American Mathematical Monthly*, vol. 62, 1955.

584. Saaty, T. L.: Coefficient Perturbation of a Constrained Extremum, *Operations Research*, vol. 7, no. 3, 1959.

585. Saaty, T. L.: "Optimization in Integers and Related Extremal Problems," McGraw-Hill Book Company, New York, 1970.

586. Saaty, T. L., and J. Bram: "Nonlinear Mathematics," McGraw-Hill Book Company, New York, 1964.

587. Saaty, T. L., and S. I. Gass: The Parametric Objective Function, Part I, *Operations Research*, vol. 2, no. 3, 1954.

588. Sakarovitch, M.: "Notes on Linear Programming," Van Nostrand Reinhold Company, New York, 1971.

589. Sakarovitch, M.: "Linear Programming," Springer-Verlag, New York, 1983.

590. Sakarovitch, M., and R. Saigal: An Extension of Generalized Upper Bounding Techniques for Structured Linear Programming Problems, *SIAM Journal of Applied Mathematics*, vol. 15, no. 4, 1967.

591. Salkin, H. M.: "Integer Programming," Addison-Wesley, Reading, MA, 1975.

592. Salkin, H. M.: The Knapsack Problem: A Survey, *Naval Research Logistics Quarterly*, vol. 22, no. 1, 1975.

593. Salkin, H. M., and W. L. Balinsky: Integer Programming Models and Codes in the Urban Environment, *Socio-Economic Planning Science*, vol. 7, pp. 739–753, 1973.

594. Salkin, H. M., and R. D. Koncal: Set Covering by an All Integer Algorithm: Computational Experience, *Journal of the ACM*, vol. 20, no. 2, 1973.

595. Salkin, H. M., and J. Saha: "Set Covering: Uses, Algorithms, Results," Department of Operations Research Technical Memo 272, Case Western Reserve University, Cleveland, OH, 1972.

596. Saunders, M. A.: A Fast Stable Implementation of the Simplex Method Using Bartels-Golub Updating, pp. 213–226 in Bunch and Rose [101].

597. Saunders, M. A.: "Large-Scale Linear Programming," Systems Optimization Laboratory, Department of Operations Research, Stanford University, Stanford, CA, 1980.

598. Scanlan, R. A.: "Flavor Quality: Objective Measurement," American Chemical Society, Washington, DC, 1977.

599. Schaible, S.: A Survey of Fractional Programming, pp. 417–440 in S. Schaible and W. T. Ziemba (eds.), "Generalized Concavity in Optimization and Economics," Academic Press, New York, 1981.

600. Schaible, S.: Fractional Programming: Applications and Algorithms, *European Journal of Operational Research*, vol. 7, no. 2, 1981.

601. Schrage, L.: Implicit Representation of Generalized Variable Upper Bounds in Linear Programming, *Mathematical Programming*, vol. 14, no. 1, 1978.

602. Schrage, L.: "Linear Programming Models with LINDO," The Scientific Press, Palo Alto, CA, 1981.

603. Sengupta, J. K.: "Stochastic Programming," North-Holland Publishing Co., Amsterdam, 1972.

604. Sengupta, J. K.: "Decision Models in Stochastic Programming," North-Holland, New York, 1982.

605. Sengupta, J. K., G. Tintner, and C. Millham: On Some Theorems of Stochastic Linear Programming with Applications, *Management Science*, vol. 10, no. 1, 1963.

606. Shanno, D. F.: The Simplex Method in Mixed Integer and Integer Programming— A Unified Computational View, *INFOR*, vol. 11, no. 2, 1973.

607. Shanno, D. F., and R. L. Weil: Linear Programming with Absolute-Value Functions, *Operations Research*, vol. 19, no. 1, 1971.

608. Shapiro, J. F.: "Mathematical Programming," John Wiley & Sons, Inc., New York, 1979.

609. Sherali, H. D., A. L. Soyster, and S. G. Baines: Nonadjacent Extreme Point Methods for Solving Linear Programs: *Naval Research Logistics Quarterly*, vol. 30, no. 1, 1983.

610. Shetty, C. M.: Solving Linear Programming Problems with Variable Parameters, *Journal of Industrial Engineering*, vol. 10, no. 6, 1959.

611. Simonnard, M. A.: "Linear Programming," Prentice-Hall, Inc., Englewood Cliffs, NJ, 1966.

612. Simonnard, M. A., and G. F. Hadley: Maximum Number of Iterations in the Transportation Problem, *Naval Research Logistics Quarterly*, vol. 6, pp. 125–129, 1959.

613. Sinha, S. M.: A Duality Theorem for Nonlinear Programming, *Management Science*, vol. 12, no. 5, 1966.

614. Sklar, M. G., and R. D. Armstrong: Least Absolute Value and Chebychev Estimation Utilizing Least Squares Results, *Mathematical Programming*, vol. 24, no. 3, 1982.

615. Slater, M.: "Lagrange Multipliers Revisited," RAND Report RM-676, The RAND Corporation, Santa Monica, CA, 1951.

616. Smale, S.: On the Average Number of Steps of the Simplex Method of Linear Programming, *Mathematical Programming*, vol. 27, no. 3, 1983.

617. Soland, R.: Optimal Facility Location with Concave Costs, *Operations Research*, vol. 22, no. 2, 1974.

618. Soland, R.: Multicriteria Optimization: A General Characterization of Efficient Solutions, *Decision Sciences*, vol. 10, pp. 26–38, 1979.

619. Sperry Univac: "Functional Mathematical Programming System—FMPS Programmer Reference," Sperry Univac Co., St. Paul, MN, 1981.

620. Spielberg, K.: Enumerative Methods in Integer Programming, *Annals of Discrete Mathematics*, vol. 5, pp. 139–183, 1979.

621. Srinivasan, V., and G. L. Thompson: An Operator Theory of Parametric Programming for the Transportation Problem—I and II, *Naval Research Logistics Quarterly*, vol. 19, pp. 205–252, 1972.

622. Stancu-Minasian, I. M., and M. J. Wets: A Research Bibliography in Stochastic Programming 1955–1975, *Operations Research*, vol. 24, no. 6, 1976.

623. Stanley, E. D., D. Honig, and L. Gainen: Linear Programming Bid Evaluation, *Naval Research Logistics Quarterly*, vol. 1, no. 1, 1954.

624. Steinberg, D. I.: On Finding an Initial Solution for the Dual Simplex Algorithm, *Decision Sciences*, vol. 7, no. 1, 1976.

625. Steuer, R. E.: Algorithms for Linear Programming Problems with Interval Objective Function Coefficients, *Mathematics of Operations Research*, vol. 6, no. 3, 1981.

626. Steuer, R. E., and A. T. Schuler: An Interactive Multiple-Objective Linear Programming Approach to a Problem in Forest Management, *Operations Research*, vol. 26, no. 2, 1978.

627. Stewart, G. W.: "Introduction to Matrix Computations," Academic Press, New York, 1973.

638. Ten Kate, A.: Decomposition of Linear Programs by Direct Distribution, Approximation, *Numerische Mathematik*, vol. 2, pp. 1–17, 1960.

629. Stiefel, E.: "An Introduction to Numerical Mathematics," Academic Press, New York, 1963.

630. Stigler, G. J.: The Cost of Subsistence, *Journal of Farm Economics*, vol. 27, 1945.

631. Stoer, J., and R. Bulirsch: "Introduction to Numerical Analysis," Springer-Verlag, New York, 1980.

632. Stoer, J., and C. Witzgall: "Convexity and Optimization in Finite Dimensions," Springer-Verlag, New York, 1970.

633. Strum, J. E.: "Introduction to Linear Programming," Holden-Day, Inc., San Francisco, 1972.

634. Symonds, G. W.: "Linear Programming: The Solution of Refinery Problems," Esso Standard Oil Company, New York, 1955.

635. Szwarc, W.: The Transportation Paradox, *Naval Research Logistics Quarterly*, vol. 18, no. 2, 1971.

636. Taha, H. A.: "Integer Programming," Academic Press, New York, 1975.

637. Telgen, J.: "Redundancy and Linear Programming," Mathematisch Centrum, Amsterdam, 1979.

638. Ten Kate, A.: Decomposition of Linear Programs by Direct Distribution, *Econometrica*, vol. 40, no. 5, 1972.

639. Tewarson, R. P.: "Sparse Matrices," Academic Press, New York, 1973.

640. Thomas, M. E., and M. Grunspan: "Fractional Programming: A Survey," Technical Report 50, Project Themis, Department of Industrial and Systems Engineering, the University of Florida, Gainesville, 1971.

641. Thuente, D. J.: Duality Theory for Generalized Linear Programs with Computational Methods, *Operations Research*, vol. 28, no. 4, 1980.

642. Tind, J., and L. A. Wolsey: An Elementary Survey of General Duality Theory in Mathematical Programming, *Mathematical Programming*, vol. 21, no. 3, 1981.

643. Tintner, G.: Stochastic Linear Programming with Applications to Agricultural Economics, pp. 197–228 in Directorate of Management Analysis [207].

644. Todd, M. J.: An Implementation of the Simplex Method for Linear Programming Problems with Variable Upper Bounds, *Mathematical Programming*, vol. 23, no. 1, 1982.

645. Todd, M. J.: Large-Scale Linear Programming: Geometry, Working Bases and Factorizations, *Mathematical Programming*, vol. 26, no. 1, 1983.

646. Tomlin, J. A.: On Pricing and Backward Transformation in Linear Programming, *Mathematical Programming*, vol. 6, no. 1, 1974.

647. Tomlin, J. A.: Generalized Upper Bounds and Triangular Decomposition in the Simplex Method, *Operations Research*, vol. 22, no. 3, 1974.

648. Tomlin, J. A.: On Scaling Linear Programming Problems, pp. 146–166 in Balinski and Hellerman [31].

649. Tomlin, J. A.: Modifying Triangular Factors of the Basis in the Simplex Method, pp. 77–85 in Rose and Willoughby [574].

650. Tomlin, J. A., and J. S. Welch: "Implementation of a Primal Simplex Network Algorithm in MPSIII," Ketron, Inc., Arlington, VA, 1983.

651. Tompkins, C.: Projection Methods in Calculation, pp. 425–447 in Directorate of Management Analysis [207].

652. Torng, H. C.: Optimization of Discrete Control Systems through Linear Programming, *Journal of the Franklin Institute*, vol. 278, no. 1, 1964.

653. Traub, J. F., and H. Woźniakowski: Complexity of Linear Programming, *Operations Research Letters*, vol. 1, no. 2, 1982.

654. Trauth, C. A., Jr., and R. E. Woolsey: "Practical Aspects of Integer Linear Programming," Monograph Sc-R-66-925, Sandia Corp., Albuquerque, NM, 1966.

655. Trotter, L. E., Jr., and C. M. Shetty: An Algorithm for the Bounded Variable Integer Programming Problem, *Journal of the ACM*, vol. 21, no. 3, 1974.

656. Tucker, A.: "Applied Combinatorics," John Wiley & Sons, Inc., New York, 1980.

657. Tucker, A. W.: Linear Programming, pp. 651–657 in "The Quality Control Conference Papers, 1953," American Society for Quality Control, Inc., New York.

658. Tucker, A. W.: Linear and Non-Linear Programming, *Operations Research*, vol. 5, no. 2, 1957.

659. Tucker, A. W.: "Linear Inequalities and Convex Polyhedral Sets," pp. 569–602 in Directorate of Management Analysis [207].

660. Tucker, A. W.: A Note on Convergence of the Ford-Fulkerson Flow Algorithm, *Mathematics of Operations Research*, vol. 2, no. 2, 1977.

661. Vajda, S.: "The Theory of Games and Linear Programming," John Wiley & Sons, Inc., New York, 1956.

662. Vajda, S.: "Mathematical Programming," Addison-Wesley Publishing Company, Inc., Reading, MA, 1961.

663. Vajda, S.: "Readings in Mathematical Programming," John Wiley & Sons, Inc., New York, 1962.

664. Vajda, S.: "Planning by Mathematics," Pitman Publishing Corporation, New York, 1969.

665. Vajda, S.: "Probabilistic Programming," Academic Press, New York, 1972.

666. Vajda, S.: "Problems in Linear and Nonlinear Programming," Hafner Press, New York, 1975.

667. Valentine, F. A.: "Convex Sets," McGraw-Hill Book Company, New York, 1964.

668. Van de Panne, C.: Parametrizing an Activity Vector in Linear Programming, *Operations Research*, vol. 21, no. 1, 1973.

669. Van de Panne, C.: "Methods for Linear and Quadratic Programming," North-Holland Publishing Co., Amsterdam, 1974.

670. Van de Panne, C.: A Node Method for Multiparametric Linear Programming, *Management Science*, vol. 21, no. 9, 1975.

671. Van de Panne, C., and A. Whinston: The Simplex and the Dual Method for Quadratic Programming, *Operational Research Quarterly*, vol. 15, no. 4, 1964.

672. Van de Panne, C., and A. Whinston: Simplicial Methods for Quadratic Programming, *Naval Research Logistics Quarterly*, vol. 11, no. 4, 1964.

673. Van de Panne, C., and A. Whinston: The Symmetric Formulation of the Simplex Method for Quadratic Programming, *Econometrica*, vol. 37, no. 3, 1969.

674. Veinott, A. F., Jr.: Minimum Concave-Cost Solution of Leontief Substitution Models of Multi-Facility Inventory Systems, *Operations Research*, vol. 17, no. 2, 1969.

675. Vorob'ev, N. N.: "Game Theory," Springer-Verlag, New York, 1977.

676. Votaw, D. F., and A. Orden: The Personnel Assignment Problem, pp. 155–163 in Directorate of Management Analysis [205].

677. Wagner, H. M.: "A Linear Programming Solution to Dynamic Leontief Type Models," RAND Report RM-1343, The RAND Corporation, Santa Monica, CA, 1954.

678. Wagner, H. M.: A Comparison of the Original and Revised Simplex Methods, *Operations Research*, vol. 5, no. 3, 1957.

679. Wagner, H. M.: On the Distribution of Solutions in Linear Programming Problems, *Journal of the American Statistical Association*, vol. 53, pp. 161–163, 1958.

680. Wagner, H. M.: The Dual Simplex Algorithm for Bounded Variables, *Naval Logistics Research Quarterly*, vol. 5, no. 3, 1958.

681. Wagner, H. M.: An Integer Linear-Programming Model for Machine Scheduling, *Naval Research Logistics Quarterly*, vol. 6, no. 2, 1959.

682. Wagner, H. M.: Linear Programming Techniques for Regression Analysis, *American Statistical Association Journal*, March 1959.

683. Wagner, H. M.: On a Class of Capacitated Transportation Problems, *Management Science*, vol. 5, no. 3, 1959.

684. Wagner, H. M.: "Principles of Operations Research," 2d ed., Prentice-Hall, Inc., Englewood Cliffs, NJ, 1975.

685. Wagner, H. M., and T. M. Whitin: Dynamic Version of the Economic Lot Size Model, *Management Science*, vol. 5, no. 1, 1958.

686. Walker, W. E.: A Heuristic Adjacent Extreme Point Algorithm for the Fixed Charge Problem, *Management Science*, vol. 22, no. 5, 1976.

687. Walkup, D. W.: The Hirsch Conjecture Fails for Triangulated 27-Spheres, *Mathematics of Operations Research*, vol. 3, no. 3, 1978.

688. Ward, L. E., Jr.: Linear Programming and Approximation Problems, *The American Mathematical Monthly*, January 1961.

689. Waren, A. D., and L. S. Lasdon: The Status of Nonlinear Programming Software, *Operations Research*, vol. 27, no. 3, 1979.

690. Waugh, F. V.: The Minimum-Cost Dairy Feed, *Journal of Farm Economics*, August 1951.

691. Webb, K. W.: Some Aspects of the Saaty Linear Programming Sensitivity Equation, *Operations Research*, vol. 10, no. 2, 1962.

692. Wendell, R. E.: A Preview of a Tolerance Approach to Sensitivity Analysis in Linear Programming, *Discrete Mathematics*, vol. 38, pp. 121–124, 1982.

693. Wendell, R. E.: "A New Perspective on Sensitivity Analysis in Linear Programming: The Tolerance Approach," Graduate School of Business Report WP-448 RE, University of Pittsburgh, Pittsburgh, PA, 1982.

694. Westlake, J. R.: "A Handbook of Numerical Matrix Inversion and Solution of Linear Equations," Krieger Publishing Co., Inc., Huntington, NY, 1968.

695. Whalen, B. H.: On Minimal-Time and Minimal-Fuel Problems, *Institute of Radio Engineers Transactions on Automatic Control*, p. 46, July 1962.

696. White, W. W.: A Status Report on Computing Algorithms for Mathematical Programming, *Computing Surveys*, vol. 5, no. 3, 1973.

697. Widder, D. V.: "Advanced Calculus," Prentice-Hall, Inc., Englewood Cliffs, NJ, 1947.

698. Wiest, J. D., and F. K. Levy: "A Management Guide to PERT/CPM," Prentice-Hall, Inc., Englewood Cliffs, NJ, 1969.

699. Williams, A. C.: Marginal Values in Linear Programming, *SIAM Journal*, vol. 11, no. 1, 1963.

700. Williams, A. C.: A Treatment of Transportation Problems by Decomposition, *SIAM Journal*, vol. 10, no. 1, 1962.

701. Williams, H. P.: The Formulation of Mathematical Programming Models, *Omega*, vol. 3, no. 5, 1975.

702. Williams, H. P.: "Model Building in Mathematical Programming," John Wiley & Sons, Inc., New York, 1978.

703. Williams, J.: "The Compleat Strategyst," McGraw-Hill Book Company, New York, 1954.

704. Williams, N: "Linear and Nonlinear Programming in Industry," Sir Isaac Pitman & Sons, Ltd., London, 1967.

705. Willner, L. B.: On Parametric Linear Programming, *SIAM Journal on Applied Mathematics*, vol. 15, no. 5, 1967.

706. Wilson, R. J.: "Introduction to Graph Theory," Longman Group Limited, London, 1972.

707. Winkler, C.: "Block-Angular Linear Programs: Unified Theory of Partitioning and Decomposition Using the Simplex Method," Technical Report SOL-74-9, Operations Research Department, Stanford University, Stanford, CA, 1974.

708. Winkler, C.: "Basis Factorization for Block-Angular Linear Programs: Unified Theory of Partitioning and Decomposition Using the Simplex Method," Technical Report SOL 74-19, Department of Operations Research, Stanford University, Stanford, CA, 1974.

709. Witzgall, C.: "On Labelling Algorithms for Determining Shortest Paths in Networks," National Bureau of Standards Report 9840, Washington, DC, May 1968.

710. Wolfe, P.: "The Simplex Method for Quadratic Programming," RAND Report P-1295, The RAND Corporation, Santa Monica, CA, 1957.

711. Wolfe, P.: The Simplex Method for Quadratic Programming, *Econometrica*, vol. 27, no. 3, 1959.

712. Wolfe, P.: "Recent Developments in Non-Linear Programming," R-401-PR, The RAND Corporation, Santa Monica, CA, 1962.

713. Wolfe, P.: Some Simplex-Like Non-Linear Programming Procedures, *Operations Research*, vol. 10, no. 4, 1962.

714. Wolfe, P.: Recent Developments in Nonlinear Programming, pp. 155–187 in F. Alt (ed.), "Advances in Computers," vol. 3, Academic Press, New York, 1962.

715. Wolfe, P.: A Technique for Resolving Degeneracy in Linear Programming, *SIAM Journal*, vol. 11, no. 2, 1963.

716. Wolfe, P.: Methods of Nonlinear Programming, pp. 66–85 in Graves and Wolfe [328].

717. Wolfe, P.: "Foundations of Nonlinear Programming," RAND Report RM-4669-PR, The RAND Corporation, Santa Monica, CA, 1965.

718. Wolfe, P.: The Composite Simplex Algorithm, *SIAM Review*, vol. 7, no. 1, 1965.

719. Wolfe, P.: Methods of Nonlinear Programming, chap. 6 in Abadie [1].

720. Wolsey, L. A.: Integer Programming Duality: Price Functions and Sensitivity Analysis, *Mathematical Programming*, vol. 20, no. 2, 1981.

721. Wüstefeld, A., and V. Zimmerman: Nonlinear One-Parametric Linear Programming and T-Norm Transportation Problems, *Naval Research Logistics Quarterly*, vol. 27, no. 2, 1980.

722. Yen, S. Y.: Finding the Lengths of All Shortest Paths in N-Node Nonnegative-Distance Complete Networks Using 1/2 N^3 Additions and N^3 Comparisons, *Journal of the ACM*, vol. 19, no. 3, 1952.

723. Yudin, D. B., and E. G. Gol'shtein: "Linear Programming," Israel Program for Scientific Translations Ltd., Jerusalem, 1965.

724. Zadeh, L. A.: An Optimal Control and Linear Programming, *Institute of Radio Engineers Transactions on Automatic Control*, pp. 45–46, July 1962.

725. Zanakis, S. H.: Heuristic 0-1 Linear Programming: An Experimental Comparison of Three Methods, *Management Science*, vol. 24, no. 1, 1977.

726. Zangwill, W. I.: The Convex Simplex Method, *Management Science*, vol. 14, no. 3, 1967.

727. Zangwill, W. I.: "Nonlinear Programming," Prentice-Hall, Inc., Englewood Cliffs, NJ, 1969.

728. Zeleny, M.: "Linear Multiobjective Programming," Springer-Verlag, New York, 1974.

729. Zimmerman, H. J.: Fuzzy Programming and Linear Programming with Several Objective Functions, *Fuzzy Sets and Systems*, vol. 1, no. 1, 1978.

730. Zionts, S.: "Linear and Integer Programming," Prentice-Hall, Inc., Englewood Cliffs, NJ, 1974.

731. Zionts, S., and J. Wallenius: An Interactive Multiple Objective Linear Programming Method for a Class of Underlying Nonlinear Utility Functions, *Management Science*, vol. 29, no. 5, 1983.

732. Zionts, S., and J. Wallenius: An Interactive Programming Method for Solving the Multiple Criteria Problem, *Management Science*, vol. 22, no. 6, 1976.

733. Zlobec, S.: Characterizing an Optimal Input in Perturbed Convex Programming, *Mathematical Programming*, vol. 25, no. 1, 1983.

734. Zoutendijk, G.: "Methods of Feasible Directions," Elsevier Publishing Company, Amsterdam, 1960.

735. Zoutendijk, G.: "Mathematical Programming Methods," North-Holland Publishing Co., Amsterdam, 1976.

736. Zukhovitskiy, S. I., and L. I. Avdeyeva: "Linear and Convex Programming," W. B. Saunders Company, Philadelphia, 1966.

ABOUT THE AUTHOR

Saul I. Gass received his B.S. in Education and M.A. in Mathematics from Boston University, and his Ph.D. in Engineering Science/Operations Research from the University of California at Berkeley. Currently Professor of Management Science and Statistics, College of Business and Management, University of Maryland, Dr. Gass began his career in linear programming with the USAF Directorate of Management Analysis—members of which originally formulated and developed the field of linear programming in the late 1940s and 1950s. While there, he helped to develop new techniques and applications to Air Force programming problems. He has taught linear programming for the U.S. Department of Agriculture Graduate School, the George Washington University, and the American University. Dr. Gass has published numerous papers in the field, and serves as a consultant to industry and government. He is a past president of the Operations Research Society of America.

Index